Conference Proceedings of the Society for Experimental Mechanics Series

Series Editor

Kristin B. Zimmerman, Ph.D.
Society for Experimental Mechanics, Inc
Bethel, CT, USA

More information about this series at http://www.springer.com/series/8922

Gary L. Cloud • Eann Patterson • David Backman
Editors

Joining Technologies for Composites and Dissimilar Materials, Volume 10

Proceedings of the 2016 Annual Conference
on Experimental and Applied Mechanics

Springer

Editors
Gary L. Cloud
Composite Vehicle Research Center
Michigan State University
Lansing, MI, USA

Eann Patterson
University of Liverpool
Liverpool, UK

David Backman
National Research Council Canada
Ottawa, ON, Canada

ISSN 2191-5644 ISSN 2191-5652 (electronic)
Conference Proceedings of the Society for Experimental Mechanics Series
ISBN 978-3-319-42425-5 ISBN 978-3-319-42426-2 (eBook)
DOI 10.1007/978-3-319-42426-2

Library of Congress Control Number: 2012943040

Printed on acid-free paper

This Springer imprint is published by Springer Nature
The registered company is Springer International Publishing AG Switzerland

Preface

Joining Technologies for Composites and Dissimilar Materials represents one of ten volumes of technical papers presented at the 2016 SEM Annual Conference & Exposition on Experimental and Applied Mechanics organized by the Society for Experimental Mechanics and held in Orlando, FL, on June 6–9, 2016. The complete Proceedings also includes volumes on *Dynamic Behavior of Materials; Challenges in Mechanics of Time-Dependent Materials; Advancement of Optical Methods in Experimental Mechanics; Experimental and Applied Mechanics; Micro and Nanomechanics; Mechanics of Biological Systems and Materials; Fracture, Fatigue, Failure and Damage Evolution; and Residual Stress, Thermomechanics & Infrared Imaging, Hybrid Techniques and Inverse Problems.*

Composite materials are being increasingly utilized at multiple levels in application areas including automotive, aerospace, marine, biomechanical, and civil infrastructure, so the need for improved joining of these materials has become critical. While the design of the composite laminate is important, it is the ability to join sections of composite to one another or to components made of dissimilar materials that is the enabling technology for creating structures that approach optimum in function, weight, durability, and cost.

Composite joining technologies have been routinely classified in the past as either mechanical or adhesive. Increasingly, joint optimization requires combinations of the two types as well as the introduction of innovative new methods, such as composite welding, that provide high strength and light weight. Hybrid composite joints that allow the joining of composites to monolithic or other classes of material comprise another important technology that will facilitate the use of composites in many new application areas. Today, developments in composite joining technologies are progressing at a rapid rate, driven by both technology and user requirements.

This symposium addresses pertinent issues relating to design, analysis, fabrication, testing, optimization, reliability, and applications of composite joints, especially as these issues relate to experimental mechanics of macroscale and microscale structures.

East Lansing, MI, USA — Gary L. Cloud
Liverpool, UK — Eann A. Patterson
Ottawa, ON, Canada — David Backman

Contents

Chapter 1
How to Join Fiber-Reinforced Composite Parts: An Experimental Investigation

Yaomin Dong, Arnaldo Mazzei, Javad Baqersad, and Azadeh Sheidaei

Abstract A coupler has been developed to prevent windshield wiper systems from being damaged by excessive loads that can occur when the normal wiping pattern is restricted. Unlike the traditional steel coupler used in wiper systems, the composite coupler will buckle at a prescribed compressive load threshold and become extremely compliant. As a result, the peak loading of the coupler and the entire wiper system can be greatly reduced. The coupler is composed of a pultruded composite rod with injection-molded plastic spherical sockets attached at either end. The sockets are used to attach the coupler to the crank and rocker of the windshield wiper linkage. Because the loads exerted on a coupler vary in magnitude and direction during a wiping cycle, the joint between the sockets and the pultruded composite rod must be robust. The paradigm for attaching sockets to steel couplers (i.e. over-molding the sockets around holes stamped into the ends of traditional steel couplers) was tested and found to produce inadequate joint strength. This paper details the methodology that was employed to produce and optimize an acceptable means to join the injection-molded sockets to the fiber glass pultruded rods. Specifically, a designed experiment based on the Robust Design Strategy of Taguchi was used to identify the process, processing parameters, and materials that yield a sufficiently strong joint at a reasonable manufacturing cost without damaging the integrity of the underlying composite structure.

Keywords Composite materials • Buckling • Wiper systems • Durability • Joint strength

1.1 Introduction

Automotive windshield wiper systems, in conjunction with washer systems, are used in vehicles to remove contaminants such as rain, sleet, snow, and dirt from the windshield. As shown in Fig. 1.1, a typical wiper system consists of an electric motor, a linkage to transform the rotational motion from the motor to oscillatory motion, and a pair of wiper arms and blades. The areas of the windshield that must be wiped by the wiper system are mandated by the federal motor vehicle safety standards FMVSS 104 [1].

Figure 1.2 shows that snow has accumulated above the cowl screen and caused the normal wiping pattern of the system to be restricted. Under such conditions, the loads in the wiper system have been observed to be approximately four times greater than those encountered under normal wiping conditions for a particular application. The elevated loads are due to two factors: (1) As the arm(s) and/or blade(s) come into contact with the restriction, the speed of the wiper system decreases and the output torque of the motor increases. (2) This factor is aggravated by the very large mechanical advantage of the wiper linkage near the reversal positions.

In extreme cases, the loads that result from restricting the wipe pattern can cause failures. For example, Fig. 1.3 depicts a 5-mm thick hardened steel rocker arm that has fractured as a result of such a loading. Clearly, such a damage is unacceptable from both safety and warranty standpoints.

Penrod and Dong [2] reviewed various methods that have been developed to solve this problem. One such solution is to use a coupler fabricated from a thermoset polyester/glass fiber pultruded composite material [3–6]. This coupler—referred to as the composite link—has the advantage of being sufficiently stiff during normal operation providing very good pattern control. However, in the event that elevated loads occur, the coupler is designed to buckle at a prescribed load level (i.e. the critical load). Once buckled, the coupler becomes extremely compliant. As a result, the peak loading of the coupler and, moreover, the entire system, can be limited and greatly reduced when compared to the same system without the composite link.

Y. Dong, Ph.D. (✉) • A. Mazzei, Ph.D. • J. Baqersad, Ph.D., P.E. • A. Sheidaei, Ph.D.
Department of Mechanical Engineering, Kettering University, 1700 University Avenue, Flint, MI 48504-4898, USA
e-mail: ydong@kettering.edu

G.L. Cloud et al. (eds.), *Joining Technologies for Composites and Dissimilar Materials, Volume 10*
Conference Proceedings of the Society for Experimental Mechanics Series, DOI 10.1007/978-3-319-42426-2_1

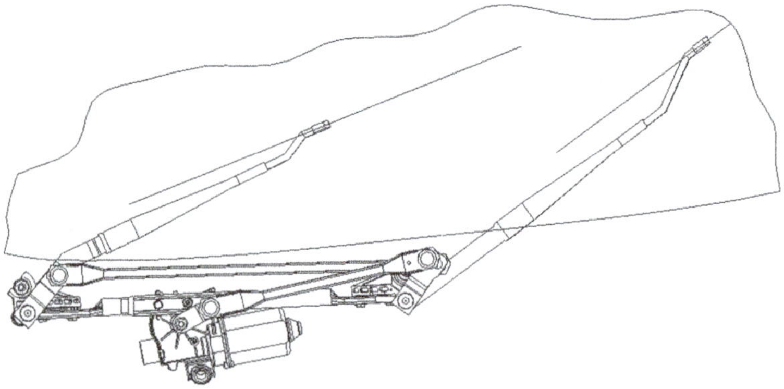

Fig. 1.1 A typical wiper system consists of wipers, motor, and a linkage, and wiper arms

Fig. 1.2 Restricted wiping pattern due to snow accumulation

Fig. 1.3 Fractured rocker arm resulting from snow accumulation

Figure 1.4 shows the operating principle of the composite link. In the illustration the wiper arms have encountered snow/ice accumulation (depicted by the hatched pattern) above the cowl screen that blocks the normal wiping pattern of the system. As a result, the wiper system loads have increased sufficiently to buckle the composite link (depicted by the arcuate member

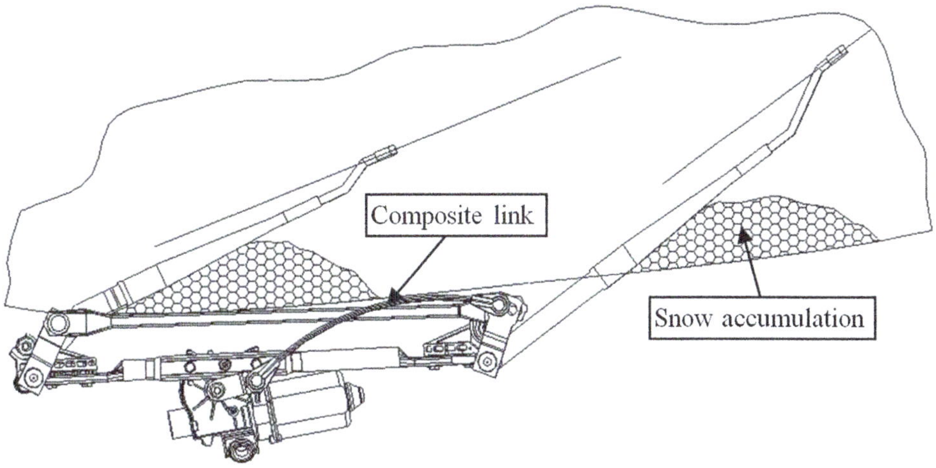

Fig. 1.4 Operating principle of the composite link

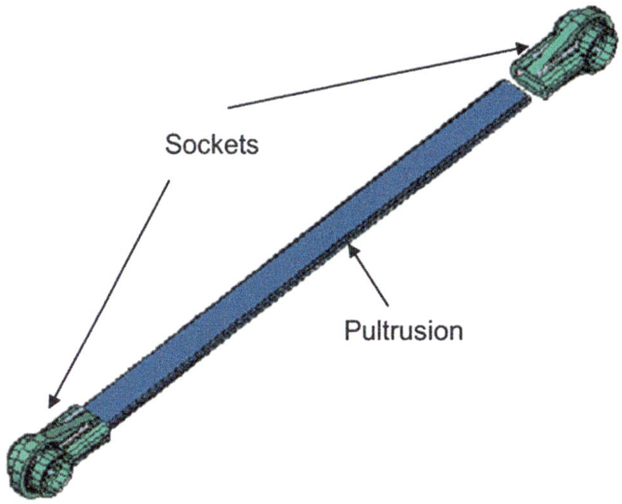

Fig. 1.5 Composite link assembly

of the wiper linkage). Once buckled, the composite link behaves nearly perfectly-plastic. As a result, the chord length of the composite link decreases significantly without a corresponding appreciable increase in axial load. Thus the crank is able to continue rotating and the wiper system is enabled to wiper at a reduced area above the restriction.

The composite link is an assembly of a 16.7 mm × 5.2 mm rectangular cross-section pultrusion and glass-filled plastic sockets attached at either end, as shown in Fig. 1.5. The design methodology, material selection, validation testing, and application guidelines for the composite link are discussed in detail by Penrod, Dong, and Buchanan [3]. Because the loads exerted on the coupler vary in magnitude and direction during each wiping cycle, the joint between the sockets and the pultrusion is critical and must be robust.

Initially, it was planned to attach the sockets to the pultrusion rod using the same approach that sockets have been traditionally joined to steel couplers (i.e. over-molding the sockets around holes stamped into the ends of the traditional steel coupler). In the case of the composite link, holes were drilled into the ends of the pultrusion and the sockets were over-molded onto the pultrusions. Unfortunately, this joining method performed dismally as it was characterized by bearing stress failures of the pultrusion at relatively low tensile loads. In retrospect, this failure mode was no surprise. The pultrusions consist of longitudinally oriented continuous glass fibers bound together by a thermosetting polyester matrix. Drilling holes on the pultruded rod cuts the continuous fibers and weakens the composite structure. As such, the material has little strength on short-transverse planes where the bearing failure was observed to occur.

Since the paradigm for attaching sockets to steel links did not work, a Greenfield approach to attaching sockets to the pultrusions was embarked upon. This paper details the approach that was employed to generate, evaluate, and optimize an acceptable means of attaching the injection-molded plastic sockets to the pultrusions.

1.2 Designed Experiment for Joining Socket to Pultrusion

How can the plastic sockets be attached to the pultruded composite rod? The traditional approach of over-molding the sockets to steel links proved inadequate as was discussed previously. In order to answer this question, a set of design specifications were generated to assess design concepts against. Qualitatively, the specifications of paramount importance were that the attachment method.

- Provide adequate joint strength in tension and compression when subjected to monotonic and cyclic loads.
- Avoid damaging the integrity of the glass fibers to the extent that the pultrusion would no longer be capable of providing a predictable critical load and adequate post-buckled axial compressive deflection.
- Be cost-effective.

Once the aforementioned specifications were established, three phases of the development were sequentially undertaken. Specifically,

- *System Design*: the first phase of the development where a variety of state-of-the-art candidate processes were generated using input from a cross-functional team of material, manufacturing, and product specialists. Once identified, the candidate processes were then narrowed to the most promising approaches via conceptual evaluation by the cross-functional team. Additionally, some ad hoc testing was completed in this phase of the development that suggested adding grooves to the ends of the pultrusion so that the over-molded socket could form a positive lock was the most viable approach.
- *Parameter Design*: the second phase of the development where a series of designed experiments were used to quantitatively ascertain the optimal combination of design and processing parameters.
- *Tolerance Design*: the third and final phase of the development where the allowable tolerance range on the critical dimensions and processing parameters were determined.

The focus of this paper is on the series of designed experiments that were used to identify the optimal combination of designs, materials, and processes during the parameter design phase of the development.

1.2.1 Parameter Selection Stage 1: Screening

The Robust Design Strategy of G. Taguchi [7] has gained broad acceptance in Japan, the United States, and elsewhere for use in product design and development. Taguchi's approach to parameter design provides a systematic and efficient method for determining near-optimum design parameters. The Taguchi method allows for the sensitivity of the design to a large number of parameters to be evaluated using a relatively small number of sample parts by utilizing orthogonal arrays from design of experiments theory. The conclusions drawn from such small scale experiments are valid over the entire experimental range spanned by the control factors and control factor levels.

As mentioned previously, adding grooves to the pultrusion ends so that the over-molded socket would be positively locked to the pultrusion was deemed the most viable approach. However, the influence of other factors on the joint strength was still unknown. Specifically, what is the influence of the surface finish of the pultrusion on joint strength? What role does the location of the grooves play? Does the socket material significantly impact the joint strength? What benefit might be derived from the use of adhesives? What is the influence of glass content in the pultrusion on joint strength? To answer these questions, an experiment was designed to screen these factors for relative importance and identify where the various factors might interact—both in a positive and negative sense.

Table 1.1 depicts the L18 Orthogonal Array that was used to screen the design candidates. Notice that the five aforementioned design parameters (A ~ F) are chosen as control factors, with each factor having three levels. Descriptions of the control levels indicated in the array are as follows.

- Surface Finish: Smooth denotes the surface finish of the pultrusion without any additional processing. Abrasive-fine, and abrasive-coarse denote that the ends of the pultrusion have been grit blast with a fine and coarse medium, respectively. Chemical etched refers to the resin-rich surface of the pultrusion ends being removed using an acidic solution. Laser etches #1 and #2 denote two power levels used to remove the resin-rich surface of the pultrusion via the use of a laser.
- Machined Grooves: Smooth denotes no grooves, top–bottom denotes grooves machined into the top and bottom surfaces of the pultrusions. Side-side denotes grooves machined into the edges of the pultrusions. Ultimately, axial grooves which

Table 1.1 L18 Orthogonal array for stage 1 (screening)

	1&2	3	4	5	6
	A	B	D	E	F
	Surface finish	Machined grooves	Socket material	Attachment/adhesives	Pultrusions
1	Smooth	Smooth	Nylon	Insert mold/hot curing	Low modulus
2	Smooth	Top-bottom	Acetal #1	Mech attach/chem. cure	Moderate modulus
3	Smooth	Side-side	Acetal #2	Insert mold/no adhesive	High modulus
4	Abrasive-fine	Smooth	Nylon	Mech attach/chem. cure	Moderate modulus
5	Abrasive-fine	Top-bottom	Acetal #1	Insert mold/no adhesive	High modulus
6	Abrasive-fine	Side-side	Acetal #2	Insert mold/hot curing	Low modulus
7	Abrasive-coarse	Smooth	Acetal #1	Insert mold/hot curing	High modulus
8	Abrasive-coarse	Top-bottom	Acetal #2	Mech attach/chem. cure	Low modulus
9	Abrasive-coarse	Side-side	Nylon	Insert mold/no adhesive	Moderate modulus
10	Chemical etch	Smooth	Acetal #2	Insert mold/no adhesive	Moderate modulus
11	Chemical etch	Top-bottom	Nylon	Insert mold/hot curing	High modulus
12	Chemical etch	Side-side	Acetal #1	Mech attach/chem. cure	Low modulus
13	Laser etch #1	Smooth	Acetal #1	Insert mold/no adhesive	Low modulus
14	Laser etch #1	Top-bottom	Acetal #2	Insert mold/hot curing	Moderate modulus
15	laser etch #1	Side-side	Nylon	Mech attach/chem. cure	High modulus
16	Laser etch #2	Smooth	Acetal #2	Mech attach/chem. cure	High modulus
17	Laser etch #2	Top-bottom	Nylon	Insert mold/no adhesive	Low modulus
18	Laser etch #2	Side-side	Acetal #1	Insert mold/hot curing	Moderate modulus

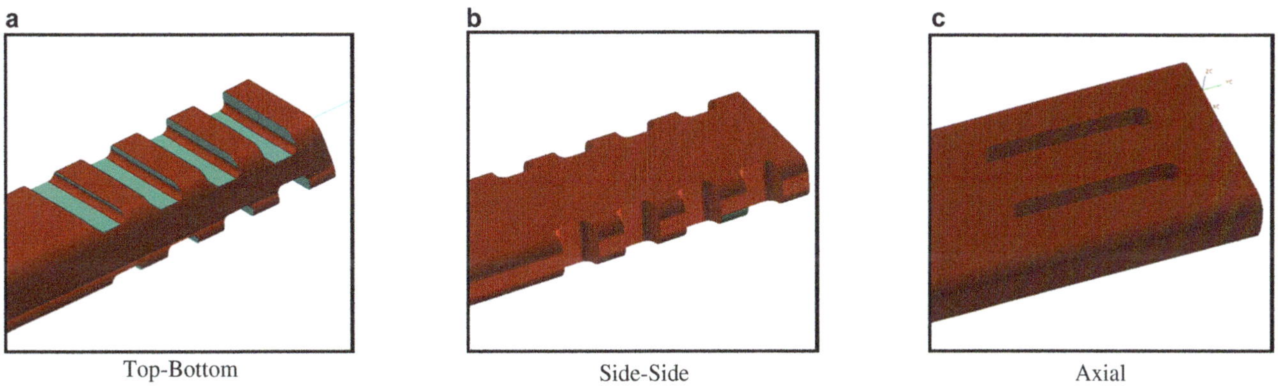

Fig. 1.6 Groove configurations

are oriented longitudinally were also considered. (Please see Sect. 2.3 for details.) All grooves were full-filleted as indicated in Fig. 1.6 and were machined by grinding.

- Socket Material: Nylon denotes a glass-filled nylon 6/6. Acetal #1 refers to 20 % glass-filled acetal, whereas acetal #2 refers to 30 % glass-filled acetal.
- Attachment/adhesives: Three possibilities were considered. Insert molding the sockets over the ends of the pultrusion was done both without adhesives (i.e. insert mold/no adhesive) and with adhesives (i.e. insert mold/hot curing). Hot curing refers to the heat available from the injection molded material promoting rapid curing of the adhesive.) Mechanical attachment of the socket to the pultrusion was also considered and used an adhesive to bond the sockets to the pultrusions. In this case an accelerant was used to promote curing of the adhesive. Thus, this approach is denoted by mech attach/chem cure in Table 1.1. In all cases, the adhesive used was a commercially available cyanoacrylate.
- Pultrusions: Low-, moderate-, and high-modulus refer to elastic modulus of the pultusions resulting from three different glass content levels being used. The higher the glass content, the higher the modulus.

It should be noted that a full-factorial experiment would require evaluating the performance of ($6\times3\times3\times3\times3=$) 486 combinations of control factors and control levels. Using the Taguchi approach, only 18 combinations need to be considered with the 18 specific combinations of control factors and control levels dictated by the method.

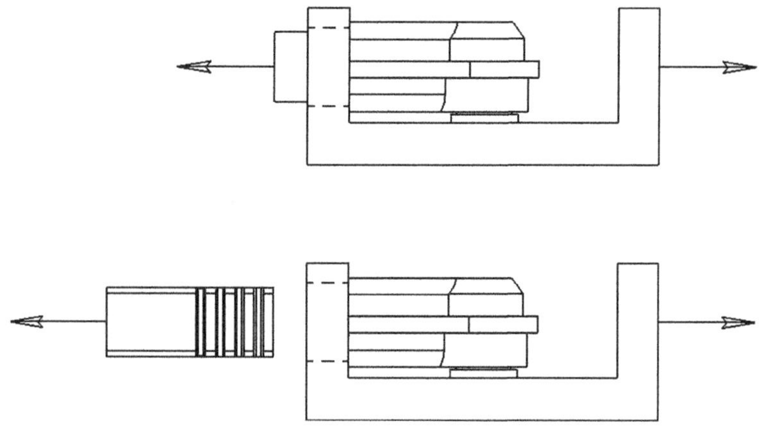

Fig. 1.7 Strip test used to determine the ultimate strength of the joint

Table 1.2 L9 Orthogonal array for stage 2 (groove design)

	1	2	3
	A	B	C
	Groove size	Groove spacing	Number of grooves
1	Small	#1	Few
2	Small	#2	Moderate
3	Small	#3	Many
4	Medium	#1	Moderate
5	Medium	#2	Many
6	Medium	#3	Few
7	Large	#1	Many
8	Large	#2	Few
9	Large	#3	Moderate

Once the 18 combinations were prototyped, a strip test was run on each configuration to determine the ultimate strength of each joint in tension. Figure 1.7 depicts the test. The assembled composite link was placed in a specially designed fixture that captured the socket only. The tensile load was increased until the assembly of the socket and pultrusion separated. Analyzing the data collected from the testing using the Taguchi approached allowed for the contribution to the joint strength of each control factor and control level, and interactions of the control factors and control levels to be quantified. From this analysis, it was found that grooves added to the top and bottom faces of the pultrusion were the single greatest contributor to joint strength. The use of adhesives to produce bonding between the socket and pultrusion was found to be the second largest contributor to strength, but was ruled out due to processing difficulties. Additionally, altering the surface finish of the pultrusion provided little benefit and was quite costly. As such, it was also disqualified from further consideration.

1.2.2 Parameter Selection Stage 2: Groove Design

Based on the results of the screening designed experiment, a second designed experiment was constructed to further optimize the groove design. Table 1.2 depicts the L9 Orthogonal Array that was constructed for this purpose. In this experiment three control factors (i.e. groove size, groove spacing, and number of grooves) were considered at three different control levels. Once again it is noteworthy that the number of configurations to be tested is only nine and that a full factorial experiment would require 27 (i.e. $3 \times 3 \times 3$) configurations.

The control levels indicated in Table 1.2 are described as follows.

- Groove Size: Small denotes 0.5 mm depth, while medium corresponds to 1.0 mm depth, and large denotes a 1.5 mm depth.
- Groove Spacing: The spacing between adjacent grooves with level #1 corresponding to 2 mm spacing, level #2 corresponding to 3 mm spacing, and level #3 corresponding to 4 mm spacing.
- Number of Grooves: Few denotes two grooves per face, moderate denotes four grooves per face; and many corresponding to six grooves per face.

Prototype parts of each configuration were fabricated and evaluated against two criteria. The first criterion was simply evaluating the tensile strength of the joint as was performed in the screening experiment. The second criterion involved examining the performance of the composite links in compression after being buckled and axially compressed to a preset deflection. Doing so revealed that under certain circumstances, the pultruded material tends to delaminate on long-transverse planes with the delamination initiating at the root of the groove.

Analyzing the data using the Taguchi approach once again allowed for the contribution to the joint strength of each control factor and control level, and interactions of the control factors and control levels to be determined. From this analysis and delamination considerations, it was found that numerous small, closely-spaced grooves provided the optimal combination of joint strength in tension and delamination resistance after buckling.

1.2.3 Parameter Selection Stage 3: Final Design

Once the groove optimization designed experiment was completed, a third and final designed experiment was constructed to examine various processing methods to produce the grooves. Additionally, surface finish was once again considered and the use of adhesives was revisited. An additional six configurations were added to the experiment for checking purposes.

Table 1.3 depicts the L18 Orthogonal Array that was constructed for this final investigation. Table 1.4 contains six additional configurations of interest that were included in the experiment. Prototype parts representing the 24 configurations specified in Tables 1.3 and 1.4 were fabricated and evaluated for tensile strength per the procedure used in the previous experiments. Notice that four different processes were considered to produce the grooves both across the grain (cross grooves) and axially—machining (i.e. cutting), grinding, grit blasting, and laser ablation, as well as, a control case of no grooves.

Table 1.3 L18 Orthogonal array for stage 3 (final design)

	1&2	3	4
	A	C	D
	Grooves	Surface finish	Attachment method
1	Smooth	Smooth	Insert mold
2	Smooth	Ground axial fiber exposure	Adhesive assy W/o surface primer
3	Smooth	Laser axial fiber exposure	Adhesive assy W/surface primer
4	Ground axial grooves	Smooth	Insert mold
5	Ground axial grooves	Ground axial fiber exposure	Adhesive assy w/o surface primer
6	Ground axial grooves	Laser axial fiber exposure	Adhesive assy W/ surface primer
7	Ground top-bottom grooves	Smooth	Adhesive assy W/o surface primer
8	Ground top-bottom grooves	Ground axial fiber exposure	Adhesive assy W/ surface primer
9	Ground top-bottom grooves	Laser axial fiber exposure	Insert mold
10	Ground side-side grooves	Smooth	Adhesive assy W/surface primer
11	Ground side-side grooves	Ground axial fiber exposure	Insert mold
12	Ground side-side grooves	Laser axial fiber exposure	Adhesive assy W/o surface primer
13	Laser axial grooves	Smooth	Adhesive assy W/o surface primer
14	Laser axial grooves	Ground axial fiber exposure	Adhesive assy W/ surface primer
15	Laser axial grooves	Laser axial fiber exposure	Insert mold
16	Cut top-bottom grooves	Smooth	Adhesive assy W/ surface primer
17	Cut top-bottom grooves	Ground axial fiber exposure	Insert mold
18	Cut top-bottom grooves	Laser axial fiber exposure	Adhesive assy W/o surface primer

Table 1.4 Six additional configurations of interest

19	Ground cross grooves	Plasma axial fiber exposure	Insert mold
20	Ground cross grooves	Plasma axial fiber exposure	Adhesive assy W/o surface primer
21	Grit blast cross grooves	Smooth	Adhesive assy W/o surface primer
22	Grit blast cross grooves	Laser axial fiber exposure	Insert mold
23	Laser cross grooves	Smooth	Adhesive assy W/o surface primer
24	Laser cross grooves	Laser axial fiber exposure	Insert mold

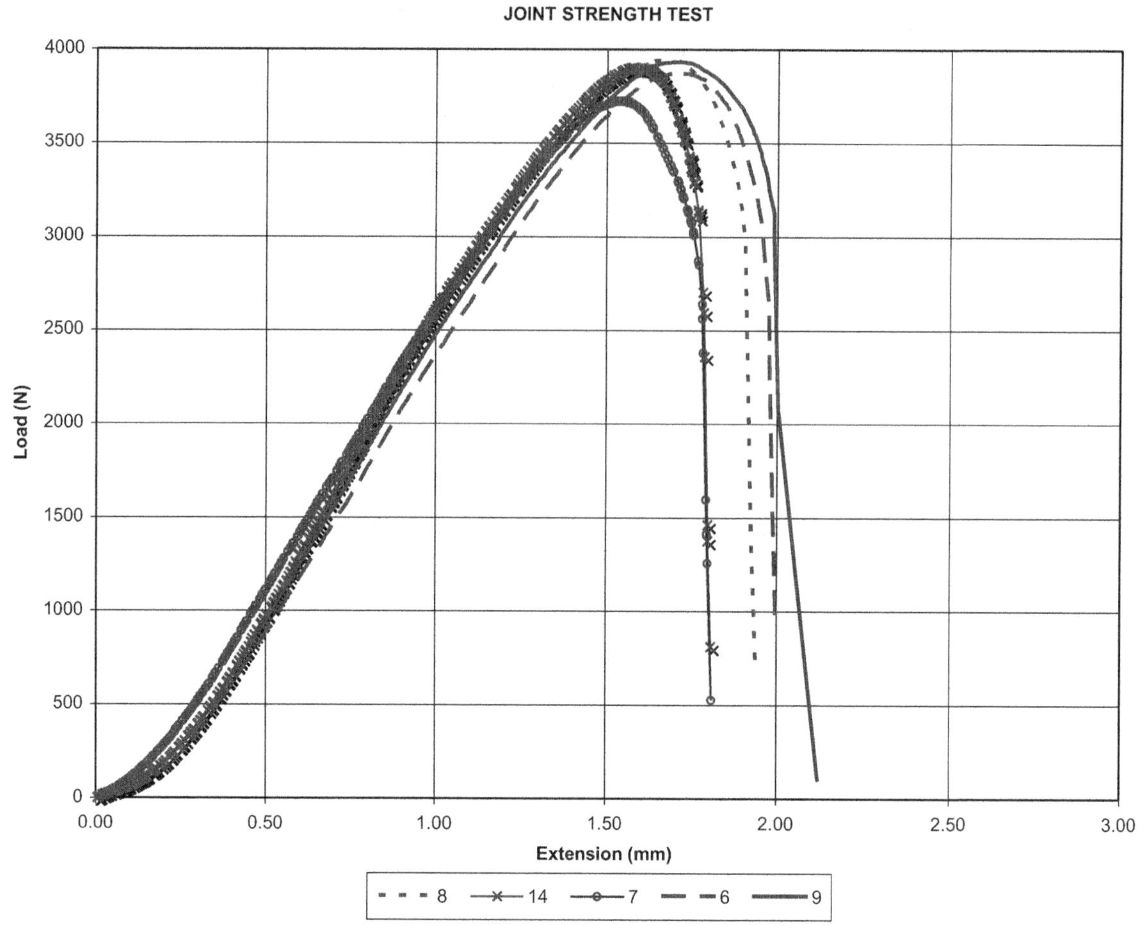

Fig. 1.8 Joint strength of the top five performing configurations

Three surface treatments were considered—grinding, laser ablation, and plasma cutting, as well as, untreated (smooth) finish. Additionally, insert-molding the sockets was compared to attaching the sockets using adhesives with and without a surface primer. It is again worth noting that a full factorial experiment would require 54 (i.e. $6\times3\times3$) configurations, whereas the designed experiment requires only 18 configurations.

Figure 1.8 depicts the test results for the top five performing configurations. In all five cases, the joint strength requirement was exceeded by a factor of two. Once again the strip test method depicted in Fig. 1.7 was used to obtain these results. The non-linear portion of the curves near the origin is attributable to the small amount of "play" in the joint due to shrinkage of the injection-molded socket material. Then, the loading noted to increase linearly with respect to deflection until the peak load-carrying capability of the joint is reached. At this point, either the socket material filling the grooves sheared off or the pultruded sheared off on a transverse plane.

1.3 Conclusions

Based on the aforementioned testing, and considering cost and quality control, the results of the investigation can be summarized as follows:

- When machining grooves into the pultrusions, grinding is superior to cutting due to the nature of the glass-filled composite structure.
- Avoid attaching parts to pultrusions via through holes as the transverse planes of the pultrusion which are sheared by tensile loading have little strength.

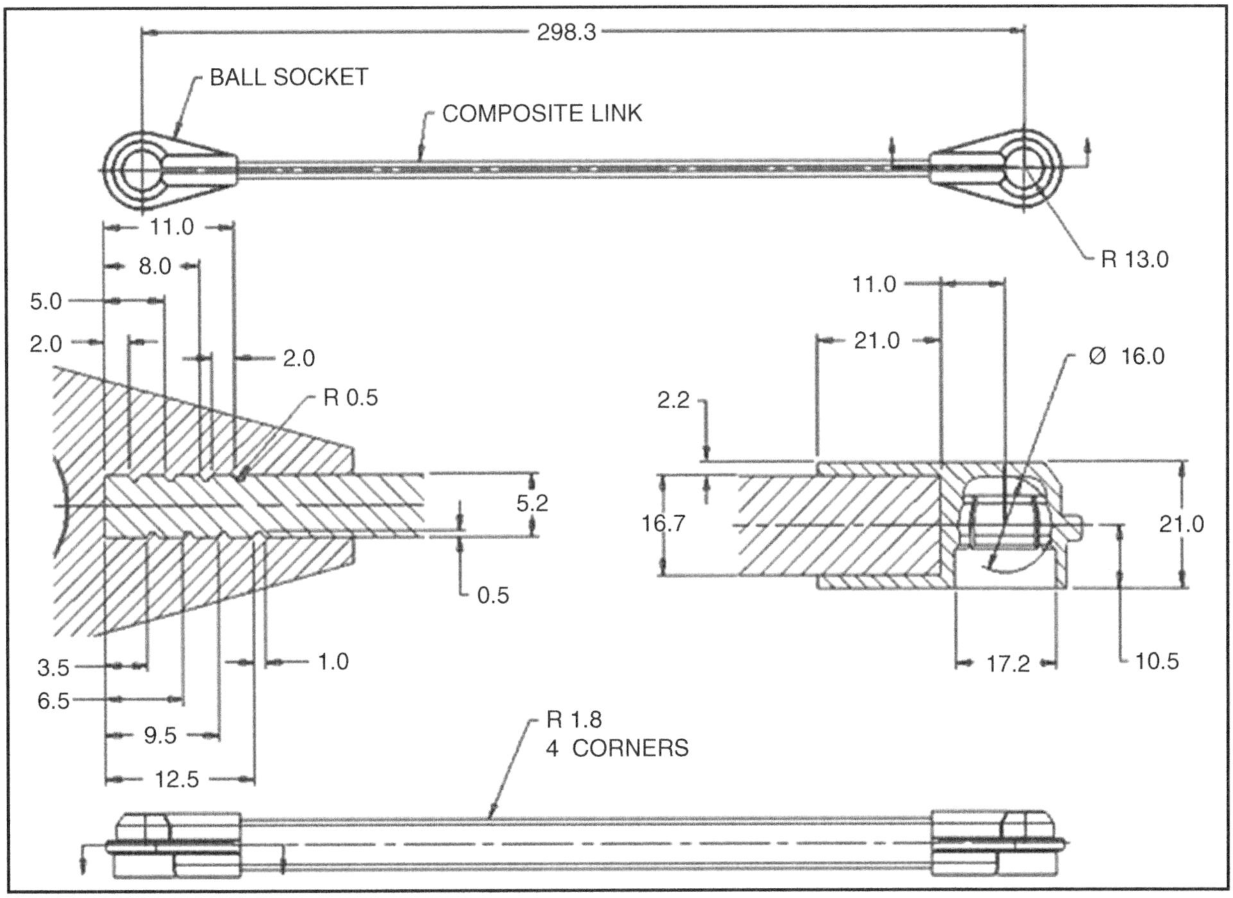

Fig. 1.9 Composite link socket attachment

- Grooves should be on the faces of the pultrusion having the greatest surface area (e.g. the top and bottom in the present design). Offsetting the grooves has been observed to make the design less susceptible to delamination.
- Numerous fine grooves are superior to a fewer coarse grooves.
- Adhesives can be used to significantly improve joint strength, but are noted to be difficult to process since they are very time-sensitive and difficult to control during the molding process.
- Surface roughness has minimal impact on joint strength. Due to the extra cost associated with this processing, it should only be considered when the load requirement cannot be satisfied otherwise.

In lieu of these assertions, the final, optimized groove configuration and dimensions of the joint between the sockets and the pultrusion is depicted in Fig. 1.9.

References

1. FMVSS 571-104, 49 CFR. Federal Motor Vehicle Safety Standards; Windshield Wiping and Washing Systems. National Highway Traffic Safety Administration, DOT, Washington, DC
2. Penrod, J., Dong, Y.: An Application Flexible Method to Limit the Loads in Windshield Wiper System. SAE Technical Paper Series, 2005-01-1835 (2005)
3. Penrod, J., Dong, Y., Buchanan, H. C.: A Novel Use of a Composite Material to Limit the Loads in Windshield Wiper System, SAE Technical Paper Series, SP-1575, 2001-01-0104 (2001)
4. Buchanan, H.C., Dong, Y.: Windshield wiping system, US Patent US 6,148,470, Valeo Electrical Systems (2000)
5. Buchanan, H.C., Dong, Y.: Windshield wiping system, US Patent US 6,381,800 B1, Valeo Electrical Systems (2002)
6. Bryson, B.A., Buchanan, H.C., Dong, Y., Penrod, J.P.: Windshield wiping system manufacturing method, US Patent US 068 81 373, Valeo Electrical Systems (2005)
7. Peace, G.: Taguchi Methods: A hands-On Approach to Quality Engineering. Addison, Reading, MA (1993)

Chapter 2
Analysis of a Composite Pi/T-Joint Using an FE Model and DIC

Chris Sebastian, Mahmoodul Haq, and Eann Patterson

Abstract An analysis of composite Pi/T-joint was performed through the comparison of a finite element simulation and experimental results obtained using digital image correlation. The finite element simulation was performed in ABAQUS using a realistically modeled adhesive layer consisting of cohesive zone elements. The composite Pi/T-joints were manufactured using vacuum assisted resin transfer molding (VARTM). The resulting Pi/T-joints were loaded in out-of-plane (web pull-out) until failure, during which three-dimensional digital image correlation (3D-DIC) was used to measure the in-plane displacements and strains.

From the experiments, it was found that the average peak pull-out force was 12.56 ± 0.82 kN, which was 10 % less than that from the FE simulation at 13.77 kN. However, this does not give much information about whether the stress and strain distribution has been accurately predicted by the simulation. To better understand this, the experimental data obtained using DIC was compared to that from the FE simulation using image decomposition. The image decomposition process reduces the large, full-field data maps to a feature vector of 130 or so elements which can be compared much more easily. In this case, there was good agreement between the experiment and simulation, when taking into account the experimental uncertainty.

Keywords Composite joint • VARTM • Digital image correlation • Numerical simulation

2.1 Introduction

Composite Pi-joints are being extensively used in the marine and aerospace industries and are called T-joints, or Tee joints in some sectors. 'T' refers to the parts being joined while 'Pi' describes the shape of the preform used to form the joint. These joints are of special importance due to their complex geometry and criticality to overall structural integrity. T-joints are found at composite bulkhead-to-skin, rib-to-skin and spar-to-skin interfaces [1]. T-shaped stiffeners are extensively used in aircraft wings [1]. Additionally, the requirement for reduced structural weight and the ever-increasing demand for efficient aerospace structures has driven the development of adhesively bonded T-joints. However, these joints have been found to be sensitive to peel and, or delamination and through-the-thickness stresses. It has been reported that composite T-joints often fail near the web/skin interface [2].

Various configurations of composite T-joints have been studied extensively, mostly in the context of marine structures [3–18], including tee-joints made of sandwich panels [9–12]. The damage and failure analysis of such marine composite joints has also been studied [13–18]. A good overview on adhesively bonded composite joints is provided by Banea and Silva [19]. Moreover most of the aforementioned T-joints were manufactured by connecting the horizontal (flange) and vertical (web) laminates with a hand-layup laminate/skin (overlaminate). Similarly, literature on T-joints using a pi-preform (hence a Pi-joint) is relatively limited [20]. The use of Pi-preforms compared to hand-layup skins offers ease of manufacturing and speedy construction. Researchers from the Wright Patterson Air Force Research Laboratory (AFRL), Ohio, US, have studied adhesively bonded Pi-preform T-joints and also incorporated them in an aircraft wing and reported enhanced performance relative to a similar wing manufactured by conventional methods [2, 21, 22]. In addition, they tested the wings with T-joints

C. Sebastian (✉) • E. Patterson
School of Engineering, University of Liverpool, Liverpool L69 3GH, UK
e-mail: c.sebastian@liverpool.ac.uk

M. Haq
Composite Vehicle Research Center, Department of Civil and Environmental Engineering, Michigan State University, Lansing, MI 48910, USA

G.L. Cloud et al. (eds.), *Joining Technologies for Composites and Dissimilar Materials, Volume 10*
Conference Proceedings of the Society for Experimental Mechanics Series, DOI 10.1007/978-3-319-42426-2_2

in fatigue and found no failure in the joints [21, 22]. They reported considerable savings in manufacturing time (days instead of weeks) and up to 75 % reduction in costs. Apart from the AFRL study [2, 21, 22], the literature on pre-form Pi-joints is relatively limited. Hence, this work focuses on evaluating the pull-out performance using both structural tests to destruction and integrated experimental and simulation approach to analyzing the strain fields in structural Pi-preform composite T—joints.

In this work, the out-of-plane behavior or pull-out performance behavior of the composite Pi-joints was experimentally studied by using three-dimensional Digital Image Correlation (3D DIC). To the best of author's knowledge, the use of DIC to study the composite Pi-joints is unique and no other work has been reported, except the preliminary results reported earlier by the authors [23, 24]. The following sections provide brief details on the manufacturing process, materials used, and the experimental and numerical results.

2.2 Experimental Method

2.2.1 *Manufacture of the Pi-joints*

The Pi-joints of dimensions shown in Fig. 2.1 were manufactured by connecting flange and web plates with a Pi-preform and infusing the resin (adhesive) using vacuum assisted resin transfer molding (VARTM). The VARTM technique was also used to manufacture the glass fiber composite adherends (base and web plates) for the Pi-joints. The reinforcement used for the adherends was Owens Corning ShieldStrand S, S2-glass plain weave fabric with areal weight of 818 g/m^2. The base and web plates had 16 and eight layers of plain weave glass fabric respectively resulting in cured thicknesses of 9.53 mm (3/8 in.) and 4.76 mm (3/16 in.), respectively. The resin used was a two part toughened epoxy (SC-15, Applied Poleramic). The base and web plates were then connected with the Pi-preform and infused with SC-15 adhesive. The Pi-preform used was a carbon-fiber 3D woven preform (Albany Engineered Composites, Inc., Rochester, NH, USA.).

A constant adhesive bond-line thickness in Pi-joints was maintained by placing steel wires of 0.127 mm (0.005 in.) diameter in the bond-line. These spacers were placed strategically such that they did not influence the resulting performance. The tensile properties of the adhesive were obtained from experimental tests following the ASTM D638 standard. The fracture properties of the resin were obtained experimentally from Mode-I and Mode-II tests. The properties of the flange and web plates were also obtained experimentally. A summary of the material properties is provided in Table 2.1.

All bonded surfaces were grit-blasted and then cleaned with acetone. Figure 2.2 provides a step-by-step overview of the Pi-joint manufacturing process. Figure 2.2b shows the aluminum mold used to obtain the desired pi-preform dimensions. The aluminum mold has inlet and outlet pipe-fittings that attach to the vacuum pump and the resin-bath, respectively. The inlet

Fig. 2.1 Pi-joint dimensions (in mm) with the adhesive thickness omitted. Not to scale

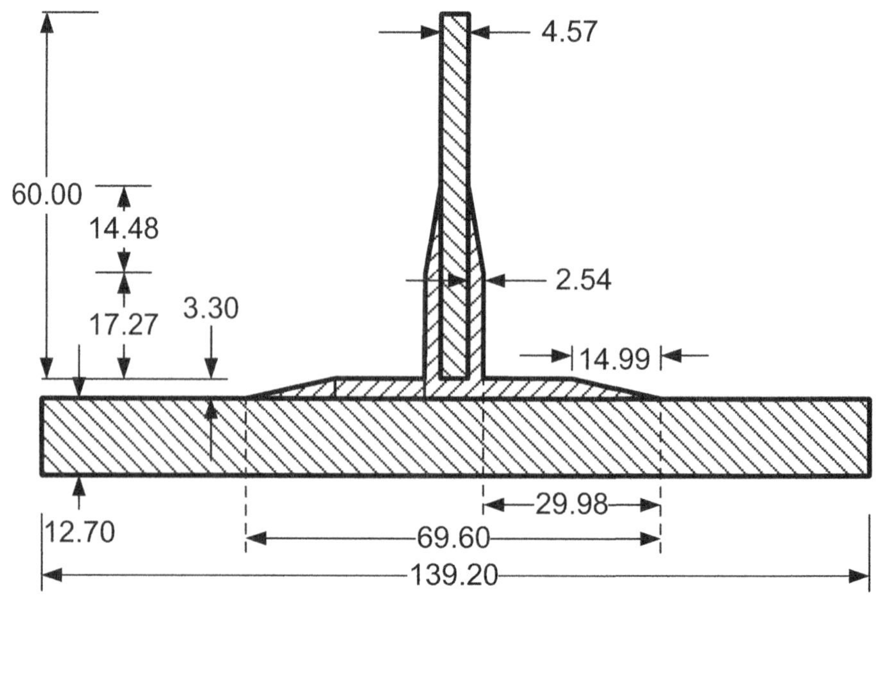

Table 2.1 Material properties used in the numerical simulation

(a)	Adhesive tensile properties (experimental)	
	Modulus (GPa)	2.69 ± 0.19
	Tensile strength (MPa)	58.10 ± 3.94
	Tensile failure elongations (%)	3.45 ± 0.37
(b)	Adhesive fracture toughness [29]	
	GI_c (N/m)	0.132
	GII_c (N/m)	0.146
(c)	Plates—GFRP (experimental)	
	Web, tensile modulus (GPa)	17.25
	Base, flexural modulus (GPa)	7.00
	Poisson's ratio [5]	0.17
(d)	3D-Preform—CFRP [30]	
	Elastic, isotroPic modulus (GPa)	32.8
	Poisson's ratio	0.17

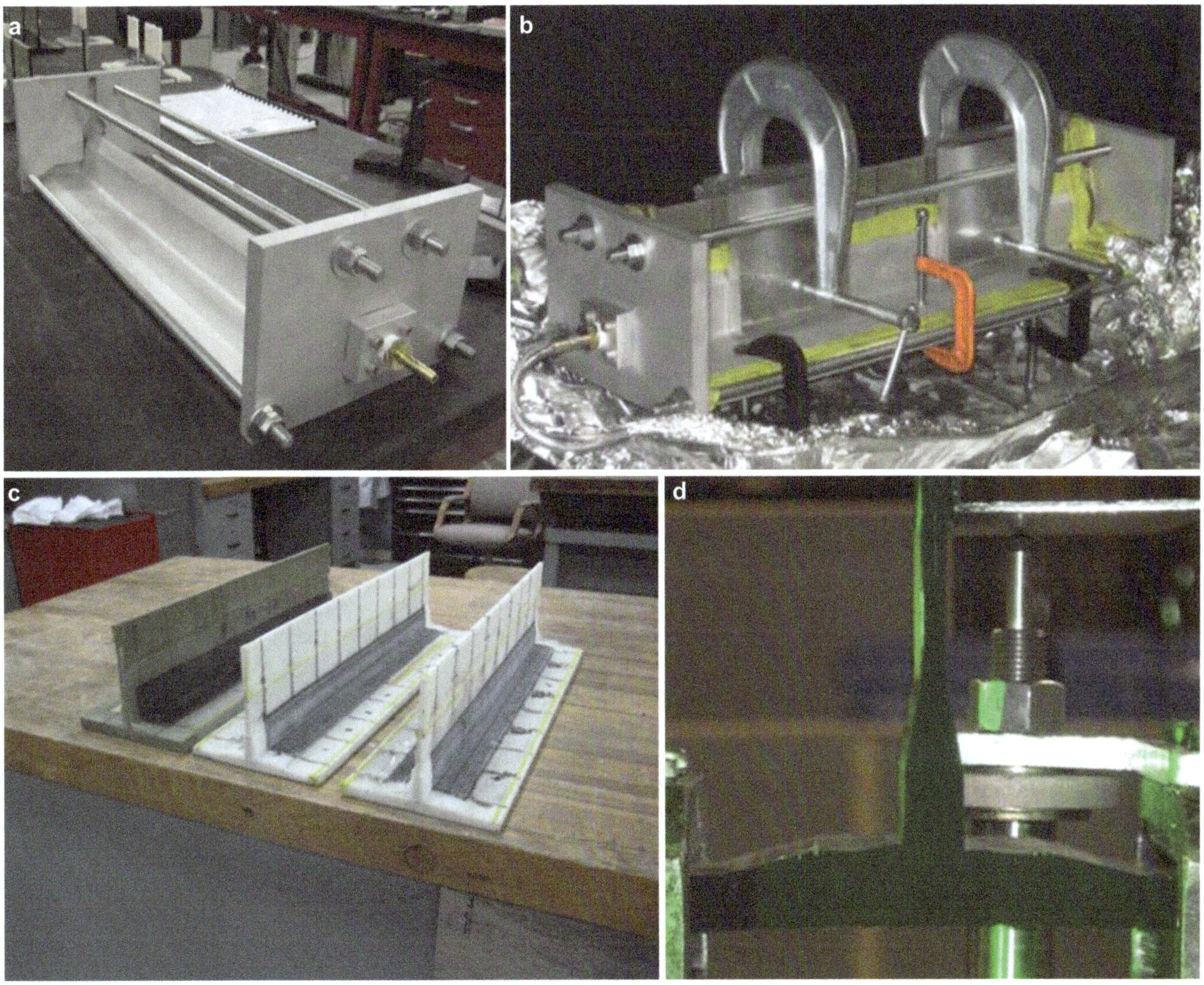

Fig. 2.2 Manufacturing of the Pi-joint using VARTM: (**a**) aluminum mold, (**b**) assembled pi-joint during VARTM, (**c**) completed pi-joints, and (**d**) a 50 mm wide section of pi-joint during a tensile pull-out test

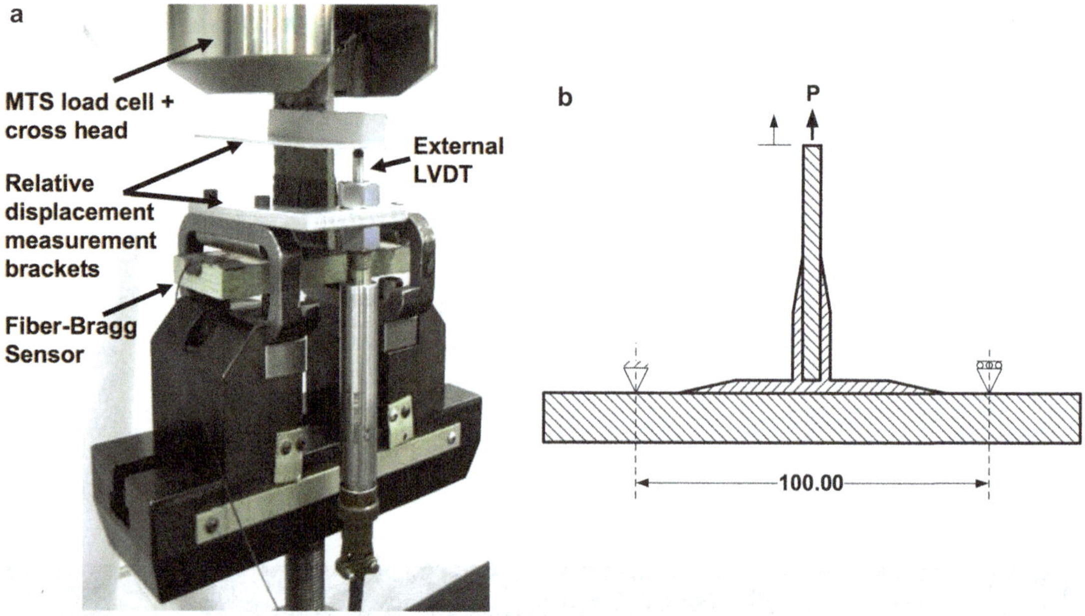

Fig. 2.3 Picture of a typical pull-out test setup (**a**); schematic of the experimental test setup (**b**)

and outlet are located just above the web-location of the Pi-preform. The permeability of the preform was found to be sufficient for successful manufacture of the Pi-joints, and hence, a distribution media was not required. The aluminum mold was fitted with end-plates and enclosed with vacuum tape prior to infusion as shown in Fig. 2.2c. The resin-infused joint was cured in a convection oven at 60 °C for 2 h and post cured at 94 °C for 4 h. The completed joints, which are shown in Fig. 2.2c, were then cut into sections of 50 mm wide using a water jet prior to being subjected to experimental testing as shown in Fig. 2.2d.

2.2.2 *Pull-Out and Damage Resistance (DIC)*

The out-of-plane behavior of the Pi-joints was evaluated by performing pull-out tests on the manufactured Pi-joints using a specially designed rig in a servo-hydraulic test machine, as shown in Fig. 2.3a. The specimens were tested at a loading rate of 1 mm/min until failure. The load and displacement from the test machine cross-head was recorded. Additionally, an external LVDT (linear voltage displacement transducer) was used to measure the relative displacement of the web with respect to the support.

In addition to the load and displacement values measured during the pull-out tests, a 3-D Digital Image Correlation system was employed to measure the full-field displacement and strains on the cut surface of the pi-joint. This was a commercially available DIC system (Dantec Dynamics Q-400) which consisted of a pair of cameras (AVT F-125B) along with a matched pair of lenses with a 12 mm focal length. Illumination was provided using a small panel of green LEDs. The cut surface of the Pi-joint was prepared by painting it first with a coat of matte white spray paint (Krylon matte white) and then misting with matte black (Krylon matte black) to provide the contrasting speckle pattern. Reference images for the correlation process were captured with a small pre-load applied to the joint, in order to remove any slack from the experimental setup. Images were then captured in 500 N increments up until failure of the joint occurred.

2.3 Numerical Simulations

An integrated approach to numerical simulations and experiments was taken with the intention of using simulations to facilitate the design of the experiments, sensor placement and also to support the interpretation of the experimental data, as well as creating a validated numerical model that could be used in future design work with a high-level of confidence. In this work, simulations of Pi-joint pull-out behavior were performed using a commercially-available finite element package, ABAQUS® [25].

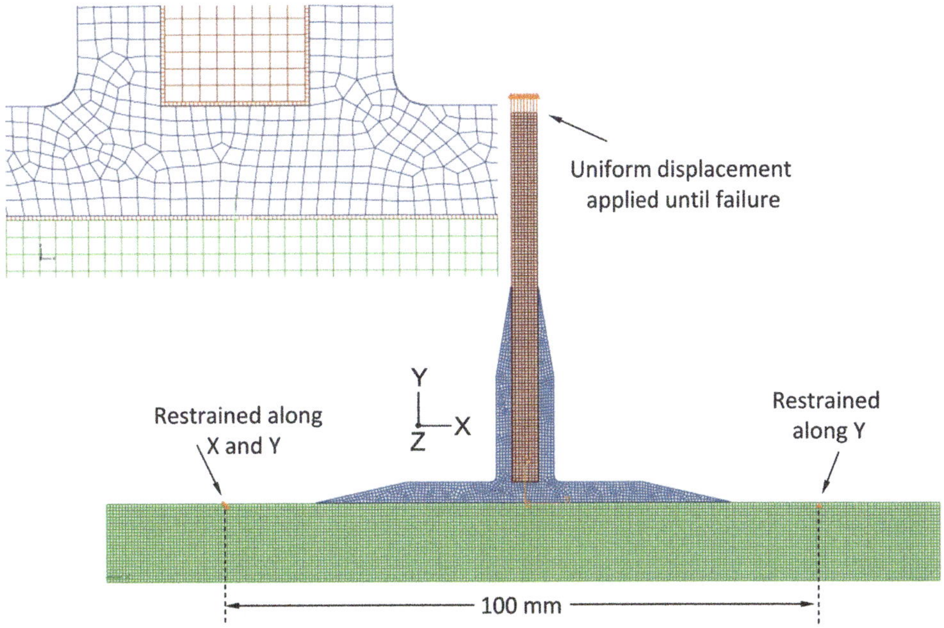

Fig. 2.4 Details of the FE model showing the boundary conditions and applied loading. The *inset* shows a more detailed view of the mesh with the quadrilateral elements and the adhesive elements

A two-dimensional plane-stress model of the Pi-joint was created in ABAQUS®. The web, flange and the pre-form were created as separate parts for ease of defining the properties and also the meshing. The adhesive was also created as a separate part but with a much finer mesh using cohesive elements. The adhesive surfaces were appropriately connected with respective surfaces of the web, flange and pre-form to assemble the model. Models that reproduced the three cases studied in the experiments were analyzed.

2.3.1 Mesh, Gap Creation and Material Models

Four-node quadrilateral plane-stress elements were used to mesh all the parts of the model in ABAQUS®. All parts were given an out-of-plane thickness of 50 mm corresponding to the width of the Pi-joint in the experiment. The flange, web and Pi-preform were modeled as linear elastic materials. An element size of 0.5 mm was used for all parts except the adhesive layers. A meshed finite element model for a Pi-joint is shown in Fig. 2.4. For adhesive layers, cohesive elements, the mesh was more refined as illustrated in the inset of Fig. 2.4. The element size of cohesive elements was equal to bond-line thickness (0.005 in., 0.127 mm), and an aspect ratio of 1 (unity) prior to load application. The adhesive was modeled in greater detail using cohesive elements and the delamination was simulated using a traction-separation material model [25]. The damage evolution was modeled in ABAQUS by defining the fracture energy as a function of the mixed modes using the analytical power law fracture criterion. The material properties used in the simulations are provided in Table 2.1.

2.3.2 Boundary Conditions and Loading

The loading pattern and boundary conditions used in the experiments were replicated in the finite element models. A schematic of the FE model is provided in Fig. 2.4. It has been reported that the number of degrees of freedom at the constraints and the angle of loading play a vital role in the accuracy of the results [6]. Kesavan et al. [6] simulated marine composite T-joints and applied the loading at an angle of 0.55° to the positive Y-axis in Fig. 2.4, to take into account manufacturing inconsistencies, and found good agreement with their experimental results. In this work, the web was assumed to be perfectly perpendicular and the angle of loading was not considered. The applied load was distributed along the top nodes in the web. The model was constrained symmetrically at two points on the flange simulating a simply-supported condition with a total span of 100 mm. The left-hand side support constraint does not permit translation in either the X- or Y-direction while the right-hand support constraint does not permit translation in the Y-direction. All simulations were performed on an eight-core 3.16 GHz machine with 64 gigabytes of RAM.

2.4 Results and Discussion

2.4.1 Experimental Pull-Out Results

Figure 2.5 provides the pull-out performance of the Pi-joints from the experiments. The general failure mode in the experiments for the Pi-joints consisted of an increasing load until failure, which led to the separation of the Pi-preform from the base. The failure was instantaneous and the direction of failure propagation could not be seen with the naked eye. A high-frame rate camera would be required to confirm the propagation path of failure. Since symmetric failure initiation was observed, the delamination propagation was also expected to be symmetric. Instead an asymmetric failure was observed. Such a discrepancy may be caused by a combination of eccentricity in load application and manufacturing flaws. The use of a brittle adhesive, like the one used in this work, makes the joint highly susceptible to manufacturing flaws and less tolerant to damage.

2.4.2 Digital Image Correlation Results

Digital Image Correlation was used to capture the displacement and strain on the cut surface of the Pi-joint as it was loaded in the tensile test. The images were processed using commercial software (Dantec Dynamics Istra 4D) to calculate the displacements and strains in the Pi-joint. A facet (or subset) size of 25 pixels was used, with an offset (or shift) of 15 pixels. The measured surface strain in the x-direction is shown in Fig. 2.7 at an applied load of 3 kN. The two red boxes indicate the areas for comparison with the FE results.

2.4.3 Numerical Simulation Results

The finite element (FE) simulations successfully modeled the complete pull-out response of the damaged and undamaged Pi-joints and the results are shown in Fig. 2.5. Figure 2.6 shows the failure mechanisms observed in the FE simulation. The initial onset of failure occurs in the web/Pi region (see inset in Fig. 2.6) and continues asymmetrically on one side of the web. This onset of failure does not reduce the load carrying capacity as the load transfer continues mostly across the other bonded

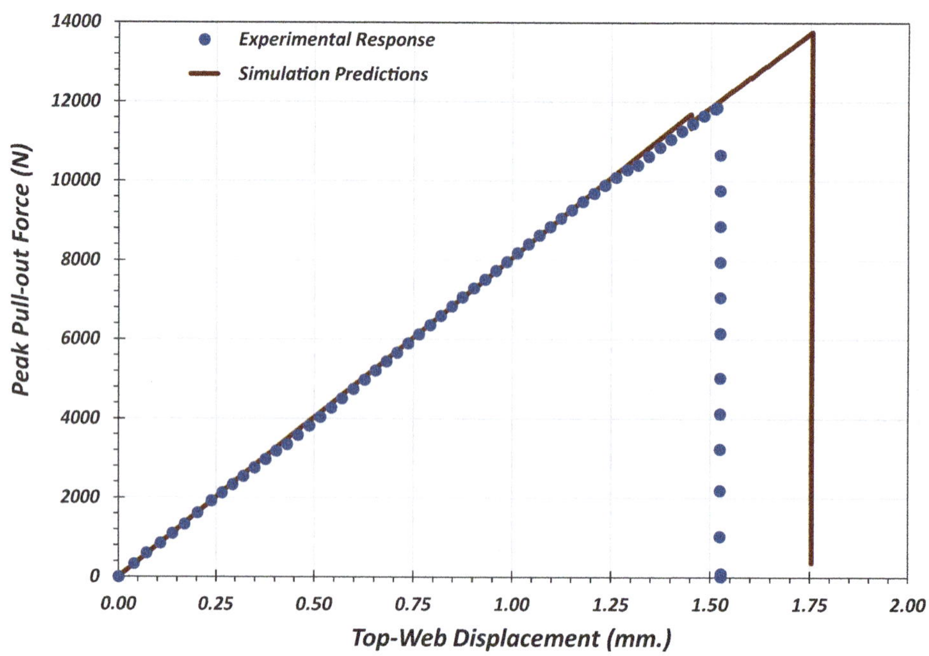

Fig. 2.5 Comparison of the complete pull-out response of the experiment and the FE simulation

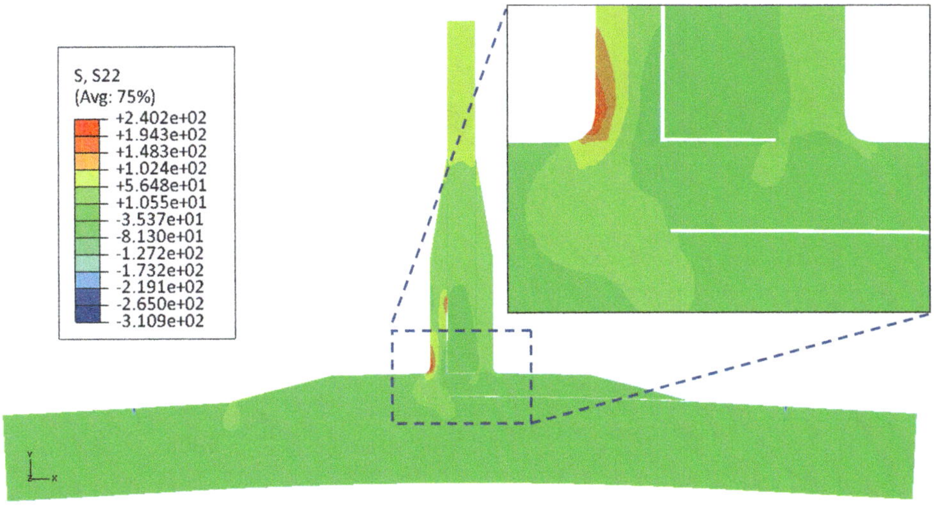

Fig. 2.6 Failure mechanism for the pi-joint in the simulation

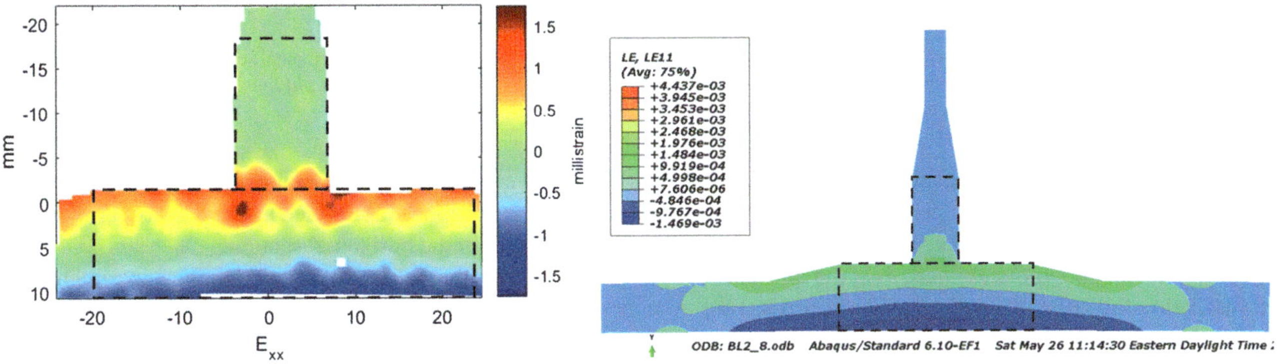

Fig. 2.7 Comparison of the principal strain from DIC (*left*) and FE (*right*). The highlighted boxes indicates where there the image decomposition comparison was performed

side of the web/Pi. As the applied load increases, the subsequent onset of failure occurs in the adhesive/bond between the preform and the base. This failure initiated from the termination of the preform with the base and continues until separation of the Pi-preform from the base.

The peak pull-out loads from the simulations agreed very well with the average experimental data as shown in Fig. 2.5. Similar load eccentricity and mis-alignment issues have been reported by Kesavan et al. [6] who applied the pull-out load at an angle of 0.55° counterclockwise to the vertical axis (web) in order to obtain better agreement between experimentation and computational results. Such modifications were not considered made in this study and all loads were applied to the web.

Additionally, the experimental data provided in Fig. 2.5 shows the standard deviation (as error bars) for each Pi-joint studied. The variation in experimental data may be due to many factors including manufacturing inconsistencies, variation in material properties of constituents, errors induced in experimental test-setup, instruments etc., Similarly, the numerical simulations assume average material properties and ideal test configurations.

2.4.4 *Comparison of DIC and FE Using Image Decomposition*

Although the pull-out force data matched well between the experiments and the simulation, it was desired to perform a more thorough validation of the simulation. Figure 2.7 shows the measured strain in the x-direction from the DIC experiment (left) compared to the x-strain found from the FE simulation (right). A recently published CEN document [26] recommends the use of image decomposition using Tchebichef polynomials to make quantitative comparisons of measured and predicted strain fields as part of a model validation process, This approach, which has been used previously to validate models of a

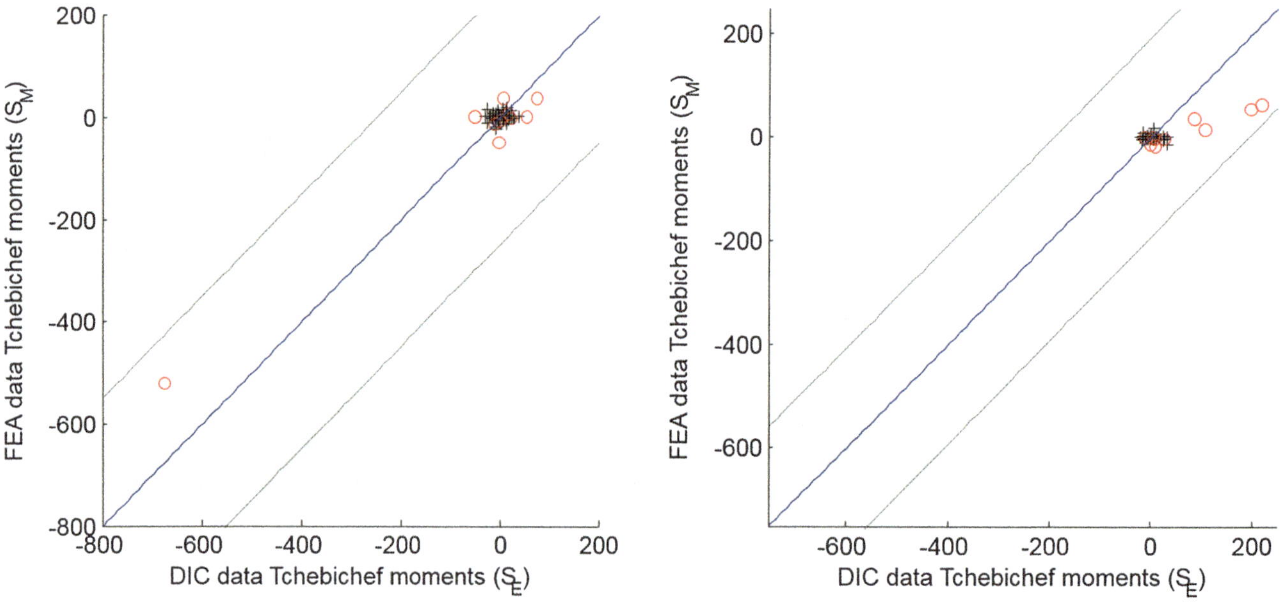

Fig. 2.8 Comparison plots of the most significant moments from the experiment and the simulation for the horizontal section (*left*) and the vertical section (*right*). If the data were in perfect agreement, all points would fall on a line with a gradient of unity (*solid line*) the dashed lines represent confidence bands equal to twice the uncertainty in the data from the experiments

vibrating plate [27] and a composite protective panel [28], was adopted here. The image decomposition using the discrete Tchebichef moments can only be performed over a rectangular domain, so the pi-joint was divided into two section for analysis. One rectangular patch was taken from the vertical section, and another from the horizontal section.

The Tchebichef polynomials were used to represent the strain component in the x-direction from both the experiment and the simulation following the procedure described in [28]. A reconstruction of the original strain image was performed from the Tchebichef coefficients to ensure that they were an accurate representation of the original data. In this fashion, 136 coefficients were used to represent the strain data from the vertical section and 176 coefficients from the horizontal section. The larger number of coefficients required to describe the horizontal section would indicate that the strain distribution was more complicated in this region.

The decomposition was performed on both the experimental and simulation data sets, producing a set of coefficients representing each strain field. These coefficients were then plotted against each other for the horizontal and vertical sections, as shown in Fig. 2.8. The experimental coefficients were plotted along the x-axis, and the simulation coefficients along the y-axis. If there were perfect agreement between the two sets of coefficients, the points would all lie along the line with gradient of unity (solid line in the figure).

Deviation from this line indicates differences between the experimental and simulation data sets. In reality, uncertainty in both the experiment and the simulation will cause deviations, and so it is unrealistic to expect all the points to fall on this line. For that reason, Fig. 2.8 has been plotted along with confidence bounds (dashed lines) which relate to the experimental uncertainty. The bands are defined by combining of the uncertainty from the experiment and from the image decomposition reconstruction process. For the case shown here, all of the moments fell within the confidence bounds. The points are primarily concentrated along the solid line, but there are a few moments close to the dashed lines, particularly in the comparison of the vertical section.

2.5 Conclusions

The performance of adhesively bonded structural composite Pi/T-joints was evaluated by experimental pull-out tests and numerical simulations. The values of the peak pull-out force recorded experimentally matched well with those from the numerical simulation (~10 % difference). The vacuum assisted resin transfer molding (VARTM) technique was found to be a successful technique to produce Pi/T-joints. 3-D digital image correlation was used to capture the displacements and strains in the pi-joint during the pull-out tests. The CEN procedure for making quantitative comparisons of measured and predicted strain fields was applied using image decomposition. The simulation was found to produce acceptable results within the confidence bounds provided by the uncertainty in the experiment. Experimentally validated simulations can be used to further understand the behavior of Pi-joints with other joint configurations and also to understand damage-induced behaviors beyond the case explored in this work, and such validated simulations have great potential in use as a design tool.

Acknowledgements This work was supported by the U.S. Army and ONR under TACOM/MSU Cooperative Agreement No. W56HZV-07-2-0001 and in collaboration with US Army TARDEC. The authors are thankful to Albany Engineered Composites Inc., NH, USA, for supplying the 3D-woven Pi-preforms. EAP was the recipient of a Royal Society Wolfson Research Merit Award.

References

1. Stickler, P.B., Ramulu, M.: Investigation of mechanical behavior of transverse stitched T-joints with PR520 resin in flexure and tension. Compos. Struct. **52**, 307–314 (2001)
2. Flansburg, B.D., Engelstad, S.P.: Robust design of composite bonded Pi joints. 50th AIAA/ASME/ASCE/AHS/ASC Structures, Structural Dynamics, and Materials Conference. Palm Springs, CA (2009)
3. Dulieu-Smith, J.M., Quinn, S., Shenoi, R.A., Read, P.J.C.L., Moy, S.S.J.: Thermoelastic stress analysis of a GRP Tee Joint. Appl. Compos. Mater. **4**, 283–303 (1997)
4. Marcadon, V., Nadot, Y., Roy, A., Gacougnolle, J.L.: Fatigue behavior of T-joints for marine applications. Int. J. Adhes. Adhes. **26**, 481–489 (2006)
5. Li, H.C.H., Dharmawan, F., Herszberg, I., John, S.: Fracture behavior of composite maritime T-joints. Compos. Struct. **75**, 339–350 (2006)
6. Kesavan, A., Deivasigamani, M., John, S., Herszberg, I.: Damage detection in T-joint composite structures. Compos. Struct. **75**, 313–320 (2006)
7. Shenoi, R.A., Violette, F.L.M.: A study of structural composite tee joints in small boats. J. Compos. Mater. **24**, 644–666 (1990)
8. Shenoi, R.A., Hawkins, G.L.: Influence of material and geometry variations on the behaviour of bonded tee connections in FRP ships. Composites **23**, 335–345 (1992)
9. Hicks, I.A., Read, P.J.C.L., Shenoi, R.A.: Tensile compressive and flexural characteristics of tee-joints in foam-cored sandwich structures. Proc. 3rd Int. Conf. Sandwich Construction, 579–590 (1996)
10. Shenoi, R.A., Read, P., Jackson, C.L.: Influence of joint geometry and load regimes on sandwich tee joint behaviour. J. Reinf. Plast. Compos. **17**, 725–740 (1998)
11. Dulieu-Barton, J.M., Earl, J.S., Shenoi, R.A.: Determination of the stress distribution in foam-cored sandwich construction composite tee joints. J. Strain Anal. Eng. Des. **36**, 545–560 (2001)
12. Toftegaard, H., Lystrup, A.: Design and test of lightweight sandwich T-joint for naval ships. Compos. Part A: Appl. Sci. Manuf. **36**, 1055–1065 (2005)
13. Theotokoglou, E.E., Moan, T.: Experimental and numerical study of composite T-joints. J. Compos. Mater. **30**, 190–209 (1996)
14. Theotokoglou, E.E.: Study of the numerical fracture mechanics analysis of compositeT-joints. J. Reinf. Plast. Compos. **18**, 215–223 (1999)
15. Dodkins, A.R., Shenoi, R.A., Hawkins, G.L.: Design of joints and attachments in FRP ships' structures. Mar. Struct. **7**, 365–398 (1994)
16. Theotokoglou, E.E.: Strength of composite T-joints under pull-out loads. J. Reinf. Plast. Compos. **16**, 503–518 (1997)
17. Rispler, A.R., Steven, G.P., Tong, L.: Failure analysis of composite T-joints including inserts. J. Reinf. Plast. Compos. **16**, 1642–1658 (1997)
18. Phillips, H.J., Shenoi, R.A.: Damage tolerance of laminated tee joints in FRP structures. Compos. Part A: Appl. Sci. Manuf. **29**, 465–478 (1998)
19. Banea, M.D., da Silva, L.F.M.: Adhesively bonded joints in composite materials: an overview. J. Mater. Des. Appl. **223**, 1–18 (2009)
20. Haq, M., Patterson, E.A., Drzal, L.T.: Pull-out behavior of adhesively bonded composite Pi-joints: damage modeling and comparison with experiments. Proceedings of the 25th Annual Technical Conference, American Society of Composites, Dayton, OH (2010)
21. Online: Air Force Print News. http://www.wpafb.af.mil/news/story_print.asp?id=123035046 (2009). Accessed 16 Sept 2009
22. Russell, J.D.: Composites affordability initiatives. Adv. Mater. Process. **165**, 29–32 (2007)
23. Haq, M., Conway, A., Patterson, E.A.: A study on damage-induced behavior of bonded Pi-/T-joints using digital image correlation (DIC) and FE modeling, 2012 SEM XII, International Congress and Exposition on Experimental and Applied Mechanics, Costa Mesa, CA (2012)
24. Sebastian, C., Haq, M., Conway, A., Patterson, E.A.: Comparison of FE and DIC measurements of a Pi-joint using feature vectors, 2012 SEM XII, International Congress and Exposition on Experimental and Applied Mechanics, Costa Mesa, CA (2012)
25. Simulia, I.: Abaqus/CAE Version 6.8-2 (2008)
26. CWA16799. Validation of Computational Solid Mechanics Models. Comite European de Normalisation, Brussels (2014)
27. Wang, W., Mottershead, J.E., Ihle, A., Siebert, T., Reinhard Schubach, H.: Finite element model updating from full-field vibration measurement using digital image correlation. J. Sound Vib. **330**(8), 1599–1620 (2011)
28. Sebastian, C.M., Patterson, E.A., Ostberg, D.: Comparison of numerical and experimental strain measurements of a composite protective panel using image decomposition. Appl. Mech. Mater. **70**, 63–68 (2011)
29. Bucinell, R.B.: Composite materials fatigue and fracture, vol. 7. STP-1330 ASTM, PA (1998)
30. Truong, T.C., Vettori, M., Lomov, S., Verpoest, I.: Carbon composites based on multi-axial multi-ply stitched preforms: part 4, mechanical properties and damage observation. Compos. Part A: Appl. Sci. Manuf. **36**, 1207–1221 (2005)

Chapter 3
5xxx Aluminum Sensitization and Application of Laminated Composite Patch Repairs

Daniel C. Hart

Abstract Sensitization can occur in 5xxx aluminum alloy utilized in welded ship structures. Cumulative exposure of plating to elevated temperatures causes magnesium in the aluminum to migrate to the grain boundaries and develop a corrosion cell. As a result, the sensitized plate is susceptible to intergranular corrosion, which when exposed to salt water and stress leads to stress-corrosion cracking (SCC), typically at stress levels below nominal plate strength. SCC typically occurs at a level of sensitization that renders welded repairs impractical, requiring an alternative methodology. One alternative to traditional welded repair of marine structures is the application of a composite patch. Composite patch repairs have seen significant use by the aerospace community on thin aluminum aircraft skins to address stress related cracking while only limited use on thicker marine structures. Patches for marine structures have demonstrated their ability to mitigate crack growth, inhibit water ingress, and reinforce damaged or degraded structure for commercial and military vessels. The use of composite patches for repair of SCC onboard U.S. Navy ships began in December of 2010. Patch design and installation are supported by ASTM material testing, environmental conditioning, large center crack tension fracture mechanics specimens, and experience with a full scale plate ductile tearing test. Patch repairs patches are installed in-situ with vacuum consolidated hand laminations and have demonstrated the ability to increase static plate strength, fatigue life, large displacement capacity of the crack plate, and durability for in-service applications.

Keywords Composite • Sensitization • Aluminum • Fatigue • Stress corrosion cracking

3.1 Background

Aluminum became a common material for use in ship structures in the 1960s with destroyers for the Netherlands Navy, frigates of the British Royal Navy, and U.S. Navy vessels from patrol boats to frigates. Early issues included galvanic corrosion of plating resulting in exfoliation and fatigue cracking of aluminum superstructures on steel hulls. These issues were corrected by updating the temper of the alloys. Recent times have seen an increasing number of Navy and maritime ships constructed of welded aluminum, either hull or topside, or in the case of high speed catamarans and the independence class of littoral combat ships (LCS) both the hulls and topside structures. Marine grade 5xxx aluminum alloys are used because of their corrosion resistant and residual weld strength characteristics. These enhanced characteristics are developed in part by the magnesium content of the alloy, which is typically greater than 3 %.

Ships and superstructures constructed of welded aluminum have generally proven to be rugged while in service, and easily modified for alterations and modernizations. Damage to these structures occurs in service from accidental causes, battle damage, fatigue fracture, corrosion, or sensitization of aluminum structure. The primary method for repair of aluminum structure requires welding and material replacement, though decreasing maintenance budgets and increased operational requirements are increasing the demand and frequency of lower cost alternatives. An additional driver for alternative repair methods is the fact that high magnesium content 5xxx alloys are susceptible to sensitization in months to years when subjected to elevated temperatures [1]. Then, starting in the early 2000s sensitization was identified as a significant issue for high speed aluminum marine vessels [2] and for the U.S. Navy's CG-47 class ship superstructures [3]. A high degree of sensitization makes welding impractical or infeasible. This is where repairs such as composite patches and bonded aluminum repairs have filled the gap to support fleet needs.

D.C. Hart (✉)
Naval Surface Warfare Center, Carderock Division, 9500 MacArthur Blvd., West Bethesda, MD 20817, USA
e-mail: daniel.c.hart@navy.mil

G.L. Cloud et al. (eds.), *Joining Technologies for Composites and Dissimilar Materials, Volume 10*
Conference Proceedings of the Society for Experimental Mechanics Series, DOI 10.1007/978-3-319-42426-2_3

Fig. 3.1 Sensitized grain structure and the results of stress corrosion cracking

Sensitization is the cumulative exposure of 5xxx aluminum to elevated temperatures causing the magnesium in the alloy to migrate to the grain boundaries. This migration eventually develops a corrosion cell when the magnesium precipitate forms a continuous network around the grains considered intergranular corrosion. Exposing the intergranular corrosion to salt water and stress, levels typically well below nominal plates strength, results in stress-corrosion cracking (SCC). SCC typically occurs when plate becomes highly sensitized, and in some cases the level of sensitization makes welding not possible. In the most severe cases SCC resulted in loss of a small section of deck plating or large bifurcating cracks adjacent to recent weld repairs as shown in Fig. 3.1.

The level of sensitization currently governs the repair method used to address damage or cracking. Several methods exist to determine the level of sensitization in the material. Detection of the sensitization level can be performed using a destructive test such as the ASTM G67 mass loss [4], or with non-destructive means such as detailed visual inspection or the Degree of Sensitization (DoS) probe, the later measures electric current due to a localized chemical reaction [5]. All methods produce a value comparable to the G67 mass loss. Aluminum repair guidance exists for three ranges of mass loss values as shown in Fig. 3.3. Starting on a plate where simple welding can be performed, as the sensitization increases pre-welding cold working of the material must be performed, and finally component replacement when the sensitization has progressed far enough welding is not permitted [5, 6]. Additionally, alternative methods are needed when welding is not permitted due to time or safety concerns, such as proximity to combustible liquid or gas or other ignitable material.

One alternative to traditional welded repair of marine structures is the application of a composite patch. Composite patch repairs of metallic aircraft structures were pioneered in the 1970s and 1980s with work in both Australia and the US [7–11]. Development of aircraft patch repairs continued through mechanical testing, design method development, and analysis efforts [12–16]. Typical repairs were performed on thin aluminum aircraft skins to address stress related cracking. Composite patch repairs of marine structures have seen increasing use, demonstrated their ability to mitigate crack growth, inhibit water ingress, and reinforce damaged or degraded structure. Allan and Clark examined the strength of carbon-epoxy patches applied to the fatigue or over stressed areas of the aluminum superstructure on British Royal Navy Type 21 Frigates. Repair performance was backed up with fatigue testing of butt-joints and welded aluminum joints with and without double sided composite patches. Cracks were weld repaired followed by application of a carbon epoxy patch for six ships in the class; patch repairs were in service for 10 years prior to sale [17]. The Royal Australian Navy experimented with carbon fiber vinyl ester composite patches on FFG-7 class frigates to address an overstressed region of the deckhouse [18, 19]. Composite patches were large, 5 m × 1 m × 1.25 cm thick, and were designed to carry longitudinal seaway loads at the top level of the deckhouse at the midship dog-bone section. Fiber, resin, and, surface preparation procedure selections were made following material characterization, environmental conditioning, Boeing wedge testing, and lap shear testing efforts. After 15 years of service, maintenance, and repairs the composite patches were considered a success [20]. Carbon composite patches designed with cross sectional stiffness matching that of the base metal were applied to the UK's Royal Navy Type 42 Destroyers, and also commercial off-shore oil platforms. Patch repairs of steel food tray lift shafts onboard the Type 42 destroyers offered a cost and time efficient repair method to address the cracking. Seven ships and 35 patches were applied using hand layup, infusion, and pre-preg installation methods performed successfully in restricted spaces [21]. Following these successes, QinetiQ installed composite patch repairs to floating production storage and offloading (FPSO) oil ship platforms, especially since a weld repair or "hot-work" is dangerous and restricted onboard FPSO's. Performing hot-work requires gas free environment and oil free tanks for a minimum of two bulkheads in any direction, which is a costly endeavor. In contrast,

Fig. 3.2 U.S. Navy composite patch repair examples

composite patch repairs do not require hot-work, and in 2002 QinetiQ repaired the steel structure of an FPSO ballast tank with an adhesively bonded pre-preg repair [21]. In the early 2000s, Det Norske Veritas (DNV) researched composite patch repairs of FPSO structure to address corrosion and fatigue cracking. This resulted in current DNV recommended practice for design, fabrication, operation, and qualification of bonded repairs on steel structure [22–24]. More recently, as shown in Fig. 3.2, U.S. Navy patches have been installed and in service on aluminum superstructure of 11 ships starting in December of 2010 [5, 25].

3.2 Sensitization and Detection

Magnesium content has been identified as the root cause of 5xxx aluminum being susceptible to sensitization. Marine grade alloys have magnesium contents that range from 2.2 to 5.5 %, those that have greater than 3 % magnesium content are super saturated and allow small amounts of magnesium to precipitate to the grain boundaries as temperature exceeds 40–65 °C [1, 5]. From a simple sail plane study reported by Greene, solar heating combined with ambient temperature and surface color can drive surface temperatures beyond 50 °C [26]. Ship structure operating in tropical environments consistently achieve surface temperatures at which sensitization begins. Enough time at temperature leads to the formation of a continuous film of beta phase (Mg_2Al_3) at the grain boundaries. Beta phase is anodic, and when exposed to moisture a corrosion cell leads to intergranular corrosion, and the plate becomes brittle [6]. Applying stress to the degraded plate results in SCC at stress levels below design limits and when subjected to fatigue stresses the crack growth rate increases [27]. Stress can be from solar or thermal loads, longitudinal bending from sea way motion, local equipment or systems, and welding. Performing a weld repair of SCC requires knowledge of the level of sensitization.

Detection of plate sensitization level can be performed using a destructive test or with non-destructive means such as detailed visual inspection or the Degree of Sensitization (DoS) probe. The primary destructive method for measuring the DoS is ASTM G67 nitric acid mass loss test (NAMLT), which measures the amount of specimen mass lost in mg/cm^2 [4]. Mass loss is measured using nitric acid to dissolve the beta phase from a 50 mm by 6 mm specimen. Specimens are typically extracted from shipboard material using a hole saw leaving 75 mm diameter or larger hole in the structure that must now be repaired. Visual techniques are being developed to optically measure the amount of beta phase particles at the plate surface, and estimate a mass loss value. Initial efforts successfully matched probabilistically processed digital image analysis results of laboratory prepared specimens to ASTM G67 measurements [28]. One of the non-destructive methods requires etching and polishing a 25–50 mm square area to a 0.5 μm finish, and utilizing an image magnification in the ×400–600 range to show clear contrast between grain and beta-phase material. Without the ability to digitally analyze the image onboard the ship, a catalog of images with known sensitization levels is being compiled for in-situ identification of the sensitization level [5]. Lastly, another non-destructive method that is being employed at the RMC's uses a portable non-destructive degree of sensitization (DoS) probe. This method requires removal of surface coatings and abrasion of a 100 × 100 mm area of the surface with 1200 grit abrasive. The DoS probe measures the electric current due to a localized chemical reaction as nitric acid dissolves the surface beta-phase [29]. The results have been calibrated for 5456 aluminum and are currently used by repair and maintenance center (RMC) personnel to determine the degree of sensitization. Continued development is underway to calibrate the results from the DoS probe for 5083 aluminum [5].

3.3 Mitigation and Repair of Sensitized Aluminum

Knowledge of the DoS governs the scope and methodology for repairing the affected structure. Typical repair methods implemented on marine structures utilize metal inert gas (MIG) welding techniques. Repair scope ranges from direct welding of a small crack to large scale crop out and complete replacement with new material. For some repairs cold working of the surrounding material prior to welding can improve residual strength of the repair. When welding is not an option, an alternative repair method such as mechanically fastened or adhesively bonded reinforcement is needed. Limitations and guidance on repair methods (Fig. 3.3) are based on the DoS from G67 test data, or alternative equivalent to mass loss testing. G67 material mass loss values below 30 mg/cm^2 are considered weldable, from 30 to 60 mg/cm^2 are weldable but require cold working, and above 60 mg/cm^2 indicates the material is not weldable and requires an alternative repair method [5].

Repair methods include cold working the aluminum, advanced welding processes to reduce the heat imparted on the plating, and adhesively bonding metallic or composite reinforcements to the affected structure. Finally, new and updated aluminum materials have been developed to reduce or eliminate sensitization such as the H128 temper.

3.3.1 Weld Repair

Sensitization levels below 30 mg/cm^2 can be repaired with welding. Basic repair of cracks can be performed by grinding the crack and re-welding. Often the search for weldable material requires cropping areas of highly sensitized plate and replacing the removed material with new plate and welding that to the existing structure. Repair material typically matches the existing alloy and is either the same thickness or thicker to mitigate sensitization. New or advanced materials developed to resist sensitization can be substituted as replacement plate.

Research into sensitization and material processing has lead to the development of a new temper, H128, registered with the Aluminum Association for 5083 alloys but can also be produced as 5456 alloys [5]. The H128 temper was developed to extend the new ASTM B928 specification, which only sets requirements for "as delivered" plate, and does not include requirements related to sensitization rates. Bushfield et al. discussed the background of establishing the ASTM B928 specification [2]. Cracking and corrosion that occurred on high speed aluminum vessels was directly linked to the use of plate manufactured to ASTM B209 specifications but delivered in a sensitized condition. The new standard, B928, requires as delivered plate to pass both ASTM G66 and G67 tests. This, however, only qualifies the as-delivered plate and does not specify sensitization rate during service. The updated temper, H128, extends ASTM B928 requirements to include a 7 day 100 °C treatment prior to the ASTM testing and also meets all of the H116 temper requirements [5]. Designing the temper to meet these strict requirements should limit the rate of sensitization under normal ship operating conditions.

Another alternative material to prevent sensitization is the recently developed Novelis Fusion plate produced by Novelis Inc. using their patented casting method (US Patent 7,472,740). The patented process produces a single cast aluminum ingot sandwich with outer layers of low magnesium content alloy and high magnesium content 5456 alloy core. This ingot can then be rolled to produce plate of the desired thickness. Using a sandwich structure isolates any beta phase that forms in the core material from the corrosive marine environment. Both the Novelis Fusion and the H128 temper plate can be welded to existing plate with existing welding methods.

3.3.2 Cold Worked Welded Repair

Mass loss measurements between 30 and 60 mg/cm^2 require cold working of the material prior to accomplishing a weld repair. One cold working method currently employed is ultrasonic peening, which uses ultrasonic impulses to drive an indenter onto the aluminum surface resulting in a refined micro structure and imparting residual compressive stress in the plate.

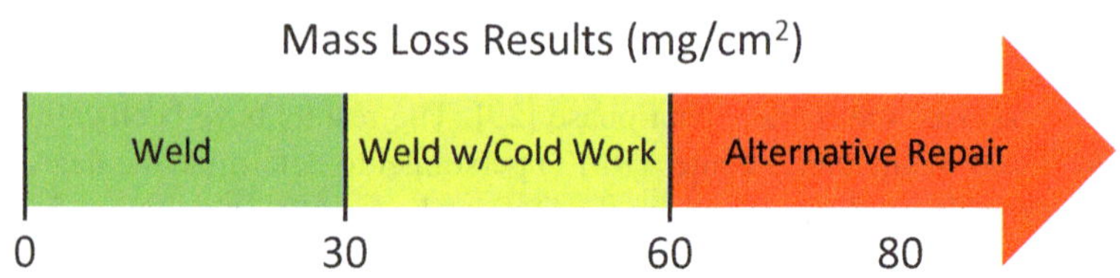

Fig. 3.3 5xxx repair recommendations based on G67 mass loss test results

Though this is typically performed only on one side of the plate, residual stress has been measured at depths from the surface of up to 10 mils. The improved micro structure and the altered stress state near the weld reduce the residual stress present after the weld cools and have been shown to prevent re-cracking [5]. From experience, this method requires careful planning and can induce cracking when not performed correctly or in locations where bending loads can reverse the stress state of the surface where cold worked stresses can add to the stress state.

3.3.3 Un-Weldable Material

Guidance for 5xxx aluminum defines the mass loss welding limit as 60 mg/cm^2. At or above this limit welding is not permitted, though through experience is often attempted and results in cracking at or in close proximity to the weld. When shipboard plating is not weldable the currently approved alternative methods are application of an E-Glass epoxy patch or a temporary adhesively bonded aluminum repair (BAR) kit applied by ship's force. Both methods are designed to reinforce the region ahead of the crack tip, carry loads across the crack plane, and restore weather tight integrity of the plate.

3.4 Alternative Repair Methods

Two primary alternative repair methods developed for U.S. Navy use since 2010 are the composite patch and bonded aluminum repair (BAR) kit. Composite patch repairs are the primary method for repairing non-weldable aluminum. Application requires trained and qualified personnel, the developed technique requires detailed surface preparation procedures, and advanced lamination techniques involving vacuum consolidation of the hand laminated composite. As experience grows there are composite patches that have seen over 5 years of service, continue to pass inspections, and receive approval for continued use. In parallel with composite patch research and implementation, BAR kits were developed to address emergent cracking while underway and can be applied by Navy sailors. These patches have been applied to areas experiencing low stresses during typical operation.

3.4.1 Bonded Aluminum Repair

The current surface preparation and lamination procedures associated with composite patch repairs are not suitable for shipboard personnel, this is due to the level of training and experience required to achieve a consistent repair quality, and the expense of hazardous material handling and storage requirements. As a result, a Bonded Aluminum Repair (BAR) kit was developed to provide a temporary repair method that effectively mitigates crack growth, restores weather-tight integrity, and allows ship's personnel to apply the repair while underway. BAR kits consist of malleable 11 gauge 5052 O-temper plates, methyl-methacrylate (MMA) adhesive cartridges with dispensing equipment, abrasive surface preparation tools, and disposables required for repair application. The steps required to install the kit are shown in Fig. 3.4. First, basic surface abrasion

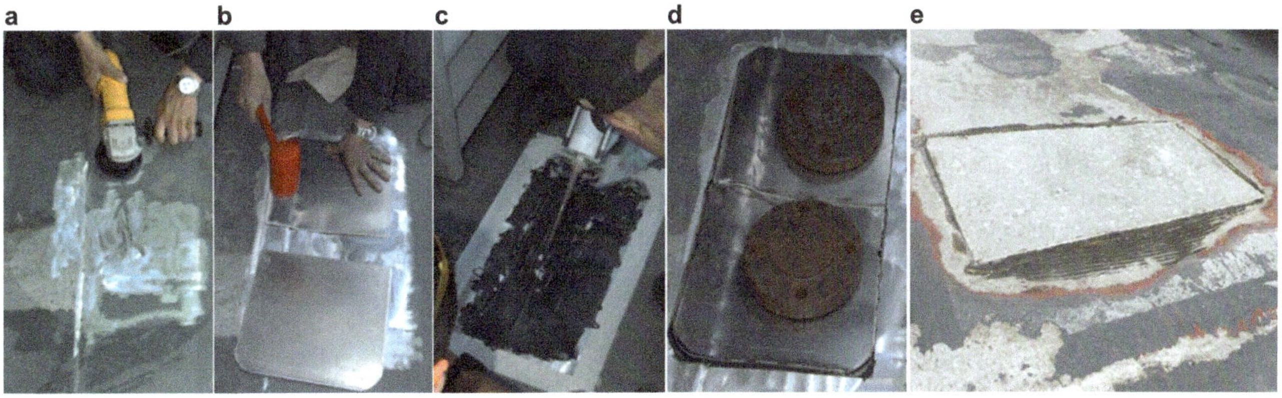

Fig. 3.4 BAR installation process: surface preparation (**a**), initial plate forming (**b**), adhesive application (**c**), final BAR repair step using weights to apply pressure (**d**), and BAR after 6 months in service (**e**)

is performed to remove paint and coatings (a). The O-temper plates are easily formed to the shape of the deck (b) and the MMA adhesive is applied to the deck (c) and then cures with simple fixturing methods (d). The MMA achieves a durable bond with minimal surface preparation. The BAR method achieved similar fatigue performance benefits to the composite repairs; However upon base plate failure, the BAR plate immediately cracked and did not carry additional load. As will be discussed below, the composite patches remained intact after the base structure aluminum failure, and continued to carry load. Though BAR is meant to be temporary, there are several repairs that have been in service for over 3 years (e).

3.4.2 Composite Patch Repair

In cases where welded repairs are not practical for immediate accomplishment such as fuel tank repairs, which must be first emptied and gas-freed, or the case of highly sensitized plate, alternative repair methods are necessary. Other potentially dangerous conditions which would preclude immediate welded repairs include ammunition spaces, locations with equipment that must be temporarily removed to permit access to the back side of the structure, or the lack of trained personnel. Deck repair using a composite patch offers a significant repair cost and safety benefit from the fact that applying the patch does not require "hot work", and can be applied from one side. Hot work refers to torch cutting and welding tasks that require a fire watch and a clear area on both sides of the repair, and when in the vicinity of a flammable substance requires the spaces to be gas free.

Strength and fatigue life benefits of the composite patch repair method have been well documented. Static testing of edge notch panels repaired with a composite patch demonstrated a significant increase in panel strength with a thin composite patch across the crack tip. Testing results showed that a single layer of unidirectional E-Glass increased ultimate strength of 6.35 mm thick specimens by 33 and 37 % for 3.18 mm inch thick aluminum plate [30]. Testing 6.35 mm thick aluminum center crack tension specimens with stiffness-matched bonded unidirectional boron patches to study effects of a composite patch dimensions on the aluminum plate fatigue life. Results showed no benefit of increased patch length perpendicular to the crack plane. A critical finding was that an elliptical crack front was developed during fatigue loading which lagged by up to 10 mm from the patched to un-patched surfaces [13]. Fatigue performance benefits were also demonstrated for three configurations of 3-ply unidirectional boron-patched, stiffened 1 mm. thick aluminum plates were tested to study the tension-tension fatigue crack growth rate: cracked plate, cracked stiffened plate with 101.6 mm spacing, and cracked stiffened plate with 152.4 spacing. FEA was utilized to calculate the Stress Intensity Factor (K) and modeled the crack growth using Paris Law coefficients for the underlying plate. Test and analysis results showed that the composite patches decreased the crack growth rate and increased the fatigue life of the repaired panels [12]. Over the past 5 years, the Navy has utilized thin low modulus E-glass epoxy composite patch repairs on 12 ships. Patches have demonstrated their ability to mitigate crack growth, inhibit water ingress, and maintain a weather tight compartment. These patches were installed in-situ with vacuum consolidated hand laminations [5, 25]. Laminate design and installation procedures were supported by laboratory lamination trials, ASTM standard material testing, environmental conditioning, large center crack tension fracture mechanics specimens, and experience with a full scale plate ductile tearing test. Patch repairs increased the fatigue life of the center crack tension (CCT) plates 4×, 6×, and 10× for three far-field stress levels [31]. Additional performance capability was demonstrated on a full scale ductile tearing specimen subjected to high stress and durability was demonstrated on patched aluminum plates subjected to dynamic impact.

3.5 Composite Patch Strength and Durability

Surface preparation, lamination procedure, and material selection were supported by laboratory lamination trials and material testing. The surface preparation method that achieved the highest lap shear strength involves mechanical abrasion of the surface followed by a phosphoric acid scrub and application of an acetic acid based coupling agent. Vacuum schedule, peel ply, perforated film, and bleeder cloth selection were based on post lamination quality assurance testing, such as fiber volume fraction and void content, and limited in-plane mechanical testing. Additionally, resin gel time studies were performed to understand behavior at various temperatures experienced at repair facilities around the world. ASTM standard tests, performed on material coupons pictured in Fig. 3.5, were used to measure elastic-plastic resin behavior, lamina and laminate in-plane mechanical properties, interface ply fracture behavior, and long term durability of environmentally exposed samples. Baseline composite patch laminates are nominally 6.44 kg/m^2 (1.32 lbs/ft^2) with a fiber volume fraction (FVF) of 32.5 %, an elastic modulus of 11.7 GPa, tensile strength of 181.3 MPa, and compressive strength of 214.4 MPa.

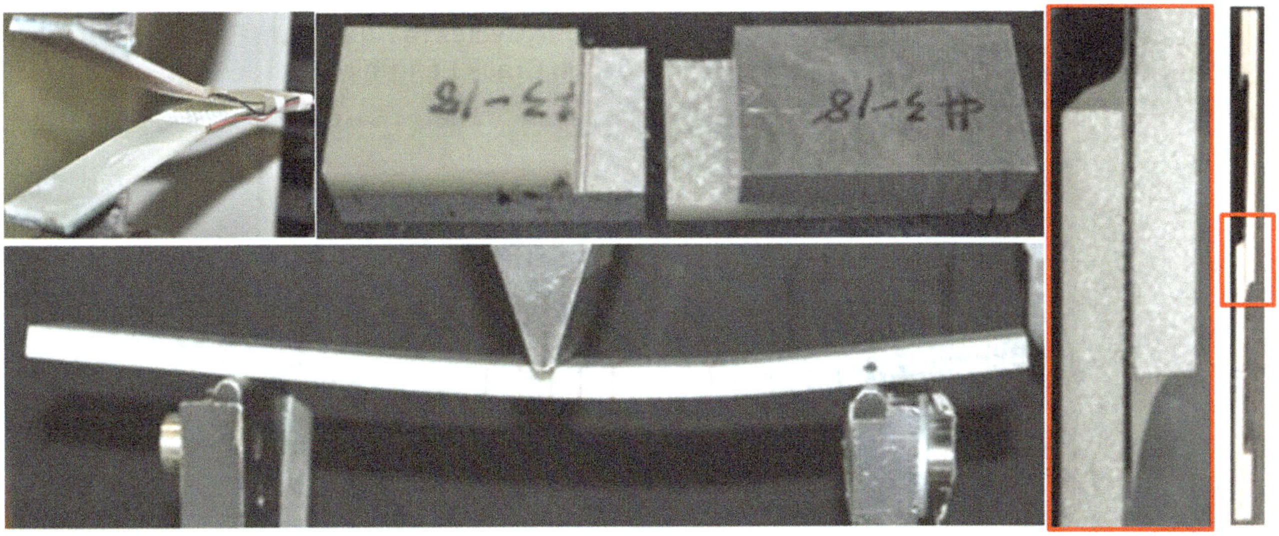

Fig. 3.5 Material testing specimens. Clockwise from upper right: mode I fracture, compressive notch shear, lap shear, and mode II fracture

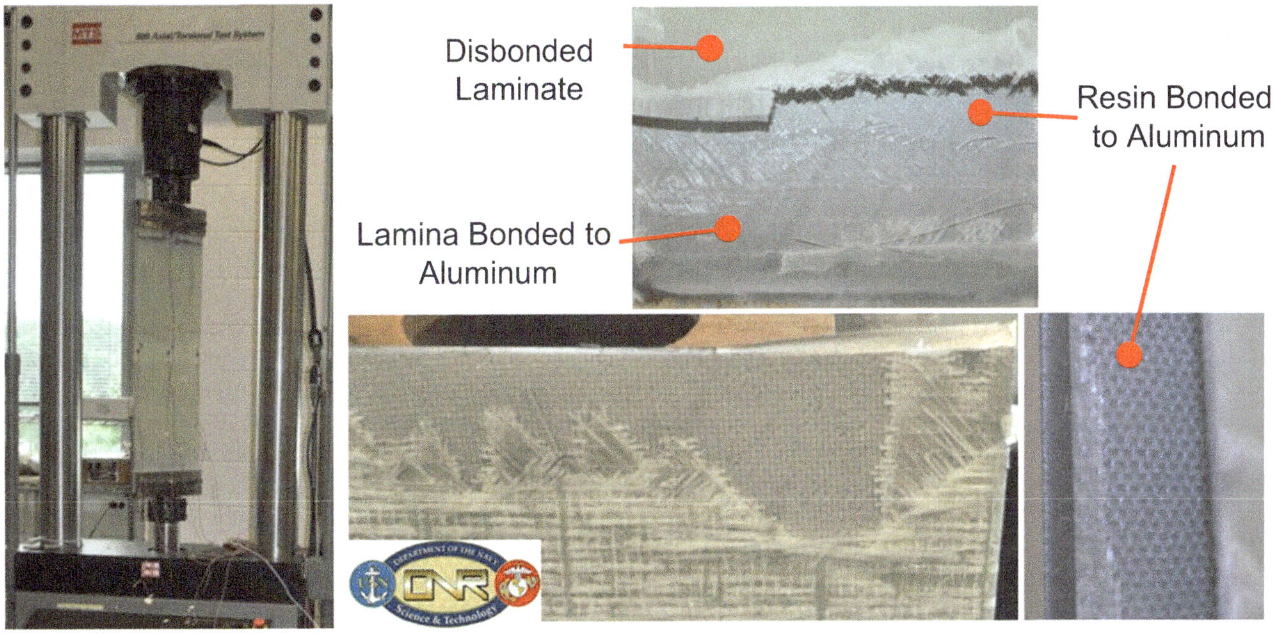

Fig. 3.6 Load frame with CCT test specimen (*left*). Failure surfaces typical of the 100.3 MPa far-field stress (*right*)

Center crack tension (CCT) tests were used to demonstrate the design method and resulting laminate configuration for the thin E-glass epoxy patch repairs performed on U.S. Navy ships. Though reinforcement length relative to the crack plane does not affect crack growth, initial CCT specimens had composite patches that extended 15 in. from the crack plane in each direction to represent the configuration of repair patches installed on ship structure. Minimum recommended patch length perpendicular to the crack is 40 cm for sensitized plate and often extend further past the nearest structural detail, such as a stiffener or frame below the deck being repaired. The CCT configuration shown in Fig. 3.6 was tested for three far-field stress levels; working stress, above the Eurocode 9 fatigue design stress limit, and a stress level resulting in plastic deformation at the crack tip. For two of three applied stress levels tested, 34.5 and 75.2 MPa, the composite patch continued to carry the applied load with only minor delamination adjacent to the crack, even after the aluminum crack extended the full width of the plate. Composite patch disbond failures, when they occurred, were limited to a small area adjacent to the crack plane and extended less than 12 mm beyond the crack tip. When the load was increased to 100.3 MPa the composite patch disbonded and delaminated from the base panel as shown in Fig. 3.6. Typical failure surfaces showed resin bonded to the aluminum with fiber tearing and inter-laminar delamination between various plies through the thickness of the composite patch.

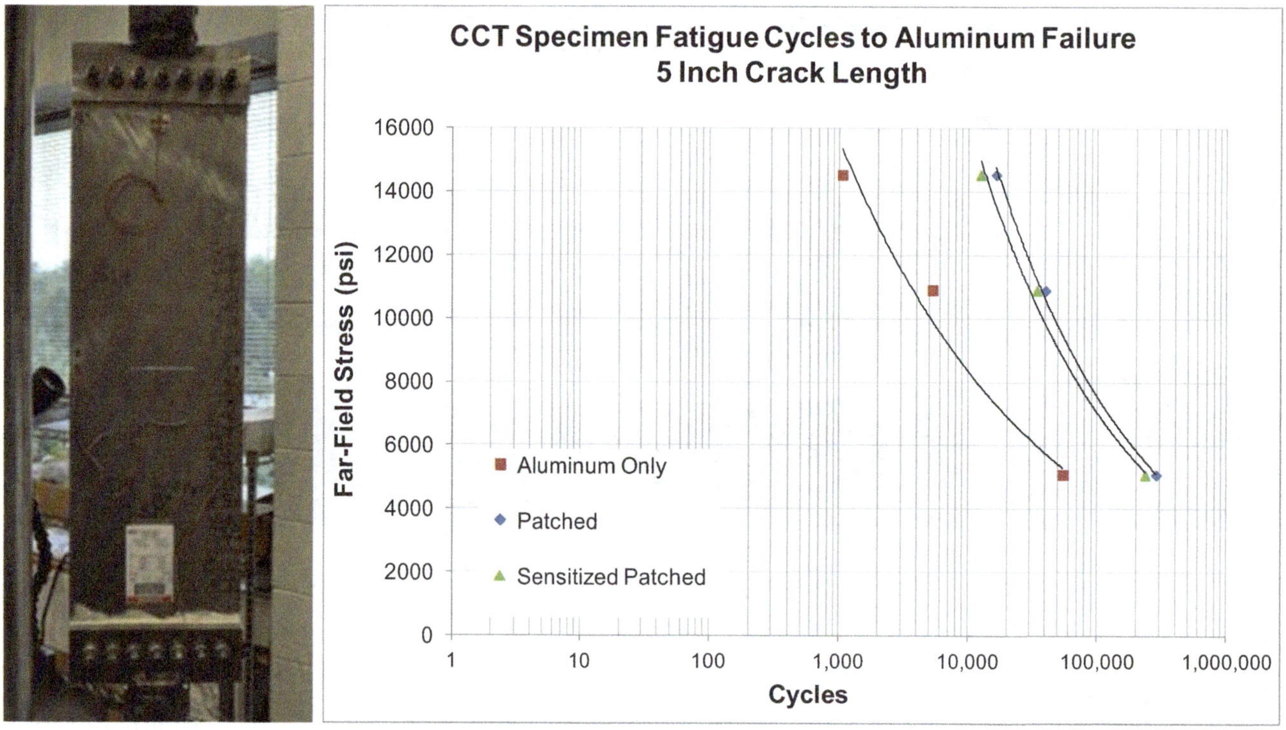

Fig. 3.7 Aluminum specimen and series far-field stress level versus cycles to failure for $2a_0=5$ in. fatigue testing

Stress level as a function of failure cycle for the un-patched aluminum, composite patch repaired aluminum, and the oven sensitized composite patch aluminum for the three stress levels are shown in Fig. 3.7. Sensitized plate was "as-delivered" aluminum conditioned at 121 °C for 10 h, which resulted in a DoS of 47.3 mg/cm^2 as measured by the G67 test method. The data points for each plate condition are fit with a power law to highlight the general trend of the data and to show the difference between the plate condition groups. Each data point represents one specimen. Application of a composite patch repair to a cracked aluminum plate subjected to a cyclic tensile load increased the cycle count to failure for the three stress levels tested. The benefit of the composite patch repair system increased with the level of applied stress. When subjected to a far-field tensile stress level of 34.5 MPa the number of load cycles increased by four to five times for the sensitized and as-delivered plate condition. Similarly, for the 75.2 MPa stress level the cycle counts increased 6.5 to 7.5 times. When the stress level was increased to 100.3 MPa the cycle count increased between 12 and 15.7 times the baseline cracked plates [31].

Composite patch repairs increase static plate strength, fatigue life, and large displacement capacity of the crack plate. Additionally, to demonstrate performance under high stress, a large composite patch reinforced aluminum edge notch test specimen (0.8 × 1.0 m) was tested in the custom designed fixture shown in Fig. 3.8. The fixture consists of four floating load arms and two compression struts. With load applied to the arms a large region of high stress forms ahead of the blunt notch that typically produces highly controlled ductile fracture of the aluminum plate [32]. The reinforced specimen consisted of a 28 ply E-Glass epoxy laminate with a nominal thickness of 15 mm and an axial elastic modulus of 18.6 GPa bonded to a 5083-H116 aluminum plate 6.35 mm thick. Uni-directional reinforcement was inserted between plies of the typical quasi isotropic laminate to increase patch modulus and strength across the crack plane. Initial quasi static analysis of the specimen indicated grip failure prior to laminate failure. Mitigation of this risk involved extending the 50.8 mm V-notch into a 127 mm blunt notch and removal of 152 mm of composite at the notch tip as shown in Fig. 3.8. Dominant failure modes observed during testing are shown along with the test fixture in Fig. 3.8. Aluminum fracture and laminate failure were observed, however disbonding of the composite did not occur. Visible aluminum bonding surfaces include multiple composite plies starting with the interface ply, which indicates the composite to aluminum bondline remained intact. Laminate failure propagated through the bottom half of the laminate until testing was terminated but the patch above the crack remained intact and securely bonded to the aluminum.

The load-deflection response of the composite reinforced plate is compared with baseline aluminum and welded aluminum plates in Fig. 3.9 [32]. Under load the aluminum crack propagated approximately 30° to the crack plane and then arrested 2 cm from the composite, which is the location where the crack direction changes in Fig. 3.8. Loading increased until aluminum plate failure, shown in Fig. 3.9, occurred within the grips. Loading continued to increase until composite

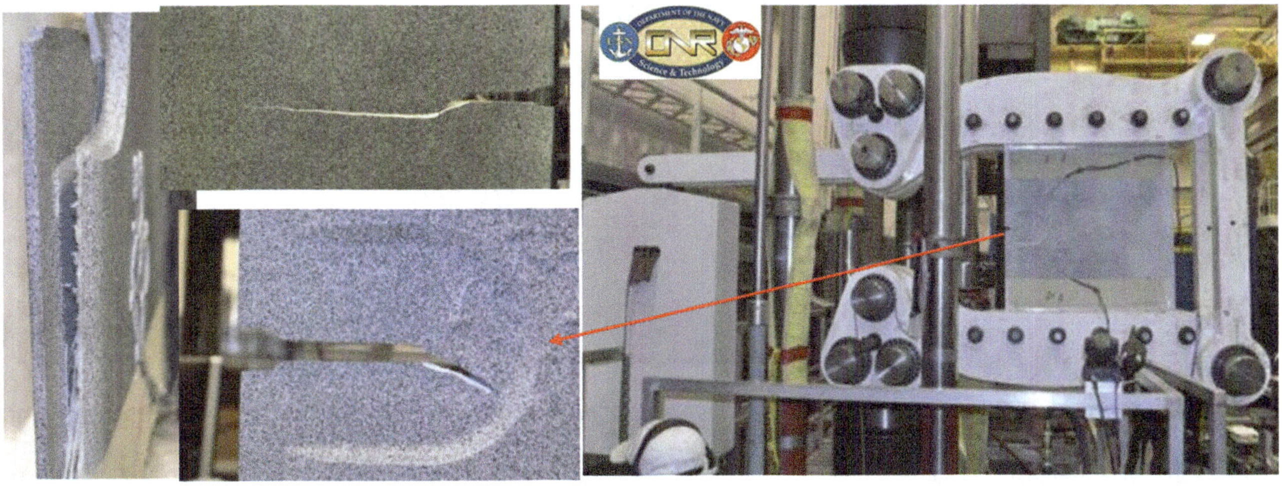

Fig. 3.8 Ductile tearing specimen failure (*left*) and test fixture (*right*)

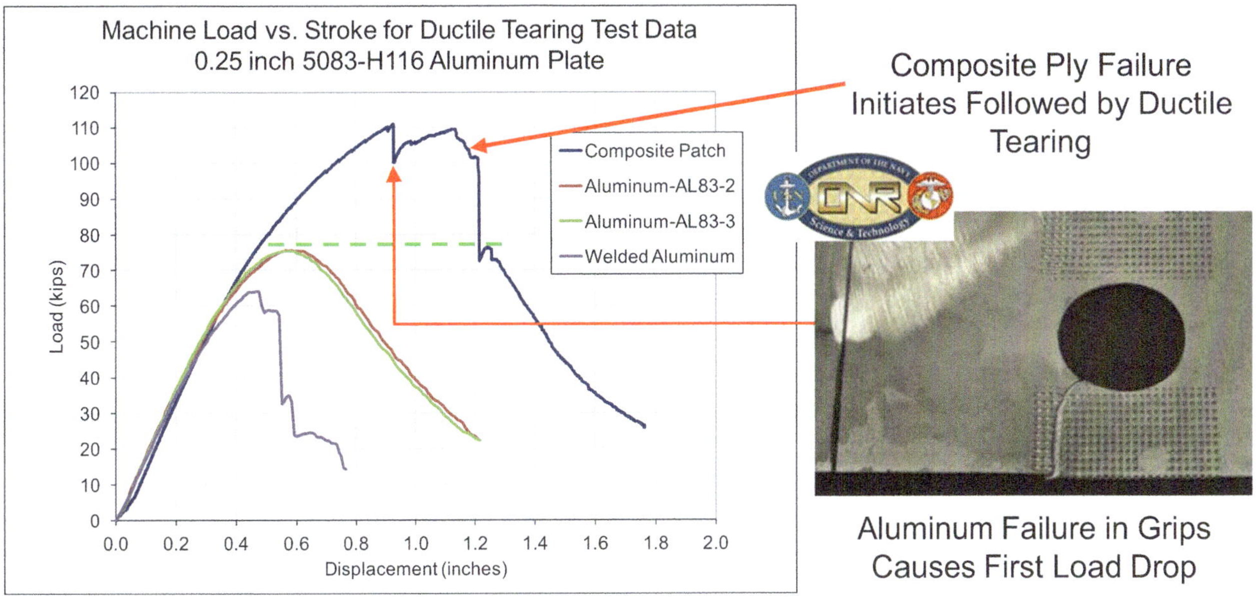

Fig. 3.9 Ductile tearing test machine load versus cross head displacement results with failure notes

failure initiated. After the load drop associated with composite failure, the aluminum plate fracture propagated following the same unloading measured for the baseline aluminum plate. The demonstrated benefit of the composite patch was additional in-plane load and displacement capacity of the plate. Though compared below, the baseline aluminum plate performance cannot be directly compared to that of the reinforced specimen. Baseline plates had electrical discharge machined (EDM) V-notches while the composite reinforced plate tested had a 76 mm crack extension cut with a bandsaw, resulting in a blunt notch. The more than 45 % of additional load capacity of the composite reinforced specimen matched that achieved with the 9.53 mm aluminum plate with the added benefit of an additional 100 % of displacement capacity. The peak load and the work (area under the curve) to fracture the hybrid panel was greater than the equivalent aluminum plate. Optimization of the composite patch could conservatively reduce composite patch weight by 15 % and still achieve equivalent performance.

Demonstration of composite patch repair durability was performed with 40.6 cm^2, and 6.35 mm thick aluminum plates, reinforced with composite patches subjected to dynamic impact. Impact energies were designed to represent tools or equipment being dropped during service or maintenance. Testing was performed using the U.S. Naval Academy's Instron Dynatup model 9250HV spring assisted drop tower with a 50 mm diameter steel cylinder and a 76 mm diameter spherical impact TUP pictured in Fig. 3.10. Two impact energies, 400 and 815 J, were used with a consistent 7.62 m/s impact velocity to impart

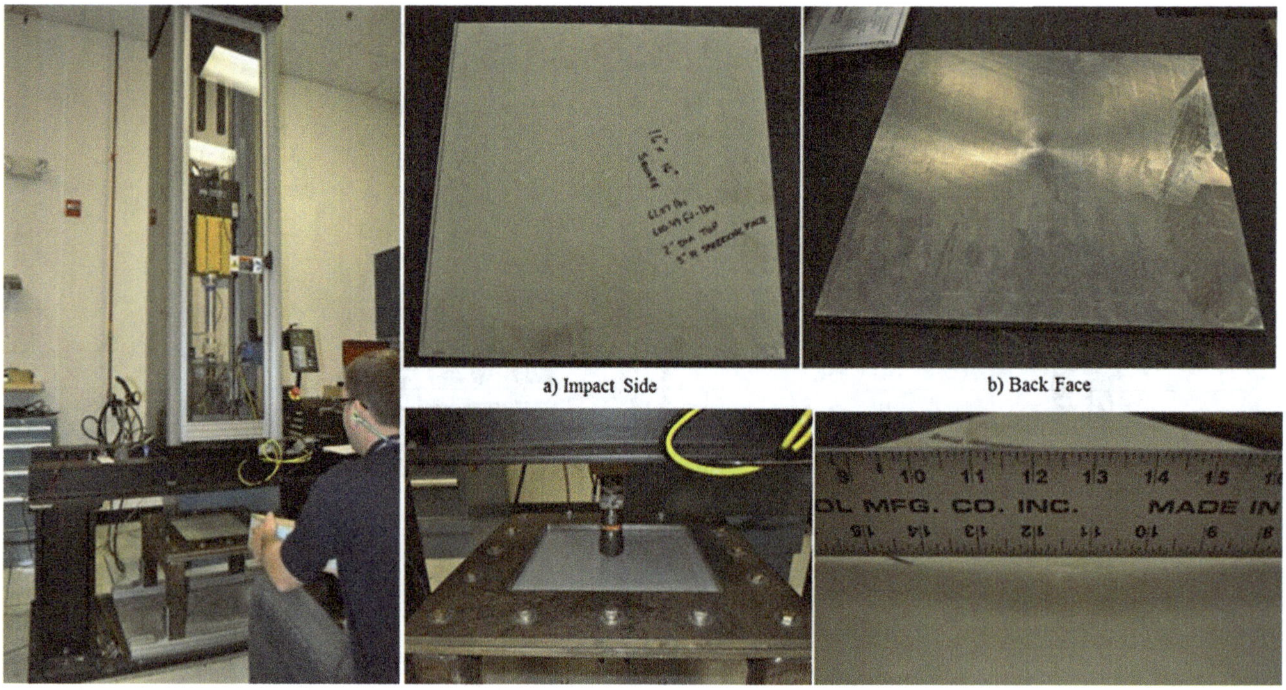

Fig. 3.10 Impact testing and residual panel deflection after 600 ft/lb impact

damage on reinforced panels with a 35.6 cm unsupported span and clamped edges. Impact produced permanent deflection of the aluminum plate, as shown in Fig. 3.10. Ultrasonic testing (UT) C-scan results showed matrix damage in the region of impact but did not include delamination of the composite from the aluminum. The area of damage for the quasi-isotropic laminate was elliptical with an area proportional to the impact energy. At 400 J the impact damage was no larger than the diameter of the impactor TUP but increased to twice the area of the TUP when impacted with 815 J.

3.6 Summary

High magnesium content 5xxx marine aluminum alloys used for welded structures are susceptible to sensitization when operating in moderate to hot marine environments. Cumulative exposure of plate to elevated temperatures causes magnesium in the aluminum to migrate to grain boundaries making the plate susceptible to intergranular corrosion and stress-corrosion cracking (SCC), which is a failure mechanism that can occur at stress levels well below nominal plate strength. Determining the repair method for sensitized plate requires knowledge of the degree of sensitization (DoS). The three methods for DoS detection used by the U.S. Navy are the ASTM G67 NAMLT, the DoS Probe, and visual analysis techniques. ASTM G67 is a destructive mass loss test, while the DoS Probe and visual analysis offer non-destructive alternatives. The resulting level of sensitization governs whether a standard weld repair, weld repair in combination with cold working, or an alternative repair method is required. Two alternative repair methods are currently approved for use, the composite patch repair and the bonded aluminum repair (BAR) kit. Though the BAR kit has demonstrated fatigue performance, the strength and durability of the composite patch repair exceeds that of the BAR. In fatigue, thin low modulus composite patches increased the cycles to failure more than four times. In addition to the cycle count increase, composite patches carried tensile load beyond aluminum plate failure for stresses up to 72.5 MPa. Under high stress associated with ductile tearing, composite reinforcement significantly increased the work required to fracture the panel by increasing the peak load and displacement achieved during testing. For in-service applications, testing up to an impact energy of 815 J, which represents a dropped toolbox or tool during maintenance, demonstrated the durability of composite patch repairs. In addition to the testing, composite patch repair of SCC onboard U.S. Navy ships began in December of 2010, with repairs that have achieved over 5 years of service, continue service, and are subject to annual inspections. Patches for marine structures have demonstrated their ability to mitigate crack growth, inhibit water ingress, and reinforce damaged or degraded aluminum structure. The U.S. Navy is continuing composite patch development with the goal of designing and implementing structural reinforcement for load carrying applications and survivability.

Acknowledgements Work described was performed by the Naval Surface Warfare Center Carderock Division's Survivability, Structures, Materials, and Environmental Department with impact testing performed at the U.S. Naval Academy in the Mechanical Engineering department's impact testing lab. Financial and technical support was provided by Dr. Paul Hess of the Office of Naval Research Code 331, the Naval Sea Systems Command Surface (NAVSEA) Surface Warfare Directorate (SEA 21), and the NAVSEA Engineering Directorate (SEA 05).

References

1. O'Shaughnessy, T.G.: ALCOA 5000-Series Alloys Suitable for Welded Structural Applications. Aluminum Company of America, Application Engineering Division, ALCOA Laboratories, ALCOA Center, Pennsylvania (1985)
2. Bushfield, H., Cruder, M., Farley, R., Towers, J.: Marine Aluminum Plate—ASTM Standard Specification B 928 and the Events Leading to its Adoption, in Society of Naval Architects and Marine Engineers, World Maritime Technology Conference and Exposition, San Francisco, CA, 2003
3. Cavas, C.P.: Cracks Plague Ticonderoga-Class Cruisers, Navy Times, 9 Dec 2010
4. ASTM G67-13: Standard Test Method for Determining the Susceptibility to Intergranular Corrosion of 5xxx Series Aluminum Alloys by Mass Loass After Exposure to Nitric Acid (NAMLT) Test. ASTM International, Conshohocken, PA (2013)
5. Golumbfskie, W.J., Tran, K.T., Noland, J.M., Park, R., Stiles, D.J., Grogan, G., Wong, C.: Survey of detection, mitigation, and repair technologies to address problems caused by sensitization of Al-Mg alloys on navy ships. Corrosion **72**(2), 314–328 (2016)
6. Golumbfskie, W.J.: Aluminum sensitization and the Navy. In: The Minerals, Metals, and Materials Society 143rd Annual Meeting and Exhibition, San Diego, CA, 2014
7. Baker, A.A.: Repair of cracked or defective metallic aircraft components with advanced fibre composites-an overview of Australian work. Compos. Struct. **2**, 153–181 (1984)
8. Baker, A.A., Rose, L.R.F., Jones, R.: In: Advances in the Bonded Composite Repair of Metallic Aircraft Structures, vol. 1. Elsevier, Amsterdam (2002)
9. Baker, A.A., Rose, L.R.F., Walker, K.E., Wilson, E.S.: Repair substantiation for a bonded composite repair to F111 lower wing skin. Appl. Compos. Mater. **6**, 251–267 (1999)
10. Christian Jr., T.F., Hammond, D.O., Cochran, J.B.: Composite material repairs to metallic airframe components. J. Aircraft **29**(3), 470–476 (1992)
11. Guyt, C.B.: Design and Installation of Bonded Repairs to Cracked Fuselage Skins on C-5A T/N 69-021, SPC 94-4050, USAF European Office of Aerospace Research and Development and the Fatigue and Fracture Group, Flight Dynamics Directorate, 1995
12. Sabelkin, V., Mall, S., Avram, J.B.: Fatigue crack growth analysis of stiffened cracked panel repaired with bonded composite patch. Eng. Fract. Mech. **73**, 1553–1567 (2006)
13. Schubbe, J.J., Mall, S.: Investigation of a cracked thick aluminum panel repaired with a bonded composite patch. Eng. Fract. Mech. **63**, 305–323 (1999)
14. Tsai, G.-C., Shen, S.B.: Fatigue analysis of cracked thick aluminum plate bonded with composite patches. Compos. Struct. **64**, 79–90 (2004)
15. Jones, R., Chiu, W.K.: Composite repairs to cracks in thick metallic components. Compos. Struct. **44**, 17–29 (1999)
16. Duong, C.N.: Design and validation of composite patch repairs to cracked metallic structures. Compos. Part A: Appl. Sci. Manuf. **40**, 1320–1330 (2009)
17. Allan, R.C., Bird, J., Clarke, J.D.: Use of adhesives in repair of cracks in ship structures. Mater. Sci. Technol. **4**, 853–859 (1988)
18. Grabovac, I., Pearce, P., Camilleri, J.A., Challis, K., Lingard, J.: Are composites suitable for reinforcement of ship structures? In: 12th International Conference on Composite Materials, Paris, 1999
19. Grabovac, I.: Bonded composite solution to ship reinforcement. Compos. Part A: Appl. Sci. Manuf. **34**, 847–854 (2003)
20. Grabovac, I., Whittaker, D.: Application of bonded composites in the repair of ships structures—a 15-year service experience. Compos. Part A: Appl. Sci. Manuf. **40**, 1381–1398 (2009)
21. Turton, T.J.D.-J., Livingstone, F.: Oil platforms, destroyers, and frigates—case studies in QinetiQ's marine composite patch repairs. Compos. Part A: Appl. Sci. Manuf. **36**, 1066–1072 (2005)
22. McGeorge, D., Echtermeyer, A.T., Leong, K.H., Melve, B., Robinson, M., Fischer, K.P.: Repair of floating offshore units using bonded fibre composite materials. Compos. Part A: Appl. Sci. Manuf. **40**(9), 1364–1380 (2009)
23. Weitzenbock, J.R., McGeorge, D.: A cold repair method for FPSOs. In: Offshore Technology Conference, Houston, TX, 2012
24. Recommended Practice DNV-RP-C301: Design, Fabrication, Operation and Qualification of Bonded Repair of Steel Structures. Det Norske Veritas, 2012
25. Noland, J.M., Hart, D.C., Udinski, E.P., Sielski, R.A.: Initiatives in bonded ship structural repairs. In: ASNE Fleet Maintenance and Modernization Symposium, American Society of Naval Engineers, San Diego, 2013
26. Greene, J.P.: Why is White So Sacred? Soaring (1975)
27. Brosi, J.K., Lewandowski, J.J.: Delamination of a sensitized commercial Al-Mg alloy during fatigue crack growth. Scr. Mater. **63**, 799–802 (2010)
28. LaMontagne, K., Stiles, D.J.: Aluminum alloy sensitization data visualization and analysis. In: Department of Defense Corrosion Conference, Gaylord National Resort and Confenvention Center, Washington, DC, 2009
29. Bumiller, E., Shedd, M., Kazerooni, B., Adedeji, W., Kelly, R.G.: The electrochemistry of intergranular corrosion and sensitization in 5XXX aluminum alloys. In: Department of Defense Corrosion Conference, Gaylord National Resort and Confenvention Center, Washington, DC, 2009
30. Bone, J.S., Karr, D., Waas, A.M.: Testing of composite patches for ship plating fracture repair. In: Naval Architecture and Marine Engineering, University of Michigan, 2003

31. Hart, D.C., Udinski, E.P., Hayden, M.J., Liu, X.: Fatigue performance and analysis of composite patch repaired cracked aluminum plates. In: Sustainability and Stewardship, Vessel Safety and Longevity Through Research, Conference, Ship Structures Committee, Linthicum Heights, 2014
32. Nahshon, K., Hoffman, W.A., Ullagaddi, C.B.: Characterization of structural scale ductile fracture of aluminum panels using digital image correlation. In: Proceedings of the 2014 Annual Conference on Experimental and Applied Mechanics, Conference, 2014

Chapter 4
Investigation and Improvement of Composite T-Joints with Metallic Arrow-Pin Reinforcement

Sebastian Heimbs, Michael Jürgens, Christoph Breu, Georg Ganzenmüller, and Johannes Wolfrum

Abstract Aiming at an increase in failure resistance and damage tolerance of composite T-joints, a novel reinforcement technique in through-thickness direction using metallic arrow-pins has been proposed by the authors in previous studies. In a recent investigation, different options for further improvement of the T-pull performance have been assessed. These include optimized arrow-pin configurations, filler and ply materials and thermoplastic interleaf layers. T-specimens have been manufactured and tested under quasi-static and high-rate dynamic loading conditions to quantify the influence of these measures. Additionally, FE models have been developed in LS-Dyna to predict the performance numerically. Model validation was conducted step by step using material coupon test data, dedicated single-pin pull-out tests and T-pull tests.

Keywords T-joint • Composite material • Metal pin reinforcement • T-pull testing • Numerical simulation

4.1 Introduction

T-joints can be found in many box-like or stringer-stiffened composite structures and they can be the weakest link for structural failure under severe loading. Hence, lots of past and current research activities are related to the assessment and improvement of the strength of composite T-joints by advanced designs [1, 2] or reinforcement techniques [3–5]. The load case in the focus of the current study is the hydrodynamic ram loading of a fluid-filled composite tank structure. The structural performance of the T-joints for this kind of load case shall not only provide maximized static strength at failure initiation but also maximized post-damage energy absorption until complete separation to ensure structural integrity.

The authors of this study have proposed a novel reinforcement technique of composite T-joints in through-thickness direction using metallic arrow-pins, which has shown good potential for damage resistance and energy absorption in first T-pull test series (Fig. 4.1) [6–9]. This type of hybrid joint makes use of the metallic plasticity of the reinforcing pins, which is beneficial for high energy absorption targets, and of the mechanical interlocking effect of the arrow-head under pull-out loads. The principle of this patented technique is to use a thin metallic sheet, in this case a 0.4 mm DC4 steel sheet, where the pin geometry is cut and bent on a special bending machine in order to obtain a reinforcement sheet with pins of various length on both sides. This sheet is intended to be placed between two laminates, in this case between skin and vertical spar of the T-joint (Fig. 4.2).

Now, the previous studies were extended in order to further improve the T-pull performance by new design solutions of the arrow-pin reinforcement architecture and by new material solutions for filler, plies and ply interfaces. The basis for these

S. Heimbs (✉) • M. Jürgens • C. Breu
Airbus Group Innovations, 81663 Munich, Germany
e-mail: sebastian.heimbs@airbus.com

G. Ganzenmüller
Fraunhofer Institute for High-Speed Dynamics (EMI), 79104 Freiburg, Germany

J. Wolfrum
Bundeswehr Research Institute for Materials, Fuels and Lubricants (WIWeB), 85435 Erding, Germany

G.L. Cloud et al. (eds.), *Joining Technologies for Composites and Dissimilar Materials, Volume 10*
Conference Proceedings of the Society for Experimental Mechanics Series, DOI 10.1007/978-3-319-42426-2_4

Fig. 4.1 Composite T-joint with metallic arrow-pin reinforcement

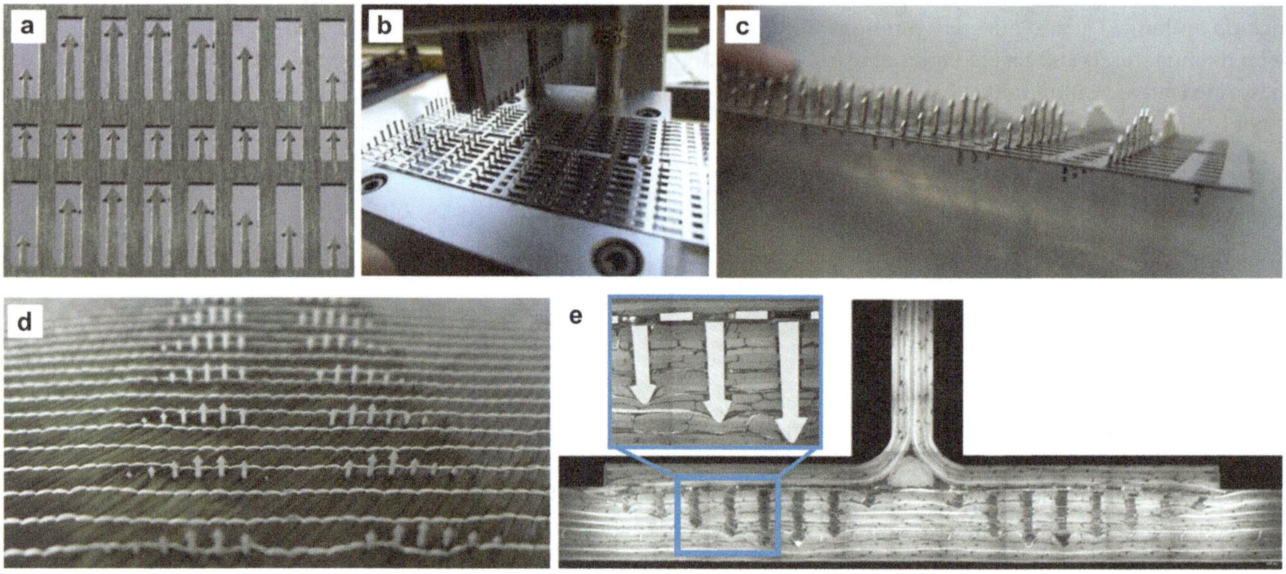

Fig. 4.2 Arrow-pin reinforcement of composite T-joint: (**a**) laser-cut steel sheet, (**b**) bending of pins, (**c**) final pin-reinforcement sheet, (**d**) pins inserted into dry carbon fiber textile, (**e**) micrograph of cured composite T-joint

considerations was a numerical T-pull simulation study with LS-Dyna, which was performed to identify the influence and optimum of different design and material parameters like the T-joint radius, the spar feet length, the filler material stiffness and the interlaminar fracture toughness [10]. The results of this study led to the decisions for the new T-specimen manufacturing and T-pull testing campaign described below.

The aim of this investigation is to assess the potential performance improvements by advanced design and material solutions using quasi-static and high-rate dynamic T-pull tests and numerical T-pull simulations with LS-Dyna.

4.2 T-Joint Design Specification

Seven different composite T-joint designs are compared in this study (Fig. 4.3). The first two designs D1 and D2 have already been investigated in a past study [6, 9] and form the basis for further improvements.

Design D1 is a conventional monolithic composite T-joint without through-thickness reinforcement acting as a reference or as a state-of-the-art in-service solution. It is made of Tenax HTS carbon fibers (CF) and Hexcel HexFlow® RTM6 epoxy resin. The stacking sequence of the 7 mm thick skin ($[0°/+45°/90°/-45°/0°/-45°/90°/+45°/0°/+45°/90°/-45°/0°/90°]_S$) and of the 4 mm thick spar ($[+45°/0°/-45°/+45°/-45°/+45°/0°/-45°]_S$) is similar for all seven T-joint designs.

Design D2 is made of the same material but has metallic arrow-pins of 2 mm length on the top side and varying length on the bottom side as through-thickness reinforcement.

Design	Illustration	Details
D1	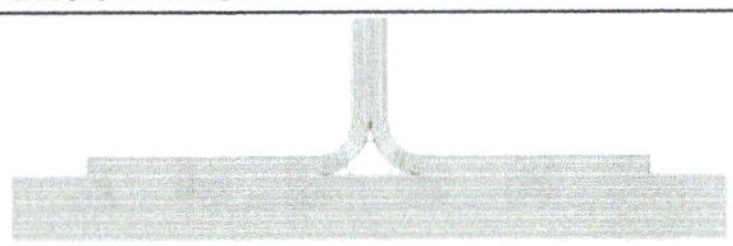	HTS-CF/RTM6; no metal pins
D2	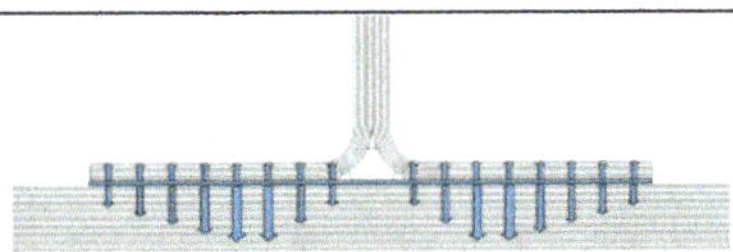	HTS-CF/RTM6; metal pins
D3	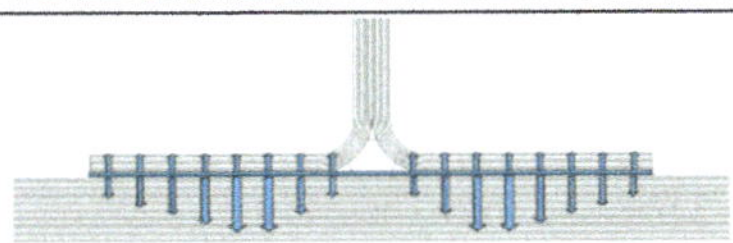	HTS-CF/EP2400; metal pins
D4	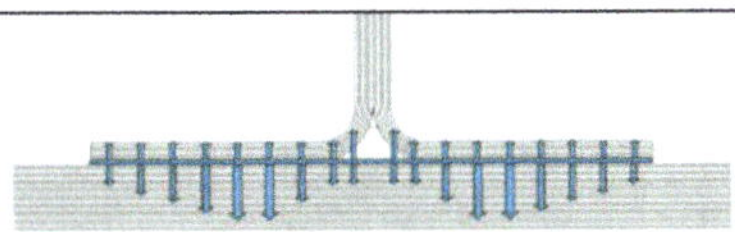	HTS-CF/EP2400; metal pins + additional center pins
D5	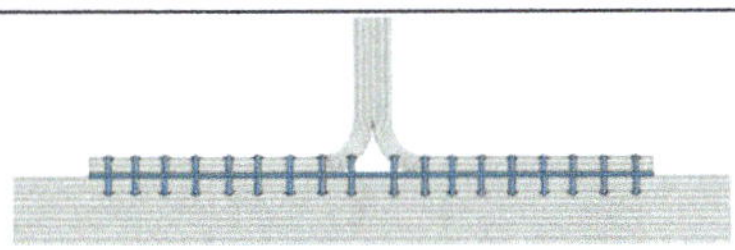	HTS-CF/EP2400; metal pins of maximum density
D6	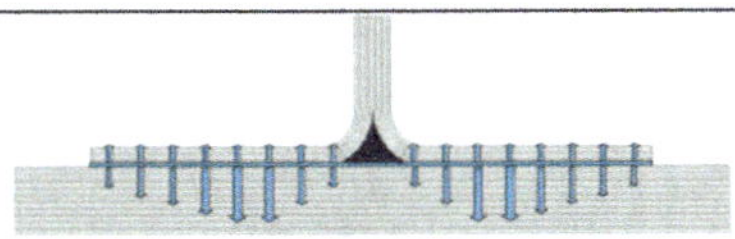	HTS-CF/EP2400; metal pins + thermoplastic filler
D7	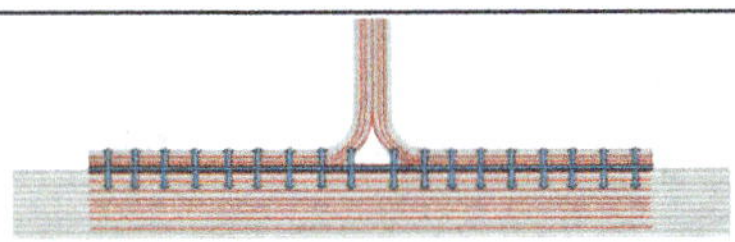	HTS-CF/EP2400; metal pins of max. density + thermoplastic binder

Fig. 4.3 Overview of composite T-joint designs investigated in this study

Design D3 has the same arrow-head reinforcement but instead of the un-toughened RTM6 the toughened epoxy resin system Cytec Prism® EP2400 was used in order to improve the fracture toughness. This toughened epoxy resin is used for all further specimens D4–D7, too.

Design D4 is comparable to D3 but has additional pins in the center region in order to investigate their influence on failure initiation and energy absorption.

An optimization of the reinforcement architecture or pin lay-out was attempted in **design D5**. This optimized design is characterized by a maximum pin density in order to investigate, if more shorter pins are superior to fewer longer pins. All pins have an equal length of 2 mm and cannot be positioned any closer than in the selected design in order to prevent undulation problems and excessive resin-rich zones.

Design D6 is similar to D3 but uses a different filler material in the T-joint center. Instead of a carbon fiber braid like in all other designs, a milled thermoplastic filler made of polyetherimide (PEI) was used, which enables higher elasticity and a more precise ply path of the spar laminate plies and avoids resin-rich zones at the corners.

In **design D7** the influence of additional thermoplastic co-polyamide (PA) binder fleece layers between all skin and spar plies was assessed, which is supposed to further increase the fracture toughness and damage resistance. The pin reinforcement pattern is similar to design D5 with the high pin density.

4.3 Experimental Characterization

The experimental test series was divided into three categories. First of all, standard coupon tests with the composite laminate and steel material of the pin sheets were performed in order to obtain mechanical properties for the numerical material modeling. These involved tensile, compression and shear tests as well as double cantilever beam (DCB) and end-notched flexure (ENF) tests for delamination properties characterization.

Fig. 4.4 Single-pin pull-out tests: (**a**) test set-up, (**b**) single-pin specimen (post-test), (**c**) 5×5 pins specimen (post-test)

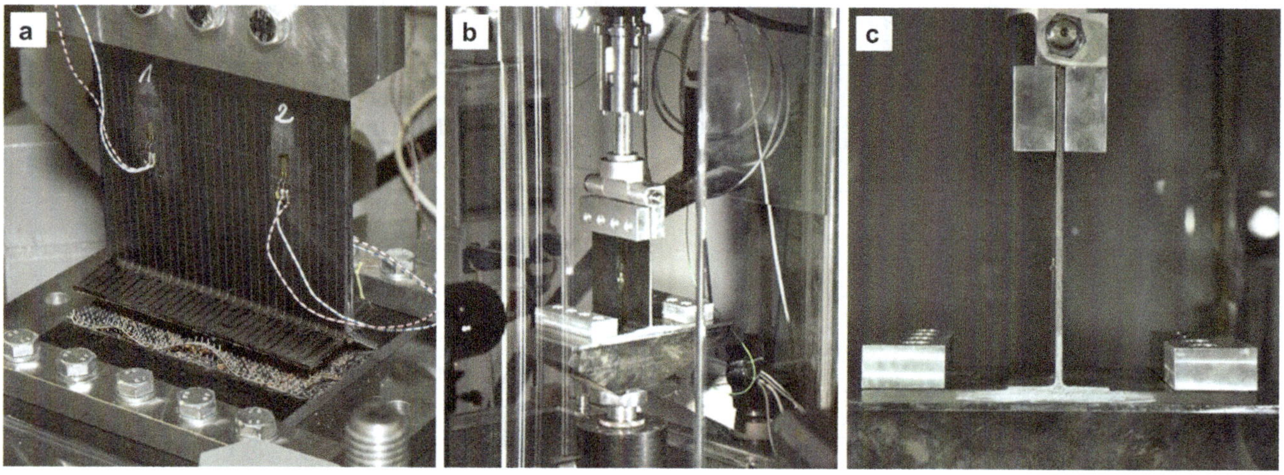

Fig. 4.5 T-pull tests: (**a**) quasi-static test set-up, (**b**) high-rate dynamic test set-up, (**c**) specimen fixation

The second test campaign was dedicated to characterizing the metal pin-to-laminate bonding strength and pin pull-out behavior in detail. For this purpose, single-pin pull-out tests were performed by manufacturing and testing specimens with just one pin in normal tension mode (Fig. 4.4). A Teflon foil was used to decouple the upper laminate from the steel sheet in order to let only the pin carry the tensile load. However, high scatter of test results and dependence of specimen quality led to the decision to go for 5×5 pins specimens, which turned out to be more robust and reproducible. The test results, which are documented in detail in a separate publication [11], gave valuable information on bonding strength and pin bending, detachment and pull-out behavior, which are helpful for later comparisons of simulation and test.

Finally, the third test campaign involved the T-pull testing under quasi-static and high-rate dynamic loading conditions. The latter are more representative of the transient hydrodynamic ram loading as the realistic target load case (Fig. 4.5). The specimen geometry was similar to previous test series [6, 9] with a width of 240 mm, depth of 150 mm and height of 188 mm. The T-specimens were clamped with metallic blocks through boreholes in the specimen and pulled in 90° direction. Digital image correlation technique was used for optical displacement and strain measurement. Additional strain gauges on both sides of the vertical spar laminate were used to identify asymmetric loading or spar bending. The strain gauge signal was also used to calculate the tensile force in the dynamic tests, as this signal was less affected by oscillations than the load cell signal itself. The high-rate dynamic tensile tests were performed on a servo-hydraulic testing machine at a loading velocity of 5 m/s and revealed an increase in stiffness and failure load, which is discussed in further detail in [6, 7]. This paper will focus on the quasi-static test results only, as these curves are less influenced by complex dynamic effects and can more easily be interpreted.

An overview of all quasi-static force–displacement curves of all seven T-joint designs is given in Fig. 4.6. Just one representative curve of each design is shown, although each test was repeated with 4–6 specimens. In order to make comparisons easier and to visualize the influence of the different design improvement options, Fig. 4.7 shows the same curves as a one to one comparison. It can be seen that the reference specimen D1 without reinforcement in through-thickness direction shows

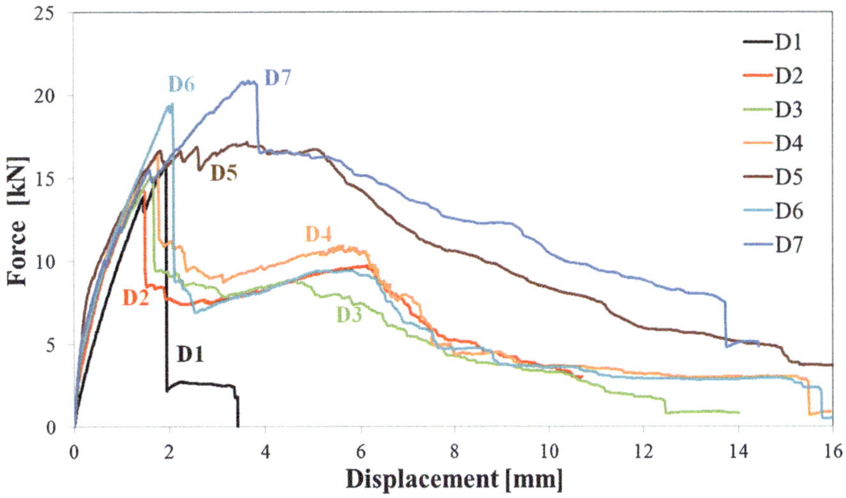

Fig. 4.6 T-pull test results: force–displacement diagram of all seven T-joint designs

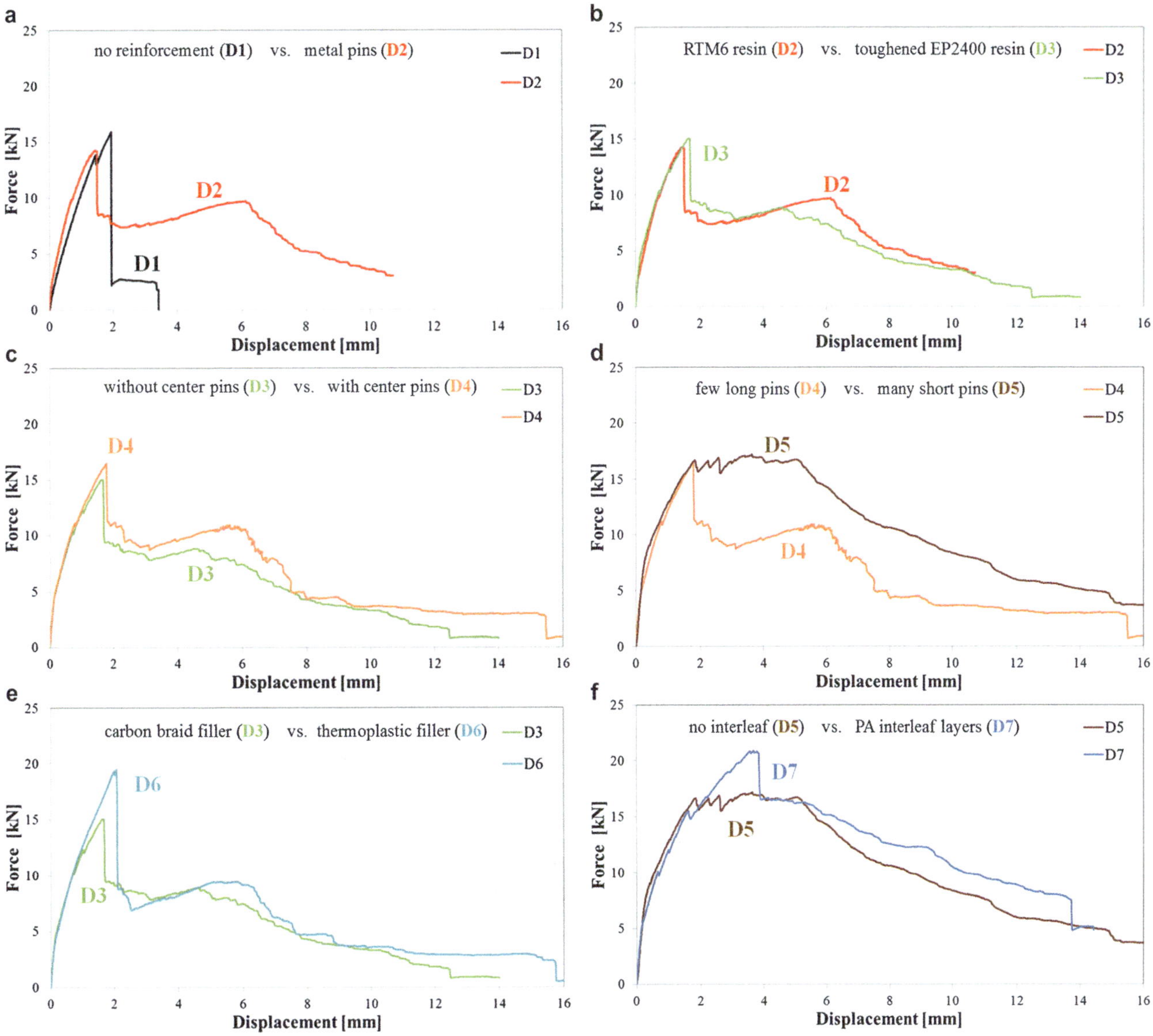

Fig. 4.7 T-pull test results: comparison of force–displacement curves

a very brittle behavior with a sudden drop of the force curve when spar-skin detachment occurs. The absorbed energy for T-joint separation was calculated as the integral of the force–displacement curve and has the lowest value for the D1 specimens with 22 J. The implementation of metallic arrow-pin-reinforcements in D2 slightly reduces the failure initiation load due to induced imperfections that enable crack initiation, but the residual post-damage force level is much higher than for specimens D1 (Fig. 4.7a). The energy absorption increases by 250 % to 78 J. Using the toughened epoxy resin EP2400 in D3 compared to un-toughened RTM6 in D2 leads to a slight increase in failure initiation load and a comparable total energy absorption of 79 J (Fig. 4.7b). The addition of further pins in the specimen center in design D4 leads to a significant improvement of failure initiation load and residual force level (Fig. 4.7c). The total absorbed energy of 109 J increases by 37 % compared to design D3.

Even more impressive is the improvement with design D5, which uses a maximized pin density with many short pins instead of fewer longer pins (Fig. 4.7d). The energy absorption increases to 170 J, which is 56 % higher than design D4 or 114 % higher than design D3. The influence of the milled thermoplastic PEI filler in design D6 compared to the carbon braid filler in design D3 with a similar pin-reinforcement is shown in Fig. 4.7e. Also this measure leads to a significant improvement with 23 % higher energy absorption of 98 J. Most obvious is the strong increase of failure initiation load. This is mainly ascribed to the ideal filler geometry without any resin-rich zones or other imperfections, which can lead to an earlier crack initiation in design D3. Finally, the influence of the PA binder fleece in design D7 is shown in Fig. 4.7f in a direct comparison to design D5 with the same reinforcement pattern. The absorbed energy rises by further 7 % to 181 J and the initial failure load is significantly increased due to the higher fracture toughness provided by the thermoplastic interleaf layers. This design shows the best performance of this whole study with an impressive improvement in energy absorption of 720 % compared to the state-of-the-art reference design D1.

4.4 Modeling and Simulation

Parallel to the specimen manufacturing and experimental testing activities, numerical simulation studies with the commercial finite element (FE) solver LS-Dyna were performed to predict the structural performance and failure behavior of the different T-joint designs. In order to increase the accuracy of the numerical predictions, the building block approach along the test pyramid was used to support the model build-up and to validate the models step by step from coupon level up to the T-specimen and even fuel tank structure level (Fig. 4.8).

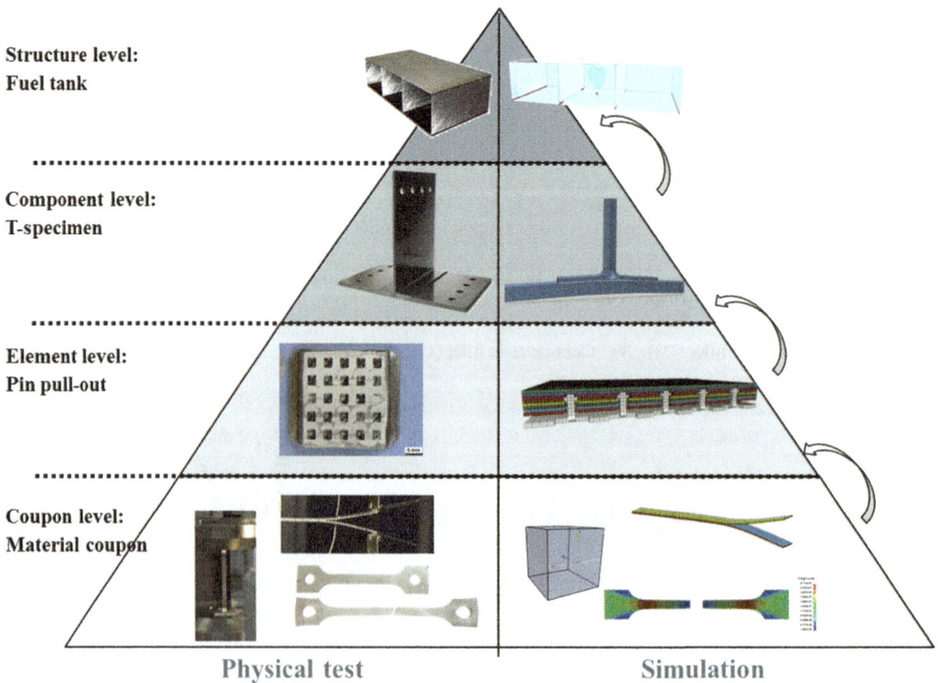

Fig. 4.8 Building block approach for T-joint FE model generation and validation

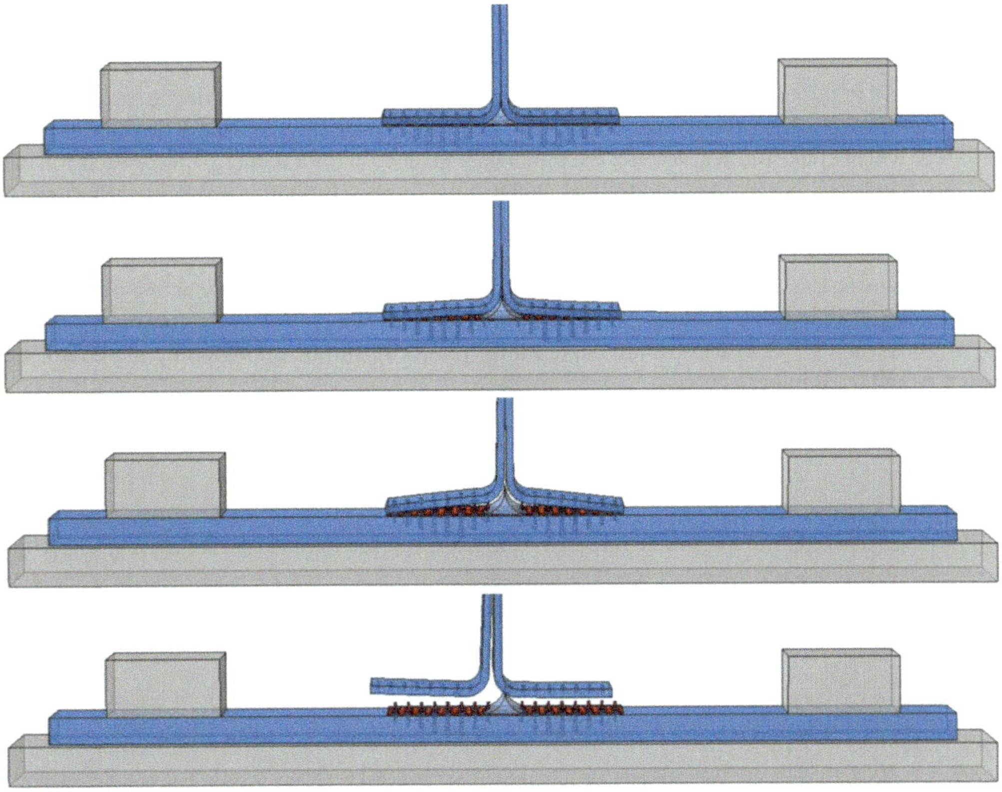

Fig. 4.9 FE simulation of T-pull loading of design D6 specimen

On coupon level, the tensile tests of the steel material were simulated to validate the metallic material law and plasticity behavior. Material model MAT24 with the tabular input of the plasticity curve in combination with solid elements was used for the metallic pin modeling. Also the composite coupon tests were simulated to validate the material model MAT59 for the modeling of each individual ply with solid elements. Special focus was also put on the simulation of the delamination tests in order to enable an accurate interlaminar failure representation and parameter calibration for the surface-based delamination tiebreak contact formulation between each ply.

The same tiebreak contact type was also used to connect the metal pins to the composite material in the T-joint model. The parameters of this metal-to-composite tiebreak contact were defined and validated in dedicated simulations of the single pin pull-out tests and 5×5 pin pull-out tests as the second level of the test pyramid (Fig. 4.8).

Finally, the T-joint models were built up using the aforementioned techniques. The model size was similar to the tested T-specimens in terms of specimen width and height, but the depth was reduced from 150 mm to just one repeating pattern of upper and lower pins with periodic boundary conditions. The realistic elasticity of the metallic clamping blocks was also covered in the models, as this has shown to have a significant influence on the slope of the force–displacement curve. With a mesh size of 0.25 mm and element numbers of 200.000–900.000—depending on the design—and tiebreak contact interfaces between each individual part and ply, the final calculation clock times in double precision mode were between 24 and 48 h on a 14 CPU workstation. An image sequence of such a T-pull simulation is shown in Fig. 4.9.

In the validation phase of the T-pull simulation models, it turned out that the models are very sensitive to parameter changes, especially with respect to the tiebreak contact parameters of the metal pin-to-composite interface. Although the initial stiffness in the resulting T-pull force–displacement diagrams could be predicted with very high accuracy compared to the test results, and the failure initiation load could be predicted with acceptable accuracy, the discrepancy between test and simulation was larger for the post-damage pin pull-out behavior. One explanation was found in the too simplified modeling of the arrow-pins and especially their pin root geometry, which leads to a different pin upward bending behavior between test and simulation. Furthermore, certain imperfections like resin-rich zones around the pin roots and pin heads as well as fiber undulations are not represented in the model. Including these details would make the already very complex models even more complex and reduce mesh size and computational efficiency.

4.5 Conclusion

The reinforcement concept in through-thickness direction of composite T-joints using metallic arrow-pin sheets was further elaborated in this study both on experimental and numerical side. More precisely, new material and design improvement options were assessed, which could further increase the damage resistance and total energy absorption for joint separation. All these options, i.e. toughened epoxy resin, additional center pins, increased pin density, milled thermoplastic filler and additional thermoplastic binder fleece layers, showed beneficial effects. Design D7, which incorporates almost all of these measures, could increase the total energy absorption by impressive 720 % compared to the state-of-the-art design D1.

The numerical studies using detailed meso-models indicated the high complexity of these arrow-pin-reinforced structures and their failure behavior. Although the models were built and validated step by step along the test pyramid, the overall predictive quality of local failure modes and overall energy absorption can still be improved by more realistic modeling of the pin sheet geometry and local imperfections.

As a next step of this study, these improved T-joint designs will be used for fuel-filled tank structures in order to verify the performance enhancements under hydrodynamic ram loads.

Acknowledgement Sincere thanks are given to Jose Luis Sandoval Murillo and Florian Troeltsch for their contributions to the numerical studies and to Christian Metzner for the support in the thermoplastic interleaf technology.

References

1. Burns, L., Mouritz, A.P., Pook, D., Feih, S.: Bio-inspired hierarchical design of composite T-joints with improved structural properties. Compos. Part B-Eng. **69**, 222–231 (2015)
2. Burns, L., Mouritz, A.P., Pook, D., Feih, S.: Strengthening of composite T-joints using novel ply design approaches. Compos. Part B-Eng. **88**, 73–84 (2016)
3. Motsch, N., Magin, M.: Influence of structural stitching on composite T-joint strength. In: 20th International Conference on Composite Materials (ICCM-20), Copenhagen, Denmark, 19–24 July 2015
4. Yan, S., Long, A., Zeng, X.: Experimental assessment and numerical analysis of 3D woven composite T-joints under tensile loading. In: 20th International Conference on Composite Materials (ICCM-20), Copenhagen, Denmark, 19–24 July 2015
5. Bianchi, F., Koh, T.M., Zhang, X., Partridge, I.K., Mouritz, A.P.: Finite element modelling of z-pinned composite T-joints. Compos. Sci. Technol. **73**, 48–56 (2012)
6. Heimbs, S., Nogueira, A.C., Hombergsmeier, E., May, M., Wolfrum, J.: Failure behaviour of composite T-joints with novel metallic arrow-pin reinforcement. Compos. Struct. **110**, 16–28 (2014)
7. May, M., Ganzenmüller, G., Wolfrum, J., Heimbs, S.: Analysis of composite T-joint designs for enhanced resistance to hydrodynamic ram. Compos. Struct. **125**, 188–194 (2015)
8. Heimbs, S., Duwensee, T., Nogueira, A.C., Wolfrum, J.: Hydrodynamic ram analysis of aircraft fuel tanks with different composite T-joint designs. In: 13th International Conference on Structures Under Shock and Impact (SUSI 2014), New Forest, UK, 3–5 June 2014
9. Heimbs, S., Mierzwa, A., Duwensee, T., Nogueira, A.C., Breu, C., May, M., Less, C., Wolfrum, J.: Investigation of static and dynamic failure behaviour of composite T-joints. In: 4th ECCOMAS Thematic Conference on the Mechanical Response of Composites (Composites 2013), Azores, Portugal, 25–27 September 2013
10. Sandoval Murillo, J.L., Ganzenmüller, G.C., Heimbs, S., May, M.: Design parameter study of a CFRP T-joint under overpressure conditions due to ballistic impact. In: 17th European Conference on Composite Materials (ECCM17), Munich, 26–30 June 2016
11. Juergens, M., Von Hafe Perez Ferreira Da Silva, M.T., Heimbs, S., Lang, H., Ladstaetter, E., Hombergsmeier, E.: Pull-out testing of multiscale structured metallic z-reinforcements for CFRP laminates. Compos. Struct. submitted for publication (2016)

Chapter 5
Review of Natural Joints and Bio-Inspired CFRP to Steel joints

Evangelos I. Avgoulas and Michael P.F. Sutcliffe

Abstract There is a great variety of joint types used in nature, from jaws, bones and tendons to root systems and tree branches. To understand how to optimise biomimetic-inspired engineering joints, first it is important to understand how biological joints work. In this paper, a review based on the functions of natural joint systems is presented. Emphasis was given to understanding natural joints from a mechanical point of view, so as to inspire engineers to find innovative methods of joining man-made structures. Nature has found many ingenious ways of joining dissimilar materials, with a transitional zone of stiffness at the insertion site commonly used. In engineering joints, one way to reduce the material stiffness mismatch is to gradually decrease the effective stiffness of the steel part of the joint by perforating it with holes. This paper investigates joining of flat perforated steel plates to a CFRP part by a co-infusion resin transfer moulding process. The joints are tested under static tensile loading. The perforated steel joints show a 175 % increase of joint strength comparing to non-perforated ones. Finite element analyses are used to interpret the experimental results.

Keywords Biomimetics • Nature • Joints/joining • CFRP • Perforated steel • Resin transfer moulding (RTM)

5.1 Introduction

Joining of composites parts, together with their damage tolerance and inserts, are of paramount importance in mechanical design because they cause structural discontinuities that are often the Achilles heel of a structural design. Adhesive bonding is a popular method for joining dissimilar materials because produces joints with higher structural efficiency, excellent fatigue life, a particularly small weight penalty and more uniform stress fields than alternatives such as fastening or riveting. Additionally, corrosion between the dissimilar materials is prevented [1–5]. However, joining composite with metal parts leads to high stress concentrations, and thus weak joints, due to the material property mismatch. Since joining of dissimilar materials, such as CFRP to steel, is widely used in high performance structures, an innovative approach is needed to meet this challenge. This paper uses the biomimetic approach to inspire a new joint design. Biomimetics is the imitation of nature's design principles applied to engineering/man-made structures. Nature has the advantage of 3.8 billions of years of research and development programme to produce an amazing range of design solutions. There is an abundance of natural world examples where two components of differing material properties are joined together [6]. The aim is to understand and analyse the functions of natural joints from a mechanical perspective and, thus, to identify characteristics that can be transferred from natural to engineering joints between metal and composite. A common solution found in nature when joining dissimilar materials is to use a transitional zone of stiffness in the insertion site [6]. One way to apply this design to engineering joints is to gradually decrease the stiffness of the joint metal part by perforating it. The concept of reducing the stiffness of a steel plate using perforations was firstly proposed and patented by Unden and Ridder [7]. Related studies have been conducted in GFRP to perforated steel joints [8] and in a sandwich structure with GFRP skins joined to a perforated steel structure [9].

In the present paper, co-infused CFRP to perforated steel joints were numerically and experimentally investigated under static mechanical testing. The structure of this paper is as follows. A brief review of natural joint systems is presented in Sect. 5.2. Section 5.3 introduces the finite element modelling and experimental methodology of the bio-inspired CFRP to perforated steel joints. The results and discussion of the outputs are analysed in Sect. 5.4, with the conclusions presented in Sect. 5.5.

E.I. Avgoulas (✉) • M.P.F. Sutcliffe
Department of Engineering, University of Cambridge, Trumpington Street, Cambridge CB2 1PZ, UK
e-mail: eia21@cam.ac.uk; mpfs@eng.cam.ac.uk

G.L. Cloud et al. (eds.), *Joining Technologies for Composites and Dissimilar Materials, Volume 10*
Conference Proceedings of the Society for Experimental Mechanics Series, DOI 10.1007/978-3-319-42426-2_5

5.2 Review of Natural Joint Systems

A literature review of natural joint systems has been carried out [6], where natural joints were classified into five different categories, based on the different joining method used by nature to create the joint. These are network structures, stiffness transitional zone, mechanisms, hooks and adhesive. In the present paper, the first two categories will be briefly mentioned as they have already inspired researchers to develop biomimetic engineering joint design concepts. An in depth review of natural joint systems from a mechanical perspective can be found in [6].

5.2.1 *Joining Method—Network Structures*

The study of the joints between tree branches and trunks has provided a source of inspiration for the design of biomimetic aerospace composite T-joints with increased damage tolerance [10]. In a study of the tree species Pinus Radiata, the branch-trunk connection was found to exhibit a ductile failure response, despite the relatively brittle nature of the underlying cellulose constituent material comprising the natural wood composite. This ductile response and high structural efficiency was achieved in nature using an efficient and complex hierarchical structure to the joint. Such a hierarchical design is common in biological structures, where the nano-, micro-, meso- and macro-length scales can interact so as to achieve a uniform strain distribution associated with high structural efficiency [11]. Burns et al. [10] applied the embedded tree-joint design to a composite T-joint and demonstrated that the ductility and the damage tolerance of the joint were increased. However, the joints showed a reduction in the load at which damage started because of the higher interlaminar stresses induced in the embedded design at the radius bend. By optimising the stacking sequence of the plies of the composite laminate at the joint this study showed that the interlaminar tension and shear stresses in the radius could be reduced [12]. This change in the local stresses led to an increase in the load at which damage initiated as well as the ultimate strength and the failure displacement of the composite T-joint.

Figure 5.1 shows a set of X-ray tomography images of the branch-trunk joint of Pinus Radiata [10]. The images highlight the key design features that give rise to the excellent mechanical properties of these branch-trunk joints. These features are embedded design, 3D fibre lay-up and variable density. Figure 5.1a illustrates the embedded design, which is comprised of a cone-shaped branch end embedded in the main trunk geometry. Such an embedded branch design has the advantages of increasing of the effective load transfer area at the joint load so as to reduce the average stress at the interface of the branch-trunk joint. Moreover, any interfacial cracks which initiate have to grow in a mode II shear mode than in a tensile mode I crack opening. This switch to shearing leads to an increased toughness. A second strengthening feature relates to the 3D fibre lay-up in which the fibres within the trunk extend forward and laterally within the joint. In this way the branch is fully enclosed in a "ball and socket" joint configuration (Fig. 5.1b). Finally there is a fibre density variation within the joint (see Fig. 5.1c) which leads to the creation of an iso-strain condition in the joint by reducing the elastic modulus mismatch between the fibres aligned along the branch and trunk directions.

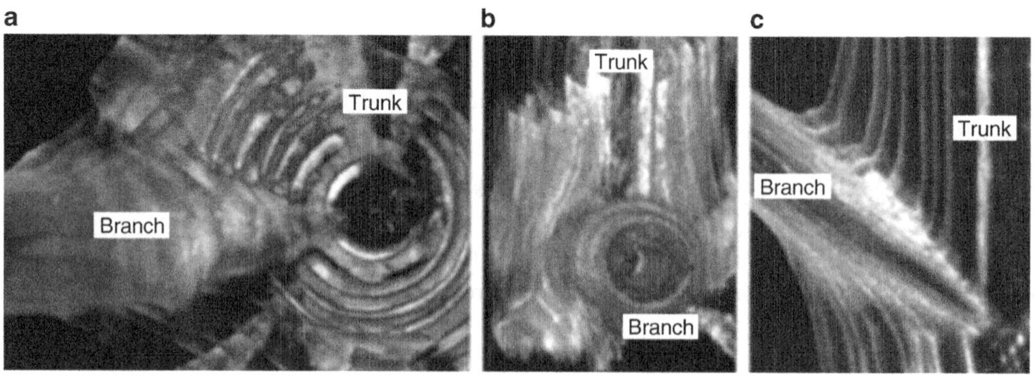

Fig. 5.1 (**a**) Internal cone-shaped branch structure embedded in the tree centre, (**b**) 3D fibre lay-up and (**c**) fibre density variation across the joint [10]

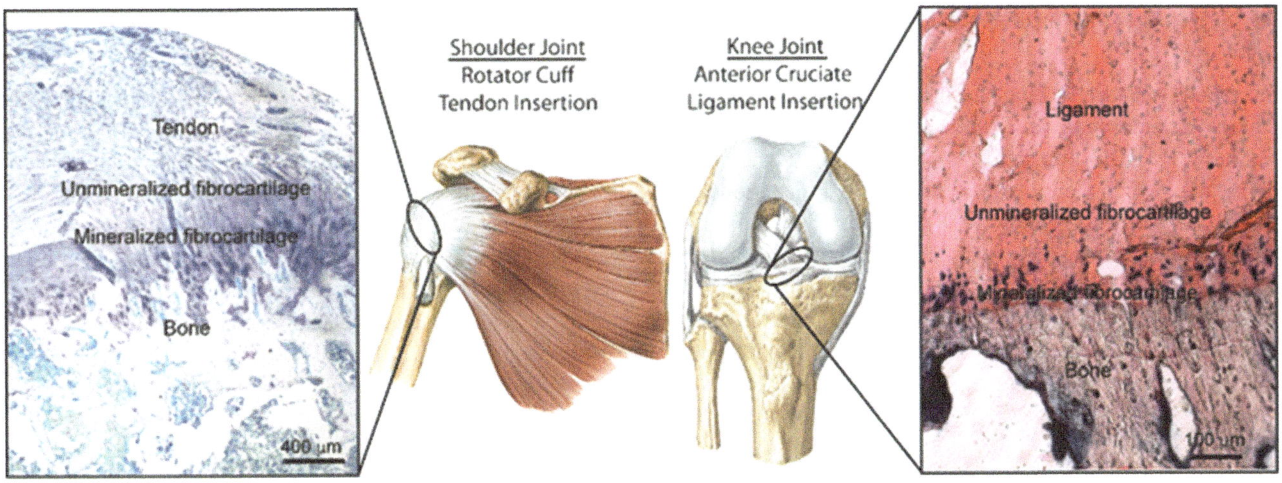

Fig. 5.2 Morphology of the supraspinatus tendon-to-bone insertion site [16]

5.2.2 *Joining Method: Transitional Zone of Stiffness*

Nature has a very effective way of joining flexible tendon to stiff bone. By introducing a complex transitional region at the interface, stress concentrations that could potentially arise at the joint are reduced or avoided. This is achieved by having a functionally-graded change in material properties at the tendon-to-bone insertion site, in which there are variations in mechanical properties, extracellular matrix composition (EMC), structural organization, and mineral content [13, 14]. Four different types of tissue can be identified in the tendon-to-bone insertion site when under an optical microscope [15]. These are tendon, fibrocartilage, mineralized fibrocartilage, and bone (see Fig. 5.2). There are two major factors which contribute to the increase in tissue stiffness from the tendon to the bone material. Firstly, the linearly increasing mineral concentration stiffens the partially mineralized collagen fibres [14]. Secondly, a decrease in the collagen fibre orientation distribution in the transition region going from the tendon to the bone gives rise to tissue stiffening. The combination of these two factors gives a variation in stiffness across the insertion site. In summary this tendon example demonstrates how nature is able to use a functionally-graded material transition region to provide an effective attachment mechanism between two mechanically different materials.

5.2.3 *Options to Increase the Strength of Engineering Joining Systems*

Joining of different materials with differing bulk properties is needed both in natural and man-made structures. In both cases the materials' mismatch leads to potential stress concentrations. So the issue in both cases is to ensure that the possibility of interfacial failure is reduced where such joints between different materials occur [17]. There are two options for engineering joints. The first option is to increase the interaction energy at the zone of contact between the two materials being joined. This can be effected by applying a surface treatment to the metal adherends prior to bonding (for example primers or a coupling agent) [18]. Although this procedure will not reduce the interfacial radial stresses, it will increase the failure stress. The second option is to avoid sharp boundaries between the different materials being joined by introducing functional gradients [19]. This method creates materials with graded mechanical properties such that they are able to resist damage more effectively than their homogeneous counterparts. In this novel procedure the radial stresses which are generated are spread over a larger surface area. This use of functional gradients in man-made structures has several advantages [19], such as reducing stress concentrations by eliminating stress singularities and by developing smooth stress distributions, increase fracture toughness and improve bonding strength.

5.3 Bio-Inspired CFRP to Perforated Steel Joints

This paper is inspired by the biomimetics approach, where nature uses a transitional zone of stiffness in the insertion site when joining dissimilar materials. One way to apply this design to engineering joints is to gradually decrease the stiffness of the joint metal part by perforating it. Figure 5.3 illustrates the proposed joint configurations, consisting of unidirectional CFRP and steel plates, which were perforated to create a transitional zone of stiffness between the steel and CFRP adherends. For comparison, reference specimens (REF) were tested with the same configuration but with a non-perforated steel part. A parametric finite element analysis was conducted to optimise the perforation patterns, while keeping all the other joint geometry the same. The analysis showed that two styles of perforation pattern performed well, a pattern with a maximum hole diameter of 4 mm (denoted "SH" for small hole), and a large hole ("LH") pattern with a maximum hole diameter of 5 mm. In both cases there were seven rows of holes, with a change in diameter between rows of holes of 0.25 mm. The only difference between the two perforated configurations is the hole diameters in the steel plates. All the other geometric variables remained the same.

5.3.1 *Numerical Modelling*

Three-dimensional finite-element models were developed with ABAQUS®. The overlap length and the free lengths of the steel and CFRP parts were all equal to 60 mm. The dimensions of the model can be seen in Fig. 5.4a. The CFRP and steel adherends were modelled with elastic orthotropic and isotropic material properties, respectively. Linear 8-node reduced integration brick elements (C3D8I) available in the ABAQUS® element library were used for both adherends.

The adhesive layer was modelled with a single layer of 8-node three-dimensional cohesive elements (COH3D8) with a thickness of 0.2 mm. A triangular cohesive zone degradation formulation was chosen because of its simplicity, widespread use for investigation purposes, especially for brittle adhesives [20], and availability in ABAQUS®. A geometrically non-linear static analysis was performed. Fixed boundary conditions were applied to the CFRP end of the joint to reproduce the testing machine gripping. A displacement *ux* was applied to the steel end of the model together with a lateral restraint (Fig. 5.4b).

5.3.2 *Specimen Manufacture and Tensile Testing*

Resin transfer moulding was used to co-infuse and bond the CFRP/steel joint. The steel parts were perforated with a water-jet cutter, sand-blasted and cleaned before moulding. The panels were then cut into specimens with a nominal width of 30 mm. It was hypothesised that the mechanical interlocking effect of the resin ingression in the holes would enhance the joint

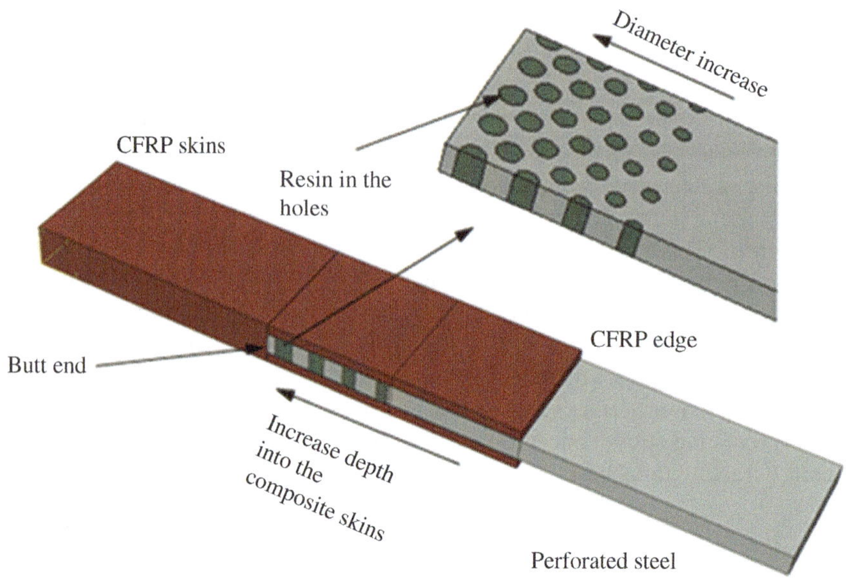

Fig. 5.3 CFRP to perforated steel model design configuration

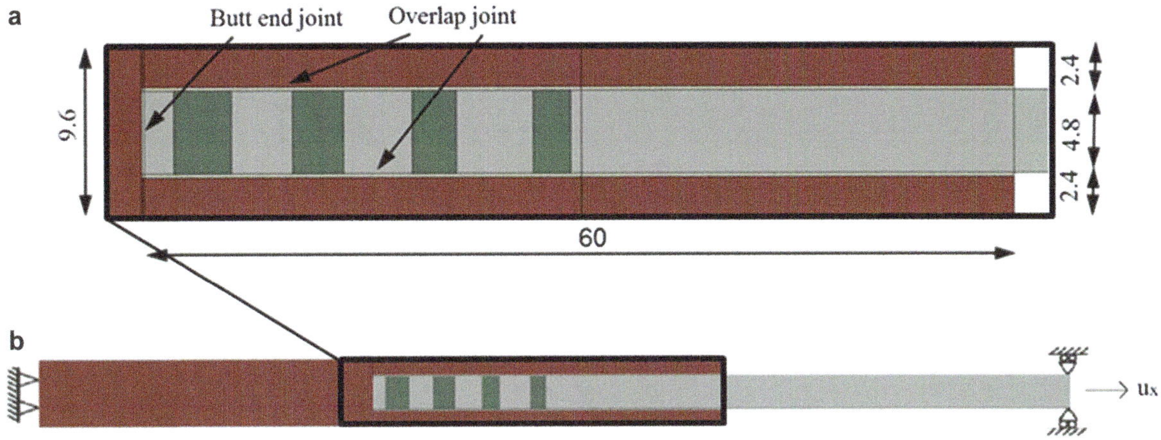

Fig. 5.4 (**a**) Joint geometry (dimensions in mm) and (**b**) loading and boundary conditions

mechanical properties. To test this all of the holes in some specimens were filled with PTFE plugs, preventing ingress of resin into the holes. Further details of the design and manufacturing can be found in [21]. Static tensile tests were performed with a fixed displacement rate of 0.2 mm/min. The load and displacement readings were obtained from the machine load cell and clip gauge extensometers, respectively. A high speed camera (16,000 frames per second) was used to image the initiation and propagation of the joint failure.

5.4 Results and Discussion of Bio-Inspired CFRP to Perforated Steel Joints

5.4.1 Load-Displacement Response

Figure 5.5 shows the load versus displacement results from experimental tests on a SH joint without PTFE inserts and on a reference joint configuration. The perforated joint has only a small drop in stiffness before it fails catastrophically. By contrast, the non-perforated reference joint failed prematurely at the butt end of the joint steel part, with a crack propagating away from there to give failure of one of the two laps. This gives the sharp change in the load–displacement curve (Fig. 5.5). After this failure, the remaining intact lap was able to carry a further increased load until final failure occurred. Included in Fig. 5.5 is a simulation for the standard reference configuration, and an additional calculation where one of the laps of the reference joint was assumed to fail at the load observed in the experimental test.

5.4.2 Maximum Load

Figure 5.6 summarises the average maximum loads obtained from the experiments. Compared to non-perforated joints, the perforated SH with PTFE configuration showed a 175 % increase of joint strength. The difference in the average maximum load for perforated joints with and without PTFE in the SH specimens was negligible. Thus, the hypothesis that resin ingress into the holes gives an increase in joint strength associated with mechanical interlocking is not observed for this perforation case. Instead it is presumed that the dominant factor which contributed to the significantly higher strength of the perforated joints, comparing to the non-perforated ones, is the transitional zone of stiffness.

5.4.3 Failure Observation Using High Speed Camera

For all the perforated joints, high-speed camera images showed that the failure initiated at the CFRP end of the joint with a crack propagating rapidly towards the butt end. The crack propagated between the steel and composite interface. Figure 5.7 shows selected frames capturing the failure initiation and the catastrophic failure of a SH joint. On the other hand, the

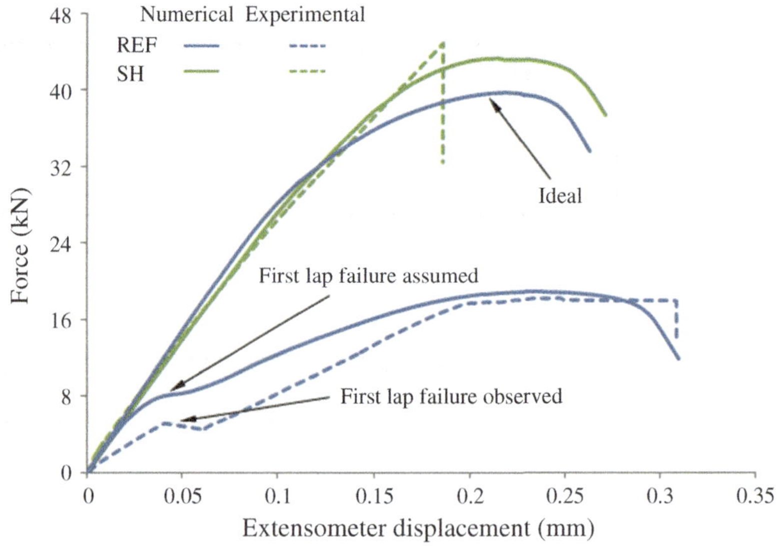

Fig. 5.5 Comparison between the numerical predictions and measured experimental load-displacement response curves, comparing non-perforated (REF) and perforated (SH) steel joints

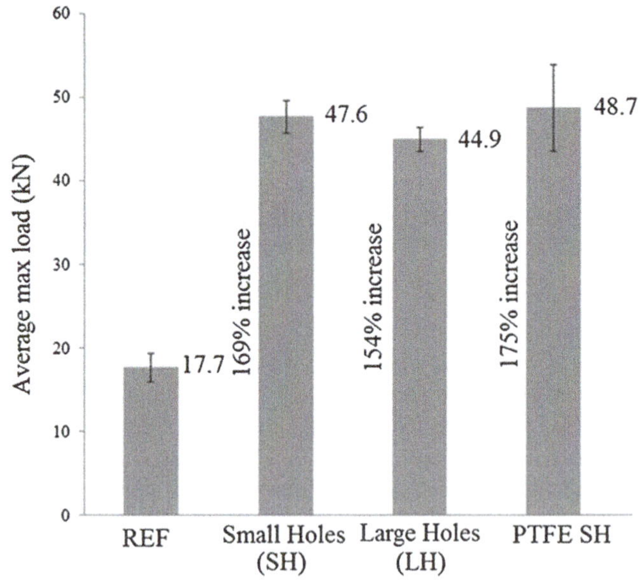

Fig. 5.6 Comparison of the average maximum load between non-perforated and perforated steel joints

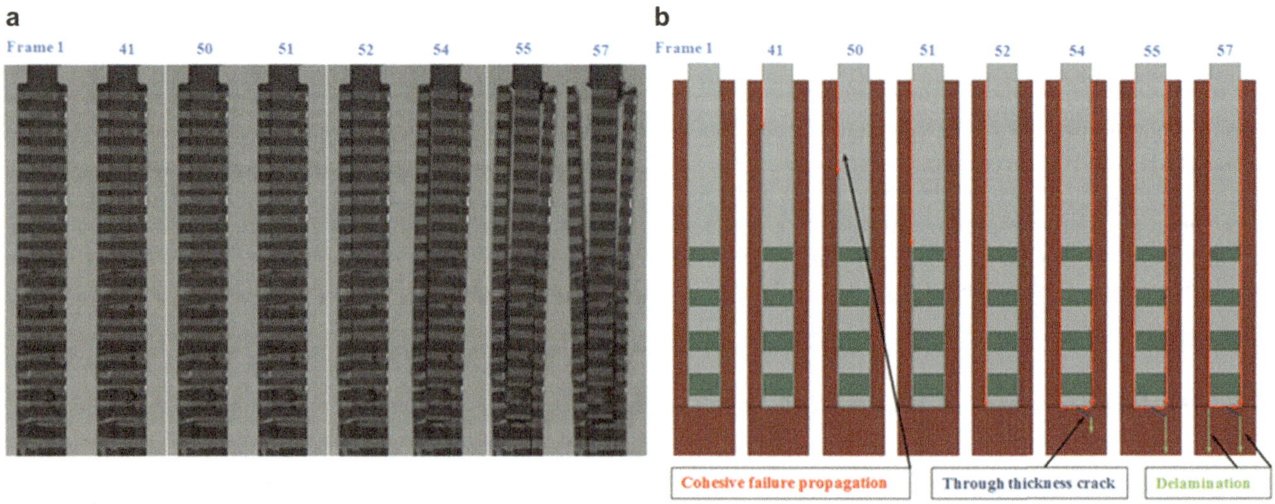

Fig. 5.7 (**a**) Catastrophic failure of a perforated SH specimen, showing selected frames from the high speed camera video (time proceeds from left to right as indicated by the frame number; the time between frames is 1/16,000 s), (**b**) schematic representation of the failure seen in the images

non-perforated joints indicated that the failure initiated at the butt end of the steel part of these joints, propagating towards the CFRP joint end until the first lap failed. From that point, the single intact lap carried a further increased load until final failure occurred. It is inferred that it is this change in failure mode associated with the change in steel stiffness which gives the observed increase in failure strength.

5.5 Conclusions

The biological principle of uniform strain found in branch-trunk joints has already inspired researchers to mimic this design to improve engineering joints. In nature, the tendon-to-bone attachment uses a gradual transition zone of stiffness to join flexible and stiff materials. One way to apply the idea of a transitional zone of stiffness to engineering joints is to gradually decrease the stiffness of the joint metal part by perforating it. A numerical and experimental investigation of bio-inspired CFRP to perforated steel joints has been presented. Compared to non-perforated joints, perforated joints showed a 175 % increase of joint strength. High speed camera images showed that the failure initiated at the CFRP end with a crack propagating abruptly towards the butt end of the joint. By contrast, non-perforated joints failed prematurely at the butt end and a crack propagated away from there to give failure of the first lap. The factor that contributed to the significantly higher strength of the perforated joints was the transitional zone of stiffness that the perforated joints offer. The resin in the holes offered no additional increase of the joint strength in the SH specimens. The numerical and experimental results were in good agreement.

Acknowledgements The authors acknowledge the financial support provided by the Engineering and Physical Sciences Research Council (EPSRC) and Dowty Propellers (part of GE Aviation).

References

1. Banea, M.D., da Silva, L.F.M.: Adhesively bonded joints in composite materials: an overview. J. Mater. Des. Appl. **223**, 1–18 (2009)
2. Campilho, R.D.S.G., Banea, M.D., Neto, J.A.B.P., da Silva, L.F.M.: Modelling of single-lap joints using cohesive zone models: effect of the cohesive parameters on the output of the simulations. J. Adhesion **88**, 513–533 (2012)
3. Anyfantis, K.N.: Finite element predictions of composite-to-metal bonded joints with ductile adhesive materials. Compos. Struct. **94**, 2632–2639 (2012)
4. Anyfantis, K.N., Tsouvalis, N.G.: Loading and fracture response of CFRP-to-steel adhesively bonded joints with thick adherents—Part I: experiments. Compos. Struct. **96**, 850–857 (2013)
5. Anyfantis, K.N., Tsouvalis, N.G.: Loading and fracture response of CFRP-to-steel adhesively bonded joints with thick adherents—Part II: numerical simulation. Compos. Struct. **96**, 858–868 (2013)
6. Avgoulas, E.I., Sutcliffe, M.P.F.: Biomimetic-inspired joining of composite with metal structures: a survey of natural joints and application to single lap joints. In: Proceedings of the SPIE Smart Structures and Materials: Bioinspiration, Biomimetics, and Bioreplication IV, San Diego, CA, 9–13 March 2014
7. Unden, H., Ridder, S.O.: Load-introducing armature as component part of a laminated structural element US Patent 4,673,606, 13 Feb 1985
8. Melogranaa, D.J., Grenestedt, L.J.: Improving joints between composites and steel using perforations. Compos. Part A: Appl. Sci. Manuf. **33**, 1253–1261 (2002)
9. Cao, J., Grenestedt, L.: Design and testing of joints for composite sandwich/steel hybridic ship hulls. Compos. Part A: Appl. Sci. Manuf. **35**, 1091–1105 (2004)
10. Burns, L.A., Mouritz, A.P., Pook, D., Feih, S.: Bio-inspired design of aerospace composite joints for improved damage tolerance. Compos. Struct. **94**, 995–1004 (2012)
11. Mattheck, C., Bethge, K.: The structural optimisation of trees. Naturwissenschaften **85**, 1–10 (1998)
12. Burns, L.A., Mouritz, A.P., Pook, D., Feih, S.: Strength improvement to composite T-joints under bending through bio-inspired design. Compos. Part A: Appl. Sci. Manuf. **43**, 1971–1980 (2012)
13. Benjamin, M., Toumi, H., Ralphs, J.R., Bydder, G., Best, T.M., Milz, S.: Where tendons and ligaments meet bone: attachment sites ('entheses') in relation to exercise and/or mechanical load. J. Anat. **208**, 471–490 (2006)
14. Thomopoulos, S., Genin, G.M., Galatz, L.M.: The development and morphogenesis of the tendon-to-bone insertion—What development can teach us about healing. J. Musculoskelet. Neuronal Interact. **10**(1), 35–45 (2010)
15. Benjamin, M., Kumai, T., Milz, S., Boszczyk, B.M., Boszczyk, A.A., Ralphs, J.R.: The skeletal attachment of tendons-tendon 'entheses'. Comp. Biochem. Physiol. A Mol. Integr. Physiol. **133**(4), 931–945 (2002)
16. Lu, H.H., Thomopoulos, S.: Functional attachment of soft tissues to bone: development, healing, and tissue engineering. Annu. Rev. Biomed. Eng. **15**, 201–226 (2013)
17. Waite, J.H., Lichtenegger, H.C., Stucky, G.D., Hansma, P.: Exploring molecular and mechanical gradients in structural bioscaffolds. Biochemistry **43**(24), 7653–7662 (2004)
18. Vincent, J.F.V.: Structural Biomaterials. Princeton University Press, Princeton, NJ (1990)
19. Suresh, S.: Graded materials for resistance to contact deformation and damage. Science **292**, 2447–2451 (2001)

20. Campilho, R.D.S.G., Banea, M.D., Pinto, A.M.G., da Silva, L.F.M., de Jesus, A.M.P.: Strength prediction of single- and double-lap joints by standard and extended finite element modelling. Int. J. Adhes. Adhes. **31**, 363–372 (2011)
21. Avgoulas, E.I., Sutcliffe, M.P.F.: Numerical and experimental investigation of CFRP to perforated steel joints. In: Proceedings of the 20th International Conference on Composite Materials, Copenhagen, Denmark, 9–24 July 2015

Chapter 6
Fabrication of 3D Thermoplastic Sandwich Structures Utilizing Ultrasonic Spot Welding

Cassandra M. Degen and Navaraj Gurung

Abstract The ultimate goal of this work was to create 3D polycarbonate sandwich structures fabricated utilizing ultrasonic spot welding (USSW). First, a study of weld strength vs. weld time, frequency, amplitude of vibration and depth of penetration was carried out. Lap shear and peel tests were used to quantitatively characterize weld strength, and from these results, a set of parameters for USSW of polycarbonate was determined. To create the sandwich structures, three layers of 1.59 mm polycarbonate sheet were stacked together, and using an alternating pattern of USSW, the middle layer was joined to each of the outer layers. Post-weld processing of the stacked layers involving heat and pressurization of the structure with air generated a 3D open cell geometry sandwich structure of polycarbonate. 3-point bend and drop impact tests were conducted to study the stiffness and the impact properties of the sandwich structures, and the results suggest that the fabricated polycarbonate sandwich structures might be suitable for applications requiring high strength and energy absorption while maintaining low weight. Additionally, the ability to easily incorporate different geometries in the structure allows the possibility to design structures with tailored mechanical properties.

Keywords Sandwich structures • Ultrasonic spot welding • Thermoforming • Polycarbonate • Thermoplastic

6.1 Introduction

As industries across the globe move towards optimization of efficiency, lightweight structural components are becoming increasingly important. Large strides have been made in high strength-to-weight ratio structures, and manufacturing techniques are needed to match the expanding material technology and the applications in which they are used.

Ultrasonic welding is a widely used technique for joining polymeric materials, and in particular, thermoplastic polymers. This welding method uses high frequency, low amplitude vibrations applied through an ultrasonic horn to create frictional heat [1–3]. Localized melting occurs at the interface of the materials during the welding process, and upon solidification, a solid-state joint is produced between polymer sheets. Equipment for ultrasonic welding is available from a number of companies, small in physical size, and easy to operate. Most ultrasonic spot welders have a constant frequency, typically either 20 or 40 kHz for polymeric materials. Adjustable parameters include: amplitude of vibration, pressure applied to the material during welding, length of time of vibration applied during welding, and depth of penetration of the tip into the material, each of which has an effect on the strength of the ultrasonic weld [4]. The ultrasonic welding system used in this work is a 20 kHz instrument from the Dukane Corporation [5] shown in Fig. 6.1.

Advantages of ultrasonic welding over other joining methods include high production rates, design freedom, energy efficiency, and ease of assembly [6]. These advantages make ultrasonic welding attractive for automation and industrial use. Disadvantages of this joining method include sample size limitations, material limitations to mainly thermoplastic polymers, noise considerations, and expensive equipment [6]. Currently, ultrasonic welding is used in industry in the automotive arena, for medical applications [6], for packaging, and for sealing of MEMS devices [7]. Ultrasonic welding has also been used for applications such as the bonding of wood [8], and for application to recent technologies such as the bonding of thermoplastic microfluidics [9]. Modeling efforts have also been used to contribute to the understanding of ultrasonic welding, capturing deformation, vibration dissipation, material flow, and heat flow [10].

C.M. Degen (✉) • N. Gurung
Department of Mechanical Engineering, South Dakota School of Mines and Technology, 501 E. Saint Joseph St., Rapid City, SD 57701, USA
e-mail: cassandra.degen@sdsmt.edu

G.L. Cloud et al. (eds.), *Joining Technologies for Composites and Dissimilar Materials, Volume 10*
Conference Proceedings of the Society for Experimental Mechanics Series, DOI 10.1007/978-3-319-42426-2_6

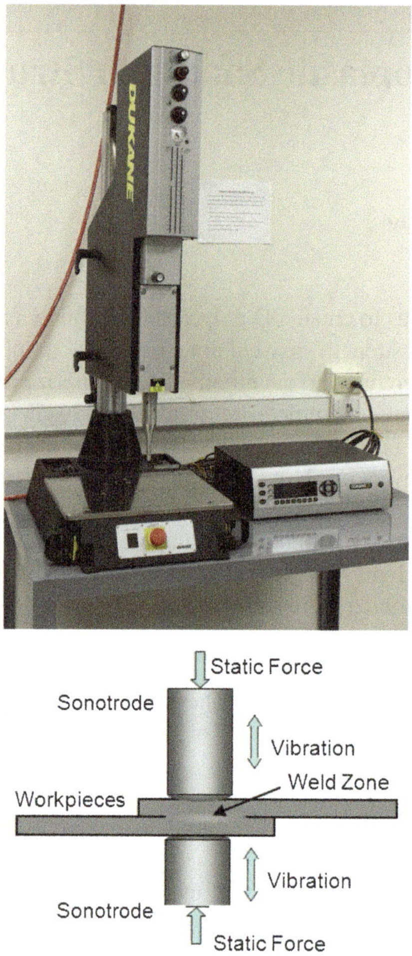

Fig. 6.1 Dukane ultrasonic welding system in PI's laboratory (*top*) and side-view schematic of ultrasonic spot welding process (*bottom*)

Polymers are extensively used for many consumer products. Low density, low electrical conductivity, high strength to weight ratio, good resistance to chemicals, wide choices of colors and transparencies, ease of manufacturing, and relatively low cost make polymers suitable for many applications [11]. The scope of applications of polymers ranges from common plastic bags to precision medical equipment. Polymers can contain a variety of additives and fillers to improve a particular property of the polymer such as processability, thermal or environmental stability, or the modulus of the final product [12]. Generally, for structural purposes, glass or carbon fibers are added to polymer resin to achieve superior strength. Carbon fiber reinforced plastic composites are one of the highly sought after materials for structural applications and the use of these composites has been growing steadily since the early 1950s [13]. Despite their superior structural efficiency, the complex processing techniques and high manufacturing cost of these reinforced plastics has restricted their use in general low budget applications.

The structural efficiency of any polymer component can be maximized by either using stronger materials or by optimizing the structure's geometry [13]. Therefore the limitation posed by the processing techniques and high manufacturing cost of reinforced polymers can be overcome by optimizing its structural geometry. One of the best methods to fabricate high strength-to-weight ratio structures is to use stiffeners, as in the cases of I-beams and T joints. These stiffeners can be used as an integral part of the component or may be attached through adhesive bonding [13].

Sandwich-structures are composite structures consisting of two face sheets and a preformed, lightweight core. In general, a sandwich layer consists of (1) a pair of thin stiff skins; (2) a thick, light weight core to separate the skins and carry the load from one skin to another; (3) an adhesive or some other form of attachment which transmits loads in between core and the skins. The separation created by the core increases moment of inertia of the skin plates, which makes them more efficient in resisting bending and buckling loads. Table 6.1 shows a comparison between the stiffness and strength of structures made out of the same thickness plate with varying thickness of core material, with the far left structure containing no core.

Table 6.1 An example of structural efficiency of sandwich panels in terms of weight [13]

	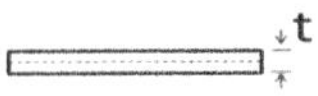	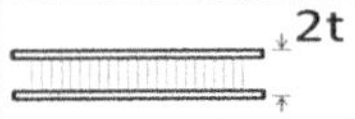	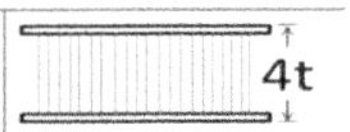
Relative bending stiffness	1	7.0	37
Relative bending strength	1	3.5	9.2
Relative strength	1	1.03	1.06

Sandwich-structures have benefits of high strength-to-weight ratios [14] and enhanced resistance to impulsive loads [15], characteristics particularly important for aerospace or automotive applications, and have received significant attention in recent years. In addition, varying the geometry of sandwich-structures has been shown to affect their mechanical response [16, 17]. Sandwich-structures with a sinusoidal unit cell have been studied for their elastic properties analytically and experimentally [18]. Analytical models have been developed to predict the mechanical behavior of corrugated paperboard [18]. The dynamic shock cushioning of sandwich-structures is also an area of recent research, and has been explored experimentally with a drop shock tester to find that the structures have superior cushioning properties, particularly useful for the packaging industry [19, 20]. The adaptability of sandwich-structures as multifunctional materials has been explored with the incorporation of polymer resin and glass spheres into the structure to enhance the fire response [21], and the incorporation of shape memory materials into the structure to create sandwich-structures to detect damage [22]. Analytical and computational modeling of the mechanical [23–27] and thermal [28, 29] responses of sandwich-structures is also an area of interest. While the benefits of sandwich-structures are well-studied, cost- and time- effective manufacturing methods remain an active research area.

The goal of this work was to utilize ultrasonic spot welding to fabricate 3D polycarbonate sandwich structures and characterize the bending stiffness and impact properties of such structures. Experiments were carried out to determine the proper thermoforming parameters for the fabrication process. 3D sandwich structures designed and fabricated with two different grid sizes were analyzed to determine the possibilities of tailoring the mechanical properties of the structure. 3-point bend tests and impact tests were conducted to characterize the sandwich structures. While the work presented here focuses on using polycarbonate panels as both outer layers and core, the flexibility of the ultrasonic spot welding technique allows for the replacement of polycarbonate panels with fiber reinforced composite material, which broadens the scope of the work.

6.2 Characterization of Ultrasonic Spot Welds in Polycarbonate

The welding parameters that dictate the quality of ultrasonic spot welds are frequency, weld amplitude, weld time, weld pressure, and weld depth. The Dukane ultrasonic spot weld system has a fixed frequency of 20 kHz, a maximum amplitude of 120 μm adjustable from 20 % (24 μm) to 100 % (120 μm), variable weld time, pressure, and depth. The depth, to which the welding tip penetrates through the material, was kept constant for this work at 27 % through the thickness of the bottom layer of material. The material used for this work was 1.59 mm thick Bayer MaterialScience Makrolon® polycarbonate sheet. To study the effect of varying parameters and determine effective parameters, welds were made with the parameters listed in Table 6.2 by varying one parameter while keeping the others constant, and the strength of those welds were measured by both lap shear and peel tests. During the welding process, the material at the welding interface melts and flows across the vicinity of spot weld. The area covered by this melted material can be visually traced from optical images of the spot weld. Image processing software (*ImageJ* 1.47 t) was used to measure each bond area.

To measure the tensile shear strength of the spot welds, lap shear tests were carried out according to ASTM D3164 [30]. Mode II failure of the welds was achieved by gluing tabs at both ends of the specimen to minimize the offsetting of uniaxial shear force. Peel tests designed to initiate Mode I failure were conducted according to ASTM D1876-08 [31], using metal hinges glued to this specimen to mount in the testing grips. A MTS 858 Mini Bionix II (Fig. 6.2) was used for both testing setups and the displacement rate used for all testing was 1.27 mm/min. During the lap shear and peel tests, the load was measured and the peak load was recorded. The peak load experienced by the sample before failure was divided by the previously measured bond area to calculate the peak stress.

Figure 6.3 shows the peak tensile shear stress achieved in ultrasonic spot welds with varying weld amplitude, weld time, and weld pressure during lap shear and peel testing. The Dukane ultrasonic welding system also measured the total energy input during the weld process by calculating the required power (voltage × current) for each 0.5 ms and summing the power over the entire weld cycle to compute the input energy (power × time). Figure 6.3 also shows the summary of weld strength measured by peak stress during lap shear and peel testing as a function of the energy input during the weld cycle. While the trend was weak and data was highly varied, the maximum weld strengths were seen when the energy input was near 100 J. The data suggests further optimization is needed for weld parameters as well as repeatability of ultrasonic spot welding to

Table 6.2 Ultrasonic spot welding parameters used for lap shear and peel samples

	Pressure (kPa)	Amplitude (um)	Time (s)	
Range of parameters	103, 172, 241	24, 30, 50.4	0.7, 1.4, 2.1,	2, 4, 6, 8
Number of samples (n)	12	9	12	3

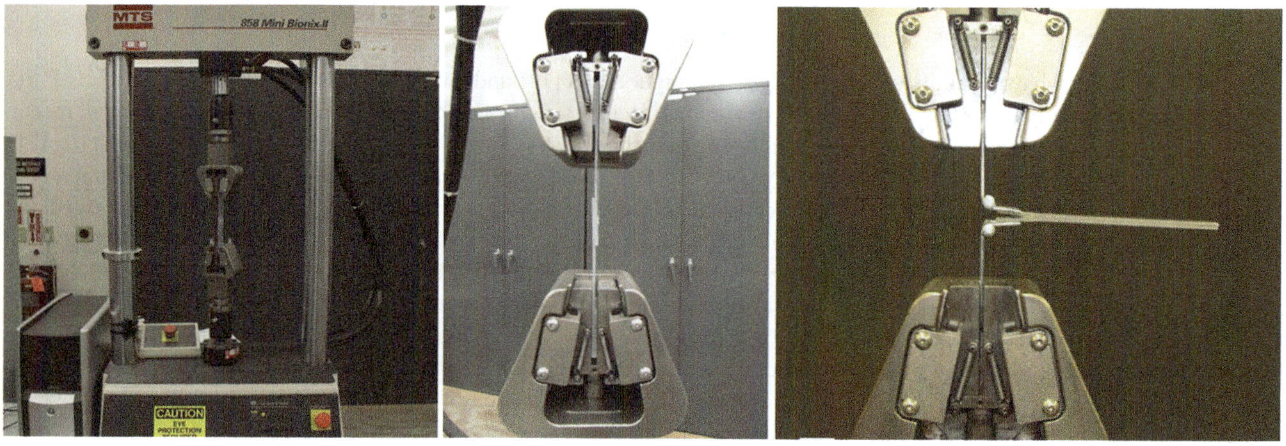

Fig. 6.2 MTS Mini Bionix II with lap shear sample (*left*), and close up of the lap shear test (*center*) and peel test (*right*)

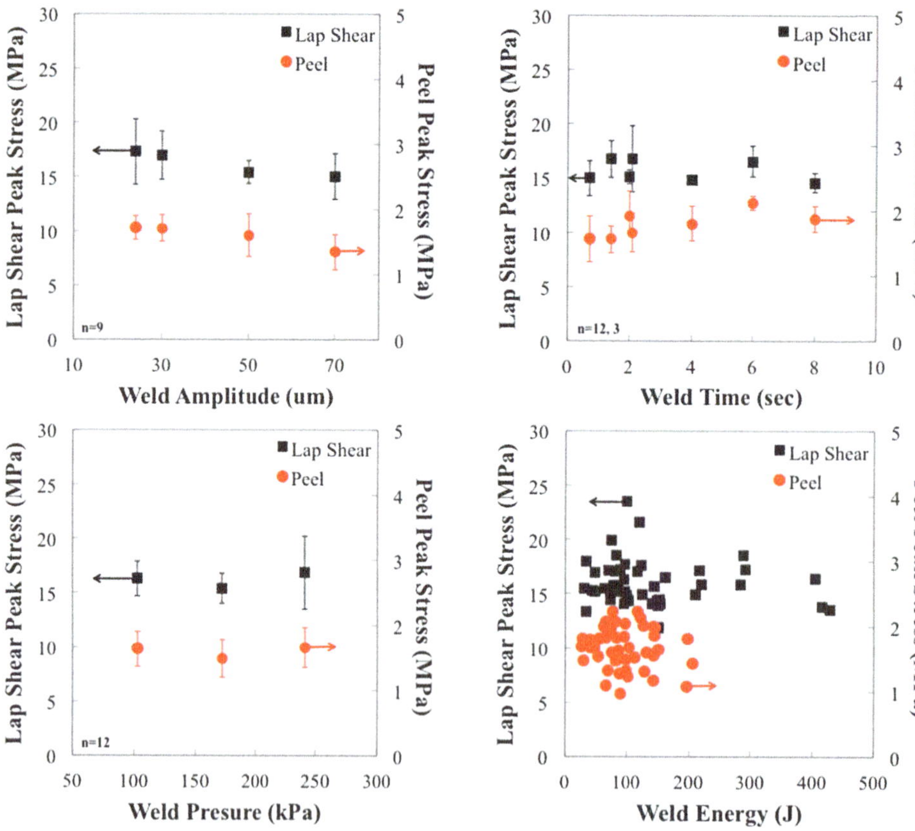

Fig. 6.3 Lap shear and peel test results for varying weld amplitude (*top left*), weld time (*top right*), weld pressure (*bottom left*), and weld energy (*bottom right*). Error bars represent one standard deviation

give clear trends for the ideal welding parameters to achieve maximum lap shear and peel strength. The results indicate the following as appropriate parameters for the ultrasonic spot welding of 1.59 mm thick polycarbonate sheet: welding amplitude of 24 μm, welding time of 6 s, and welding pressure of 241 psi. These parameters were used throughout the remainder of the work.

6.3 Fabrication of Sandwich Structures

To create sandwich structures, three stacked layers of polycarbonate sheets were prepared for ultrasonic spot welding by marking the grid size and spot weld location, drilling a hold on the corner of the lower layer and cutting a slot in the middle layer to serve as an air channel for later thermoforming, and spraying a thin layer of urethane mold release on the surfaces that come into contact with each other. Using the weld parameters discussed above (weld time = 6 s, amplitude = 24 μm (20 %), weld pressure = 241 kPa, depth of weld = 1.89 mm), ultrasonic spot welds were placed in a weld pattern as shown in Fig. 6.4. The welding process involved creating ultrasonic spot welds in a grid pattern on one side of the polycarbonate stack, flipping the stack, and welding an alternate grid pattern of welds. The weld depth was changed by 0.6 mm when the stack was flipped over to account for the increased thickness of the stack due to the first set of welds. In Fig. 6.4, the dark black spots represent the spot welds between the top and middle layer, and the light grey spots represent the welds between middle and lower layer. Two structures with differing grid sizes were designed, and depending on the structure desired, the distance between the weld (x in Fig. 6.4) was either 50.8 or 76.2 mm.

A thermoforming process was used to expand the welded stack of polycarbonate sheets into a sandwich structure. A 6061 aluminum frame (Fig. 6.5) consisting of a base frame and a cover was fabricated to fit the polycarbonate stack. The protruded part of the cover fit in the cavity of the base frame but did not touch the bottom of the base frame. The welded polycarbonate stack sat in between the gaps so that all four edges were clamped evenly. An air inlet was drilled on one corner of the mold, which was later connected to a pressurized air supply. A 30 T CARVER hot press was used to heat the whole assembly (Fig. 6.5). From DMA results, the glass transition temperature was found to be 160 °C for the polycarbonate material used. Several experiments showed that forming the polycarbonate above 160 °C resulted in a degraded polycarbonate, while temperatures lower than 160 °C caused the ultrasonic spot welds to fail before thermoforming was complete.

The forming procedure was divided into five steps: (1) pre-drying, (2) clamping, (3) heating, (4) pressure forming, and (5) cooling. Because water vapor trapped in the polycarbonate sheets created bubbles and was detrimental to the thermoforming process, the polycarbonate sheets were pre-dried before thermoforming. As recommended by the material supplier, the polycarbonate was kept in a dehumidified air circulated oven at 120 °C for 4 h [32]. Within 10 h of pre-drying, the polycarbonate

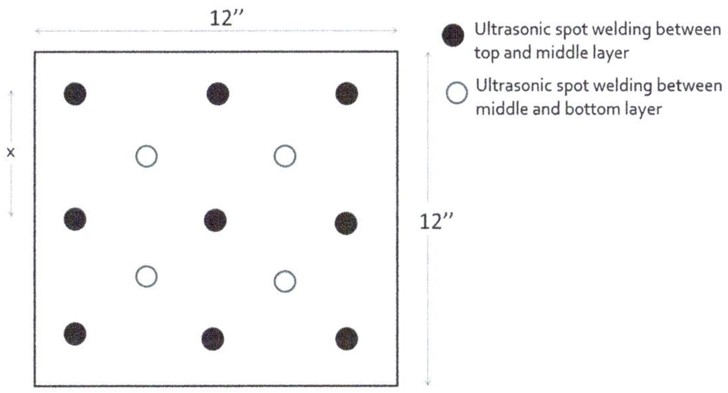

Fig. 6.4 Ultrasonic spot weld placement for sandwich structure fabrication

Fig. 6.5 Aluminum frame (*left*) to hold polycarbonate stack in hot press (*center*), and a side view schematic of the polycarbonate stack in the molds within the hot press (*right*)

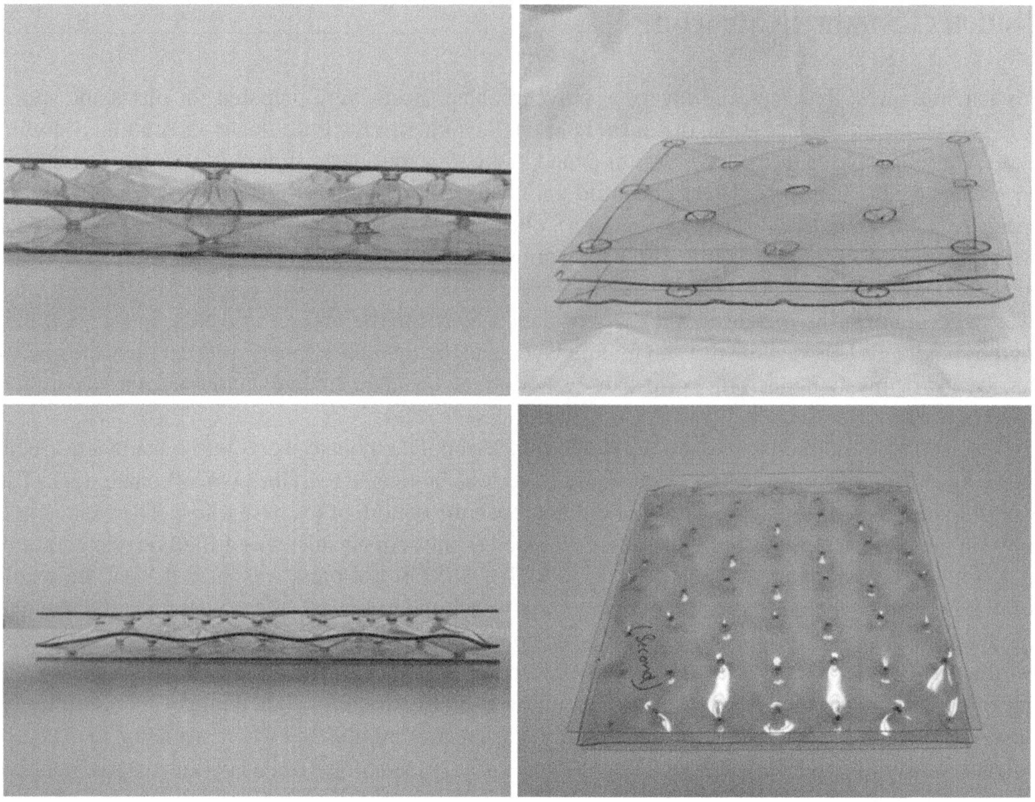

Fig. 6.6 Side (*left*) and top (*right*) views of the 50.8 mm (*top*) and 76.2 mm (*bottom*) grid size sandwich structures. The thickness of each structure is 25.4 mm

stack was placed in the aluminum mold with the air inlet port in the mold aligned with the hole drilled in the stack. The surface of the aluminum mold that came into contact with the polycarbonate sheet was packed with silicone gasket maker to ensure air tightness of the assembly. The molds were then clamped together and left to partially cure for about 30 min, so that any leaks would be further sealed when the assembly was clamped by the hydraulic press. A very low pressure was then applied to the assembly to check for air leaks. The entire assembly is shown schematically in Fig. 6.5.

A thin layer of urethane mold release was sprayed on the outer exposed surface of polycarbonate stack, which prevented the polycarbonate sheet from sticking to the surface of hot platens. The assembly was then pressed with a hydraulic pressure of 17.8 kN in between the hot platens, allowing the temperature of the polycarbonate to rise to the thermoforming temperature of 160 °C. Experiments showed that 96.5 kPa was a suitable pressure to properly form the structure without causing weld failure. When the assembly reached the desired forming temperature, the pressure was regulated to form the polycarbonate sheets into the designed shape. After forming for 30 min, the temperature of the hot press was turned off and the sandwich structure was cooled slowly in the hot press while the air pressure and clamping pressure were still applied. A representative fully formed 76.2 mm grid size sandwich structure and 50.8 mm grid size sandwich structure are shown in Fig. 6.6.

6.4 Mechanical Characterization of Sandwich Structures

6.4.1 3-Point Bend Testing

To investigate the bending behavior of the fabricated 3D structures, 3-point bend tests were carried out on both the 50.8 and 76.2 mm grid size polycarbonate sandwich structures. Under the 3-point bending loading scheme, each of the three layers of the sandwich structure experienced different loading stresses. The outer two layers acted as the flanges and carried the bending stress in response to the normal loading. The top layer was under compression while the bottom layer experienced expansion. The middle corrugated layer resisted the bending of the structure and held the outer layers apart such that the stiffness of the overall structure was increased. The ultrasonic spot welds, in between the outer and middle layer, held these layers together and allowed the structure to act as a single unit, providing bending rigidity.

Table 6.3 Welding parameters used for initial studies of ultrasonic spot welding of 1.59 mm thick polycarbonate

Specimen type	Sandwich structure peak load (kN)	Deflection of top surface (mm)	Deflection of bottom surface (mm)	Load required for solid polycarbonate sheet (kN)
50.8 mm grid sandwich structure	0.82	7.37	4.06	0.11
	1.0	10.9	7.62	0.20
	1.0	11.9	8.13	0.22
76.2 mm grid sandwich structure	0.44	11.2	3.56	0.10
	0.52	5.59	3.81	0.10
	0.32	11.2	6.60	0.18

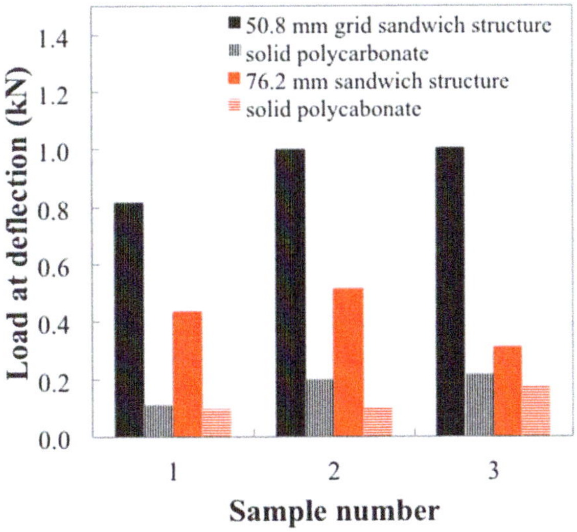

Fig. 6.7 Comparison of the 3-point bending load required to produce the same deflection in the 50.8 mm grid sandwich structure and solid polycarbonate of the same weight (*black*), and in the 76.2 mm grid sandwich structure and solid polycarbonate of the same weight (*red*)

ASTM D790-07 [33] was adapted to perform the 3-point bend test, and the rate of the cross-head motion was maintained at 14.4 mm/min. A MTS 810 was used as the test machine and special fixtures were designed and fabricated as recommended by the ASTM guidelines. Due to the compressible nature of the sandwich structures, the specimen was slightly compressed during the test. As a result, the top layer and the bottom layer did not experience the same strain. The deflection on the top layer is recorded by the machine itself and a dial gauge was used to measure the deflection of the lower layer. Three specimens of each 50.8 and 76.2 mm grid size sandwich structures were tested. The peak load is taken as the load that the structure can withstand before failing. For this work, failure of a weld is considered to be failure of the structure. These peak values occur just before the welds begin to fail. The peak load and the corresponding deflections of the top surface and bottom surface are listed below in Table 6.3.

To compare the bending properties of the sandwich structures to solid polycarbonate, the maximum deflection of the sandwich structures is used to calculate the load required to produce the same deflection in a solid polycarbonate material of the same weight. The thickness of polycarbonate sheet was taken as the combined thickness of all three layers, i.e. 4.76 mm, and the flexural modulus of polycarbonate was obtained from published values [34]. The required load to deflect a solid polycarbonate sheet of the same weight was calculated from ASTM guidelines using:

$$P = \frac{4EDbd^3}{L^3} \tag{6.1}$$

where, P = required load E = flexural modulus of polycarbonate (2380 MPa [34]), D = deflection of outer surface, b = width of the specimen (241 mm), d = thickness of specimen (25.4 mm), L = length of support span (210 mm). The results from the 3-point bend test of the sandwich structures and the calculated loads for solid polycarbonate are listed in Table 6.3 and compared graphically in Fig. 6.7.

The load bearing capacity of the sandwich structures averaged approximately five times that of plain polycarbonate sheets of the same weight when deflected by the same amount, indicating the effectiveness of sandwich structures for holding bending loads. This result shows that for applications where the weight of the structure plays a critical role, the sandwich structures can be a viable option for structural components.

6.4.2 *Impact Testing*

The behavior of the sandwich structures was also characterized with impact loading. Using an Instron CEAST 9350, a weight was vertically dropped along guide rails to strike the specimen. The 0.5″ radius spherical head striker of the machine was fitted with a force transducer to measure the impact force during the test. The force-time data was used to extract information regarding each specimen. The energy of impact was determined by the mass of the falling weight (5.18 kg) and the initial height (75 mm). Three specimens of each of the 50.8 and 76.2 mm grid size structures as well as a solid polycarbonate sheet of the same thickness (25.4 mm) were tested under impact.

Figure 6.8 shows the force experienced by the striker for representative samples of the 50.8 mm grid size, 76.2 mm grid size, and the solid polycarbonate, and shows the peak force experienced by the striker with the 25.4 mm thick polycarbonate sheet was significantly higher than of with the sandwich structures and occurred over a shorter time.

The time over which the impact occurs indicates the cushioning ability of the material and the higher the time of impact, the better the material can mitigate the detrimental effects of impact loads. The results from the impact test showed the 76.2 mm grid size sandwich structure had a higher time of impact as compared to the 50.8 mm grid size sandwich structure. Similarly, the 50.8 mm grid size sandwich structure had a higher time of impact compared to the solid polycarbonate sheet. The impact properties of the 3D sandwich structures show their potential in impact and shock mitigation applications. The impact depth, associated with the compressible nature of the structure, was also measured. The results suggest that the 50.8 mm grid size sandwich structure is the most compressible among all three specimens.

Because energy absorption is a major criteria used to characterize many structural components, the amount of energy absorbed by each specimen during the impact test was calculated by a plate impact theory developed by Jang, Huang, Hsieh and Kowbel [35]. Here,

$$(t) = \int_0^t P(t')v(t')dt' \tag{6.2}$$

where $\Gamma(t)$ is the total absorbed energy, $P(t')$ is the load as a function of time, $v(t')$ is the velocity as a function of time, and dt' is the size of the time interval. The calculated absorbed energy per weight for each sample is shown in Fig. 6.9, showing the advantage of the sandwich structures in energy absorption per weight over solid polycarbonate. The results of the impact tests suggest that the 3D polycarbonate sandwich structures might be suitable for applications where a large energy absorption and low weight are desired.

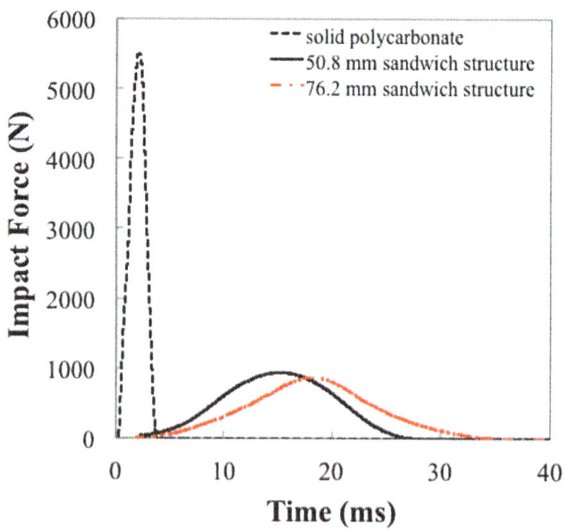

Fig. 6.8 Force read by the impact striker vs. time for a 50.8 mm grid sandwich structure, a 76.2 mm grid sandwich structure, and a solid polycarbonate of the same thickness

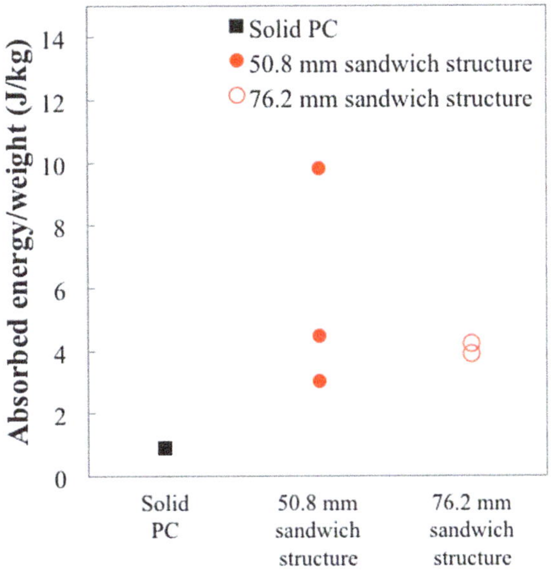

Fig. 6.9 The calculated absorbed energy per weight for a 50.8 mm grid sandwich structure, a 76.2 mm grid sandwich structure and a solid polycarbonate of the same thickness

6.5 Conclusions

The results of this work showed that 3D sandwich structures of polycarbonate could be fabricated using ultrasonic spot welding and demonstrated the flexibility in the fabrication technique. Tailoring of the mechanical properties of the sandwich structures was possible and was demonstrated with different grid size structures. Here, the ultrasonic spot welding process was used as the attachment method of different layers in a 3D sandwich structure. The strength of the ultrasonic spot welds was tested experimentally and was used to hold the layers of polycarbonate together while exposed to the thermoforming process. Both pressure and temperature proved to be important parameters of the thermoforming process. The 3D sandwich structures were tested for their structural properties in 3-point bend testing and impact testing. Experimental results revealed that the sandwich structures could withstand loads five times higher than solid polycarbonate of the same weight. Impact tests showed that the sandwich structures were superior in mitigating impact loads, making the sandwich structures appropriate for shock impact load resisting applications.

There are many aspects of the fabrication technique that could be improved in order to enhance the 3D sandwich structures, and the incorporation of other features to the polycarbonate sandwich structures is promising. The weld quality can be improved by investigations on the microscopic level regarding material interaction at the weld zone. Structures with desired properties can be designed and fabricated by altering not only the grid sizes, but also the thickness of the structure and the thickness of the individual layers. Moreover, the ultrasonic spot welding technique is compatible with dissimilar thermoplastic materials, which enables the use of different polymers for each layer of structure. This adaptability could help to integrate the properties of different materials into a single structure. Fiber reinforced composite materials are also a candidate for use on the outer layer of the structure to increase the stiffness and impact resistance of the overall structure.

References

1. Gutnik, V.G., Gorbach, N.V., Dashkov, A.V.: Some characteristics of ultrasonic welding polymers. Fibre Chem. **34**, 426–432 (2002)
2. Rani, M.R., Prakasan, K., Rudramoorthy, R.: Study of different joints for ultrasonic welding of semicrystalline polymers. Exp. Tech. **33**, 36–42 (2009)
3. Roopa Rani, M., Prakasan, K., Rudramoorthy, R.: Designing joints for ultrasonic welding of plastics. Weld. J. **84**, 50 (2005)
4. Liu, S.J., Chang, I.T., Hung, S.W.: Factors affecting the joint strength of ultrasonically welded polypropylene composites. Polym. Compos. **22**(1), 132–141 (2001)
5. Dukane iQ Series Ultrasonic Press System Specification Sheet, 2013
6. Amancip-Filho, S.T., Dos Santos, J.F.: Joining of polymers and polymer–metal hybrid structures: recent developments and trends. Polym. Eng. Sci. **49**, 1461–1476 (2009)
7. Kim, J.B., Jeong, B., Chiao, M., Lin, L.: Ultrasonic bonding for MEMS sealing and packaging. IEEE Trans. Adv. Pack. **32**, 461–467 (2009)

8. Gfeller, B., Zanetti, M., Properzi, M., Pizzi, A., Pichelin, F., Lehmann, M., Delmotte, L.: Wood bonding by vibrational welding. J. Adhes. Sci. Technol. **17**, 1573–1589 (2003)
9. Tsao, C.-W., DeVoe, D.L.: Bonding of thermoplastic polymer microfluidics. Microfluid. Nanofluid. **6**, 1–16 (2008)
10. Nonhof, C.J., Luiten, G.A.: Estimates for process conditions during the ultrasonic welding of thermoplastics. Polym. Eng. Sci. **36**, 1177–1183 (1996)
11. Singh, R., Mattoo, A., Saigal, A.: Optimizing the design and impact behavior of a polymeric enclosure. Mater. Des. **27**, 955–967 (2006)
12. Fried, J.R.: Polymer Science & Technology, 2nd edn. Prentice Hall, Upper Saddle River, NJ (2003).
13. Petras, A.: Design of sandwich structures (Doctoral dissertation), Cambridge University Engineering Department, Cambridge, England (1998)
14. Besse, C.: Development and optimization of a formable sandwich sheet (Doctoral dissertation), Ecole Polytechnique, Palaiseau, France (2012)
15. Rabczuk, T., Kim, J.Y., Samaniego, E., Belytschko, T.: A homogenization method is proposed for sandwich structures. Int. J. Numer. Meth. Eng. **61**, 1009–1027 (2004)
16. Kocher, C., Watson, W., Gomez, M., Gonzalez, I., Birman, V.: Integrity of sandwich panels and beams with truss-reinforced cores. J. Aerospace Eng. **15**, 111–117 (2002)
17. Xin, F.X., Lu, T.: Effects of core topology on sound insulation performance of lightweight all-metallic sandwich panels. Mater. Manuf. Process. **26**, 1213–1221 (2011)
18. Qiao, P., Wang, J.: Mechanics of composite sinusoidal honeycomb cores. J. Aerospace Eng. **18**, 42–50 (2005)
19. Krusper, A., Isaksson, P., Gradin, P.: Modeling of out-of-plane compression loading of corrugated paper board structures. J. Eng. Mech. **133**, 1171 (2007)
20. Guo, Y., Xu, W., Fu, Y., Wang, H.: Dynamic shock cushioning characteristics and vibration transmissibility of X-PLY corrugated paperboard. Shock Vib. **18**, 525–535 (2011)
21. Giancaspro, J., Balaguru, P.N., Lyon, R.E.: Use of inorganic polymer to improve the fire response of balsa sandwich structures. J. Mater. Civil Eng. **18**, 390–397 (2006)
22. Okabe, Y., Minakuchi, S., Shiraishi, N., Murakami, K., Takeda, N.: Smart honeycomb sandwich panels with damage detection and shape recovery functions. Adv. Compos. Mater. **17**, 41 (2008)
23. Doumanidis, C., Gao, Y.: Mechanical modeling of ultrasonic welding. Weld. J. **83**, 140–146 (2002)
24. Gao, Y., Doumanidis, C.: Mechanical analysis of ultrasonic bonding for rapid prototyping. J. Manuf. Sci. Eng. **124**, 426–434 (2002)
25. Besse, C.C., Mohr, D.: Optimization of the effective shear properties of a bidirectionally corrugated sandwich core structure. J. Appl. Mech. **80**, 011012 (2012)
26. Brauner, C., Block, T.B., Herrmann, A.S.: Meso-level manufacturing process simulation of sandwich structures to analyze viscoelastic-dependent residual stresses. J. Compos. Mater. **46**, 783–799 (2011)
27. Thomsen, O.T., Bozhevolnaya, E., Lyckegaard, A., Mortensen, F.: Characterisation and assessment of load response, failure and fatigue phenomena in sandwich structures induced by localized effects—a review. Strain **44**, 85–101 (2008)
28. Arao, Y., Koyanagi, J., Utsunomiya, S., Kawada, H.: Analysis of thermal deformation on a honeycomb sandwich CFRP mirror. Mech. Adv. Mater. Struct. **17**, 328–334 (2010)
29. Song, W., Ma, Z.: Behavior of honeycomb FRP sandwich structure under combined effects of service load and low temperatures. J. Compos. Constr. **15**, 985–991 (2011)
30. ASTM D3164. In: Standard Test Method for Strength Properties of Adhesively Bonded Plastic Lap-Shear Sandwich Joints in Shear by Tension Loading. ASTM International (2008)
31. ASTM D1876-08. In: Standard Test Method for Peel Resistance of Adhesives (T-Peel Test). ASTM International (2008)
32. Technology, B.S.: Bayer Guide on Thermoforming Makrolon
33. ASTM D790-07. In: Standard Test Methods for Flexural Properties of Unreinforced and Reinforced Plastics and Electrical Insulating Materials. ASTM International (2008)
34. LLC (BMS). Makrolon GP (Polycarbonate). Plastics International (2010)
35. Jang, B.P., Huang, C.T., Hsieh, C.Y., Kowbel, W., Jang, B.Z.: Repeated impact failure of continuous fiber reinforced thermoplastic and thermoset composites. J. Compos. Mater. **25**, 1171–1203 (1991)

Chapter 7
Impact and Lap Shear Properties of Ultrasonically Spot Welded Composite Lap Joints

Cassandra M. Degen, Lidvin Kjerengtroen, Eirik Valseth, and Joseph R. Newkirk

Abstract Ultrasonic spot welding (USSW) is a widely used technique for joining thermoplastics where high frequency, low amplitude vibrations are applied through an ultrasonic horn resting on the polymer surface to create frictional heat, producing a solid state joint between polymer sheets. Advantages such as short weld cycle time, fewer moving components and reproducibility make this technique attractive for automation and industrial use. The goal of this work was to evaluate the feasibility and analyze the lap shear and impact strength of a composite material joint created using ultrasonic spot welding. The base material used for the joints was a composite consisting of a polycarbonate matrix with chopped glass fibers. The strength of the lap joints was determined through experimental lap shear and impact testing. A finite element analysis was conducted for more thorough insight into the stress patterns in the lap joints. Experiments showed that the ultrasonically spot welded joints tested in tensile lap shear loading carried a load 2.3 times higher than adhesive joints and the impact tested joints had an impact strength 3.5 times higher than adhesive joints. The results of this work suggest ultrasonic spot welding as a viable joining method for thermoplastic composite materials.

Keywords Composite joining • Ultrasonic spot welding • Thermoplastic • Composites • Impact testing

7.1 Introduction

The use of composite materials has seen an explosive growth over the last 20 years as the need for lightweight materials pushes to the forefront of many industries [1]. While the design of a single-piece composite part is ideal, molding considerations and other mechanical limitations make joining of composite parts necessary to complete assembly and a critical step in the manufacture of composite parts [2]. In general, composite parts can be joined using mechanical fastening, adhesive bonding, or fusion welding processes. The main limitations of mechanical fastening include increased weight and the presence of large stress concentrations around the fastening locations. Adhesive bonding has material dependent limitations and requires proper mating of the material surface to the adherent [3, 4]. Composite fusion welding relies on the melting, intermolecular diffusion across the joint, and re-solidification of the matrix material. The molecular details of thermoplastic materials allows for melting due to applied heat. For this reason, thermoplastics matrix composites are candidate materials for fusion welding techniques.

Fiber-reinforced thermoplastic composites are one of the fastest growing material choices of the composites industry [5, 6]. Fueled largely by the automotive industry, composite materials with a thermoplastic matrix offer several advantages of traditional composites with a thermoset matrix, including: increased toughness, longer material lifespan, non-toxic and energy-saving production, recyclability, lower manufacturing cost, and the ability to reshape and reform [7]. As thermoplastic composite use increases, manufacturing techniques such as molding and joining for such materials must also progress, and these topics have been the focus of many recent investigations [5, 8–12].

Fusion welding may be achieved through a variety of welding techniques, and the technique studied here is ultrasonic spot welding (USSW). USSW is a mechanical welding process where low amplitude, high frequency vibrations are used to locally melt thermoplastic materials or metals thus creating a bond. This welding technique was developed in the 1940s, but was not patented or adapted in the US until the 1960s. The process of ultrasonically welding thermoplastics was started in 1963 by Sonics & Materials Inc [13]. USSW is now one of the most commonly used welding methods to join thermoplastic materials [14].

C.M. Degen (✉) • L. Kjerengtroen • E. Valseth • J.R. Newkirk
Department of Mechanical Engineering, South Dakota School of Mines and Technology,
501 E. Saint Joseph St., Rapid City, SD 57701, USA
e-mail: cassandra.degen@sdsmt.edu

G.L. Cloud et al. (eds.), *Joining Technologies for Composites and Dissimilar Materials, Volume 10*
Conference Proceedings of the Society for Experimental Mechanics Series, DOI 10.1007/978-3-319-42426-2_7

The energy transferred to a thermoplastic material subjected to ultrasonic vibrations is dissipated through intermolecular friction, leading to heating of the material and localized melting [14]. An ultrasonic welder typically consists of six main components: a pneumatic press, generator, transducer, booster, welding horn and the base of the machine [15]. The ultrasonic spot welder used in this work is a 20 kHz Dukane iQ series Ultrasonic pneumatic press 43Q-220. The process of ultrasonic welding is typically separated into five steps [16]: (1) the two parts to be welded are fixed at the base in the fixture, (2) the welding horn is moved into contact with the upper part in the fixture, (3) a force (in the form of a pressure) is applied to hold the two parts to clamp the parts together during welding, (4) the ultrasonic vibrations are induced by vibrating the welding horn vertically at the desired frequency, thus melting the plastic material locally at the interface of the two pieces of material, (5) the ultrasonic vibrations are shut off after a preset time, and the horn is held in place by continuing to apply a force. The force is maintained until the melted material has solidified, and the horn is returned to its original position. USSW has several advantages over other welding techniques, such as fast cycle times, and is faster than other comparable welding or joining methods [14]. The process is energy efficient since the energy needed to induce local melting of the plastic material is modest, in the order of 2–3 kJ for this work. Minimal preparation of material before welding is required, compared to adhesive bonding or bonding using mechanical fasteners [14].

This work examines the feasibility of USSW of a composite consisting of a polycarbonate matrix with chopped glass fibers by analyzing the lap shear and impact strength joints created using ultrasonic spot welding. Experimental results coupled with a finite element analysis (FEA) give insight into the stress patterns of lap joints. Experiments showed that the ultrasonically spot welded joints tested in tensile lap shear loading carried a load 2.3 times higher than adhesive joints and the impact tested joints had an impact strength 3.5 times higher than adhesive joints. The results of this work suggest ultrasonic spot welding as a viable joining method for thermoplastic composite materials.

7.2 Ultrasonic Spot Welding of a Composite Material

Ultrasonic spot welding was used to join a thermoplastic fiber reinforced polymer (Tecanat GF20, produced by Ensinger Inc). This composite consists of a polycarbonate matrix and 20 wt% chopped glass fibers, and the mechanical properties are shown in Table 7.1. Tecanat GF20 is available in pre-compounded, extruded sheets, 6.35 mm thick, with 4.76 mm long fibers. The glass fibers are chopped and randomly oriented in three dimensions. However, the extrusion process in which the material is produced does align the fibers within the matrix, and thus the material is not fully isotropic due to this alignment of the fibers, and will have different properties in each direction. No data on the anisotropic properties of Tecanat GF20 was given by the supplier.

Since the welding parameters were expected to influence the strength of the welds, a study was carried out to determine the effect of changing amplitude, time, and pressure during the ultrasonic spot welding process. Because the Tecanat GF20 composite is 80 wt% polycarbonate, pure polycarbonate material was used to study and estimate appropriate ultrasonic spot welding parameters. Lap shear samples of 6.35 mm thick polycarbonate material were ultrasonically spot welded and subsequently analyzed by both optical observation and mechanical testing.

To find the acceptable amplitude, welding time, and welding pressure of polycarbonate, and to estimate the parameters to weld Tecanat GF20, a study of the effect of each of the three parameters was carried out, keeping the other parameters constant. The welding parameters studied are shown in Table 7.2. For each weld, the hold pressure and the hold time were kept constant at 276 kPa and 2.0 s, respectively.

The criteria used to find the appropriate parameters were the total bonded area (shear area), the maximum load carried by the samples during a lap shear test, and the standard deviation (σ) of each set of sample set. The shear area was measured optically prior to lap shear testing. To measure the tensile lap shear strength of ultrasonic spot welds in polycarbonate, lap shear samples were prepared. Tabs on either end of the samples were glued to the lap shear specimen to reduce the offsetting of the applied force to reduce the bending moment acting on the joint. The lap shear samples were tested at a rate of 1.3 mm/min according to ASTM 3164 using an MTS 810 load frame. Four lap shear samples of each welding parameter combination were tested and the average maximum load and corresponding standard deviation are listed in Table 7.2. The failure type observed in most of the lap shear samples was almost immediate fracture after the peak load was reached. The fracture took

Table 7.1 Mechanical properties of Tecanat GF20 [17]

Property	Metric units
Young's modulus	4826 MPa
Ultimate tensile strength	110 MPa
Elongation at break	4 %
Impact strength (IZOD, unnotched)	1.01 J/mm
Poisson's ratio	0.37
Compressive strength	76 MPa

Table 7.2 Welding parameter combinations and results of lap shear testing of 6.35 mm thick pure polycarbonate ultrasonic spot welds

Sample number	Welding amplitude [μm]	Weld pressure [kPa]	Weld time [s]	Average maximum load [N]	Standard deviation of maximum load [N]
A1	46.5	207	12.5	3075	2038
A2	46.95			4057	1014
A3	47.4			3703	418
A4	47.85			6192	1301
T1	47.4	207	12.2	5457	1473
T2			12.5	3816	1206
T3			12.7	6213	1493
T4			13.0	3612	1243
P1	47.4	172	13.0	2739	413
P2		207		2642	1091
P3		241		3099	2412

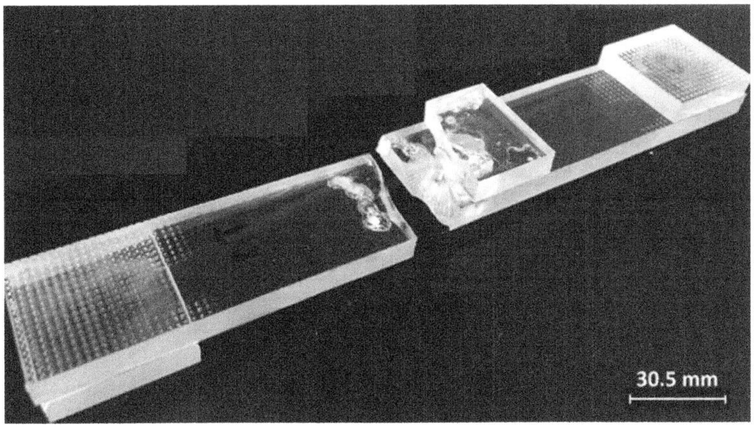

Fig. 7.1 Failed polycarbonate lap shear sample after tensile testing

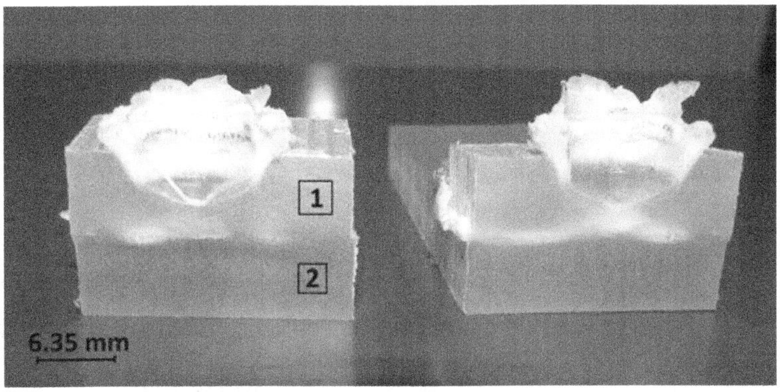

Fig. 7.2 Sectioned view of an ultrasonically spot welded composite material

place close to the point where the welded area begins, indicating a bending moment and a large stress concentration in this area. A failed sample can be seen in Fig. 7.1. This fracture was likely due to bending stresses and stress concentrations at the interface between the weld and polycarbonate similar to those observed by Hart-Smith [18].

It was clear that varying welding parameters influenced the strength of the joints and it was noted that a more thorough analysis is required for full optimization. For this study, the following parameters produced welds that carry high loads in polycarbonate, and were used when ultrasonically spot welding the composite material, Tecanat GF20: amplitude of 47.85 μm, weld time of 12.7 s, and weld pressure of 241 kPa. The other adjustable parameters of hold time and hold pressure were kept constant at 2.0 s and 276 kPa, respectively. A sectioned view of an ultrasonically spot welded Tecanat GF 20 composite welded with these parameters is shown in Fig. 7.2. In this weld, the tip did not penetrate all the way through the top layer of the composite labeled 1 in Fig. 7.2, but the ultrasonic vibrations still induced melting at the interface between the top and bottom layer, labeled 2 in Fig. 7.2 to create a bond.

7.3 Characterization of Ultrasonically Spot Welded Tecanat GF 20

7.3.1 Ultrasonically Spot Welded Lap Joints

The ultrasonically spot welded lap joints of Tecanat GF20 were characterized experimentally with a tensile lap shear test (based on ASTM 3164) and an impact test (based on ASTM 950) to assess their load bearing capacity. For impact testing, the lap joints used were unconventional without notches or other modifications to create stress concentrations to reduce the amount of material for each sample (Fig. 7.3).

Lap shear samples were tested according to the procedure described for pure polycarbonate above. To impact test unconventional lap joints, a test rig, shown in Fig. 7.4a, was designed and built to hold the lap joints in the correct position such that they were impacted in a repetitive fashion to yield useful results. The design of the test rig was based on the design specified in ASTM D950 to impact test adhesively bonded lap joints. The impact tester used was an Instron SI-1D3 (Fig. 7.4b), and had the ability to impact samples with an energy of up to 407.6 J.

The absorbed energy was recorded from the impact tests and was analyzed by normalizing to both the shear area of the weld and the shear length of the weld. The fibers in Tecanat GF20 cause the material to be opaque, and the weld area and weld length could not be determined before the experiments using optical techniques, therefore, the area and length were found optically after the welds were tested to failure. The impact test was set up as seen in Fig. 7.4 where the lap joints were loaded in the test rig and mounted to the test machine. A total of seven lap shear and six impact samples were tested, with results shown in Tables 7.3 and 7.4, respectively. Representative failed samples are shown in Fig. 7.5.

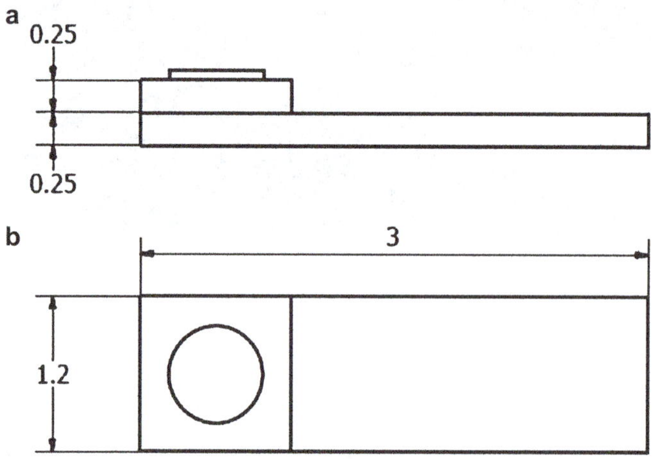

Fig. 7.3 Unconventional lap joint, (**a**) side view (**b**) top view

Fig. 7.4 CAD model of impact test rig, isometric view (*left*), the thickness of side walls are 3.175 mm, the Instron SI-1D3 impact tester (*center*), and test rig with mounted impact sample (*right*). The direction of the striker is shown by the *arrow*

Table 7.3 Results from lap shear test, maximum load, and elongation at maximum load

Sample	Maximum load [N]	Elongation at maximum load [mm]
LS-1	5552	1.2
LS-2	4745	1.0
LS-3	5703	1.1
LS-4	6169	1.2
LS-5	5352	1.0
LS-6	6039	1.2
LS-7	5589	1.2
Average	5593	1.1

Table 7.4 Results from impact tests, measured weld width, weld area, calculated joint strength in terms of weld area and weld length

Sample	Absorbed energy [J]	Weld length [mm]	Weld area [mm^2]	Impact strength [J/mm]	Impact strength [J/mm^2]
I-1	51	30	722	1.7	0.07
I-2	66	29	775	2.2	0.08
I-3	88	28	789	3.1	0.11
I-4	44	29	641	1.5	0.07
I-5	66	29	596	2.2	0.11
I-6	65	30	794	2.2	0.08
Average	63	29	719	2.2	0.09

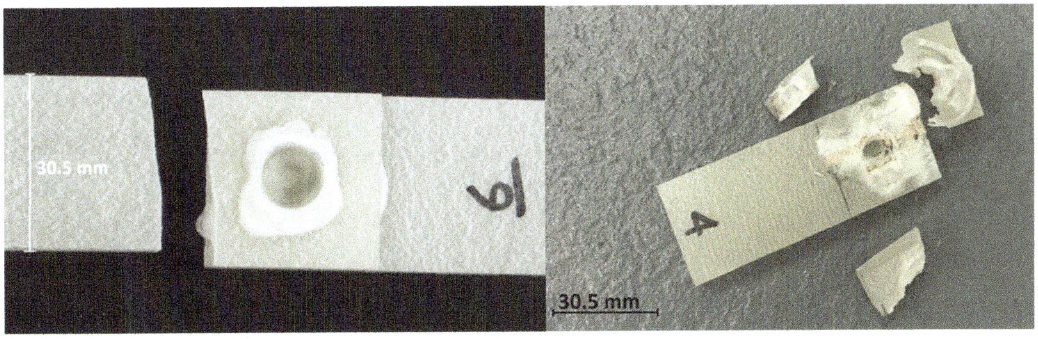

Fig. 7.5 Representative failed lap shear sample (*left*) and failed impact sample (*right*)

All seven lap shear samples failed in a similar manner to that shown in Fig. 7.5 (left). The samples failed in a near linear region of the load-deformation profile and not undergo significant plastic deformation, suggesting the presence of bending stresses and stress concentrations contributing to the failure of the joint. Additionally, the thickness of the welds induced a bending moment in the joints observed in videos of the tests. Lap joints subjected to a load that is not perfectly aligned, as in these experiments, tend to fail in the base material due to the bending moment and stress concentrations that arise due to the offsetting of the force as observed by Hart-Smith [18]. The stresses in the lap joints were further investigated by conducting a finite element analysis (FEA). All impact samples failed in a consistently brittle manner, as shown in Fig. 7.5 (right), attributed to the high strain rate of the impact test. This type of behavior under impact loading was also observed by Barlow et al. [19].

For each lap shear sample, the normal stresses in the base material and the shear stresses in the weld were calculated and used to determine a stress concentration factor to compare with the finite element model. The nominal normal stress in the material was calculated using:

$$\sigma_{x_q} = \frac{P}{A} \tag{7.1}$$

In Eq. (7.1), P is the peak load of the applied tensile force and will be defined as such for the remainder of the calculations, and A is the area of the base material. The average shear stress in the weld was defined as the applied tensile force divided by the shear area:

$$\tau_{x_1 x_2} = \frac{P}{A_S} \tag{7.2}$$

In Eq. (7.2), A_s is the shear area. This definition of average shear stress assumes that the shear stresses are uniform throughout the shear area. It was shown by Goland and Reissner [20], and Hart-Smith [18] that the shear stress distribution in adhesive lap joints has a significant distribution and that shear stress increases near the edges of the adhesive area. However, the experiments showed that the lap joints did not fail due to the shear stresses in the welds, and as such the average shear stress definition was used. The calculated average shear stresses are all lower than the published shear strength of Tecanat GF20, and as expected none of the lap joints failed in the welds. The bending stresses in the base material were calculated using:

$$\sigma_{bending} = \frac{My}{I} \tag{7.3}$$

In Eq. (7.3), M is the bending moment at the root of the weld, y is the distance from the neutral axis, equal to one half of the thickness of the base material, and I is the area moment of inertia of the base material. The values of y and I were calculated for each sample. The bending moment M was calculated using:

$$M = Pk\frac{t_{weld} + t_{base\,material}}{2} \tag{7.4}$$

In Eq. (7.4), t_{weld} is the thickness of the welds, $t_{base\ material}$ is the thickness of the base material, equal to 3.4 mm, and k is the bending moment factor defined by Hart-Smith [23]. The sum of the bending stress and normal stress was also calculated and this total stress in the x_1 direction at the experimental peak load were all lower than the published ultimate tensile stress for Tecanat GF20 of 110 MPa, suggesting the presence of a stress concentration at the root of the weld. The stress concentration factor at the weld was defined as:

$$K_f = \frac{\sigma_f}{\sigma_{total}} \tag{7.5}$$

where σ_f is the ultimate tensile stress of Tecanat GF20, and σ_{total} is the total stress in Table 7.5.

7.3.2 Adhesively Joined Lap Joints

To compare ultrasonically spot welded lap joints to adhesively joined lap joints, Pace Technologies ultrathin 2 low viscosity epoxy resin [21] was used to join Tecanat GF20. Here, the surfaces were wiped clean to remove large surface contaminations, the joints were clamped together to ensure the correct geometry, and cured for 24 h minimum. The published tensile strength of the adhesive is 55.1 MPa [21]. The lap joints failed in the adhesive, as expected since Tecanat GF20 has a significantly higher failure load than the epoxy. Two lap shear and two impact samples were tested using the same methods as described above, and the results are shown in Table 7.6. The load–displacement curves for the lap shear tests showed that the adhesive underwent significant plastic deformation as compared to the curves for ultrasonic spot welds. The adhesive samples failed in a mixed type of fracture between adhesive failure and cohesive failure, highlighted in Fig. 7.6 (top). The absorbed energy from the impact tests and the areas measured optically after testing are also listed in Table 7.6, along with the calculated impact strength per area and per length. The impact tested samples failed cohesively as shown by the outlined epoxy on one side of the joints in Fig. 7.6 (bottom), suggesting less than ideal bonding [22].

Table 7.5 Average shear stress at peak load, nominal stress, bending stress, total stress, and stress concentration factor for all seven lap shear tested samples

Sample	Average shear stress [MPa]	Nominal stress [MPa]	Bending stress [MPa]	Total stress [MPa]	*Kf*
LS-1	13.2	27.5	16.6	44.1	2.5
LS-2	10.2	23.5	19.1	42.6	2.6
LS-3	12.1	28.1	17.3	45.4	2.4
LS-4	11.8	30.8	18.1	48.9	2.3
LS-5	12.8	26.6	23.8	50.4	2.2
LS-6	14.0	29.3	18.7	48.0	2.3
LS-7	11.4	27.7	17.2	44.9	2.5
Average	12.2	27.6	18.7	46.3	2.4

Table 7.6 Results from adhesive lap shear and impact tests

Sample	Maximum load [N]	Elongation at maximum load [mm]	Absorbed energy [J]	Adhesive width [mm]	Adhesive area [mm²]	Impact strength [J/mm]	Impact strength [J/mm²]
LS-A-1	2242	0.6					
LS-A-2	2445	0.7					
LS-A-Average	2343	0.65					
I-A-1			13.6	24	579	0.6	0.02
I-A-2			15.6	24	582	0.6	0.03
I-A-Average			14.6	24	581	0.6	0.025

Fig. 7.6 Failed adhesive lap shear (*top*) and impact samples (*bottom*) with remaining epoxy highlighted

7.3.3 *Comparison of Adhesively Joined and USSW Joints*

In Fig. 7.7, LS-USSW and LS-Adhesive are the results from the lap shear tested USSW and adhesively joined Tecanat GF20, respectively, and I-USSW and I-Adhesive are the impact tested USSW and adhesively joined Tecanat GF20, respectively. The results show that the USSW joints tested in lap shear has a maximum load 2.3 times higher than adhesively joined samples (Fig. 7.7). The adhesive failure mode suggests that the Tecanat GF20 would benefit from a surface treatment before adhesion, and the cohesive mode suggests that the epoxy resin used was not sufficiently strong to bond Tecanat GF20 in lap shear loading conditions [23]. Increasing the thickness of the adhesive layer could reduce the normal stresses in the adhesive and increase the strength of the adhesive joints. The experimental impact results had an average impact strength 3.5 times higher than the adhesively joined samples (Fig. 7.7). This may be due to the low impact strength of epoxy resins as suggested by LeGrand [24].

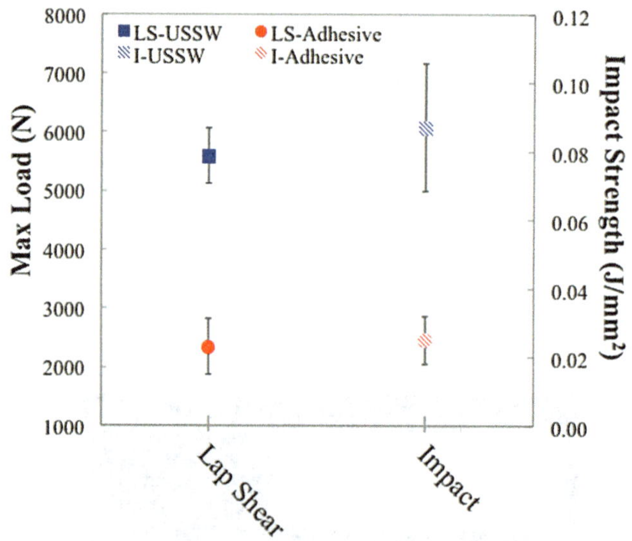

Fig. 7.7 Comparison of ultrasonically spot welded and adhesively joined Tecanat GF20 in both loading scenarios

7.4 Finite Element Analysis

A finite element analysis (FEA) of a lap joint in a loading situation comparable to the lap shear test was conducted using the commercial software ABAQUS CAE. The lap joint was modeled in the software with geometric simplifications where the weld was modeled as perfectly circular while the hole left by the pilot tip was omitted. The load was applied as a prescribed displacement of tabbed ends of 1.1 mm (the average experimental displacement at the maximum load). The goal of the FEA analysis was to investigate the distribution of stresses in the lap joints and assess how they affect the failure of the lap joints. Tecanat GF20 was assumed to be isotropic and linearly elastic as the material has very little plastic deformation before failure. Generalized Hooke's law was used as the constitutive relationship and Young's Modulus and Poisson's ratio were assumed to be 4826 MPa and 0.366, respectively.

The mesh used consisted of two types of elements from the ABAQUS three-dimensional solid element library, the C3D8R element and C3D6 element. In this case, the C3D6 element, was used to ease meshing around the spot weld which was modeled as perfectly circular. The number of these elements in the mesh was low compared to the number of hexahedral elements (<0.4 % of the total number of elements in the mesh used), and it was assumed that their effect on the results was negligible. To ensure that the error in the finite element model was as low as possible, a convergence study was conducted. The maximum Von Mises stress and the maximum displacement magnitude were calculated for a varying number of nodes. The mesh used had a change in maximum stress of less than 10 %, which was deemed sufficient for this application.

Figure 7.8 shows the relative stresses throughout the entire sample, the normal stresses in the z direction for half of the model, and a side view of the stress distributions. In Fig. 7.8 (center), high stresses are present at the transition between weld and base material. To calculate the stress concentration created by the weld, the ratio between the maximum stress at the root of the weld and the stresses far away from the weld were used. A stress concentration of 2.6 was found using this method.

The FEA also showed that the shear stresses in the weld were low. From the analysis, the maximum shear stress in the sample was 24.7 MPa, which is well below the shear strength of Tecanat GF20. This results was expected as the experiments showed that the lap joints did not fail due to shear forces. To analyze the bending stresses, a small portion of the base material was investigated and the stresses in the z direction were plotted through a cross section of the y-z plane (Fig. 7.9). The stress distribution in Fig. 7.9 shows that bending stresses are present in the z direction, as the stresses change from tensile to compressive over the cross section.

The results from the finite element analysis confirmed that the presence of bending in the base material, and the presence of a stress concentration of 2.6 at the root of the weld. The stress concentration factor of 2.6 found in the finite element model agrees with the calculated results for each sample. The error in the calculated K_f and the K_f from FEA may be attributed to the geometric simplifications, the boundary conditions used, and the linear assumptions.

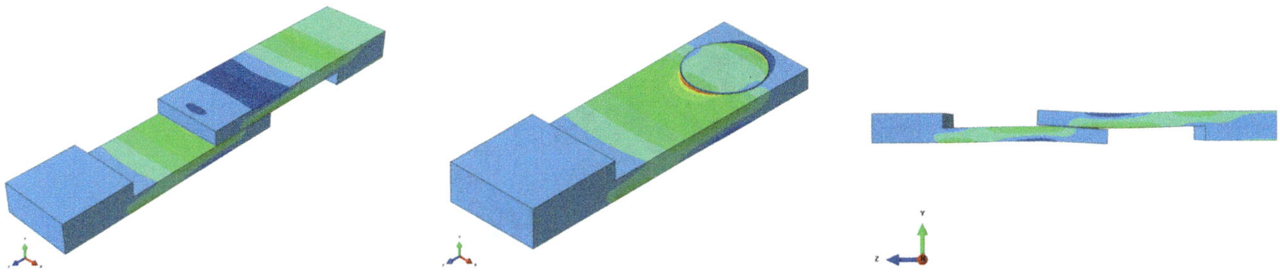

Fig. 7.8 Isometric view of lap joint (*left*), isometric view of half of the joint (*center*), and side view (*right*) of the joint showing plot of normal stress in the z direction. *Dark colors* indicate compressive stress, *light colors* indicate tensile stress, *red colored areas* are subjected to the highest tensile stress, and the thickness of material is 6.35 mm

Fig. 7.9 Stresses in z direction plotted on the y-z cross section. *Arrows* in the negative z-direction are tensile stresses, *arrows* in the positive z-direction are compressive stresses, and the thickness of material is 6.35 mm

7.5 Microscopy of Welds

To optically assess the fractures in the lap shear samples, scanning electron microscopy of the fracture was conducted with a Zeiss Supra 40 scanning electron microscope. This was done to compare with the results of Suzuki et al. [24], who investigated the fracture surface of a composite material subjected to bending stresses. A stitched image of the fracture surface of sample LS-1 at ×60 magnification can be seen in Fig. 7.10.

The highlighted areas indicated in Fig. 7.10 show areas that have been studied more closely. The top area is close to where the highest tensile stresses were due to the bending and stress concentration from the results of the FEA. The lower area is near an area with the lowest stresses due to the bending stress distribution.

In Fig. 7.10 (top), the dark spots (circled) are voids left after fiber pullout, which is expected in high tensile stress regions [25]. In Fig. 7.10 (bottom), a region near lower stresses is shown, where fewer fiber pullout areas are present, suggesting that this region failed due to crack propagation through this region of the composite [25].

The scanning electron microscopy of the fracture surface supports the results from the FEA, showing a region of high tensile stresses close to the weld due to the bending stresses and stress concentration. These stresses were the driving force of the failure of the lap joints [18].

7.6 Conclusions

The results of this work have successfully shown the use of ultrasonic spot welding to join chopped fiber thermoplastic composite materials. The results from the experiments showed that welding composite materials are similar to welding the thermoplastic material of the composite matrix, and welding parameters used to join thermoplastics may be used to join composites with the same thermoplastic matrix.

Lap joints were manufactured by ultrasonically spot welding 6.35 mm thick Tecanat GF20 and characterized by lap shear and impact testing to determine their mechanical strength. A high repeatability suggests that the ultrasonic spot welding process may be well suited for high volume production. The lap shear samples all failed in a consistent manner, due to bending stress and a stress concentration at the interface between the weld and the base material. This theory was further confirmed by conducting FEA and scanning electron microscopy of the fracture surfaces. The stresses in the experimentally tested lap joints were calculated, and agreed with the results from the finite element model.

Fig. 7.10 Fracture surface of a lap shear sample (*center*) and enlarged region subjected to lowest tensile stresses (*top*) and region subjected to highest tensile stresses (*bottom*)

To compare the performance of ultrasonic spot welds to adhesives, lap joints joined using an epoxy resin were fabricated. These adhesively bonded joints were subjected to the same tests as the ultrasonically spot welded joints. The ultrasonically spot welded samples outperformed the adhesively joined samples in both lap shear and impact tests. The adhesively joined samples required 24 h to fabricate due to the curing time of epoxy resin, while the ultrasonically spot welded samples took 14.7 s to weld plus time taken to fix the samples correctly in the spot welder, typically less than one minute total. The significantly shorter fabrication time, in addition to the high strength of the ultrasonically spot welded joints are the benefits of ultrasonic spot welding when used to join thermoplastic composite materials. Future work includes further assessment of the strength of ultrasonic spot welds, manufacturing and testing of double lap joints to reduce the bending stresses, and a more detailed study of the microscopic details of the joint.

References

1. Hull, D., Clyne, T.W.: An Introduction to Composite Materials. Cambridge University Press, Cambridge (1996)
2. Yousefpour, A., Hojjati, M., Immarigeon, J.P.: Fusion bonding/welding of thermoplastic composites. J. Thermoplast. Compos. Mater. **17**(4), 303–341 (2004)
3. Vinson, J.R.: Adhesive bonding of polymer composites. Polym. Eng. Sci. **29**(19), 1325–1331 (1989)
4. Banea, M.D., da Silva, L.F.: Adhesively bonded joints in composite materials: an overview. Proc. Inst. Mech. Eng. Part L: J. Mater. Des. Appl. **223**(1), 1–18 (2009)

5. Ageorges, C., Ye, L., Hou, M.: Advances in fusion bonding techniques for joining thermoplastic matrix composites: a review. Compos. A: Appl. Sci. Manuf. **32**(6), 839–857 (2001)
6. Ageorges, C., Ye, L.: Fusion Bonding of Polymer Composites: From Basic Mechanisms to Process Optimization. Springer, London (2002)
7. Friedrich, K., Almajid, A.A.: Manufacturing aspects of advanced polymer composites for automotive applications. Appl. Compos. Mater. **20**(2), 107–128 (2013)
8. Stavrov, D., Bersee, H.E.N.: Resistance welding of thermoplastic composites—an overview. Compos. A: Appl. Sci. Manuf. **36**(1), 39–54 (2005)
9. Ahmed, T.J., Stavrov, D., Bersee, H.E.N., Beukers, A.: Induction welding of thermoplastic composites—an overview. Compos. A: Appl. Sci. Manuf. **37**(10), 1638–1651 (2006)
10. Benatar, A.: Ultrasonic welding of advanced thermoplastic composites. Doctoral dissertation, Massachusetts Institute of Technology (1987)
11. Bates, P., Couzens, D., Kendall, J.: Vibration welding of continuously reinforced thermoplastic composites. J. Thermoplast. Compos. Mater. **14**(4), 344–354 (2001)
12. Stokes, V.K.: Joining methods for plastics and plastic composites: an overview. Polym. Eng. Sci. **29**(19), 1310–1324 (1989)
13. Weber, A.: Welding Still Ensures High-Strength Joints (2007, November)
14. Troughton, M.J.: Handbook of Plastics Joining: A Practical Guide, 2nd edn. William Andrew, Norwich (2008)
15. The Welding Institute. "Ultrasonic welding of injection moulded components." TWI. http://www.twiglobal.com/technical-knowledge/job knowledge/ultrasonic-welding-of-injection-moulded-components-part-1-process-and-equipment-061/. Accessed 21 Jul 2016.
16. Dukane Intelligent Assembly Solutions. "What is Ultrasonic Welding?" Dukane. http://www.dukane.com/us/PPL_whatisUPA.htm. Accessed 21 Jul 2016.
17. Ensinger Inc. "TECANAT GF20 glass filled PC." Ensinger. Accessed 21 July 2016. http://www.ensingerinc.com/products.cfm?page=product &product=tecanat+gf20+glass+filled+pc. Accessed 21 Jul 2016.
18. Hart-Smith, L.: Adhesive-bonded single-lap joints. Technical Report NASA CR 112236 (1973)
19. Barlow, C., Kumar, V., Flinn, B., Bordia, R., Weller, J.: Impact strength of high density solid-state microcellular polycarbonate foams. J. Eng. Mater. Technol. **123**, 229–233 (2001)
20. Goland, M., Reissner, E.: The stresses in cemented joints. J. Appl. Mech. **11**, A17–A27 (1944)
21. Pace Technologies. "Metallographic Castable Mounting." Pace Technologies. http://www.metallographic.com/Technical/Castable%20Mount.htm. Accessed 21 Jul 2016.
22. LeGrand, D.: Handbook of Polycarbonate Science and Technology. Marcel Dekker, New York (2000)
23. Davis, M.J., Bond, D.A.: The importance of failure mode identification in adhesive bonded aircraft structures and repairs. In: ICCM 12, Paris, France (1999)
24. Suzuki, K., Ohsawa, I., Uzawa, K., Nagata, K., Matsuo, T., Yamane, M., Takahashi, J. (2012): Joint Efficiency of Ultrasonic Welding of CFRTP (Carbon Fiber Reinforced Thermoplastics) for Structural Applications
25. Srinivasa, V., Shivakumar, V., Nayaka, V., Jagadeeshaiaih, S., Seethram, M., Shenoy, R., Nafidi, A.: Fracture morphology of carbon fiber reinforced plastic composite laminates. Mater. Res. **13**(3), 417–424 (2010)

Chapter 8
Numerical and Experimental Characterization of Hybrid Fastening System in Composite Joints

Ermias G. Koricho, Mahmoodul Haq, and Gary L. Cloud

Abstract Drilling of holes in continuous fiber composites for fastening reduces the load carrying capacities, introduces delamination and creates sites for failure initiation and progression. Recent work on a novel hybrid fastening system, which introduces a structural resin insert through a channel in the bolt shaft, overcomes the effects of drilling, eliminates slip, reduces delamination and increases the load-carrying capacities. While the benefits of this system are experimentally observed, the phenomena that govern these improvements are not fully understood. In this work, numerical simulations and experimental tests of hybrid fastening systems comprised of glass fiber reinforced polymer (GFRP) composite substrates fastened using a ½″ Grade 5 bolt, and with a preload torque of 35 N-m were performed. The results were compared one to the other for stiffness and strength. The simulations captured the damage initiation and progress around the hole, and showed the reduction in bolt tilt ascribed to the incorporation of the adhesive insert. Overall, numerical simulations show promise in providing quantitative information about these complex phenomena, and such post-experimental validations can be used as powerful design tools.

Keywords Hybrid bolted joint • Bolt • Composite • FEM • Damage

8.1 Introduction

The use of composite materials is getting significant attention in the transport industries such as in automotive, aerospace, and naval applications in order to achieve lightweight structures. Bolted joints are one of the most frequently used methods of fastening in composite structures where there is a desire for repairing/replacing structural components or increasing the reliability of the adhesively bonded joints.

However, composite bolted joints have their own disadvantages, since composite substrates need to be drilled to insert the bolt. Drilling and machining of holes in composites cause stress concentrations and delamination that reduce the resistance of the joint construction to applied loads, and thus its efficiency. Besides, gaps usually occur at wearing surfaces during assembly of large scale structures. This problem can be exacerbated for composite structures as it is comparably more difficult to control dimensional tolerances of composite components than those of conventional materials such as steel and aluminum. Consequently, pre-stresses could be developed by these gaps and the assembly forces, which could result in initiation of cracks and delaminations that would adversely affect the performance of the structures [1–3]. To overcome these problems, inserts, such as shims, have often been used to fill the clearance to avoid possible structural damage caused by pre-stress and delamination [4–7]. Huhne et al. [5] carried out quasi-static tests to investigate the influence of liquid–shim thickness on the strength and structural behavior of carbon fiber laminate single-lap bolted joints, and they found that the joint stiffness reduced with shim thickness but ultimate joint load was unaffected up to a liquid–shim thickness of 1.5 mm. However, the Mil-Handbook [6] holds a different point of view: it states the usage of shims will cause reduction of bearing strength in bolted composite joints even if the shim is very thin, and, as the thickness of the shim increases, the reduction will be greater. To clearly understand the effect of shims on the mechanical behavior of bolted joints, Liu et al. [7] investigated both experimentally and numerically the performance of composite-to-titanium bolted joints with liquid shims. Results showed that both the failure load and stiffness decrease with the thickness of the shim, however, the effect of the shim becomes negligible as the thickness is diminished between 0.5 and 1 mm.

E.G. Koricho (✉) • G.L. Cloud
Composite Vehicle Research Center, Michigan State University, 2727 Alliance Drive, Lansing, MI 48910, USA
e-mail: koricho@msu.edu

M. Haq
Composite Vehicle Research Center, Michigan State University, 2727 Alliance Drive, Lansing, MI 48910, USA

Department of Civil and Environmental Engineering, Michigan State University, Engineering Building, East Lansing, MI 48824, USA

G.L. Cloud et al. (eds.), *Joining Technologies for Composites and Dissimilar Materials, Volume 10*
Conference Proceedings of the Society for Experimental Mechanics Series, DOI 10.1007/978-3-319-42426-2_8

On the other hand, Haq et al. [8, 9] have studied the effects on the quasi-static behavior of hybrid bolted joints of resin inserts with various reinforcements (nano-fillers and glass/carbon sleeves) that are similar to the joints studied in this work. They have observed zero slip, considerable improvements in static load carrying capacities, increased strengths, and delayed onset of delamination in the resulting joints. However, no numerical study on the mechanical behavior of such novel hybrid bolted joints is presented in their works.

In this work, numerical simulations and experimental tensile tests of conventional and hybrid fastening systems comprised of glass fiber reinforced polymer (GFRP) composite substrates fastened using a ½″ Grade 5 bolt, and with a preload torque of 35 N-m were performed. The results were compared one to the other for stiffness and strength. The damage initiation and evolution were also predicted at particular points of interest using the Hashin failure criteria available in ABAQUS® and compared with the experimental results.

8.2 Materials

In this study, the vacuum-assisted resin transfer molding (VARTM) technique was used to manufacture the glass fiber reinforced plastic (GFRP) composite laminates used for the single-lap composite bolted joints. The reinforcement was S2-glass plain weave fabric (Owens Corning ShieldStrand S) with a real weight of 818 g/m^2. The resin used was a two part toughened epoxy, namely SC-15 (Applied Poleramics Inc., CA). The SC-15/glass-fiber laminates were cured in a convection oven at 60 °C for 2 h followed by a post cure of 4 h at 94 °C. Sixteen layers of plain-weave glass reinforcement were used for each adherend in this study. The thickness of the cured laminate was found to be 9.6 mm. The laminate coupons for the bolted joints were cut by a water-cooled, diamond-coated-disc cutting machine to achieve final dimensions of 100 mm width by 150 mm length.

The bolt was a nominal ½in. grade five, medium-strength steel hex head (head width: 19.05 mm or 3/4 in.; head height 7.94 mm or 5/16 in.) bolt with a right-hand thread having a length of 38.1 mm (1.5 in.), fully threaded with a 12.7 mm-13 (1/2 in.-13) thread. The bolts were drilled with 2 mm diameter passageways for resin injection.

8.3 Fabrication of Bolted Joints

Two approaches were used to develop the composite single lap bolted joints: conventional and hybrid. The schematic diagram of the hybrid joint bolt (see Fig. 8.1a) shows that, in its simplest form, it basically consists of a bolt that has a channel through the shank that allows the injection of an insert that fills the clearance space between the bolt and work-pieces and acts as a structural component [10, 11]. One of major advantages of this invention is that the presence of the adhesive/insert compensates for the variation of hole clearance that typically occurs during conventional machining processes. Additionally, in layered composites, the delaminations and irregularities that are introduced by hole-drilling become filled with the resin insert, thereby repairing any damage and creating a monolithic part. Finish reaming of the holes is thereby eliminated at considerable reduction of fabrication cost. Many variations of the insert concept exist, but only the simplest form was treated in this study.

The GFRP substrates were joined using the bolts described above and flanged sleeve washers. A 15.7 mm diameter hole was drilled in the work pieces 50.5 mm away from three edges. The sleeve washer has the following dimensions: flange diameter 25.4 mm, ID of 13.46 mm, sleeve OD of 15.88 mm, flange thickness of 3.18 mm, overall height of 4.75 mm. The radial clearance between the bolt shank and the hole was ~1.5 mm. This gap, which probably exceeds the optimum, was to be filled with resin to create the hybrid joints. The control specimens that lacked the resin insert were manufactured to the

Fig. 8.1 (**a**) Schematic of basic hybrid composite joint, (**b**) Cross section of hybrid joint with resin insert exaggerated

same specifications, and the resulting clearance is not consistent with typical manufacturing practice for critical structures although it is not unusually high for non-critical factory or field assembly. The tightening torque of 35 N-m was applied by a digital torque wrench.

After assembly, the samples that were selected for the hybrid joint were injected with SC-15 resin through the channels in the bolts so as to fill the clearance space between the threaded bolt shank and the work pieces. The injected inserts were cured by a convection oven at 60 °C for 2 h followed by post curing at 94 °C for 4 h. Figure 8.1b shows a cross-section of the bolted joint, wherein the channels are exaggerated for better visibility. As can be seen, the resin fills all the intricate areas including the threads of the bolts and, upon curing, creates a monolithic structure.

8.4 Experimental Setup

All quasi-static testing was performed using a 100 kN load cell servo-hydraulic testing machine (MTS 810), and a typical test setup is shown in Fig. 8.2. The machine is equipped with a standard load cell and an internal crosshead displacement measuring extensometer. An external laser extensometer was also used to measure the joint longitudinal displacement, as shown in the figure.

The tests were performed according to ASTM STP1455-EB. In this work, five tests were carried out for statistical significance. The average shear strength of the joint was calculated by dividing the maximum load reached in each test divided by the cross-sectional area of the bolt.

8.5 FEM Modeling

8.5.1 Modeling and Meshing

In this study, the commercially available finite element software ABAQUS® was used to obtain the stress distribution and damage initiation near the holes in the bolted composite coupons. The overall dimensions for modeling of the GFRP with a drilled hole were obtained from the specimens used for the experimental tests.

Eight-node quadrilateral in-plane general-purpose continuum shell elements with reduced integration and hourglass control were used to model the 16 ply plain-weave composite plates, $[0]_{s8}$. A minimum of three integration points using the Simpson integration rule were assigned at each layer of continuum shell elements. As is shown in Fig. 8.3a, the specimen was partitioned at specific sections to apply boundary conditions and localized meshes, especially around the hole, where they need to be refined to capture more accurately the steep stress gradients there. During the meshing process, the bias method was implemented to allow the element size to be varied with respect to the areas near the holes: the nearest element to the hole has the smallest element size, as shown in Fig. 8.3a. Figure 8.3b shows the completely assembled lap joint as modeled.

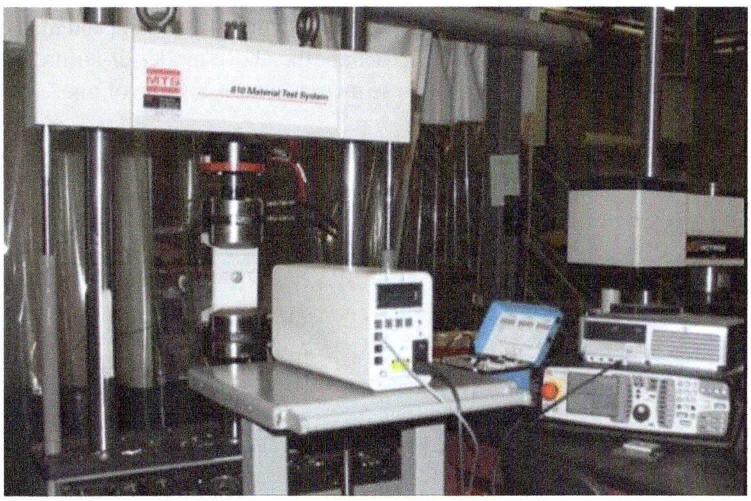

Fig. 8.2 Experimental setup

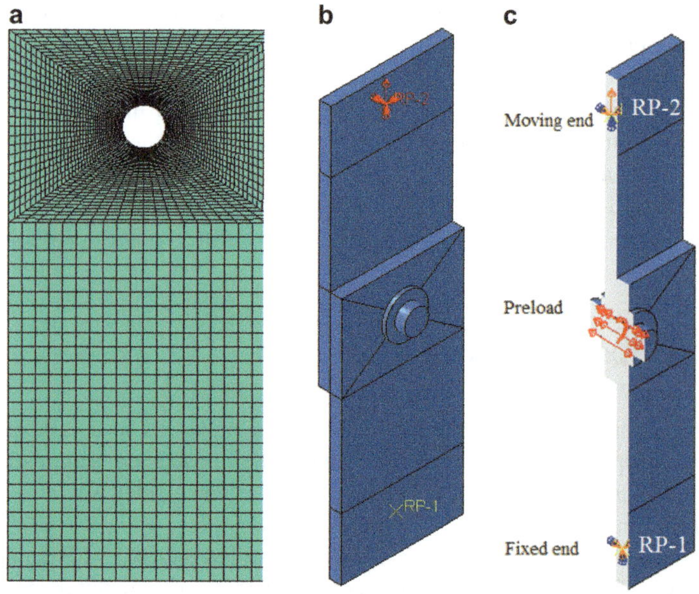

Fig. 8.3 FEM model: (**a**) mesh, (**b**) assembled joint, (**c**) sectional view of loading and boundary conditions

The bolt and the nut, which were made of a steel, were considered to be a single part in order to reduce computational time by avoiding modeling the contact between the bolt and nut. A ten-node quadratic tetrahedron three-dimensional element was used to model the bolt and the resin insert. The "surface-surface contact" model was incorporated to define the interaction in the joint. The tangential and normal contact properties were defined using the "penalty" friction formulation and "hard" contact pressure overclosure methods, respectively. For the hybrid bolted joint, the resin insert was placed between the bolt and the composite hole wall and constrained using the "Tie option" available in ABAQUS®.

The joint was subjected to tensile loading by constraining all degrees of freedom at the fixed end (RP-1 in Fig. 8.3c) of the joint and allow a constant displacement rate of 2 mm/min at the moving end (RP-2 in Fig. 8.3c). The preload, which was developed by the bolt tightening torque, was incorporated in the bolt shank through the "Bolt load" module in ABAQUS CAE, as shown in Fig. 8.3c. It was set to be 10 kN, which is the preload that corresponds to 35 N-m tightening torque according to the usual empirical formula.

8.5.2 *Material Modeling*

The damage initiation and evolution model used in this work was based on the Hashin failure criteria in ABAQUS®. In this model, four different modes of failure were considered, namely, fiber rupture in tension, fiber buckling and kinking in compression, matrix cracking under transverse tension and shearing, and matrix crushing under transverse compression and shear.

The Hashin failure initiation metrics for the above modes of failures take the following general forms, and failure is predicted to occur when any one of these metrics attains a value of one:

$$F_f^t = \left(\frac{\sigma_{11}}{X^T}\right)^2 + \alpha\left(\frac{\tau_{12}}{S^T}\right)^2, \text{for fiber tension} \left(\sigma_{11} \ge 0\right) \qquad (\text{HSNFTCRT}) \tag{8.1}$$

$$F_f^c = \left(\frac{\sigma_{11}}{X^C}\right)^2, \text{for fiber compression} \left(\sigma_{11} < 0\right) \qquad (\text{HSNFCCRT}) \tag{8.2}$$

$$F_m^t = \left(\frac{\sigma_{22}}{Y^T}\right)^2 + \alpha\left(\frac{\tau_{12}}{S^L}\right)^2, \text{for matrix tension} \left(\sigma_{22} \ge 0\right) \qquad (\text{HSNMTCRT}) \tag{8.3}$$

$$F_m^c = \left(\frac{\sigma_{22}}{2S^T}\right)^2 + \left[\left(\frac{Y^C}{2S^T}\right)^2 - 1\right]\frac{\sigma_{22}}{Y^C} + \left(\frac{\tau_{12}}{S^L}\right)^2, \text{for matrix compression} \left(\sigma_{11} < 0\right) \qquad (\text{HSNMCCRT}) \tag{8.4}$$

Table 8.1 Laminate strength data

Composite	Weave type	Fiber fraction (wt%)	Tensile strength (MPa)	E_x (Gpa)	G_{xy} (Gpa)	ν_{xy}
0/90 S-glass/epoxy	Plain	0.61	476.264	23.18	4.0	0.08

where: X^T *and* X^C are longitudinal tensile and compressive strengths, respectively;
Y^T *and* Y^C are transverse tensile and compressive strengths, respectively;
S^L *and* S^T are longitudinal and transverse shear strengths, respectively;
α is a coefficient that determines the contribution of the shear stress to the fiber tensile failure initiation criterion; and
σ_{11}, σ_{22}, and τ_{12} are components of the effective stress tensor, σ, that is used to evaluate the initiation criteria.

The response of the material during the post-damage initiation phase is computed as,

$$\sigma = S\varepsilon \tag{8.5}$$

where ε is strain and S is the damaged elastic matrix

$$S = \frac{1}{D}\begin{bmatrix} (1-d_f)E_1 & (1-d_f)\nu_{21}E_1 & 0 \\ (1-d_f)(1-d_m)\nu_{12}E_2 & (1-d_m)E_1 & 0 \\ 0 & 0 & (1-d_s)GD \end{bmatrix} \tag{8.6}$$

$$D = 1-(1-d_f)(1-d_f)\nu_{12}\nu_{21} \tag{8.7}$$

The three scalar parameters, d_f, d_m, and d_s represent the current state of fiber damage, matrix damage, and shear damage, respectively. The damage parameters have $0 \le d_i \le 1$ values, with monotonically increasing values up to $d_i = 1$, where complete failure occurs.

Some of the materials properties that were used to model the GFRP were experimentally found and are summarized in Table 8.1 [12]. To incorporate the damage evolution parameters in ABAQUS®, additional experimental data was obtained from the literature [13]. An elastic–plastic multi-linear isotropic hardening material model was used to represent the stress–strain relationship of the resin insert material that was used as the interface in the hybrid bolted joint. The resin material properties were taken from [14].

8.6 Results and Discussion

Figure 8.4 shows the load vs. deformation characteristics of conventional and basic hybrid bolted joints that have been reported earlier and were used for comparisons with the FEM study [8, 9, 12]. Unlike the conventional bolted joint, the hybrid joint exhibits a monotonic increase of load without slippage. Onset of delamination occurs at approximately 32 kN for the hybrid, while the conventional joint begins to delaminate at approximately 12 kN. Once delamination initiates, the stiffness of the hybrid joint reduces gradually until it reaches its maximum load carrying capacity, ~ 51.5 kN. If the stiffness of the conventional bolted joint is calculated between points 1 and 2, the hybrid bolted joint improves the stiffness of the joint by ~62 % prior to the onset of delamination. Furthermore, if slip is critical, then the hybrid bolt can be considered infinitely better than the conventional one. If failure is defined as the onset of delamination, then the hybrid bolt is roughly three times better than the conventional one.

Figure 8.5 compares the force-displacement curves from the experimental and numerical studies of the hybrid joint. The stiffness of the numerical model is seen to be in good agreement with the experimental one within a range of 0–7 kN. However, as the load increases further, the slope (joint stiffness) of the experimental curve decreases more quickly than does that of the analytical model. As mentioned above, the onset of delamination from the experiments, labeled as point *a* in Fig. 8.5, occurs at 32 kN. The FEM curve shows delamination onset, identified by the slight plateau labeled as point *b* in Fig. 8.5, occurring at about 30 kN. These findings agree to within 6 %, which seems very good considering the complexities of the problem and the unknown factors.

Practically, the experimental results could be influenced by several factors including the quality of the infusion process, geometry variations, joint assembly issues, distribution of the resin insert around the bolt, gripping pressure, and alignment of the specimen with the loading direction. The cumulative effect of all those factors could adversely or favorably affect the stiffness of the joint, first damage initiation point, maximum failure load, and mode of failure. None of these factors could be taken into account in the idealized FEM model. In summary, it seems that the model is quite well validated by the experiments.

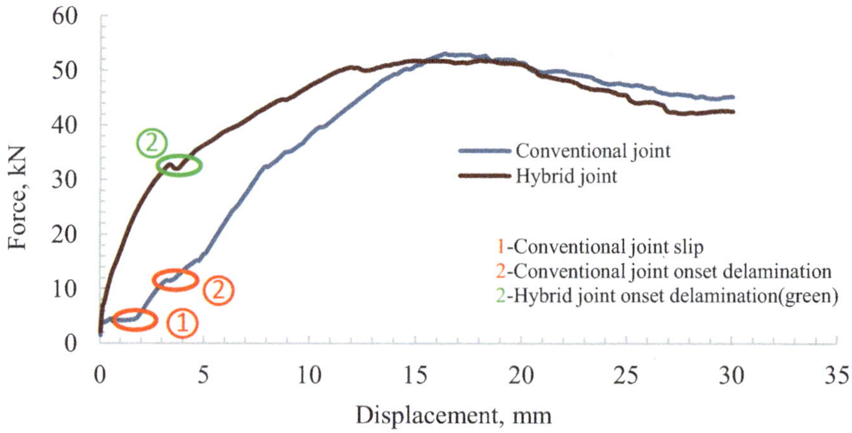

Fig. 8.4 Tensile tests of bolted lap joint with conventional and hybrid bolts

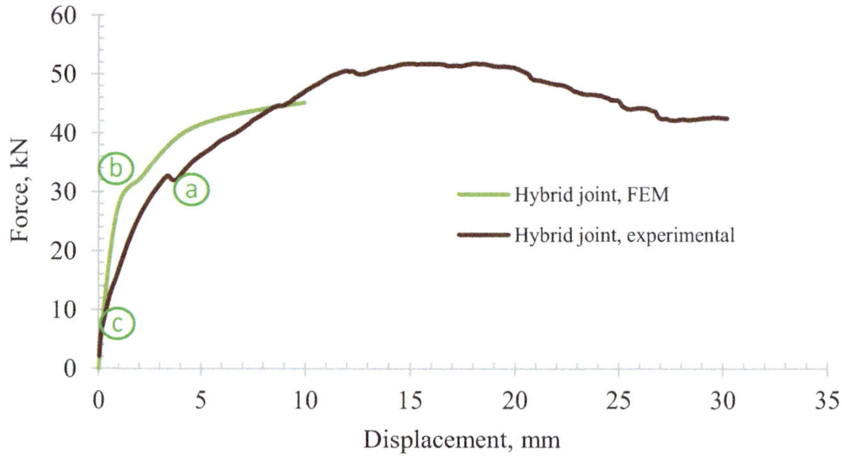

Fig. 8.5 Comparison of results of experimental and FEM modeled tensile test of hybrid bolted lap joint

One of the critical aspects in the design of composite structures and joining is understanding damage initiation and evolution at critical locations such as at hole, near abrupt change in cross-section, and at interfaces. Those kinds of features affect directly or indirectly the performance of a joint designed for a particular application. Investigation in depth of the interactions of the materials as well as in-plane and through-thickness stress distributions could help in the design of robust structural composite bolted joints. The main failure modes in a hybrid bolted joint can be categorized into three groups, namely: yielding of the bolt, deformation and fracture of the washers, and excessive damage of composite substrates characterized by matrix and/or fiber failure. Figure 8.6a shows a hybrid bolted joint that has been tested to failure and then exposed to UV light after dye penetration. For qualitative comparison, Fig. 8.6b shows the Von Mises stress distribution predicted by FEM at failure. Notice that the experimentally tested composite substrate is severely damaged around the hole due to tensile loading, bearing pressure, and localized compressive load beneath the washer. The FEM predicts well the fractured surface beneath the washer. A further interesting comparison is presented by the figure following. Figure 8.7a is a photograph of the hybrid composite bolted joint at failure under tensile loading. Figure 8.7b shows the Von Mises stress distribution in the bolt and washer from the FEM model. These results show that the tensile shear loading and secondary bending effects cause the bolt to become noticeably tilted and to develop plastic zones at points 1 and 2 of the washer, which ultimately leads to deformation and fracture. As expected, the maximum stress in the bolt shank is developed at the shear plane of the joint, that is, at its intersection with the interface of the joined components.

Reference is made to two points in the force-displacement curve obtained from the FEM model and shown in Fig. 8.5 to interpret their significances and to aid in visualizing the corresponding stress distributions and damage initiations in composite substrates. The first of these is point *c*, where, evidently, initial matrix failure occurs; and the second is point *b*, which demonstrates significant stiffness reduction at, evidently, delamination. Figure 8.8a–d show the four failure metrics at damage initiation. As shown in these figures (look inside the hole), out of the four Hashin failure criteria, the first onset of damage

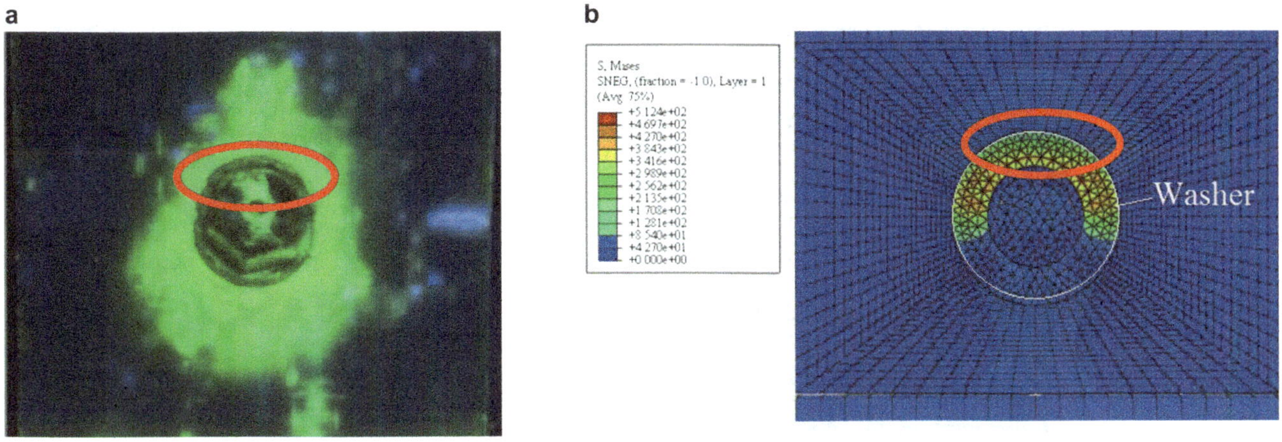

Fig. 8.6 Qualitative comparison of experimental and FEM results. (**a**) Photo from dye penetrant test of surface of hybrid joint tested to failure. (**b**) FEM prediction of Von Mises stess beneath the washer

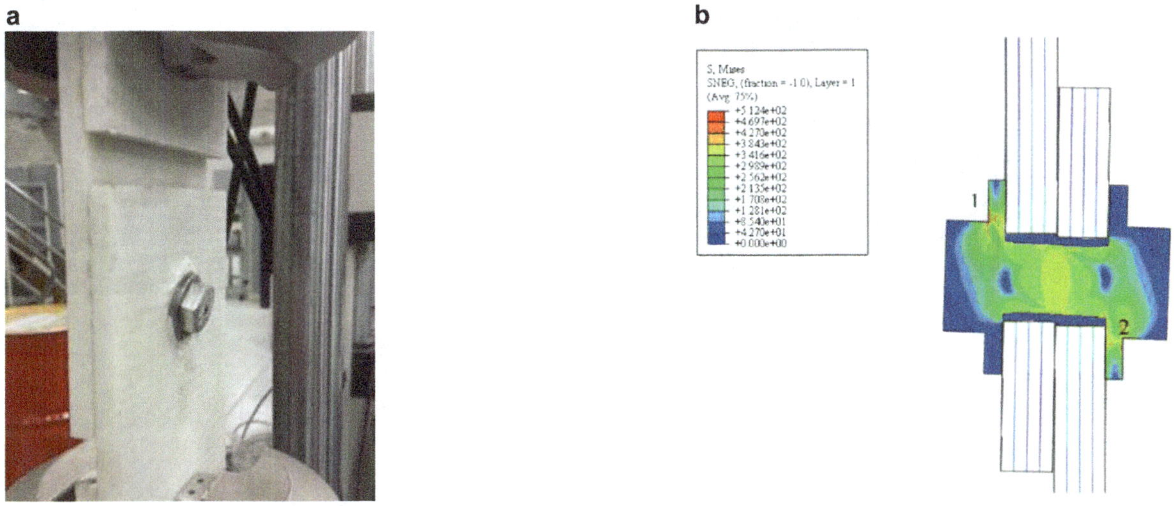

Fig. 8.7 Hybrid joint: (**a**) photograph of a hybrid composite bolted joint at failure under tensile loading, (**b**) FEM prediction of Von Mises stess in the bolt and washer

initiation due to matrix failure was observed when the failure index, R, that is based on matrix tension (see Fig. 8.8d and Eq. 8.3) is equal to one $\left(R \equiv F_m^t = \left(\frac{\sigma_{22}}{Y^T}\right)^2 + \alpha\left(\frac{\tau_{12}}{S^L}\right)^2 = 1 \right)$. The remaining failure metrics from Eqs. 8.1, 8.2, and 8.4 show low values at this stage, which implies that their influences are minimal in causing any considerable damage within the composite specimen. The maximum principal stress field was also captured in order to observe the stress gradient around the hole, and this is shown in Fig. 8.8e. At this stage, the stress field indicates that the load tends to develop a significant bearing stress in the composite substrate.

Fiber failure is a usually considered to be the critical failure mode in composite structures as compared with matrix failure. In the absence of fiber failure, the loading of the joint can be increased even though some matrix failure has occurred. However, as matrix failure becomes more dominant, it will detrimentally affect the stiffness of the joint.

Figure 8.9 shows the FEM damage predictions according to each of the four Hashin criteria for the composite substrate of the hybrid bolted joint at the onset of delamination, which corresponds to point *b* of Fig. 8.5. Figure 8.9a, b show that the predicted failure begins with the near-simultaneous compressive and tensile matrix failures at separate locations and progresses with broadening damage areas when the corresponding failure indexes $R \geq 1$. At this point, the stiffness of the

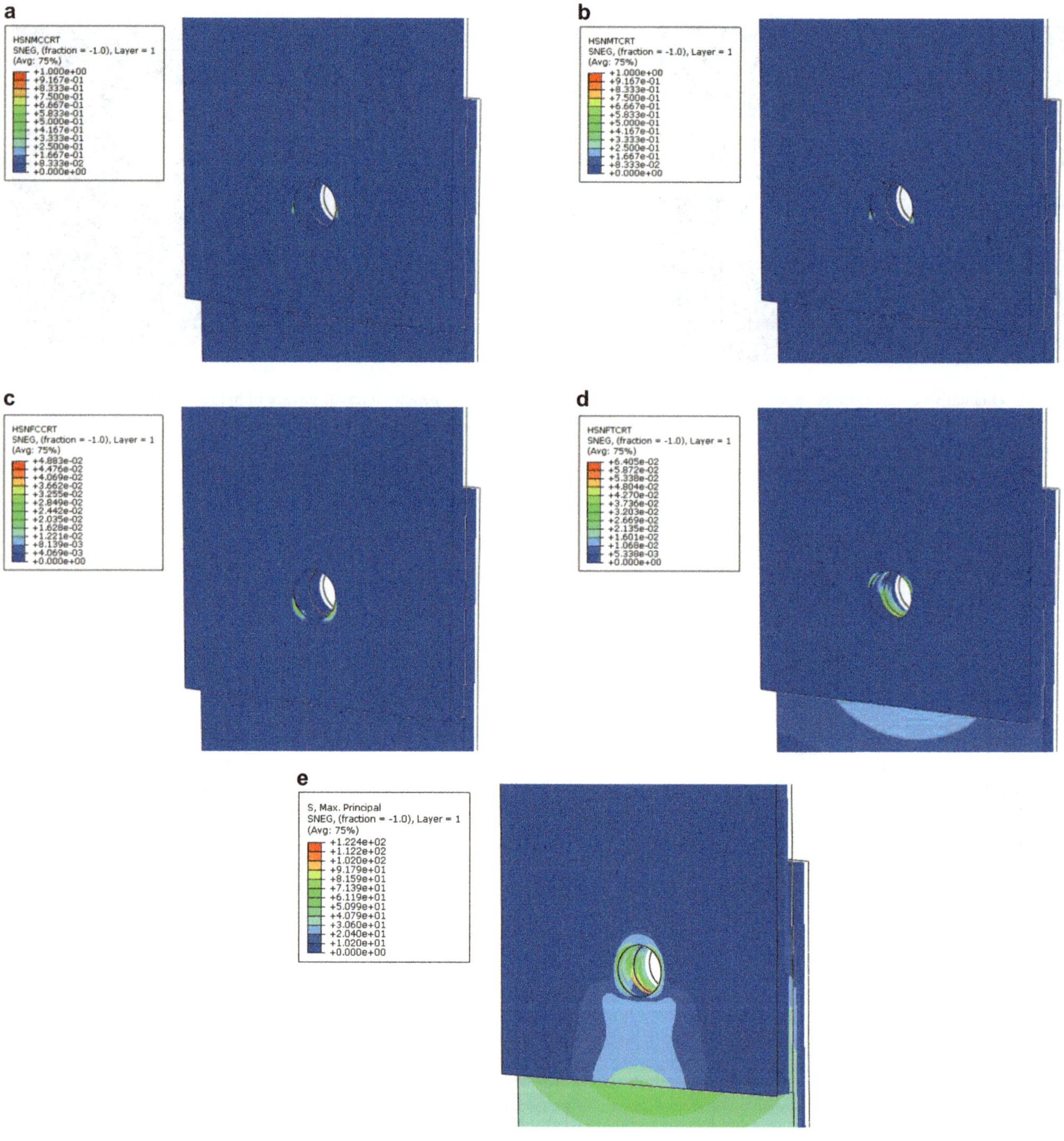

Fig. 8.8 Hashin damage metrics and stress prediction from FEM model of hybrid bolted joint at first onset of matrix failure: (**a**) Hashin matrix compression metric; (**b**) Hashin matrix tension metric; (**c**) Hashin fiber compression metric; (**d**) Hashin fiber tension metric; (**e**) Maximum principal stress distribution

joint declines to its lowest value prior to ultimate load. On the other hand, the tensile fiber failure [HSNFTCRT] and compressive fiber failure [HSNFCCRT] metrics, represented in Fig. 8.9c, d respectively, show values that are less than one, specifically, $R \equiv F_f^t = \left(\frac{\sigma_{11}}{X^T}\right)^2 + \alpha\left(\frac{\tau_{12}}{S^T}\right)^2 = 0.7244$, $R \equiv F_f^c = \left(\frac{\sigma_{11}}{X^C}\right)^2 + \alpha\left(\frac{\tau_{12}}{S^T}\right)^2 = 0.799$. These results confirm that in composite bolted joints, even though the fibers might remain intact at elevated load, the performance of the joint is degraded owing to the presence of matrix failure. Matrix failure is one of the main factors driving delamination between the layers. Usually, delamination can also be caused by interlaminar stresses, which are developed due to mismatches in material properties, discontinuities in the structure (e.g. holes), and high deformation gradients [15].

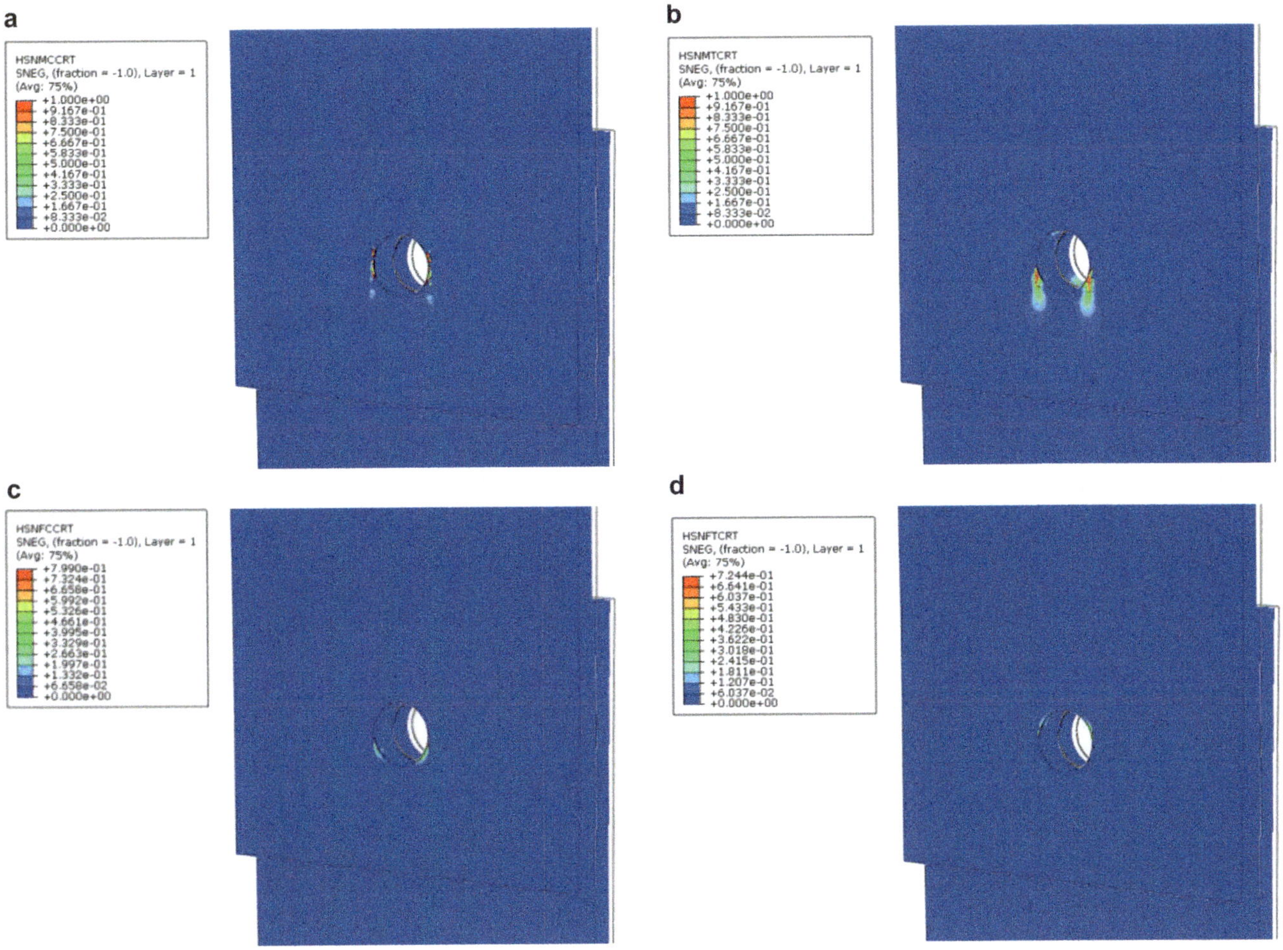

Fig. 8.9 Hashin damage metrics from FEM model of hybrid bolted joint at first onset of delamination: (**a**) matrix compression metric; (**b**) matrix tension metric; (**c**) fiber compression metric; (**d**) fiber tension metric

8.7 Conclusions

In this work, experimental and numerical methods were adopted to investigate the mechanical properties of hybrid composite bolted joints. The numerical model was developed in Abaqus/Standard. By comparing the numerical results with the experimental results, the validity of the numerical model was verified in terms of load–displacement and damage patterns. The observed results can be summarized as follow:

- Generally, this study targets the development of a numerical model of a single-lap hybrid bolted composite joint that was extensively studied experimentally by the same group in the past.
- As reported in the previous works, the hybrid joint eliminates the presence of slippage and delays the onset of delamination from occurring at low load level.
- The stiffness of the hybrid joint predicted by the numerical method is shown to be in a good agreement with the experimental result, particularly at the initial stage of the load–displacement curve.
- Major phenomenon, such as onset of delamination and maximum load carrying capacity of the hybrid joint, are also predicted well using the numerical model along with the Hashin failure criteria.
- The damage of the composite beneath the washer is well predicted.
- The deformation of the bolt and washer was predicted, and the results agree qualitatively with experimental results.
- Significant stiffness variations and absences of detailed load fluctuations are observed as the applied load increases. These behaviors can be explained by the fact that, in the current numerical model, the interfaces between the individual plies within the laminate were assumed to be ideal.

- A delamination model, which is usually represented by cohesive elements between the plies, was intentionally excluded in order to minimize computational time.
- Future work includes detailed modeling of the hybrid bolted joint including a delamination model within the interface, tailored inserts that contain particulates and fibers, and multi-material joints.

References

1. Dhôte, J.X., Comer, A.J., Stanley, W.F., Young, T.M.: Investigation into compressive properties of liquid shim for aerospace bolted joints. Compos. Struct. **109**, 224–230 (2014)
2. Walker, T.H., Minguet, P.J., Flynn, W.F., Cargery, D.J., Swanson, G.D., Ilcewicz, L.B.: Advanced technology composite fuselage-structural performance. NASA Con-tractor report 4732 (1997)
3. Lu, X., Kim, K.Y., Sachse, W.: In situ determination of elastic stiffness constants of thick composites. Compos. Sci. Technol. **57**, 753–762 (1997)
4. Goering, J., Bohlmann, R.E., Wanthal, S., Kautz, E., Neri, L.: Assembly induced de-laminations in composite structures. In: Proceedings of Ninth DoD/NASA/FAA Conference on Fibrous Composites in Structural Design, Lake Tahoe (1991)
5. Hühne, C., Zerbst, A.K., Kuhlmann, G.: Progressive damage analysis of composite bolted joints with liquid shim layers using constant and continuous degradation models. Compos. Struct. **92**, 189–200 (2010)
6. Military Handbook - MIL-HDBK-17-1F: Composite Materials Handbook, Volume 1 - Polymer Matrix Composites Guidelines for Characterization of Structural Materials. U.S. Department of Defense, ELECTRONIC ISBN 978-1-59124-509-4, (2002)
7. Liu, L., Zhang, J., Chena, K., Wanga, H., Liub, M.: Experimental and numerical analysis of the mechanical behavior of composite-to-titanium bolted joints with liquid shim. Aerosp. Sci. Technol. **49**, 167–172 (2016)
8. Haq, M., Khomenko, A., Cloud, G.: Novel hybrid fastening system with nano-additive reinforced adhesive inserts. In: SEM 2014 Annual Conference & Exposition on Experimental and Applied Mechanics, Greenville, 2–5 June 2014
9. Haq, M., Cloud, G.: Flexible hybrid fastening system. In: SEM 2013 Annual Conference & Exposition on Experimental and Applied Mechanics, Chicago, 3–5 June 2013
10. Cloud, G.: 2012 William M. Murray Lecture: some curious unresolved problems, speculations and advances in mechanical fastening. Exp. Mech. **53**(7), 1073–1104 (2013). doi:10.1007/s11340-013-9736-3
11. Cloud, G.L.: Hybrid fastener. Provisional Patent 6550-000226-US-PS1-1 (2012). General patent application filed (2013)
12. Koricho, E.G., Khomenko, A., Haq, M., Cloud, G.L.: Fatigue behavior of novel hybrid fastening system with adhesive inserts. In: Proceeding of SEM 2015 Annual Conference & Exposition on Experimental and Applied Mechanics, Costa Mesa, 7–11 June 2015
13. Belingardi, G., Koricho, G.K., Beyene, A.T.: Characterization and failure modeling of notched cross-ply and angle-ply fabric GFRP composite material. J. Compos. Struct. **112**, 237–2049 (2013)
14. Khomenko, A., Koricho, E.G., Haq, M.: Fiber Bragg-grating sensors for SHM and FEM of in service bonded multi-material Pi-joints. In: Proceeding of American Society for Composites 29th Technical Conference, La Jolla, 8–11 Sept 2014
15. Gara, A.C.: Delamination—a damage mode in composite structures. Eng. Fract. Mech. **29**, 557–587 (1988)

Chapter 9
Application of Digital Image Correlation to the Thick Adherend Shear Test

Jared Van Blitterswyk, David Backman, Jeremy Laliberté, and Richard Cole

Abstract The purpose of this study was to develop and validate a novel method for measuring shear deformation during the Thick Adherend Shear Test (TAST), to determine *in situ* mechanical properties of an adhesive under tensile shear loading, using two-dimensional (2D) Digital Image Correlation (DIC). Shear strains were optically measured using DIC from the bond line and also from adherend deformations; both were compared against measurements made using the ASTM D5656 standard (KGR-1 extensometers). The results from 17 TAST specimens showed that all techniques were in good agreement, however, both DIC techniques had significantly lower variance on measured mechanical properties compared to the KGR-1 extensometers. The use of correlated adherend deformations was the preferred technique for deriving plastic shear strains, and overall produced the lowest scatter on mechanical properties. This study demonstrates the potential for the use of 2D DIC as a more precise, and time-efficient alternative to the KGR-1 extensometers for room temperature *in situ* characterization of adhesives in shear.

Keywords Adhesives • Mechanical properties • Single-lap joint • Digital image correlation (DIC) • KGR-1 extensometer • Experimental test methods

1 Introduction

Bonded joints are seeing increasing application in the aircraft industry for their many advantages over mechanically fastened joints. One of the main challenges facing aircraft designers relates to demonstrating compliance with certification requirements. Certification of highly-loaded adhesive bonded joints typically requires a detailed characterization of the stress–strain response of the adhesive and a statistically-based set of material strength allowables. The large volume of testing required to develop such bases for material design allowables necessitates an experimental methodology that is credible, precise, repeatable, and time and cost efficient.

The thick adherend single-lap joint specimen specified in the ASTM D5656 standard [1], is the accepted methodology for measuring *in situ* adhesive shear properties. The use of thick aluminum adherends is most effective for measuring the response of an adhesive in shear due to enhanced bending rigidity along the bonded overlap, which reduces peel stresses and joint rotation. These considerations restrict the use of other thin adherend single-lap joint test methods, such as ASTM D1002 [2] and ASTM D3165 [3], to "apparent" strength tests of bonded joints, as peel stresses having significant influence on the overall joint strength [2–5]. ASTM D5656 [1] requires the use of the KGR-1 extensometer system to measure relative adherend deformation across the bond line of Thick Adherend Shear Test (TAST) specimens that are subjected to a tensile shear load. The KGR-1 extensometer is a specialized mechanical device, tailored for the TAST specimen, which can resolve sub-micron, relative, uniaxial displacements. The KGR-1 mounting pins span the bond line and measure relative shear displacements of the adjacent adherends. The measured displacements consist of both adhesive shear and adherend elongation. To isolate the strain in the adhesive, the adherend deformations are approximated based on KGR-1-measured displacements on a solid aluminum specimen of the same geometry as the bonded joint, commonly referred to as a "dummy specimen".

J. Van Blitterswyk • J. Laliberté
Department of Mechanical and Aerospace Engineering, Carleton University, 1125 Colonel By Drive, Ottawa, ON, Canada K1S 5B6

D. Backman (✉) • R. Cole
National Research Council Canada, Aerospace Portfolio, 1200 Montreal Road, Ottawa, ON, Canada, K1A 0R6
e-mail: david.backman@nrc.gc.ca

G.L. Cloud et al. (eds.), *Joining Technologies for Composites and Dissimilar Materials, Volume 10*
Conference Proceedings of the Society for Experimental Mechanics Series, DOI 10.1007/978-3-319-42426-2_9

The KGR-1 system is tedious and time consuming to operate, and is susceptible to numerous experimental errors which can significantly affect the level of data scatter. Furthermore, the quality of measurements is heavily dependent on placement repeatability and operator skill, and the resulting experimental errors are difficult to quantify. Some of the experimental errors include: mounting pin slippage, rotation, and poor placement repeatability both across the bond line, and along the joint overlap [6–8]. Quantifying the experimental errors associated with the KGR-1 and the ASTM D5656 test method has received much attention over the past 20 years [4, 6–9]. In addition, a small number of studies have explored modifications to the KGR-1 device and other experimental techniques for characterizing structural adhesives including pilot holes and additional mounting pins [4, 7], clip gauge extensometers and digital video microscopy [10, 11], and stacked strain gauge rosettes [12].

Since none of the proposed techniques have been properly developed, have limited working ranges, require correction factors, or are time consuming or expensive, the KGR-1 remains the accepted instrument for measuring adhesive properties in shear. With the development of full-field optical techniques, several researchers have considered the use of Digital Image Correlation (DIC) as an alternative technique for characterizing adhesives using single lap-joint specimens.

Lemmen *et al.* [13] characterized the shear stress-shear strain behaviour of an adhesive using a TAST specimen and an optical high magnification digital grid deformation technique. This allowed strains to be extracted using bond line deformations, thus eliminating the need for correction factors to compensate for adherend deformations. Unfortunately, the high magnification created issues with image focusing and field of view, as the camera required repositioning mid-test to capture the non-linear adhesive deformations. The combination of the two problems created large scatter in the data. Moreover, only two specimens were tested and no comparison was provided to a reference standard methodology, thus leaving the repeatability and credibility of this technique in question. Several other studies have been conducted using two-dimensional (2D) and three-dimensional (3D) DIC on single and double lap-joints [14–20], however these efforts used thin adherend specimens, which do not create the proper stress field for adhesive characterization [2–5]. In the case where material properties were estimated [14], a comparison was not provided to a reference standard and several assumptions were required, which limited the technique to linear strains.

To the authors' knowledge there has yet to be a rigorous comparison of DIC to the accepted ASTM D5656 methodology in terms of measurement precision, repeatability, cost and testing time. The purpose of this research is to validate 2D DIC as a new, faster, more precise and cost effective method for characterizing the *in situ* properties of adhesives in shear using the ASTM D5656 test specimen and procedure. This study compares the KGR-1 in ASTM D 5656, to two DIC data extraction approaches, using 17 ($n = 17$) TAST specimens bonded with aerospace-grade AF 163-2K epoxy structural film adhesive.

2 Experimental

2.1 KGR-1 Extensometer

The KGR-1 extensometer, from Cytec Engineered Materials, is a mechanical instrument for precisely measuring relative uniaxial deformation. The KGR-1 extensometer system consists of two Linear Variable Differential Transformers (LVDTs), each housed in a two-piece (front/back) rigid frame, which span the bonded joint as shown in Fig. 9.1. Each front frame has one steel mounting pin while the back frame has two. The front and rear frame mounting pins are held in plane, parallel to the specimen surface, using an extension arm attached to the back side of the frame. This configuration allows the two sets of mounting pins to displace independently from one another, causing the LVDT core to move with respect to the LVDT coils, which outputs a voltage from the transformer. The instrument is mounted on the specimen using a self-supporting spring-clamp mechanism (Fig. 9.2) and suspends itself on the specimen via the mounting pins, mechanically restraining the assembly to displace with the adherends. The mounting pins have sharp points that indent into the adherends to minimize slippage.

Calibration of the KGR-1 output voltage was performed using an extensometer calibrator (MTS Calibrator Model 650.03). The calibrator has two aluminum mounting blocks that are spanned by the KGR-1 mounting pins and a micrometer dial that enables precise relative displacements to be applied to the aluminum blocks, which was then correlated to the output voltage.

2.2 Digital Image Correlation

Digital Image Correlation (DIC) is a non-contact optical method for measuring surface deformation and strain. The technique requires at least two images of the surface of interest; one in a state with no applied load (reference), and one under applied load. A computer algorithm discretizes the images into regions referred to as "subsets", and characterizes the transformation of each subset, between images, using shape functions based on optimizing a cross-correlation coefficient of

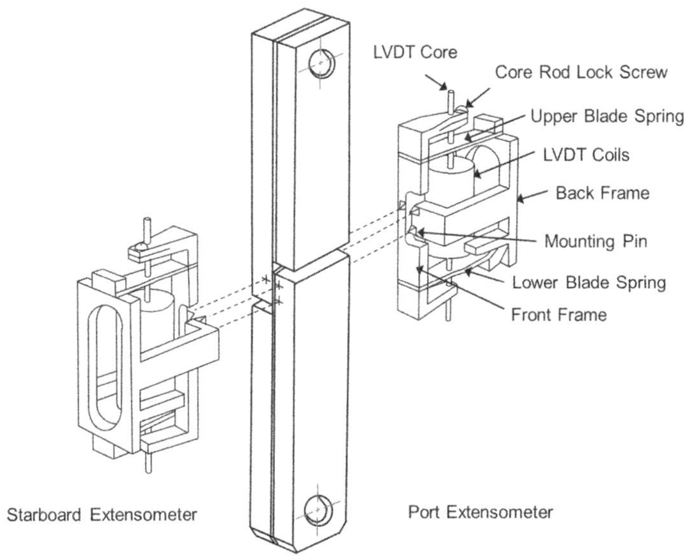

Fig. 9.1 KGR-1 extensometer schematic layout (based on [21])

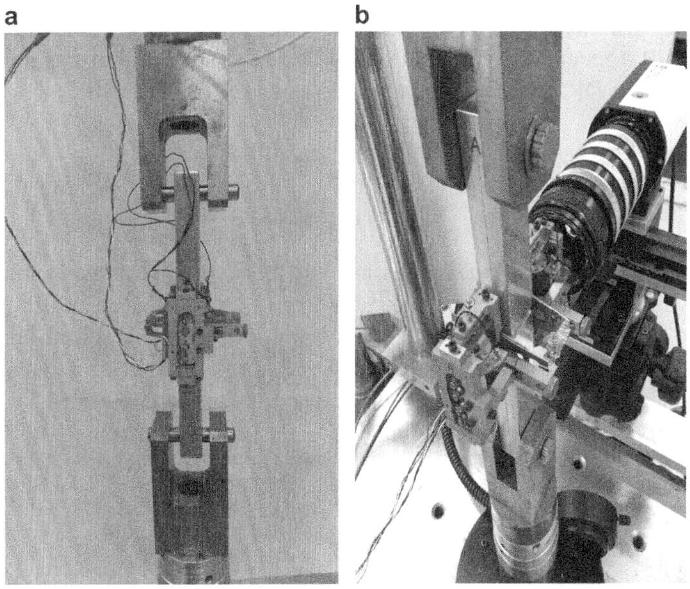

Fig. 9.2 Experimental setup for (**a**) KGR-1 and (**b**) DIC with KGR-1 in place to show stand-off distance

subset pixel intensities. The 2D DIC camera system consisted of a 1.3 M pixel QImaging Retiga 1300i CCD monochrome video camera, outfitted with a Canon 50 mm macro zoom lens with three extension tubes. The system had a field of view of approximately 6.5 mm × 5.5 mm with a corresponding spatial resolution of 5.30 μm/pixel (189 pixels/mm). The camera was mounted on a set of high precision vernier cross-stages and fixed to a tripod head. The entire assembly was fixed to a custom frame, supported on the upright supports of the load frame, as shown in Fig. 9.2b. The tripod enabled three degrees of freedom for the camera orientation, with fine focal adjustments achieved via the linear vernier stages.

A ring light was mounted on the lens to uniformly illuminate the bonded overlap region. Images of the gauge region and synchronized recordings of applied load and displacement from the load frame were acquired using the VicSnap software (v. 2010) from Correlated Solutions Inc. The image processing and data extractions were performed via post-processing with Vic2D (v. 2009 1.0) from Correlated Solutions Inc. The strain and displacement resolution was estimated using a series of correlated images of the stationary specimen surface under no applied load. The corresponding maximum strain and maximum displacement provides an indication of the level of Gaussian camera noise, lighting noise, and lack of contrast or focus. For the current setup, the strain and displacement noise levels did not exceed 100 microstrain and 35 μm, respectively. For the purposes of this comparative study, image correlation was only performed on the port side of the specimens, rather than both sides, as was done with the KGR-1 extensometer system.

2.3 Manufacturing

Eighteen single-lap shear joints were prepared from two sets of two 254 mm×305 mm×9.5 mm (10 in.×12 in.×0.375 in.) MIC-6 aluminum plates bonded together with AF 163-2K-0.06 epoxy film adhesive. Plate flatness was determined to be within 0.025 mm (0.001 in.) as measured with a dial gauge prior to bonding. The plates were prepared by cleaning with Methyl-Ethyl-Keytone followed by grit blasting, and a surface pre-treatment with Sol–gel AC130-2. The bond line thickness was regulated using four 0.203 mm (0.008 in.) spacer wires. The panels were vacuum bagged and cured in an autoclave according to the adhesive suppliers' suggested cure cycle (100 min at 121 °C (250 °F) and 674 kPa (83 psig) with a temperature ramp rate of 1 °C (3.5 °F) per minute). The bond line temperature was monitored by embedding a thermocouple at the plate-to-edge dam interface. An ultrasonic C-scan was performed on each bonded plate to verify acceptable bond quality and uniformity. One bonded plate was labeled 'J' and the other 'K'. The specimens were rough cut from the bonded plates and finished to the dimensions specified by ASTM D5656 [1] (25.4 mm in width and 228.6 mm in length) with an end mill. The notches were cut with a slitting saw; however, the adhesive was not removed from the notches to avoid any possibility of damaging the bond line at the ends of the gauge region. Only 17 specimens were tested as one was damaged during manufacturing. Prior to tests utilizing either of the DIC techniques, the port side of the gauge region was lightly abraded to generate a random surface texture.

Bond line metrology was performed using an optical microscope and an in-house program which extracted the coordinates of the microscope crosshairs. A calibrated stage was used to align the specimen bond line with the microscope crosshairs and the bond line thickness was manually calculated using extracted coordinates from each side of the bond line. The overall bond line thickness was calculated using an average of five measurements, equally spaced along the overlap. All of the specimens were measured to have the same bond line thickness on either side to within 0.013 mm (0.0005 in.). For the KGR-1 extensometers, strain was calculated using the average bond line thickness between both sides of the specimen, whereas DIC strains were calculated using the bond line thickness measured on the port side of each specimen.

2.4 Testing

Since simultaneous measurements with the KGR-1 extensometers and DIC was not possible, an alternative testing approach was developed, with measurements taken with DIC and the KGR-1 on the same specimen, in separate loading sequences up to 4482 N (1000 lbf) (within adhesive linear range). This permits direct comparisons of the measured shear modulus between the techniques. Following these measurements, and without removing the specimen from the load frame, each specimen was loaded to failure. Randomly selected specimens were allocated to the DIC (n=8) and to the KGR-1 (n=9) systems, respectively for tests to failure.

3 Data Processing

The images, along with the file containing synchronized analog outputs from the test frame's load and displacement sensors, were uploaded into the Vic2D software. A reference image was captured with the specimen subjected to a small tensile load of 222 N (50 lbf) to ensure proper self-alignment of the universal joint and the bushings within the mounting holes. The Vic2D software correlates the deformed images to the reference image within a user-defined Area of Interest (AOI). The selected AOI enclosed the entire bond line within the available field of view, and outwards on the adherends to the locations of the KGR-1 mounting pin indents (Fig. 9.3). The images were processed using a correlation algorithm based upon the normalized squared differences criterion, with sub-pixel interpolations performed using an optimized eight-tap spline; the most efficient interpolation method in terms of minimizing Gaussian noise [22]. To achieve maximum strain and displacement resolution, a smoothing window of 5 pixels (step=1 pixel, filter=5) or 27 μm was selected. During large plastic deformations, the image correlation algorithm could not correlate the deformed images to the reference image. Therefore, beyond the onset of plasticity incremental correlation was applied, which correlates each successive image with the previous image.

Two data extraction techniques were explored for the DIC measurements. The first used three small area extractions of adherend displacement (here forth referred to as 'point' extractions), in the same general arrangement as the KGR-1 mounting pins, but much closer to the bond line, and the second technique utilized a shear-strain extraction directly from the bond line (see Fig. 9.3). Adherend displacement was extracted from a small area, rather than a point, to minimize measurement noise. For consistent DIC displacement extractions, in line with the KGR-1 mounting pins, the KGR-1 was mounted on the speci-

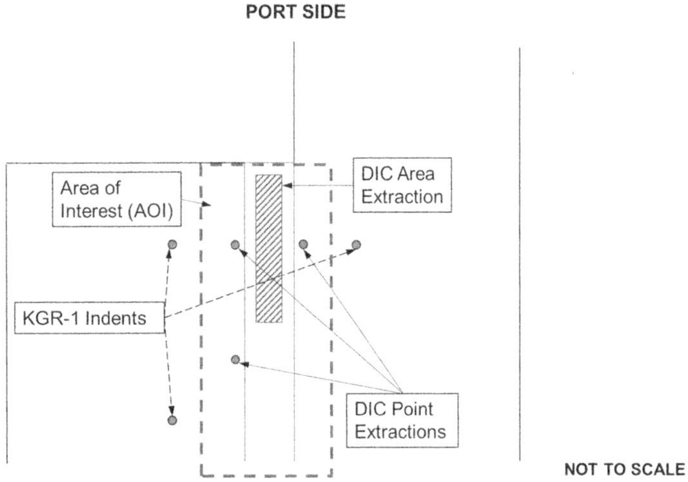

Fig. 9.3 Locations of DIC data extractions on the port side of a TAST specimen

men prior to DIC measurements to indent the adherends. A custom mounting fixture was used to place the KGR-1 mounting pins in the same position on each specimen to within 0.5 mm. The two upper DIC points were taken in-line with the upper KGR-1 indents with each extraction centered 50 pixels (265 μm) away from the bond line. The bond line area extraction covered the centerline of the bond and was positioned such that it extended equal distances on either side of the upper KGR-1 indents (Fig. 9.3). The area extraction did not extend to the edge of the overlap to avoid bias from the strain concentration.

The ability to place the DIC data extraction points closer to the bond interface than the KGR-1 mounting pins greatly reduced the influence of adherend deformation on the measured relative displacements. To simplify the analysis, corrections for adherend deformation were neglected for DIC measurements as they were estimated to account for a mere 0.3 % of the shear strain at 4482 N (1000 lbf).

3.1 Correcting DIC Displacement Measurements for Rigid Body Rotation

Although the TAST specimen is designed to mitigate adherend bending effects and thus, gauge region rotation, this cannot be completely eliminated due to the inherent loading eccentricity across the bond line. The three contact points force the extensometers to rotate with the specimen and resolve only relative shear displacements in a local reference frame, whereas DIC measures deformation of the specimen in a global reference frame. Therefore, DIC-measured displacements capture aggregate motion of rigid body rotation and relative shear displacement (Fig. 9.4a), with rigid body rotation increasing with increasing tensile load. In much of the same way as the KGR-1 mounting pins, three extractions of displacement allow one to resolve the rigid-body rotation, effectively characterizing the transformation between an equivalent local reference frame and the global reference frame. This process will be described next.

Let the original position of the three points be denoted lo, ro and bo. As the specimen deforms, the gauge region rotates by an angle θ and the extraction points move to their respective deformed positions denoted as l', r', b'. This motion is captured through displacements in the x-direction (ur, ul, ub) and y-direction (vl, vr, vb). Embedded in this motion is relative shear displacements, which when added to the rotation, results in the final deformed positions at l'', r'', b''. Therefore, with the relative orientation of the points known, the reference frame rotation angle can be calculated for each image using Eq. (9.1).

$$\theta = \tan^{-1}\left(u_b - u_r / h\right) \tag{9.1}$$

Neglecting relative displacements in the x-direction between point l and r, the relative shear displacement (Δv) can then be found by transforming the deformed coordinate frame back to the reference orientation. This is equivalent to subtracting the vertical displacement due to pure rotation from the relative vertical displacements between the two points spanning the bond line, as measured by DIC ($\Delta vDIC$) (Eq. (9.2)). The resulting relative displacement is that caused by pure shear deformation in the adhesive (Eq. (9.3)).

$$v_{DIC} = v_r - v_l \tag{9.2}$$

$$v_{resultant} = v_{DIC} - l\tan\theta \tag{9.3}$$

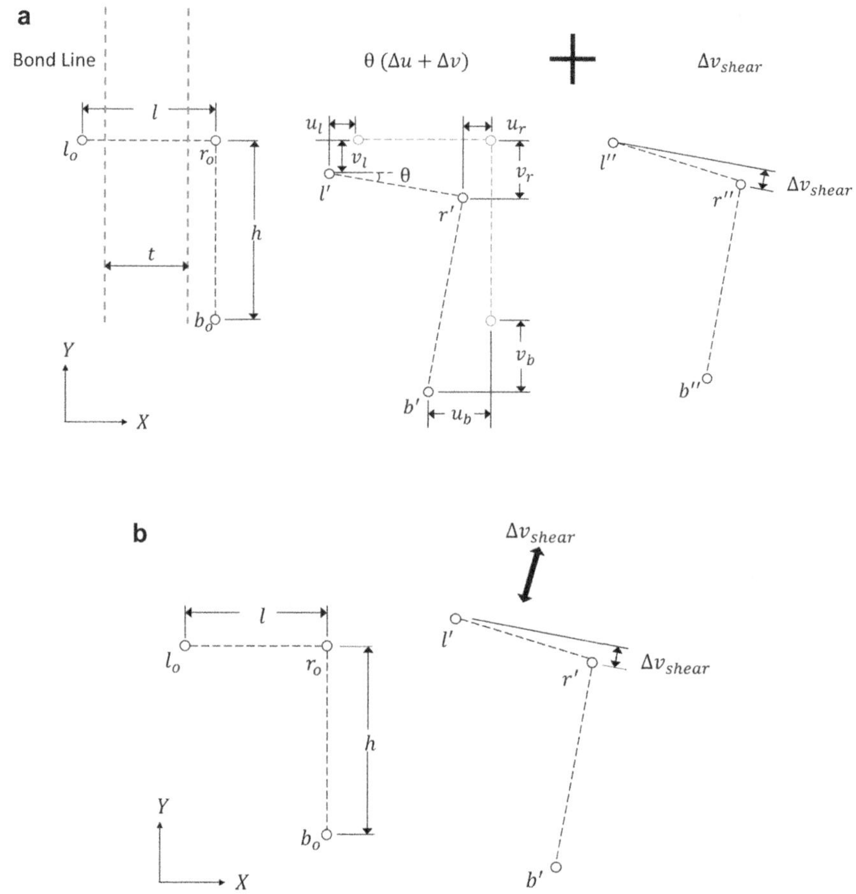

Fig. 9.4 Effect of reference frame on measurements of relative shear displacements using (**a**) a DIC global frame, consisting of rigid body rotation and shear deformation and, (**b**) the KGR-1 in a body-fixed frame of reference which measures only relative shear deformations

Shear strain is then calculated by dividing Δ*vresultant* by the bond line thickness, *t*. This correction was not required for the strain measurements directly from the bond line as the Vic 2D software accounts for rigid-body motion between subsets. The KGR-1 data was corrected for adherend deformation using measurements from the solid 'dummy' specimen, as per ASTM D5656 [1]. The *in situ* shear stress-shear strain curves and mechanical properties were then computed for all data sets using the graphical process outlined in ASTM D5656 [1].

4 Results

4.1 Shear Stress-Shear Strain Curves

The upper and lower bounds of all shear stress–strain curves measured with the KGR-1 system, DIC three point extractions of adherend displacement, and DIC area extractions of bond line strains, are shown in Fig. 9.5a–c, respectively, with a comparison of the three measurement methods shown in Fig. 9.5d.

4.2 Statistical Testing

A statistical analysis (student's *T*-test) was performed to detect significant differences in average measured mechanical properties on a per-plate, and overall (or 'aggregate') basis. The average mechanical properties, $\bar{x}$, and standard deviation, σ, along with results from the statistical analysis are detailed in Table 9.1. In Table 9.1, 'SD' and 'NSD', denote that a significant difference between techniques exists, or does not exist, respectively. Furthermore, 'DIC-P' denotes the DIC technique using three extractions of adherend displacement, and 'DIC-A' denotes DIC area extractions of strain from the bond line.

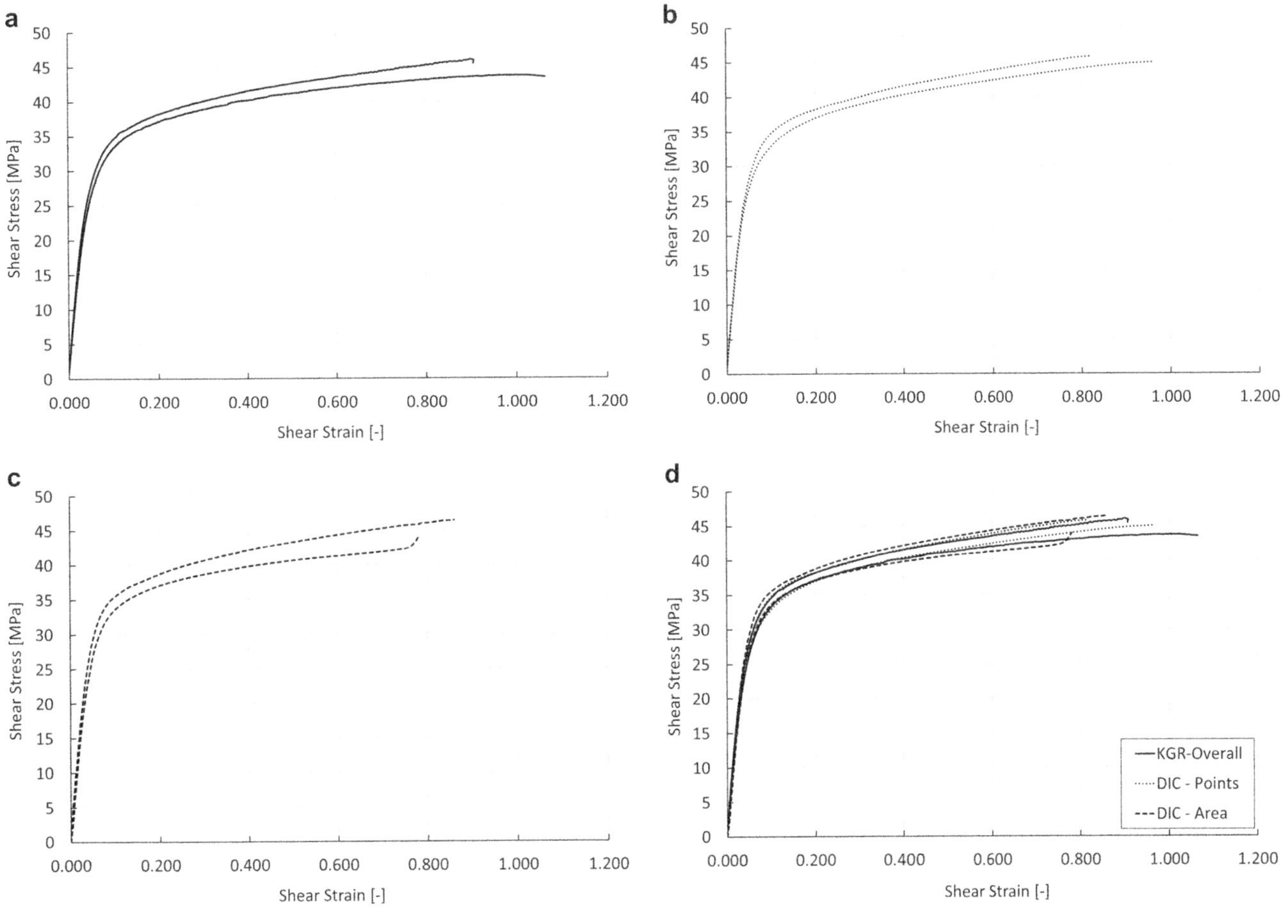

Fig. 9.5 Upper and lower bound shear stress-shear strain curves for AF 163-2K film adhesive measured using: (**a**) the KGR-1 extensometers, (**b**) DIC-measured displacements on the adherends, (**c**) DIC-measured strains extracted directly from the bond line, and (**d**) an overall comparison of three methods

Table 9.1 Statistical significance testing for differences in average measured properties between the KGR-1 and DIC data sets

Mech. property		Shear modulus, MPa	Prop. limit stress, MPa	Prop. limit strain	Knee stress, MPa	Knee strain	Ultimate strain
Plate J							
KGR-1	$\bar{x}(\sigma)$	696.8 (35.4)	11.5 (2.6)	0.014 (0.004)	35.5 (0.5)	0.147 (0.008)	0.998 (0.066)
DIC–P	$\bar{x}(\sigma)$	669.6 (7.7)	17.3 (1.2)	0.024 (0.001)	35.4 (0.2)	0.124 (0.013)	0.888 (0.043)
DIC-A	$\bar{x}(\sigma)$	720.0 (12.5)	16.5 (0.6)	0.022 (0.003)	35.1 (1.3)	0.115 (0.013)	0.840 (0.052)
KGR-1 vs. DIC-P		NSD	SD	SD	NSD	SD	NSD
KGR-1 vs. DIC-A		NSD	SD	SD	NSD	SD	NSD
DIC-P vs. DIC-A		SD	NSD	NSD	NSD	NSD	NSD
Plate K							
KGR-1	$\bar{x}(\sigma)$	685.4 (90.7)	13.9 (2.1)	0.018 (0.005)	35.6 (0.3)	0.140 (0.015)	1.021 (0.075)
DIC-P	$\bar{x}(\sigma)$	653.6 (21.1)	17.2 (0.7)	0.024 (0.001)	35.4 (0.5)	0.127 (0.012)	0.925 (0.061)
DIC-A	$\bar{x}(\sigma)$	713.7 (22.7)	16.8 (1.0)	0.023 (0.001)	35.4 (0.6)	0.110 (0.003)	0.842 (0.054)
KGR-1 vs. DIC-P		NSD	SD	NSD	NSD	NSD	NSD
KGR-1 vs. DIC-A		NSD	SD	SD	NSD	SD	SD
DIC-P vs. DIC-A		SD	NSD	NSD	NSD	SD	NSD

(continued)

Table 9.1 (continued)

Mech. property		Shear modulus, MPa	Prop. limit stress, MPa	Prop. limit strain	Knee stress, MPa	Knee strain	Ultimate strain
Aggregate							
KGR-1	$\bar{x}(\sigma)$	690.8 (68.5)	12.6 (2.6)	0.016 (0.005)	35.5 (0.4)	0.144 (0.011)	1.008 (0.067)
DIC-P	$\bar{x}(\sigma)$	661.1 (22.9)	17.2 (0.8)	0.024 (0.001)	35.4 (0.4)	0.126 (0.011)	0.911 (0.055)
DIC-A	$\bar{x}(\sigma)$	716.7 (29.8)	16.7 (0.8)	0.023 (0.002)	35.3 (0.8)	0.112 (0.007)	0.842 (0.050)
KGR-1 vs. DIC-P		NSD	SD	SD	NSD	SD	SD
KGR-1 vs. DIC-A		NSD	SD	SD	NSD	SD	SD
DIC-P vs. DIC-A		SD	SD	NSD	NSD	SD	SD

5 Discussion

Despite having a small sample size, several useful observations were made from the comparisons between DIC and the KGR-1 regarding the influence of instrumentation on the degree of data scatter and mechanical properties derived from the measured adhesive response.

Generally speaking, the shear stress–strain curves obtained with the KGR-1 system and DIC, shown in Fig. 9.5, are in excellent agreement and have low scatter. With this experimental test, some scatter can always be expected due to the inherent variability in the epoxy adhesive, degree of cure within the bond, bond line thickness between specimens, residual stresses within the specimens, and defects within the bond line [23]. However, any biases imposed by these variations were mitigated through random assignment of the specimens and repeated measurements of the shear modulus with different measurement techniques on each specimen. DIC measurements have lower variance compared the KGR-1 measurements for all mechanical properties, with exception to knee stress and knee strain. In the case of the shear modulus, the aggregate statistics show that the variance associated with the KGR-1 is up to 300 % greater than DIC. This may be caused by progressive pin slippage or rotation, which was observed in almost all post-test inspections. The significant difference detected for measured knee strain, and proportional limit stress and strain between all techniques suggests that the shape of the stress–strain curve may have an appreciable dependence on strain measurement technique.

A closer examination of the elastic region (Fig. 9.6) shows that, in addition to measuring lower strains, the DIC-area extractions had higher noise levels compared to the point extractions of adherend displacement. This is a result of reduced contrast in the bond line, strain resolution limits imposed by system noise (100 microstrain) and spatial resolution within the bond line. The issue of spatial resolution, within and along the bond line, is an inherent compromise when using DIC on this specimen due to the large aspect ratio of the gauge region. The resolution chosen for this work (5.30 μm/pixel) was such that one half of the overlap region was captured, from which strain extractions were likely to be more representative of the average response of the whole overlap, compared to a setup with higher resolution across the bond line. Further studies are required to investigate the sensitivity to spatial resolution within the bond line. A separate test with a high contrast, fine speckle pattern applied to the surface resulted in a slight reduction in measurement noise, but had no appreciable effect on measured mechanical properties. The strains resolved using the DIC point extraction technique were not affected by the speckle pattern. The minimal improvements from the speckle pattern do not justify the time required to paint each specimen, especially in the context of adhesive characterization for certification due to the extensive amount of testing required. However, this requires further study before definitive claims regarding the effects of surface contrast can be made.

With increasing shear strain, all techniques show larger scatter, likely due to inter-specimen variations (varying bond line thickness, bond line imperfections, etc.). The KGR-1 measured slightly higher ultimate strains compared to either DIC technique, however the validity of these ultimate strains was brought into question as post-test inspection showed substantial pin slippage on four of the nine specimens tested to failure. Similar issues were encountered when collecting data from the dummy specimen, as 16 tests were required to capture 10 quality data sets. The ultimate shear strains measured within the bond line were consistently lower due to large relative displacements within the bond line, which caused problems for the correlation algorithm. This problem was first observed just beyond the onset of plasticity, and incremental correlation of images was required to capture the response. However, near failure the surface features changed so drastically that correlation was not possible. The omission of these regions of large deformation from the average shear strain measurement, causes the strain attenuation. This issue was not resolved using a speckled pattern as the paint did not adhere well to the surface under such large plastic deformations.

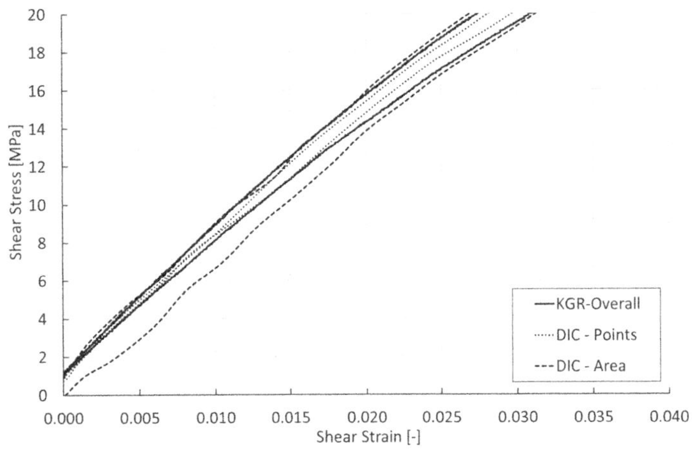

Fig. 9.6 Comparison of the bounds on the initial linear response region as measured using KGR-1 and DIC techniques

The KGR-1 does offer some advantages over DIC, especially in regards to measuring material allowables in less favourable environmental conditions. Adhesives prescribed for aerospace applications must be capable of sustaining ultimate design loads in the most adverse conditions, i.e. hot and wet environments. The KGR-1 can be mounted to the specimen in an environmental chamber, whereas these tests become far more difficult with DIC as the cameras may require specialized enclosures and heating can cause mirage effects and artificial strains. Further, the onset of poor quality data (i.e. due to pin slippage) is typically easier to detect with the KGR-1 by comparing the output voltages from the two extensometers. The results also suggest that the adherend correction from the dummy specimen, the validity of which has disputed in the literature [6, 7], should not necessarily be discredited as the KGR-1 measurements are in good agreement with DIC. This is surprising given the inherent assumption that the adherend deformation is insensitive to adhesive compliance and bond line thickness. That said, any errors in this correction are difficult to resolve with the current data set since the true mechanical properties for a given specimen are not known.

This study has shown that the most reliable and repeatable DIC technique is that which uses measured deformations in the adherends adjacent to the bond line. The lower relative adherend displacements are optimal for correlation right up to failure and the rotation correction is simple to apply. Further, the ability to perform correlations on the bare specimen surface offers significant time savings. Additional time savings are offered in pre-test setup time as the KGR-1 requires consistent recalibration to ensure that the shock loading at failure does not create bias or error in subsequent measurements. Frequent pin slippage can also double or triple specimen testing times [24]. The post-processing time was slightly higher for DIC as the image correlation routine had to be first performed before extracting the data, however this expense could be mitigated with the use of more powerful computing resources. As a final point, an additional advantage of DIC is the availability of formalized calibration procedures, such as that developed by Patterson et al. [25], whereas no such procedures exist for the KGR-1 extensometers.

6 Conclusions

From this study of 17 thick adherend shear tests, it was found that DIC measurements, using both adherend point displacements adjacent to the bond line, and direct strain extractions from the bond line, were in good agreement with the KGR-1 measurements. However, material properties measured using DIC measurements had up to 300 % lower variance than those from the KGR-1, and the failure strains recorded by the KGR-1 system were brought into question due to frequent pin slippage at large plastic strains.

Correlating images based solely on surface texture was found to be a viable technique for strain measurement. Of the two DIC techniques trialed, the most reliable technique in this study was that using DIC-measured relative adherend displacements (three-point technique), since the adherends undergo much smaller and more controlled relative displacement, and can be reliably correlated right up to failure. DIC also offers significant time savings in pre-test setup and regular calibrations. The statistical analysis performed on the KGR-1 and DIC measurements found significant differences in the knee strain, and proportional limit stress and strain values, suggesting that the shape of the shear stress-shear strain curve may have an appreciable dependence on measurement technique.

Future work includes further testing to develop a larger statistical data base to compare DIC to the KGR-1 extensometers, and detailed studies on the sensitivity of material properties to the location of data extraction and spatial resolution within the bond line.

Acknowledgement The technical support of Richard Desnoyers and Matthieu Harrison of NRC-Aerospace and Dr. Steven Philips of McGill University are gratefully acknowledged.

References

1. ASTM D5656-10: Standard Test Method for Thick-Adherend Metal Lap-Shear Joints for Determination of the Stress-Strain Behaviour of Adhesives in Shear by Tension Loading. ASTM International, West Conshohocken (2010). doi:10.1520/D5656-10. http://www.astm.org
2. ASTM D1002-10: Standard Test Method for Apparent Shear Strength of Single-Lap-Joint Adhesively Bonded Metal Specimens by Tension Loading (Metal-to-Metal). ASTM International, West Conshohocken (2010). doi:10.1520/D1002-10. http://www.astm.org
3. ASTM D3165-07: Standard Test Method for Strength Properties of Adhesives in Shear by Tension Loading of Single-Lap-Joint Laminated Assemblies. ASTM International, West Conshohocken (2007). doi:10.1520/D3165-07. http://www.astm.org
4. Tomblin, J.S., Yang, C., Harter, P.: Investigation of thick bondline adhesive joints. U.S. Department of Transportation Federal Aviation Administration, Report: DOT/FAA/AR-01/33 (2001)
5. Vijaya Kumar, R.L., Bhat, M.R., Murthy, C.R.L.: Experimental analysis of composite single-lap joints using digital image correlation and comparison with theoretical models. J. Reinf. Plast. Compos. **32**(23), 1858–1876 (2013)
6. Kassapogluo, C., Adelmann, J.: KGR-1 thick adherend specimen evaluation for the determination of adhesive mechanical properties. In: 23rd International SAMPE Technical Conference, Society for the Advancement of Material and Process Engineering, pp. 162–176 (1991)
7. Yang, C., Huang, H., Tomblin, J.S., Oplinger, D.W.: Evaluation and adjustments for ASTM D5656 standard test method for thick-adherend metal lap-shear joints for determination of the stress-strain behaviour of adhesives in shear by tension loading. J. Test. Eval. **29**(1), 36–43 (2001)
8. Tsai, M.Y., Morton, J., Krieger, R.B., Oplinger, D.W.: Experimental investigation of the thick-adherend lap shear test. J. Adv. Mater. **27**(3), 28–36 (1996)
9. Tomblin, J.S., Seneviratne, W., Escobar, P., Yoon-Khian, Y.: Shear stress-strain data for structural adhesives. U.S. Department of Transportation Federal Aviation Administration, Report: DOT/FAA/AR-02/97 (2002)
10. da Silva, L.F.M., da Silva, R.A.M., Chousal, J.A.G., Pinto, A.M.G.: Alternative methods to measure the adhesive shear displacement in the thick adherend shear test. J. Adhes. Sci. Technol. **22**(1), 15–29 (2008)
11. da Silva, L.F.M., Adams, R.D.: Measurement of the mechanical properties of structural adhesives in tension and shear over a wide range of temperatures. J. Adhes. Sci. Technol. **19**(2), 109–141 (2005)
12. Tsai, M.Y., Morton, J., Oplinger, D.W.: In situ determination of adhesive shear moduli using strain gages. Exp. Mech. **36**(4), 297–304 (1996)
13. Lemmen, H.J.K., Alderliesten, R.C., Benedictus, R., Hofstede, J.C.J., Rodi, R.: The power of Digital Image Correlation for detailed elastic-plastic strain measurements. In: WSEAS International Conference on Engineering Mechanics, Structures, Engineering Geology, Creta, 22–24 July 2008
14. Nunes, L.C.S.: Shear modulus estimation of the polymer polydimethylsiloxane (PDMS) using digital image correlation. Mater. Des. **31**(1), 583–588 (2009)
15. Comer, A.J., Katnam, K.B., Stanley, W.F., Young, T.M.: Characterising the behaviour of composite single lap bonded joints using digital image correlation. Int. J. Adhes. Adhes. **40**, 215–223 (2013)
16. Moutrille, M.P., Derrien, K., Baptiste, D., Balandraud, X., Grediac, M.: Through-thickness strain field measurement in a composite/aluminum adhesive joint. Compos. Part A **40**(8), 985–996 (2008)
17. Adams, R.D., Wake, W.C.: Structural Adhesive Joints in Engineering. Elsevier Applied Science, New York (1984)
18. Colavito, K.W., Gorman, J., Madenci, E.: Refinements in digital image correlation technique to extract adhesive strains in lap joints. In: Proceedings of the 50th AIAA/ASME/ASCE/AHS/ASC Structures, Structural Dynamics, and Materials Conference, American Institute of Aeronautics and Astronautics, Palm Springs, 4–7 May 2009
19. Cognard, J.Y., Davies, P., Sohier, L.: Design and evaluation of bonded composite assemblies. In: European Congress of Computational Methods in Applied Sciences and Engineering (ECCOMAS), Jyväskylä, 24–28 July 2004
20. Wang, Z.Y., Wang, L., Guo, W., Deng, H., Tong, J.W., Aymerich, F.: An investigation on strain/stress distribution around the overlap end of laminated composite single-lap joints. Compos. Struct. **89**(4), 589–595 (2008)
21. Cytec Engineered Materials, Extensometer Operating Manual (2000)
22. Sutton, M., Orteu, J., Schreier, H.: Image Correlation for Shape, Motion and Deformation Measurements. Springer, New York (2009)
23. Baker, A., Dutton, S., Kelly, D.: Composite Materials for Aircraft Structures, 2nd Edition. AIAA Education Series (2004). ISBN 978-1-56347-540-5
24. Seneviratne, W.P., Tomblin, J.S.: Adhesive characterization using NIAR-modified KGR extensometer. Soc. Adv. Mater. Process Eng. **47**(5), 37–44 (2011)
25. Patterson, E.A., Hack, E., Brailly, P., Burguete, R., Saleem, Q., Siebert, T., Tomlinson, R., Whelan, M.: Calibration and evaluation of optical systems for full-field strain measurement. Opt. Laser Eng. **45**, 550–564 (2007)

Chapter 10
Interfacial Strength of Thin Film Measurement by Laser-Spallation

Leila Seyed Faraji, Dale Teeters, and Michel W. Keller

Abstract Improving adhesion properties and strength of thin films used in wide range of important application is the key factor in successfully manufacturing micro-electro-mechanical devices, multilayer micro-electronic and optical devices. Mechanical performance and reliably of thin film on a substrate depends on strength of interfacial adhesion. Understanding the role of process parameters like deposition condition and design parameters such as substrate thickness and film thickness, on mechanical performance of the layer is essential in optimizing manufacturing. However, there are limited techniques for characterizing intrinsic interfacial strength of thin films. The use of laser-generated stress pulses has shown the ability to provide quantitative evaluation of the adhesion strength of thin films. In the current work, we use the laser-spallation approach for the investigation of the adhesion strength of thin tin oxide films to alumina substrates.

Keywords Laser-spallation • Interfacial strength • Aluminum thin film • Tin oxide thin film • X-ray diffraction

10.1 Introduction

The mechanical strength of thin films used in electronic component such as electrodes for batteries, solar cells, or batteries is dependent on many factors. Critical issues include deposition technique and the chemical and electrical environment in which the films exist. Probing the mechanical strength of thin films and their interfaces is challenging and frequently requires complex micromachining and processing to produce samples. In order to reduce the time required for iterative designs that may eventually result in an acceptable structure, fundamental investigations of the thin film properties and interfaces are essential. These studies are expected to improve the efficiency of the design process and speed the optimization of the thin film process conditions. A critical point in any multi-layer structure is the interface region between two dissimilar materials. This interface is commonly a location for mechanical stress concentration, which can lead to failure of the interface and eventually the device [1, 2]. Due to lattice mismatch or thermal expansion differences between the layers, there are also residual stresses at interface. Depending on the magnitude and sign of the residual stress at the interface, this location may become a preferred site of the film debonding. Many approaches have been adopted to probe these interfaces, such as scratching by sharp tip [3, 4], AFM nano indentation [5], and laser spallation [6, 7]. All mentioned interfacial strength measurement methods are destructive tests. For example in scratching method, spalling and buckling failure modes occur as a result of the compressive stress field preceding the moving tip scratching the surface [8]. There are many other failure modes of films that depend on the measurement method. Debonding, spallation, cracking, channeling are some of these failure modes that could be used for attachment strength of a film [9].

One application for this technique is in the understanding of adhesion in solid state battery systems. There is high demand for advanced rechargeable lithium-ion batteries with high energy densities, small dimensions and ultra-light weight for portable electronic devices. Fabrication of small-scale solid state batteries requires optimization of fabrication method, microstructure, crystal structure and electrochemical properties of both anode and cathode thin films. The interfaces in a solid state battery system consisting of the current collector/anode/solid electrolyte/cathode/current collector system are all susceptible to severe stress during charge/discharge cycles. One promising anode material for these batteries is tin oxide, SnO_2 [10, 11].

L.S. Faraji • M.W. Keller (✉)
Department of Mechanical Engineering, The University of Tulsa, Tulsa, OK, USA
e-mail: michael-keller@utulsa.edu

D. Teeters
College of Engineering & Natural Sciences, Chemistry and Biochemistry, The University of Tulsa, Tulsa, OK, USA

G.L. Cloud et al. (eds.), *Joining Technologies for Composites and Dissimilar Materials, Volume 10*
Conference Proceedings of the Society for Experimental Mechanics Series, DOI 10.1007/978-3-319-42426-2_10

SnO_2 has a theoretical capacity of more than twice that of graphite, the commonly used anode material in commercial batteries [12] hence batteries using this material as an anode would have very high energy densities that is very desirable for the heavy demand placed on batteries today. However, when lithium is intercalated into SnO_2 during the charging process, SnO_2 can swell almost 300 %, [13, 14] which cause stresses that can debond the attached current collector or separate the anode from a solid electrolyte. Nam et al., reported on enhancing electrochemical properties of thin film tin oxide negative electrodes for lithium batteries [15] and nano-structuring of the SnO_2 anode thin films can help in the management of interfacial stresses [16]. Yet interfacial strength is still an issue and gaining an understanding of interface properties by measuring the adhesion strength at battery interfaces would be invaluable in designing battery systems. Battery electrode systems such as SnO_2 and other electrode materials that have excellent capacities, but adhesion problems could be better understood and processing conditions changed to enhance adhesion mechanisms in these new, high capacity battery systems.

This paper presents the experimental results for spallation of tin oxide layer from Alumina substrate [17]. Back side of the alumina substrate was coated by Aluminum thin film and spalled by laser. After propagation of the pressure pulse throughout the sample thickness, it reaches to the thin layer and either forms the second spallation on tin oxide side or delamination of tin oxide happens. The primary goal of the current work is to optimize the power of the laser to preferentially detach the tin oxide layer instead of bulk substrate spallation. The residual stresses in the heat-affected zone of laser spalled Aluminum layer deposited on alumina are also measured.

10.2 Material and Experimental Setup

10.2.1 SnO_2 Thin Film Deposition

To model a battery anode, a thin film of SnO_2 was coated on surface with an RF magnetron sputter coater. Film thicknesses of approximately 100 nm were deposited. A quartz crystal deposition monitor was used to determine deposition thickness. The sputtering chamber was evacuated and refilled with pure argon gas three times to remove reactive gases. Sputtering plasma was maintained by adjusting the power supply tune and load and maximized power to improve deposition. The substrate was 99.6 % purity Alumina with 1 mm +/− 0.05 mm thickness. The Alumina was lapped both sides and was polished to optical degree.

10.2.2 Aluminum Thin Film Back Layer

Aluminum thin films were deposited by a DC off axis magnetron sputtering system (AJA Orion) onto polished alumina substrates. The metallic target (purity 99.99 %) was 2 in. in diameter and 0.25 in. thick. The working distance between the cathode and the substrate holder was ~100 mm and the base pressure of 5×10^{-6} Torr was achieved by a turbo molecular pump prior to the introduction of argon into the deposition chamber. The final film thickness was measured using a surface profiler (Alpha Step 500). The process parameters for thin film deposition are given in Table 10.1.

10.2.3 Laser Spallation

A 1064 nm wavelength Litron Nano S 130 mJ YAG Pulsed laser with focusing lens was used for laser spallation. The laser produces a Gaussian laser energy pulse with a rise time of 4.4 ns. The peak laser energy could be adjusted between 1 and 130 mJ. A confining layer of waterglass was deposited on an absorbing layer of aluminum as described in [18].

Table 10.1 Process parameters used for Aluminum deposition

	DC power	Process pressure	Ar flow rate	Deposition rate	Final thickness
Al	350 W	3 mT	12 sccm	~8.3 nm/min	500 nm

10.2.4 Thin Film Characterization

The surface topography of the spallation sites were studied by scanning electron microscopy with an FEI Helios Nanolab 600 Scanning Electron-Focused Ion dual-beam microscope (SEM/FIB). Due to the micro-scale diameter sizes (500–900 μm) of the damage, the SEM images were captured at relatively low magnification (×54) and relatively high (~10–14 mm) working distance. However, the sufficiently high energy of the electron beam, at 5.0 kV accelerating voltage and 43 pA electron beam current, produced very fine, detailed images of the topography of the substrate.

X-ray diffraction was performed a Bruker's D8 Discover XRD^2 micro-Diffractometer equipped with Pilot software and VANTEC-500 area detector. Tube parameters of 40 kV/40 mA using Cu-Kα radiation at the detector distance of 30 cm which covers approximately the area of 20° in 2θ and 20° in γ with 0.02° resolution was used. For sample positioning, a motorized five axis (X, Y, Z (translation), ψ (tilt), φ (rotation)) stage was used for positioning which can move the measurement spot to the instrument center within 12.5 μm position accuracy and 5 μm repeatability. A photograph of experimental set up is shown in Fig. 10.1a, b.

Figure 10.2 shows the laser-camera focusing method used for sample alignment. 2D XRD frames from center of spallation sites and the surface that is coated with aluminum were collected using 800 μm pinhole collimator while oscillating the sample 1 mm in both x and y direction. Sample was also rotated during exposure more grains. X-ray exposure time per frame was 120 s.

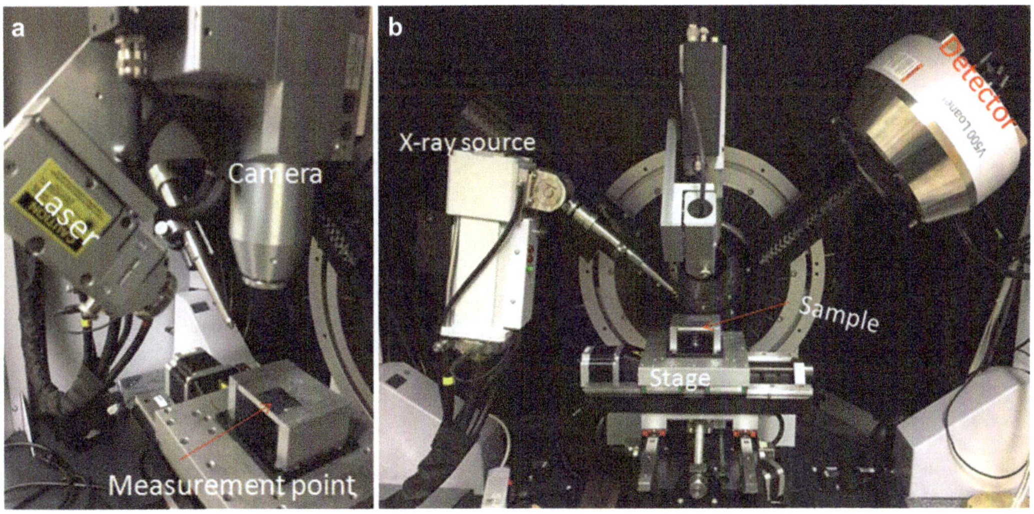

Fig. 10.1 (**a**) Sample orientation on the XRD stage (**b**) Photograph of the experimental setup

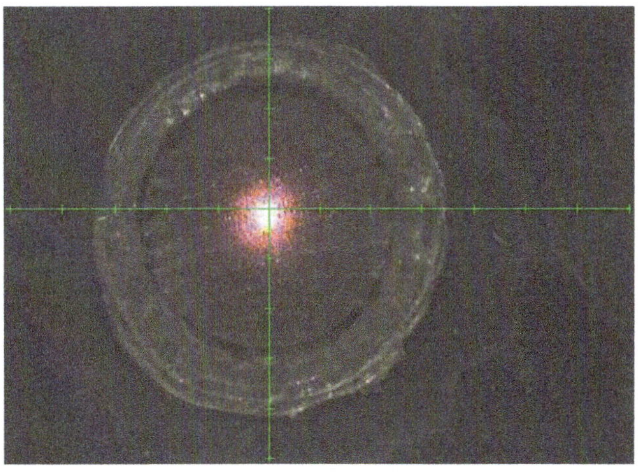

Fig. 10.2 *Top view* optical microscopy image of Aluminum sample, superimposed with a target points. Camera cross hair and laser point and X-ray are coincide at measurement points which is the laser spallation site

10.3 Result and Discussion

In laser spallation, a YAG laser pulse is incident on a thin aluminum layer which vaporizes and induces a large localized temperature increase due to the dissipation within the shock front [19]. In practice, spallation happens in very short time and material goes back to ambient in order of approximately a nano second.

Figure 10.3 shows SEM micrograph of spallation point on the Aluminum side for four different laser powers. For 15 and 20 mJ laser energy large circumferential ring cracks were observed around spallation point. When a short and intense laser pulse irradiated on the Aluminum thin film, high pressure plasma is formed on the surface [20]. An elastic–plastic wave propagates trough the Alumina substrate and hits the tin oxide layer. This shock wave has created a plastic strains behind on the surface of alumina. Most of the Aluminum layer has been evaporated but very thin layer aluminum left behind is solidified with wave marks on the surface as labeled with arrows on Fig. 10.3b.

The YAG laser beam energy causes the aluminum absorbing layer to vaporize. The thermal expansion of the aluminum causes a compressive stress wave to form that travels longitudinally towards the film/substrate interface. When this compression wave reflects off the free surface of the test film, the wave is reflected back towards the substrate and becomes a tensile wave that loads the film/substrate interface in tension. The tensile stress wave pulls the test film from the substrate. If the tensile stress is stronger than the film/substrate adhesion strength, the film is spalled from the substrate.

In order to determine if spallation occurred at a lower power, the test film spots on SnO_2 were then viewed with a Scanning Electron Microscope as shown in Fig. 10.4. The power increases as we move from top left to bottom right. The lowest power level at which the film spalled is the point at which the test film to substrate adhesion fails. For the current tests we see that at all powers investigated that some substrate spallation occurred for this SnO_2 system. In ongoing studies we are modifying the test specimen to include a metal layer that will simulate the current-collector that is present in these battery systems.

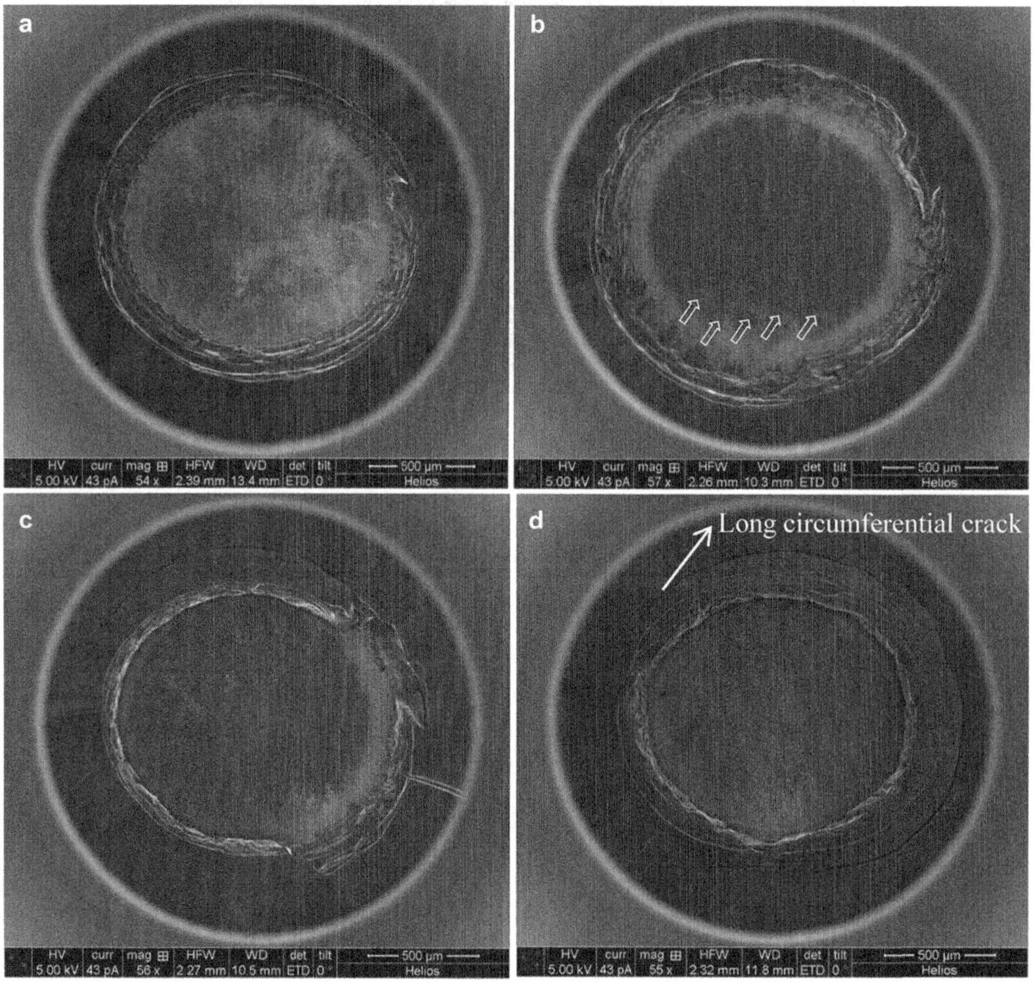

Fig. 10.3 Spalled spot of Aluminum coating from an Alumina Substrate (**a**) 5 mJ, (**b**) 10 mJ, (**c**) 15 mJ, (**d**) 20 mJ

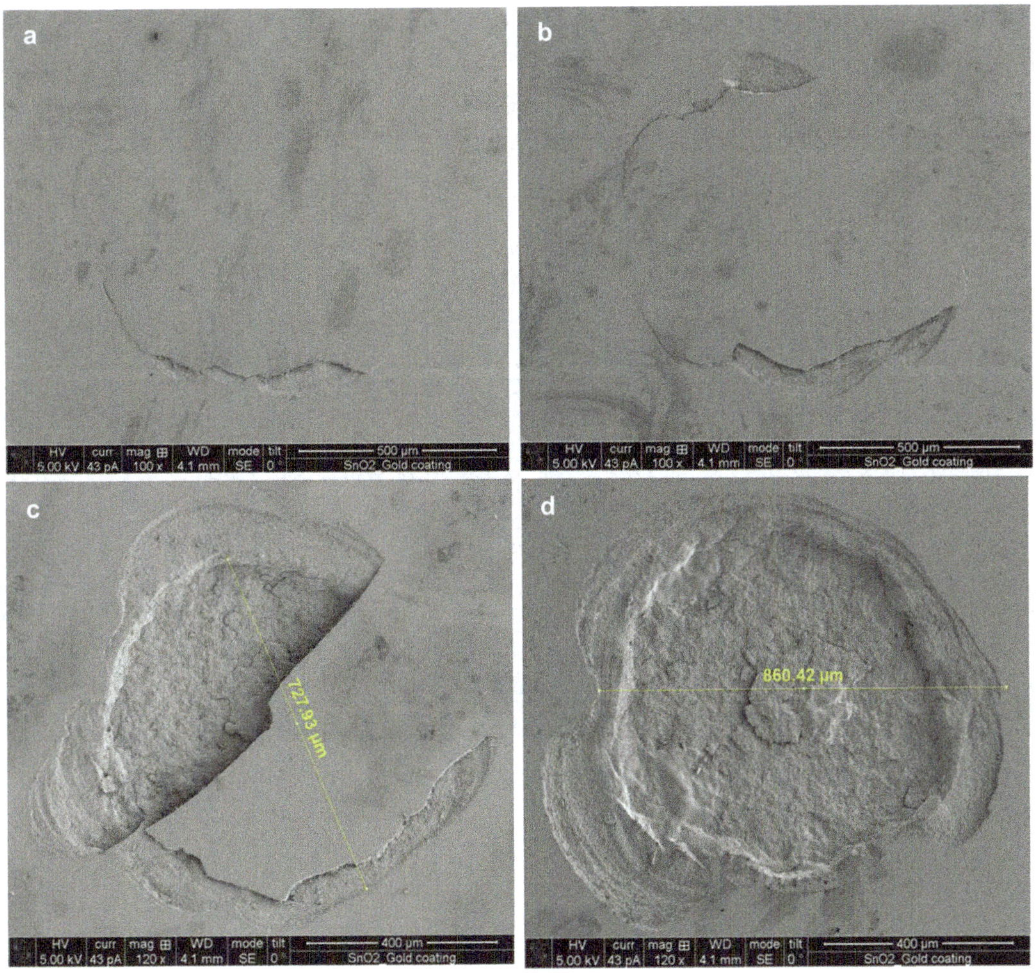

Fig. 10.4 SEM micrograph of delaminated SnO_2 (**a**) 5 mJ, (**b**) 10 mJ, (**c**) 15 mJ, (**d**) 20 mJ

10.4 Residual Stress Measurement After Laser Spallation Using Two Dimensional μXRD Data

Laser spallation can generate heterogeneous plastic deformation and microstructure modifications, leading to high residual stresses appearance. Sin2ψ method was used to measure the residual stress state induced by the laser interaction with surface with Bruker's LEPTOS software [21]. Laser spallation spots on alumina substrate with four different values of the laser intensity 5, 10, 15 and 20 (mJ) were used for X-ray diffraction measurement.

Figure 10.5 shows indexed rings on frames for Alumina substrate only and Aluminum coating on Alumina. Diffraction rings are indexed according to 01-081-2266 PDF card for AL2O3 and 03-065-2869 PDF card for Aluminum. The aluminum layer is very thin and while exposing the Aluminum layer, X-ray penetrates all the way through the Aluminum layer and diffracted beam contains information from both Aluminum layer and surface of Alumina substrate. Using low angle incident beam 5° we were able to see aluminum related rings. Aluminum (111), (220) and (311) rings at 38.47, 44.719, and 78.225 and 2θ degrees respectively are shown in Figs. 10.5 and 10.6.

10.5 Peak Selection for Residual Stress Measurement

Diffraction rings at high 2θ angles are appropriate for residual stress measurement. There is a larger peak shift Δ2θ°, and less effect from sample displacement error for high 2θ rings. The selected peak for residual stress measurement is better to have sufficient intensity and preferentially not in vicinity of other diffraction peaks. Therefore, the (311) ring could be the best

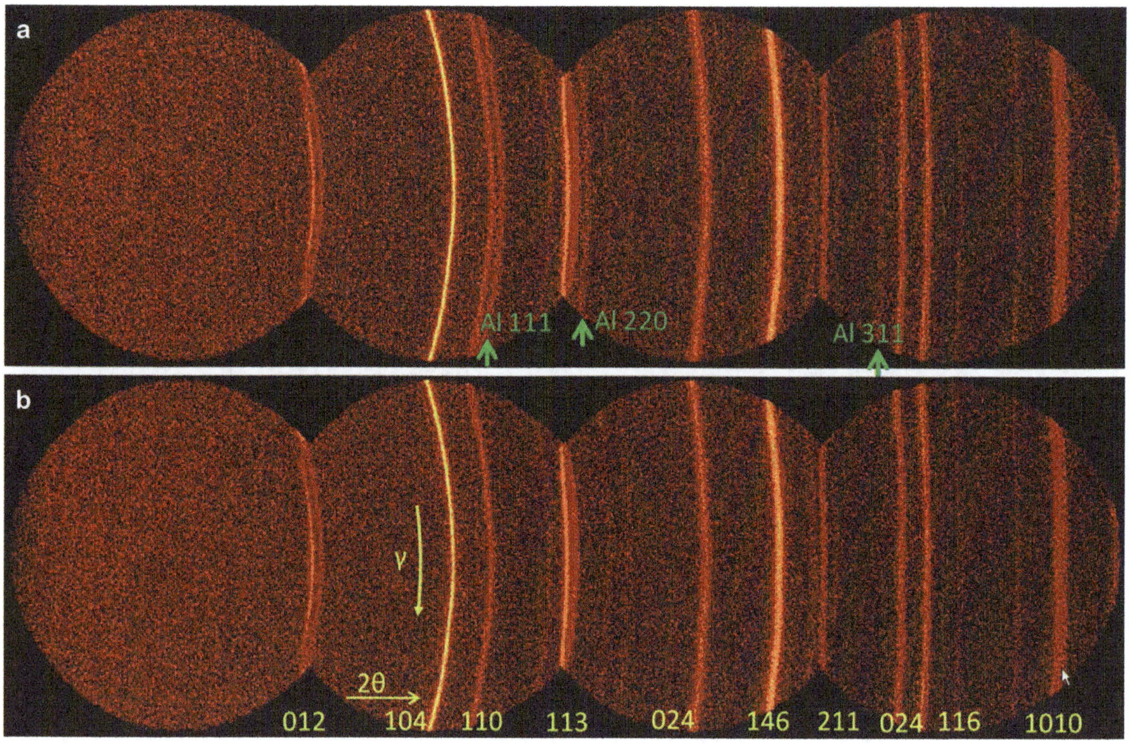

Fig. 10.5 (**a**) 2D X-ray diffraction pattern of Aluminum coating on Alumina (**b**) Alumina surface without coating

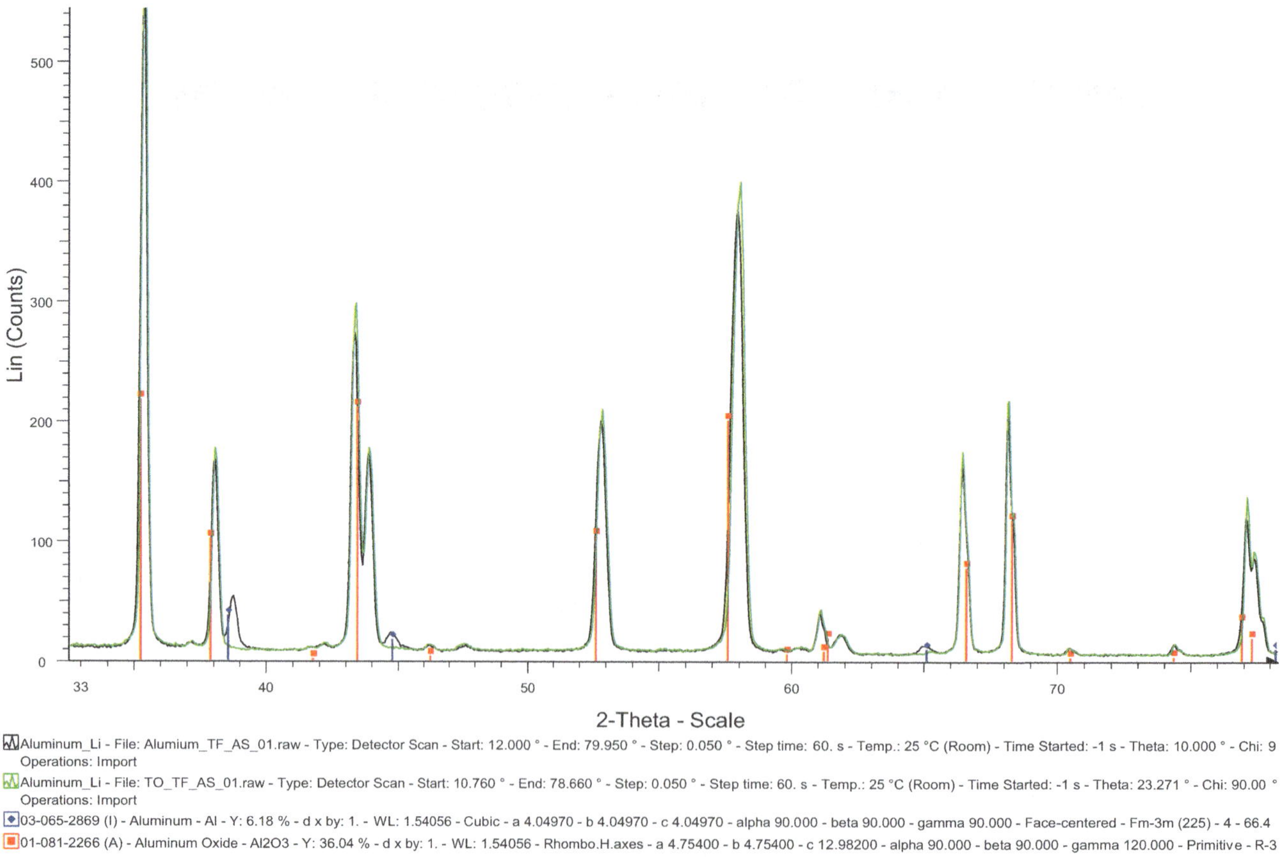

Fig. 10.6 1D X-ray diffraction pattern of Aluminum coating on Alumina compared with Alumina sample before coating

option which could be selected for our residual stress measurements in order to include material isotropy and the possibility of obtaining high angle measurements with higher stress sensitivity [22, 23]. But only 111 peak at $2\theta = 34.4$ has acceptable intensity for peak fitting.

Using Bragg's law for $\lambda = 1.54178\text{Å}$ and given $\theta_0 = 38.47$ and 111 ring,

$$\frac{1}{d_{hkl}^2} = \frac{h^2 + k^2 + l^2}{a^2} \rightarrow \frac{1}{d_{111}^2} = \frac{1^2 + 1^2 + 1^2}{(4.04970)^2} \rightarrow d_{111} = 2.3381 \tag{10.1}$$

The peak shift is expressed in degrees [24],

$$\Delta(2\theta)^\circ = -2\frac{180}{\pi}\frac{1+v}{E}\sigma_\phi \tan\theta \sin^2(\psi) \Rightarrow \sigma_\phi = \frac{^\circ\Delta(2\theta)\pi}{-360\tan\theta\sin^2(\psi)}\frac{E}{1+v} \tag{10.2}$$

For given resolution of 0.01°, $E = 69$ MPa, $v = 0.33$, minimum σ_ϕ as a function of 2θ and ψ can be obtained. For a typical $2\theta = 34.4°$ and $\psi = 12°$, a residual stress with magnitude of ±23.34 MPa is measurable.

Residual stress was measured on four laser exposed points and one point on a sample without spallation as reference for zero residual stress sample. Two dimensional frame data were imported to Bruker's LEPTOS software for residual stress evaluation. Each frame was divided to sub regions and locally integrated along the γ direction to obtain 1D diffraction pattern for each sub region. Correction on polarization, background, $K_{\alpha 2}$ was performed. The peak positions were determined by fitting Pearson VII type peak. Due to the related large number of points available of a diffraction peak, the Pearson VII function was statistically reliable. The stress was calculated assuming the stress is measured as uniform averaged value and stress has a scalar nature.

We would like to control the energy of laser to create laser spallation on aluminum surface, transfer energy trough alumina substrate and delaminate the tin oxide layer. When the energy is high (10–20 mJ) the second spallation also happens in SnO_2 layer. Aluminum grains around the spallation region act like strain gauges recording plastic deformation due to laser shock wave around the spallation point. Residual stresses were obtained at each exposure location on the sample are shown in Fig. 10.7. Location in which low energy laser is used (1–5 mJ) has smaller residual stress values (around −80 Mpa). On this sample SnO_2 layer has started to delaminate from the substrate. But higher energy laser (10–20 mJ) used for laser spallation in addition to laser spallation on Aluminum side have created a second spallation on the SnO_2 layer. An interesting observation is spallation from high energy laser beam have left more residual stress (around −145 Mpa) in the Aluminum layer.

Fig. 10.7 Residual stress vs. spallation power

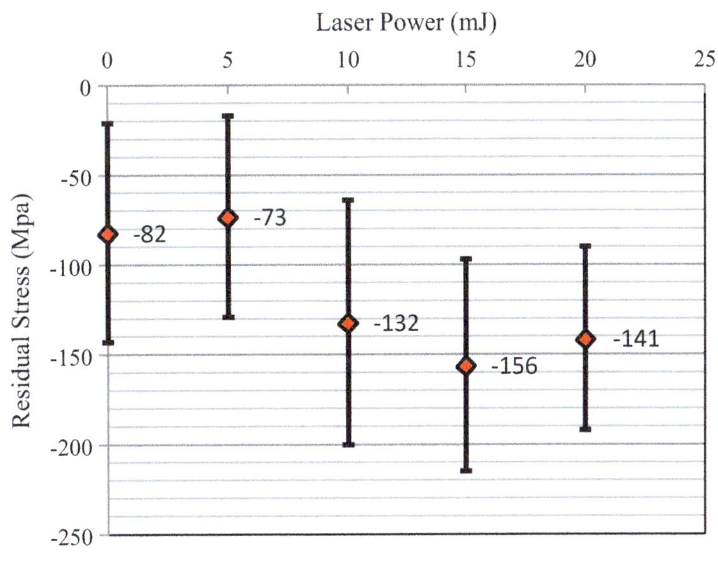

10.6 Conclusion

Tin oxide and Aluminum thin films were deposited on both sides of optically polished alumina substrate with an RF magnetron sputter coater and a DC off axis magnetron sputtering system. A pulsed YAG laser (wavelength of 1064 nm) was used for laser spallation technique in order to determine threshold of energy of laser beam required to study adhesion properties of tin oxide thin film on Alumina substrate. It was found that if the energy of laser pulse is less than 10 mJ the thin oxide thin film starts to delaminate. Higher energy up to 20 mJ was tested and found to create spallation in the SnO_2 layer as well. Laser spallation implies heterogeneous plastic deformation and microstructure modifications, leading to high residual stresses appearance. $Sin^2\psi$ method was used to measure the residual stress. For the first time residual stresses of (−80 to −145 MPa) was recorded in laser spalled samples caused by individual laser interaction with film and substrate surface. Spallation from high energy laser beam generates large residual stresses (around −145Mpa) in the Aluminum layer while low energy laser pulse (1–5 mJ) creates smaller residual stress values.

References

1. Kilinc, Y., Unal, U., Alaca, B.E.: Residual stress gradients in electroplated nickel thin films. Microelectron. Eng. **134**(20), 60–67 (2015)
2. Allahkarami, M., Bale, H.A., Hanan, J.C.: Analytical model for prediction of residual stress in zirconia porcelain bi-layer. In: Narayan, R., Colombo, P. (eds.) Advances in Bioceramics and Porous Ceramics III: Ceramic Engineering and Science Proceedings, vol. 31, pp. 19–26 (2010)
3. Larsson, M., Olsson, M., Hedenqvist, P., Hogmark, S.: Mechanisms of coating failure as demonstrated by scratch and indentation testing of TiN coated HSS on the influence of coating thickness, substrate hardness and surface topography. Surf. Eng. **16**, 436–444 (2000)
4. Beake, B.D., Harris, A.J., Liskiewicz, T.W.: Review of recent progress in nanoscratch testing. Tribol. Mater. Surf. Interfaces **7**(2), 87–96 (2013)
5. Seyed Faraji, L.: Nanoscale carbon fiber-matrix interphase characterization with atomic force microscopy indentation. Oklahoma State University, ProQuest Dissertations Publishing (2014)
6. Gupta, V., Yuan, J., Pronin, A.: Recent developments in the laser spallation technique to measure the interface strength and its relationship to interface toughness with applications to metal/ceramic, ceramic/ceramic and ceramic/polymer interfaces. J. Adhes. Sci. Technol. **8**(6), 713–747 (1994)
7. Gupta, V., Yuan, J.: Measurement of interface strength by the modified laser spallation technique II Applications to metal/ceramic interfaces. J. Appl. Phys. **74**(4), 2397–2399 (1993)
8. Rats, D., Hajek, V., Martinu, L.: Micro-scratch analysis and mechanical properties of plasma-deposited silicon-based coatings on polymer substrates. Thin Solid Films **340**(1–2), 33–39 (1999)
9. Hutchinson, J.W.: Stresses and Failure Modes in Thin Films and Multilayers. Notes for Academic Course. Technical University of Denmark, Lyngby (1996)
10. Idota, Y., Matsufuji, A., Maekawa, Y., Miyasaki, T.: Tin-based amorphous oxide: a high-capacity lithium-ion-storage material. Science **276**, 1395 (1997)
11. Courtney, I.A., Dahn, J.R.: Electrochemical and in situ X-ray diffraction studies of the reaction of lithium with tin oxide composites. J. Electrochem. Soc. **144**, 2045 (1997)
12. Kim, T.J., Son, D., Cho, J., Park, B., Yang, H.: Enhanced electrochemical properties of SnO_2 anode by $AlPO_4$ coating. Electrochim. Acta **49**, 4405–4410 (2004)
13. Beaulieu, L.Y., Eberman, K.W., Turner, R.L., Krause, L.J., Dahna, J.R.: Colossal reversible volume changes in lithium alloys. Electrochem. Solid-State Lett. **4**, A137–A140 (2001)
14. Robert, F., Lippens, P.E., Olivier Fourcade, J., Jumas, J.C., Gillot, F., Morcrette, M., Tarascon, J.M.: Mössbauer spectra as a fingerprint in tin-lithium compounds: applications to Li-ion batteries. J. Solid State Chem. **180**, 339–348 (2007)
15. Nam, S.C., Yoon, Y.S., Cho, W.I., Cho, B.W., Chun, H.S., Yun, K.S.: Enhancement of thin film tin oxide negative electrodes for lithium batteries. Electrochem. Commun. **3**(1), 6–10 (2001)
16. Smith, M.R., Johnson, P.L., Teeters, D.: Interfacial storage of lithium in the nanostructure of SnO_2 nanobaskets for capacities exceeding theoretical values. Solid State Ion. **225**, 680–684 (2012)
17. Wang, J., Sottos, N.R., Weaver, R.L.: Mixed-mode failure of thin films using laser-generated shear waves. J. Mech. Phys. Solids **57**, 51–66 (2009)
18. Wang, J.: Thin-film adhesion measurement by laser-induced stress waves. Ph.D. thesis, Theoretical and Applied Mechanics, University of Illinois at Urbana Champaign (2002)
19. De Rességuier, T., Loison, D., Dragon, A., Lescoute, E.: Laser driven compression to investigate shock-induced melting of metals. Metals **4**(4), 490–502 (2014)
20. Ballard, P., Fournier, J., Fabbro, R.: Study of residual stresses induced by laser-shocks. J. Phys. IV **01**(C7), 132 (1991)
21. He, B.B.: Two-Dimensional X-Ray Diffraction. Wiley, Hoboken (2009)
22. Allahkarami, M., Jayakumar, B., Hanan, J.C.: Residual stress of individual aluminum grains from three dimensional X-ray diffraction. Exp. Appl. Mech **6**, 123–130 (2015)
23. Benedetti, M., Fontanari, V., Allahkarami, M., Hanan, J.C.: Notch fatigue behavior of shot peened high-strength aluminium alloys: role of the residual stress field ahead of the notch root. In: Experimental and Applied Mechanics, vol. 6, pp. 113–121. Springer (2015)
24. Krawitz, A.D.: Introduction to Diffraction in Materials, Science, and Engineering. Wiley, New York (2001)

Chapter 11
Joining of UHTC Composites Using Metallic Interlayer

Noritaka Saito, Laura Esposito, Toshio Yoneima, Koichi Hayashi, and Kunihiko Nakashima

Abstract Ultra-high temperatures ceramics (UHTCs) are the subject of intense worldwide research effort, and their stability in severe environments makes them candidates for aerospace, nuclear and solar energy applications. Widespread usage UHTCs requires the development of effective and reliable joining methods that facilitate the fabrication of large, complex-shaped, and potentially multimaterial components and devices. Joining of HfB_2 and ZrB_2, UHTC diborides, which exhibit outstanding thermo-mechanical and thermochemical properties and good erosion and corrosion resistance, was the focus of the present study. $MoSi_2$ is an effective sintering aid and a composite component for both HfB_2 and ZrB_2, resulting in dense bulk materials with excellent mechanical properties. HfB_2–10 vol.% $MoSi_2$ composites were joined at 1500 °C with a Ni/Nb/Ni interlayer that forms a thin liquid film. Joint-region characterization revealed well-bonded interfaces with interfacial reaction products with the $MoSi_2$. Well-bonded interfaces were also obtained for a ZrB_2–10 vol.% $MoSi_2$ composite bonded at 1500 °C with both Ti and Zr interlayers. It was found that the Ti interlayer exhibited more intensive interfacial reaction with ZrB_2 composite than the Zr interlayer. Additionally, well-bonded interfaces were also found for a ZrB_2–10 vol.% $MoSi_2$ composite bonded at 1500 °C with ZrB_2-X vol.% Ni (X = 20, and 40) powder-based interlayer. Joint-region characterization revealed well-bonded interfaces with microstructures strongly dependent on the Ni content.

Keywords Ultra-high temperature ceramics • Joining • Transient liquid phase • Interfacial reaction • Four-point bending test

11.1 Introduction

The attractive and unusual properties that can be exhibited by ultrahigh temperature ceramics (UHTCs) have made them appealing candidate materials for a wide range of applications, including ones in the nuclear, solar, and aerospace sectors. The ultrarefractory borides, (e.g., ZrB_2, TiB_2 and HfB_2), like the ultrarefractory carbides, are difficult to sinter in undated form, and for both types of UHTCs, the use of modest amounts of an appropriate sintering aid has played a key role in facilitating densification (even in the absence of an applied pressure), refining the microstructure, and improving properties. For the specific case of ZrB_2, additions of $MoSi_2$ have produced the desired processing and microstructural benefits, and thereby, have drastically improved the mechanical performance [1–4].

Applications often require that components be joined. This has many practical advantages. It allows the fabrication of complex-shaped structures, and for multimaterial structures, can allow costly high-performance materials to be used selectively and judiciously, where they are most needed. Prior work [5–7] has demonstrated the feasibility of a transient-liquid-phase (TLP) inspired approach to joining undoped ZrC, HfC, and TaC. The introduction of sintering aids, while beneficial to the properties of the material, may create additional challenges when joining is required. When joining undoped carbide UHTCs with a multilayer Ni/Nb/Ni interlayer, a Ni-Nb alloy liquid forms at reduced temperature that wets and partially penetrates some of the undoped carbides examined. Well-bonded, crack-free joints were produced. The dissolution and incorporation of the sintering aid into the liquid will affect its composition, likely change its total volume, and may alter the

N. Saito (✉) • K. Hayashi • K. Nakashima
Kyushu University,
744, Motooka, Nishi, Fukuoka 819-0395, Japan
e-mail: saito@zaiko.kyushu-u.ac.jp

L. Esposito
CNR ISTEC National Research Council of Italy–Institute of Science and Technology for Ceramics,
via Granarolo, 64, Faenza RA 48018, Italy

T. Yoneima
Shoei Chemical Inc.,
5-3, Aza Wakasakura, Fujinoki-machi, Tosu 841-0048, Japan

G.L. Cloud et al. (eds.), *Joining Technologies for Composites and Dissimilar Materials, Volume 10*
Conference Proceedings of the Society for Experimental Mechanics Series, DOI 10.1007/978-3-319-42426-2_11

wetting behavior, potentially deleteriously. In addition, as a result of the change in composition and volume, both new and increased amounts of phases can develop either due to reactions occurring at the bonding temperature or during cooling (especially if residual liquid is present at the onset of cooling). The potential duality in the role of sintering aids is explored in the present work.

We have examined the bonding of HfB_2- and ZrB_2-based composite with $MoSi_2$, by employing three joining techniques, which are transient liquid phase (TLP) bonding using a Nb-dominated Ni/Nb/Ni interlayer, reactive bonding using group IV transition metal monolayer, and TLP bonding using ZrB_2-Ni powder-based interlayer, to assess the potential for making reliable bonds using the more easily fabricated $MoSi_2$-containing UHTC boride materials. A thorough microstructural characterization of the bond region has been performed intended to identify the nature of the phases formed, and the extent of the reaction/interaction zone. The four-point bending strength of the joints has also been measured.

11.2 Materials and Experimental Procedures

11.2.1 HfB_2- and ZrB_2-Based Composites

Powders of HfB_2 (99.5 %, −325 mesh, Materion, Mayfield Heights, OH, USA) and ZrB_2 (98.0 %, 1.5–2.5 microns, Japan New Metals Co., Ltd., Toyonaka, Osaka, Japan) as principal material and $MoSi_2$ (99.0 %, Sigma-Aldrich Japan, Tokyo, Japan) as a sintering aid were employed. HfB_2 + 10 vol.% $MoSi_2$ and ZrB_2 + 10 vol.% $MoSi_2$ mixture powders were ball milled for 24 h using zirconia milling media (2 mm in diameter) in ethanol as dispersion medium, dried in a rotary evaporator, and passed through a −50 × 47 mesh polyethylene sieve. The resulting powders were uniaxially pressed into 12 mm diameter pellets, and then cold isostatically pressed (CPA50-300, Sansho Industry, Osaka, Japan) under 300 MPa for 3 min. The pellets were placed in a BN crucible (N1, Denki Kagaku Kogyo, Tokyo, Japan), sintered under 0.1 MPa argon atmosphere at 1950 °C for 60 min for HfB_2-$MoSi_2$ composite, and sintered under 0.1 MPa argon atmosphere at 2000 °C for 60 min for ZrB_2-$MoSi_2$, using a graphite element furnace (FVPHP-R-3, Fuji Dempa Kogyo, Osaka, Japan) with a heating rate of 5 °C/min, and furnace cooled by shutting off the power.

The cross-section of the sintered bodies was polished and investigated using a typical field emission scanning electron microscope (FESEM, Ultra55, Carl Zeiss, Oberkochen, Germany) with energy dispersive spectroscopy (EDS). Mean relative density of the HfB_2 and ZrB_2 composite was measured by Archimedean method. Mean bending strength was measured by four-point bending test explained in later Sect. 11.2.5. One surface of each composite pellet was polished with 9, 6, 1, and 0.25 μm diamond suspension solutions by using an automatic polisher (Automet250, Buehler, Lake Bluff, IL, USA). The polished pellets were ultrasonically cleansed in mixed solution of ultrapure water and soap, ultrapure water, acetone, and 2-propanol in the order listed before the bonding process.

11.2.2 Transient Liquid Phase (TLP) Bonding Using Ni–Nb–Ni Interlayer

In principle, TLP bonding of ceramics has the potential to join high-temperature, high-performance materials at significantly reduced temperatures relative to brazing and solid-state diffusion bonding, in processing cycles that are short in duration (comparable to those in brazing). The extension of TLP bonding to ceramics employs a multilayer (A/B/A) interlayer. A minor portion of the interlayer (A) generates the TLP, and the major portion (B) remains solid during joining, and provides the sink for the liquid former. By choosing a major component, the core layer, with a high melting point and a good thermal expansion match with the ceramic, a refractory alloy that minimizes residual stresses is expected to evolve during processing and use. In order for the approach to be successful, there needs to be a synergistic interaction between several features or characteristics of the interlayer system. The outer cladding layer material must promote the formation of a liquid at reduced temperature. Phase diagrams can provide guidance on this issue. The liquid that forms will generally include some of the core layer due to dissolution, and can similarly interact with the ceramic being joined. The resulting liquid is thus at least a binary and can include additional components. Ideally the wetting characteristics and volume of the liquid are such that all of the interfacial gaps are completely filled, producing a well-bonded interface. Extensive reaction with the ceramic or sintering aids therein is not necessarily desirable, particularly if it leads to the formation of brittle intermetallics or substantially increases the amount of liquid. Ideally, once the liquid has formed and filled interfacial gaps, the liquid former (A) diffuses rapidly into the core layer (and/or adjoining ceramic) causing a progressive decrease in liquid volume at the joining temperature, and ultimately complete isothermal solidification of the joint. As a result, and particularly if extensive redistribution and dilution of the melting point depressant occurs, the remelt temperature of the joint can be several 100° above the original joining temperature, which is obviously advantageous for high-temperature applications. Creating an interlayer that meets all of these criteria simultaneously is challenging, and experiments that are intended to examine individual aspects of the process can often be difficult.

Here, a schematic illustration of HfB_2 composite TLPB joint is shown in Fig.11.1a. The interlayer sandwiched between being joined composites was consisted of core Nb layer of 127 μm (99.9 %, Nilaco, Tokyo, Japan) with cladding Ni layer of 2 μm (99.9 %, Goodfellow). The Ni and Nb foils were cleaned in the same manner as the HfB_2 composites and immersed in 1:1 H_2O:HNO_3 aqueous solution for 10 s to remove any oxide layer. The Ni/Nb/Ni interlayer was placed between the HfB_2 composites as shown in Fig. 11.1a. The joining assembly was heat-treated using a graphite element hot press furnace (FVPHP-R-3, Fuji Dempa Kogyo, Osaka, Japan) under 20 Pa vacuum at 1500 °C with holding time of 30 min and a constant 8.5 MPa applied pressure. The furnace was heated at a constant heating rate of 15 °C/min and cooled by cutting the furnace power off. The joining samples were embedded in epoxy and sectioned across the interfacial region using a diamond saw (Isomet4000, Buehler, Lake Bluff, IL, USA). The sectioned samples were then polished and cleaned in the same manner as mentioned in Sect. 11.2.1. The interfacial regions of the samples were examined using FESEM with EDS and an electron probe microanalysis (EPMA-1610, Shimadzu, Kyoto, Japan) with a wavelength-dispersive X-ray spectrometer (WDS).

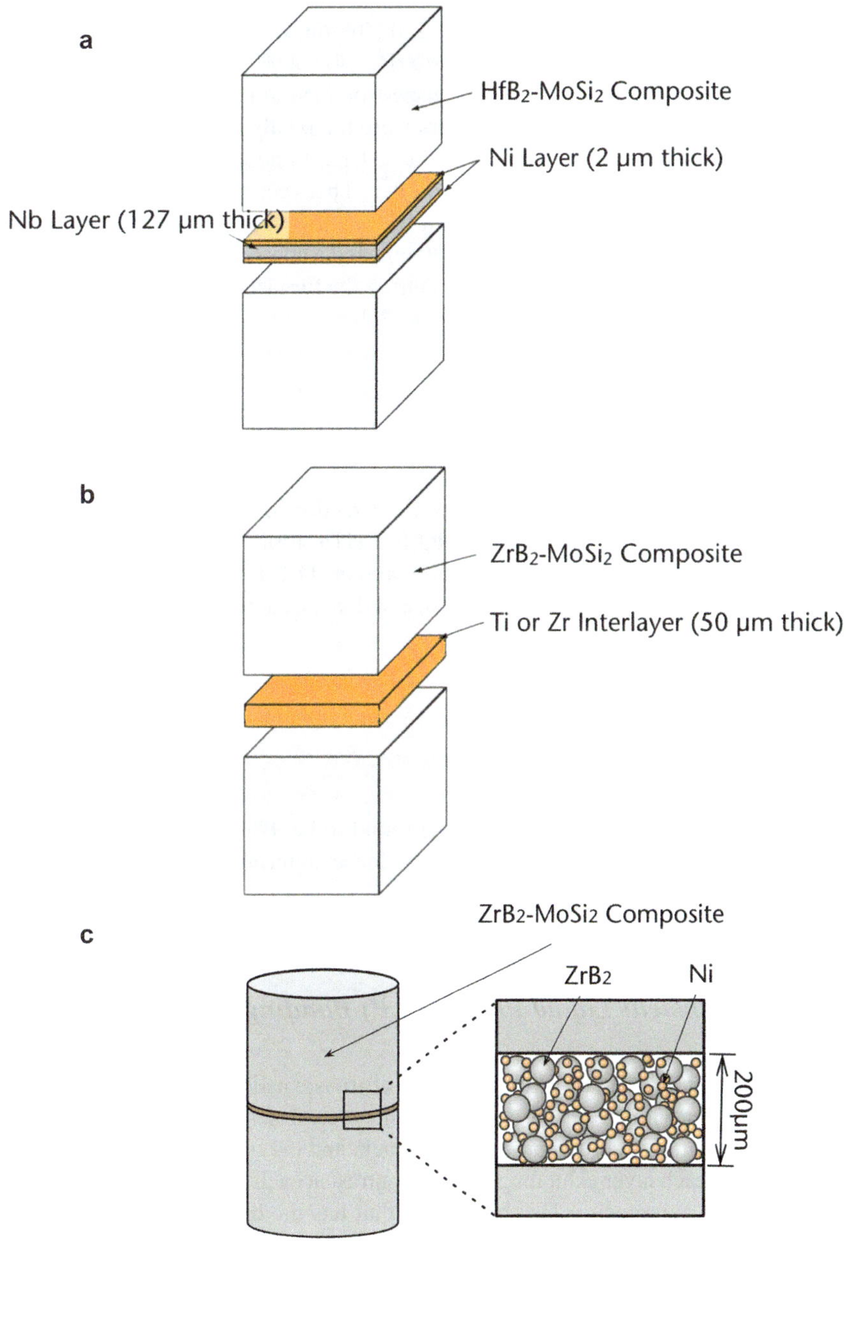

Fig. 11.1 Schematic illustrations of joining assemblies, (**a**) Transient liquid phase (TLP) bonding using Ni-Nb interlayer, (**b**) Reactive bonding using Ti or Zr interlayer, and (**c**) Transient liquid phase (TLP) bonding using ZrB_2-Ni powder-based interlayer

11.2.3 Reactive Bonding Using Ti or Zr Interlayer

A schematic of ZrB_2 composite reactive bonding joint is shown in Fig. 11.1b. The interlayer was consisted of monolayer of 50 μm thick Ti (99.9 %, Nilaco, Tokyo, Japan) or Zr (99.9 %, Nilaco, Tokyo, Japan). The Ti and Zr foils were cleaned in the same manner as the ZrB_2 composites and immersed in 1:1 H_2O:HNO_3 aqueous solution for 10 s to remove any oxide layer. The Ti or Zr interlayer was placed between the ZrB_2 composites as shown in Fig. 11.1b. The joining assembly was heat-treated using a graphite element hot press furnace (FVPHP-R-3, Fuji Dempa Kogyo, Osaka, Japan) under 20 Pa vacuum at 1500 °C with holding time of 30 min and a constant 13 MPa applied pressure. The furnace was heated at a constant heating rate of 15 °C/min and cooled by cutting the furnace power off. The interfacial regions of the samples were examined in the same manner as mentioned above.

11.2.4 Transient Liquid Phase (TLP) Bonding Using ZrB_2-Ni Powder-Based Interlayer

A schematic of ZrB_2 composite TLPB is shown in Fig. 11.1c. The interlayer was consisted of powder mixture of ZrB_2 (98.0 %, 1.5–2.5 microns, Japan New Metals Co., Ltd., Toyonaka, Osaka, Japan) and Ni (99.9 %, 0.2 microns, Shoei Chemical Inc., Tosu, Saga, Japan). ZrB_2+20 vol.% Ni and ZrB_2+40 vol.% Ni mixture powders were ball milled for 24 h using zirconia milling media (2 mm in diameter) in ethanol as dispersion medium, dried in a rotary evaporator, and passed through a −50×47 mesh polyethylene sieve. The resulting powders were uniaxially pressed into 12 mm diameter pellets, and then cold isostatically pressed (CPA50-300, Sansho Industry, Osaka, Japan) under 300 MPa for 3 min, to form 200 micron thick interlayer of the mixed powder compact. The interlayer was placed between the ZrB_2 composites as shown in Fig. 11.1b. The joining assembly was heat-treated using a graphite element hot press furnace (FVPHP-R-3, Fuji Dempa Kogyo, Osaka, Japan) under 20 Pa vacuum at 1500 °C with holding time of 30 min and a constant 8.5 MPa applied pressure. The furnace was heated at a constant heating rate of 15 °C/min and cooled by cutting the furnace power off. The interfacial regions of the samples were examined in the same manner as mentioned in Sect. 11.2.2.

11.2.5 Four-Point Bending Test of the Joints

Following bonding, the joining samples were sectioned into 25 mm×2.5 mm×2 mm (length×width×thickness) beams using a electrical discharge machine (ROBOCUTα-400iC, Fanuc, Yamanashi, Japan). The beams were beveled and then polished in the same manner as mentioned in Sect. 11.2.1. 10 bending beams were tested at room temperature. In addition, HfB_2 and ZrB_2 composite beams were processed in the same manner and tested for 12 times at room temperature to compare the strength of the joints.

11.3 Results and Discussion

The relative densities of composites were found to be 97.6 % for HfB_2 composite and 98.2 % for ZrB_2 composite, respectively. Both composites revealed nearly full-dense sintering characteristics by utilizing $MoSi_2$ addition as the sintering aid and composite component.

11.3.1 Transient Liquid Phase (TLP) Bonding Using Ni-Nb Interlayer

Back-scattered electron (BSE) micrographs of transient liquid phase (TLP) bonding joint of HfB_2-$MoSi_2$ composite using Ni-Nb interlayer with the results of WDS chemical analyses are shown in Fig. 11.2. As can be seen at the lower magnification, it was found that no macroscopic void and crack, and the original Ni/Nb/Ni interlayer structure has evolved into a multilayer structure with each layer exhibiting distinct contrast at each interface as a result of melting, dissolution, and interfacial interaction with HfB_2 composite. The thick layer that has the brightest contrast was identified as Nb using the WDS analysis.

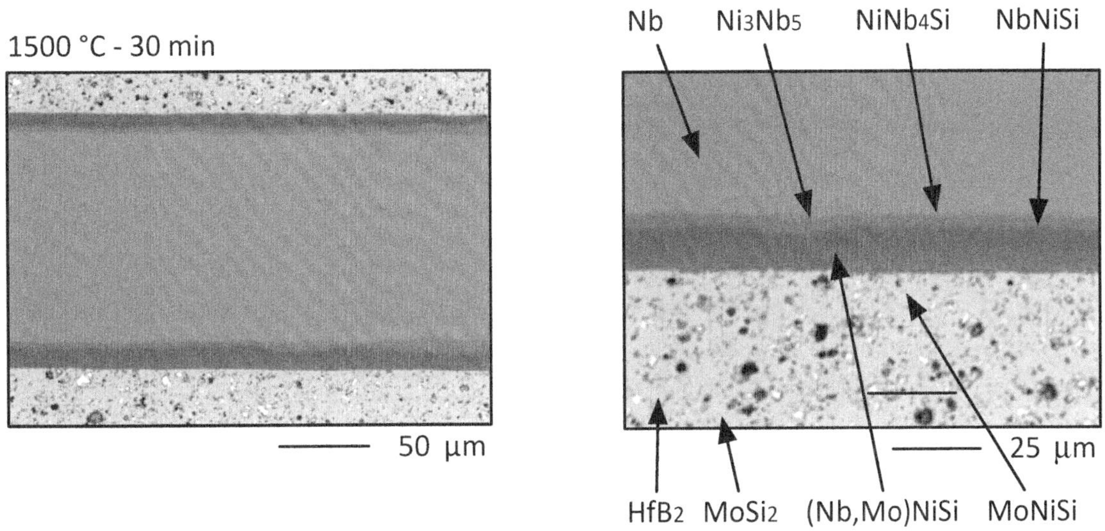

Fig. 11.2 Microstructural and microchemical analyses results of the polished images of the reaction layer and related phases identified on the basis cross section of HfB_2-$MoSi_2$ composite joint

The total "Nb" thickness decreased from 125 μm to ≈110 μm. The decrease in the thickness of the Nb layer indicates enhanced dissolution of Nb because, at the bonding temperature, 1500 °C, the solubility limit of Nb into molten Ni is expected to ≈ 55 at.%. Since the Ni layer is only 2 μm thick, one would only expect a comparable thickness of Nb to dissolve on each side in the absence of an effect of HfB_2 dissolution.

The thicker reaction layer, ≈10–20 μm depending on the position measured, is a consequence of the enhanced dissolution of both the core layer and the HfB_2 composite, and is made up of multiple layers. From the HfB_2 side of the reaction layer, (Nb,Mo)NiSi, NbNiSi, $NiNb_4Si$, and Ni_3Nb_5 inter metallic phases were identified by WDS analysis. The Mo and Si found in these intermetallics are originally from the $MoSi_2$ sintering aid. Although reaction between the Ni/Nb/Ni interlayer and the HfB_2 composite was extensive, the interface was relatively flat, suggesting that the dissolution of HfB_2 composite was uniform along the interface. Furthermore, the HfB_2 composite well maintained the microstructure of the as-processed composite, which is a good indication for less degradation of the bulk properties during the TLP joining process. The four-point bending test on the joints revealed 412 ± 53.9 MPa of fracture strength at the room temperature, besides 462 ± 45.7 MPa for the HfB_2 composite itself, which suggests the TLP joints of HfB_2 composite with Ni-Nb interlayer well-maintains the strength of the composite.

11.3.2 Reactive Bonding Using Ti or Zr Interlayer

As described in the Sect. 11.3.1, UHTC boride was found to be reactive against metallic phase. In the current section, we would like to utilize interfacial reaction between UHTC boride composite and metallic interlayer for the bonding. Back-scattered electron (BSE) micrographs of reactive bonding joint of ZrB_2-$MoSi_2$ composite using (a) Ti or (b) Zr interlayer with the results of WDS chemical analyses are shown in Fig. 11.3. As can be seen in Fig. 11.3a, b, no void and crack were observed in the current magnification, which indicates the reactive bonding of ZrB_2 composite have successfully done. However, the interfacial microstructures of ZrB_2 composite joints were found to significantly differ from each other. In the case of using Ti interlayer, cf. Fig. 11.3a, the interfacial reaction between the ZrB_2 composite and Ti interlayer was tremendously intensive, and resulted in the formation of reaction products as needle like TiB grains, (Ti,Zr,Mo)B, (Ti,Zr,Mo) and (Ti-Zr-Si) solid solutions. These reactions were attributed to accelerated diffusion via Ti-Zr-Si liquid phase at the bonding temperature of 1500 °C. By contrast, shown in Fig. 11.3b, the ZrB_2 composite joint with Zr interlayer revealed modest interfacial reaction against the ZrB_2 composite, and ended up to generate (Zr,Mo) solid solution, Zr_4Si and Zr_2Si silicides, because of the solid phase bonding. The four-point bending test on the joints revealed 477.7 ± 88 MPa for the Ti interlayer and 439.3 ± 70 MPa for the Zr interlayer, respectively at the room temperature, besides 376.8 ± 77.0 MPa for the ZrB_2 composite itself, which suggests the RB joints of ZrB_2 composite with Ni-Nb interlayer do not degrade the strength of the composite.

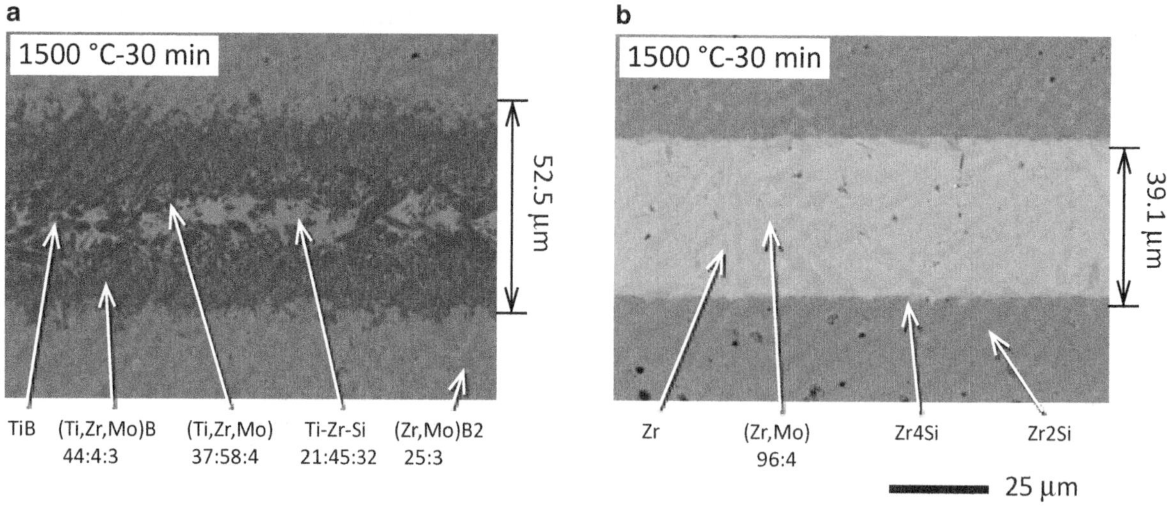

Fig. 11.3 Microstructural and microchemical analyses results of the polished images of the reaction layer and related phases identified on the basis cross section of ZrB_2-$MoSi_2$ composite joints, (**a**) with Ti interlayer, and (**b**) with Zr interlayer

11.3.3 Transient Liquid Phase (TLP) Bonding Using ZrB_2-Ni Powder-Based Interlayer

Here, the other effort on TLP bonding of ZrB_2 composite employing ZrB_2-Ni powder-based interlayer, which can provide us more rapid interfacial reaction of TLP interlayer because of larger surface area that could be involved into the intensive reaction. Back-scattered electron (BSE) micrographs of reactive bonding joint of ZrB_2-$MoSi_2$ composite utilizing (a) ZrB_2-20 vol.% Ni (ZN20) interlayer and (b) ZrB_2-40 vol.% Ni (ZN40) interlayer with the results of WDS chemical analyses are shown in Fig. 11.4. As can be seen in Fig. 11.4a, b, no void and crack were observed, which indicates the joining of ZrB_2 composite have successfully done by using ZrB_2-Ni powder-based interlayer. Similar interfacial microstructures were found in both joints, however, the porosity in the ZN40 interlayer seemed to be slightly smaller than that in the Zn20 interlayer. Additionally, B-Ni-Si-Zr phases with the different compositions were identified in both the composite and interlayer side. The four-point bending test on the joints revealed 423.0 ± 83.2 MPa for the ZN20 interlayer and 512.5 ± 54.7 MPa for the ZN40 interlayer, respectively at the room temperature, besides 376.8 ± 77.0 MPa for the ZrB_2 composite itself, which suggests the TLP joints of ZrB_2 composite with ZrB_2-Ni interlayer enhances the strength of the composite.

11.4 Summary

The bonding of HfB_2- and ZrB_2-based composite with $MoSi_2$, by employing three joining techniques have been explored, which were transient liquid phase (TLP) bonding using a Nb-dominated Ni/Nb/Ni interlayer, reactive bonding using group IV transition metal monolayer, and TLP bonding using ZrB_2-Ni powder-based interlayer. Major findings were as follows.

1. HfB_2-10 vol.% $MoSi_2$ and ZrB_2-10 vol.% $MoSi_2$ composites were successfully processed with 97.6 % and 98.2 % in relative densities, and 462 ± 45.7 MPa and 376.8 ± 77.0 MPa in four-point bending strength.
2. Micro-structural and -chemical analyses at the joining interface region of the joints fabricated in the present study revealed well-bonded features without any voids and cracks along the interface.
3. For TLP bonding of HfB_2 composite using Ni–Nb–Ni interlayer, the interlayer showed interfacial reaction against the composite at the bonding temperature, and resulted in the formation of intermetallic silicides. The four-point bending test on the joints revealed 412 ± 53.9 MPa of fracture strength at the room temperature, besides 462 ± 45.7 MPa for the HfB_2 composite itself, which suggests the TLP joints of HfB_2 composite with Ni-Nb interlayer well-maintains the strength of the composite.
4. For reactive bonding of ZrB_2 composite using Ti or Zr interlayer, the interfacial reactions were found to be significantly dependent on the formation of liquid phase. The Ti interlayer indicated more intensive and active interfacial reaction against ZrB_2 composite than the Zr interlayer. The four-point bending test on the joints revealed 477.7 ± 88 MPa for the Ti interlayer and 439.3 ± 70 MPa for the Zr interlayer, respectively at the room temperature, besides 376.8 ± 77.0 MPa for

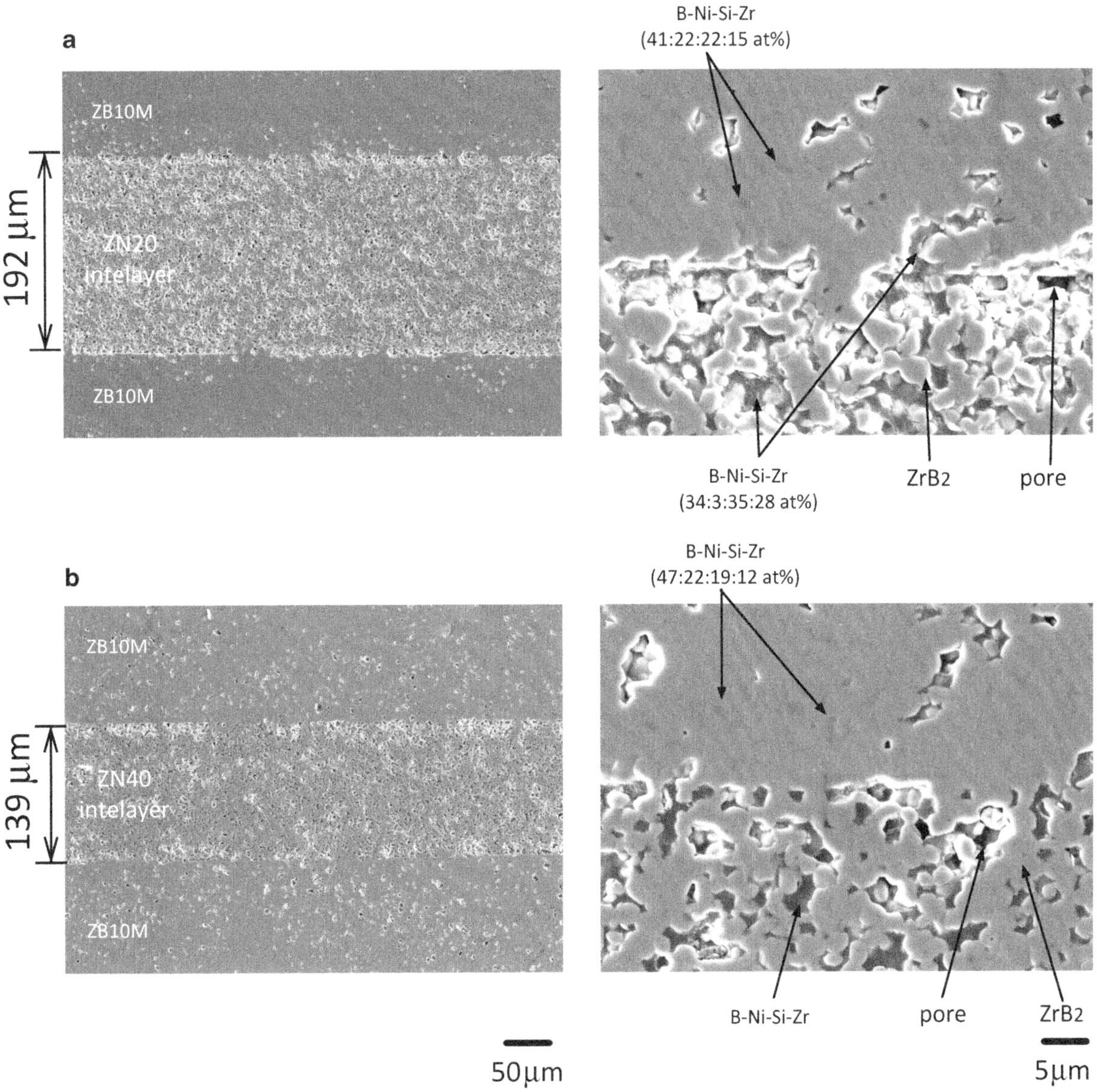

Fig. 11.4 Microstructural and microchemical analyses results of the polished images of the reaction layer and related phases identified on the basis cross section of ZrB_2-$MoSi_2$ composite joints, (**a**) with ZrB_2-20 vol.% Ni interlayer, and (**b**) with ZrB_2-40 vol.% Ni interlayer

the ZrB_2 composite itself, which suggests the RB joints of ZrB_2 composite with Ni-Nb interlayer do not degrade the strength of the composite.

5. For TLP bonding of ZrB_2 composite using ZrB_2-Ni powder-based interlayer, the similar interfacial microstructures were observed except thickness of the interlayer and porosity along the joining interface. The four-point bending test on the joints revealed 423.0 ± 83.2 MPa for the ZrB_2-20 vol.% Ni interlayer and 512.5 ± 54.7 MPa for the ZrB_2-40 vol.% Ni interlayer, respectively at the room temperature, besides 376.8 ± 77.0 MPa for the ZrB_2 composite itself, which suggests the TLP joints of ZrB_2 composite with ZrB_2-Ni interlayer enhances the strength of the composite.

References

1. Sciti, D., Brach, M., Bellosi, A.: Oxidation behavior of a pressureless sintered ZrB_2-$MoSi_2$ ceramic composite. J. Mater. Res. **56**(3), 149–158 (2005)
2. Sciti, D., Brach, M., Bellosi, A.: Long-term oxidation behavior and mechanical strength degradation of a pressurelessly sintered ZrB_2-$MoSi_2$ ceramic. Scr. Mater. **53**(11), 1297–1302 (2005)
3. Sciti, D., Guicciardi, S., Bellosi, A., Pezzotti, G.: Properties of a pressureless-sintered ZrB_2-$MoSi_2$ ceramic composite. J. Am. Ceram. Soc. **89**(7), 2320–2322 (2006)

4. Silvestroni, L., Kleebe, H.J., Lauterbach, S., Muller, M., Sciti, D.: Transmission electron microscopy on Zr- and Hf-borides with $MoSi_2$ addition: densification mechanisms. J. Mater. Res. **25**(5), 828–834 (2010)
5. Saito, N., Ikeda, H., Yamaoka, Y., Glaeser, A.M., Nakashima, K.: Wettability and transient liquid phase bonding of hafnium diboride composite with Ni-Nb alloys. J. Mater. Sci. **47**(24), 8454–8463 (2012)
6. Silvestroni, L., Sciti, D., Esposito, L., Glaeser, A.M.: Joining of ultra-refractory carbides. J. Eur. Ceram. Soc. **32**(16), 4469–4479 (2012)
7. Esposito, L., Sciti, D., Silvestroni, L., Melandri, C., Guicciardi, S., Saito, N., Nakashima, K., Glaeser, A.M.: Transient liquid phase bonding of HfC-based ceramics. J. Mater. Sci. **49**(2), 654–664 (2014)

Chapter 12
Metal-to-Composite Structural Joining for Drivetrain Applications

Peter J. Fritz, Kelly A. Williams, and Javed A. Mapkar

Abstract Dissimilar material components in structural applications require advanced joining geometries and processes that permit the interface of these components to have mechanical behavior equivalent to the lesser of the base metal material behavior and the base composite material behavior. In volume-constrained applications, such as drivetrain parts, the interface must have the strength of the base materials, and the envelope of space of the current drivetrain componentry. The interface surfaces of the composite and metal are demonstrated here to achieve base material capability for the thickness, width, and length directions, resulting in conservation of mechanical properties throughout the interfaces, for all loading and deformation at material transitions. The method of material forming, modification, combining, and product operation is explained here, for optimization of materials usage and for processing equipment.

In designs to date, the metal to composite interface has been the weak link of the assembly, and product performance has been limited to this characteristic. This new method of interface build, and theory of base-to-joint equivalent strength, results in a new capability for drivetrain component design, and will start a new era of dissimilar materials combinations to offer the consumer improved product functionality with enhanced interface design, manufacture, envelope, and operational improvements.

Keywords Drivetrain • Joining • Dissimilar • Structural • Composite

12.1 Introduction

Metal parts in vehicle drivetrains dominate the materials selection for cars and trucks. The mostly steel, cast iron, and aluminum parts of the drivetrain are reliable, and are low-cost, but these metal parts result in drivetrain components that are heavy. As drivetrain components are generally very heavily loaded, composite material substitutions require the higher-strength composite selections, and joining between metal and composite must be sufficient to equal the strength of the base materials. To achieve such high interfacial strength capabilities, metal-to-composite interfaces need to be tailored by composite-fiber orientation, loading, geometry, and metal surface preparation.

Basic research has been completed here, to characterize joining between dissimilar materials, such as metal and composite, to determine the capability of standard joining within the framework of application requirements, and the special geometries necessary to attach metals to composites in drivetrains. Fundamental interfacial shear strengths for rotational component interfaces will be demonstrated here, and compared to the minimal necessary shear strengths of the base materials. Basic information on metal surface preparation, cylindrical interfaces, classical analysis, and thermal limitations will be verified here, so that the metal-to-composite interface capability of drivetrain applications can be compared to the base material of component metal shear capability and component composite interlaminate capability, leading to the design of joints that are stronger than the base materials.

The Eaton Vehicle Group provides automotive, aerospace, and commercial vehicle manufacturers worldwide with products and systems that are designed to improve vehicle efficiency, performance and power—such as emission control components, engine valves, valve-train systems and superchargers—as well as offering driveline expertise through its transmission, clutch and torque management products and systems. The testing and verification in this report was done for Eaton on driveline parts including transmission, differential, supercharger, and clutch.

P.J. Fritz (✉) • K.A. Williams • J.A. Mapkar
Eaton Corporate Research and Technology, 26201 Northwestern Highway, Southfield, MI 48076, USA
e-mail: peterfritz@eaton.com

G.L. Cloud et al. (eds.), *Joining Technologies for Composites and Dissimilar Materials, Volume 10*
Conference Proceedings of the Society for Experimental Mechanics Series, DOI 10.1007/978-3-319-42426-2_12

12.2 Shear Strength

12.2.1 Shear Strength Samples

To determine the effect of the metal surface preparation, the metal surface of coupon parts for bonding attachment were prepared as plain, textured (grit blasted), dimpled, and holes (penetrated). These preparations were all cleaned by acetone wipe, and then adhesively bonded or overmolded to a polymer, such as a thermoplastic, and cooled to strength. In tensile testing, the shear coupons were subjected to the highest loading to lead to failure, and that load was divided by the contact area, such as both sides summed, leading to the average shear stress of the interface at failure. In general, these coupons were loaded with an increasing load until debond occurred. At coupon debond, the peak load was recorded, and the interface area with the thermoplastic polymer was recorded, to calculate the peak shear stress. The schematic of Fig. 12.1 shows a standard lap-shear coupon indicating two green areas where the composite base material (d_1) on the left is bonded by overmolding to the metal component on the right (d_0). The bond area A_b then, is

$$A_b = 4lz_0 \tag{12.1}$$

where z_0 is the width of the metal specimen near the green area, but in a direction z that is not shown, but is the direction perpendicular to both directions x and y. This geometry is useful for symmetrical loading of the bonded area so that the shear strength can be measured with reduced peeling of the bond area.

The specimens used a composite thermoplastic, as shown on the left in Fig. 12.2, where the reinforced thermoplastic was overmolded on steel half-dogbone-shaped base material. Unlike Fig. 12.1, the reinforced thermoplastic was bonded on five sides of the steel. As the "end" of the steel was small and the end area offered only tensile bonding in all specimens, its area was ignored. But the other four sides did provide shear strength so all were utilized in the measurement of the shear area A_s as

$$A_s = 4l\left(z_0 + d_0\right) \tag{12.2}$$

This geometry of testing follows the method of ASTM D3528 [1], and uses symmetrical loading to prevent wrenches (force couples) in the interlaminar bond area. This deviates from the method of Watson in AFWAL-TR-82-4013 where steel plates are used on both ends of the coupon [2], but allows for symmetrical overmolding of the composite component.

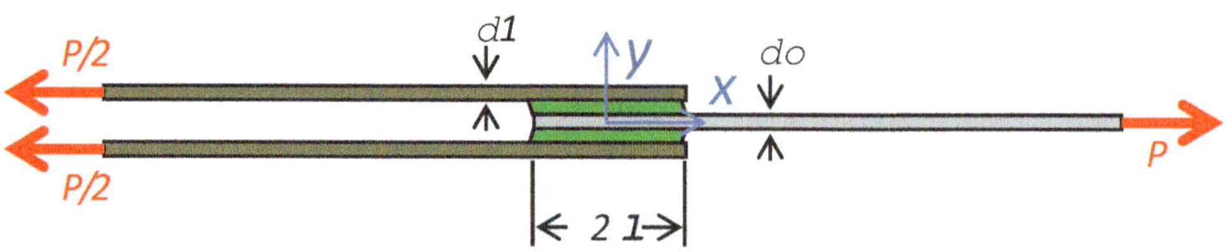

Fig. 12.1 Schematic of composite material surrounding metal sheet specimens for shear loading

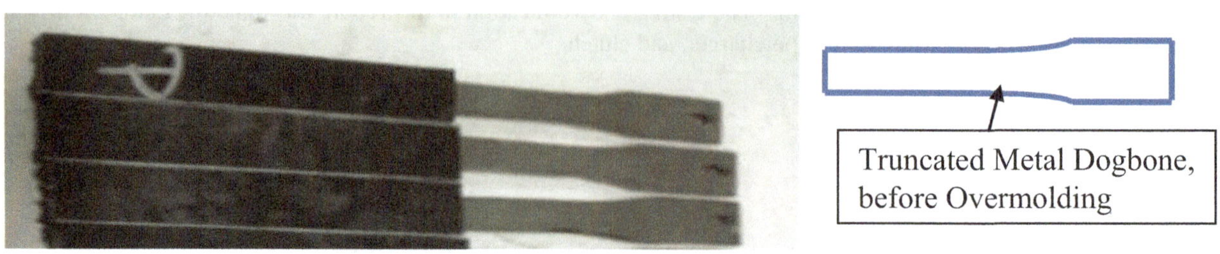

Fig. 12.2 Photograph of composite material bonded to truncated metal dogbone specimens for shear loading

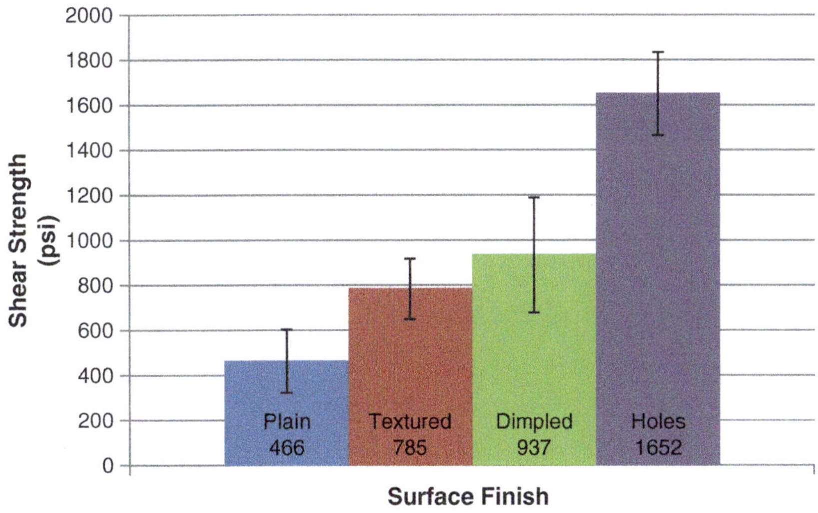

Fig. 12.3 Shear strength of metal-to-composite interfaces with application of metal surface preparation, and overmolding of the composite coupons, with the metal as an insert

12.2.2 Shear Strength Verification

The base metal surface preparation, per the test plan, resulted in significant variations in interfacial shear strength. For these samples, all of the composite was overmolded onto the metal extended half-dogbones, so that results would only indicate the relative performance of the surface preparation. The weakest was the metal with flat faces, and no preparation. When texturing (grit blasted) was added, the interfacial shear strength increased with the additional geometry and increase in surface area, because of the texturing. The dimpling was a more severe case of geometry alteration, so more surface area was generated, and more shear strength was measured. Finally, with "holes," an additional material interlock of the metal and composite was introduced, so significant additional shear strength was obtained, because all of the bonded surface was loaded and the composite material that extruded into the holes formed a base material mechanical "interlock," increasing the maximum strength of the interface. The interlock was independently about as strong as the overmold (flat surfaces van der Waals) bond, so that strength of the "holes" samples was about double that of the other textured and dimpled attachment methods. Unfortunately, the holes interface strength also only achieve under 10% of that of the shear-strength of the steel base material (Fig. 12.3).

12.2.3 Shear Strength with Metal Surface Preparation

For metal preparation, the more severe the treatment is, and as the additional surface area is created by disturbing the metal surface, the higher the apparent shear-bond strength is attained. Where mechanical interlocks (holes) are used in the preparation of the base material, strength, beyond that of the chemical bonding, whether van der Waals or covalent, can be readily achieved.

12.3 Adhesive Bonding

12.3.1 Adhesive Bond Samples

To determine the interfacial strength of a rotational-part subassembly-interface, a set of two metal components were made with a cylindrical interface. The components were fitted to allow a 0.25 mm gap, as recommended by the adhesive manufacturer, between the metal parts for the joining. The bonding surfaces were all cleaned with an acetone wipe to remove grease and oil residue. Two-part epoxy adhesive was mixed by hand and applied to all surfaces on both the metal parts where adhesion was desired, as shown in Fig. 12.4, on a cylindrical interface. The parts were slid together on a fixture that assured concentricity. Excess adhesive flowed on both sides of the cylindrical interface during assembly, assuring complete coverage. After green strength was achieved in an hour, the bonded part was removed from the fixture. The adhesively bonded parts were cured by manufacturer specification, and then inspected for tolerance stack. All parts were found to be within 0.025-mm of

Fig. 12.4 Rotational parts slathered with cyan-colored adhesive, before assembly

Fig. 12.5 Rotational parts assembled together with cylinder-shaped adhesive dollops

circumferential and radial dimensioned positions, well within tolerance of rotational equipment. One assembled part is shown in Fig. 12.5, where the cyan-colored band is the thermoset adhesive.

12.3.2 Adhesive Bond Verification

The bonded parts were subjected to rotational loading as pure torsion, and torqued to failure. The smooth surface adhesive-bonded area failed by a combination of cohesive and adhesive failure. All of the 57 square centimeters of adhesive bond failed simultaneously, and the failure occurred at a load of 2400 Nm. For design purposes, this corresponds to a shear load on the specimen of 15 MPa, that is average for a structural epoxy adhesive. Compared to a flat interface, such as with a symmetrical shear-load coupon, this cylindrical shaped strength is slightly more than can be achieved by the same adhesive in a flat area. This shear strength of the adhesive bond is less than 10 % of that of the base material steel, so is significantly low for joining.

12.3.3 *Adhesive Bond of Rotational Parts*

In the situation of a cylindrical shaped adhesive bond interface (rotational part), the shear loading is uniform, and not dependent on a symmetrical and flat surface (e.g. lap-shear coupon), that has stress concentrations at the ends of the bonded intervals. For a symmetrical shear flat-adhesive coupon, the ends are subject to stress concentration, lowering the measured strength of the complete adhesive dollop. So, although the strength of the adhesive may be the same at all sites, in practice, the stresses will appear higher where there are starts and ends to the bond area, as is the situation for a lap-shear coupon. For the situation of a cylindrical bond shape, the strength will be slightly higher, as the load is distributed evenly.

12.4 Thermal Range

12.4.1 *Thermal Range Samples*

Rotational components need to operate at high and low temperatures. To determine the range of temperature that interface joining can survive, samples were constructed with cylindrical interfaces to represent rotational machinery. The samples were built with metal and composite parts, and these parts were attached by adhesive and by interference fit. To create a mechanical load, the subassemblies were shaken with an acceleration of up to 50 g's to represent a 10,000 N load, as rotational components encounter as their highest known loading. Simultaneously, the assemblies were subjected through a 24-h cycle to achieve a high dwell temperature of 175 °C and a low dwell temperature of −45 °C, repeatedly. Various combinations of base materials were tested, but all had a 0.25 mm adhesive gap, and were assembled with coaxial alignment. The temperatures and accelerations were ramped up with a slow gradient, so that if an intermediate failure occurred, the maximum achievable temperature and acceleration could be recorded as the joint capability. The shake was applied in the radial direction of the rotational component, the component was constrained to a central axis, and to prevent circumferential movement, the component was further constrained at a line near the outer radius. The shake was applied at a rate lower than one cycle per second, and hundreds of cycles were conducted at each temperature extreme, as traced in Fig. 12.7. To establish capability, these parts were disassembled from the fixture, visually inspected for cracks and disassembly, at every temperature extreme shown in Fig. 12.7. If no cracks or disassembly could be observed by visual, tactile, CMM, thermography, and NDE, then the assembly was labeled to be structurally intact, and capable of that loading at temperature.

12.4.2 *Thermal Range Verification*

The rotational components were subjected to ramped-up temperature extremes for exposed automotive driveline components, and ramped-up accelerations. All samples were inspected after shaking, for each temperature low extreme and each temperature high extreme, for microcracking, relative movement between parts, and catastropic separation of parts. All parts passed this testing, showing no deterioration, during all of the temperature extremes, to the last cycle, with no microcracking, relative movement, or part separation. This is a significant observation, as both the base materials and the interfaces survived readily within the temperature bounds.

In Fig. 12.6, three drivetrain components (samples) that are each rotationally symmetric were tested simultaneously in the thermal chamber. At each temperature extreme, the components were disassembled from the fixture and inspected for survivability. All bonded components passed completely, with no visible damage. These parts were subjected to additional fatigue testing to indicate that there was no operational damage to all of these components.

In other testing of rotational components, a reinforced thermoplastic part was overmolded onto a metal part, and subjected to the thermal excursions of Fig. 12.7, without external loading. All of the adhesively-bonded parts passed this test, and when inspected by ultrasonic, tactile, and thermography methods, indicated that no opening occurred, and the bonding by van der Waals force of overmolding had no gapping.

Fig. 12.6 The hydraulic actuator is behind the step-ladder and the rotational components to be tested are in the thermal chamber. The rotational components are shaken in the stationary thermal chamber

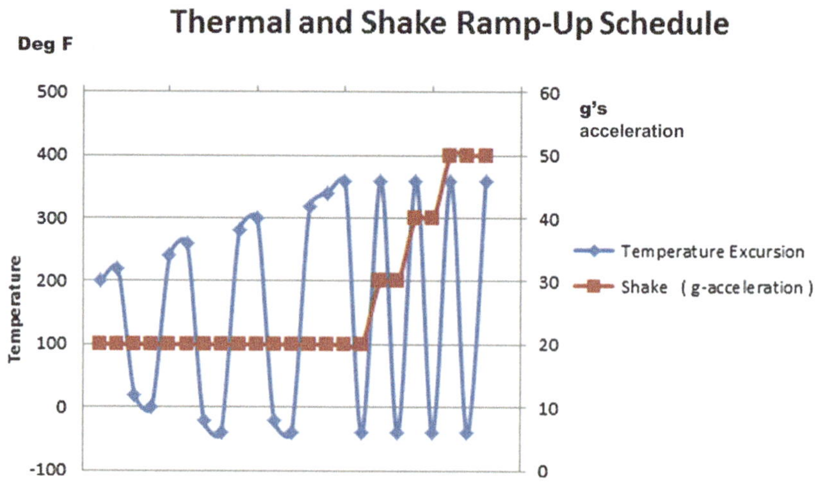

Fig. 12.7 Ramped loading and thermal excursions for the shake testing of the metal-to-composite interfaces

12.4.3 Thermal Capability of Composite to Metal Interfaces

Rotational parts bonded with thermoset adhesive were subjected to forces of over 1000 N. No damage (microcracking or disassembly) was observed by visual, tactile, CMM, or NDE. Multiple "baseline" rotational components that were interference-fit assembled (no van der Waals or covalent bond) failed by disassembly in this testing procedure as shown in Fig. 12.6, indicating that the thermoset adhesive and overmold bond did have a significant role. Although this testing indicates significant capability for thermoset adhesive interfaces, it is well known that both the base composite material, and the thermoset adhesive material can be weaker at temperatures at our use of 175 °C (350 F), compared to room temperature structural performance [3].

In another situation, with a thermoplastic composite overmolded to metal, the van der Waals (polymer to metal) bond was sufficient for the component to survive multiple cycles of extreme hot and cold with no exterior structural loading, successfully, as proven by ultrasonic and thermographic testing to assure metal-to-composite bonding without gapping.

Fig. 12.8 Metal-and-composite components are prototyped full-scale and tested inside existing subassemblies as shown in the dynamometer test cell here

12.5 Structural Behavior

12.5.1 Structural Behavior Samples

Samples of rotational equipment were prototyped and loaded into drivetrain assemblies to test duty cycles to failure. The metal-to-composite interfaces had machnical interlocks, and were attached by overmolding in some parts, interference-fit in some parts, and adhesive bonding in some parts. The rotational components were tested for noise, torsional strength, and fatigue life in dynamometers. The rotational parts were made of metal components and composite components. The composite components were attached to the metal by van der Waals forces (overmolding), friction (interference fit), and by covalent bonds (two-component adhesive bonding), depending on how these were assembled. All were built to current standards for such equipment, and tested in existing housings that are readily available in the marketplace.

12.5.2 Structural Behavior Verification

The components were prototyped to full scale, assembled into a drivetrain subassembly, and mounted on dynamometers that were customized to perform the duty cycles of vehicles, in a stationary test cell. For the rotational components that were overmolded, as reinforced thermoplastic on textured metal, the parts failed after conventional all-metal subassembly components failed and were replaced, but before the duty cycle was completed, for verification testing. The metal-and-composite rotational part assembled by overmolding failed just before achieving 20 % of the duty cycle, where the baseline is an all-cast-iron part that achieves five-times more cycling to failure. This metal-and-composite rotational part assembled by overmolding showed in testing that it was significantly weaker than the all-metal current production part.

In other verification testing, for significantly different rotational components made with thermoset adhesive to bond the composite to the textured metal, all of the components survived the entire duty cycle for that component, as did the current all-steel part. This thermoset adhesive method of assembly was fully successful, and indicates a significant method for successful joining of metal and composite. A dynamometer test cell for rotational eqipment is shown in Fig. 12.8, with the dissimilar material component inside the blue housing.

12.5.3 Full Prototype Dynamometer Duty Cycles

For full-scale testing, on various drivetrain components, the material choice, processing, surface preparation, and component loading for the duty cycle, influence the outcome of the verification testing. Although there were mixed reports in the described results, the outcome is dependent on multiple inputs, making broad conclusions difficult. But it is clear that certain designs and materials in drivetrains, with cylindrical geometries, can be made successfully, by use of large interface-areas with mechanical-interlocks, with the appropriate choices of materials, process, and interface bonding method.

12.6 Conclusion: Utility of Composite-and-Metal Components for Use in Drivetrains

Drivetrains represent the highest demand of performance for materials used in the Eaton vehicle and aerospace group, with regard to structural and thermal capability, and indicate potential capabilities of primary-structural components, such as control arms, for vehicles. Verification testing here for metal and composite two-component parts have shown that metal-to-reinforced-thermoplastic and metal-to-reinforced-thermoset driveline components are achievable in current vehicle configurations and envelopes. Metal-to-composite components have the capability for base-material-attachment, cylindrical interfaces, thermal excursions, and full duty-cycle durability. Furthermore, the interfaces can be enhanced by any metal surface preparation that leads to additional surface area and mechanical interlock of base materials. Driveline applications for lightweghting can greatly influence vehicle dymanics, as rotational components tend to be unsprung components, components to add to the rotational inertial of the vehicle, components that add to the translational inertia of the vehicle, and components that generate undesired vibration, harshness, or noises. Metal-and-composite components have now been built, analyzed and verification tested, so are available for new designs where lightweighting, and NVH need to be enhanced.

References

1. ASTM D3528-96(2008): Standard Test Method for Strength Properties of Double Lap Shear Adhesive Joints by Tension Loading, p. 249. ASTM International, West Conshohocken (2008)
2. Watson, D.C.: Mechanical Properties of E293/1581 Fiberglass-Epoxy Composite and of Several Adhesive Systems, p. 30. Final Report, May 1982, AFWAL-TR-82-4013
3. Ruggles-Wrenn, M.B.: Effects of Temperature and Environment on Mechanical Properties of Two Continuous Carbon-Fiber Automotive Structural Composites. DOE DE-AC05-00OR22725, Oak Ridge National Laboratory, Oak Ridge, TN (2003)

Chapter 13
Short-Term Preload Relaxation in Composite Bolted Joints Monitored with Reusable Optical Sensors

Anton Khomenko, Ermias G. Koricho, Mahmoodul Haq, and Gary L. Cloud

Abstract Bolted fastening is one of the oldest joining techniques, and it is widely used in vast number of applications. Although such fasteners provide many advantages, bolted joint assemblies and in-service behaviors demonstrate complex phenomena and stress distributions. Among these factors, bolt tension affects joint reliability and residual lifetime most significantly. One of the major concerns in bolted joints with composite panels is the effect of creep in the through-the-thickness direction of the material, which leads to preload reduction and premature failure of the loosened bolted joint. Thus, it is essential to monitor the preload after joint assembly. In this work, a previously developed bolt tension monitor that is instrumented with a reusable optical sensor was used to evaluate the short-term relaxation of preload in similar and dissimilar composite bolted joints. A single-parameter model was used to fit experimental data and obtain a relaxation parameter, which was assumed to be an extensive property. Preload reductions in lap joints with steel and glass fiber reinforced plastic (GFRP), and GFRP with GFRP were measured for ca. 9 min after the application of two levels of initial preloads. It was found that, depending on the joint configuration, the preload reduction varied considerably; and an increase in initial preload tends to reduce the preload relaxation. Overall, this technique provides robust and cost-efficient health monitoring in composite bolted joints.

Keywords Bolt tension monitor • FBG sensor • Preload relaxation • Bolted joint • Composite creep

1 Introduction

Bolted fastening is one of the most-used joining techniques in vast numbers of applications including manufacturing, construction, machine design, transport and aerospace sectors, etc. While these fasteners exhibit many advantages such as ease of assembly and repair, in-service behavior is defined by many complex phenomena that depend on various parameters affecting serviceability and durability. One of the major reasons for failure of bolted joints is the preload relaxation that can be caused by rotational self-loosening or creep in the through-the-thickness direction of one or both of the joined components. This issue is especially acute in structures containing plastics or composites that are increasingly being used to reach mandated strength/weight goals. In service self-loosening of components have been reported frequently; and a survey of automobile service managers in the United States reported that one-quarter of all service problems were traced back to

---

A. Khomenko (✉)
General Photonics Corporation, 5228 Edison Avenue, Chino, CA 91710, USA

Composite Vehicle Research Center, Michigan State University, 2727 Alliance Drive, Lansing, MI 48910, USA
e-mail: khomenka@msu.edu

E.G. Koricho • G.L. Cloud
Composite Vehicle Research Center, Michigan State University, 2727 Alliance Drive, Lansing, MI 48910, USA

M. Haq
Composite Vehicle Research Center, Michigan State University, 2727 Alliance Drive, Lansing, MI 48910, USA

Department of Civil and Environmental Engineering, Michigan State University,
Engineering Building, East Lansing, MI 48824, USA

G.L. Cloud et al. (eds.), *Joining Technologies for Composites and Dissimilar Materials, Volume 10*
Conference Proceedings of the Society for Experimental Mechanics Series, DOI 10.1007/978-3-319-42426-2_13

lose fasteners [1]. Hence, bolt loosening has become an important research area as ways are sought to prevent failures in a variety of applications, particularly in similar and dissimilar composite joining [2]. The loss of preload induced by creep has a detrimental effect on the strength and the residual lifetime of the connection. Thus, it is essential to monitor the bolt preload in service conditions to prevent catastrophic failure of the bolted joint. Many studies on composite creep have been performed, but in all of these investigations the clamp-up forces were randomly selected and neither the influences of different clamp-up forces nor the effects of external static and dynamic loads on relaxation were investigated in detail [3]. In most of the previous studies, the tools for preload measurements were used ‘as is’ without any systematic evaluation of instrumentation errors on the preload measurements, a critical omission for long-term measurements. Moreover, since there is not any standard or single procedure, monitoring times have been randomly selected and the results have been presented in different formats, meaning that the various results cannot be compared directly. Lastly, the results for initial preload effect on composite creep display some contradictions [3, 4]. Comprehensive research on the creep of similar and dissimilar composite joints, including proper selection of the NDE technique employed, is badly needed in order that designers and manufacturers might meet weight and performance criteria.

Numerous techniques for bolt tightening control and structural health monitoring exist and can be classified into several categories as follows: external miniature load cells [2], acoustic inspection [5, 6], optical inspection [7], internally embedded strain gages and fiber Bragg grating (FBG) sensors [8, 9], and techniques based on other physical principles. Each of these techniques has its own merits and limitations. Optical inspection is complex and is not suitable for continuous monitoring under field conditions [1]. Acoustic inspection is sensitive to external vibrations [5]. Sensing washers have to be incorporated into the structure, and the load cells are quite expensive [10]. Moreover, sensing washers and load cells add weight to the structure. Installation of electrical resistance strain gages (RSGs) in the bolt shaft is cumbersome, they are temperature sensitive, and the drilled channel has to be several mm in diameter, which is detrimental to the integrity of the bolt. Moreover, RSGs are not compatible with some types of bolts. FBG sensors are very small in diameter (up to 100 μm) and easy to install in any type of the bolt, but permanent embedding is not practical because the cost of the sensor is much higher than that of traditional RSG’s. Previously, we developed a novel technique to monitor bolt clamping force utilizing removable and reusable embedded FBG sensors, and we validated the technique for static preload measurements in multi-material bolted joint configurations [11]. The “bolt tension monitor” uses a temporary adhesive for embedding an FBG sensor into the bolt shaft, thereby converting the bolt into a tension measuring device that is a reliable tool for evaluating the behavior of the bolted joint. The FBG sensor can be embedded at any given time into any type of bolt, then be removed and reused later in the same or another bolt. The bolt tension monitor provides real-time precise clamping force measurements and does not require complicated manufacturing or setup preparation. Significantly, the bolt tension monitor is resistant to most external environmental effects, the major exception being changes in gross temperature of the bolt.

In this work, bolt tension monitors utilizing hot melt adhesive or polylactide (PLA) plastic as the medium for embedding FBG sensors were compared and used for the monitoring of preload relaxation caused by creep in similar and dissimilar composite bolted joints. Initially, instrumented bolts were calibrated using an external load cell, and the creep in the embedding thermoplastic was evaluated using a single steel plate. Afterwards, short-term preload reductions that occurred within the first 9 min after bolt tightening in lap joints with two material combinations, i.e., steel with glass fiber reinforced plastic (GFRP) and GFRP with GFRP, were studied for two levels of initial preload. A single-parameter model was used to fit experimental data and obtain a relaxation parameter, which was assumed to be an extensive property. This means that the value of the relaxation parameter for the system would be the sum of the relaxation parameters for each subsystem. Finally, the thru-hole load cell used for bolt tension monitor calibration was evaluated for preload relaxation measurements in a composite material.

2 Specimen Preparation

2.1 Bolt Tension Monitor Fabrication

Two bolt tension monitors were manufactured as is described in [9]. The bolts were grade five medium-strength steel hex head (head width: 19.05 mm or 3/4″; head height 7.94 mm or 5/16″) bolts with a right-hand thread having a length of 38.1 mm (1.5″), threaded with a 12.7 mm-13 (0.5″-13) thread. A 1.8 mm diameter hole was drilled along the axis of each bolt through one-half of its length (23 mm). A conventional heat gun was employed to embed an FBG sensor inside each bolt using in one case 3M™ Scotch-Weld™ Hot Melt Adhesive 3779 PG Amber and in the second case polylactide (PLA) plastic. The embedded FBG sensors used were Micron Optics os1100 with a gage length of 10 mm. The spectral responses of the FBG sensors were acquired using a Micron Optics sm125 700 Optical Sensing Interrogator. Both bolt tension monitors are illustrated in Fig. 13.1.

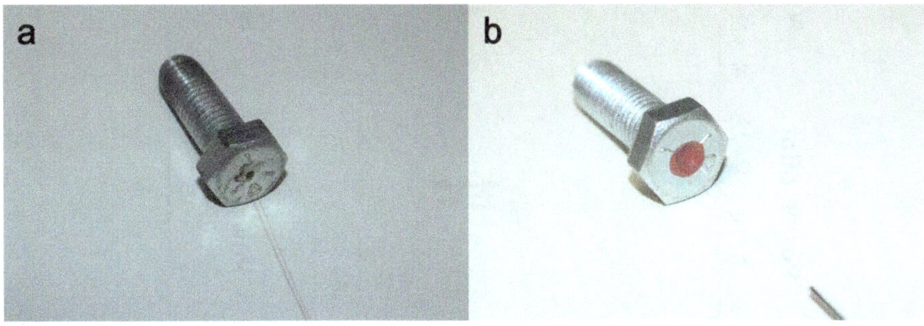

Fig. 13.1 Bolt tension monitors: (**a**) with hot melt adhesive embedding medium, (**b**) with PLA plastic embedding medium

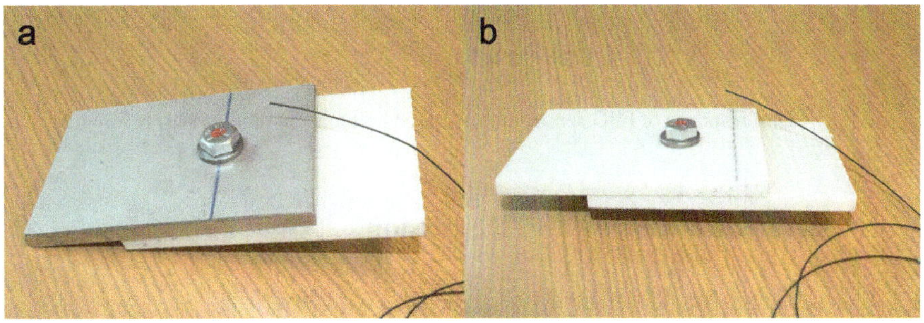

Fig. 13.2 Assembled bolted joints: (**a**) steel-GFRP configuration, (**b**) GFRP-GFRP configuration

The operating principle of the bolt tension monitor is based on the spectrum monitoring of the embedded FBG sensor. Such a sensor acts as an optical filter that transmits the entire spectrum of the light source and reflects back the resonant Bragg wavelength [11]. Any perturbation of the Bragg grating period results in a shift of Bragg wavelength. The strain response arises from both the physical elongation of the sensor and the change in fiber effective refractive index caused by stress-birefringence, whereas a thermal response arises from the inherent thermal expansion of the fiber material and the temperature dependence of the refractive index. Because FBG sensors require only very small stresses to induce appreciable strains, temporary adhesives (e.g., thermoplastics) suffice for attachment of these sensors well enough so that, within certain limits, they accurately indicate strain in the bolt body. Within the linear elastic regime, the axial tension load in the bolt is directly proportional to the measured elongation strains and can be easily calculated if the bolt material properties are known. However, as with all load transducers, direct calibration with known loads is the only truly valid way to accurately establish the relationship between force and system output for each transducer.

2.2 *Bolted Joints Assembly*

GFRP composite panels for the bolted joints were manufactured using a liquid molding process. The reinforcement used was Owens Corning ShieldStrand® S, S2-glass plain weave fabric with areal weight of 818 g/m^2. Sixteen layers of fabric were used. The distribution medium was Airtech Advanced Materials Group Resinflow 60 LDPE/HDPE blend fabric. The resin used was a two-part toughened epoxy, namely SC-15, obtained from Applied Poleramic. A steel mold having dimension of 609.6 mm × 914.4 mm with point injection and point venting was used to fabricate 508.0 mm × 609.6 mm plates. After the materials were placed, the mold was sealed using a vacuum bag and sealant tape. The mold was then infused under vacuum at 29 in Hg. The resin-infused panel was cured in a convection oven at 60 °C for 2 h and post-cured at 94 °C for 4 h.

The steel and GFRP coupons used in the joint assembly had dimensions of 152 mm × 101 mm × 9.5 mm and 152 mm × 101 mm × 9 mm, respectively. A 15.7 mm diameter hole was drilled in the plates 50.5 mm away from three edges, and a steel sleeve washer was inserted. The sleeve washer had the following dimensions: flange diameter 25.4 mm, sleeve ID of 13.46 mm, sleeve OD of 15.88 mm, flange thickness of 3.18 mm, overall height of 4.75 mm. The resulting assemblies for steel-GFRP and GFRP-GFRP configurations with the PLA-based bolt tension monitor are illustrated in Fig. 13.2.

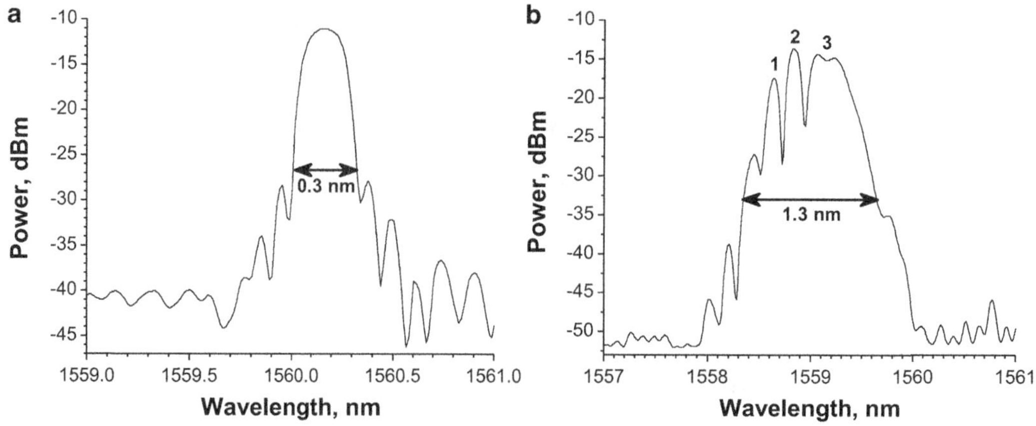

Fig. 13.3 Spectral response of FBG sensor in PLA-based bolt tension monitor: (**a**) before installation, (**b**) after installation

3 Experimental Results and Discussion

3.1 Bolt Tension Monitor Features and Calibration

Spectral responses of FBG sensors embedded in hot melt adhesive and PLA plastic were compared before and after FBG sensor installation. The spectral changes were not pronounced for the sensor that was embedded in hot melt adhesive; however, in the case of PLA as the embedding medium, the spectral response the FBG sensor was distorted and exhibited multi-peaks, as shown in Fig. 13.3.

The spectrum broadening from ~0.3 to ~1.3 nm and the significant distortion were probably caused by non-uniform shrinkage of the PLA thermoplastic during cooling. Although it would seem to be detrimental to the FBG sensor, this multi-peak feature can be considered as beneficial. Indeed, instead of a single gage, the various peaks serve as several randomly distributed gages, each of which can be used to monitor strains [12]. This system then provides redundant measurements and the ability to monitor strain uniformity along the bolt shaft [11]. In this work, peaks 1, 2, and 3 shown in Fig. 13.3b were used to measure the preload with the PLA-based bolt tension monitor.

Both bolt tension monitors were calibrated using a thru-hole load cell model THC-1K-T manufactured by Transducer Techniques® with a maximum load capacity of 4448 N (1000 lb). During the manual tightening of the bolt, the spectral responses of the FBG sensor embedded into the bolt were acquired using the interrogator and synchronized with the load cell data. The calibration curves, showing typical dependences of Bragg wavelength shift of the sensors on the applied preload as measured by the load cell, are reproduced in Fig. 13.4. The relationship between applied load and wavelength shift was found to be linear and the calibration coefficient was found to be 37.1 ± 0.8 N/pm for the hot melt adhesive-based bolt tension monitor. For the PLA-based bolt tension monitor, the calibration factors are and 23.7 ± 0.8 N/pm, 22.8 ± 0.6 N/pm, and 21.7 ± 0.6 N/pm for peak 1, peak 2, and peak 3, respectively. The differences in calibration factors indicate that sensor installation has a notable effect on calibration factor, thus each transducer has to be calibrated prior to actual measurements.

3.2 Data Analysis Using a Single-Parameter Model for Preload Relaxation

After the measurements were completed for each case, the preload reduction was approximated using a modified version of the equation according to Caccese et al. [2]:

$$\frac{P_t}{P_0} = p\left(t + t_0 + 1\right)^{-\alpha}, \tag{13.1}$$

where P_t is the measured preload, P_0 is the initial preload, $p \sim 1$ is the coefficient that takes into account the continually changing load value at the start of the tests which is affected by the primary creep response [2], t is time in seconds, $t_0 = 2$ is the time interval between the maximum preload and the first fitted data point, and α is the relaxation parameter. Coefficient

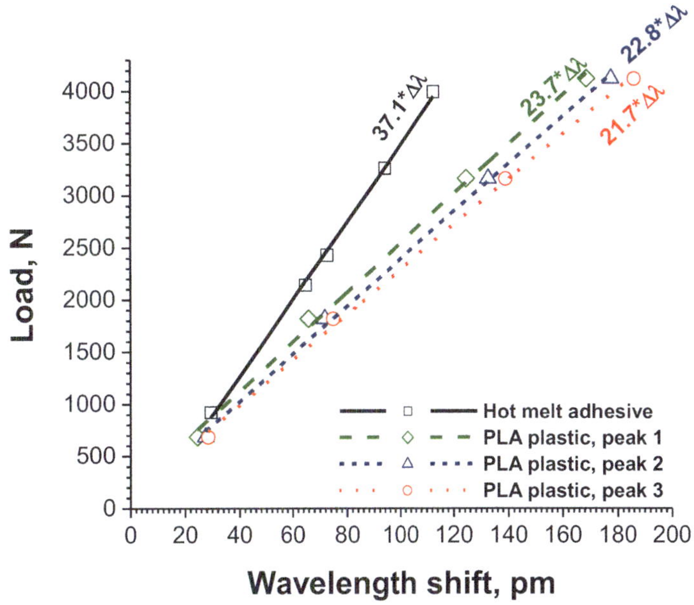

Fig. 13.4 Calibration curves for hot melt adhesive- and PLA-based bolt tension monitors

t_0 was introduced for proper data fitting, since the measurements were taken 2 s after the maximum preload. An important hypothesis was adopted, namely that the relaxation parameter is an extensive property. Thus, the relaxation parameter of the selected configuration can be represented as the sum of the relaxation parameters of individual components. For example, if a transducer has its own relaxation parameter α_T, the relaxation parameter of studied configuration α_C can be found by a simple subtraction of α_T from measured relaxation parameter α, i.e., $\alpha_C = \alpha - \alpha_T$. Hence, Eq. (13.1) can be rewritten as:

$$\frac{P_t\left(t+t_0+1\right)^{\alpha_T}}{P_0} = p\left(t+t_0+1\right)^{-\alpha_C}, \tag{13.2}$$

A singular beneficial characteristic of this approach is that the preload relaxation contributed by the transducer itself is eliminated by scaling of the measured preload to the value which would be expected in the absence of the transducer. The results are thereby caused to be relatively independent of test conditions so that they become physical rather than phenomenological, and, if properly interpreted, they have broad application.

The use the single parameter α allows simple case-by-case comparison, whereas a two-parameter model is not quite practical and suitable for this purpose [2, 4, 13]. The value of α may depend upon such variables as material creep properties, temperature, humidity, moisture content, stress and thickness. Preload reduction in a bolted joint is a complex phenomenon, which is caused by continuously changing material creep and stress relaxation. After initial preload, the material begins to creep causing a small reduction in material thickness which in turn leads to contraction of the bolt and decrease of reaction force. Thus, the more the creep in the material the faster the preload decreases (relaxation parameter increases).

3.3 Short-Term Preload Relaxation Measurements by Hot Melt Adhesive-Based Bolt Tension Monitor for 4500 N Initial Preload

In the first set of experiments, short-term preload relaxation measurements using the bolt tension monitor that incorporated the hot melt adhesive as the embedding medium were performed on the steel-GFRP and GFRP-GFRP bolted joint configurations for an initial preload of 4500 N and for periods of ca. 9 min after bolt tightening. At least three tests were performed at $45 \pm 5\,\%$ humidity and room temperature, which was $24 \pm 2°$, for each case; and approximately 2700 measurement data points were obtained during each test. Before every test run, the system (bolted joint and bolt tension monitor) was verified to be in its initial state. In order to achieve that, the system was kept untightened for at least 15 min so it could achieve temperature equilibrium and return to its initial state with no residual creep in the embedding thermoplastic or the GFRP plate(s). After the manual tightening, the short-term preload relaxation was monitored for about 9 min. Preload measurements were

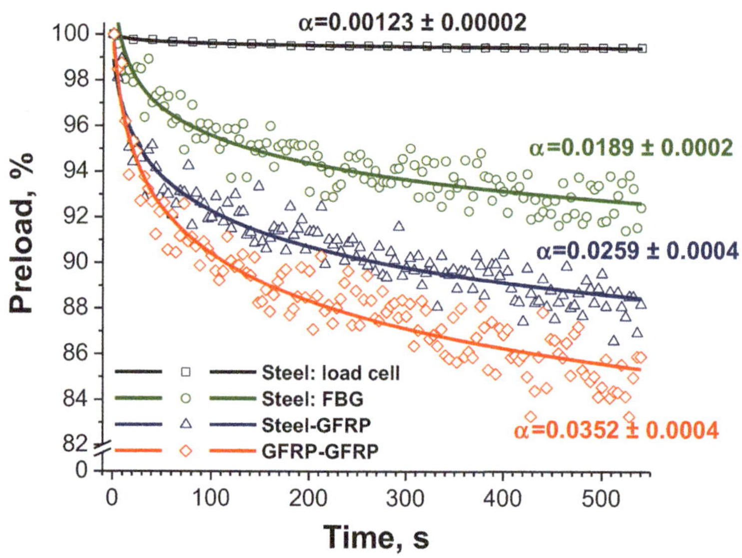

Fig. 13.5 Typical preload relaxation measured (every 20th point plotted) in steel-GFRP and GFRP-GFRP bolted joint configurations with hot melt adhesive-based bolt tension monitor and in steel plate measured by thru-hole load cell and monitored with FBG sensor embedded with hot melt adhesive for 4500 N initial preload

begun 2 s after the bolt was tightened to eliminate the effect of tightening rate. Typical preload relaxation measurements from one test are presented in Fig. 13.5.

When a single steel plate was used, the average relaxation parameters α from the series of tests were found to be 0.0013 ± 0.0002 and 0.0158 ± 0.0029 measured by the thru-hole load cell and the embedded FBG, respectively. In other words, assuming that the steel did not creep, the hot melt adhesive-based bolt tension monitor had a creep in the embedding thermoplastic itself, which is not desired for long-term measurements. The average relaxation parameters for a steel-GFRP and GFRP-GFRP bolted joint configurations measured using the hot melt adhesive-based bolt tension monitor were found to be 0.0268 ± 0.0024 and 0.0346 ± 0.0046, respectively. These values, however, are contaminated by the creep-induced relaxation in the embedding thermoplastic mentioned above. Since the relaxation parameter was assumed to be an extensive property, the relaxation parameter for the studied configuration can be easily obtained by simple subtraction of the thermoplastic relaxation parameter from the measured configuration relaxation parameter, as described above. Thus, the average relaxation parameters were 0.0110 ± 0.0038 and 0.0188 ± 0.0054 for the steel-GFRP and GFRP-GFRP configurations.

3.4 *Short-Term Preload Relaxation Measurements by PLA-Based Bolt Tension Monitor for 4500 N Initial Preload*

Following the procedure described previously but using the bolt tension monitor with the PLA thermoplastic embedding medium, short-term preload relaxation was measured again for 9 min in steel-GFRP and GFRP-GFRP bolted joint combinations for an initial preload of 4500 N. Typical preload relaxation measurements are shown in Fig. 13.6.

When a single steel plate was used, the average relaxation parameter α was found to be 0.0013 ± 0.0001 and 0.0021 ± 0.0008 measured by the thru-hole load cell and the PLA-embedded FBG, respectively. The average relaxation parameters for a steel-GFRP and GFRP-GFRP bolted joint configurations measured by the PLA-based bolt tension monitor were found to be 0.0123 ± 0.0025 and 0.0241 ± 0.0023, respectively. The measured relaxation in the PLA thermoplastic was much lower than that obtained for the hot melt adhesive, which means that PLA is a better candidate for embedment of the FBG in the bolt tension monitor. Subtracting the previously obtained relaxation parameter found for PLA plastic, the average relaxation parameters for steel-GFRP and GFRP-GFRP bolted joint configurations are found to be 0.0102 ± 0.0026 and 0.0220 ± 0.0024, respectively. Since the creep-induced relaxation of the PLA plastic was insignificant, preload relaxation measurements would seem to yield higher precision. The measured relaxation parameter for a bolted joint with two composite panels was approximately twice as great as that for a bolted joint with a single composite plate. This observation can serve as an indication that the relaxation parameter is, indeed, an extensive property.

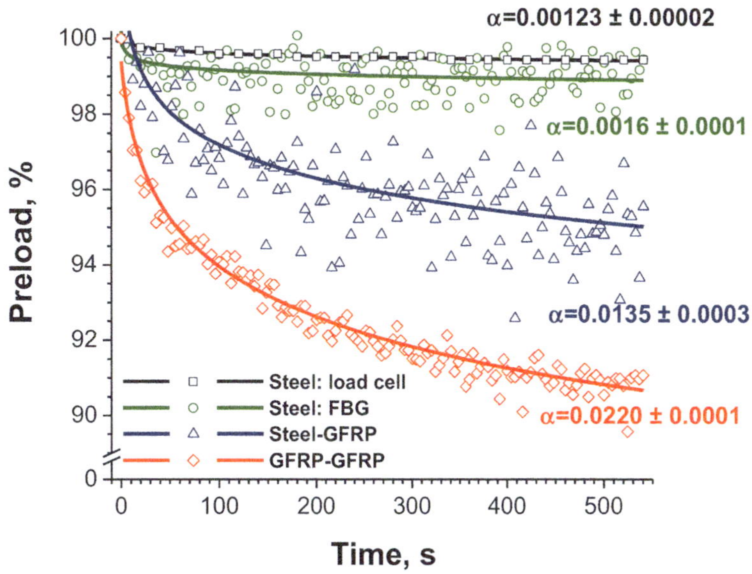

Fig. 13.6 Typical preload relaxation measured (every 20th point plotted) in steel-GFRP and GFRP-GFRP bolted joint configurations with PLA-based bolt tension monitor and in steel plate measured by both the thru-hole load cell and by FBG sensor embedded with PLA plastic for 4500 N initial preload

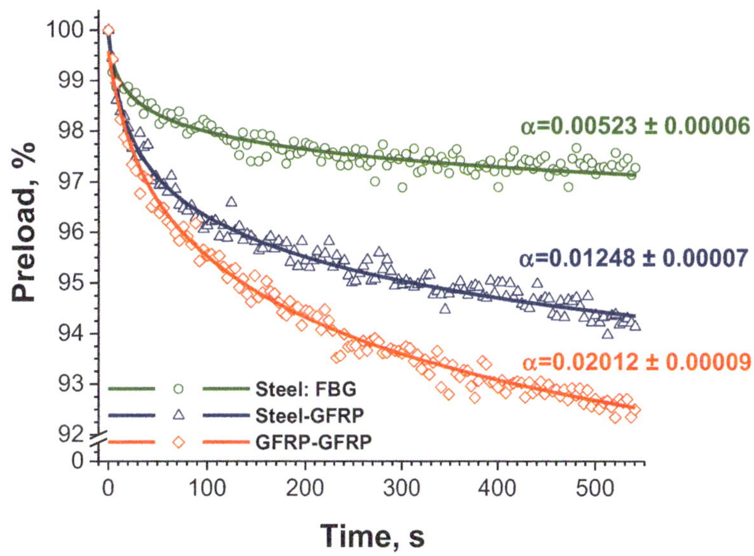

Fig. 13.7 Typical preload relaxation measured (every 20th point plotted) in steel-GFRP and GFRP-GFRP bolted joint configurations and in steel plate measured with PLA-based bolt tension monitor for 8000 N initial preload

3.5 Short-Term Preload Relaxation Measurements by PLA-Based Bolt Tension Monitor for 8000 N Initial Preload

Short-term preload relaxation measurements in the PLA thermoplastic embedding medium, the steel-GFRP and the GFRP-GFRP bolted joint configurations were performed for 9 min using an initial preload of 8000 N. The procedures were the same as those outlined above. Typical preload reduction measurements are reported in Fig. 13.7.

The average relaxation parameter α for the PLA plastic measured by testing the embedded FBG sensor with the steel plate was found to be 0.0050 ± 0.0007, which was higher than that for the 4500 N preload. The average relaxation parameter for the steel-GFRP and GFRP-GFRP bolted joint configurations measured by the PLA-based bolt tension monitor for this higher initial preload was found to be 0.0127 ± 0.0005 and 0.0204 ± 0.0003, respectively. Taking into account the previously obtained relaxation parameter in PLA plastic, the average relaxation parameters for the steel-GFRP and GFRP-GFRP bolted joint configurations can be found to be 0.0077 ± 0.0008 and 0.0154 ± 0.0008, respectively. As with the 4500 N initial preload case, the bolted joint with two composite panels had a relaxation parameter nearly twice that of a bolted joint with a single composite plate. This agreement can be taken as further validation of the single parameter approximation and proposed hypothesis. Although the average values showed a reduction of the relaxation parameter with increased preload, more experiments with higher preloads and longer monitoring times are required to confirm this dependency.

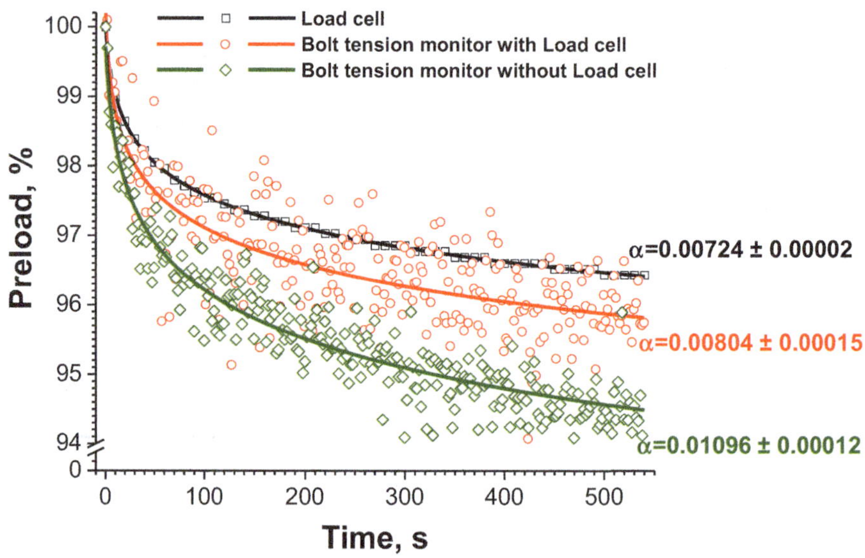

Fig. 13.8 Typical reload relaxation measured (every tenth point plotted) in GFRP plate by thru-hole load cell and by PLA-based bolt tension monitor with and without thru-hole load cell in the assembly for 4500 N initial preload

3.6 Evaluation of Thru-Hole Load Cell for Preload Relaxation Measurements

In the final set of experiments the thru-hole load cell was evaluated for use in composite creep/preload relaxation measurements. For this purpose a single GFRP plate in combination with the PLA-based bolt tension monitor and the thru-hole load cell was tightened to 4500 N, and the preload reduction was monitored. In the first case, the system consisted of a GFRP panel with a steel washer, the PLA-based bolt tension monitor, the thru-hole load cell, and a nut. In the second case, the thru-hole load cell was replaced by a steel block of similar thickness and size. All performed measurements followed the procedure described previously. Typical preload reduction measurements are illustrated in Fig. 13.8.

The average relaxation parameters α measured by the thru-hole load cell, the PLA-based bolt tension monitor with and without the load cell present in the assembly were found to be 0.0077 ± 0.0007, 0.0081 ± 0.0009, and 0.0108 ± 0.0005, respectively. If the PLA relaxation parameter obtained previously is taken into account, the average relaxation parameter for the GFRP plate can be obtained, giving values of 0.0060 ± 0.0012 and 0.0087 ± 0.0009 for the GFRP plate and PLA-based bolt tension monitor assembly with and without the thru-hole load cell, respectively. The average GFRP plate relaxation parameter measured in the assembly without the load cell shows good agreement with the steel-GFRP bolted joint relaxation parameter obtained earlier. This indicates that the majority of the preload relaxation in the steel-GFRP bolted joint combination was caused by the GFRP plate, as expected. Note also that the presence of the thru-hole load cell in the assembly tends to reduce the measured relaxation parameter of the GFRP plate. Such behavior can be explained if the thru-hole load cell acts as a spring washer, which partially offsets the reduction of the preload, hence artificially decreasing the actual preload relaxation. This example demonstrates that any tool for preload measurements cannot simply be used 'as is'; and more rigorous study is required to fully evaluate the use of such a tool for creep monitoring.

4 Conclusions

In this work, bolt tension monitors utilizing hot melt adhesive and PLA plastic as an embedding medium were compared and used for the monitoring of short-term preload relaxation in similar and dissimilar GFRP composite bolted joints. It was found that if the sensor is embedded in hot melt adhesive, the spectral changes in the FBG output were not pronounced; while, in the case of PLA as the embedding medium, the spectral response of the FBG sensor was distorted and exhibited multiple peaks, which were caused by non-uniform shrinkage of the PLA thermoplastic as it cooled and solidified. Although it would seem to be detrimental to the proper functioning of the FBG sensor, this multipeak feature can be viewed as beneficial and afforded the opportunity to obtain several redundant preload measurements. The relationship between applied load and wavelength shift (or calibration curve) was found to be linear for both hot melt adhesive- and PLA-based bolt tension monitors.

The results demonstrated that for a 4500 N initial bolt preload the hot melt adhesive-based bolt tension monitor exhibited the creep in the embedding thermoplastic itself, which is not desirable for long-term measurements. However, if a single-parameter approximation (see Eq. 13.1) is used and the relaxation parameter is assumed to be extensive property, the relaxation parameter of the studied configuration can be obtained by subtraction of the embedding thermoplastic relaxation parameter from the measured one.

It was found that for the PLA-based bolt tension monitor, the PLA relaxation parameter slightly increased for higher initial preload. An important finding is that the PLA relaxation parameter was considerably lower than that of hot melt adhesive. This allowed more precise evaluation of the relaxation parameters for steel-GFRP and GFRP-GFRP bolted joint configurations. It was demonstrated that for both the 4500 N and the 8000 N initial preload cases, the relaxation parameter for the bolted joint with two composite panels was nearly twice as large as the relaxation parameter compared with a bolted joint with a single composite plate. This observation can serve as an indication that the relaxation parameter is an extensive property. Although the average values showed reduction of the relaxation parameter with increased preload, more experiments with higher preloads and longer monitoring times are required to confirm this trend.

Finally, it was demonstrated that the presence of the thru-hole load cell in the assembly tends to reduce the measured relaxation parameter of the GFRP composite. This can be explained if the thru-hole load cell acts as a spring washer, which partially offsets the reduction of the preload, hence artificially decreasing the actual relaxation parameter.

The results of these tests demonstrate that the single-parameter approximation used in the study provided good fitting of the experimental data and allowed simple case-by-case comparison, provided the relaxation parameter is proven to be an extensive property. The comparison of relaxation parameters for various configurations further confirms that the chosen approximation was valid. The next step is further validation of the proposed hypothesis for multi-material configurations under various conditions, and more detailed research on the limits of the chosen approximation.

Overall, this study aims to assist in the development of a single procedure for simple, robust and cost-efficient characterization of preload relaxation that is caused by creep in bolted joints. Future work includes the use of distributed fiber optic sensors and long-term strain measurements, which can be accomplished if temperature and strain effects are discriminated. This step can be achieved in several ways, e.g., use of temperature insensitive fiber optic sensors, temperature compensation with secondary fiber optic sensors, and the use of a fiber optic sensor which provide separate measurements for strain and temperature.

References

1. Li, C., Xu, C.-N., Adachi, Y., Ueno, N.: Real-time detection of axial force for reliable tightening control. Proc. SPIE **7522**, 75223G (2010)
2. Caccese, V., Berube, K.A., Fernandez, M., Melo, J.D., Kabche, J.P.: Influence of stress relaxation on clamp-up force in hybrid composite-to-metal bolted joints. Compos. Struct. **89**, 285–293 (2009)
3. Thoppul, S.D., Finegan, J., Gibson, R.F.: Mechanics of mechanically fastened joints in polymer-matrix composite structures—a review. Compos Sci. Technol. **69**(3-4), 301–329 (2009)
4. Thoppul, S.D., Gibson, R.F., Ibrahim, R.A.: Phenomenological modeling and numerical simulation of relaxation in bolted composite joints. J. Compos. Mater. **42**(17), 1709–1729 (2008)
5. Coelho, C.K., Das, S., Chattopadhyay, A., Papandreou-Suppappola, A., Peralta, P.: Detection of fatigue cracks and torque loss in bolted joints. Proc. SPIE **6532**, 653204 (2007)
6. Amerini, F., Meo, M.: Structural health monitoring of bolted joints using linear and nonlinear acoustic/ultrasound methods. Struct. Health Monit. **10**(6), 659–672 (2011)
7. Huang, Y.H., Liu, L., Yeung, T.W., Hung, Y.Y.: Real-time monitoring of clamping force of a bolted joint by use of automatic digital image correlation. Opt. Laser Technol. **41**(4), 408–414 (2009)
8. Industrial and research news: miniature load cells. Strain **23**, 196 (1987)
9. Hay, A.D.: Bolt, stud or fastener having an embedded fiber optic Bragg grating sensor for sensing tensioning strain. US Patent 5,945,665, 31 Aug 1999
10. Jiang, Y., Zhang, M., Park, T.-W., Lee, C.-H.: An experimental study of self-loosening of bolted joints. J. Mech. Des. **126**(5), 925–931 (2004)
11. Khomenko, A., Koricho, E.G., Haq, M., Cloud, G.L.: Bolt tension monitoring with reusable fiber Bragg-grating sensors. J. Strain Anal. Eng. Des. (2015). doi:10.1177/0309324715598265
12. Haq, M., Khomenko, A., Udpa, L., Udpa, S.: Fiber Bragg-grating sensor array for health monitoring of bonded composite lap-joints. In: Proceedings of the 2013 Annual Conference on Experimental and Applied Mechanics: Experimental Mechanics of Composite, Hybrid, and Multifunctional Materials, vol. 6, pp. 189–195 (2014)
13. Shivakumar, K.N., Crews, J.H.: Bolt clampup relaxation in a graphite/epoxy laminate. In: O'Brien, T.K. (ed.) Long-Term Behavior of Composites, ASTM STP, vol. 813, pp. 5–22. American Society for Testing and Materials, Philadelphia (1983)

CPI Antony Rowe
Eastbourne, UK
April 03, 2019

excavation projects in the Tucson Basin have yielded excellent data for multiple settlements arrayed across the Santa Cruz Valley floodplain (Mabry 1998, 2008; Mabry and Doolittle 2008; Vint 2015; Whittlesey et al. 2010). Excavations have yielded hundreds of storage pits, many dozens of houses, middens, high rates of maize recovery, numerous grinding tools, and burials that date to the San Pedro (1200 to 800 cal BC) and Cienega phases (800 cal BC to cal AD 50). Settlements were frequently abandoned and moved every few years due to flooding and the archaeological signature of short-term sedentism, flood-induced abandonment, and reoccupation results in clusters of archaeological sites termed districts (Mabry 2008c).

By the San Pedro phase, Tucson Basin settlements were organizationally committed to maize agriculture, yet foraging remained a key part of their subsistence. People invested in building and maintaining irrigation canals as early as 1500 cal BC (Mabry 2002, 2006, 2008c; Mabry and Stevens 2014; Vint 2015). By 1200 cal BC, Tucson Basin farmers engineered, built, and maintained field systems with complex canal networks. The Las Capas site system included diversion dams, main and secondary canals, and field laterals which provided water to an estimated 4000 constructed small planting basins (Nials 2015). Labor estimates for building and maintaining the system are modeled as 200 to 400 person days for construction plus another 400 person days per year for maintenance requiring cooperative multifamily labor groups (Mabry 2008c; Nials 2015:455). Mabry models Las Capas population levels as containing a work force of 78 to 234 persons, with Nials (2015) favoring the lower end of this range. By the Cienega phase, populations in the Santa Cruz River valley further increased (Gregory and Diehl 2002; Mabry 2005b; Whittlesey and Ciolek-Torrello 1996; Wills and Huckell 1994).

The scale of these Santa Cruz Valley Early Agricultural period settlements and the degree of sedentism they represent remain key issues. Determining if houses and features on a site represent a series of reoccupations or if they constitute a single continuously occupied settlement remains a fundamental methodological challenge and source of debate (Diehl 2005c; Kelly 1992; Rafferty 1985; Roth 2016b). Indications that these occupations were more sedentary than many had believed include sites with pithouses, internal spatial organization, formal midden deposits, numerous large storage pits, burials, irrigation systems, and archaeobotanical remains representing most seasons of the year. However, some argue that these sites are not fully sedentary locations but the accumulations of repeated, short duration occupations by smaller groups (Gregory and Diehl 2002; Roth 2016b; Schurr and Gregory 2002; Whittlesey 2016; Whittlesey and Ciolek-Torrello 1996). Other investigators argue that by the Early Agricultural period settlements represent substantial, largely sedentary, maize-dependent villages (B. Huckell 1995; Mabry 1998, 2008c; Vint 2015a). The term "short-term sedentism" implies that these largely sedentary settlements were abandoned every few years due to flooding, only to be reoccupied at a later time (Mabry 2008c).

Today lively debates concern the role of agriculture, population size, the degree of sedentism, duration and continuity of occupation, nature of storage, the pace of change, and nature of the social organization at these Early Agricultural period southern Arizona communities, continuing to draw upon a rich body of data, method, and theory (e.g., Diehl 1997; Diehl and Davis 2016; Gregory and Diehl 2002; Mabry 1998, 2008c, 2008; Schurr and Gregory 2002; Whittlesey 2016; Whittlesey et al. 2010). In any case, Early Agricultural period settlements in the Tucson Basin are not consistent with traditional models of a gradual incremental change from small mobile hunting and gathering societies to the large, sedentary villages with heavy dependence on agriculture that emerged alongside ceramic technology. However, the impact of maize during the millennium between its introduction and the formation of San Pedro phase settlements is poorly understood.

There has also been controversy over the basic mechanisms by which early agriculture spread into the southwestern United States and northwestern Mexico. Most researchers assumed local foragers adopted domesticates and farming technology from the south and then developed agricultural economies in situ (Mabry et al. 2008a; Minnis 1992; Wills 1988). Other investigators have argued that in at least some cases agriculture was introduced by migrant farmers who either moved in to occupy empty niches or who actually displaced local foragers (Bellwood 1997; Berry and Berry 1986; Berry 1982; Hill 2001; LeBlanc 2008; Matson 1991, 2002). The presence of precocious large farming communities in southern Arizona and northwestern Chihuahua has been cited as support of southern immigrant models (Bellwood 2005; LeBlanc 2002; Matson 2002). However, the large Early Agricultural period settlements are also explicable with the in situ model as local foragers may have found greater investment in farming an attractive option given local ecological, demographic, and farming conditions (Mabry 2008c; Minnis 1985). We have argued that the long-distance migration of Proto-Uto-Aztecan (PUA) speaking farmers from Mesoamerica or others into the Southwest is not supported by the linguistic, genetic, demographic, and archaeological evidence (Merrill et al. 2009). While long-distance migrations by PUA famers do not appear to be a tenable explanation for the introduction of maize, short distance migrations and diffusion may have indeed been a part of the mechanisms of maize dispersal involving ancestral Uto-Aztecan speakers as well as indigenous people (Mabry et al. 2008a). Merrill (2012:233) shows that the Proto-Southern Uto-Aztecan speech community had not yet dispersed and that this population was exploiting wild resources in northeastern Sonora or southeastern Arizona when it obtained maize, placing it near to some of the earliest maize in the borderlands.

Investigations at La Playa in northwestern Sonora have also contributed to substantial rethinking about the nature of the Early Agricultural period and this project has been summarized by Carpenter and colleagues (2015). Most of the archaeological materials at La Playa date to the Early Agricultural period, including a substantial San Pedro phase occupation (1200 to 800 BC), and by the latter portion of the Cienega phase (800 BC to AD 50) the site encompassed 10 km^2, making it the largest known Early Agricultural period settlement. La Playa is located on a floodplain along a perennial section of the Río Boquillas, where the bedrock of the adjacent hills forces groundwater to the surface providing reliable moisture for irrigating the alluvial plain. Over 3,000 features have been recorded on the site and the vast majority are roasting features. Erosional processes have exposed 345 human burials and hundreds more are likely present (Carpenter et al. 2015; McLaurin et al. 2012). The vast majority are inhumations and most of the 54 dated remains belong to the Cienega phase although San Pedro phase remains are also well represented and five burials are associated with the earlier unnamed phase. An extensive irrigation system watered at least 3 km^2 during the Cienega phase (Cajigas 2016, 2018), and according to unverified suggestions, it was operational by the San Pedro phase (Copeland et al. 2012; Schott 2017). Erosional processes account for the fact that few domestic structures have been identified, although midden deposits and thousands of manos and metates attest to domestic life. La Playa was a fully sedentary occupation with substantial reliance on maize agriculture as well as wide range of wild plants. Tremendous quantities of marine shell from the Sea of Cortez, 120 km distant, were manufactured into bracelets and other items with specialized tools in shell-working areas. In addition to settlements along the Santa Cruz River and Río Casas Grandes, La Playa represents another key location where Early Agricultural period farmers dramatically modified their environment by organizing multifamily labor units, taking advantage of readily accessible subsurface moisture to farm along a reach boundary of a perennial river.

We argue in this volume that the Early Agricultural period Río Casas Grandes cerros de trincheras were constructed as defensive settlements during periods of intense raiding and warfare, and that these sites represent some of the earliest evidence for organized warfare in the southwestern United States and northwestern Mexico (Hard and Roney 2007; LeBlanc 1999). However, evidence of Early Agricultural period warfare and violence is not confined to northwest Chihuahua but is also found at La Playa and in the Tucson Basin. In the Tucson Basin the early cerro de trincheras Tumamoc Hill, with both a Cienega phase (800 BC to AD 50) and Early Ceramic period Tortolita phase occupations (AD 475 to 700) has long been argued to have served as a defensive settlement (Wallace et al. 2007; Wilcox 1979). However, Paul and Suzanne Fish and others have argued that Tumamoc Hill and other cerros de trincheras are locations of specialized ritual practices and social aggregation (Fish et al. 2011; Wallace et al. 2007). Indeed, these explanations are not mutually exclusive as aggregation is a fundamental element of defense.

Bioarchaeological data provide direct evidence of persistent intercommunity violence during both the San Pedro and Cienega phases of the Early Agricultural period from both the Tucson Basin and La Playa (Watson and Phelps 2016). Sixteen percent of 172 individuals showed evidence of violent trauma including healed cranial blunt-force trauma, perimortem cranial blunt-force trauma, and projectile point trauma (Watson and Phelps 2016:594). The combined evidence from northwestern Chihuahua, northwestern Sonora, and southern Arizona indicate that competition and conflict was endemic to the formation of Early Agricultural period floodplain communities.

The research detailed in this book, along with investigations in the Tucson Basin and at La Playa, has stimulated new discussions of the Early Agricultural period. The Río Casas Grandes sites show that within a thousand years of the arrival of maize, groups formed settlements that were dramatically different from contemporary hunter and gatherer sites in other parts of the Southwest but shared parallels with communities along the Santa Cruz River and the Río Boquillas. All are situated on stream reach boundaries with access to flood waters and subsurface flows where Early Agricultural period farmers extensively modified their environment by participating in substantial cooperative building projects and developed economies that were organized around agriculture and included substantial foraging. Some of the communities along these rivers represented large aggregations of 100 or more people, and substantial levels of intercommunity violence were a part of life.

The book describes our investigation of Cerro Juanaqueña and related sites in northwestern Chihuahua along the Río Casas Grandes. Cerro Juanaqueña and the other 13 sites discussed are cerros de trincheras. With their complexes of terraces, walls, and other rock features, they are among the most impressive prehistoric sites in the southwestern United States and northwestern Mexico. In Arizona and Sonora, investigators have recognized cerros de trincheras for decades. However, preceramic, large-scale cerros de trincheras sites were totally unanticipated, since these imposing sites have consistently yielded evidence of dating to the late ceramic periods, generally between AD 1000 and 1350 (e.g., Downum 1986; Downum et al. 1994; Fish et al. 2007b; McGuire and Villalpando 2007, 2015; O'Donovan 2002, 2004; Sauer and Brand 1931; Stacy 1974; Wilcox 1979). Early Agricultural period components are recognized at Tumamoc Hill (Fish et al. 1986; Fish et al. 2013; Wallace et al. 2007) and now at cerros de trincheras on the Upper Gila River in southeastern Arizona, including Round Mountain, the DotMon site, and the Sanchez site (Hard et al. 2016a; Hard et al. 2016b; Hard et al. 2018; Roney 1999; Roney et al. 2014, 2015a, 2015b). It is now apparent that Early Agricultural period cerros de trincheras are a broadly expressed cultural phenomenon across the borderlands.

Systematics

Archaeologists recognize time as a continuous variable, but it is often convenient to divide time into periods or phases, in order to facilitate comparison and communication, and to describe general sequences of development. In this book we do this using several different systems, depending on purpose and context. At the most general level we recognize a dichotomy between the Prehistoric and Historic periods, sometimes recognizing the Protohistoric period as a transitional interval following the initial arrival of Europeans, but before written records provide much information about indigenous peoples of the region.

Prehistoric time is often divided into the Paleo-Indian, Archaic (or Preceramic), and Ceramic periods. As our understanding of prehistory becomes more detailed, these distinctions are increasingly blurred. Initial use of ceramics is now recognized as early as 2100 cal BC in the Tucson Basin (Heidke 2006), but in northwestern Chihuahua ceramic containers did not come into general use until around AD 200. Along with the change in material culture, we take AD 200 as the division between the Archaic and Ceramic periods.

A general division of the Archaic period into an Early Archaic period (6500 BC to 3500 BC), Middle Archaic period (3500 BC to 1200 BC), and Late Archaic period (1200 BC to AD 50) is widely recognized (B. Huckell 1996a; Mabry 2008a). The difference between the Late Archaic period and the temporally overlapping Early Agricultural period is sometimes blurred and the distinction based on presence or absence of agriculture remains controversial (B. Huckell 1995, 1996a; Whittlesey 2016). We think of the Early Agricultural period as the subset of Middle and Late Archaic contexts which includes evidence of agriculture. Because most discussion in this book pertains to adaptations that clearly included agriculture, we mostly use the term "Early Agricultural period," but do occasionally reference the "Late Archaic," especially when a more general connotation is intended. As discussed below, we believe that the most recent information suggests 1500 cal BC as a better division between the Middle Archaic period and the Late Archaic period. We place the beginning of the Early Agricultural period at 2200 cal BC with the first definitive evidence of maize in the American Southwest and northwestern Mexico.

Many years ago Archaic period remains in a large portion of the southern Southwest were designated as the Cochise Culture (Sayles and Antevs 1941), contrasting with the Armagosa Culture to the west, the Oshara Tradition to the north, and the Jornada area to the east. Today, the term "Cochise Culture" is not used much, but the geographically bounded sequence of changes in material culture which it implies remains valid, especially in the latter part of the Archaic period. The research described in this book deals with a series of sites in northwestern Chihuahua which are clearly most closely related to the Cochise Culture sequence (Chapter 6).

The Cochise sequence was originally described as a series of stages: the Sulphur Springs stage, the Chiricahua stage, and the San Pedro stage (Sayles and Antevs 1941). This sequence has been replaced by the Early, Middle, and Late Archaic periods (B. Huckell 1996a). Through a series of major archaeological projects in the Tucson Basin investigators have further refined the Early Agricultural period. As currently understood, this sequence begins with the poorly known Silverbell interval, a time preceding the San Pedro phase when incipient agriculture was practiced (Whittlesey et al. 2010:79). Its beginning date is not specified, but current literature places its end around 1200 cal BC.

The interval between 1200 cal BC and 800 cal BC is the formally recognized San Pedro phase, defined on the basis of a distinctive suite of artifacts and feature configurations (Mabry 2008a:73). As will be shown in Chapter 5, many of the archaeological remains that we discuss in this book date between 1300 cal BC and 1100 cal BC. Technically, this chronological placement implies that they should be interpreted as Silverbell interval manifestations. However, in economy, scale of settlement, projectile point types, and many other details of material culture, they clearly correspond to the San Pedro phase. In the Tucson Basin itself, canal systems have recently been documented at ca. 1500 cal BC (Thiel and Mabry 2006), and aside from projectile points, assemblages from contexts dating to this time also resemble those of the San Pedro phase (J. Adams 2006; Sliva 2006). For these reasons, we suggest that 1500 cal BC should be recognized as the beginning date for the San Pedro phase.

Some of the sites discussed in this book evidence an occupation or reoccupation dating between 300 cal BC and cal AD 100. Although the sample of material which we can confidently attribute to this interval is smaller, we are comfortable discussing these remains as manifestations of the Cienega phase. Elsewhere in the southern Southwest the Cienega phase is dated between 800 cal BC and cal AD 50 (Mabry 2008a). The artifacts recovered from sites in northwestern Chihuahua with occupations in this range are generally consistent with Cienega phase assemblages, and the construction and use of cerros de trincheras firmly ties the Chihuahuan sites to similar developments in the Tucson Basin, such as the early occupation on the summit of Tumamoc Hill.

In recent years AMS radiocarbon dating, Bayesian statistical analyses, and generally improved archaeological practices have resulted in much more precise dating of archaeological assemblages from Early Agricultural times. These factors, combined with a rapidly expanding sample of dated contexts, frequently allow us to compare and discuss contexts that were actually contemporaneous, or that were separated from one another by specified amounts of time, rather than relying on broad, generalized constructs such as periods and phases. This allows more nuanced interpretations, as much of the recent literature reflects.

Organization of this Volume

In the following chapters we present the results of our study regarding age, environment, subsistence, demography, and technology, and we address some of the ecological and social conditions that promoted the formation of these settlements. Chapter 2 reviews the modern environment and integrates the project's paleoenvironmental data with the regional literature. The sites are in the Semidesert Grassland biotic community between the Plains Grasslands to the west and Chihuahuan Desert to the east (Brown and Lowe 1980). Modern rainfall here is barely sufficient for successful rainfed farming. However, the Río Casas Grandes provides a series of wide, fertile floodplains with moisture potentially available from a high water table and receding floodwaters. Regional paleoecological studies suggest that increased moisture was underway prior to the primary period of occupation.

Chapter 3 focuses on the description of Cerro Juanaqueña and the excavations we conducted there. The residential terraces cover an area of about 10 ha divided between a group of terraces on the lower toe of the hill and the larger, more heavily occupied group on the upper slopes and summit. The spatial structure of the terraces suggests an integrated, planned building effort under the direction of organized leadership. There is a 400 m long perimeter berm encircling three-fourths of the site. In order to better appreciate the degree of effort invested in these features, we constructed an experimental terrace, calculated the volume in all of the terrace features, and estimated the number of person days required for their construction. Results showed that the occupants spent about 30 person years of labor to construct the site. This expenditure of effort is roughly equivalent to that of a 550-room pueblo. In addition to the terraces, about 100 rock rings are either built on the terraces, surfaces or elevated slightly above them. Excavations revealed a number with unprepared compacted floors, a hearth inside one rock ring, and a metate resting on the floor of two others. These findings show that at least some of the rock rings are footings for brush structures used as dwellings.

Chapter 4 describes the 13 other cerros de trincheras in the region that are included in the study. We took aerial photographs, mapped, and excavated a few units at the sites of Cerro Vidal, Cerro los Torres, and Cerro el Canelo. The project also made surface observations at the other 10 cerros de trincheras. All of these cerros de trincheras are located in similar ecological settings, on high, isolated basalt hills adjacent to the floodplain. All but one are south of Cerro Juanaqueña, distributed mostly along the Río Casas Grandes. Three sites are in the next major drainage to the east and south, the Río Santa Maria. The sites vary in size and content. Cerro el Canelo, at about half the size of Cerro Juanaqueña, is the largest of these sites with a similar range of artifacts. Cerro el Canelo possesses one particularly striking feature, a 70 m diameter rock ring that we argue is associated with the Late Archaic occupation. If we are correct, it is one of the earliest examples of communal architecture in northwest Mexico or the American Southwest. We suggest that eight of the 13 sites date to the Late Archaic period, based on a combination of radiocarbon dates, artifact associations, and overall layout and construction. The other five sites may well belong to the Late Archaic period based only on site layout and construction. Further, virtually all have evidence of large-scale spatial organization involving multiple conjoined terraces and/or substantial perimeter berms that define exterior boundaries. In sum, these sites suggest that a relatively large population occupied the landscape and that their inhabitants were involved in a similar range of social and economic strategies during this period. As we later argue in Chapter 7, these populations lived on high hills to maintain a defensive advantage during a time of intensive raiding and warfare.

Chapter 5 presents the analysis of 33 AMS radiocarbon dates, most on annuals or other short-lived species. The occupation of Cerro Juanaqueña was underway no later than ca. 1350 to 1300 cal BC, and the site was abandoned by 1100 cal BC. During this period indigenous people also inhabited Cerro los Torres and Cerro el Canelo. Statistical analysis of the dates indicates that people inhabited the upper terraces of Cerro Juanaqueña for more than 30 years but less than 200 years. Cerro Vidal and two of the lower terraces of Cerro Juanaqueña were reoccupied by about 300–400 cal BC and abandoned by cal AD 1. The two discrete, synchronous uses of multiple cerros de trincheras sites, separated by eight centuries, suggests a similar response to similar conditions, most likely the return of warfare within the context of ecological conditions appropriate for farming.

Chapter 6 presents a descriptive analysis of the tools and miscellaneous artifacts including 500 projectile points and bifaces, plus numerous hammerstones and cores and other chipped stone items. The chapter describes miscellaneous stone, bone, and shell artifacts such as cruciforms, micro-eccentrics, stone pipes, spindle whorls, shaped abraders, awls, beads, and a pendant. We use ample photographs and drawings to describe these items and compare them with known objects. We suggest that at least some of these artifacts are social artifacts, used in activities which helped create and reinforce group solidarity. The few ceramics recovered from the surface are presented here as well. These clearly pertain to later, ephemeral use of the sites.

Chapter 7 argues that Cerro Juanaqueña and most of the other cerros de trincheras were constructed to serve a defensive intent. They are on the tops of high, steep, isolated, easily defended hills. The overall spatial organization of most sites includes 360-degree orientations and perimeter berms. In some cases concentric berms provide for multiple layers of defense. The inter-site visibility among many of the sites is also consistent with a defensive function. The punctuated and synchronic occupation of the cerros de trincheras sites also suggests multiple groups were responding in similar ways to similar pressures. The aggregation of such a sizable population is also most likely a product of its defensive character, particularly given its early period. A global ethnographic survey suggests that people in traditional societies rarely maintain residential settlements on elevated landforms. The primary exceptions are for defense or when there are no other places to live. Finally, a number of the general conditions that promote warfare appear to be present.

In Chapter 8, Karen R. Adams and J. Kevin Hanselka report on the ethnobotanical work. The recovery of charred botanical remains was a key goal of the fieldwork to address questions related to plant use and to recover datable samples. The botanical remains came from the examination of 198 flotation samples plus 203 separate bags of macrofossils recovered from excavation screens. Adams identified 19 separate taxa from charred seeds and other reproductive parts and 13 separate wood charcoal types. Maize was in 51% of the flotation samples Adams examined, with cheno-ams (undistinguished chenopodium or amaranth) the next most common. Both the high recovery rates of maize and the high diversity of wild species suggest that maize was prominent in the diet but wild foods were used as well.

Gayle J. Fritz, in Chapter 9, presents a detailed analysis of the charred cheno-am seeds. One of our original hypotheses was that domesticated indigenous plants may have played a prominent role in the diet. Fritz concludes that the charred amaranth seeds possess one of the key characteristics of domesticated varieties, a thin seed coat. However, the generally poor preservation of these tiny, charred seeds precluded the observation of a number of other key attributes of domestication. She concludes that amaranth was likely an important element of the diet and there is some evidence that it was domesticated or at least significantly manipulated. Further studies will be required to say conclusively if the utilized amaranth was a domesticated, nondomesticated, or transitional variety. In contrast, chenopodium was important as well but there is no evidence of its possible domestication, and apparently it was simply harvested wild.

In Chapter 10 we summarize results of limited pollen and phytolith studies as well as the stable isotope analyses of a small number of human and jackrabbit bone samples.

In Chapter 11 Kari M. Schmidt, through the analysis of thousands of bone fragments, addresses the occupants' use of animals. Hunting was a major part of the subsistence economy as indicated by the recovery of 30,600 bone fragments. Schmidt identified approximately 2,400 of these to at least the level of class. Jackrabbits and cottontails comprise 88% of the identifiable bone fragments. Deer, pronghorn antelope, and rodents made up the bulk of the remaining mammals. A few bighorn sheep bones and a single bison bone suggest limited hunting at more distant locations. The people also harvested fish. A small percentage of the identified bone recovered from screens is from small, minnow-sized fish, but fish bones are far more common in samples recovered from the heavy fraction of flotation samples, suggesting they were an important component of the diet. Chapter 11 also includes discussion of 13 small fragments of human bone recovered from Cerro Juanaqueña. These fragments were not recovered from burials but instead they are isolated fragments found during excavation. It is not clear if these remains are simply post-depositional disturbances of burials that were not located during excavations, or if they represent poorly treated human remains (sensu LeBlanc 1999).

Lee C. Nordt presents a detailed geomorphological analysis of the floodplain below Cerro Juanaqueña in Chapter 12. Nordt dug 64 backhoe trenches, described trench profiles, obtained 29 radiocarbon dates, and processed stable carbon isotope samples (see also Nordt 2003). The alluvial sequence indicates erosional events continually lowered the floodplain surfaces beginning after 15,050 cal BC and continuing until after 1250 cal BC, after which substantial vertical floodplain building began. Two of five periods of landscape stability and soil formation correlate with the two occupations of Cerro Juanaqueña. The Trincheras paleosol was under formation by 1400 cal BC and correlates with the primary occupation of the Cerro Juanaqueña. This paleosol is now buried 2 m below the modern surface. The Janos paleosol dates to 400 cal BC to cal AD 200, and its formation correlates with the second occupation of the site. The aggrading floodplain at these intervals provided an attractive niche for farming at a time when grasslands dominated the landscape. The termination of both the San Pedro and Cienega phase occupations of Cerro Juanaqueña correlate with erosional events that destroyed large segments of the arable Trincheras and Janos paleosurfaces and degraded the agricultural potential of the floodplain, likely contributing to both abandonments.

The Cerro Juanaqueña ground stone plays a significant role in our understanding of the site and is detailed in Chapter 13. The 1,600 pieces of documented ground stone, including basin metates, grinding slabs, small manos, and stone mortars or bowls, plus a few other items, allow us to address a variety of issues. Based on detailed surface recording the metates, grinding slabs, manos, and stone mortars are concentrated around the upper terraces and indicate that the terraces were the locus of substantial domestic activity. The small manos are similar to handstones or "one-hand" manos but are actually somewhat larger than typical Archaic manos. The basin metates show an unusual deeply-worn, "V"-shape elongated wear pattern resulting from years of maize grinding with reciprocal strokes. The basin metates and manos represent specialized tools used to crush and reciprocally grind large quantities of hard, flinty maize as well as wild seeds. The elongated, deep basin metates may represent a technology that is transitional between that of typical Archaic basin metates and Ceramic period trough metates. These heavily worn basin metates also provide key data that advanced our understanding of the intensity of the occupation.

Bradley J. Vierra presents his detailed analysis of the chipped stone in Chapter 14. He describes the lithic reduction that occurred at the site, the raw materials and their sources, and the tools manufactured. Vierra compares the lithic technology at Cerro Juanaqueña with that of Archaic hunter-gatherers and sedentary pueblo occupations. Typically, Archaic campsites are characterized by extensive biface reduction and pueblo sites by core reduction. Vierra analyzes over 15,000 chipped stone artifacts from excavated contexts plus a sample of cores and hammerstones from surface contexts. He finds that the people emphasized the reduction of raw material available in river gravels and outcrops near the site including rhyolite, chalcedony, chert, and basalt. The occupants reduced the coarser rhyolite and basalt cores for simple flake tools. Obsidian is the only material not locally available. The occupants reduced higher quality core materials including chalcedony, chert, and obsidian for tools, particularly projectile points and bifaces. The presence of early, middle, and late stage bifaces indicates that all stages of manufacture occurred at the site and demonstrates the relatively significant importance of this activity. Vierra compares the particular mixture of biface production and core reduction at Cerro Juanaqueña, and finds that both technologies are well represented in the assemblage. He concludes, that therefore, that Cerro Juanaqueña is not similar to either Archaic campsites or Puebloan sites. Instead, it is more analogous to other Archaic habitation sites. Based on comparisons between the chipped stone and ground stone assemblages, Vierra suggests that patterns of male–female labor organization were undergoing change.

Chapter 15 summarizes the major arguments of this study. We assert that Cerro Juanaqueña represents a large settlement characterized by low residential mobility, large population, and substantial reliance on maize, and that it was constructed rapidly for defensive purposes. The Early Agricultural period Río Casas Grandes settlements share parallels with those on the Santa Cruz River and Río Boquillas. Our overall goal in this volume is to describe primary data with enough detail to be useful to future researchers, while still placing the study within a current theoretical framework.

Notes

1. Cal indicates a date is tree-ring calibrated, that is the date is corrected for the variation in the amount of ^{14}C in the atmosphere.
2. AMS dating allows a single small sample of charred maize to be directly dated, eliminating most of the traditional problems of association by attempting to date maize via a quantity of associated charred wood or other material.
3. This date is calibrated at two standard deviations.

CHAPTER 2

The Environment of the Río Casas Grandes Region of Northwestern Chihuahua

Introduction

Our early training as archaeologists and as anthropologists taught us the importance of environment in understanding cultural change. The natural environment does not determine the course of human developments, but it does present a range of opportunities, and it can limit options and impose constraints on human behavior. It makes some outcomes more likely than others. Because important aspects of the natural environment, such as climate and vegetation, are dynamic, coupled with interactions between humans and their environment, environmental change is often implicated as an important factor driving cultural change.

Physical Setting

One element of the natural environment which usually changes very slowly in human terms is topography. Regional geomorphology and geology are discussed in detail in Chapters 12 and 14. In this more general discussion, we note that the 14 cerros de trincheras which we documented in northwestern Chihuahua are all located in a transition zone between the Mexican Basin and Range physiographic province to the east and the Sierra Madre Occidental to the west (Figure 2.1). East of these Early Agricultural settlements, topography consists of numerous small mountain ranges aligned north–northwest by south–southeast, typically less than 20 km long and 10 km wide, with peaks that are usually 2000 m or less above mean sea level. These mountain ranges are separated by vast interconnected basins lying at 1200 to 1500 m in elevation. Seasonal playas are found on the floors of many of these basins. Collectively about two-thirds of the Basin and Range province consists of basins, while one-third is mountainous (Brand 1937; Hawley 1969). The small mountains of the basin and range country near our project area are important partly because they contribute to environmental diversity, but also because they often consist of rhyolite and basalt bedrock with detrital slopes and alluvial fans that are sources of nodular chert, chalcedony, obsidian, rhyolite, and basalt used for the manufacture of chipped and ground stone tools.

To the west is the Sierra Madre Occidental, part of the great spine of the North American continent. In the vicinity of our project area these mountains consist of ash-flow tuffs and rhyolite. They rise from the Basin and Range province in a series of north–south trending ridges, each higher than the last, eventually culminating in peaks as high as 2600 m (Swanson et al. 2006; Swanson and McDowell 1984).

All of the cerros de trincheras we investigated in northwestern Chihuahua are associated with floodplains along rivers which originate in the Sierra Madre and terminate in the seasonal playas of the basin and range country. Three main rivers drain the northeastern slopes of the Sierra Madre. From east to west these are the Río Santa Clara/Río del Carmen, the Río Santa María, and the Río Casas Grandes. The Río Casas Grandes is by far the largest of the three and is one of the major rivers of Northwest Mexico and the American Southwest. Its watershed has an area of 16,600 km^2, making it similar in scale to Arizona's Verde River and Salt River (Schmidt 1973:21–22; Whalen and Minnis 2001). The Río Santa María drains a watershed of about 11,000 km^2, and the Río del Carmen is smaller still. For comparison, the Mimbres River, which empties into the same series of interconnected basins, has a watershed about 13,000 km^2 in size. Despite its geographic proximity, the Mimbres River is fundamentally different from the Chihuahuan rivers because it is fed by winter storms, as well as by summer monsoons. In contrast, the Río Casas Grandes, Río Santa María, and Río del Carmen derive their peak flows primarily from midsummer monsoonal precipitation.

Ten of the cerros de trincheras discussed here are located along the main course of the Río Casas Grandes. One is on a major tributary of the Río Casas Grandes and three others are located along the Río Santa María.

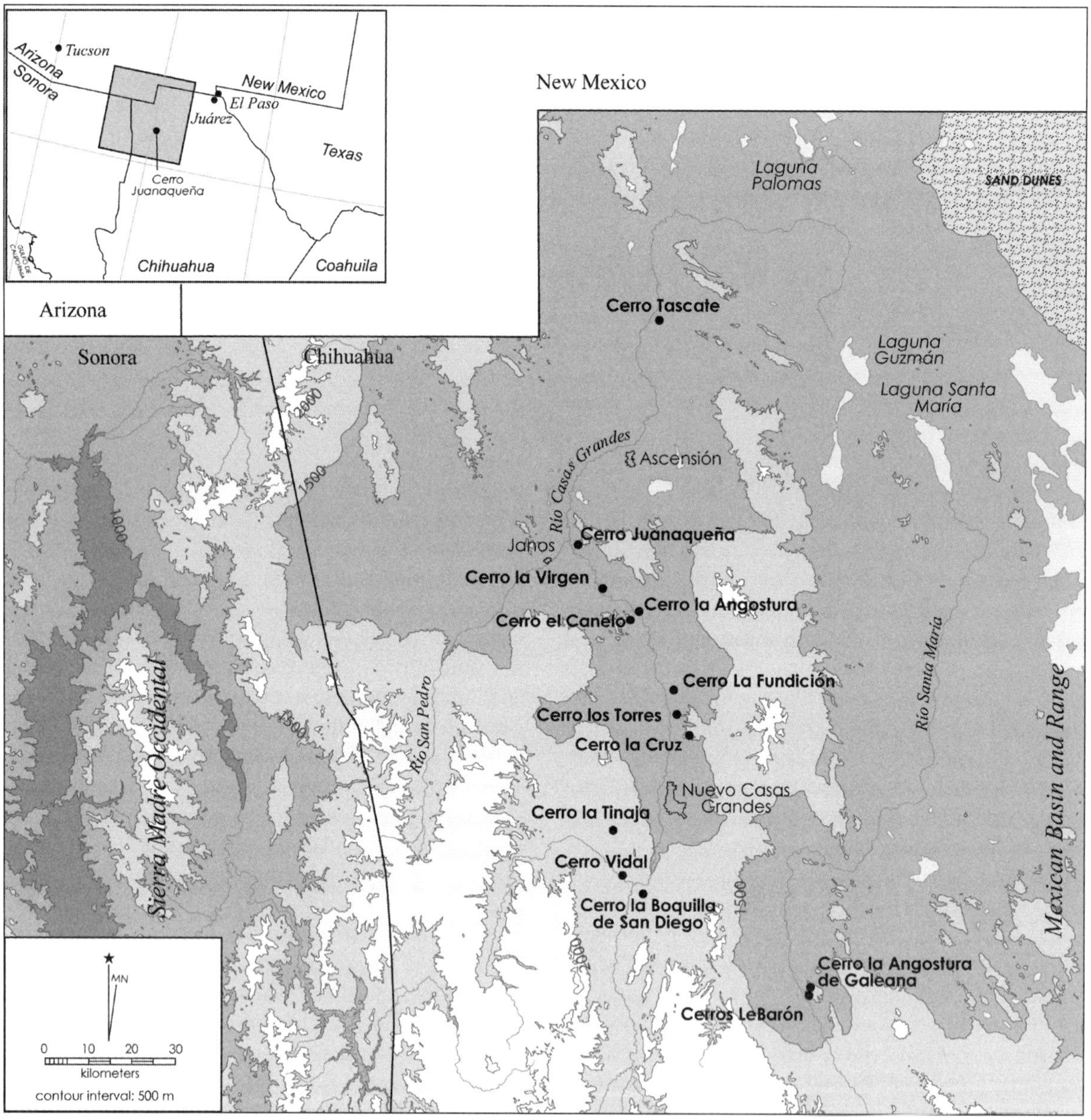

FIGURE 2.1. Archaeological sites and major landscape features in northwestern Chihuahua. Elevation contours indicated by dark gray 1000–1500 m, light gray 1500–2000 m, white 2000–2500 m.

All 14 are associated with broad, fertile floodplains. More specifically, all of these sites are associated with constrictions of the floodplain which form river reach boundaries, and they have particular geomorphic properties relevant to ancient Southwestern farmers (Gregory and Nials 2005; Nials et al. 2011). Reach boundaries occur where bedrock constrains the course of the river or where tributaries meet the main channel. These constrictions slow water velocity, enhance alluvial deposition, and force ground water to the surface. In this way they help form productive farmland and increase resilience to agricultural shortfalls in the face of fluctuating climate. All of the cerros de trincheras in northwestern Chihuahua are associated with reach boundaries. We argue that this observation indicates that Early Agricultural period people were establishing defensive settlements in key agricultural settings that have the best potential to enhance productivity, reduce risk, and increase resilience to both aridity and social competition.

Climate

Climate is clearly another important environmental variable for both foraging and farming peoples. Table 2.1 summarizes nearby Janos, Chihuahua weather data (Servicio Meteorológico y Geográfico 1997). The July average temperature is hot, 76.6°F, and winters are mild with an average January temperature of 44.6°F. The available data from the Janos weather station recorded a frost-free period of 219 and 230 days for 2000 and 2001 respectively. Based on temperature sensors we placed on and near Cerro Juanaqueña from 1999 to 2001, the average frost-free period is 177 days in the Río Casas Grandes floodplain and increases to 258 days on the upper terraces, above the cold air that settles in the floodplain (Sidebar 2.1). Average annual Janos precipitation is 334 mm ($s = 91$ mm, $n = 32$, 1961 to 1993), of which two-thirds falls between July and October and less than one-fourth falls in the winter between November and February. The spring season is dry, with less than one-tenth of the annual total falling between March and

TABLE 2.1. Precipitation and Temperature, Janos, Chihuahua, 1961–1993.

	Jan	Feb	Mar	Apr	May	Jun	Jul	Aug	Sep	Oct	Nov	Dec	Annual
Mean Precipitation (mm)	19.5	12.4	9.5	6.8	6.6	16.8	84.2	65.3	40.6	27.5	16.2	28.1	334.0
Standard Deviation (mm)	19.7	11.1	10.1	10.5	10.1	23.4	36.2	41.4	26.0	27.7	18.0	28.8	90.5
Minimum Precipitation (mm)	0.0	0.0	0.0	0.0	0.0	0.0	35.0	4.0	0.1	0.0	0.0	0.0	151.1
Maximum Precipitation (mm)	89.0	45.0	36.0	41.5	38.0	80.0	191.0	189.5	88.0	119.0	65.0	129.8	513.5
Mean Temperature (F)	44.6	47.7	52.0	57.6	67.5	76.1	76.6	75.2	67.1	58.6	49.6	44.8	59.8

SIDEBAR 2.1. TEMPERATURE SENSORS

In 1998, we wanted to know if temperatures and growing seasons on the slope of Cerro Juanaqueña were greater than in the floodplain (Fish et al. 1984). In order to explore this issue, we installed eight electronic thermometers (Hobo Model H08-001-02) in a rough east–west transect from the Río Casas Grandes floodplain, up the western slope of Cerro Juanaqueña to its summit. The thermometers were sealed in clear weatherproof cylinders, and each one was wired to a shrub about two feet above the ground in broken shade. These instruments collected temperatures at four-hour intervals between 1998 and 2002.

The data they collected show that maximum temperatures on the lower terraces are systematically higher than temperatures on either the floodplain or the upper terraces. Minimum temperatures in the floodplain are 10° to 20°F lower than minimum temperatures on the terraces, demonstrating a dramatic difference in growing season between these two settings.

The thermal cells recorded three complete growing seasons of temperature data, 1999–2001. The growing season is defined as continuous days without frost (> 32°F). In the floodplain, the last frost occurred between April 21 and May 3, and the first frost occurred between October 12 and October 28. The average frost-free period was 177 days in the floodplain but increased steadily further upslope. On the lower terraces, at Rock Ring 353, it was 211 days. At the lower edge of the upper terrace group (Terrace 183) it was 245 days; and at the middle of the upper terraces (Terrace 215) it was 258 days.

The dramatic lengthening of the frost-free period, progressing from 177 to 258 days up the west slope, is clearly due to the cold-air drainage effect. In comparison, we obtained the available data from the Janos weather station, which is located in the town of Janos at a residence situated above the Río San Pedro floodplain. There, frost-free periods of 219 and 230 days are reported for 2000 and 2001 respectively. These values are comparable to the frost-free period measured by the temperature cells on the lower terraces of Cerro Juanaqueña of 220 and 218 days for 2000 and 2001 respectively.

The results of this temperature study potentially supported our initial hypothesis that the constructed terraces on Cerro Juanaqueña were built as farming plots which served to extend the growing season (Fish et al. 1984). The small size of the terraces, their poor water-holding capacity, high labor investment, and the fact that length of growing season is not an important constraint on farming in this environment allowed us to soundly reject this notion as is discussed in Chapter 3.

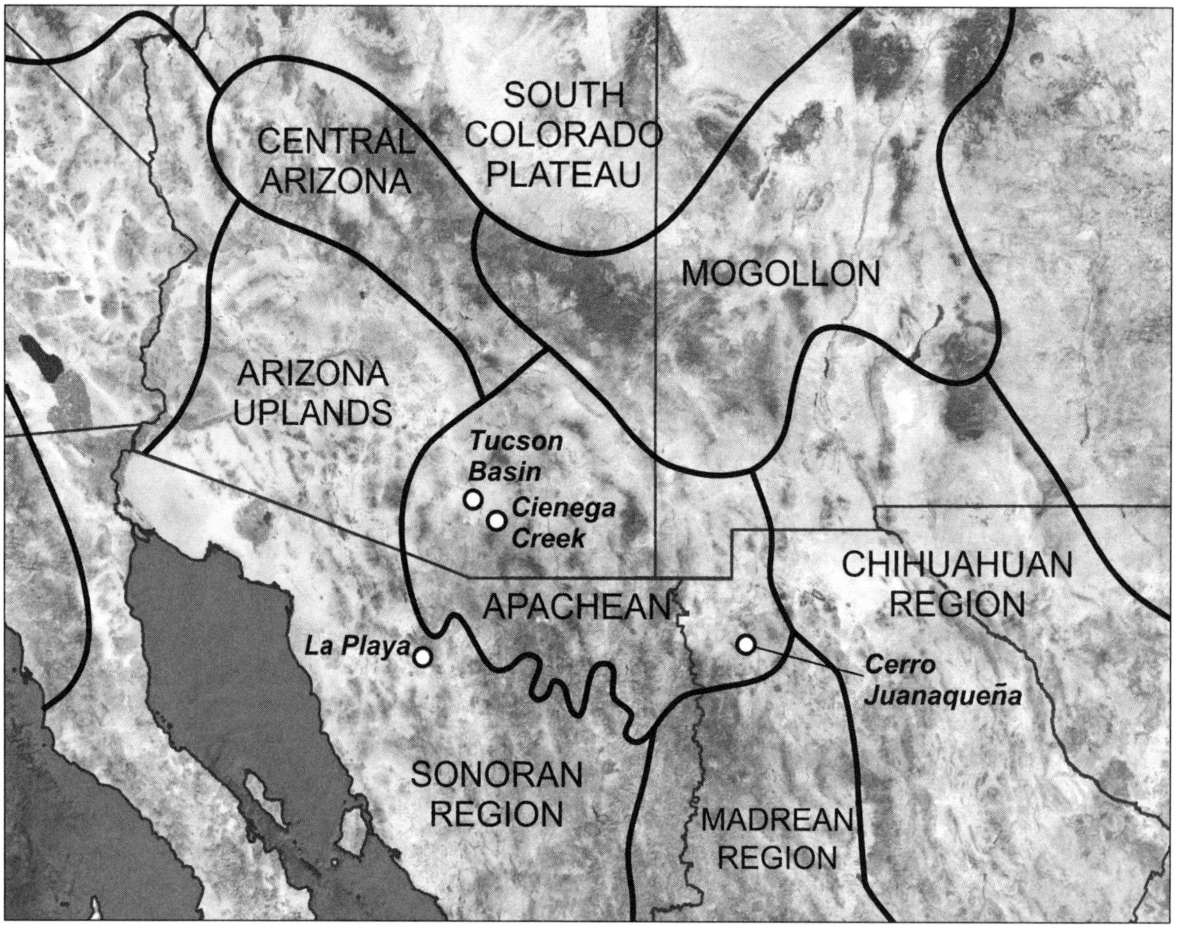

FIGURE 2.2. Floristic regions mapped on the basis of degree of similarity of flora by Bowers and McLaughlin as cited in text and four Early Agricultural period locations in the International Four Corners.

May. However, rainfall in the upper watershed of the Río Casas Grandes, above ca. 2000 m, averages over 400 mm per year (García et al. 1973), supporting its perennial river flow.

Critical summer rains fall in July as part of the North American Monsoon System. Moisture from the northwestern tropical Pacific enters Chihuahua from the southwest bringing moisture from a number of ocean sources including the Gulf of California and Gulf of Mexico (Barron et al. 2012; Cavazos et al. 2002; Douglas et al. 1993). Movement of the Pacific and Gulf of Mexico air masses induces high year-to-year variability in the summer monsoons.

Apachean Floristic Area

Floristic communities in the Southwestern United States and Northwestern Mexico have been mapped at a regional scale and provide frameworks to consider past human environmental interactions. University of Arizona ecologists mapped floristic regions based on statistical similarity of flora using data from at least 100 different study locations (Bowers and McLaughlin 1982, 1987, 1996; McLaughlin 1986, 1989, 1992, 1994, 2007). They defined the Apachean floristic area (Figure 2.2) that incorporates northeastern Sonora, northwestern Chihuahua, southeastern Arizona, southwestern New Mexico, and extends along the Mogollon Rim (Bowers and McLaughlin 1987; McLaughlin 1986; McLaughlin and Bowers 1982). It includes the Río Casas Grandes area as well as the Tucson Basin and the Cienega Valley (B. Huckell 1995) of southeastern Arizona, all locations where Early Agricultural period occupants made surprisingly early substantial investments in farming, although La Playa lies just outside the mapped boundaries of the Apachean floristic area.

The Apachean floristic area is an area of mild temperatures, high summer rainfall, and variable spring rainfall. These climatic factors stimulate growth of annual and perennial herbs and result in unusually high species diversity. Topographic factors, including elevation gradients associated with mountain ranges and the presence of riparian zones and other mesic habitats, also contribute to significant species diversity. Within the Apachean floristic area, desert grasslands and mid-elevation woodlands are particularly high in species diversity (Bowers and McLaughlin 1982; McLaughlin 1992), a point that runs counter to modern impressions of overgrazed desert grasslands. Comprehending the dynamic interactions between the ecological characteristics of the Apachean floristic area and humans will be important to understanding why foragers made such early investments in agriculture in this region. For example, grasslands, with their dense ground cover, retain soil, thereby reducing sediment discharged into rivers. Low discharge fosters stable stream morphologies that form deep alluvial terraces and productive agricultural niches.

Modern Biotic Communities

Brown (ed. 1982) and his contributing authors based their descriptions of biotic communities in the Southwestern United States and Northwestern Mexico on a nested hierarchy consisting of hydrologic regimes (upland and wetland), plant formations (e.g. desert, grassland, woodland, and forest), climatic zones (e.g. cold temperature and warm temperature), and broad geographically defined biotic provinces (e.g. Chihuahuan, Madrean, and Great Basin). The resulting system of classification includes 27 biotic communities which they mapped across a large area of the Southwestern United States and Northwestern Mexico (Brown and Lowe 1980). Boundaries among these communities are neither sharp nor static, but grade one into another in dynamic ways subject to land use practices and paleoclimate. Over much of its length the Río Casas Grandes flows through the Semidesert Grassland biotic community, but there are six biotic communities within 60 km to the east and west of this river, underscoring the wide range of potentially useful plants and animals which may have been available to inhabitants of the hill-top settlements. These biotic communities are summarized below using descriptions provided by Brown and Lowe, supplemented by other sources (Brand 1937; Lesueur 1945; Rzedowski and Huerta 1981; SPP 1980, 1981a, 1981b, 1985). All six of these biotic communities fall within the Apachean floristic region discussed above.

Chihuahuan Desertscrub

This biotic community dominates lowland areas of the vast basin and range province east of the Río Casas Grandes. Below elevations of about 1250 m, this community is more or less continuous, while between 1250 and 1500 m in elevation it forms a complex mosaic with the Semidesert Grassland biotic community described next. The pattern of Chihuahuan Desertscrub vegetation is conditioned by elevation, soils, moisture, drainage features, and historic land use. Desertscrub species prefer lower elevations, lower rainfall, saline soils, well-drained soils, alluvial fans, rocky slopes, and overgrazed areas (Brown 1982a). In contrast, Semidesert Grassland species prefer higher moisture, deeper soils, floodplains, areas adjacent to playas, and increasing elevations found on mountain slopes.

The principal shrubs of the Chihuahuan Desertscrub biotic community include saltbush, tarbush, mesquite, acacia, creosote, ocotillo, and lippia. Cacti and succulents are not widespread, but in some locations are common, and these include cholla, prickly pear, agave, sotol, and yucca. Following the rainy season numerous forbs can germinate rapidly in the unvegetated spaces between shrubs on alluvial fans and playa edges. Bunch grasses and drop-seed grasses are found on dunes, basin floors, playa margins, and other locations where soils are deep and moisture retention is improved (Brown 1982a).

Economic plants include diverse shrubs, succulents, forbs, and food plants. Among these, succulents, particularly the leaf succulents such as agave and yucca can be eaten in early spring. Most annuals grow rapidly following summer rains and so are available in mid-to-late summer, while some shrubs such as mesquite produce substantial crops in late summer. Mule deer, jackrabbits, cottontail rabbits, and a variety of rodents could be hunted in this community. Additional information about plant resources is provided in Chapter 8, while faunal resources are detailed in Chapter 11.

Semidesert Grassland

The Semidesert Grassland biotic community is transitional between the Chihuahuan Desertscrub biotic community to the east and the Plains Grassland biotic community to the west. It is typically found between about 1250 and 1500 m in elevation, and it is dominant in the low mountain ranges of the basin and range provinces, as well as a broad region along the base of the Sierra Madre. The more detailed vegetation maps prepared by the Mexican federal government (SPP 1980, 1981a, 1981b, 1985) accurately reflect an even more complex mosaic of grasses and shrubs within this biotic community. It is difficult to label any particular location as "grassland" or "scrubland" because the modern landscape is a fluctuating continuum. Year-to-year variability in precipitation and overgrazing exacerbate conditions where shrubs, half-shrubs, cacti, and forbs outnumber and even replace grasses. Under low summer rainfall conditions, annuals increase at the expense of perennial grasses (Brown 1982d:124). Under improved conditions perennial bunch grasses are interspersed with bare ground and shrubs are less common (Brown 1982d). Indeed, in some locations, adequate rainfall, minimal erosion, deep soils, and minimal shrub competition have allowed perennial bunch grasses to cover large expanses leaving few shrubs and annuals. Grasses are also predominant at higher elevations, on mountain slopes, in and around playas, and on floodplains. Given the variability in lifeforms, species diversity in semidesert grasslands tends to be high. A detailed, comparative, quantitative study of numerous plant inventories in Arizona demonstrates that semidesert grasslands, as well as Madrean woodlands have significantly higher species diversity than many other vegetation communities, including Sonoran Desert communities (Bowers and McLaughlin 1982).

Some of the major species found in the Semidesert Grassland include tobosa grass, black grama, hairy tridens, curly mesquite, and fluffgrass, among others. Succulents including sotol, agaves, and yuccas are conspicuous in the area. Shrubs include mesquite, one-seed juniper, allthorn, Mormon tea, Wright's lippia, catclaw acacia, and ocotillo. Typical invader shrubs from the desert scrub include tarbush, whitethorn, and creosote bush, as well as mesquite (Brown 1982d). Barrel cactus, cholla, and hedgehog cactus are important as well. Oaks and willow can be found along drainages. Numerous forbs, especially annuals, grow quickly in response to moisture. Mammals found in the Semidesert Grassland are not specific to this biotic community, but tend to be found in the Chihuahuan Desertscrub, as well as Plains Grasslands, including jackrabbit, cottontail, pronghorn antelope, mule deer, and an array of rodent species (Chapter 11).

Plains Grassland

Above the Semidesert Grasslands but below the woodlands of the Sierra Madre is a 20 to 40 km wide strip of the Plains variant of the Plains and Great Basin Grassland biotic community. This large, contiguous area of Plains Grassland is far removed from its home biotic province in eastern New Mexico and western Texas due to long-term environmental changes and plant migrations (Brown 1982c). Below 1400 to 1500 m, the eastern border of the Plains Grassland biotic community intermingles with the Semidesert Grassland, while above 1700 to 2000 m it merges into the Madrean Evergreen Woodland (Brown 1982c; Brown and Lowe 1980). Grazing has resulted in some shrub invasion, but today this zone presents relatively uncluttered grass-dominated landscapes. Generally grazing, fire suppression, and perhaps declines in rainfall have resulted in shrub invasion and the expansion of Semidesert Grassland at the expense of Plains Grasslands (Brown 1982c; Dick-Peddie 1993; Heerwagen 1956; Van Auken 2000; York and Dick-Peddie 1969).

The grasses of the Plains Grassland biotic province tend to be short-grass communities dominated by grama, including black grama, hairy grama, sideoats grama, slender grama, blue grama, tanglehead, and three-awn. Shrubs and trees include mesquite, one-seed juniper, oaks, and ocotillo. Among the succulents are cholla, bear grass, and agave (SPP 1985).

This biotic community offers food resources from trees and shrubs, especially oak and mesquite which grow along drainages, as well as succulents and some forbs. Important mammals include pronghorn antelope, mule deer, jackrabbit, and cottontail. Today some of the largest remaining prairie dog communities in North America are found in this zone and are the focus of important conservation efforts (Ceballos González et al. 2010; Dinerstein et al. 2000; Hruska et al. 2017; List et al. 2010).

Madrean Evergreen Woodland

This biotic community dominates the foothills and lower slopes of the Sierra Madre at elevations from 1700 to 2000 m, up to ca. 2300 m, and is generally intermediate between the Plains Grasslands and the Madrean Conifer Forest. Like the Semidesert Grassland biotic community, the Madrean Evergreen Woodland is particularly rich in species relative to many other biotic communities, in part due to its forb understory that proliferates following spring and summer rains (Bowers and McLaughlin 1982).

Oak-dominated communities are widespread (Brown 1982b). The major species are oaks and pines, including Emory oak, Arizona white oak, Chihuahua oak, silverleaf oak, Arizona pine, and one-seed juniper. These trees often form an open canopy that allows a well-developed grass and forb understory. Succulents include agaves, yuccas, bear grass, barrel cactus, and cholla.

More than any other community, the Madrean Evergreen Woodland offers substantial tree resources, particularly pinyon pine nuts and acorns from the many oaks. Shrubs, forbs, and succulents remain important as well. Vegetation maps show a number of small rain-fed (temporal) agricultural fields in this biotic community, indicating that moisture is sufficient at this elevation for farming (SPP 1985). Important mammals include mule deer and white-tail deer.

Madrean Montane Conifer Forest

Beginning at elevations of 2000 to 2400 m, the Evergreen Woodland grades into the Madrean Montane Conifer Forest biotic community depending upon moisture, slope, exposure, and soils. Because the highest elevations in the region are only around 2500 to 2600 m, this biotic community occupies only small isolated areas on the uppermost ridges and summits, although its extent is more expansive further south in the Sierra Madre (Brown and Pase 1982).

The major tree species include Chihuahua pine,

pinyon pine, Arizona pine, gray oak, silverleaf oak, and Arizona white oak. Spruce and fir generally occur above 2,600 m, but probably exist in this biotic community in particularly moist settings. A grass understory is common and includes muhly and love grass.

A shrub understory is not especially well developed in the Madrean Montane Conifer Forest, but it does include sumac, a potential economic resource. Pinyon pine, an important source of nuts, occurs in this biotic community, but its density and distribution are unknown. Small rain-fed (temporal) agricultural fields are also present in this community (SPP 1985).

Wetlands

Probably more than any other biotic community, wetlands have been severely altered in historic times, due to the combined effects of water diversion, groundwater pumping, grazing and other human uses, as well as recurrent drought. Regional vegetation maps do not systematically map mesic communities, but these were undoubtedly critical to human adaptations, especially at lower elevations. Riverine floodplains supported wetlands along with backwater meanders, cut-off channels, and active channel margins. Mid-nineteenth century historical accounts describe a string of cottonwood trees along the course of the Río Casas Grandes (Bartlett 1854 [1965]). Donald Brand (1937), working primarily in the early 1930s, remarked upon the numerous springs present along the Río Casas Grandes and noted that many important spring-fed wetlands were present in the lowlands of the basin and range country.

Plants collected by paleoethnobotanist Karen Adams at a warm spring marsh area known as La Palotada located about 10 km west of Janos, Chihuahua are listed in Table 8.1. Such mesic habitats greatly enhance species richness and provide important resources such as bulrush.

Paleoecology

The modern environment described above is an essential starting point for thinking about early maize agriculture and the lifeways of people who occupied early hilltop villages in northwestern Chihuahua. However, the biotic communities described above, and to some extent even some key aspects of topography and geomorphology along the main river courses, are the products of a long and complex sequence of climatic change reaching back to the last ice age. In this section, we summarize some recent advances in this sequence with particular reference to northwestern Chihuahua. The temporal sequence used below is a modification of Van Devender's chronology (1990). Uncalibrated dates and chronologies from the literature are calibrated (Bronk Ramsey 2017; Reimer et al. 2013) and the uncalibrated age range is included for each temporal unit to facilitate comparisons.

Late Wisconsin, ca. 26,000 to 11,500 cal BP

Modern environmental conditions have not always prevailed, and it is important to understand how temperature, precipitation, and patterns of climatic variability have changed through time. During the earth's latest ice age, the Late Wisconsin (26,000 to 11,500 cal BP or 22,000 to 10,000 uncalibrated radiocarbon years BP), the overall climate was cooler than today and the precipitation regime was much different. Ice sheets covered much of Canada and the northern tier of the United States. Pleistocene conditions suppressed the summer tropical monsoonal pattern so the North American Monsoon, which brings summer precipitation, had not yet developed. The average latitude of westerly storm tracks was much further south than today so winter storms brought rain inland with greater frequency. Cooler and drier summers and milder and wetter winters made for a more equable climate than modern conditions (Metcalfe 1997).

A modern visitor to the region during the Late Wisconsin period would recognize most of the plants and animals but would find them significantly displaced from their modern habitats. Lowlands that today support only cactus and desert shrubs were covered with grasslands, while pine, juniper, and oak woodlands extended lower down mountain slopes and much further south than today. By 12,850 cal BP, oak-juniper woodlands had largely replaced once widespread pinyon-dominated communities (Holmgren et al. 2003). One of the most dramatic differences, in comparison to modern conditions, was the presence of deep and extensive lakes in many of the basins of the basin and range country. Today the basins are mostly dry playas, but during the Late Wisconsin period the combination of higher winter moisture and lower evaporation rates maintained natural lakes that were astonishingly large, deep, and cold. For example, Lake Palomas (Figure 2.3) reached a depth of 25 m, connected four major sub-basins in northwestern Chihuahua, and covered 7,000 km^2 (Castiglia 2002; Castiglia and Fawcett 2006). By the end of this interval, marked by the Younger Dryas cooling interval (12,850 to 11,650 cal BP), most of the large fauna including mammoths, camels, dire wolves, and saber-tooth tigers were extinct.

Early Holocene, 11,500 to 9500 cal BP

During the early Holocene (11,500 to 9500 cal BP or 10,000 to 8500 uncalibrated radiocarbon years BP), global temperatures began to rise and the glaciers covering Canada and the northern tier of the United States began to retreat. Winter precipitation, although greater

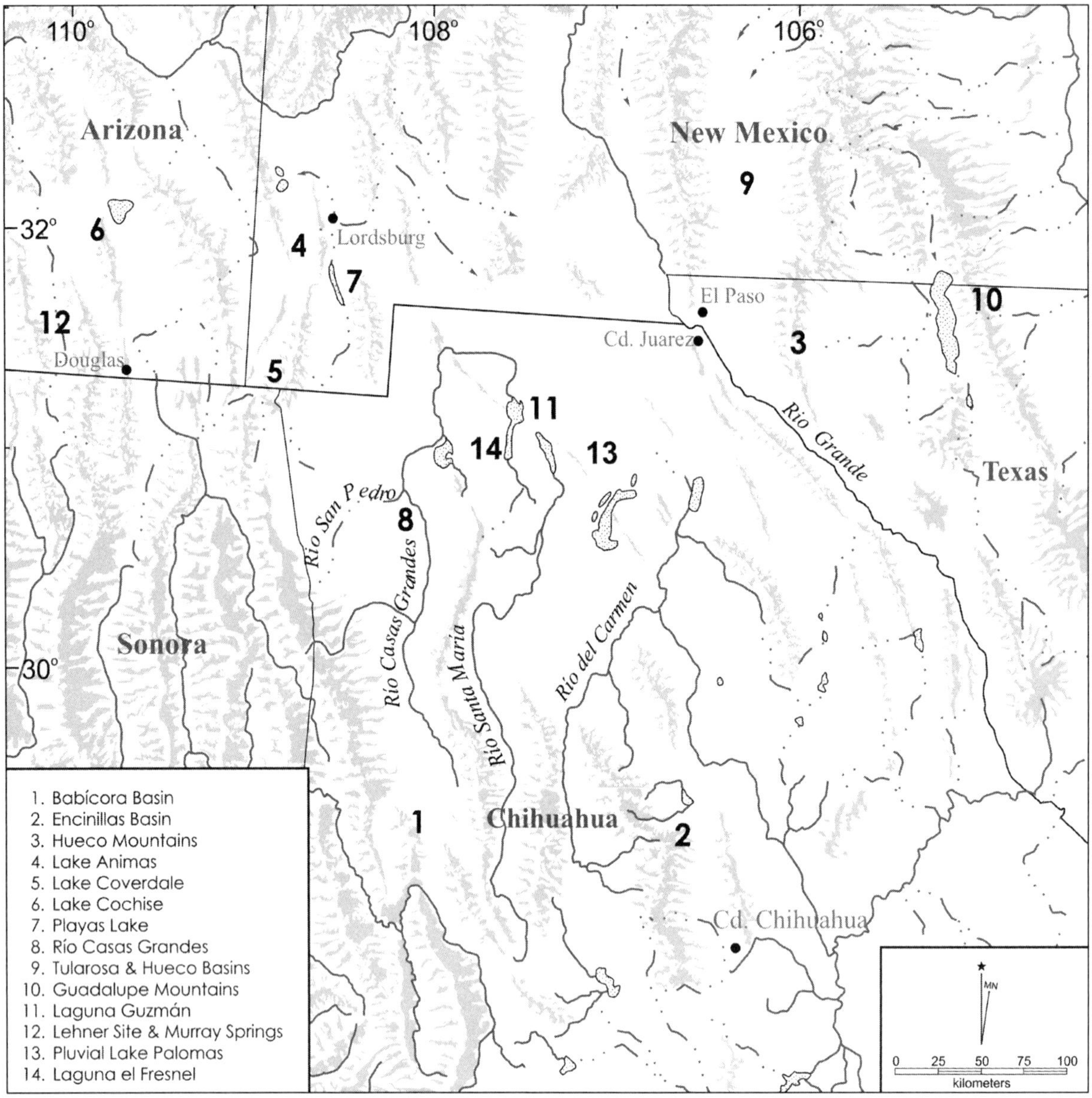

FIGURE 2.3. Paleoenvironmental proxy locations discussed in this chapter.

than today, declined as winter storm tracks shifted northwards. Toward the end of this period, ca. 9000 cal BP, there was a return to more mesic conditions related to increased El Niño activity (Conroy et al. 2008). Grasslands dominated many basins, while mountain slopes supported oak-juniper woodlands. In other settings, such as the Babícora Basin on the eastern slopes of the Sierra Madre, grasslands slowly replaced woodlands. Reduced precipitation, warmer temperatures, and higher evaporation rates dried or shallowed pluvial lakes.

Some detailed proxy data from regional paleoenvironmental studies support these patterns. The playa Lake Cochise experienced a refilling event at 10,260 to 9270 cal BP (Waters 1989) and Laguna El Fresnal had a high stand at 9550 to 9130 cal BP (Castiglia and Fawcett 2006). Speleothem data from Pink Panther Cave in the Guadalupe Mountains also reflected a wet period from 9950 to 9400 cal BP (Asmerom et al. 2007).

Middle Holocene (early), 9500 to 7500 cal BP

By the early part of the middle Holocene (9500 to 7500 cal BP or 8500 to 6600 uncalibrated radiocarbon years BP), average winter storm tracks approximated their modern positions and winter precipitation was sharply lower. Winter precipitation declined also due to diminished El Niño–Southern Oscillation (ENSO) conditions in the eastern Pacific (Conroy et al. 2008). At the same time, atmospheric conditions that promote summer monsoons had yet to develop, so summer precipitation remained low. In short, during the early part of the middle Holocene, the United States–Mexico borderlands experienced an unusually long and pronounced time of reduced precipitation.

Environmentally, this interval had no documented lacustrine high stands in northwestern Chihuahua and adjacent portions of the United States. The Guadalupe Mountains speleothem data support this model, reflecting a long period of aridity from 9400 to 7500 cal BP (Asmerom et al. 2007). Consistent with these observations, the Estancia Basin dried at 7840 cal BP (Menking and Anderson 2003). A global cooling event is also reflected in the speleothem data at 8000 cal BP (Alley et al. 1997; Asmerom et al. 2007). At higher elevations grasses continued to replace oak-juniper woodlands (Van Devender 1990), while at lower elevations shrubby vegetation with C_3 photosynthetic pathways replaced C_4 grasslands (Buck and Monger 1999). In the Río Casas Grandes valley proxy records indicate a warming and drying trend was underway (Nordt 2003).

Middle Holocene (late), 7500 to 4500 cal BP

Recent data indicate that the latter part of the middle Holocene (7500 to 4500 cal BP or 6600 to 4000 uncalibrated radiocarbon years BP) marked a regime shift, as the aridity of the early middle Holocene ended with the establishment of the North American Monsoon pattern. For some extended periods of time the monsoon may have actually been 50% to 100% stronger than today bringing moisture from both the Gulf of California and the Gulf of Mexico into the Southwest (Barron et al. 2012; Huckleberry 2015; McAuliffe and Van Devender 1998). However, relatively weak ENSO events continued to provide low levels of winter precipitation (Conroy et al. 2008). Following 7500/7000 cal BP, sea-surface temperatures altered, narrowing the geographic extent of the monsoonal patterns, bringing them close to their modern distribution, and ending the early middle Holocene dry period (Barron et al. 2012; Poore et al. 2003; Poore et al. 2005). The monsoon, ENSO, and other atmospheric processes interacted with one another to produce complex cycles of dry and mesic conditions. Modern associations of plant species and their modern distribution patterns were established by 5000 cal BP (McAuliffe and Van Devender 1998).

Details of proxy data support the view that the late middle Holocene in the southern Southwest was a period of increasing moisture, although there were periods of aridity. Gulf of Mexico sediment cores indicate the North American Monsoon peaked between 7000 and 4800 cal BP and diminished following 4500 cal BP (Poore et al. 2003; Poore et al. 2005). North American Monsoon contributions from the Gulf of California had a somewhat similar, but more variable temporal trajectory (Barron et al. 2012). The Guadalupe Mountains Pink Panther speleothem data reflected increased moisture between 7300 and 4600 cal BP, but included periods of aridity at 7000 to 6500, 5900 to 5800, and 5300 to 5200 cal BP (Asmerom et al. 2007).

Proxy records from ancient lakes indicate variations in effective moisture with multiple refilling and drying events. Laguna El Fresnal filled between 7650 and 6900 cal BP and again between 4900 and 4100 cal BP (Castiglia 2002; Castiglia and Fawcett 2006). Lake Cochise refilled at 6600 to 5900 cal BP and a second time after 5800 cal BP. The latter high stand correlated with increased moisture seen in the Murray Springs pollen record between 5730 and 4490 cal BP (Mehringer et al. 1967; Waters 1989). Lake Cloverdale also filled, but the timing is not well constrained, dating between 5700 and 2000 cal BP (Krider 1998). Lake Playas contained perennial water until 4500 cal BP and grasslands thrived in the Playas Valley (Van Devender 1995; Van Devender and Worthington 1977). In central New Mexico Lake Estancia had refilled after 7400 cal BP but experienced a second drying event at 6250 cal BP (Menking and Anderson 2003).

Packrat midden data from the eastern margins of the Sonoran Desert in Arizona also offered data supporting a late middle Holocene North American monsoon 50% to 100% greater than modern. Blue palo verde (*Cercidium floridium*) was present in a series of middens dating 7320 to 4490 cal BP but was absent from earlier and later middens and is absent on nearby modern slopes (McAuliffe and Van Devender 1998). Other species responsive to summer moisture, including bristlegrass (*Setaria leucophila*), showed similar late middle Holocene distributions. Holmgren and her colleagues (2007) also used packrat midden data to argue for greater than modern summer moisture and temperature, and cold, dry winters during the middle Holocene (8000 to 4000 cal BP) in the northern Chihuahuan Desert.

Harden and her colleagues (2010) accumulated radiocarbon dates from valley flood events across the Southwest, examining both bedrock and alluvial valley settings (cf. Ely 1997). Across the Holocene, alluvial flood chronologies tended not to correlate with bedrock flood events or with the Guadalupe Mountains speleothem data because bedrock flood events may destroy alluvial flood records (Harden et al. 2010). However, these investigators did find a pronounced peak in the radiocarbon distribution from bedrock flood events from 6730 to 4930 cal BP, and this peak correlated with the record of increased moisture (6500 to 4500 cal BP) in the Guadalupe Mountains speleothem data (Asmerom et al. 2007).

There was contradictory evidence for some of these patterns. For instance, Huckleberry (2015) reviewed evidence for a dry, rather than a wet middle Holocene dated to between 8000 and 4000 cal BP. The Babícora Basin was portrayed as reaching peak aridity at 6840 cal BP (Ortega-Ramírez et al. 1998), and pollen data from Animas Creek in southwestern New Mexico suggested maximum aridity at 5730 cal BP (Davis 1994:44). Indeed our reconstruction appears inconsistent with the warm and dry middle Holocene Altithermal proposed for the Great Basin and applied to the Southwest (Antevs 1948). However much of the evidence for a dry middle Holocene is from the northerly portions of the Southwest which may have had diminished winter and spring moisture. Other evidence is from prior to 7500/7000 cal BP and falls in our dry early middle Holocene (Huckleberry 2015).[1]

The bulk of speleothem, lacustrine, flood chronologies, packrat midden data, ocean cores, and climate models suggested at the millennial scale the existence of a dry early middle Holocene (9400 to 7500 cal BP) that contrasted with a wet late middle Holocene (7500 to 4500 cal BP), punctuated by dry periods, as the summer monsoonal pattern emerged. These changes ended a long period of aridity and established modern vegetation patterns. However, there is much we do not understand about this critical period and more well-dated proxies that allow fine grained comparisons are needed.

The late middle Holocene set the stage for the northward diffusion of maize from tropical Mexico into northwestern Mexico and the southwestern United States. Maize was initially domesticated in Mesoamerica between 8700 and 6250 cal BP (Piperno and Flannery 2001; Piperno et al. 2009; Smith 2005). Maize occurred in multiple settings in the Southwest by 4200 cal BP and was therefore likely present even earlier (Merrill et al. 2009). As indicated in Chapter 1, direct AMS dates on maize found out of stratigraphic context raised the possibility of maize being in southern Arizona as early as 5600 cal BP (Vint 2015a). The development of the North American Monsoon and its expansion into the southwestern United States during this climatic period would have facilitated the northward spread of maize during the late middle Holocene.

Late Holocene, 4500 cal BP to Present

During the late Holocene the amplitude of millennial scale temperature and moisture variability declined while interannual variability increased. There was an overall decline in intensity of the North American Monsoon and an increase in the influence of ENSO. The declining monsoon reduced both the amount and intensity of summer precipitation while changes in the ENSO phenomenon generally increased winter precipitation, but its cyclical nature brought increased variability. A third major pattern of the late Holocene was a dramatic reduction in flood frequency and severity.

Several proxies from both outside and inside the Southwest provide support for these three patterns. Deep sea cores from the Gulf of Mexico indicate the North American Monsoon sharply dropped in intensity following 5200 cal BP and remained at an intermediate level until 3500 to 3300 cal BP when it again declined (Poore et al. 2003; Poore et al. 2005). Other than a spike at 3000 cal BP, it remained diminished, until 2000 cal BP when it returned to an intermediate level (Barron et al. 2012; Poore et al. 2003; Poore et al. 2005). Late Holocene climate also shifted when the ENSO combined with warm phases of Pacific Decadal Oscillation. This combination of factors increased winter precipitation as warm Pacific waters fostered increased moisture from California to the Southwest (Onken et al. 2017). These trends are visible after 4200 cal BP and increased at 3200 cal BP; but peak ENSO activity occurred from 2300 until 1500 cal BP after which it declined again (Barron and Anderson 2011; Conroy et al. 2008; Onken et al. 2017).

During the late Holocene, a period of increased flooding occurred between 4550 and 3320 cal BP followed by a pronounced drop in frequency of large floods. After a slight interval of renewed flooding at ca. 3000 cal BP, episodes of flooding declined sharply between 3000 and 2000 cal BP. Flood events again increased in frequency between 2000 cal BP and the present (Harden et al. 2010).

The details of a number of regional paleoecological proxies suggested that portions of the Early Agricultural period experienced greater effective moisture than the present. However, given the nature of the data, these reconstructions are necessarily incomplete and contain inconsistencies. Increases in early late Holocene moisture are in north-central Sonoran Desert packrat middens as they reflect ongoing arid conditions (McAuliffe and Van Devender 1998), but evidence of increasing moisture is in the northern Chihuahuan Desert proxies. Soil stable carbon isotope studies from the Tularosa and Hueco Basins indicate C_4 grasslands dominated the basin floor and alluvial fans from 4500 to 2200 cal BP, and returned briefly to late Pleistocene levels (Buck and Monger 1999). Following 2200 cal BP the vegetation reverted to C_3 shrubs (Buck and Monger 1999). From the adjacent Hueco Mountains, packrat midden data also reflect a late Holocene mesic interval with pinyon, oak, and juniper appearing while desert shrubs and succulents exhibit a variable decline between 3200 and 1380 cal BP. Desert shrubs returned again after 1380 cal BP (Mauldin 1995; Van Devender 1990:Figure 7.9). A moist interval was visible in packrat midden data from the Guadalupe Mountains dated to 3670 cal BP (Wilkens and Currey 1999). A meta-analysis of northern Chihuahuan Desert packrat midden data indicate that warmer winter temperatures and ENSO winter rainfall at 5000 to 4000 cal BP facilitated the northward spread of subtropical desert shrubs including whitethorn acacia (*Acacia constricta*), ocotillo (*Fouquieria splendens*), and creosote bush (*Larrea tridentata*) (Holmgren et al. 2007).

Speleothem data from Pink Panther Cave in the Guadalupe Mountains provide a continuous 12,000 year record of changing moisture levels. These data indicated dry conditions between 4600 and 3800 cal BP, with wet conditions at 3800 cal BP. The speleothem data indicate extremely wet conditions at 3400 to 3100 cal BP with the wettest conditions from 2800 until 2400 cal BP, likely due to enhanced winter moisture (Asmerom et al. 2007; Polyak and Asmerom 2001). However, this timing is inconsistent with the proposed earlier onset of increased ENSO activity at 4200 cal BP. The spelothem data indicate dry conditions followed from 2400 until 1900 cal BP succeeded by another wet period (Asmerom et al. 2007).

A number of sequences from regional playas provide important complimentary data. Moisture at Playas Lake in southwestern New Mexico was sufficient to support chub fish and salamanders, in raptor deposits dating 3520 to 2410 cal BP (Van Devender and Worthington 1977). High effective moisture was recorded in Babícora Basin marsh deposits and paleosols dating between 4360 and 2900 cal BP followed by an arid period (Barron et al. 2012; Ortega-Ramírez 1995; Ortega-Ramírez et al. 1998: 1177). Diatoms and ostracods indicate that the basin refilled, perhaps intermittently, and for a brief period the lake resembled cold (46° F) late Pleistocene conditions. Unfortunately, the dating of the refilling was imprecise at 4900 to 1260 cal BP. A shallowing occurred before 2550 cal BP (Metcalfe et al. 1997; Ortega-Ramírez 1995), but cold winters persisted until 1260 cal BP (Palacios-Fest et al. 2002:202). Lake Cochise refilled beginning at 5400 cal BP and lacustrine conditions may have persisted until 3200 cal BP after which the lake dried (Waters 1989). A ca. 5000 to 4100 cal BP beach ridge at Laguna el Fresnal also marked a refilling event (Castiglia 2002; Castiglia and Fawcett 2006).

Pollen profiles from a series of borderland cienegas, including Animas Creek, indicate expansion of aquatic vegetation at about 3200 cal BP (Davis 1994:33). Evidence of a shift to less effective moisture is in the pollen record from the Lehner site in southeastern Arizona dating to about 2570 cal BP and at Lehner and Murray Springs dating to about 1400 cal BP (Mehringer 1967).

The longest tree-ring sequence in the Southwest comes from the El Malpais National Monument in west-central New Mexico. This record showed a period of increased moisture extending from 1869 until 1693 cal BP (AD 81 to 257), and the beginning of a long arid period at 1692 cal BP (AD 258) (Grissino-Mayer 1996).

In summary, these records suggested that during the Early Agricultural period (4150 to 1750 cal BP), greater effective moisture than present was driven by increased winter precipitation and reduced monsoons, but interannual variability was likely high. Decreased flooding followed 3340 cal BP and increased flooding returned after 2000 cal BP (Ely 1997; Harden et al. 2010; Onken et al. 2017). By 1700 cal BP, a shift to drought conditions emerged (Grissino-Mayer 1996). Thus, the Early Agricultural period, despite reduced North American Monsoons, was generally a period of higher effective moisture than present, accompanied by a reduction in the frequency of large floods.

Winter precipitation tended to be more widespread, falling in lower intensities for longer duration when vegetation was dormant, resulting in decreased runoff and less loss from evaporation and evapotranspiration (Onken et al. 2017:13). More water infiltrated the soil, providing moisture for spring germination and for recharge of subsurface water. As summer monsoonal precipitation declined, locations where subsurface water was accessible became some of the most predictable moisture sources on the landscape.

Hunter and gatherers who were beginning to utilize maize responded to these increasingly variable conditions by seeking the most reliable sources of moisture for farming, which included accessible subsurface water along drainage reach boundaries, springs, and marshes. Higher water tables, spring-fed marshes, and river reach boundaries enhanced the value of farming locations where subsurface water was available. Reduced flood frequencies and severity created stable floodplains and depositional environments further enhanced floodplains as productive and reliable farming niches (Harden et al. 2010; Huckleberry 1999). This ecological context partly explains the close association between early cerro de trincheras locations and drainage reach boundaries.

Agricultural Potential

Given a meandering floodplain, deep, loamy sediments, higher than modern rainfall, and a high water table, what maize farming options would have been available to people who lived here 3000 years ago? To help answer this question, we turned to the modern local experts, the people who farm the area today. Although they use tractors, chemical fertilizers, and herbicides, local farmers in Colonia Oaxaca at the base of Cerro Juanaqueña farm the same silt basin that prehistoric people farmed 3000 years ago. They told us that prior to the drought of the 1990s, their preferred method of farming was to plant maize in their floodplain fields during the month of May. Despite the irrigation canals and arid climate, we were surprised to learn that they preferred to rely solely on summer rains for moisture. They minimized dependence on irrigation, but when rainfall was not adequate they used irrigation with water from springs or pumped water from the Río Casas Grandes, as the river was too downcut for gravity flow irrigation.

In the 1930s, the archaeologist Donald Brand (1937: 20) also noted that throughout the Río Casas Grandes valley the best farm land was in the silt basins that formed upstream from valley constrictions, and that irrigation from the numerous springs in the valley was conducted here. These constrictions also likely helped force subsurface water to the surface. Winter moisture played an important role in recharging subsurface moisture. Akimel O'odham (Riverine Pima) and Maricopa farmers took advantage of a similar geological situation to farm in silt basins along the Gila River of Arizona in the early twentieth century (Doolittle 2000; Nials et al. 2011).

Along the Río Casas Grandes, a drought that endured throughout much of the 1990s altered that pattern in Colonia Oaxaca. Against all precedent, the water level in the spring at the base of Cerro Juanaqueña fell so far that water no longer rose to the level of the floodplain. The acequias ran dry, and the stately cottonwood trees that once lined their banks passed into memory. During the years of our fieldwork, 1997 until 2000, farmers depended on water pumped from wells to irrigate their fields.

Rain-Fed Farming

Knowing that modern farmers prefer to rely on rain-fed farming, what was its potential 3000 years ago? The main factor limiting maize production in this arid environment is water. Generally, precipitation of 150 mm during the growing season is considered the lower limit needed to produce a maize crop without irrigation. Modern hybrid maize has a longer growing season and requires substantial precipitation, as much as 450 to 600 mm (Shaw 1988) for optimal growth. However, the growing season for indigenous varieties of maize is 90 to 130 days from planting to maturity (Muenchrath 1995; Muenchrath et al. 2002). In one modern experiment, Muenchrath (1995) produced a crop of 2100 kg/ha of Tohono O'odham maize in unirrigated plots with only 160 mm of precipitation.

With this minimum requirement in mind, we compiled precipitation records from a weather station only a few kilometers from Cerro Juanaqueña (Janos, Chihuahua) covering the years between 1961 and 1993 (Figure 2.4). This record shows highly variable growing season precipitation, with a mean of 221 mm between May and September. During multiple years, growing season precipitation was at or below the nominal 150 mm minimum for temporal farming. The elevated precipitation during the 1980s sharply contrasts with the depressed levels for the early 1990s. Clearly rain-fed farming is risky in Janos. Droughts are typical here, as well as in the greater region (e.g., Díaz et al. 2002).

Modern agriculture data for the region reflect these realities. By the late 1990s farmers had ceased planting maize, many fields belonging to farmers in Colonia Oaxaca were fallow, and most who did plant fields relied on irrigation water pumped from wells. Differences in yield between irrigated and rain-fed fields were dramatic. For example, official Mexican records compiled by the Secretaría de Programación y Presupuesto, Instituto Nacional de Estadística, Geografía e Informática allow a quick analysis of the risks of rain-fed maize farming in the modern era (Yale University Library: Social Science Research Services 2004). Data for the 1994 growing season indicated that in the municipio of Janos 2450 ha were planted in temporal maize fields for grain production, but less than 20% (479 ha) was harvested, and yields were only 1500 kg/ha (718 metric tons). In contrast, 1647 ha of maize were planted in irrigated fields, 100% was harvested, and the yield was 6520 kg/ha (10,750 metric tons), or nine times the yield of the rain-fed fields. This example illustrates the high risk and low yield of rain-fed farming in this setting in the late twentieth century.

The risk of rain-fed farming is further heightened by the critical need for correct timing of growing season moisture. At planting, soil moisture must be sufficient for germination. Winter and spring precipitation provides soil moisture for spring germination prior to the onset of summer rains. Hypothesized increased winter precipitation would have also enhanced soil moisture recharge. The plants must also have moisture during the pivotal four weeks surrounding silking, which occurs about ten weeks after planting. Precipitation received too late in the growing season will do little good, even though total moisture is above the minimum growing season requirement (K. Adams et al. 1999; Muenchrath and Salvador 1995).

Flood Recession Farming

In Chapter 12 Lee Nordt provides a soil water balance model using average modern rainfall and temperature data and shows that soil water recharge occurs from November to January. Soil moisture utilization occurs during the following three months. High evapotranspiration and low rainfall results in soil moisture deficits during the rest of the year, thus the potential success of rain-fed farming is limited. However, because the soil has good water-holding capacity, over-bank flooding, also known as flood recession agriculture, can also deliver water to crops in an effective manner for this setting. Overbank floodwater farming was known among historic indigenous groups along the lower Colorado River (Castetter and Bell 1951) and the lower Gila River

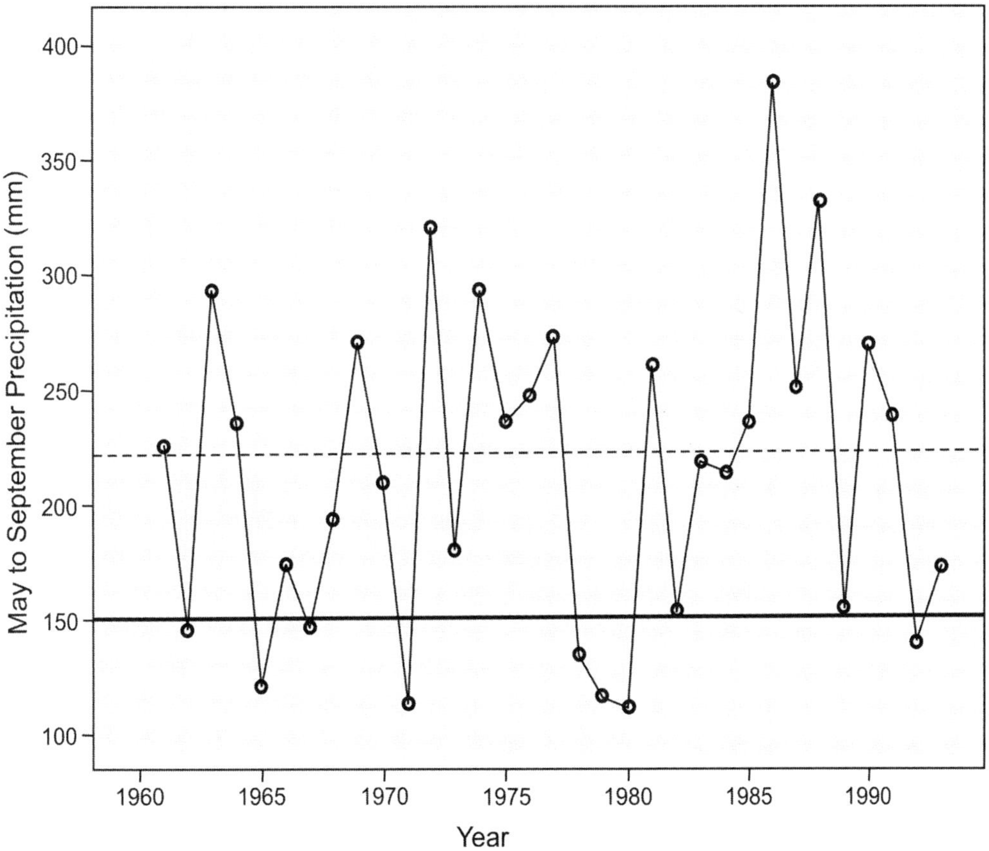

FIGURE 2.4. Janos growing season precipitation, May through September, 1961 to 1993. Horizontal lines mark minimum growing season precipitation for maize (bold line at 150 mm) and actual mean growing season precipitation at Janos, Chihuahua (dashed line at 221 mm).

(Spier 1970:58–61). In northern Mexico this form of agriculture has been reported from the La Junta de los Ríos area at the confluence of the Rio Grande and the Río de los Conchos (Kelley 1952, 1953), as well as in the Comarca Lagunera district of present-day Coahuila (Gerhard 1982:220).

Lee Nordt goes on to describe the multiple alluvial terraces and paleosols in the river valley and provides details about their variable water-holding capacities. Three thousand years ago the residents of Cerro Juanaqueña likely farmed on the Trincheras paleosol, formed on an alluvial surface that is now buried two meters below the modern surface. This paleosol has a high water-holding capacity due to its loamy texture and can hold 150 to 200 mm of water before becoming saturated.

A brief example illustrates the importance of this property. Evapotranspiration rates for modern maize in arid climates are about 5 mm per day (Muenchrath and Salvador 1995; Shaw 1988). If the full amount of average July precipitation, about 80 mm, fell in one rainfall event, it would only provide soil moisture for 16 days, due to loss through evapotranspiration. However, if the river left its channel in one flood event and covered the floodplain with 150 to 200 mm of water, it would saturate the loamy sediments and provide moisture for 30 to 40 days, perhaps even longer for native varieties of maize (Chapter 12).

Based on this analysis, it is likely that the modern rain-fed farming practices discussed above include the use of flood recession as a component of this strategy. The wide silt basins that form above the channel constrictions allow floodwaters to slow and spread, greatly reducing their destructive potential, as we witnessed on more than one occasion during July floods from our vantage point on Cerro Juanaqueña. In addition, rainfall amounts are much greater in the higher elevations of the Río Casas Grandes and Río Santa María watersheds, in comparison to settings in lowland basins. However, a significant portion of the higher elevation precipitation is transported to lowland settings along the major drainages as runoff. Floods originating in mountains can even result in flood recession moisture being delivered to fields in Janos without any local precipitation.

As we describe above, the timing of geomorphological processes, including landscape stability and low energy floods and the occupation and reoccupation of the cerros de trincheras, correlate with an agricultural strategy that used flood recession agriculture. The reduction in high magnitude flooding is followed by a depositional environment with sediments that indicate low energy flooding along a sandy, mixed-load meander belt stream. Landscape stability, including low energy flood regimes, allowed for the formation of the Trincheras paleosol, around 3200 cal BP (1250 BC), at the time when Cerro Juanaqueña was initially occupied. The abandonment of Cerro Juanaqueña correlates with an erosional episode at about 3000 cal BP. The process repeats itself a millennium later with the return of a depositional environment, landscape stability, and the formation of the Janos paleosol. These changes correlate with the smaller scale reoccupation of Cerro Juanaqueña at about 2000 cal BP. Finally, an erosional event and the resumption of high magnitude flooding correlated with the final abandonment of Cerro Juanaqueña.

Overbank floodwater farming was likely a key strategy for the occupants of the fortified hilltop settlements in northwestern Chihuahua. However, other strategies, including water table farming and potentially irrigation, may have been used as well.

Water Table Farming

The floodplain can also be a source of subsurface water for crops. Such fields are referred to as seepage, subirrigation, or water table fields (Doolittle 2000:234; Mabry 2005a). Water table fields are typically thought of as small areas that benefit from continual moisture from a nearby spring or other water source. For water table farming to be effective, roots must be able to reach subsurface moisture, but the root zone cannot be saturated for lengthy periods. Ideally, the upper capillary fringe of the water table must reach no further than the lower depth of the primary maize root system. The deep root system of some maize varieties can penetrate to 2 m, allowing it to tap subsurface water. Some investigators believe that this strategy could have provided higher yields and lower risks than many other agricultural strategies used in the southwestern United States and northwestern Mexico (Mabry 2005a). However, Wilken (1987) argues that it is difficult to achieve a balance between root systems having access to subsurface moisture but not being continually saturated. Perhaps for this reason, such fields were not common in Latin America nor in the southwestern United States. Although today local residents routinely dig wells to penetrate the water table and in the recent past used springs as sources of irrigation water, they do not practice water table farming, nor do they seem to have a clear idea how water table depths are patterned across the valley.

Nordt's geomorphologic studies of the floodplain (Chapter 12) suggest that the water table was sufficiently high for water table farming. Our backhoe trenches in the river valley located the modern water table at 50 to 300 cm below the modern surface with depths correlating with the different alluvial terraces. A high water table was also indicated by the numerous springs in the valley observed in the 1930s by Donald Brand.

Pollen and phytolith results, described in Chapter 10, suggest that damp conditions were present on the floodplain at the time Cerro Juanaqueña was occupied. Data from the Trincheras paleosol indicate the rare presence of cattail and sedge pollen, as well as trilete and monolete fern spores, and suggest greater levels of soil moisture than present today. Cattail requires constantly damp conditions while sedge and fern thrive under conditions of high soil moisture. These data suggest springs or a high water table were part of the paleoenvironment. Also grass and tree pollen and Nordt's soil stable isotope data indicate that the Trincheras paleosol, which was contemporaneous with the initial occupation of Cerro Juanaqueña, was a mixed desert grassland community with disturbance plants and included habitats with damp soils as well as locations with continual moisture.

Nordt indicates that the Trincheras paleosol exhibited redoximorphic features that form under saturated, anaerobic soil conditions. However, it is difficult to differentiate between ancient and more recent redoximorphic soil features. Assuming an ancient water table that was high, but not too high, it is possible that water table farming would have been feasible 3000 years ago.

Irrigation Farming

We did not identify any irrigation features, but we cannot rule out the possibility that 3000 years ago irrigation was practiced. In fact, we would not be surprised at the future discovery of ancient irrigation canals. Irrigation technology on the Santa Cruz River in the Tucson Basin was underway by 1500 cal BC and was well developed by 1200 cal BC with multiple communities constructing and maintaining irrigation canals (Mabry 2008c; Mabry and Stevens 2014; Vint 2015a). At the site of Las Capas, a complex, hierarchical ditch system included the construction of thousands of small planting basins (Nials 2015). For these reasons, we believe that irrigation farming was a strong possibility during the Early Agricultural period in northwestern Chihuahua. Certainly, the construction of Cerro Juanaqueña and other cerros de trincheras shows that people had the capacity to organize and direct demanding labor projects. We have argued that farming could readily have been practiced in the absence of irrigation, and that 3000 years ago moisture regimes were higher than today. However, inevitably, dry years occurred and irrigation canals would have decreased the risk and increased the yield of maize. The Río Casas Grandes was a meandering stream and not deeply entrenched, making water diversions easier to build and maintain.

Geomorphic Context

Stream Stability in Grassland Settings

In the final section of this chapter, we highlight stream stability, one additional aspect of the Río Casas Grandes that promoted its use for farming with relatively low labor investment and high productivity despite generally arid conditions. In preceding sections, we have described the importance of floodplain constrictions, availability of subsurface moisture, a low velocity flood regime, and soils with high water-holding capacity. Here we emphasize the fact that much of the course of the Río Casas Grandes flows through a grassland biotic community. In this, the Río Casas Grandes contrasts with many other watercourses of the Southwest, where potential agricultural lands are located in desert scrub communities. Furthermore, it is likely that C_4 grasslands have dominated much of western Chihuahua since at least the end of the Wisconsin epoch.

As it passes through the grasslands, the Río Casas Grandes is a meandering perennial stream with a mixed sediment load, deep channels, a high water table, wide floodplains, and deep soils composed of medium to fine grain sediments. These characteristics are conducive to agriculture, and they have remained remarkably stable over the past 5000 years or so. Even though the river has experienced many episodes of channel migration and cycles of aggradation, followed by periods of landscape stability and new channel erosion, these key characteristics have been remarkably resilient (Chapter 12).

In contrast, some of the streams of central and southern Arizona have been much more unstable, alternating between meandering and braided forms, and experiencing cycles of destructive channel widening and massive deposition. For example, the middle Gila River in central Arizona became a highly unstable, braided stream at several points in its geomorphologic history (Huckleberry 1995). This inherent instability results in rapid evolution of channels and steam geometry. Braided streams have numerous interwoven channels that transport and redeposit large volumes of coarse grain sediments that clog their channels. They experience frequent periods of high flows and carry large-grain sediments, while fine-grain materials are minor constituents of the sediment load (Waters 1992). Such streams can make farming quite difficult.

A single stream can change from meandering to transitional to braided along different segments of its course or through time, due to changes in tectonics, rainfall regime, water volume, water table, gradient, climate, vegetation, land use, watershed size, and transported sediment load (grain size and amount) and sediment inputs from the landscape (Schumm and Brackenridge 1987; Waters 1992). One of the most critical variables affecting channel morphology is increased sediment transport. Increased sediment transport can occur when overland flow strips sediment from a sparsely vegetated landscape. This occurrence leads to stream instability and can transform a meandering, stable stream with a wide floodplain into an unstable, braided stream choked with its own alluvium (Schumm and Brackenridge 1987: 223). Subsequent erosional events can then result in massive channel widening.

Sediment yield, in turn, is strongly controlled by vegetative cover. This consequence occurs because during high volume precipitation events some types of vegetative cover help retain surface sediments, while other types of vegetative cover leave the ground surface much more vulnerable to erosion. Of all vegetation types in the United States, the peak sediment yields are in shrub- and bunch grass-dominated semiarid environments (Schumm and Brackenridge 1987:223–224). In grasslands with denser ground cover, sediment yields drop precipitously.

At various times, both Arizona and northwestern Chihuahua experienced cycles of aridity followed by wetter conditions that resulted in erosional events (Nordt 2003; Waters and Haynes 2001). In Arizona, when wetter conditions followed drier, the prevalence of desert scrub vegetation encouraged massive sediment transport and deposition (Waters and Haynes 2001). Huge erosional events expanded channel widths by many tens of meters and changed meandering streams to braided stream forms. Alternatively, in the persistent grassland environment of northwestern Chihuahua, similar cycles of wetting and drying only resulted in erosional events such as channel downcutting and lateral channel migration within a continued meandering stream morphology. Following these erosional events, slow aggradation and floodplain building followed.

Cerro Juanaqueña and all but one of the Early Agricultural period cerros de trincheras are near the transition between Plains Grassland and Semiarid Grassland biotic communities. The combination of grassland setting and riverine constrictions are fundamental to Río Casas Grandes agroecology. The constrictions formed silt basins, slowed stream velocity, and forced ground water to the surface while grasslands limited sediment transport and contributed to stream stability and the long-term maintenance of a meandering geometry.

Floodplain Evolution and Occupation of the Cerros de Trincheras

In Chapter 12 Lee Nordt describes specific responses of the Río Casas Grandes floodplain to the sequence of climate changes outlined earlier in this chapter. In broad terms, the region-wide reduction in high magnitude flood events that occurred at around 3300 cal BP resulted in a fundamental change in depositional regime. Prior to 3300 cal BP, the dominant process on the floodplain was erosion. The river was carving a broad, shifting channel through older alluvial deposits, resulting in a valley over a kilometer wide in places, bounded on the east and west by high terraces composed of older alluvium. From time to time, sediments may have accumulated in this setting but most were subsequently flushed away by high magnitude flooding.

After 3300 cal BP, floodplain geomorphic processes were dominated by deposition, rather than erosion, and the valley began to fill with sediments. Episodes of erosion and channel cutting continued to occur from time to time, but inevitably channels refilled, and the floodplain surface rose. In Chapter 15, we will consider this sequence in more detail and will relate it to periods of occupation and abandonment at Cerro Juanaqueña. We argue that the two episodes of occupation correspond to periods of floodplain stability and high potential agricultural productivity. Surprisingly, the ends of both occupation periods coincide with major erosional episodes that destroyed much of the arable alluvial terraces and substantially reduced the agricultural potential of the Río Casas Grandes floodplain.

Summary

All 14 cerros de trincheras that we investigated are associated with floodplains along major rivers fed by primarily monsoonal precipitation falling in the Sierra Madre Occidental. In broad terms these archaeological sites are located in the Apachean floristic area, characterized by an unusually high diversity of plant species. In more specific terms, all are located in either Plains or Semidesert Grassland biotic communities and have access to Chihuahuan Desertscrub, Madrean Evergreen Woodland, and Madrean Conifer Forest biotic communities, as well as wetlands. They are all associated with stream reach boundaries that would have enhanced access to moisture, agricultural productivity, and reliability of the alluvial terraces along the floodplain.

Since the end of the latest ice age, about 11,500 years ago, the climate of northwestern Chihuahua has grown both warmer and more arid. Equally significant have been important changes in the broad patterns of atmospheric circulation that partly determine climate. By the end of the early Holocene, winter storm tracks were shifting northwards, reducing winter precipitation. The early part of the middle Holocene was arid in the US–Mexico borderlands because average winter storm tracks were in their modern (northerly) position, but the summer monsoon pattern had not yet developed. However, between 7500 and 4500 years ago, the summer monsoons were in place. The monsoons may have actually been significantly stronger than today, but the ENSO phenomena which brings winter precipitation to the area was still weak. Modern plant associations likely achieved their modern distribution patterns at this time, and maize was introduced from further south in Mexico. During the late Holocene, monsoons declined in intensity, while the variable ENSO phenomena increased in importance. The result has been a climatic pattern with variable patterns of precipitation, including shifts in the intensity of flood regimes.

Large sections of the Río Casas Grandes and Río Santa María flow through settings which were grasslands throughout the Holocene. Because of this, the character of these drainages has been more stable than many other river systems in the southwestern United States and northwestern Mexico. For most of this time, the Río Casas Grandes has been a mixed load, meandering stream. An important change did occur around 3300 years ago, when high magnitude flooding became much less common. Instead, the river became subject to frequent low energy overbank floods, which initiated a sustained period of floodplain aggradation and set up conditions ideal for early maize agriculture. Within this context, prehistoric peoples established major hilltop settlements near drainage reach boundaries, places where bedrock or major tributaries created constrictions in the floodplains. Water tables are consistently higher in these settings, stream gradients and velocities are lower, and overbank flooding is more pronounced. Higher winter moisture of the late Holocene recharged water tables enhancing access to water, particularly along stream reach boundaries. Theoretical considerations, as well as information provided by modern farmers, suggest that floodplain agriculture depended on moisture from both direct winter and monsoonal rainfall and overbank flooding or irrigation in locations where soils with good water-holding capacity were present. Eventually, erosional events recorded in the geomorphological record disrupted this setting and Cerro Juanaqueña was abandoned. The cycle repeated itself a millennium later with the return of a depositional setting, low velocity overbank flooding, and the reoccupation of Cerro Juanaqueña, only to be terminated a second time by renewed high magnitude flooding and erosion.

Notes

1. Hall (2018), in a thorough review published after we completed this chapter, also argues for a dry middle Holocene. His discussion encompasses the entire Southwest while ours focuses on the borderlands. In brief, Hall devalues the lacustrine and packrat midden data supportive of a wet late middle Holocene, and he does not consider changes in the North American Monsoon. We are in agreement that more well-dated proxies are needed for this critical period.

CHAPTER 3

Cerro Juanaqueña and Its Features

Introduction

The first time we visited Cerro Juanaqueña, in 1994, we were looking for an *atalaya*, a hilltop communication site associated with the fourteenth century Casas Grandes regional system. The location was indicated as just a dot on a map (DiPeso et al. 1974:Figure 284–285) and we really did not know what to expect. We thought that perhaps there would be a cairn or a pile of burned rock, or at most a stone circle and traces of an ancient road or trail. Instead, as we approached the summit, we found ourselves crossing massive cobble berms resembling nothing we had ever seen before. In fact, our first reaction was that there must be some geologic explanation—a system of dykes perhaps or some obscure geomorphic process. As the scale and configuration of the terraces gradually became clear, we began to realize that we were looking at features constructed by human beings, and that they were like nothing else reported in the region.

Over the next six years, including four intensive field seasons (1997 to 2000), we spent thousands of person-hours trying to define and understand these constructions. Almost immediately we began to make a map. At first, when resources were limited, we used simple tape and compass techniques. As time went by and funding increased, we graduated to methods using electronic surveying equipment, and finally to photogrammetric mapping. This progression is described in more detail in Appendix 3.1. Ultimately, we found that the terraces cover an area of over 10 ha, and that they have a combined total length of over 8 km. In addition, there are almost 100 rock rings, and a number of other cultural features described in this chapter.

Our investigations also included intensive excavations, and one of our important goals was to find out how the terraces were constructed. Using careful trowel and whisk broom techniques, we exposed cross sections through a number of terraces, as well as large surface areas of others. Following each field season, we submitted annual reports to Instituto Nacional de Antropología e Historia (INAH) which contained additional descriptions of the excavated features and are included as electronic appendices as follows: 1997 season [Appendix 3.2 (Hard and Roney 1998b)]; 1998 season [Appendix 3.3 (Hard and Roney 1999a)]; 1999 season [Appendix 3.4 (Roney and Hard 2000)]; 2000 season [Appendix 3.5 (Hard et al. 2001)]. Table 3.1 lists the excavated features and respective field seasons.

Our excavations amounted to about 52 m^3 (Table 3.1), and the methods and strategies that we used are described in Sidebar 3.1. Ultimately, we applied our understanding of terrace construction to a replication experiment, an exercise that leads to estimates of the amount of labor needed to build the constructions on Cerro Juanaqueña.

As we began to understand something of their construction, and as radiocarbon dates and other excavation results came in, we realized that we also needed a better understanding of taphonomy—that is, knowledge of how the sediments and cultural materials we excavated came to be in the contexts where we found them. Geomorphologist Lee Nordt assisted us in the taphonomic research by providing well-grounded technical descriptions of the sediments and expert interpretations about their formation processes. This understanding helped us better reconstruct the ancient methods of terrace construction and helped us interpret the results of our excavations and surface documentation.

This chapter begins with a brief description of the topographic setting of Cerro Juanaqueña, then proceeds to a detailed discussion of the terraces themselves. Next various other cultural features found on and around Cerro Juanaqueña are described. Finally, we recount a replication experiment in which we built a terrace resembling as closely as possible the ones we excavated. By keeping careful track of the amount of time required for each phase of construction we were able to develop an authoritative estimate of the total labor investment at Cerro Juanaqueña.

Topographic Setting

Cerro Juanaqueña was constructed on the summit and slopes of a steep, 140 m high basalt hill that juts above the Río Casas Grandes floodplain (Plate 1). The hill is actually a highpoint on a 4 km long, north-south trending basalt ridge (Figure 3.1). The western and southwestern sides of the hill slope steeply from the floodplain to the summit, with slopes as great as 40%. The northern and eastern sides of the hill are also steep, but the slopes are not as long in these directions. The gentlest approach is from the southeast, where a long, easy slope (ca. 25%) leads from the summit down to a saddle and drainage, which separate the basalt ridge from a larger range of hills further east. The summit of Cerro Juanaqueña is a relatively level area about 60 m by 25 m in size.

Bedrock consists of dark gray, somewhat vesicular

Sidebar 3.1. Excavation Methods

In four seasons of fieldwork we completed excavations in 30 separate features on Cerro Juanaqueña, including eight rock rings, 21 terraces, and one burned rock feature (Table 3.1). Selection of features for excavation was guided by several different objectives.

1. First, we placed a great deal of emphasis on recovery of faunal materials for subsistence information and on recovery of macrobotanical remains for both AMS radiocarbon dating and economic information. We soon discovered that both bone and charred macrobotanical remains were most likely to be preserved at depths of 30 cm or more below the surface. For this reason, we often tried to select large terraces and terraces built on steep slopes where depth to bedrock would be greatest. We also carefully inspected rodent burrows and other areas of disturbance, looking for ashy soil and small bits of charcoal. Occasionally we excavated small 30 cm by 30 cm test units to depths of 30 cm or more, hoping to find indications of faunal and macrobotanical remains. When charcoal was found in these test units, the feature was usually selected for excavation. The four terraces and two rock rings that were only tested are identified in Table 3.1.
2. We also wanted to understand details of terrace construction and the formation processes which affect the artifacts and other archaeological materials. For this objective, we needed to expose complete stratigraphic profiles through the terraces.
3. We wanted to obtain radiocarbon dates from different portions of the site in hopes of recognizing a sequence of construction and use.
4. Finally, we hoped to find, document, and interpret as wide a range of features as possible. Excavations at cerros de trincheras in southern Arizona revealed that features very similar to our rock rings were sometimes pithouses. In Sonora, other researchers have found postholes and other evidence of ramadas and jacal structures built on terrace surfaces. Also, storage pits have been identified as an important aspect of Early Agricultural period adaptations in southern Arizona, and we thought that similar storage features might be present at Cerro Juanaqueña. This objective required exposure of large areas of potential occupational surfaces, both within the rock rings and on the terrace surfaces.

These excavation objectives required two nonexclusive strategies, deep excavations for exposure of stratigraphic profiles and for recovery of bone and charcoal, and large horizontal exposures which might lead to discovery of post holes, hearths, or other features.

After we selected a terrace or other feature to excavate, our first step was to draw a detailed plan and cross section including all artifacts, except debitage (Figures 3.4 and 3.5). A base line laid out with string perpendicular to the long axis of the terrace facilitated mapping and orientation of excavation units. The initial excavation units were usually two 1 m × 1 m units that were placed to find charred plant material and to reveal the terrace formation sequence. If ashy soil, bits of charcoal, or burned animal bone fragments were revealed by ant or rodent disturbance or by our 30 cm by 30 cm probes, then we would initiate excavations there. More than one ant bite was awarded for our efforts! The terrace construction sequence was most clearly revealed when we could view the juxtaposition of the outer berm wall, the terrace fill, and bedrock. Also, charcoal was preserved best at greater depths. Placing the two adjacent excavation units at the outer edge olivine basalt with an aphanitic texture. As this basalt weathers, it breaks into angular or subangular clasts, 0.2 to 1.5 m or more in diameter, which cover the hill continuously, forming a talus-like surface. In its original, unmodified condition, this surface is difficult to walk across and is unsuitable for most human activities. Much of the construction which occurred on Cerro Juanaqueña consisted of piling and rearranging these rocks to form terraces with flat surfaces that could be used for a variety of activities.

Terraces

Distribution

The great age, number, scale, and organization of the constructed terraces on Cerro Juanaqueña impress anyone who visits the site. We defined 550 individual terraces with a total combined length of 8 km. The terraces occur in two distinct zones, one loosely centered on the summit of the hill and the other near the base of its western slope (Figures 3.1 and 3.2). The upper zone measures 325 m by 250 m and has an area of 6 ha (14.8 acres).

of the terrace and perpendicular to its long axis usually worked the best.

A modified version of the Level/Locus recording system was used throughout the project (LeBlanc 1976). With this system we were not constrained to only 1 m × 1 m squares or arbitrary 10 cm levels, although frequently these standard units worked the best. If a natural level, such as the surface aeolian deposit, was encountered, we could vary the depth of the levels. Each provenience was described with its feature number such as T(errace)167 or R(ing) 286, a sequential unit number, and a sequential level number (e.g., Terrace 67, Unit 2, Level 3). The depth of each level was recorded below datum (BD). Depths were measured using a line level and string attached to a datum stake. The vertical datum for each feature was arbitrarily set by the height of the line-level string, so that the modern surface was fixed at approximately 100–150 cm BD. The spatial relationships between the various features and datum stakes were recorded with the use of the electronic distance measuring instrument (Appendix 3.1). If a feature such as a posthole, a buried surface, or hearth, etc. was encountered, it was designated with a sequential subfeature letter and descriptive term (e.g., Subfeature posthole A). Excavation notes were written on Unit-Level forms with plan views generally drawn at the bottom of each 10 cm level.

All excavations were done by hand, primarily using trowels and whisk brooms. Trenches or test units were excavated to bedrock in all of the features, using arbitrary levels, which were usually 10 cm thick. During the first two seasons (1997 and 1998), we knew we had to establish the antiquity of the site and evaluate the role of maize in the economy, so more than half of the feature excavations were vertical exposures intended to recover radiocarbon and botanical samples as well as basic stratigraphic information from a diverse sample of features. In the last two years of fieldwork we placed more emphasis on horizontal exposures, hoping to better understand activities that took place on the hill. In many cases, the actual prehistoric occupation surfaces appeared to be only a centimeter or two below the present surface. For this reason, we attempted to follow natural compaction surfaces, removing only a few centimeters of soil and exposing larger areas. More features were excavated in this fashion in the latter two seasons.

Excavators took frequent large soil samples for flotation to obtain charred plant remains. Pollen samples were taken from recognizable surfaces during excavation and from stratigraphic profiles after completion of excavation units. In addition, we collected pollen samples from modern surface contexts for comparative purposes. Whenever possible, radiocarbon samples were taken in situ. Excavated fill was screened through ⅛ inch mesh hardware cloth, and all animal bone and artifacts, including lithic debitage, were retrieved. All charred maize or other recognizable plant remains, including most wood charcoal fragments, were collected from the screens. Archaeological specimens and samples were assigned "Bag Numbers" as they were collected. Bag numbers are sequential within each feature, and excavators maintained a running list of bag numbers, their contents, and proveniences on Feature Summary Forms. With this system, only a feature number and a bag number are needed to identify an artifact's unique provenience. After completion of all excavation units on a particular terrace or stone circle, excavators wrote detailed summaries of the results of their excavations and prepared stratigraphic drawings and descriptions. Additional details of these excavations are published in our annual field reports (Hard and Roney 1998b, 1999a, 1999b; Hard et al. 2001; Roney and Hard 2000) and are reprinted here as electronic Appendices 3.2 to 3.5.

TABLE 3.1. Features Excavated at Cerro Juanaqueña.

Feature[1]	Year Excavated	Zone	Total Volume Excavated (m^3)	Surface Area Exposed (m^2)[2]	Maximum Depth Below Surface (m)
R1A	1997	Upper	0.6	2.1	0.6
T6	1997	Upper	2.2	20.0	0.8
T10	1998	Upper	2.0	2.0	1.3
R28	1997	Upper	1.0	5.5	0.7
R51	1999	Upper	0.1	1.0	0.2
T97[3]	1998	Upper	2.0	2.0	1.0
T126	1998	Upper	1.8	1.9	1.0
T163	1998–99	Upper	5.4	23.5	0.7
T167	1997	Upper	6.6	5.1	1.4
T175	1999	Upper	1.3	8.3	0.5
T195	1998	Upper	0.2	0.3	0.8
T222	1997	Upper	2.3	15.6	1.4
R234	1998	Upper	1.3	2.0	0.9
R239	1999	Upper	0.9	5.2	0.3
R250	1997	Upper	0.9	3.5	0.9
T273	1998	Upper	0.7	2.0	0.6
R286	1999	Upper	1.9	11.0	0.8
T287	1999	Upper	2.2	22.0	1.1
T290	1998	Upper	1.4	2.0	0.7
T297	1998	Upper	1.9	2.0	1.3
R307A[3]	2000	Upper	0.2	0.3	0.6
T387	1998	Lower	2.2	2.0	1.2
T413	1998	Lower	2.4	2.0	1.4
T415[3]	2000	Lower	0.2	0.3	0.6
T463[3]	2000	Lower	0.2	0.3	0.6
T465[3]	2000	Lower	0.2	0.3	0.8
T487	2000	Lower	6.4	10.5	1.0
T508	2000	Lower	1.6	1.3	1.4
T537	1997	Upper	1.8	4.0	0.6
BR1	1999	Upper	0.7	3.0	0.5
Total			52.6	161.0	

[1] R=Rock Ring; T=Terrace; BR=Burned Rock Feature
[2] Includes only terrace or rock ring surface; excludes surface areas of terrace berms.
[3] Small test unit only.

Within this area prehistoric people modified literally every square meter by clearing cobbles from the surface and by constructing terraces.

Near the base of the hill is a 600 m long zone of contiguous terraces with an area of 4 ha (9.9 acres). Whereas bedrock is very shallow in the upper terrace area, the lower zone of terraces was constructed along the upper margin of an apron of colluvium at the base of the hill. By comparison with the upper slopes, this setting provides a gentler slope and a more generous supply of cobbles and fine sediments for terrace construction. Terraces in the lower zone occur between 10 and 40 m above the modern floodplain.

About 40 individual terraces do not fall into these two main zones. One group of about eight related terraces occurs about 50 m north of the lower terrace zone, in a similar setting at the upper edge of the colluvial deposit. The others are scattered across the hillslope between the upper and lower terrace zones in the southeastern quadrant of the site. We sometimes refer to these scattered features collectively as the middle terraces, in contrast to those of the upper and lower zones.

Terrace Construction

Prehistoric peoples used two related techniques to create level surfaces at Cerro Juanaqueña. In areas where the hill slope was gentle (ca. 12% or less), the builders simply removed the loose basalt cobbles from the surface, exposing the underlying bedrock and leaving a residue of fine sediments (Plate 1). The resulting surface

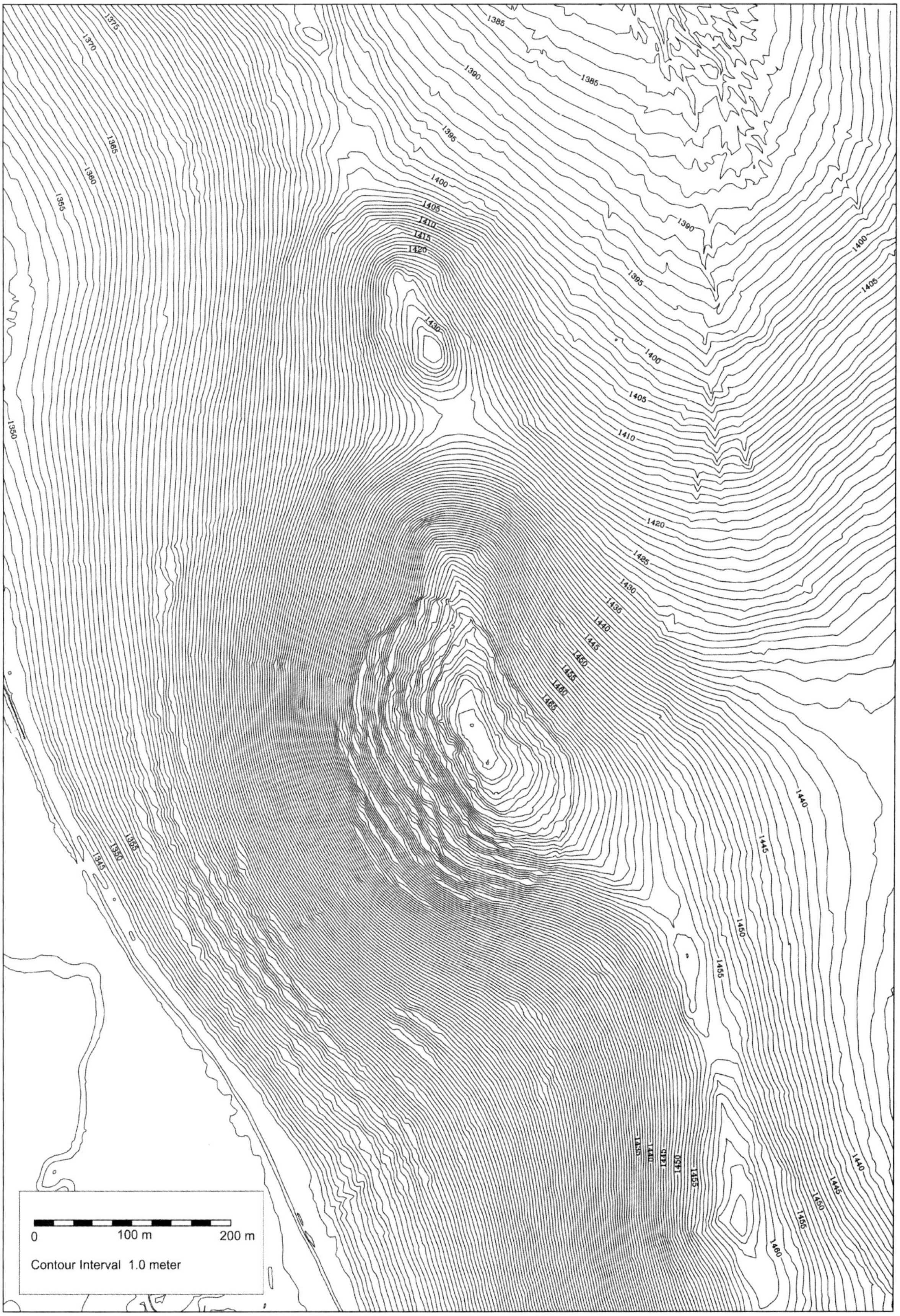

FIGURE 3.1. Photogrammetric map of Cerro Juanaqueña.

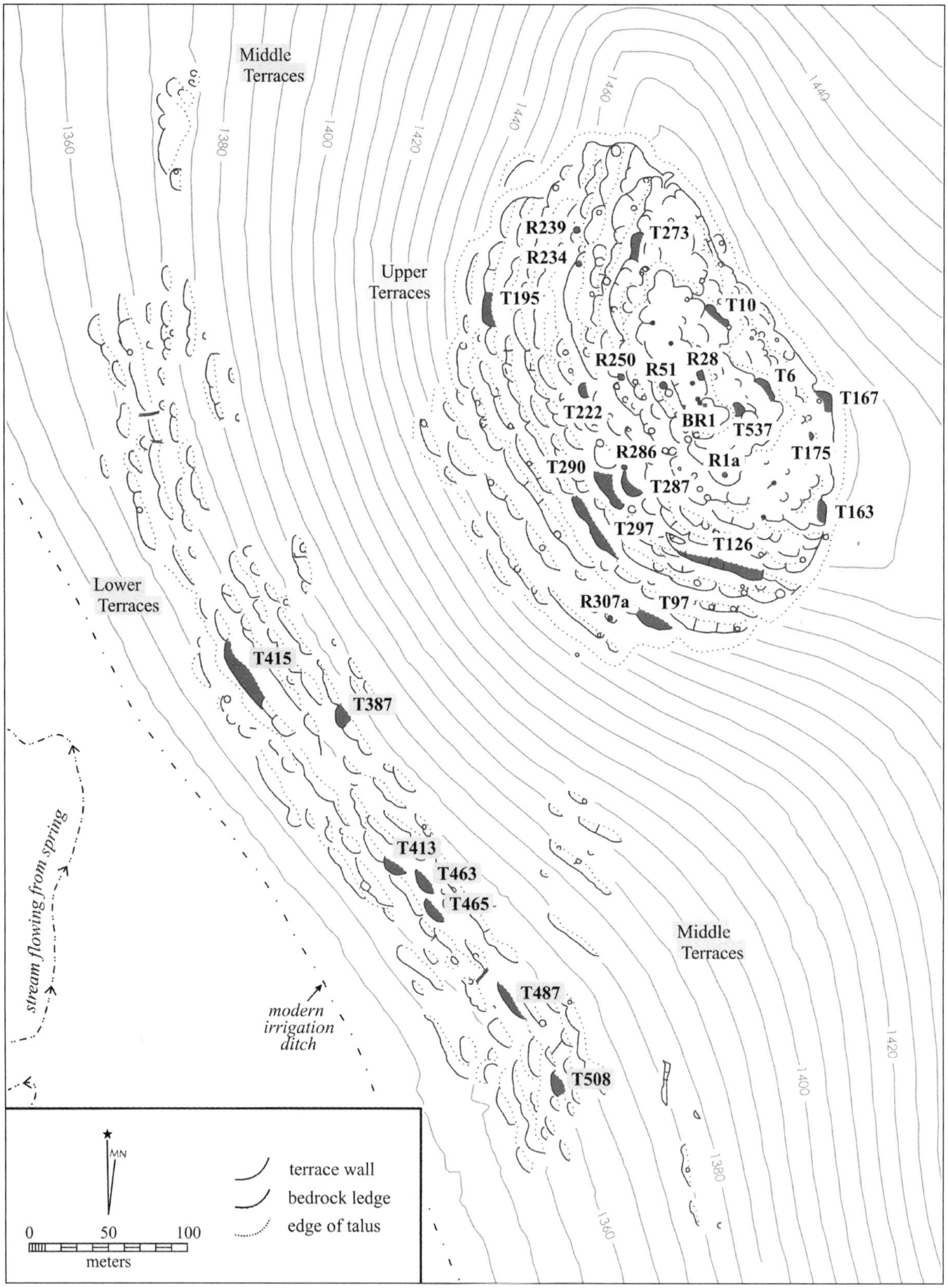

FIGURE 3.2. Locations of excavated features on Cerro Juanaqueña.

was relatively level and smooth and could be used for a variety of purposes. A typical area of this nature is illustrated in Figure 3.3. The small shaded areas are artificial terrace surfaces (discussed below) which contrast with the extensive areas of gently sloping hill surface where cobbles have been removed. The loose cobbles from the original surface were sometimes piled on bedrock outcrops near the summit, but more often they were tossed down slope and used to construct widely spaced terraces. This kind of construction was most common near the summit of the hill, where naturally level surfaces occur. At a handful of other places on the hill, small natural bedrock ledges were cleared off and used as terraces.

Actual constructed terraces are much more numerous (Plate 2). We believe that construction of these features involved several steps. First local basalt cobbles were piled into large berms that bow out in the center and pinch in at the ends against the slope (Figure 3.4). The rocks forming these berms were usually tossed carelessly into place, with no internal organization. We did find several examples where rocks had been stacked so they would hold a steeper grade, but there were no instances of the formal vertical masonry characteristic of cerros de trincheras in later time periods. Terrace cross sections (Figure 3.5 and Figure 3.6) show cobble berms reaching from 60 to 160 cm high above the original ground surface. The pocket between the apex of the terrace and the natural slope behind them was filled in with smaller rocks lower down and fine sediment near the top to form a level or slightly sloping surface. These surfaces are usually cleared of cobbles and have an average surface area of 51.5 m^2 (s = 44.1 m^2, range 2.4 to 290.6 m^2, n = 486). Sometimes the apex of the berm rises as much as 50 cm above the platform surface to form a parapet around the outer edge of the terrace. These parapets serve as retaining structures that prevent surface erosion and allow colluvial sediments to accumulate.

The principal depositional characteristics resulting from this construction procedure can all be seen in Figure 3.7 showing the horizontal excavation of Terrace 287. The larger cobbles which form the berm or retaining wall are visible at the outer edge of the terrace. Fine sediments on the surface of the terrace have been stripped away by excavation to expose the smaller cobbles that make up the underlying terrace fill. Finally, the fine sediments themselves are visible in unexcavated areas and as dirt between the cobbles that comprise the terrace fill and berm wall.

Terrace Stratigraphy

Although details varied from location to location, the stratigraphic profiles exposed by excavation were basically similar in all of the terraces that we excavated. Figure 3.8 is an idealized example which is described below.

A. *Bedrock* was basalt, often with a white carbonate coating.
B. Sometimes we found pockets of darker, more compact soil resting directly on the bedrock, a deposit that we believe represents traces of the *original surface colluvium* that covered the hillslope before terrace construction.
C. *Berm walls* rested on bedrock and/or the original colluvium. The berms consisted of up to 70% blocky cobbles with a matrix of dark grayish brown sandy clay loam.
D. *Construction fill* could usually be easily distinguished from the berm because it included much more matrix and far fewer small cobbles. Typically, the matrix was dark grayish brown sandy clay loam with 35% or 40% angular to platy pebbles, 0.3 to 3.0 cm in diameter.
E. On some terraces *fine sediment* overlay the construction fill. This was a deposit of brown, fine sandy loam with fewer pebbles. In several cases we were able to follow compaction surfaces within this deposit.
F. The uppermost deposits, labeled *colluvial deposits* in Figure 3.8, were usually yellowish brown and consisted of fine sandy loam, often poorly consolidated. We believe that the lighter color of these deposits results from an eolian component in their composition, and they are a post-occupational deposit.

Formal, technical descriptions of the sediments in specific terraces accompany a series of figures, presented here. Terrace 167 is located in the upper terrace complex (Figure 3.3), in an area where the hillslope is relatively gentle (Hard and Roney 1998b:16–20). The profile of Terrace 167 is shown in Figure 3.6 and two profile descriptions are provided, one for the terrace wall and the other for fill behind the terrace. In this case the terrace wall is actually an especially massive feature described below as the perimeter berm. Terrace 6 (Figure 3.9) is also located in the upper terrace complex near the summit of the hill, in an area where the slope is relatively gentle (Hard and Roney 1998b:10–16). The profile describes the terrace fill just behind the berm wall. Terrace 297 (Figure 3.10) is about half way down the western slope of Cerro Juanaqueña, in an area where the slope is quite steep (Figure 3.2; Hard and Roney 1999a:45–46). Finally, Terrace 387 (Figure 3.11) is located in the lower terrace complex at the base of the hill (Figure 3.2) (Hard and Roney 1999a:46–49). Despite its difference in setting, the deposits are very similar to those in terraces of the upper terrace complex.

The basic stratigraphy on the terraces seemed to be intact, and we could easily recognize the six generalized stratigraphic units discussed above in most terrace excavations. However, two important sources

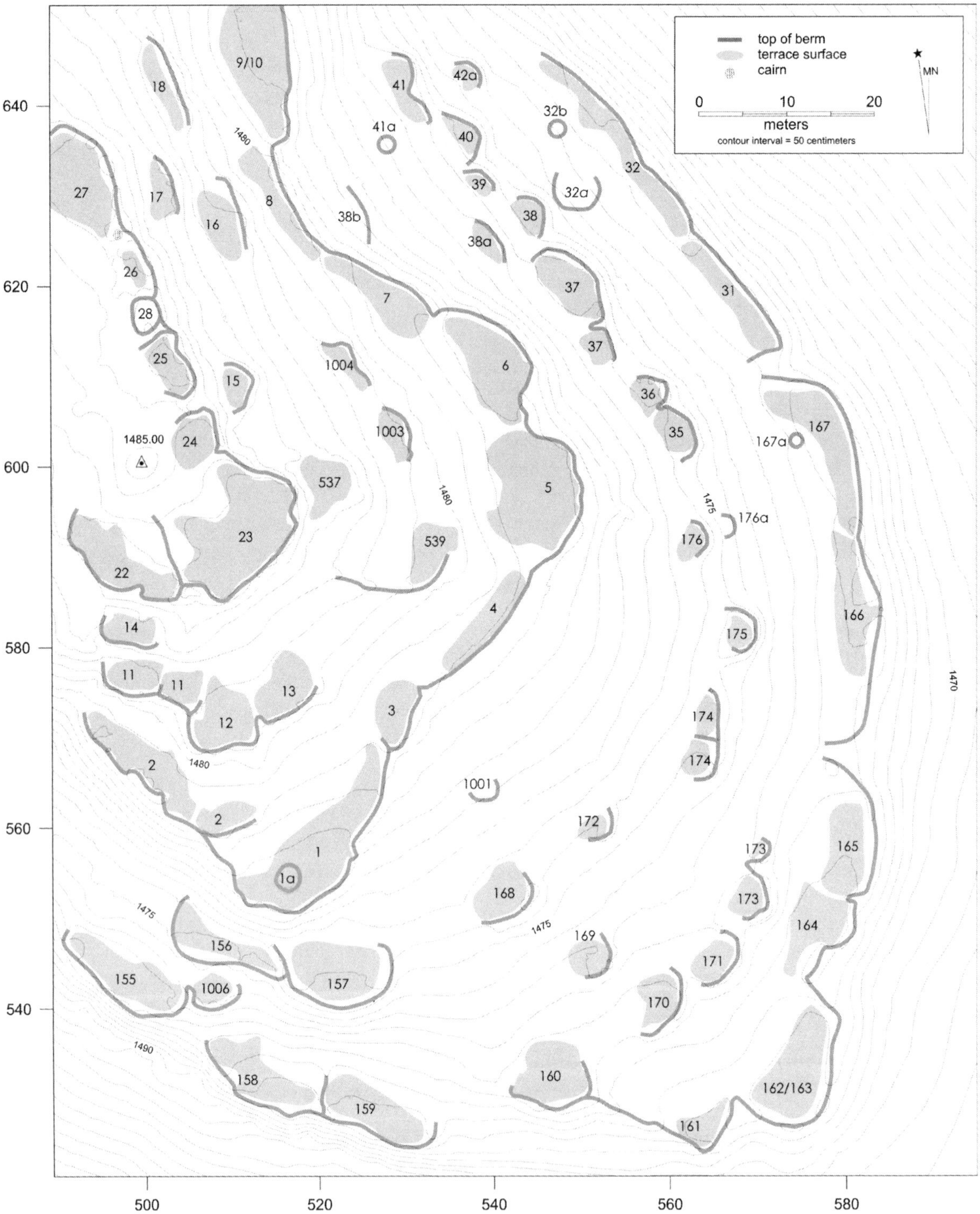

FIGURE 3.3. Features built on a gentle slope. Southeastern edge of Cerro Juanaqueña.

of post-occupational disturbance require comment. Rock squirrels live on the site today, and their widely scattered burrows are indicated by mounds of dirt peppered with pieces of white caliche, brought to the surface from deeper levels. Burrows of smaller rodents were also sometimes recognized in our excavation units, especially in deposits where fine sediments predominate. In addition to rodents, small to medium castings from biological organisms were common throughout the deposits, and uncharred plant material, as well as insect chitin were found in flotation samples from all depths.

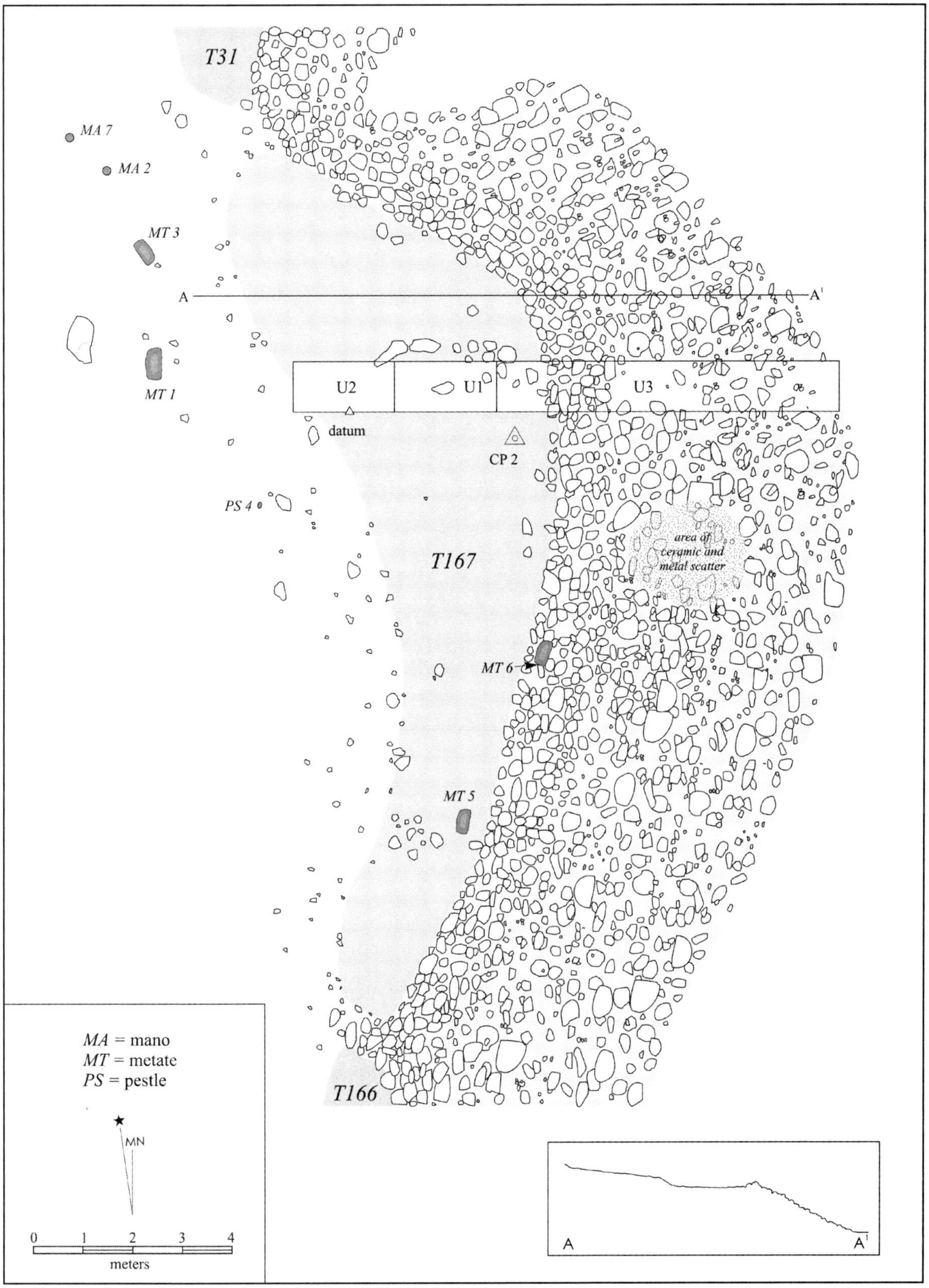

FIGURE 3.4. Terrace 167, illustrating typical terrace construction. Southeastern edge of Cerro Juanaqueña.

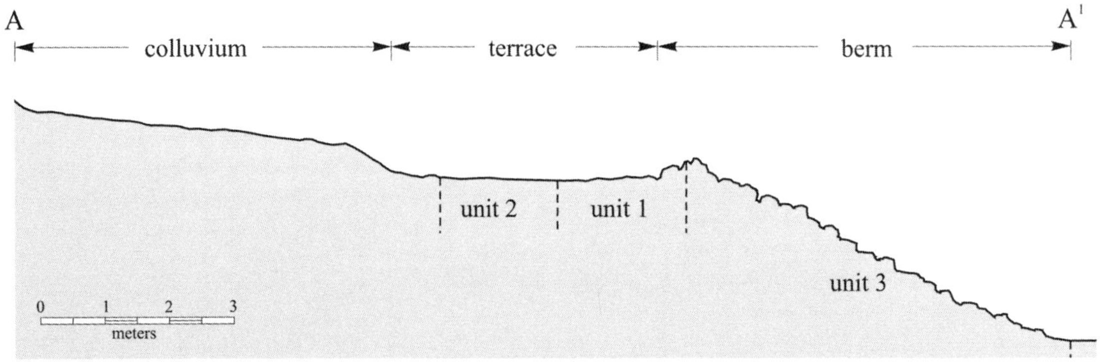

FIGURE 3.5. Cross section of Terrace 167. View towards north.

These observations indicate ongoing disturbance of the cultural deposits, but also suggest that the effects of this disturbance are selective. Specifically, there are few mechanisms at Cerro Juanaqueña capable of vertical movement of large objects such as manos, metates, cores, and hammerstones. In fact, these items are almost entirely limited to surface and near-surface contexts. Smaller items such as lithic artifacts and debitage are more susceptible to rodent disturbance. These items occur throughout the deposits but tend to be more common closer to the surface. Finally, small items such as bone fragments and flecks of charcoal seem to move freely through the deposits. We found no relationship between age and depth of our radiocarbon samples, and dates from the same stratigraphic column were often inverted. Both bone and charcoal were best preserved at depths of 30 cm or more below ground surface.

Excavation of Terrace Surfaces

Because terrace surfaces are well preserved, it seemed possible that we might find hearths, postholes relating to ephemeral structures, or other features associated with the terrace surfaces. To explore this possibility, we excavated large horizontal exposures on four terraces, clearing away the upper 5 or 10 cm of deposit over areas of up to 22 m^2, attempting to expose the more compact use surfaces discussed above. In addition, compaction surfaces were exposed on a number of other terraces prior to deep test excavations (see Table 3.1). For the most part, the results of this work were disappointing, but in several instances probable cultural features were found.

One of the most interesting features was a possible storage pit found in Terrace 487, in the lower terrace complex. This feature was apparently dug into the terrace sometime after original construction. The pit was recognized as an area of softer sediment packing a void in the rubble fill behind the terrace wall, and by a lining of 10 to 20 cm diameter cobbles. It appears to have been a meter or so in diameter and perhaps 60 cm deep (Figure 3.12). A large quantity of macrobotanical material, including maize, was recovered from its fill. There was also an unusual concentration of freshwater mollusk shell on this terrace.

On Terrace 6, excavators were able to follow extensive compaction zones over an area of about 16 m^2. They defined a shallow, informal pit dug into the terrace sediments, approximately 1.0 m in diameter and 18 cm deep, as well as two postholes. The postholes were 17 to 21 cm in diameter and 7 to 10 cm deep, with pebbles situated around their circumferences. A third possible posthole was found, but none of these features formed a clear pattern. Terrace 6 also produced the only example of burned daub found at Cerro Juanaqueña, reinforcing the impression that a structure of some sort once existed at this location. Finally, a small shallow hearth was found in Ring 239. This feature is described in more detail later in this chapter.

Comparison of Terrace Zones

In the introduction to this section, we described the overall organization of the terraces into three zones: Upper, Lower, and Middle (Figure 3.2). The upper and lower terrace zones are both compact, well-defined clusters of features, but the difference in setting is striking. This observation, along with differences in frequencies of certain features and artifact classes, raised the possibility of functional differences between the terraces in these two settings. This analytical theme is one which we will explore at various points in our discussion of Cerro Juanaqueña. It is worthwhile to note that several of the other cerros de trincheras in northwestern Chihuahua have analogous upper and lower terrace zones (Chapter 4). The distribution of terraces at Cerro la Cruz is very similar, with two well-defined and compact zones of features, one centered on the summit and the other taking the form of a long band near the base of the hill on the slope facing the Río

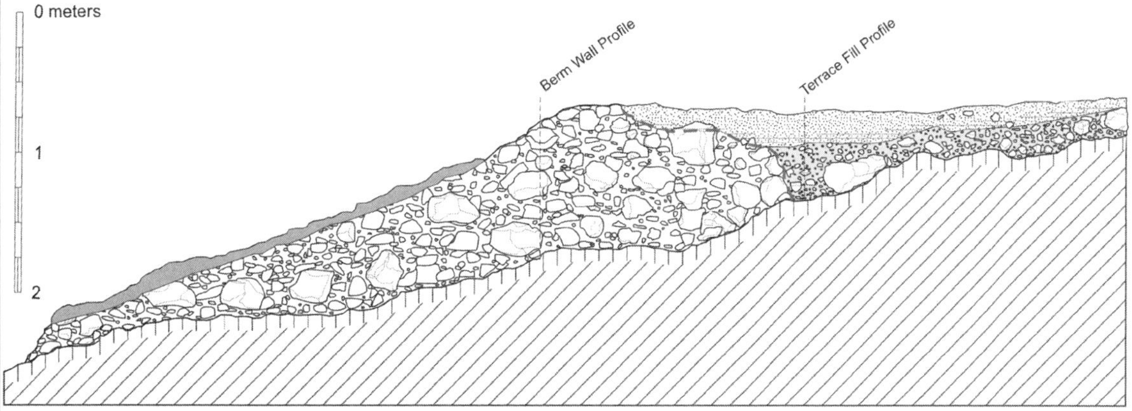

Berm Wall

0-6 cm below surface: Dark grayish brown sandy clay loam (very dark grayish brown when moist). Moderate fine platy structure, slightly hard. 5% angular basaltic pebbles, .2 to .4 cm in diameter, matrix supported, Moderately well sorted. Noncalcareous. Clear wavy boundary.

6-15 cm: Dark grayish brown sandy clay loam (very dark grayish brown when moist). Moderate medium subangular blocky structure, slightly hard. 50% angular basaltic cobbles, 10 to 20 cm in diameter; matrix supported; poorly sorted. 15% angular basaltic pebbles, .5 to 1.5 cm in diameter; matrix supported; poorly sorted. Many dark grayish brown cylindrical biocasts .4 cm in diameter. Noncalcareous. Clear wavy boundary.

15-48 cm: Very dark gray sandy clay loam (black when moist). Moderate medium subangular blocky structure, hard. 50% angular to blocky basaltic cobbles, 10 to 20 cm in diameter; matrix supported, poorly sorted. 15% angular basaltic pebbles .5 to 1.5 cm in diameter, matrix supported; poorly sorted. Many dark grayish brown cylindrical biocasts, .4 cm in diameter. Nonalcareous. Gradual wavy boundary.

48-89 cm: Black sandy clay loam (black when moist). Moderate course subangular blocky structure, hard. 50% angular to blocky basaltic cobbles; 10 to 20 cm in diameter; matrix supported, poorly sorted. 15% angular basaltic pebbles .5 to 1.5 cm in diameter, matrix supported; poorly sorted. Many dark grayish brown cylindrical biocasts, .4 cm in diameter. Calcareous.

Terrace Fill Deposits

Colluvium

0-7 cm below surface: Brown fine sandy loam (very dark brown when moist). Moderate fine platy structure, hard. 12% angular basaltic pebbles, .2 to .4 cm in diameter, matrix supported, poorly sorted. 2% angular basaltic pebbles, 2 to 3 cm in diameter, matrix supported, moderately well sorted. Few strong brown sandy inclusions. Noncalcareous. Clear wavy boundary.

Fine Sediment

7-15 cm: Brown sandy clay loam (very dark brown when moist). Moderate medium subangular blocky structure, hard. 20% angular basaltic pebbles, .2 to .4 cm in diameter, matrix supported, poorly sorted. 5% angular basaltic pebbles, 2 to 3 cm in diameter, matrix supported, moderately well sorted. 10% very dark grayish brown cylindrical biocasts. Noncalcareous. Clear wavy boundary.

15-21 cm: Brown sandy clay loam (very dark brown when moist). Massive structure with faint laminae, very hard. 5% angular basaltic pebbles, .2 to .4 cm in diameter, matrix supported, poorly sorted. Few fine faint yellowish brown soft iron masses. 3% very dark grayish brown cylindrical biocasts. Noncalcareous. Clear wavy boundary.

Construction Fill

21-45 cm: Dark gray sandy clay loam (black when moist). Massive structure, hard. 40% angular basaltic pebbles, .2 to .4 cm in diameter, matrix supported, moderately well sorted. 20% angular basaltic pebbles .5 to 3.0 cm in diameter, matrix supported, poorly sorted. 8% brown sandy patches. 10% very dark grayish brown cylindrical biocasts. Noncalcareous. Abrupt wavy boundary.

45-74 cm: Very dark gray sandy clay loam (black when moist). 35% angular basaltic pebbles, .2 to .4 cm in diameter, matrix supported, poorly sorted. 35% angular basaltic pebbles, 2 to 15 cm in diameter, matrix supported, poorly sorted. 3% brown sandy patches. 10% very dark grayish brown cylindrical biocasts. Noncalcareous.

FIGURE 3.6. Profile description of Terrace 167. View towards south.

FIGURE 3.7. Photograph of Terrace 287, partially excavated.

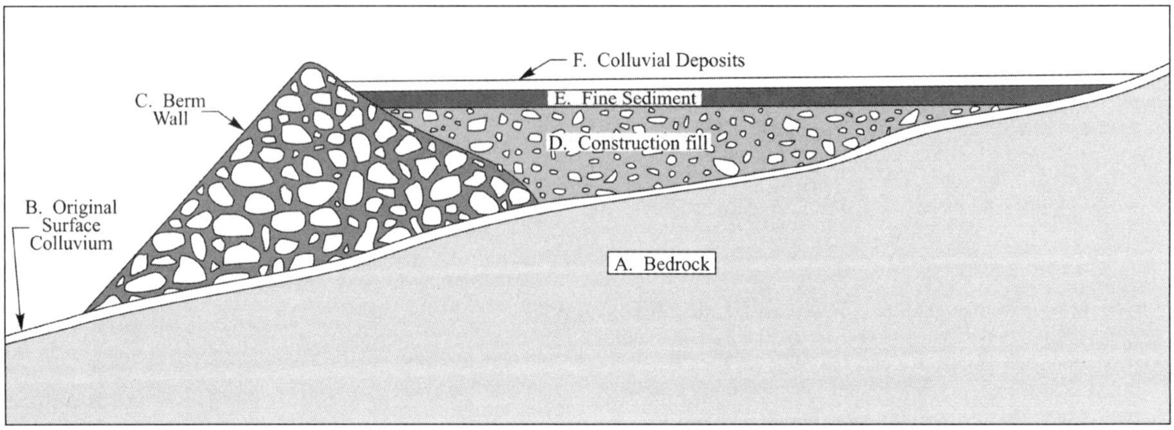

FIGURE 3.8. Idealized terrace profile.

Casas Grandes floodplain. A zone of lower terraces may also occur at Cerro la Boquilla de San Diego. Terraces in the middle zone do not form a compact group, and no clear analogues to this zone have been recognized at other sites.

The scale of the three terrace zones can be compared in several different ways. Table 3.2 tabulates statistics relevant to the scale of these constructions. Several preliminary observations are warranted at this point. Referring to Table 3.2, differences in useable surface area are interesting. Looking at columns 1 and 2, we see that slightly more than half of the terraces (measured by number or by total length) occur in the upper terrace zone. However, almost 63% of the artificial surface area is in this zone. This is true because at the summit of the hill large areas of useable surface were created by clearing cobbles from relatively level surfaces, and construction of terraces was unnecessary. As we will discuss later on, clearing cobbles from a naturally level surface is considerably less difficult than construction of a terrace on a slope. This finding implies that creation of useable surface area was less expensive at the summit of the hill

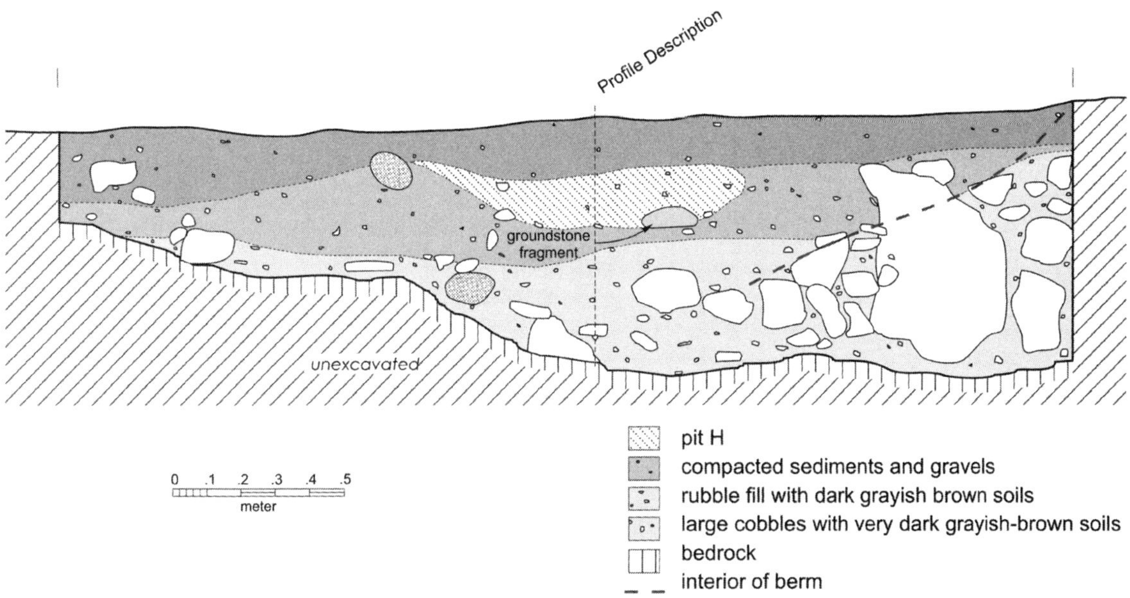

0-2 cm below surface: Yellowish brown fine sandy loam (dark grayish brown when moist). Moderate fine platy structure, hard. Few black burned streaks. Noncalcareous. Abrupt wavy boundary. Eolian or colluvium.

2-9 cm: Grayish brown sandy clay loam (very dark brown when moist). Moderate fine to medium platy structure, hard. 3% angular basaltic pebbles .5 to 1.0 cm in diameter; matrix supported; moderately well sorted. Noncalcareous. Clear wavy boundary. Fill material.

9-22 cm: Very dark gray sandy clay loam (black when moist). Moderate fine to medium subangular blocky structure, hard. 20% angular to blocky basaltic pebbles; 10 to 20 cm in diameter; matrix supported, poorly sorted. 15% angular basaltic pebbles .5 to 1.5 cm in diameter; matrix supported; poorly sorted. 5% angular basaltic pebbles .2 to .4 cm in diameter, matrix supported, poorly sorted. Noncalcareous. Clear wavy boundary. Fill material.

22-47 cm: Dark gray sandy clay loam (black when moist). Moderate medium subangular blocky structure, hard. 40% angular basaltic pebbles; 1 to 4 cm in diameter; matrix supported, poorly sorted. Calcareous. 2% carbonate filaments on gravel bottoms. Clear wavy boundary. Fill material or original ground surface.

47-72 cm: Grayish brown fine sandy loam (very dark grayish brown when moist). Weak medium subangular blocky structure, friable. 20% angular basaltic pebbles; 2 to 6 cm in diameter; matrix supported, poorly sorted. Calcareous. 2% carbonate filaments on gravel bottoms. Fill material or orginal ground surface.

FIGURE 3.9. Profile description of Terrace 6, near the summit of Cerro Juanaqueña. View towards northwest.

than on its slopes or in the lower terrace zone at the base of the hill.

The principal physical variable related to individual terraces which we monitored was terrace length, and with respect to this variable, we found no difference among the three terrace zones. Average terrace lengths in the upper, middle, and lower terrace zones are 15.4 m, 12.6 m and 14.4 m respectively. Researchers working with cerros de trincheras in later time periods in Sonora and Arizona have suggested that narrow terraces were used for agriculture while broader terraces were used for other day-to-day village activities (Fish et al. 1984). Although we did not formally monitor terrace width, no striking differences in this variable were found among the upper, middle, and lower terraces.

Macrofeatures and Perimeter Berms

In many cases individual terraces are discrete entities which can be easily distinguished from other nearby terraces. However, about 40% of the terraces articulate with one another to create linear groups consisting of two to five terraces, forming continuous surfaces. In these instances, definition of individual terraces is sometimes subjective, based on indentations in the line of the terrace berm or on cross-walls which sometimes subdivide the terraces (see later section on cobble alignments). We sometimes refer to these groups of adjacent, contiguous terraces as *macrofeatures* (Plate 2).

At Cerro Juanaqueña and many of the other cerros de trincheras in northwestern Chihuahua, the exterior boundaries of the terraced areas are defined by

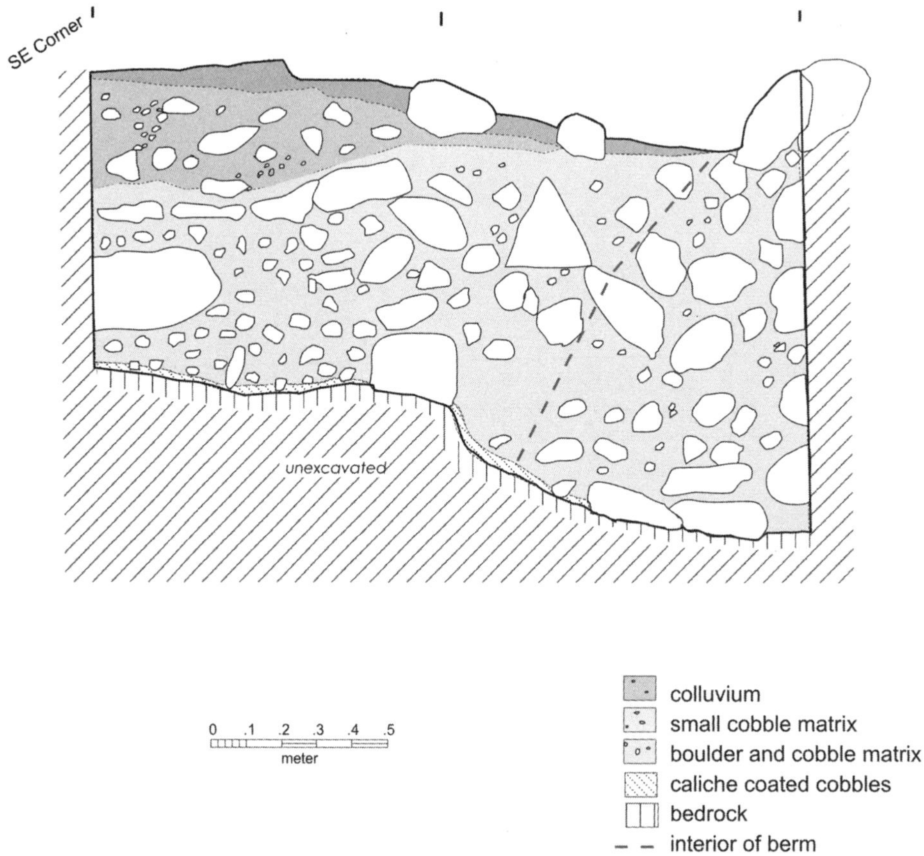

0-2 cm below surface: Grayish brown sandy clay loam (very dark grayish brown when moist). Moderate fine platy structure, slightly hard. 8% platy pebbles .3 to 1 cm in diameter, matrix supported. Noncalcareous. Clear wavy boundary.

2-9 cm: Grayish brown sandy clay loam (very dark grayish brown when moist). Moderate medium subangular blocky structure, hard. 15% angular to platy pebbles, .3 to 1 cm in diameter, matrix supported. Noncalcareous. Gradual wavy boundary. (Slightly compacted.)

9-29 cm: Grayish brown sandy clay loam (very dark brown when moist). Moderate fine and medium subangular blocky structure, hard. 20% angular to platy pebbles, .4 to 3 cm in diameter, matrix supported. Weakly calcareous. Clear wavy boundary.

29-57 cm: Dark gray sandy clay loam (very dark grey when moist). Moderate fine subangular blocky structure, slightly hard. 70% angular to platy pebbles, 2 to 6 cm in diameter. Matrix and grain supported. Calcareous. Clear wavy boundary.

57-91 cm: Dark gray sandy clay locam (very dark grey when moist). Moderate fine and medium subangular blocky structure, slightly hard. 80% angular to platy cobbles, 4 to 12 cm in diameter. Grain supported. Calcareous. Abrupt wavy boundary.

91+ cm: Basalt.

FIGURE 3.10. Profile description of Terrace 297 on the steeply sloping western slope of Cerro Juanaqueña. View towards south.

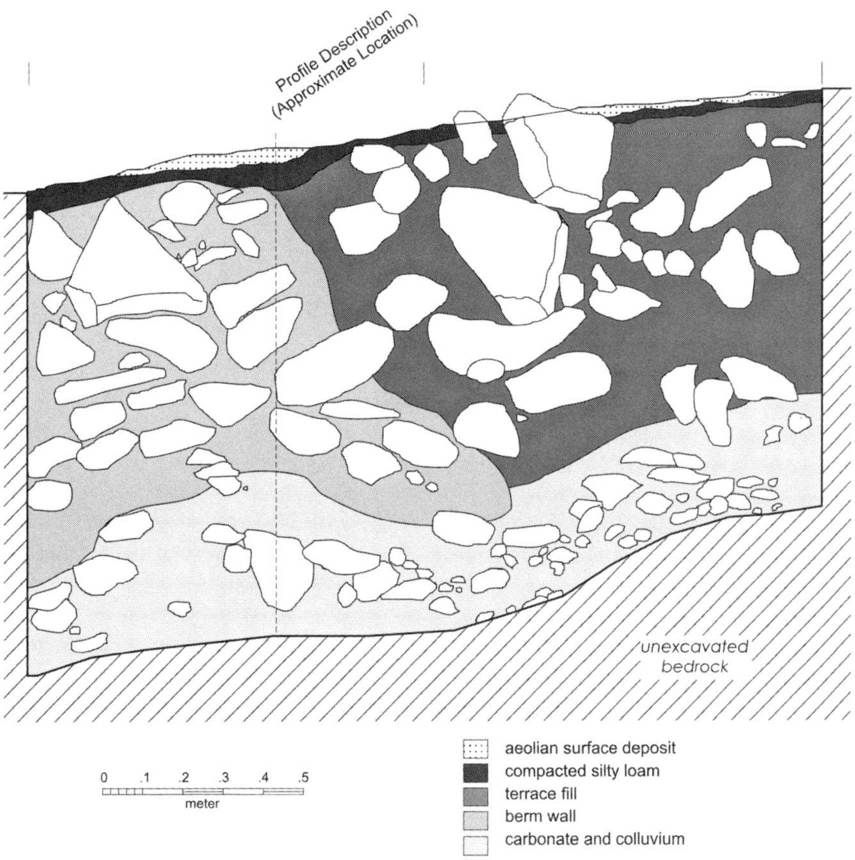

0-3 cm below surface: Brown fine sandy loam (dark brown when moist). Moderate fine and medium platy structure, slightly hard. 20% angular to platy pebbles .3 to 1.5 cm in diameter, matrix supported. Common fine biocasts, few 1 cm angular carbonate clasts. Weakly calcareous. Clear wavy boundary.

3-9 cm: Brown fine sandy loam (dark brown when moist). Moderate medium subangular blocky structure, hard. 25% angular to platy pebbles .3 to 1.5 cm in diameter; matrix supported. Common fine and medium biocasts. Few 1 cm angular carbonate clasts. Calcareous. Clear wavy boundary.

9-17 cm: Dark grayish brown sandy clay loam (very dark grayish brown when moist). Moderate fine and medium subangular blocky structure, slightly hard. 35% angular to platy pebbles .3 to 2.5 cm in diameter; matrix supported, poorly sorted. 15% angular basaltic pebbles .5 to 1.5 cm in diameter; matrix supported. Many fine and medium biocasts. Few 1 cm angular carbonate clasts. Calcareous. Clear wavy boundary.

17-33 cm: Dark grayish brown sandy clay loam (very dark brown when moist). Moderate fine and medium subangular blocky structure, slightly hard. 40% angular to platy pebbles; .5 to 3 cm in diameter; matrix supported. Many fine and medium biocasts. Few 1 cm angular carbonate clasts. Calcareous. Clear wavy boundary.

33-59 cm: Dark grayish brown sandy clay loam (very dark brown when moist). Moderate fine and medium subangular blocky structure, slightly hard. 70% blocky cobbles 4 to 12 cm in diameter. Common fine and medium biocasts. Few 1 cm angular carbonate clasts. Calcareous. Clear wavy boundary.

59-90 cm: Dark grayish brown sandy clay loam (very dark brown when moist). Moderate fine and medium subangular blocky structure, slightly hard. 80% blocky cobbles 4 to 15 cm in diameter. Common fine and medium biocasts. Few 1 cm angular carbonate clasts. Calcareous. Clear wavy boundary.

90-111 cm: Dark grayish brown clay loam (very dark brown when moist). Moderate fine and medium subangular blocky structure, slightly hard. 50% angular pebbles, 1 to 4 cm in diameter; matrix supported. Few fine and medium biocasts. Few 1 cm angular carbonate clasts. Calcareous. Clear wavy boundary.

111+ cm: Basalt with 1 cm thick white carbonate layer on top.

FIGURE 3.11. Profile description of Terrace 387, near the base of Cerro Juanaqueña. View towards north.

FIGURE 3.12. Photograph of Terrace 487, Pit B.

TABLE 3.2. Comparison of Terrace Zones.

	Number of Terraces [1]		Length of Terraces [2]		Total Area [3]		Total Artificial Surface Area [4]	
Zone	*n*	%	m	%	ha	%	ha	%
Upper	309	56.2	4754	58.3	6.0	58.7	2.27	62.9
Middle	40	7.3	502	6.1	—	—	0.20	0.6
Lower	201	36.5	2900	35.6	4.2	41.3	1.14	31.6
Total	550	100.0	8156	100.0	10.2	100.0	3.61	100.1

[1] Raw counts of the number of terraces in each setting. Distinguishing the end of one terrace and the beginning of another is sometimes subjective, but in most cases the terraces are either distinct constructions, or else pronounced indentations in their berm walls provide a basis for separation.

[2] Length of the terraces in each setting was calculated from the rectified, photo-interpreted map using ArcView GIS software.

[3] Total area is based on a simple outline of the entire zone, calculated using GIS. This measure is useful for the upper and lower terraces, which occur in compact, well-defined zones. However, the middle terraces are widely scattered, and the area in which they occur cannot be meaningfully compared to the total area of the upper and lower terrace zones.

[4] Total artificial surface area refers to the level surfaces created by terrace construction or created by clearing cobbles from naturally level surfaces. These figures were also calculated using GIS software. The map was superimposed over the rectified, scaled aerial photograph of the site, and a polygon was created for each individual surface. In general, terrace surfaces were defined as extending from the terrace wall to the talus slope behind it. Where this distinction could not be made on the basis of the map, photointerpretation of the underlaying photograph was used instead. Areas of the individual surfaces were then calculated using GIS and summed to provide the figures given in this table.

very large macrofeatures, which we refer to as *perimeter berms*. The macrofeature which defines the northern, eastern, and southern perimeter of Cerro Juanaqueña consists of 22 adjoined terraces forming a 400 m long alignment and can be seen clearly in Plates 1 and 2. Where this macrofeature runs perpendicular to the hillslope, it resembles other macrofeatures composed of individual terraces, but on the northern and southern sides of the site, it runs parallel to the slope and forms a large cobble berm, rather than a terrace. This feature is also different from most other macrofeatures in that the cobble berm extends well above terrace surfaces in many places. This cobble berm defines the exterior boundaries of the site, an observation which implies that a large portion of the site did not grow through accretionary, uncoordinated construction episodes. Instead the features appear to have been planned to accommodate definite spatial needs and constructed in a single, coordinated episode. In Chapter 4 we describe similar features from other Early Agricultural period cerros de trincheras in northwestern Chihuahua. These features are key pieces of evidence indicating aggregation and a degree of organized decision-making which were unexpected in this period.

Rock Rings

During initial inspections of Cerro Juanaqueña we identified 99 circular rock alignments, either built on the terrace surfaces or elevated slightly above them, which we refer to as *rock rings*. Similar features are standard elements of cerros de trincheras from later time periods in Sonora and Arizona. Their purpose is not well understood, although upon excavation some have proven to be shallow pithouses (Downum 1993; Fish et al. 1984). At Cerro Juanaqueña we suspected that some of these might be dwellings and others could have been storage features. In an effort to better understand them, we excavated several and made detailed surface observations for all of the others.

The rock rings at Cerro Juanaqueña encompass a considerable amount of variability. About 39 of them are very distinct and are clearly cultural features. Another 40 are reasonably well-defined and are very likely to be cultural features rather than fortuitous cobble alignments. The remaining 20 rock rings are less distinct and some of these may not actually have been deliberate constructions. Site maps presented in this report (e.g., Figure 3.2) generally show all 99 of these features, but the description which follows here is based on the 79 rock rings which are most confidently identified as cultural features. Summary information about these examples is presented in Table 3.3 and the detailed data are in Appendix 3.6.

Most of the rock rings are relatively small, with interior diameters between 0.9 and 4.3 m (but one example

TABLE 3.3. Summary Data for Rock Rings.

Interior Diameter	Number	Definition	Number
0.9 to 2.0 m	25	Discontinuous cobble alignment	10
2.05 to 3.0 m	37	Single course of cobbles	45
3.05 to 4.0 m	13	Multiple courses of cobbles	19
4.05 to 6.0 m	4	Pocket in talus	4
		Unknown	1

Shape (complete and near complete only)	Number	Circumference	Number
Round	31	Complete (360°)	27
Oval	30	Near Complete (230° to 355°)	35
Semicircular	17	Semicircular (90° to 200°)	17
Rectangular	1		

Horizontal Relationship to Terraces	Number	No. Subset
Isolated, not associated with any terrace	12	
Separate feature overlooking terrace	8	
On terrace	58	
Near middle of terrace		(18)
Offset to left or right of middle		(14)
In extreme right or left corner of terrace, between terrace wall and talus		(23)
Terrace edge		(3)
Unknown	1	
Total	79	

Vertical Relationship to Terraces	Number
On terrace, slightly elevated above surface [1]	34
On or slightly below terrace surface	32
Unknown	13

[1] Range = .1 to 1.5 m, median = 40 cm

Terrace Zones	Number
Upper terraces	59
Lower terraces	16
Middle terraces	4

Opening Direction	Count
N	8
NE	1
E	6
SE	12
S	7
SW	8
W	6
NW	7
No opening	22
Unknown	2

FIGURE 3.13. Photograph of Ring 239, showing bisected hearth. View towards west.

FIGURE 3.14. Photograph of metate resting on compaction surface in Ring 286.

had a diameter of 6.0 m). In addition to this outlier, the diameters of the rock rings form two modes; the small group is normally distributed with a mean of 2.3 m and the medium group has a mean of 4.1 m (small group: 0.9 m to 3.6 m, $s = 0.61$ m, $n = 71$; medium group: 4.0 to 4.25 m, $s = 0.1$ m, $n = 7$). The data do not reveal any patterns in construction, shape, orientation, positioning, arc, or upper and lower terrace locations that correlate with this size class distinction in a meaningful way.

A single continuous course of cobbles at ground level defines most of the rock rings, but a significant number did include multiple courses. Even among those with multiple courses, mean height of the walls is only about 26 cm. Circumferences of a majority of the rock rings were defined around an arc of 230 degrees or more, and many of these were defined around a full 360-degree circumference. Proportions of round and oval-shaped rings are approximately equal. One ring, Ring 28, was characterized as subrectangular in shape. This feature was located at the summit of the site and appeared to be unique, but excavation revealed no special features or artifact associations. There were 55 rock rings with discernable openings and these openings were oriented in all directions. There is a slight bias towards southerly and easterly orientations rather than northerly (Table 3.3). Most of the rock rings were associated with terrace surfaces, and we found no strong pattern in their position on the terrace. A majority were built directly on the terrace surface, but a significant number were elevated somewhat above the terrace surface (see section on platforms below). For those that were elevated above the terrace surface, elevations ranged from 0.1 m to 1.5 m, with a median of 40 cm ($n = 34$, interquartile range = 48 cm).

The final comparison in Table 3.3 suggests that rock rings are concentrated in the upper terrace zone and are underrepresented in the lower terrace zone. We tested this observation for statistical significance using the chi-square test, basing expected values on four different characterizations of the terrace complex: total number of terraces in each zone, total length of the terraces in each zone, total area, and total area of artificial surface in each zone (Table 3.2). The rock rings are overrepresented in the upper terrace zone and underrepresented in the lower terrace zones, with levels of significance (p values) ranging from 0.05 to 0.001. The middle terrace zone was excluded from this analysis because the number of rock rings expected in this zone was too low to test meaningfully. Within the upper terrace area, rock rings are concentrated in the central part of the zone, below the crest, but near the top of the west-facing slope of the hill.

In an attempt to better understand the rock rings, we excavated eight of these features from the upper terrace zone, although two of these received only small test units (Table 3.1). None of the small number of rock rings on the lower terraces appeared likely to provide useable excavation data and none were explored. In most of the rock ring excavations we found a compact surface only a few centimeters below the modern ground surface which we believe was used prehistorically. Terrace deposits beneath the rock rings were essentially identical to the deposits excavated on terraces without rock rings.

We found no interior features in six of the eight excavated rock rings. However, in Rock Ring 239, we found evidence of a simple hearth in the middle of the structure and two metates resting on the compaction surface near the perimeter of the feature (Figure 3.13). The hearth was a shallow depression about 70 cm in diameter and 8 cm deep, with a few burned rocks and ashy deposit in its uppermost sediments, and discoloration caused by thermally induced oxidation toward its bottom (Roney and Hard 2000). Unfortunately, because of its shallow depth, no charred plant material was preserved in this feature, so we were unable to date the hearth directly.

A second example, Rock Ring 286 did not include interior features, but it is interesting because two metates were found on its floor (Figure 3.14), and because its construction details turned out to be different than we had expected. Initially we believed that this was a large, oval-shaped rock ring overlooking Terrace 287. However, excavation results suggested that the entire feature was actually a platform, and that the structure itself was a semicircle of dry laid cobbles at least two courses high facing onto the platform. Despite careful excavation, no features were found within either the semicircle or on the platform (Roney and Hard 2000).

We did not find evidence that floors of the rock rings were placed notably below the surface in the fashion of a pithouse; it appears that the structures were constructed on, or a few centimeters below, the occupation surface. There was no evidence of wall postholes in the thin, rocky sediments. Perhaps construction posts were secured within the rock ring perimeter, although no direct evidence was found of this method. Physical characteristics of the rock rings, such as their diameter, a circular to oval form, and minimal construction effort suggest that they were brush structures, although smaller examples could have been storage structures instead. We could not identify patterns in the data that allowed us to discriminate between dwellings and storage features. Details of the chipped stone, fauna, and plant remains found during the excavation of the rock rings are found in the appropriate chapters, and these materials are generally consistent with domestic activities.

Other Surface Features

Cobble Alignments

In several instances we found linear alignments of cobbles on the terrace surfaces. Typically, these were perpendicular to the terrace berm and extend from the berm to the talus slope behind the terrace. We call these features *cross-walls*. These usually consisted of a single row of small (20 cm diameter) cobbles, but sometimes they were composed of small boulders (ca. 60 cm in diameter). They often divided individual terraces into halves or thirds. We mapped approximately 52 of these features at Cerro Juanaqueña, 38 in the upper terrace zone, three on the middle terraces, and 11 in the lower terrace zone. In three instances cross-walls were intersected by an additional cobble alignment at a right angle, and in two instances there were cobble alignments that were parallel to the terrace berm. At Cerro Vidal, described in Chapter 4, cross-walls occurred in considerable numbers. We do not understand the purpose of these cobble alignments.

Walls

At several locations we found berms of cobbles running up and down the slope between different terrace levels. These were much more massive than cross-walls, and they were found on the slopes between terraces, rather than on terrace surfaces. Walls were sometimes impressive features, 14 m in length, 2 or 3 m wide at the base, and close to a meter in height. The purpose which these served is not clear, but similar features were also documented at Cerro Vidal.

Platforms

At Cerro Juanaqueña we observed that one or both ends of the crescent-shaped terraces were elevated slightly above the rest of the terrace surface. As described above, rock rings were sometimes built on these elevated surfaces, which we call *platforms*. Usually platforms were not defined by cobble alignments, and they were subtle and subjectively recognized. In other cases the platforms were very definite, and there were also examples in which small separate terraces overlook larger terraces in a similar physical relationship. Because definition of platforms was so subjective, we did not attempt to include them on our map of the site. As noted above, 22 rock rings were built on platforms, averaging 37 cm above the main terrace surface. Rock Ring 286, described above, is an example of an excavated rock ring-platform complex.

Burned Rock Features

The native basalt at Cerro Juanaqueña fractures with an angular pattern, and we seldom observed rocks that we could confidently identify as "fire-cracked" or "fire-affected." However, in six locations near the summit of the hill there were low mounds of obviously burned rock, distinguished by relatively uniform size (diameters ca. 5 to 8 cm), by an unusually high frequency of angular breakage, and by a large number of spalls. These features were circular or oval in shape and averaged about 5.5 m in diameter, with approximately 25 cm of relief. One of them (BR 2) included an unusual concentration of mano fragments and hammerstone fragments, but this was not true of the others. Two examples (BR 5 and 6) are draped over terrace berms, indicating that they post-date construction of the features. However, radiocarbon dating shows that the burned rock features are approximately contemporaneous with the principal occupation of the site. We excavated one of the burned rock features (BR 1) located near the summit of the hill. This feature was a small (2.8 m diameter) concentration of heat-fractured rock with about 40 cm of relief. A light surficial scatter of burned rock extended downslope for 6 or 7 m. Excavation showed that there was no internal pit. The feature rested on original ground surface, although the configuration of weathering bedrock may have created a slight natural depression at this location. Intense ash deposits were found in the center of the feature, but heat spalling and fracturing was most pronounced among rocks from the upper half of the deposit.

Preservation of charred botanical materials from this feature was poor, probably because the feature is shallow. However, four pieces of charcoal from BR 1 could be identified as fragments of mesquite wood, and one of these yielded a radiocarbon age of 3210 ± 50 BP (INSTAAR, NSRL-11428; see Table 5.1). This date is comparable to others from the site and confirms that the burned rock feature was used during the primary occupation of Cerro Juanaqueña. The other five burned rock features located nearby are similar in appearance, and there is no reason to doubt that they are contemporaneous with the main occupation of the site.

Cairns

Six cairns were identified on Cerro Juanaqueña. Two of these are at or near the summit of the hill and consist of areas where cobbles have been piled over natural boulder outcrops. The resulting cairns are about 5 m in diameter and perhaps 1.5 m high. Despite their size, these features have an informal appearance and may have resulted from incidental activities rather than purposeful construction. Four small cairns consisting of 10 to 30 cobbles stacked or grouped into small piles occur elsewhere on the site, three on the summit and one on the west-facing slope of the hill. It is impossible to say whether these features pertain to the primary occupation at Cerro Juanaqueña or if they are the result of post-occupational activities.

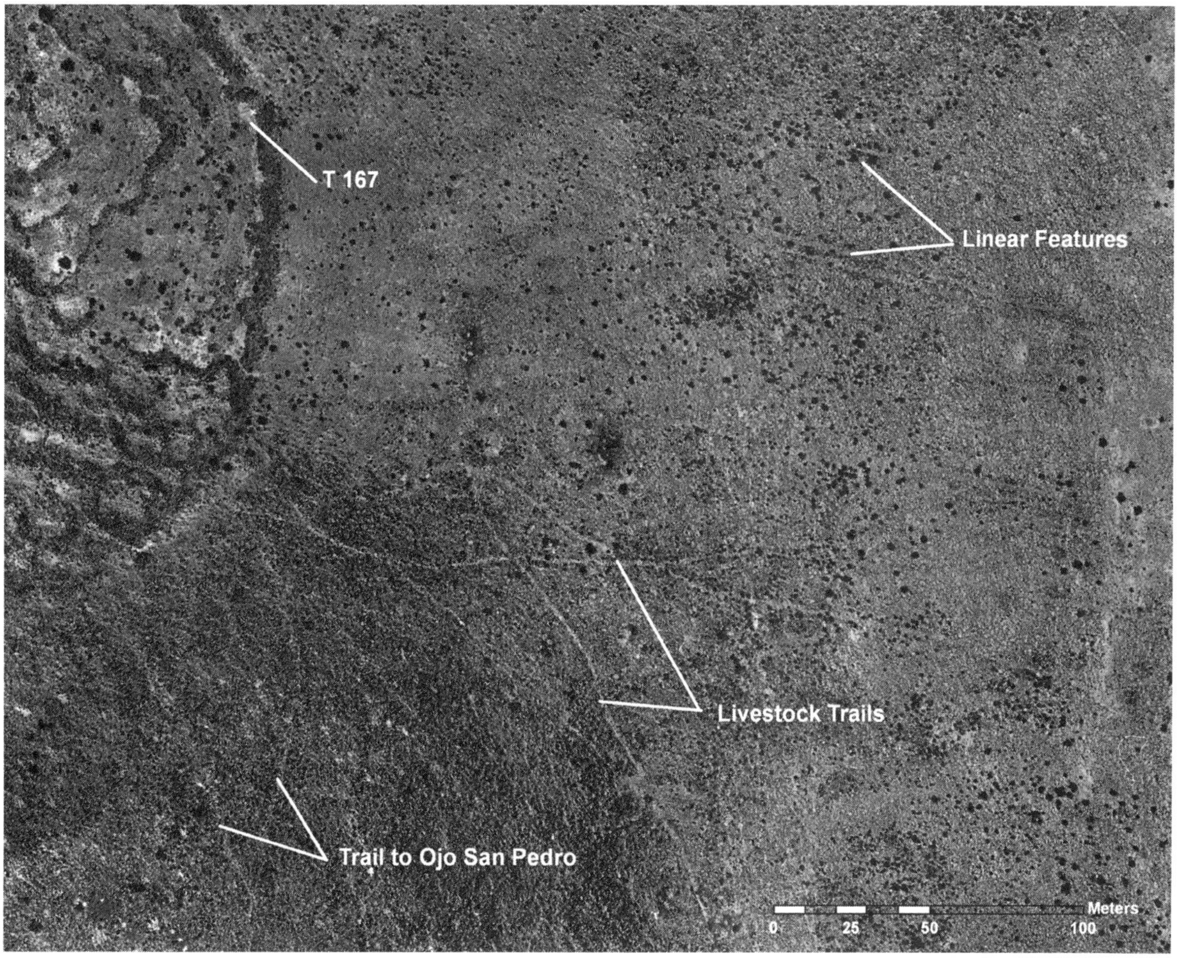

FIGURE 3.15. Aerial photograph showing linear features associated with Cerro Juanaqueña.

Trails and Linear Features

Several trails ascend Cerro Juanaqueña, providing relatively easy access from the Río Casas Grandes floodplain to the summit. One begins on a colluvial slope 600 m to the south of Cerro Juanaqueña near a major spring, Ojo San Pedro, and ascends diagonally across the slope of the ridge, reaching the crest at a point very near the southeastern corner of the Cerro Juanaqueña terrace complex (Figure 3.15). Another begins 500 m away near the southern end of the lower terrace complex and ascends to more or less the same point near the southeastern corner of the terrace complex. Both of these trails have branches that lead into the terrace complex. Several other livestock trails lead to the southeastern corner of the upper terrace complex across relatively gentle slopes to the southeast and east of the site. Another major trail, one that is used most intensively by livestock today, begins near the floodplain a little north of the lower terraces and wraps around the northern slope of the hill, ascending steadily until it reaches the saddle about 250 m west of the upper terrace complex. From this point the terraces can be reached by walking up a gentle slope that is relatively clear of cobbles.

Some of these trails may have been created and maintained by livestock use, but others, including the trail leading towards Ojo San Pedro and the one that wraps around the northern end of the site, show evidence of casual construction. Large cobbles and small boulders were moved aside at various points to make the route easier to use. However, we should point out that while working at Cerro Juanaqueña we seldom used these trails. Unless we were carrying very heavy loads, it was usually more expedient to climb directly up the hill slope.

Another set of curious linear features appear to focus on Terrace 167, on the eastern edge of the upper terrace complex. These features are clearly visible in aerial photography as two sets of parallel lines that extend eastwards from the site for at least 300 m (Figure 3.15). The paired lines are 3 or 4 m apart. One pair runs approximately 106° east of north, and the other approximately 101° east of north, leading toward the head of a small drainage immediately east of Cerro Juanaqueña.

On the ground, these features resemble the trails described above. Cobbles and boulders have been moved aside to define a swale with the appearance of a footpath. However, in comparison to the other trails, these features are much more straight and formal in appearance, they are not used by livestock today, and they could be prehistoric foot paths. The head of the drainage east of the site is a potential source of fine sediments for terrace construction, and a modern earthen livestock tank located 1.5 km east of the site might have been a source of water in prehistoric times as well.

TABLE 3.4. Grinding Slicks.

	Grinding surface length (cm)	Grinding surface width (cm)
n	13	13
Mean	38.4	22.9
Median	30.0	16.0
Maximum	99.0	64.0
Minimum	16.0	8.0
Standard Deviation	25.9	14.8
Interquartile Range	21.0	15.0

Grinding Slicks

Ground surfaces on basalt boulders and outcrops are labeled "grinding slicks," and 26 were recorded on the surface of Cerro Juanaqueña. As discussed below, these features resemble metates and might have been discussed in Chapter 6, along with other artifacts. Instead, they are included in this chapter because they are fixed locations on the site and are therefore conceptualized as features rather than artifacts. They occur on both terrace surfaces and on talus slopes of the terraces; some occur in clusters but others occur as isolates. Two clusters of grinding slicks account for over half of the total, and both are close to the top of the hill. One cluster of six occurs in the northeast sector, in a large cleared area. Here one slick formed over a likely Archaic-age petroglyph (see Rock Art Locus 19, below). A second cluster of about 10 grinding slicks occurs about 50 m to the south of the first, in a cleared area, in the summit sector. Most of the remaining slicks are more widely scattered, but in the northwest sector.

We measured 13 of the 26 grinding slicks following the same sampling procedures used for the whole metates (Chapter 13). All slicks were recorded and every other one encountered was measured. Since these grinding features are poorly understood, a comparison of their dimensions with those of the grinding slabs and basin metates may shed light on their function. Their dimensions are described in Table 3.4. The smallest slick is only 16 cm long and 8 cm wide, but the largest slick is 99 cm long and 64 cm wide, providing a very large working surface. The medians of both grinding surface length and width are intermediate between those of the basin metates and grinding slabs, but overlap the size distribution of both (see Table 13.11). As was observed in both the grinding slabs and basin metates, there is a linear correlation between grinding surface length and width among the grinding slicks. These surfaces exhibit light, medium, and heavy wear, with medium being the most common. As wear increases, both grinding surface length and width increase. The two features with the largest grinding surfaces are the only two grinding slicks with heavy wear, and they both had a measurable depth of wear of 0.5 cm.

The advantages and disadvantages of the slicks relative to the basin metates and grinding slabs are difficult to know. Given the slight amount of wear, the slicks may have served when the grinding slabs with their small surfaces were inadequate for a task. One interpretation of the clustered nature of many of these features is their use involved a social group activity. The grinding slicks are consistent with many other lines of evidence indicating the wide array of domestic activities that occurred on this extensively modified site.

Bedrock Mortars

Only six bedrock mortar holes were recorded on Cerro Juanaqueña, and four of these are clustered in the southeast quadrant of the site, within a 50 by 50 m area that includes one cupule location and a grinding slick. One is associated with Rock Art Locus 2. The mortar holes occur both on terrace surfaces and talus slopes. They are cylindrical with diameters ranging from 7 to 18.5 cm (7, 7.5, 9, 14, 18.5 cm), plus one that is an irregular 8 × 14 cm. Depths were only recorded for four, and three of these were quite shallow, 1 to 2 cm in depth, with one measuring 6 cm deep. Given the shallow nature of these, most do not appear to be typical mortar holes. In fact, some of these, particularly the two smallest ones, could be considered cupules as described below.

ROCK ART

Cupules

Cupules (not to be confused with corn cupules) are small, circular depressions ground into the faces of boulders and outcrops (Figure 3.16). We found 76 cupules at 10 different loci on the site (Table 3.5) and measured a sample of 45 of these. At Cerro Juanaqueña the cupules are usually 3 to 8 cm in diameter (mean = 5.2 cm, s = 1.2 cm, n = 45) and average about 0.7 cm in depth (s = 0.5 cm, n = 45), with rounded, smooth bottoms that appeared to have grinding wear.

The degree of patination on the cupules suggests that all pertain to the Early Agricultural period at Cerro Juanaqueña. The cupules are patinated to the same degree as the boulders and outcrops upon which they were

FIGURE 3.16. Boulder covered with cupules at Cerro Juanaqueña, Locus A, Panel 1.

created. In contrast, most petroglyphs, including those that could be confidently attributed to the AD 1200 to 1450 period on stylistic grounds, have a relatively fresh appearance, with good contrast between the pecked design and the native rock face. These observations imply that the cupules are considerably older than most of the petroglyphs at this site. Because of the disturbance associated with construction of the terraces and because of their relatively nondescript location, it is unlikely that the cupules predate the Early Agricultural period occupation at Cerro Juanaqueña.

In two cases multiple cupules are closely spaced on boulders or outcrops that are about one meter in length. In many cases the distance between adjacent cupules is less than the diameter of a cupule. In these two cases most of the cupules are generally oriented in a horizontal plane, but a few are near the sloping edge of an outcrop and are slightly angled. The surface between the cupules appears to have been smoothed by pecking.

Cupules are found all around the world, and in the western United States, they are well-documented in the Great Basin, California, the Southwest, and in southwestern Texas. In the western United States they range widely in age. Those in western Nevada are thought to be some of the oldest examples of rock art, perhaps dating as early as 5000 BC (Heizer and Baumhoff 1962). In contrast, some in southern California may have been made in recent times (Minor 1975; Nissen and Ritter 1986). Functional interpretations also vary widely. Ethnographic evidence is often cited as linking cupules to ritual activity (see Stoney 1993). On the other hand, Wallace (1983) notes that prehistoric cupules in Arizona are often associated with bedrock mortars and proposes that the features were used to sharpen pestles.

Petroglyphs

With help from Michael and J. J. Golio, we documented petroglyphs at 26 loci on Cerro Juanaqueña. Each locus usually consists of a single boulder, and although they are widely scattered, all are located toward the highest part of the upper terrace complex. Loci were subdivided into 43 panels, each usually consisting of one face of a boulder, and within these panels we recorded a total of 69 individual elements, which are characterized briefly in Table 3.6.

Only one element was tentatively attributed to the Early Agricultural period, Locus 19.A.1 (Figure 3.17a). Like the cupules discussed above, this element seemed to be fully repatinated. All of the other petroglyphs had a relatively fresh appearance, and unless they included historic iconography (horses, Christian crosses, Latin letters, numbers, and brands) they are assumed to be prehistoric, most likely pertaining to the latter part of the Ceramic period, AD 950 to 1450.

TABLE 3.5. Cupules.

Locus	Panel	Number	Comments
A	1	23	5 or 6 m from Loci A2 and A3
	2	1	About 1 m from Locus A3
	3	1	
B	1	9	
C	1	3	
D	1	9	Two adjacent boulders
E		1	
F	1	1	
G	1	6	Two separate boulders about 1 m apart
	2	4	
H	1	2	On same panel as Petroglyph Locus 7
I	1	5	On same panel as Petroglyph Locus 19
J	1	11	On same panel as Petroglyph Locus 26

TABLE 3.6. Rock Art Locations.

Locus	Panel	Element	Description	Age	Figure No.
1	A	1	Face	Prehistoric	
1	A	2	Curved line	Prehistoric	
1	A	3	Curved line	Prehistoric	
1	B	1	Complex design	Prehistoric	
1	C	1	Parallel lines	Prehistoric	
2	A	1	Initials	Historic	
2	A	2	Date	Historic	
2	B	1	Date	Historic	
3	A	1	Rabbit	Prehistoric	
4	A	1	Brand	Historic	
5	A	1	Fish	Prehistoric	
6	A	1	Anthropomorph	Prehistoric	3.17 C
6	B	1	Name	Historic	
6	C	1	Name	Historic	
6	C	2	Brand	Historic	
6	D	1	Brand	Historic	
6	D	2	Brand	Historic	
7	A	1	Interlocking lines	Prehistoric	
8	A	1	Initials	Historic	
8	B	1	Letter	Historic	
8	B	2	Church	Protohistoric/Historic	
9	A	1	Date	Historic	
9	B	1	Letters	Historic	

TABLE 3.6. (cont'd.) Rock Art Locations.

Locus	Panel	Element	Description	Age	Figure No.
10	A	1	Stick figure	Prehistoric	
10	A	2	Concentric circles	Prehistoric	
10	A	3	Concentric semicircle	Prehistoric	
10	A	4	Curvilinear	Prehistoric	
10	C	1	Concentric circles	Prehistoric	
10	B	1	Random lines	Unknown	
11	A	1	Indistinct	Prehistoric	
12	A	1	Interlocking frets	Prehistoric	
12	A	2	Interlocking frets	Prehistoric	
13	A	1	Maltese cross	Prehistoric	
14	A	1	Footprint	Prehistoric	
14	A	2	Footprint	Prehistoric	
14	A	3	Footprint?	Prehistoric	
14	A	4	Footprint?	Prehistoric	
14	A	5	Footprint?	Prehistoric	
14	A	6	Spot	Prehistoric	
15	A	1	Cuadro	Prehistoric	
15	A	2	Complex design	Prehistoric	
15	A	3	Stick figure	Prehistoric	
15	A	4	Curvilinear	Prehistoric	
16	A	1	Church	Protohistoric/Historic	
17	A	1	Large X	Prehistoric	
17	A	2	Small x	Prehistoric	
17	A	3	Small x	Prehistoric	
17	A	4	Small x	Prehistoric	
17	A	5	Rectangle?	Prehistoric	
17	B	1	Rectilinear	Prehistoric	
17	B	2	Rectilinear	Prehistoric	
17	C	1	Stick figure?	Prehistoric	
17	D	1	Small x	Prehistoric	
18	A	1	Cuadro	Prehistoric	
19	A	1	Curvilinear	Archaic?	3.17 A
20	A	1	Concentric circles	Prehistoric	
20	A	2	Circle	Prehistoric	
20	B	1	Face	Prehistoric	
20	C	1	Complex design	Prehistoric	
20	C	2	Complex design	Prehistoric	
21	A	1	Anthropomorph	Prehistoric	3.17 D
21	B	1	Man on horse	Protohistoric/Historic	3.17 B
22	A	1	Letter	Historic	
22	A	2	Letter	Historic	
22	B	1	Letter	Historic	
23	A	1	Circle with line	Prehistoric	
24	A	1	Curvilinear	Prehistoric	
25	A	1	Rectilinear	Prehistoric	
26	A	1	Curvilinear	Prehistoric	

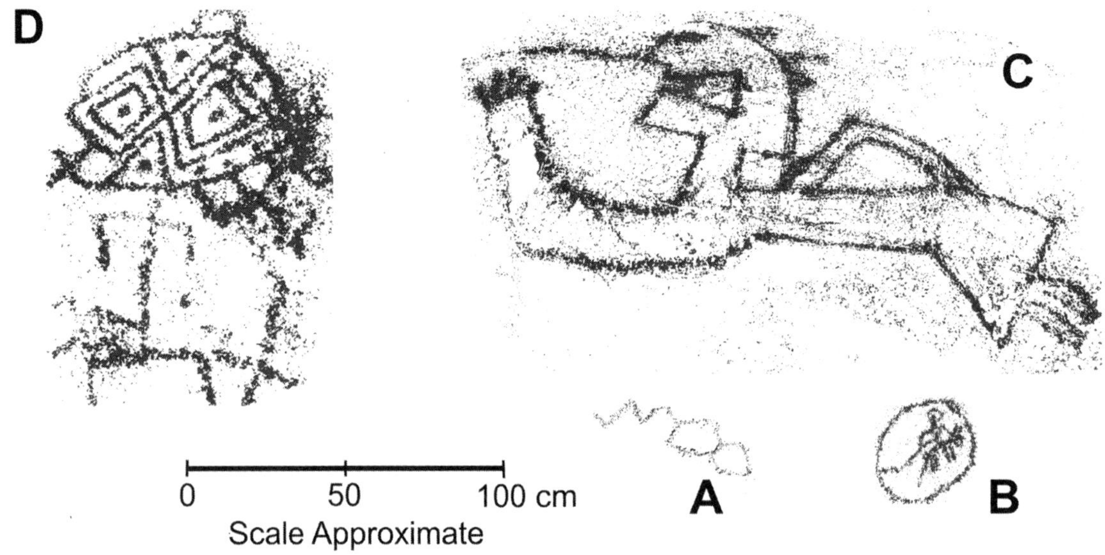

FIGURE 3.17. Examples of petroglyph elements at Cerro Juanaqueña.

The largest set included 48 elements which as a group conform to the Paquimé style (Schaafsma 1997). About half of the elements are relatively simple designs such as circles, "X"s, curvilinear, and rectilinear designs. The rest are more complex, and include footprints, stick figures, faces, cuadros or cartuchos (Mendiola Galván 2002; Schaafsma 1997), and a Maltese cross. One figure, Locus 6.A.1, is especially notable (Figure 3.17c). This distinctive and complex anthropomorphic figure is well-known from the Medio period pottery of the Casas Grandes culture (VanPool 2003). Another complex anthropomorphic figure is depicted in Locus 21.A (Figure 3.17d).

Three petroglyphs are tentatively attributed to protohistoric or historic times based on style and iconography. Two seem to represent churches, with rectangular or stepped elements surmounted by a Christian cross. The third example is a stick figure representation of a man on a horse, contained within a circle that may depict a lasso (Figure 3.17b).

A much larger group of 16 elements are clearly historic in age. These include Latin letters, dates, and brands. In general, elements in this group are either scratched onto the rock or are more lightly pecked in comparison to those attributed to protohistoric and prehistoric times.

Replication Experiment

When we began our work at Cerro Juanaqueña, most archaeologists believed that lifestyles in this part of North America during the Early Agricultural period were little changed from those of preceding Archaic times, and until the discovery of Cerro Juanaqueña the archaeological record in northern Chihuahua and southern New Mexico seemed consistent with this model. Preceramic sites tended to consist of extensive artifact scatters that accumulated gradually over millennial time scales. Architecture, when it was found at all, consisted of small, casually constructed brush shelters, easily built and easily abandoned when people were ready to move on to their next camp site.

The constructions at Cerro Juanaqueña are clearly something very different. Even casual visitors at Cerro Juanaqueña are overwhelmed by the amount of labor that must have been involved in its construction. This labor investment far exceeds anything expected to have been built by small groups of mobile hunter and gatherers. Because of this, one important objective of our project was to better quantify the amount of labor actually represented by these features.

In order to do this, we designed a replication experiment, drawing upon the results of our excavations and upon our detailed maps of the site (Hard et al. 1999). Our questions focused on the construction methods, volume of fill, labor costs, labor organization and ultimately terrace function. We decided early in the project that actually constructing a terrace ourselves would be an effective way to approach some of these issues. For many decades investigators, particularly in Europe, have built replicas of ancient monuments to gain a better understanding of the work, techniques, and materials involved in their construction (Abrams 1994; Callahan 1981; Craig et al. 1998; Erasmus 1977; Jewell 1963; Peebles and Kus 1977). Identified as an "architectural energetic" approach, this has the potential of quantifying labor investment, an important element in understanding the behavioral and social context of a site (e.g. Abrams 1989). However, architectural energetic studies have rarely been applied to Archaic period or even egalitarian societies, as most applications have focused on substantial public architecture related to complex societies. Our approach was to build a terrace feature in a manner similar to a prehistoric one and then extrapolate our experimentally derived labor rates to estimate the labor costs involved in the construction of the entire site (Hard et al. 1999). Sidebar 3.2 describes the methods we used in this experiment, and in extrapolating results to the larger site.

The results of the experiment are presented in Table 3.7. We estimate that construction of a 17 m long terrace would have required approximately 65 person hours and would have involved moving almost 34 m^3 of rock and dirt. Extrapolating this result, we estimate that a total of 31,000 m^3 of materials were used for the entire site, and that the total effort required 60,000 person hours, or about 30 person years. This estimate is equivalent to the construction of a 550-room pueblo and may be low as the replication experiment did not completely fill the voids between the cobbles in the berm and terrace fill. When we revisited the experimental terrace after several

Sidebar 3.2. Terrace Construction Experiment

Before we tackled the construction of a full-size terrace we first built a miniature one that was about 10% of the volume of a full-size terrace. Through this effort we tested and developed a construction sequence and appropriate labor, time, and volume documentation methods that allowed the work to go smoothly and the necessary data to be collected. This effort was successfully completed at the end of one day without mishap and we were encouraged to pursue building a full-scale terrace.

Based on this experience, team-member José Zapata worked with the assistance of two local ranchers constructing a second full-size terrace at a location on a hill immediately north of Cerro Juanaqueña (Hard et al. 1999). The location was selected because it had no archaeological features and because it resembled Cerro Juanaqueña with a slope of about 20%, thin soil, and abundant basalt cobbles on the surface. Zapata participated in the work, maintained a journal logging all activities and time expenditures, and documented the effort through video and still photography. Tools included wooden digging sticks, a metal pick, and metal buckets.

The first step was to clear vegetation from the hillslope where the terrace was to be constructed by pulling and stomping, without using metal tools. This required about half an hour to complete. The next step was to outline the crescentic shape of the berm using 20 to 40 cm diameter cobbles, essentially creating the first layer of the berm wall. The footprint of the berm was to be 17 m from tip to tip. Many in situ rocks were used, and this step required only 45 minutes.

The third step was to complete construction of the berm wall, using materials taken from a measured 350 m^2 area surrounding the feature. Rocks measuring between 6 and 40 cm in diameter were tossed into the outlined feature. Larger rocks were rolled into place, and two team members working together had to carry a few of the very largest ones to the growing berm wall. All of the rocks were taken from within 15 m of the replication terrace, and transport distances of 6 to 10 m were typical. After two days of work, all of the loose rocks on the surface had been used, and it was necessary to pry out partially buried rocks using digging sticks. This effort slowed the rate of progress considerably, and various strategies were utilized to reduce fatigue. In the end, the berm was about 95% complete and would have required 21 person hours to have fully finished.

Based on the results of terrace excavations, we determined that filling in behind the berm was the fourth phase of terrace construction. During the preceding phase, prying out large cobbles had left a large number of small cobbles loose on the surface. Because of this, filling in the terrace proceeded rapidly, and after only 13 person hours of work a level terrace surface had been created behind the cobble berm.

The final step was to cap the terrace surface with fine sediment, and surprisingly this turned out to be the most costly phase of the construction effort. In the end, only half of the replicated terrace was capped. The construction team spent about two hours using digging sticks to loosen the silty, hard-packed dirt within the 350 m^2 source area. They scooped the dirt up by hand and transported it in 12-liter metal buckets, a concession to contemporary technology that probably did not significantly affect the labor rate. After two hours, enough information had been collected to estimate the number of bucket loads per hour that could be obtained and transported. For the remaining three hours the team years, most of the fine sediment had migrated down into the terrace fill, and new fine sediment would have been required to maintain a useable terrace surface. As can be seen in Table 3.7, production and transport of fine sediment is the most costly phase of terrace construction.

Summary and Discussion

Cerro Juanaqueña is the largest of a series of fortified villages built in northwestern Chihuahua at about 1300 cal BC. These sites are clearly among the earliest intensive farming communities in North America, and they compare favorably with coeval developments further south in Mesoamerica. Cerro Juanaqueña alone includes 8 km of residential terraces, constructions that cover an area of over 10 ha. Based on replication experiments, we estimate that construction of the terraces involved something on the order of 60,000 person-hours of effort. A series of 23 AMS radiocarbon dates reported in Chapter 5 suggest that the terraces were not built incrementally over a long period of time. Instead, they almost certainly represent less than two centuries of intensive occupation. The spatial organization of the terraces and the occurrence of massive perimeter berms at Cerro Juanaqueña and several other cerros de trincheras

used a handpick to loosen the dirt and kept careful records about the number of full buckets that were moved. Use of a metal pick increased efficiency by about 30%. Applying this factor to the effort required by this phase and taking into account that only half of the terrace was covered with fine sediment, we estimate that this final step would have required 30 person hours to complete.

In addition to documenting the amount of construction labor required by terrace construction, we also wanted to estimate the volume of stone and soil used for different phases of construction. Estimation of the volume of soil was accomplished by counting the number of 12-liter pails full of dirt that were required to cap half of the terrace and multiplying the result by two. This exercise indicates that 4.2 m^3 of dirt were required to cover an average terrace. To estimate the volume of rock, a sample area 3 m by 3 m in size was defined within the rock collection area. All rocks in the sample area were tabulated according to size categories (e.g., 6 cm to 10 cm diameter, 11 cm to 15 cm, etc.). The formula for volume of a sphere was used to calculate the volume of rock in each size category. Based on this analysis, each square meter of hillslope includes 18.44 cobbles and .081 m^3 of stone. Thus the 350 m^2 materials collection area used in the experiment yielded about 28 m^3 of stone.

Taking into account the volume of materials and the amount of labor expended, we are able to estimate that rock can be gathered and transported at an average rate of about 1.15 hours per cubic meter. In contrast, gathering and transporting soil requires 7.1 hours per cubic meter, or more than seven times the amount of labor required to produce stone. The final figures show that 65 person hours would be required to fully construct the replicated terrace, which had a volume of 34 m^3. Therefore, the overall construction rate is 1.9 hours per cubic meter. Other documented experiments that involved procuring rocks or dirt range from 1.8 to 2.1 person hours per cubic meter (Abrams 1994:Table 3; Erasmus 1977:61; United Nations Economic Commission for Asia and the Far East 1961:Table 1), consistent with the results we obtained from our experiment.

A final set of calculations are needed to apply these results to Cerro Juanaqueña as a whole. We began by estimating the total volume of rock and sediment in the 550 terraces. Our photogrammetric map and the detailed instrument map of 68 terraces on the southeast quadrant of the site allowed us to document a series of terrace cross sections, and to estimate the mean area of a terrace cross section. A standard civil engineering method allows estimation of irregular volumes by multiplying average cross sections by length (O'Rourke 1940). For our sample of 24 terraces with detailed metric data, the mean cross section area of a terrace on a steep slope (slope > 10°) is 4.5 m^2 (sd=2.3 m^2, n=21) and the mean cross section area of terraces on more gentle slopes (slope ≤ 10°) is 1.3 m^2 (sd=.22, n=3). Using a Geographic Information System (ArcView 3.2), we determined that there are 1.4 km of berm wall on gentle slopes and 6.7 km on steeper slopes, a total of 8.1 km of berm walls on the site as a whole. Multiplying these figures by the appropriate cross section area, we estimate a total of 32,000 m^3 of rock and sediment in the 550 terraces on Cerro Juanaqueña.

At a rate of 1.9 person hours per cubic meter, it would require 60,000 person hours of labor to construct the 550 terraces. Assuming a common traditional work pattern of six-hour days, 300 days per year (Barlett 1980; Erasmus 1977; Turner 1983a, 1983b), the constructions on Cerro Juanaqueña represent about 30 person years of work, and the average terrace required about 18 person days to construct.

TABLE 3.7. Experimental Terrace Labor Data.

Task	% Completed	Person hours	Person hours extrapolated for 100% completion	Materials (m^3)	Materials extrapolated for 100% completion (m^3)	Person hours per m^3
1. Site clearing	100	0.50	0.50	—	—	—
2. Layout terrace	100	0.75	0.75	Included with rock berm below		
3. Rock berm	95	20.00	21.10	20.4	21.5	1.0
4. Rock platform	100	12.80	12.80	8.0	8.0	1.6
5. Sediment cap [1]	50	14.90	29.80	2.1	4.2	7.1
Total			65.00		33.7	1.9

[1] Assuming 100% work with digging sticks.

imply occupation by a large and cohesive social unit, as well as a degree of planning and leadership not necessarily expected at this early date. Arguments developed in Chapter 13 suggest a more or less permanent population of around 200 people.

In addition to the terraces themselves, major features include as many as 100 stone rings. Some of these were clearly residential while others may have been aboveground storage facilities. The range and numbers of stone tools, as well as vast quantities of burned animal bone and plant material indicate intensive occupation. Aside from the stone rings, evidence of structures is limited to a few isolated postholes, a single fragment of burned daub, and one probable underground storage pit. Other features attributed to the Early Agricultural period occupation at Cerro Juanaqueña include five roasting features represented by mounds of fire-cracked rock, cobble cross-walls on the terrace surfaces, a small number of walls built between terraces, an undetermined number of low elevated platforms at the ends of some terraces, and 26 bedrock grinding features. Several trails and linear features are immediately outside the area of terrace construction. There are a few cairns on Cerro Juanaqueña, some of which seem to be incidental by-products of terrace construction, while others may be much more recent. A substantial amount of rock art is also on the site. Cupules and a small number of abstract petroglyph elements are probably associated with the Early Agricultural period occupation, based on relative degree of patination. A much larger number of elements date to the Ceramic period, and a small number seem to be Protohistoric or Historic in age.

In addition to providing a physical description of Cerro Juanaqueña, this chapter develops a model of formation processes on the site. The stone constructions themselves seem to be remarkably resilient and have changed little over the past three millennia. In contrast, the fine sediments on the site are subjected to disturbance from a variety of sources, including rodents and insects. Cultural materials are affected differently, depending primarily upon their size. Large items such as manos, metates, cores, and hammerstones seem to remain on the surfaces where they were deposited. Smaller materials such as debitage, chipped stone artifacts, and larger pieces of bone are more vulnerable to mixing, but mostly remain relatively near the surface. The smallest items of archaeological interest, such as small bone fragments and charred plant pieces, including maize fragments, are found throughout the deposits and have little stratigraphic integrity.

CHAPTER 4

Other Sites

Introduction

By the beginning of our second field season in 1998 we knew that Cerro Juanaqueña was a massive Late Archaic period settlement unlike anything previously identified in the region. The site clearly held important implications for understanding the past, but was it truly unique or was it only one element in a pattern of similar sites? The ramifications for each possibility would be quite different. Not far from our minds too was the inherent skepticism of the archaeological community. A pattern of multiple sites offers a far stronger argument than a case of one.

Over the decades previous archaeologists, such as Charles DiPeso (1974) and Rex Gerald (1990), reported other cerros de trincheras (or sites that sounded much like cerros de trincheras) in the Río Casas Grandes drainage. After the discovery of Cerro Juanaqueña, Paul Minnis found several more by inspecting aerial photography, and we located still others through reconnaissance, aerial photo interpretation, and informants. By 1996 Roney had mapped Cerro Vidal and Cerro los Torres as well as Cerro Juanaqueña. We continued this program of recording, mapping, and limited surface collections, and obtained new large-scale aerial photography for other cerros de trincheras in Chihuahua. In addition to Cerro Juanaqueña we documented 13 similar sites in this manner (Table 4.1), and we selected three of the most promising for limited excavations—Cerro Vidal, Cerro los Torres, and Cerro el Canelo. Our objectives were to obtain radiocarbon, botanical, faunal, architectural, and artifactual data. This chapter describes these sites and excavations.

Cerros de Trincheras

The major sites discussed in this chapter are *cerros de trincheras* consisting of extensive complexes of terraces built on the summits and upper slopes of steep-sided, isolated hills. Radiocarbon dating and artifact assemblages suggest that many of these sites were used during the Late Archaic period and based on current evidence they appear to be among the earliest agricultural sites in the region. Many, but not all of the sites include evidence of intensive residential occupation. All are smaller than Cerro Juanaqueña and were occupied less intensively, but they are constructed in a similar way and many exhibit a range of artifacts comparable to that found on Cerro Juanaqueña. Radiocarbon dates from three of the sites (Cerro el Canelo, Cerro los Torres, and Cerro Vidal) show contemporaneous occupation with Cerro Juanaqueña, three others have yielded dart points which are characteristic of the Late Archaic period (Cerro la Boquilla, Cerro la Tinaja, and Cerro Galeana), and aceramic sites at the base of three more (Cerro la Virgen, Cerro la Fundición, and Cerro Le Barón Sur) suggest that they may also date to this period, ca. 1500 BC to AD 300.

The sections below describe 13 cerros de trincheras, with varying levels of detail. The sites are described in geographic order, beginning with Cerro Tascate, the northernmost of the group and the one nearest to the mouth of the Río Casas Grandes (Figure 1.1). After proceeding up the Río Casas Grandes to Cerro la Tinaja, we conclude with descriptions of Cerro Galeana,[1] Cerro Le Barón Norte, and Cerro Le Barón Sur, located on the Río Santa María (Figure 1.1). Some of the important characteristics of the sites are tabulated in Table 4.1 All sites are illustrated with Adriel Heisey's striking photographs in Plates 1 to 16.

Cerro Tascate

Cerro Tascate (Plate 4) is a low, 50 m high hill adjacent to the Río Casas Grandes about 50 km downstream from Cerro Juanaqueña. Two, continuous, parallel berm/terraces encircle the summit, enclosing a 200 m by 100 m area on three sides (Figure 4.1). Areas on the southwestern side of the hill, where these two encircling

TABLE 4.1. Cerros de Trincheras in Northwestern Chihuahua.[1]

Name of Site	Distance to Floodplain	Height of Hill	Area	Terrace Length	No. of Rock Rings	Period
Tascate	0.2 km	50 m	1.2 ha	550 m	3	Unknown
Juanaqueña	0.4 km	140 m	11.2 ha	8350 m	114	Late Archaic
Upper			(6.0 ha)	(4980 m)	(100)	
Lower			(5.2 ha)	(3370 m)	(14)	
La Virgen	1.5 km	80 m	1.8 ha	790 m	11	Late Archaic?
Angostura	0.8 km	80 m	1.3 ha[2]	450 m[2]	22+	Unknown
Canelo[3]	1.9 km	160 m	5.8 ha	3900 m	50	Late Archaic
Fundición	0.3 km	20 m	1.8 ha	1000 m	4	Late Archaic?
Torres	0.9 km	80 m	5.2 ha	2300 m	8	Late Archaic
La Cruz	2.0 km	140 m	2.6 ha[2]	Unknown	Unknown	Unknown
Upper			(1.8 ha)			
Lower			(0.8 ha)			
Boquilla	0.3 km	100 m	1.9 ha[2]	450 m[2]	Unknown	Unknown
Vidal	0.2 km	120 m	4.0 ha	2190 m	25	Late Archaic
Tinaja	0.2 km	80 m	0.6 ha	245 m	5	Late Archaic?
Galeana	0.8 km	120 m	5.4 ha[2]	1500 m[2]	7+	Unknown
Le Baron N	0.6 km	80 m	0.3 ha	170 m	8	Late Archaic?
Le Baron S	0.4 km	40 m	0.3 ha	110 m	2	Late Archaic?

[1] Figures in parentheses are subtotals.
[2] Estimate based on photointerpretation without ground truthing.
[3] Excludes some outlying features.

walls appear discontinuous, are broken by steep bedrock outcrops. On the eastern slope of the hill, within the area enclosed by the parallel berm/terraces are two shorter 30 or 40 m long terraces. At the summit are two contiguous subrectangular enclosures 5 or 6 m on a side, defined by low, dry-laid masonry walls, now mostly collapsed. Several vaguely defined circular areas near the summit are subjectively identified as rock rings. Near the summit of the hill and in several other places within the large enclosed area, there are poorly organized concentrations of rock, but we did not see evidence that loose rock had been systematically cleared from large areas, as is the case at Cerro Juanaqueña. Instead, most of the ground surface inside the feature complex appeared unmodified.

The entire artifact assemblage from this site consisted of the tip of a biface, a core, two flakes, and a mano fragment, as well as a few sherds found at the very summit of the hill. A wooden crucifix had also been constructed on the summit, supported by a stone cairn. There is a substantial Ceramic period occupation in the immediate vicinity of this site, and several Ceramic period sites, as well as ruins of early to mid-twentieth century historic buildings, are found at the base of the hill. Because of the paucity of cultural material on Cerro Tascate, we cannot confidently attribute it to any particular archaeological period.

Cerro la Virgen

Cerro la Virgen is located about 7 km upstream from Cerro Juanaqueña (Plate 5). It is approximately 1.5 km east of the floodplain, separated from it by a broad, gently sloping alluvial apron. The hill rises about 80 m above the surrounding terrain. This site closely resembles Cerro Tascate in that its principal features are two parallel rubble berms which encircle the top of the hill, enclosing an area about 100 m by 200 m in size (Figure 4.2). Inside this enclosure are 11 poorly defined rock rings, and an additional ring was found on a natural terrace near the base of the hill. The summit of the hill is virtually devoid of artifacts. During several visits to the site fewer than 10 flakes and a single sherd have been noted.

There is an extensive aceramic lithic scatter at the base of the hill on its southern side. This artifact scatter is about 100 m by 200 m in size and is being dissected by runoff from the hillslope. It includes debitage, cores, hammerstones, basin metates, small one-hand manos, and at least one stone bowl or small mortar comparable to those found at Cerro Juanaqueña and other Late Archaic cerros de trincheras. The single projectile point found at this site was a corner-notched dart point, with an expanding stem and convex base. It is similar to Late Archaic specimens from other sites in the region. Aside from the single sherd noted above, no pottery has been

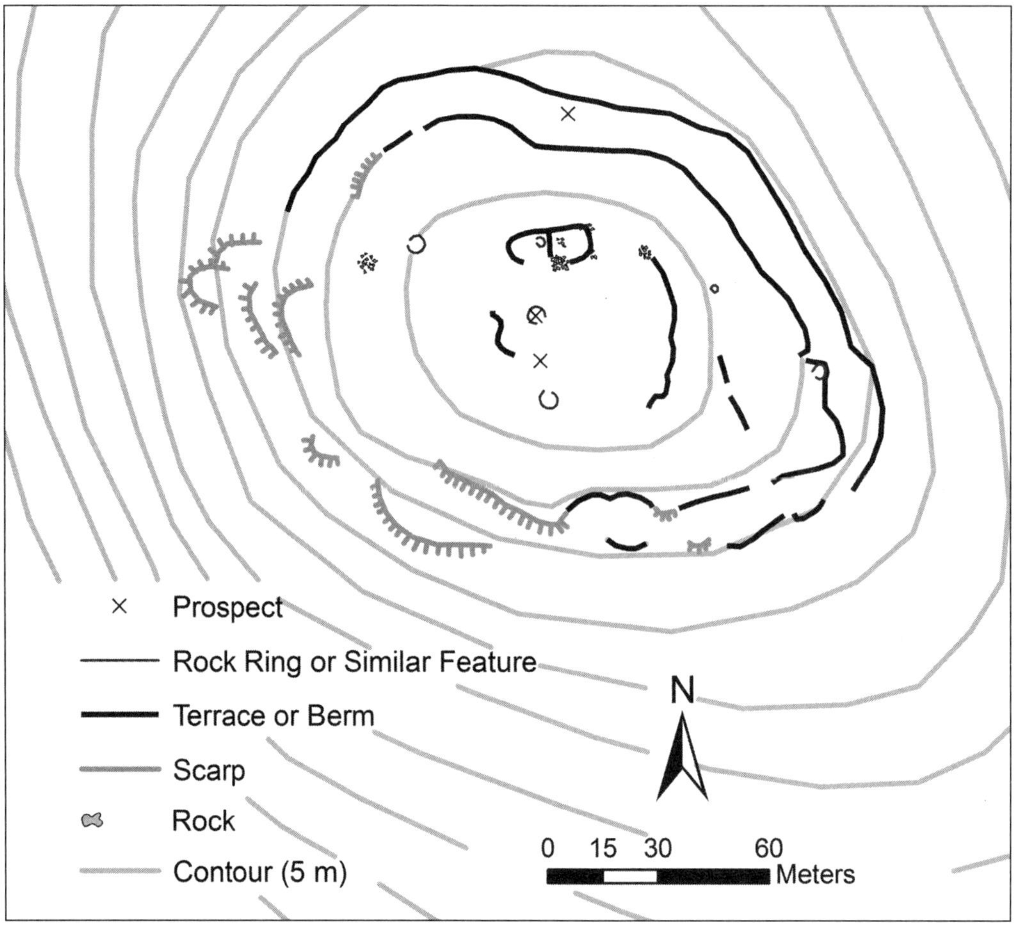

FIGURE 4.1. Cerro Tascate. Also see Plate 4.

found in the immediate vicinity of Cerro la Virgen. However, an intensive Ceramic period occupation associated with the floodplain 1.5 km west of the site includes numerous residential buildings and artifact scatters. Because of the absence of ceramics, and because of its proximity to a site which is clearly Late Archaic in age, we suspect that this site dates to the Late Archaic period.

Cerro la Angostura

Cerro la Angostura is a large, 80 m high hill overlooking the Río Casas Grandes at a major constriction in the floodplain (Plate 6). The hill is actually the southern end of a large ridge with steep sides and an open, rolling, or gently sloping crest. The principal constructions at this site consist of a 150 m long wall of dry-laid masonry, as well as several shorter wall segments, now almost completely reduced. These walls or berms enclose an area 100 m by 85 m in size (Figure 4.3). The site has not been intensively recorded, but initial impressions suggest that as many as 22 rock rings or related features were built within this enclosure. The largest wall at Cerro la Angostura defines an arc centered on a high point on the ridge, enclosing an area about 85 m by 40 m in size. On the north, west, and south, the enclosure is bounded by the wall or rubble berm. On the southeast, additional unmapped terraces extend about halfway down the steeply sloping side of the ridge. Only the upper part of the feature complex has been inspected, but in this area there are very few artifacts. Just a single core and three flakes have been noted in this part of the site. There is very little fine sediment on Cerro la Angostura, and most of the features mentioned here are above ground, rather than being actual terraces. This site has not been sufficiently evaluated, and we do not attribute it to any specific archaeological period.

Cerro el Canelo

Cerro el Canelo is the second largest of the cerros de trincheras in northwestern Chihuahua, after Cerro Juanaqueña (Plate 7). With 3.9 km of terrace berms and an area of about 5.8 ha, Cerro el Canelo is only about half the size of Cerro Juanaqueña, but it is comparable in terms of setting, construction, overall layout, and the range of activities represented. The site is located on a 160 m high hill about 25 km south of Cerro Juanaqueña. Although it overlooks the Río Casas Grandes

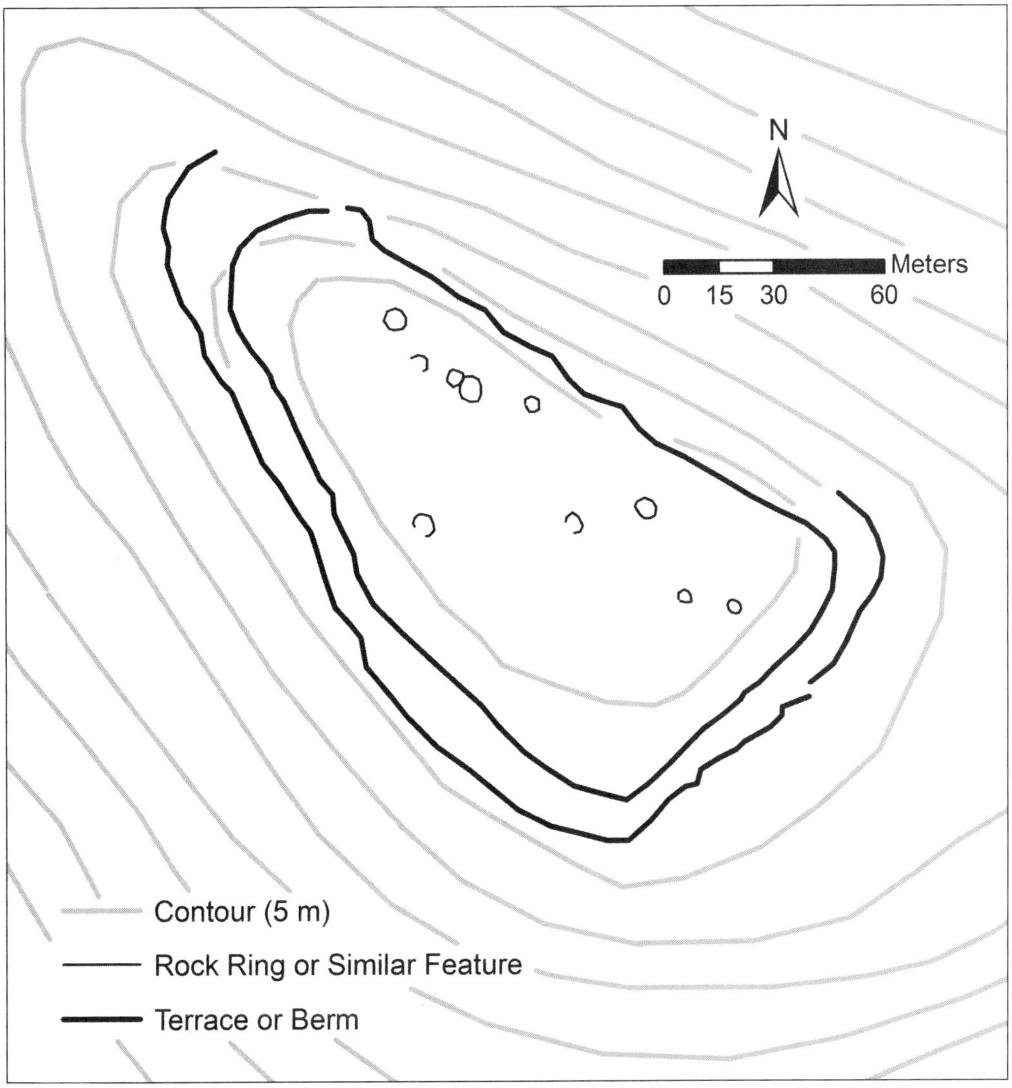

FIGURE 4.2. Cerro la Virgen. Also see Plate 5.

floodplain, the summit is almost two kilometers from the river. Cerro el Canelo includes approximately 250 arc-shaped terraces with construction identical to that at Cerro Juanaqueña, as well as 50 rock rings (Figure 4.4). As at Cerro Juanaqueña, individual terraces are often joined together to form coherent macrofeatures up to 350 m in length (Plate 7). These features are concentrated around the summit of the hill, but a number of stone circles and small terraces are scattered down a ridge which runs toward the river.

In a low saddle near the base of the hill is a stone circle which is 70 m in diameter defined by a rubble berm (Plate 8). It was made by clearing cobbles from its interior and piling them around its circumference to form a well-shaped circle with a large opening towards the southwest (Figure 4.5). Although there is still some question about its dating (see below), this feature is our best candidate for a communal or integrative structure among the cerros de trincheras of northwestern Chihuahua.

Artifacts found on the surface in the main feature complex are identical to those at the Late Archaic cerros de trincheras in the Casas Grandes region. These include side-notched and corner-notched dart points with expanding stems and convex bases, large numbers of cores, large quantities of lithic debitage, hammerstones, slab and basin metates, oval one-hand manos, and stone bowls. A small number of cupules similar to those at Cerro Juanaqueña also occur on this site. Based on these observations we were confident that the principal complex of features at Cerro el Canelo was built and occupied during the Late Archaic period.

In hopes of dating the site more precisely, we excavated four test pits, one in the main terrace complex near the summit of the hill, two on a single terrace found in the middle zone of the north slope, and one in the

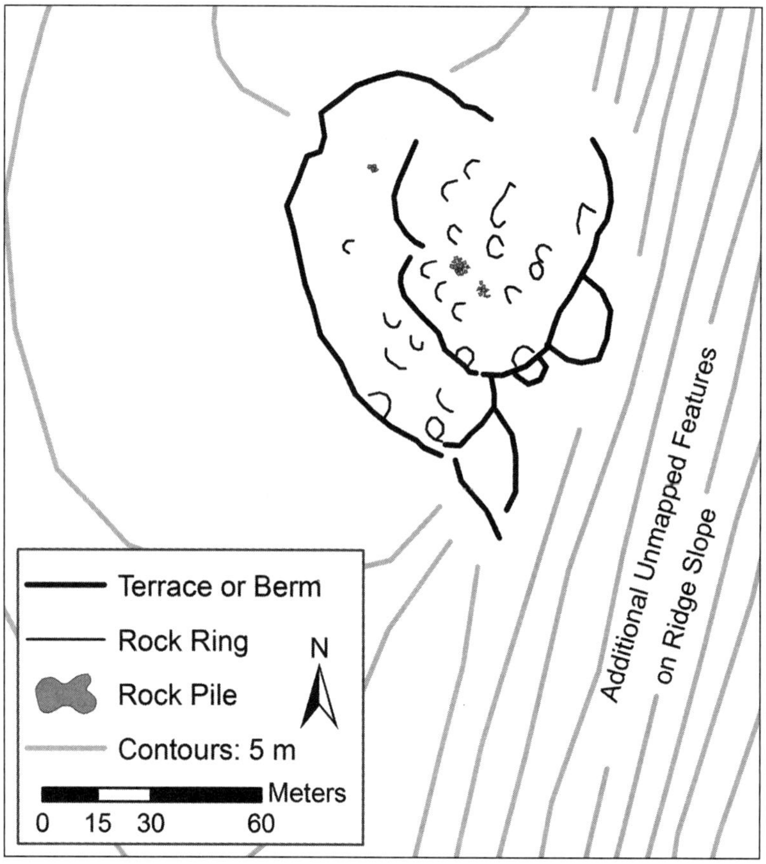

FIGURE 4.3. Cerro la Angostura. Also see Plate 6.

berm of the 70 m diameter stone circle closer to the base of the hill (Hard et al. 2001; Roney and Hard 2000; see Appendices 3.5 and 3.4). The units in the main terrace complex (Terraces 41 and 94) yielded abundant burned animal bone and charred plant remains, reinforcing our impression that this site was used for residential occupation. In addition to mesquite (*Prosopis* sp.) and saltbrush (*Atriplex* sp.) charcoal, charred remains of ocotillo (*Fouquieria* sp.), cheno-am (*Chenopodium* sp. or *Amaranthus* sp.), monocot, *Portulaca* sp., and sunflower (*Helianthus* sp.) were recovered from the terraces. Although there was quite a bit of charcoal, no maize was found in the two terraces. The only identifiable plant remains from the large stone circle were partially charred fragments of ash (*Fraxinus* sp.). (See Chapter 8 for a more complete description of plant remains from this site.) Faunal remains from Cerro el Canelo were similar to those from Cerro Juanaqueña, and included box turtle, pocket mouse, deer mouse, cottontail, jackrabbit, antelope, and other unidentified artiodactyls, snake, lizard, amphibians, rodents, and small birds. Small fish remains were especially abundant (Chapter 11).

We obtained three radiocarbon dates from this site (Chapter 5). The oldest is from a fragment of charred ocotillo, 2990 ± 45 BP which calibrates between 1300 and 1130 BC at one sigma. We accept this as the age of the initial construction and principal occupation at Cerro el Canelo because it is consistent with the material culture found on the site, including both artifacts and features. This date is also virtually identical to that of the principal occupations at Cerro Juanaqueña and at Cerro los Torres. However, a fragment of monocot stem from the same terrace gave a date of 330 ± 60 BP (uncalibrated) and a fragment of large mammal bone recovered from the large stone circle yielded a conventional radiocarbon date of 630 ± 50 BP (uncalibrated). These later radiocarbon dates may indicate that activities continued here long after the initial occupation of the site. They are not corroborated by ceramics or other artifact associations that are diagnostic of late prehistoric or protohistoric times, and we suspect that these two dates reflect more ephemeral activities at Cerro el Canelo.

Age of the large stone circle (Figure 4.5) is an especially crucial issue because it may represent a formalization of integrative activities, which in the Late Archaic period is without precedent in the prehistory of the southwestern United States and northwestern Mexico. We do believe that this feature dates to preceramic times, contemporaneous with the other terraces at Cerro el Canelo. The mode of construction, patination on the

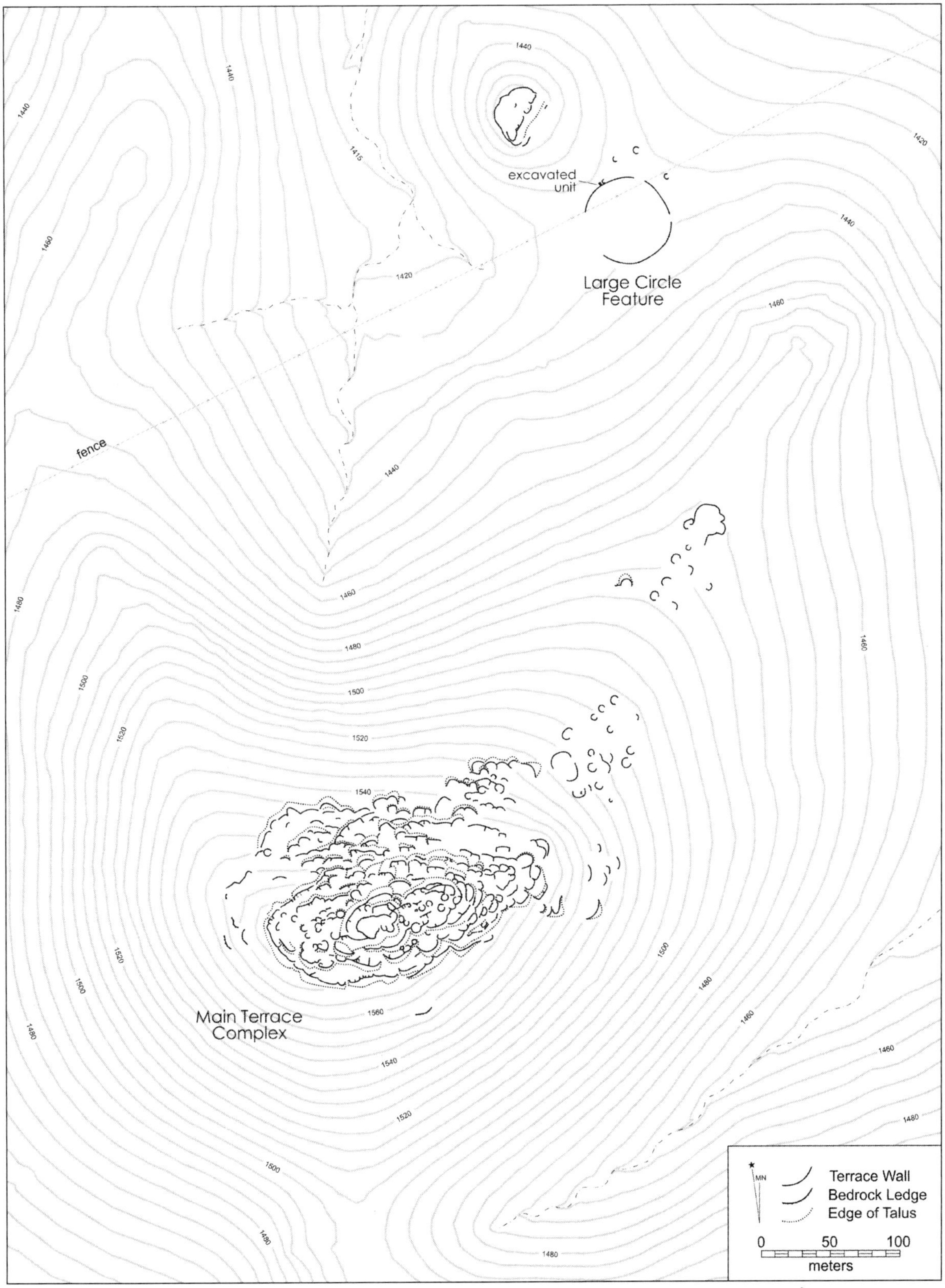

FIGURE 4.4. Cerro el Canelo. Also see Plate 7.

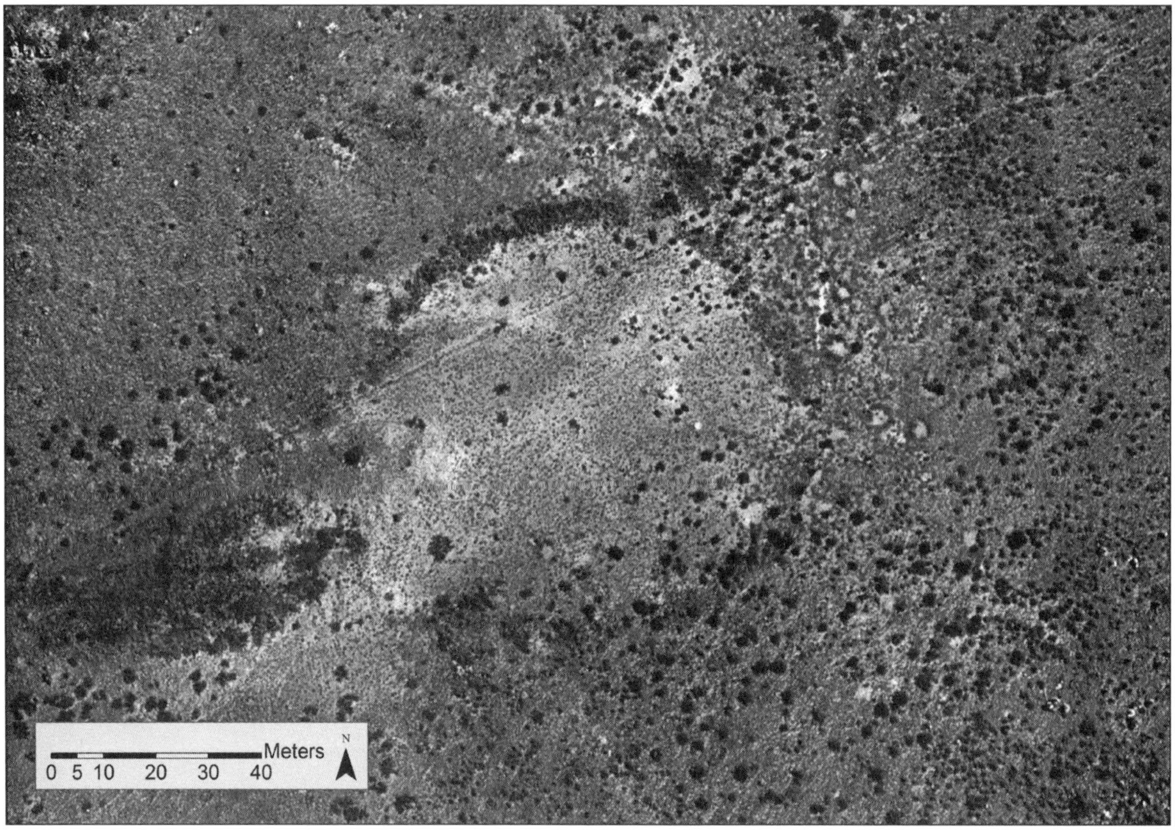

FIGURE 4.5. Possible communal or integrative structure at Cerro el Canelo. Also see Plate 8.

rocks, and vegetation on the berm are identical to those in the main terrace complex. The site has comparatively little evidence of Ceramic period occupation, and no ceramic or other artifact associations which suggest Ceramic period use of the feature itself.

Cerro la Fundición

Cerro la Fundición is a low hill adjacent to the Río Casas Grandes (Plate 9). In this area the river flows through a broad basin, and Cerro la Fundición stands as an isolated hill of basalt near the middle of the basin. The hill takes its name from a large industrial facility, rumored to have been a smelter, which is located at the base of the hill. No machinery remains, but large footings, bricks, and mounds of slag are present. This facility could have been used into the late nineteenth century. In addition, there are remnants of several historic residences at the northern and northeastern base of the hill.

The prehistoric features on the summit of Cerro la Fundición cover an area 150 by 125 m in size (Figure 4.6). At the very summit of the hill is a large stone circle about 5 m in diameter, and ceramics occur in the immediate vicinity of this feature. It is not clear if this circle is part of the original construction, a later Ceramic period feature, or a feature created by treasure hunters in more recent times. Other terraces are created by rubble berms like those at many other cerros de trincheras sites. Some are arc-shaped like those observed at Cerro Juanaqueña and Cerro el Canelo, while others are larger-scale coherent features. Most of the terraces at Cerro la Fundición are components of large, lobe-shaped macrofeatures which encircle the summit. Only four rock rings were found at this site, and they do not seem to be integral features.

Relatively few artifacts were found within the terrace complex. Ceramics are restricted to the very summit of the hill. Elsewhere multidirectional rhyolite cores and debitage are found. We also noted one small mortar and several oval, one-hand manos. A more extensive, largely aceramic lithic scatter is located at the base of the hill. This site includes roasting features and scatters of burned rock, abundant lithic debitage, cores, hammerstones, a basin metate, and several of the corner-notched dart points with expanding stems and convex bases which characterize the Late Archaic period in northwestern Chihuahua. Although some ceramics are present on this cerro, the other artifact associations and the similarity of this site to other preceramic cerros de trincheras lead us to tentatively attribute this site to the Late Archaic period.

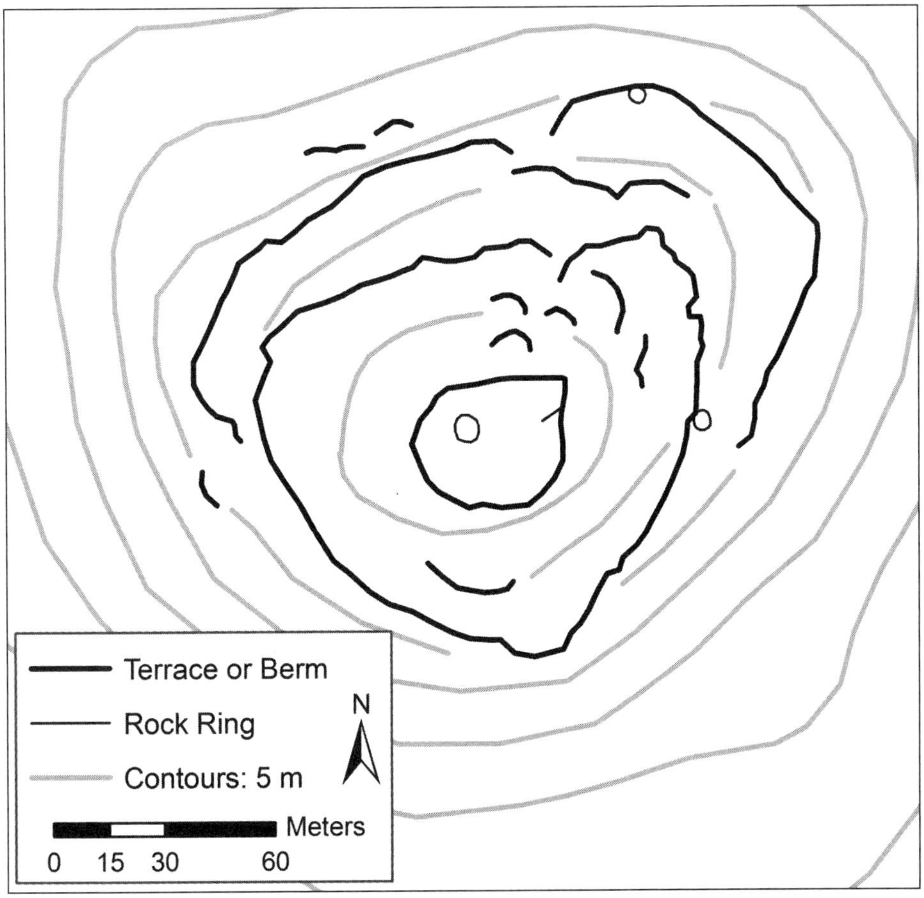

FIGURE 4.6. Cerro la Fundición. Also see Plate 9.

Cerro los Torres

Cerro los Torres is an isolated, 80 m high basaltic hill located near the eastern margin of the Río Casas Grandes floodplain, almost a kilometer from the present course of the river (Plate 10). This site is likely the one described by Schwatka (1895:61-67) as Davis Hill. At any rate, we discovered it while following Schwatka's narrative. A majority of the terraces on this site are the arc-shaped terraces with the same rubble berm constructions which characterize Cerro Juanaqueña and Cerro el Canelo (Figure 4.7). Like those two sites, Cerro los Torres also includes rock rings. The outermost terraces on the northern and eastern slopes of the hill coalesce to form a macrofeature around the perimeter of the site. Even more remarkable is the 500 m long continuous rubble berm which defines the western and southern edge of the site. Like perimeter berms on other cerros de trincheras in the region, this feature crosses the contours of the hill, rather than forming level terraces. Elsewhere on the site, eight well-defined rock rings occur among the terraces.

Cultural materials found on this site include cores, lithic debitage, and hammerstones. Both Late Archaic dart points and smaller arrow points usually attributed to the Ceramic period occur on Cerro los Torres. Seven sherds located in a restricted area at the summit of the hill constitute the entire ceramic assemblage. One small mortar, a few slab and basin metate fragments, and expedient, one-hand manos have also been noted.

We excavated three adjacent units in a single terrace near the summit on the eastern side of the hill (Terrace 1), creating a trench 2.5 m long and 1.0 m wide (Hard and Roney 1999a, 1999b). The excavation revealed a stratigraphic profile closely resembling those at Cerro Juanaqueña and other cerros de trincheras in northwestern Chihuahua (Figure 4.8). Animal bone from the test units included a large percentage of jackrabbit and cottontail, as well as smaller numbers of pocket mouse, coyote, pronghorn antelope, snake, and fish (Chapter 11). Charred plant remains included a variety of wild plant species, as well as maize (Chapter 8). An AMS radiocarbon date of 2920 ± 55 BP (uncalibrated) was obtained from one of the maize cupules (Chapter 5).

Although a small amount of pottery was present near the summit of Cerro los Torres, a much greater part of the artifact assemblage is more consistent with a

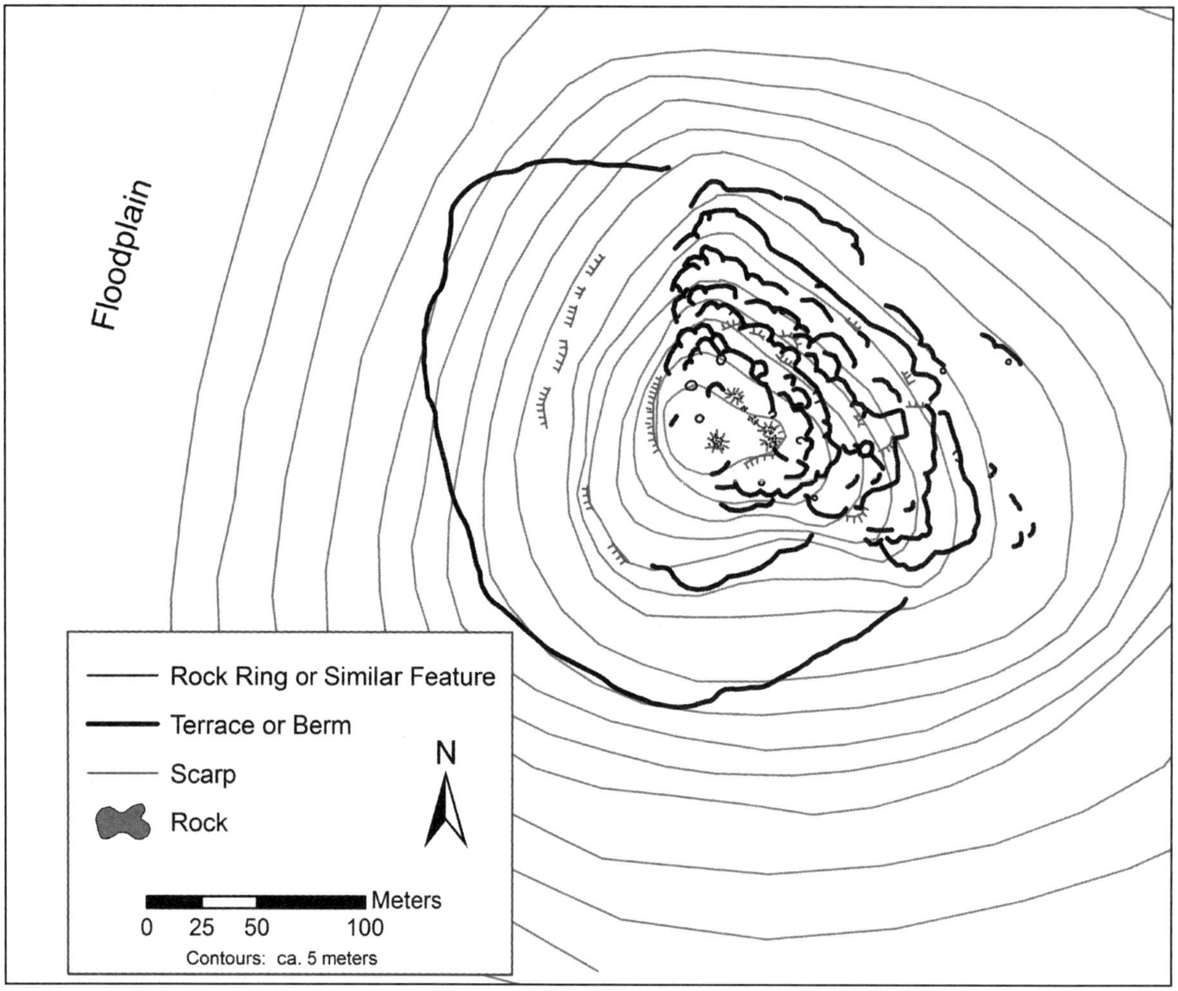

FIGURE 4.7. Cerro los Torres. Also see Plate 10.

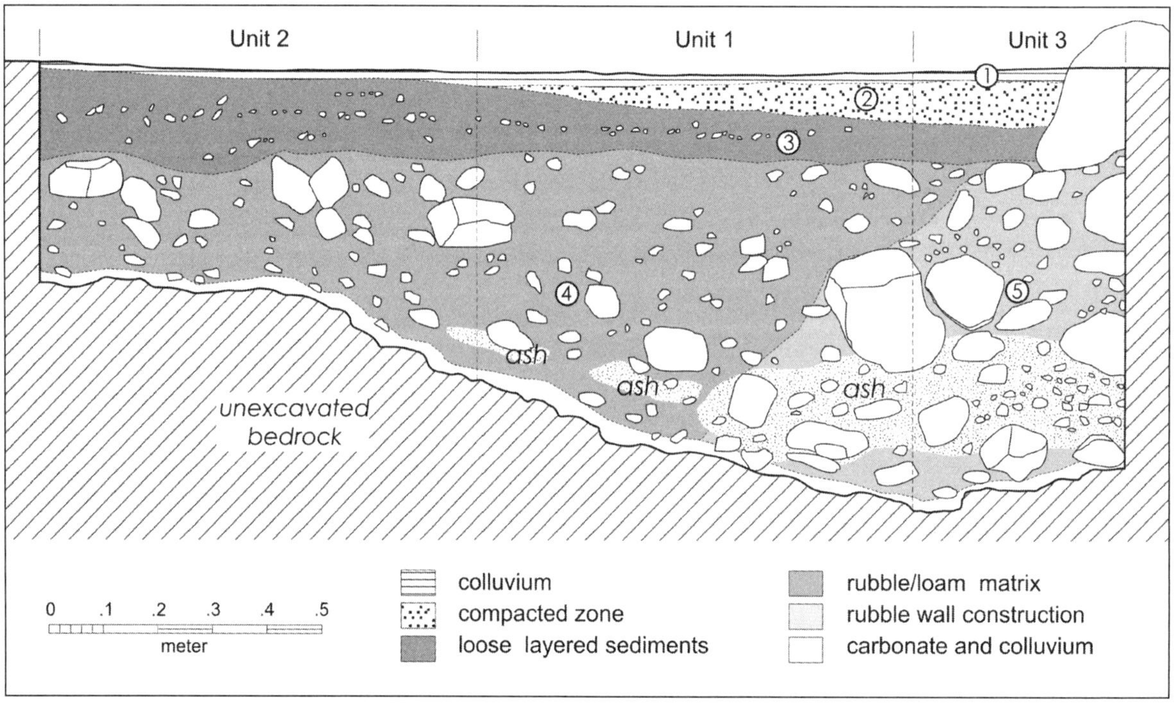

FIGURE 4.8. Profile of Terrace 1 at Cerro los Torres. View towards northwest.

Late Archaic occupation. This observation is supported by a single AMS radiocarbon date, which is virtually identical to dates from Cerro Juanaqueña and Cerro el Canelo. On this evidence we are confident that construction and major occupation of Cerro los Torres was during Late Archaic times.

Cerro la Cruz

Cerro la Cruz is one of the few cerros de trincheras, apart from Cerro Juanaqueña, that include a terrace complex at the base of the hill, as well as on the summit (Plate 11). The hill itself is 140 m high and is located 2 km from the floodplain. Arc-shaped terraces cover an area of about .8 ha. on its lower slopes and on a natural bench 10 m or so above the base of the hill. A larger complex of terraces covers an area of about 1.8 ha. centered on the summit. Detailed observations have not been made at this site, and only a small number of sherds and a light scatter of lithic artifacts were observed in the upper terrace complex.

Cerro la Boquilla de San Diego

Cerro la Boquilla has only been visited briefly. The site is located on the summit of a 100 m high hill which overlooks a constriction in the floodplain of the Río Casas Grandes (Plate 12). There is a large cleared area on the summit and five or six levels of relatively short terraces on both the eastern and western slopes of the hill. There is evidence of disturbance near the base of the western slope of the hill. This disturbance could represent an additional terraced area, but requires closer inspection. Cultural materials are relatively abundant on the summit, and in places the soil is dark and ashy, implying that midden deposits are present. There are a few plain brownware sherds on the summit, and we documented one large dart point. In addition to the prehistoric remains, there is a crucifix supported by a cairn on the summit, and a one-room rectangular dry-laid masonry building of historic origin on an outcrop just south of the summit.

Cerro Vidal

Near Hacienda San Diego, the Río Palanganas and the Río Piedras Verdes unite to form the Río Casas Grandes. Cerro Vidal is a 120 m high hill which overlooks this confluence (Plate 13). Terrace construction at this site resembles that documented on other cerros de trincheras in northwestern Chihuahua (Figure 4.9). However, at Cerro Vidal, individual arcs are often difficult to define. At this site the terraces form a series of macrofeatures which describe two concentric rings around the summit, a large lobe appended to these, and an outer berm/terrace which encircles the entire complex. Several cerros de trincheras in northwestern Chihuahua exhibit short cobble alignments oriented perpendicular to the terrace walls, which we call crosswalls. The function of these features is not understood, but they are especially common at Cerro Vidal. There are 25 rock rings inside the terrace complex at Cerro Vidal and another notable group of 12 or 13 located about 30 m south of the terrace complex.

The density of cultural material at Cerro Vidal is comparable to that at Cerro el Canelo and Cerro los Torres, and includes the same range of artifacts. Cores, hammerstones, and abundant debitage show that lithic reduction was an important activity. Metates and metate fragments are rare, but there are a number of expedient one-hand manos. Projectile points include Late Archaic dart points as well as specimens which seem small enough to have been used with the bow and arrow. No ceramics have been found. We excavated one test unit at Cerro Vidal (Hard and Roney 1999a, 1999b; see Appendix 3.3). Plant remains included maize and amaranth, as well as *Chenopodium* and cheno-am (Chapters 8 and 9). The very meager faunal assemblage from this excavation included jackrabbit and cottontail (Chapter 11). Maize cupules from this site have yielded AMS radiocarbon dates of 2100 ± 40 BP (uncalibrated) and 2340 ± 55 BP (uncalibrated), essentially identical to results from the later occupation at Cerro Juanaqueña (Chapter 5).

Cerro la Tinaja

Cerro la Tinaja is a small cerro de trincheras situated on an 80 m high hill adjacent to the Río Tapicitas, a major tributary of the Río Piedras Verdes/Río Casas Grandes (Plate 14). Brief field visits indicated that the terrace complex occupies an area 90 m by 40 m in size, centered on the summit of the hill (Figure 4.10). In addition to the terraces, at least five rock rings and several crosswalls are also present. The artifact assemblage exposed on the surface is estimated to include well over 500 lithics, perhaps half a dozen brownware sherds, one sherd of red-on-brown, and an unidentified sherd, perhaps from a polychrome vessel. Formal artifacts included several broken metates, two expedient manos, a hammerstone, and several cores. A single San Pedro style projectile point was found. The site is in an area of intensive Ceramic period occupation. Although a small number of sherds indicate some Ceramic period activity on this site, its construction and major period of use remain undated.

Cerro la Angostura de Galeana

Cerro la Angostura de Galeana (referred to here for convenience as Cerro Galeana) overlooks the floodplain of the Río Santa María (Plate 15). The hill is a long ridge, with a crest which slopes upward from northwest to a maximum height of 120 m at its southeastern end. Long

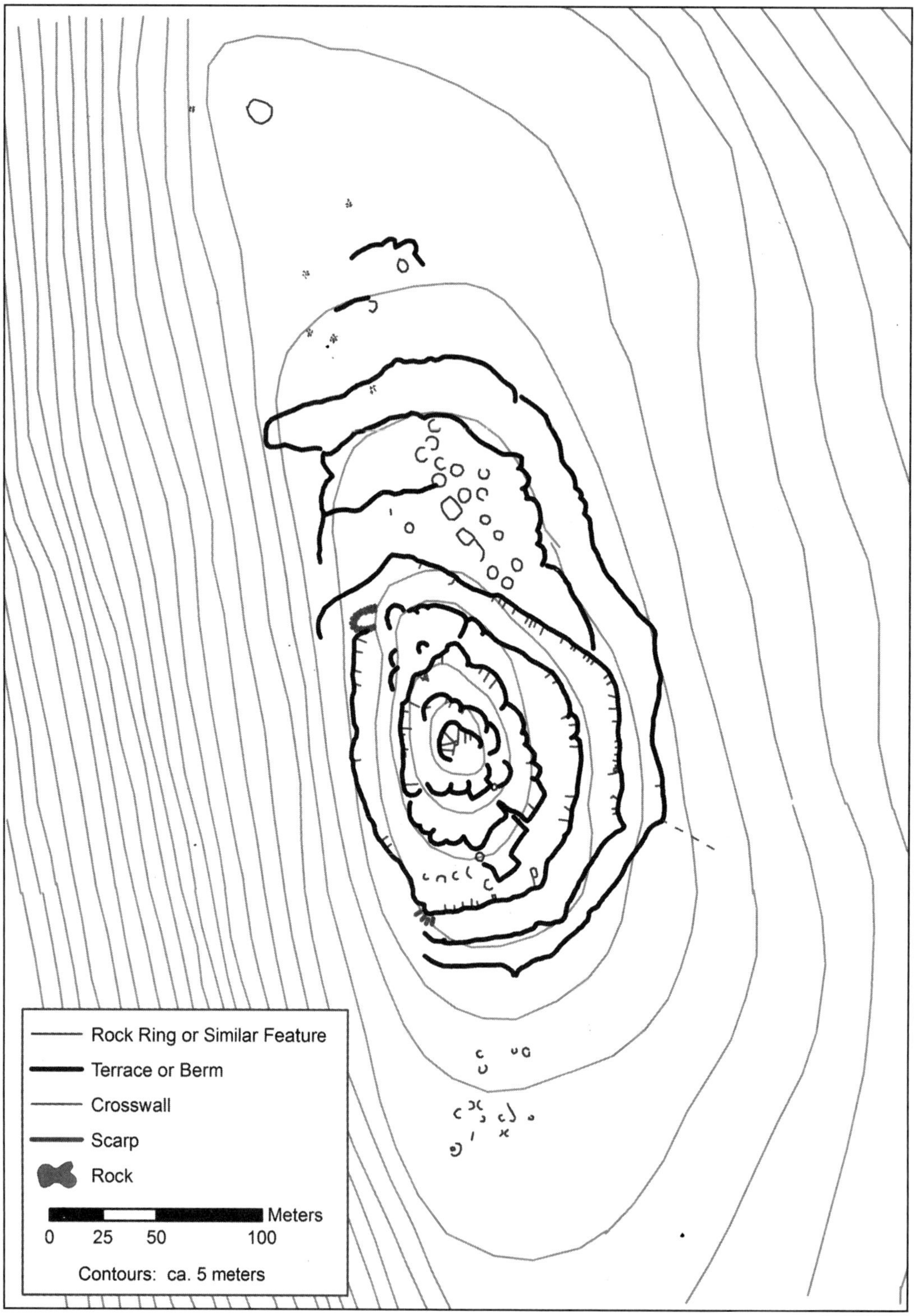

FIGURE 4.9. Cerro Vidal. Also see Plate 13.

terraces are constructed on the crest and southwest facing slope. The northeastern and southeastern slopes of the ridge are sheer outcrops where construction was not practical. The map of the site presented here (Figure 4.11) is derived entirely from photointerpretation with minimal ground truthing. It conveys the general scale and layout of the site, but may include significant errors and omissions in its details.

This extensive terrace complex has only been visited briefly, but preliminary inspection suggests that relatively little cultural material is present. Only 400 to 500 lithic artifacts were observed, and a substantial number

Plate 1. Cerro Juanaqueña. Cerro Juanaqueña is the largest cerro de trincheras in northwestern Chihuahua, and during this project it was the focus of intensive investigation. This view towards the setting sun shows the perimeter berm in the foreground and wide floodplain in the background. For another overall view of the site see S. Fish et al. (2007b), Plate 2.

PLATE 2. Cerro Juanaqueña. Level residential terraces are encompassed by a massive perimeter berm. This view is of the northwestern side of the site.

PLATE 3. Cerro Juanaqueña. This close view of terraces on the western face of Cerro Juanaqueña gives some sense of the amount of labor involved in their construction.

Plate 4. Cerro Tascate. A double perimeter berm stretches unbroken along the northeastern side of the hill and continues around the two adjacent sides of this low cerro de trincheras. Its age is uncertain. The Río Casas Grandes is visible in the background.

PLATE 5. Cerro la Virgen. This aceramic, triangular-shaped cerro de trincheras is encircled by a double perimeter berm on two sides while the third side is defined by a berm and a natural scarp. Its interior included 11 poorly defined rock rings. This view is towards the northwest.

PLATE 6. Cerro la Angostura. This small cerro de trincheras has not been fully investigated. It appears to include a number of rock rings, but almost no artifacts, and its age is unknown. This view is looking towards the southwest, across a constriction of the Rio Casas Grandes. Cerro el Canelo (Plate 7) is on the west side of the river on the most prominent hill visible in the center midground.

PLATE 7. Cerro el Canelo. After Cerro Juanaqueña, Cerro el Canelo is the second largest cerro de trincheras reported here. Excavations show that it was contemporaneous with the principal occupation at Cerro Juanaqueña. This view is towards the northeast overlooking a Río Casas Grandes reach boundary and wide floodplain. Cerro la Angostura (Plate 6) rests on the far right (southern) end of the small range on the opposite side of the river. The large cleared circle (Plate 8) lies between the two small hills in the midground.

PLATE 8. Cerro el Canelo. This large cleared circle is 70 m in diameter. It may be the earliest integrative structure yet documented in northwestern Mexico and the southwestern United States.

PLATE 9. Cerro la Fundición. Built on a small hill that rises out of the Río Casas Grandes floodplain, this small cerro de trincheras includes some pottery, as well as projectile points characteristic of the Early Agricultural period. This view is looking towards the northeast.

Plate 14. Cerro la Tinaja. This small cerro de trincheras included some brownware pottery, as well as a San Pedro style dart point. Although not visible in this view, the terrace complex is centered on the summit of the hill. This view is towards the southeast and the hill overlooks the Río Tapicitas, a major tributary of the Río Piedras Verdes/Río Casas Grandes.

PLATE 15. Cerro la Angostura de Galeana. This cerro de trincheras lies on the eastern side of the Río Santa María. In this image, extensive walls and terraces are visible on the crest and southwest-facing slope. The few artifacts include Archaic-age dart points as well as some pottery.

PLATE 16. Cerro Le Barón Norte and Cerro Le Barón Sur. Cerro Le Barón Sur is visible in the foreground of this photograph; Cerro Le Barón Norte is located on the hill beyond the house. The sharply pointed peak in the background is Cerro la Angostura de Galeana (Plate 15). This view is towards the north.

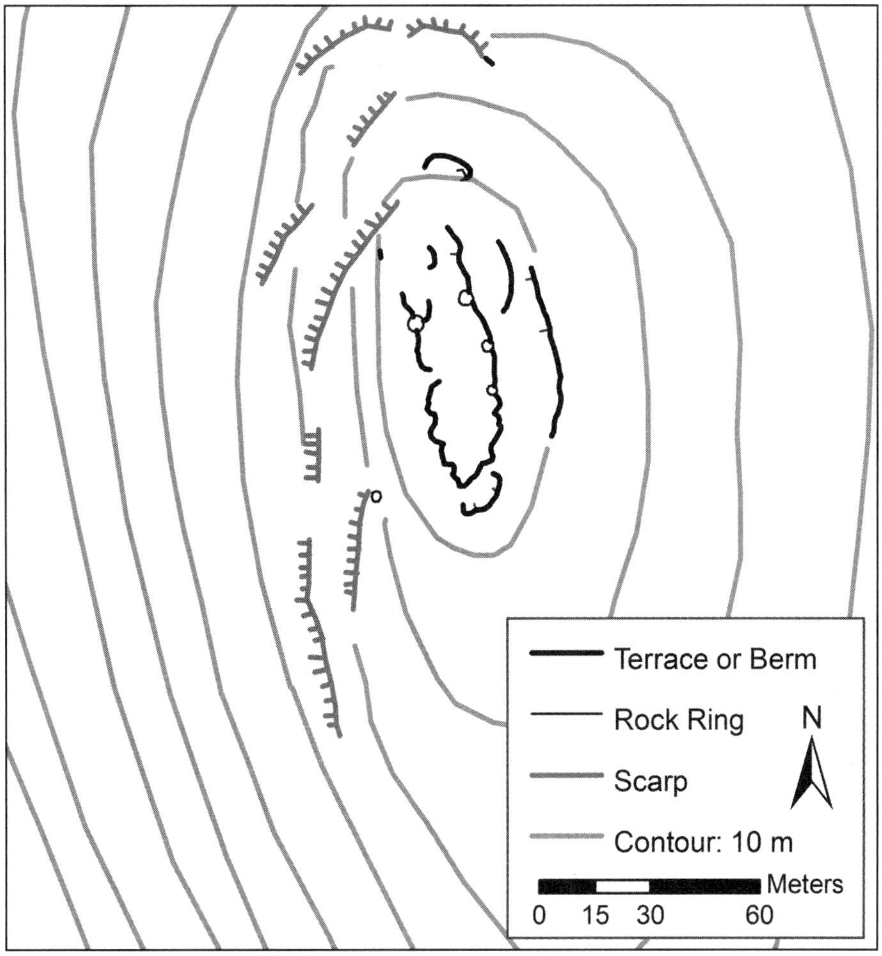

FIGURE 4.10. Cerro la Tinaja. Also see Plate 14.

of these appear to be related to a single reduction event. Projectile points included a Hatch or Hueco type and two Cienega points, as well as one that was not identifiable. Other lithic artifacts included seven or eight cores. The ground stone assemblage consisted of fragments of two basin metates. A dozen sherds were documented in various parts of the terrace complex, including examples from both prehistoric and historic times. Petroglyphs that seem stylistically related to the Ceramic period are also widely scattered on this site. A large number of additional petroglyphs, as well as a large Ceramic period ruin, are located near the base of Cerro Galeana.

Cerro Le Barón Norte

About a kilometer south of Cerro Galeana are two lower peaks which intrude on the floodplain. The two peaks are separated by a saddle, and terrace-like constructions occur on both (Plate 16, mid-ground). The northernmost hill, Cerro Le Barón Norte, is 80 m high. At this site there is a 110 m long rubble berm, several smaller walls, and 9 or 10 stone circles (Figure 4.12). Construction of the long rubble berm is similar to that observed at other prehistoric cerros de trincheras in northwestern Chihauhua. However, many of the other cultural features on this site appear to be more recent. The rocks used in their construction have not repatinated, the rocks seem more precariously balanced, and they support little vegetation. These observations raise the possibility that multiple use episodes are represented here, possibly including constructions built in historic times. We found a few sherds of historic green glass, and more recent trash at this location, but no prehistoric artifacts.

At the western base of the hill, on natural terraces overlooking the floodplain of the Río Santa María, we documented an extensive, largely aceramic lithic scatter with dark, midden stained soils. Bedrock mortars are common, and there are a series of rockshelters on the adjacent hillslope. A substantial Ceramic period occupation is also in this vicinity, and the area has been used extensively by Europeans from the colonial era until the present.

Cerro Le Barón Sur

The second group of features, Cerro Le Barón Sur, is on a 40 m high hill about 250 m south of the complex described above (Plate 16, foreground). This is a small

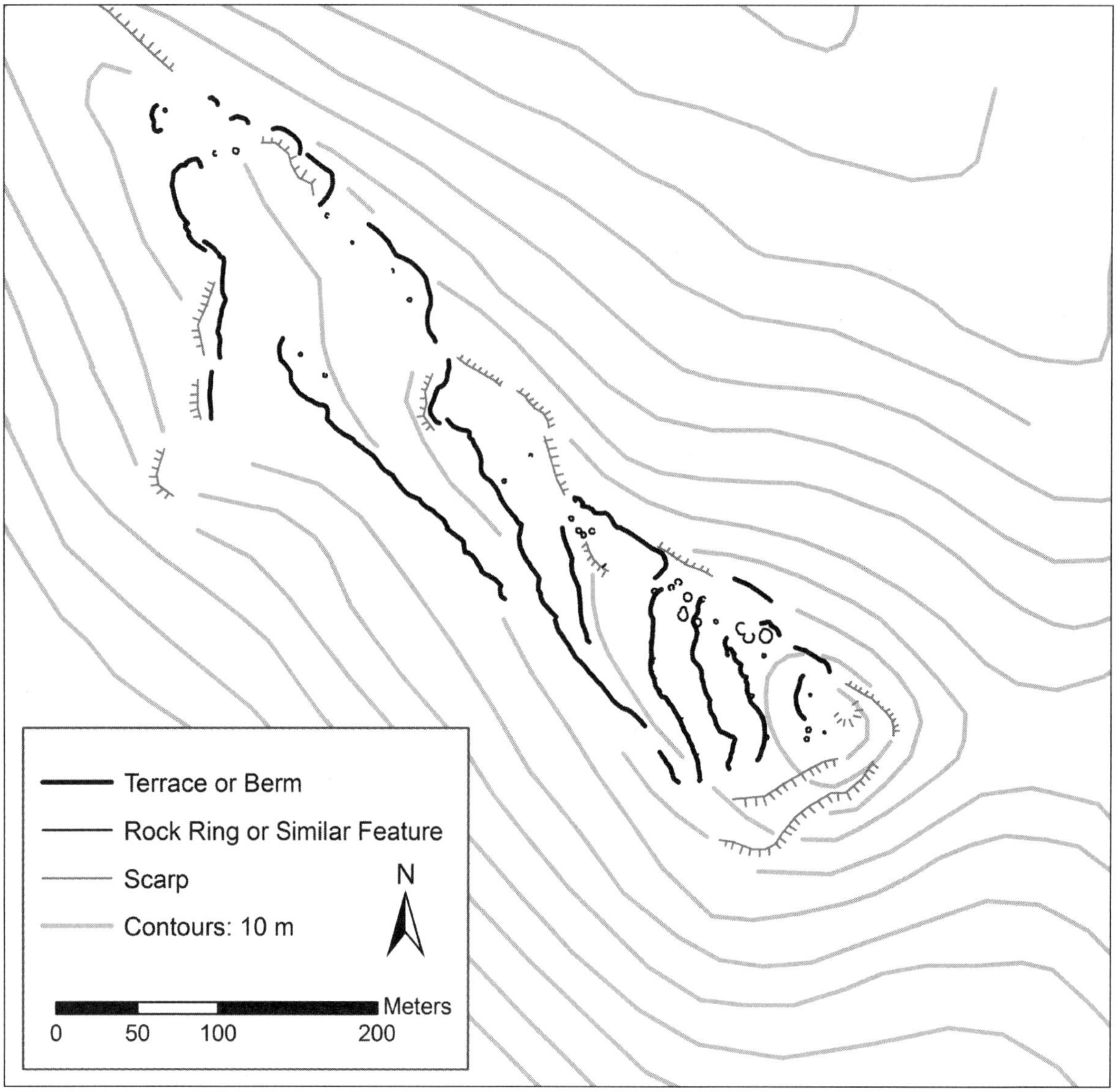

FIGURE 4.11. Cerro Angostura de Galeana. Also see Plate 15.

complex, consisting of a 50 m long rubble berm, several shorter alignments, and two rock rings (Figure 4.13). These features are located primarily on the western face of the north-south trending ridge, while a near vertical outcrop defines the eastern margin. We did not find any prehistoric artifacts at this site. Some features (not mapped) on this hill are clearly very recent, perhaps the result of children playing, and others could be of historic origin. However, as at the northern complex, some of the larger features appear subjectively to be prehistoric.

Minor Investigations at Sites on the Floodplain Near Cerro Juanaqueña

In addition to work on cerros de trincheras, we conducted limited investigations at five other sites on or near the floodplain below Cerro Juanaqueña. One of the most important of these was a buried Early Agricultural period site named Ophelia's Well. This site is buried a meter below the modern floodplain surface at the base of a Pleistocene terrace near the confluence of the Río Casas Grandes and the Río San Pedro. Cultural deposits consisted of a thin (ca. 15 cm thick) stratum of gray, ashy soil with flecks of charcoal, several fire-cracked rocks, and a few lithic flakes found within an area 17 m by 8 m in size. Three possible pits up to 30 cm in depth and 25 to 30 cm in diameter were also defined, but all three were quite indistinct. A single radiocarbon date on wood charcoal indicates that this location was used between 800 and 200 cal BC. This date is well after the principal occupation on Cerro Juanaqueña, but before the secondary occupation. These observations show that intensive Early Agricultural period use of the Río Casas Grandes floodplain continued after the early period of cerro de trincheras use. They also imply that if

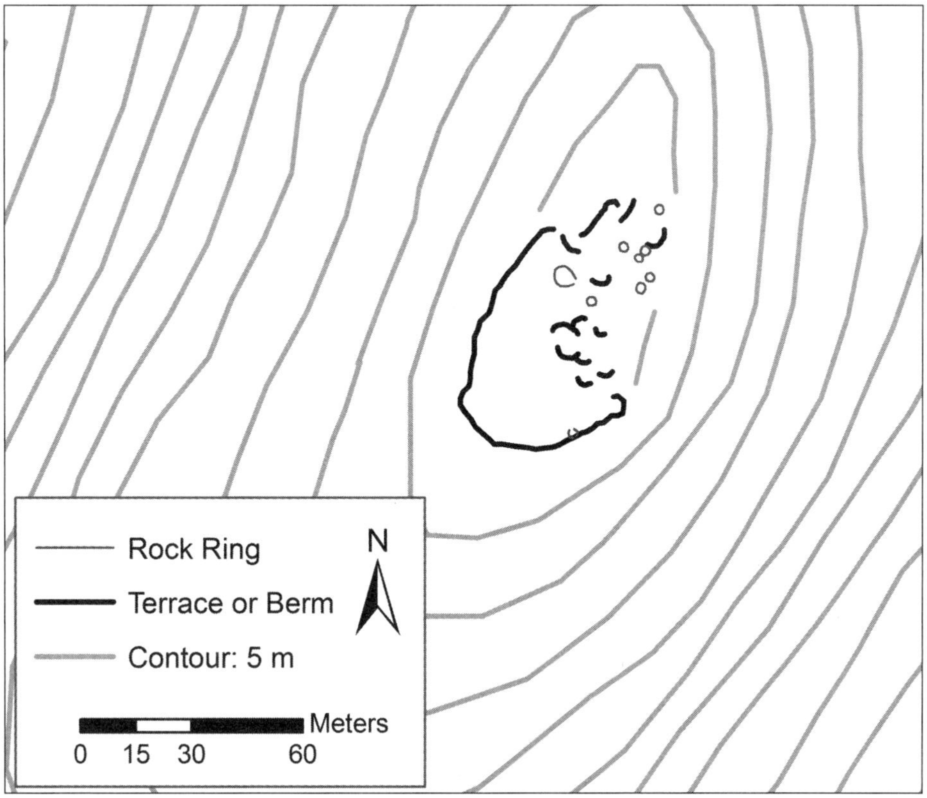

FIGURE 4.12. Cerro Le Barón Norte. Also see Plate 16.

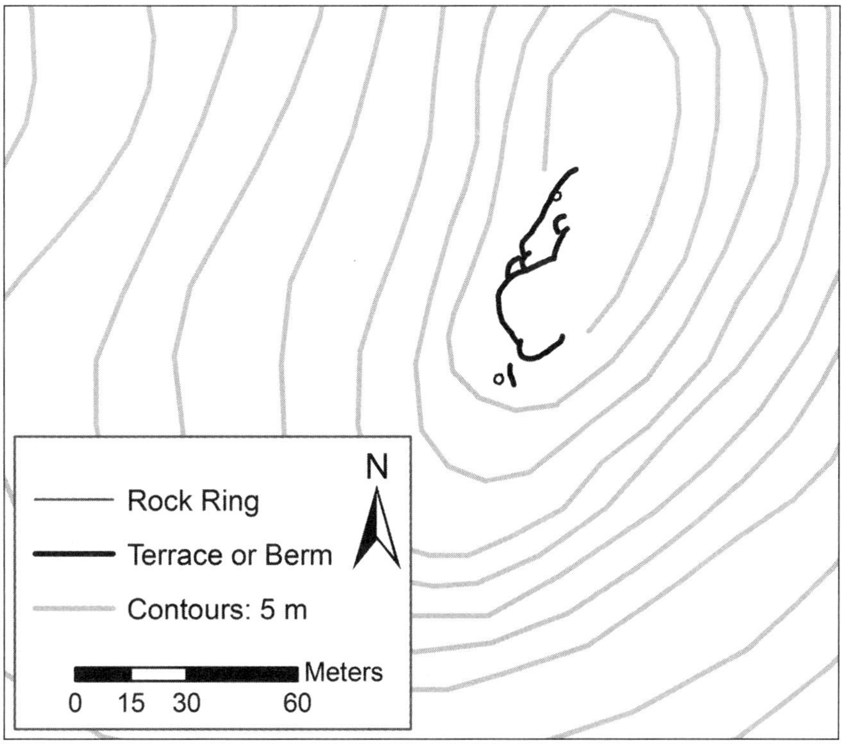

FIGURE 4.13. Cerro Le Barón Sur. Also see Plate 16.

other floodplain sites from this chronological interval exist, they are likely to be deeply buried and not visible from the surface today.

The Floodplain Site was located about four km south of Cerro Juanaqueña. It was 4 ha in size and included worked stone, fragments of ground stone, and various burned rock features. An absence of pottery suggested that it might be Late Archaic in age. Unfortunately, limited testing suggested that these remains rested on a deflated surface and no materials suitable for radiocarbon dating were recovered.

Finally, we screened backdirt from looters' pits at three Ceramic period sites near Cerro Juanaqueña in order to recover samples of lithic debitage and fauna for comparative purposes. These included Sitio Janos, a Viejo period site located just west of Janos, Sitio Lucero, a Viejo period site below Cerro Juanaqueña, and Sitio Antonio, a Medio period site also located near the floodplain below Cerro Juanaqueña.

A more complete description of work conducted at these five sites is included here as Appendix 4.1. Materials recovered from these sites are described in the appropriate analytical chapters: Chapter 8 for macrobotanical remains, Chapter 11 for faunal remains, and Chapter 14 for lithic materials. Radiocarbon dating results from Ophelia's Well are formally presented in Chapter 5, and its position in the quaternary alluvial sequence is described in Chapter 12.

Conclusions

The data presented above show that Cerro Juanaqueña is one of a series of similar cerros de trincheras found in northwestern Chihuahua. At least two of these sites were occupied contemporaneously with Cerro Juanaqueña. A suite of 14 radiocarbon dates from Cerro Juanaqueña's upper terraces shows that the principal period of occupation was a span of about two centuries that was underway no later than 1300 cal BC, with the upper terraces abandoned by 1100 cal BC. A less intensive occupation of the lower terraces occurred between 400 cal BC and cal AD 1. The calibrated AMS radiocarbon date from Cerro los Torres is ca. 1130 cal BC, which is statistically indistinguishable from dates of the principal occupation at Cerro Juanaqueña. Cerro el Canelo was also occupied during this same interval, as indicated by an AMS radiocarbon date of ca. 1215 cal BC. The two AMS dates from Cerro Vidal are ca. 400 cal BC and ca. 140 cal BC, remarkably close to the date of the later occupation at Cerro Juanaqueña. These findings, which are presented in more detail in Chapter 5, raise the possibility that occupation and use of cerros de trincheras in northwestern Chihuahua was episodic and synchronous throughout the region. That is, at least some of the cerros de trincheras may have only been used only during specific and relatively brief intervals in the Late Archaic period. We hypothesize that during lengthy intervals between use of the cerros de trincheras, Early Agricultural period peoples occupied settlements in more ordinary settings on or near the floodplains. Limited investigations at Ophelia's Well provide some preliminary support for this position.

Although radiocarbon dates are only available from four of the cerros de trincheras in northwestern Chihuahua, circumstantial evidence suggests that many of the other sites may also date to the Late Archaic period (see Table 4.1):

1. Artifacts diagnostic of this period occur on several of the undated cerros de trincheras. Other undated sites which have yielded side- and corner-notched dart points with expanding stems and convex bases include Cerro la Boquilla de San Diego, Cerro la Tinaja, and Cerro Galeana.
2. Three of the undated cerros de trincheras, Cerro la Virgen, Cerro la Fundición, and Cerro Le Barón Norte have few artifacts on their terraces, but do have substantial artifact scatters at their bases which are either completely or largely aceramic. Aceramic artifact scatters are not common in this region, and their frequent association with cerros de trincheras is striking.
3. None of the cerros de trincheras documented here can be unambiguously attributed to the Ceramic period. Pottery has been found on nine of the sites (Cerro Tascate, Cerro Juanaqueña, Cerro la Virgen, Cerro la Fundición, Cerro los Torres, Cerro la Cruz, Cerro la Boquilla de San Diego, Cerro la Tinaja, and Cerro la Angostura de Galeana). However, in each of these cases the amount of ceramic material is small and its distribution within the site is usually limited to the summit of the hill. As discussed above, two of these sites have yielded radiocarbon dates which show that periods of intensive occupation occurred well before the advent of pottery, and we suspect that this is true at others as well. At least two of the sites which included pottery have also yielded side- or corner-notched dart points with the expanding stems and convex bases characteristic of the Late Archaic period (Cerro la Boquilla de San Diego and Cerro la Tinaja). All of these sites occur in areas that were occupied intensively during an interval between approximately AD 1000 and 1450 (e.g., DiPeso 1974), and it is not surprising that small numbers of sherds are found on more ancient sites.
4. Similarities in construction among many of the sites are consistent with the hypothesis that they are contemporaneous. Except for portions of Cerro la Angostura, Cerro Le Barón Norte, and Cerro Le Barón Sur, all of the sites described here exhibit similar

construction, which we have documented through both excavation and replication (Hard and Roney 1998a, 1998b, 1999a, 1999b; Hard et al. 1999; Roney and Hard 2000).[2] Large cobbles and boulders were gathered from the surface and tossed downslope to form a cobble berm. Smaller rocks were thrown into the pocket between the hillslope and the berm to form a level surface, and fine sediment was added to finish the terrace. In rare instances rocks were stacked to hold a steeper grade, but formal, vertical masonry is absent. This mode of construction contrasts with the more formal, vertical, dry-laid masonry common in later cerros de trincheras in Sonora and southern Arizona (Downum 1993: 80; McGuire et al. 1993:25; O'Donovan 1997:101). These constructions are also very different from the cobble alignments which characterize agricultural trincheras during the Ceramic period in this part of Chihuahua (Herold 1965; Howard and Griffiths 1966; Schmidt and Gerald 1988).

5. Similarity in overall layout of the sites is also consistent with a hypothesis of contemporaneity. Inspection of the site maps shows that the terraces are often arcs approximately 15 m in length. Individual terraces are typically joined together to form larger continuous terraces following contours of the hill which we call *macrofeatures*. Most of the sites which we have mapped also have perimeter berms which define their exterior boundaries. Perimeter berms are constructed in a manner similar to ordinary terraces, although they are often more massive. They closely resemble terraces in areas where they run perpendicular to slope, but unlike terraces, perimeter berms are not consistently perpendicular to slope. When they run diagonal or parallel to slope, it is apparent that they are actually cobble berms which serve to define the site boundary rather than to create level space. At sites such as Cerro Tascate, Cerro la Virgen, and the two Cerros Le Barón, perimeter berms are the principal constructed features. At other sites, such as Cerro Juanaqueña, Cerro los Torres, and Cerro Vidal the perimeter berms are very conspicuous.

In summary, radiocarbon dates show that at least three of the 14 cerros de trincheras in northwestern Chihuahua were built and occupied intensively during the Late Archaic period, 1500 BC to AD 50.[3] Direct and indirect artifact associations provide more circumstantial evidence that six others were also built during this early period. Despite a substantial Ceramic period occupation in this region, none of the cerros de trincheras can be unambiguously attributed to the Ceramic period. Finally, construction details and similarities in layout are also consistent with the hypothesis that all of these sites were initially built in the Late Archaic/Early Agricultural period.

Notes

1. Cerro la Angostura de Galeana, which we refer to as Cerro Galeana, is a different site than Cerro la Angostura (Figure 1.1).
2. The cited annual field reports are included as electronic Appendices 3.2 to 3.5.
3. In 2017 we became aware of two additional cerros de trincheras in the Río Casas Grandes valley north of Cerro Tascate, one of which has been documented. Emiliano Gallaga (personal communication 2018) recorded Alamos Altos I, located 5 km north of Tascate. It is on the east side of the Río Casas Grandes overlooking a floodplain constriction. The site dimensions are 150 m by 500 m and include two small peaks. One peak has a perimeter berm that is 200 m by 100 m with additional interior walls. A light lithic scatter, rock rings, at least two Late Archaic projectile point forms, mortar holes, and a pot break are present.

CHAPTER 5

Radiocarbon Dating

Introduction

Cerro Juanaqueña is remarkable largely because of its great age. Although archaeologists have known about this site for some time, its early age was not appreciated and the site received little attention prior to our own investigations. As early as the mid-1960s, Charles DiPeso included Cerro Juanaqueña on a regional map of northwestern Chihuahua, marking it as the location of an atalaya or signaling site associated with the Ceramic period Casas Grandes culture (DiPeso et al. 1974:Figure 284-5). Later Rex Gerald (1990) spotted the terraces during an aerial reconnaissance and then visited the site in 1988, thinking that it might be a Ceramic period fortification. Instead, he found Archaic-style basin metates, late-appearing petroglyphs, and a complete absence of pottery. Gerald (1990) concluded that the site could date to the time of Casas Grandes or the early Historic period or the Archaic period.

In 1992, John Roney visited Cerro Juanaqueña while pursuing an interest in atalaya sites. Instead of a small pyrocommunication facility, he found an astounding series of massively constructed terraces, and instead of Casas Grandes pottery, he found large numbers of late Archaic dart points. Over the next several years Roney mapped artifacts and features at Cerro Juanaqueña, Cerro Vidal, and Cerro los Torres. He continued to find Archaic projectile points on the terraces, confirmed the virtual absence of ceramics, and noted the high densities of Archaic style metates and manos, particularly at Cerro Juanaqueña. Based on these observations, Roney (1996a, 1996b) argued that these three sites and their constructed terraces dated between 1000 BC and AD 500.

These conclusions challenged longstanding wisdom concerning the introduction of agriculture into the Southwest as well as the fundamental nature of Late Archaic cultures. We quickly realized, however, that without radiocarbon dates most archaeologists would remain skeptical of these claims. We submitted a National Science Foundation (NSF) proposal for one season of funding for a modest program of test excavations in hopes of recovering samples for radiocarbon dating. Initially, funding for that first proposal was in doubt, but in late April 1997 we learned that NSF would fund the project. We could now pursue our effort to obtain dates and other basic information. From that point on the project moved fast.

On May 30, 1997, we received our first archaeological permit while visiting the Instituto Nacional de Antropología y Historia (INAH) offices in Mexico City. An entry in Hard's field journal, made during the return trip to Janos reads: "Project Goals: 1) obtain ^{14}C samples; 2) determine if houses are present; 3) obtain macrobotanical samples." Barren, rocky hillslopes are not an ideal place to look for 3,000-year-old charcoal, and at the outset it was not at all clear that we would succeed. But we ultimately had success in all three goals. By Sunday, June 1, 1997, the first excavation units were defined and mapped at Cerro Juanaqueña. Our team had also discovered on the surface 50 more Archaic projectile points! We broke ground on June 2 and flotation samples taken that day contained datable wood charcoal and evidence of ashy soil. Our hopes soared.

However, for reasons discussed below, wood charcoal is not the best material for radiocarbon dating. Therefore we needed samples of charcoal from maize or other short-lived plant species. Graduate student team members Kari Schmidt and José Zapata did not disappoint us. On June 7, 1997, their excavation unit produced the first of many pieces of charred maize from Cerro Juanaqueña. By the end of that memorable first week of fieldwork we had also recovered an intact pipe, a cruciform, and a polished pestle. We had also identified an occupation surface, recovered a substantial number of animal bones, examined terrace construction techniques, and identified buried midden deposits. What we did not find was perhaps of equal significance—no ceramics were present in any of our excavation units. The upper portion of Cerro Juanaqueña—that massive, intensively occupied, single-component residential site with only a few centuries of occupation—was just beginning to reveal itself.

After the field season we sent the first batch of charred plant samples to paleoethnobotanist Karen Adams. Working through a binocular microscope, she was able to identify the plants that formed the tiny fragments of charcoal we had recovered. This crucial step allowed us to select the very best samples for dating, and we followed this protocol throughout the project.

Through the fall of 1997, anticipation grew as we waited for the dating results from the University of Colorado's Laboratory for AMS Radiocarbon Research. Finally, a fax came to The University of Texas at San Antonio on November 20 reporting that the first four dates were all clearly Late Archaic in age, falling around ca. 3000 BP (Table 5.1, Samples NSRL 3983, 3985, 3986, 3995). Our expectations were confirmed. The site indeed dated to the Late Archaic period, and we reported these initial results in an article in the journal *Science* (Hard and Roney 1998a).

As the project continued, we began to work at other similar sites in the area, and we were able to ask more and more precise questions about the age of the occupations at Cerro Juanaqueña, Cerro los Torres, Cerro Vidal, and Cerro el Canelo. Were there multiple occupations at Cerro Juanaqueña? Could we detect a temporal growth sequence for the site? Did the lower terraces date to the same period as the upper terraces? Were the other three sites of the same age? How long was the occupation? When did the occupations begin and end? In order to pursue these questions we maintained a rigorous and aggressive approach to radiocarbon dating.

Methods and Samples

Accelerator Mass Spectrometry (AMS) dating has brought about a revolution in archaeological dating methods. This method became widely available in the late 1980s and it allows archaeologists to date very small samples of charcoal and other material (Taylor 1997). Conventional radiocarbon dating typically uses charcoal samples weighing 10 grams or more—a large handful. However, AMS dating can analyze samples less than 0.1% of that—less than half the weight of a maize kernel. In many archaeological contexts, including the exposed conditions at Cerro Juanaqueña, it is often difficult to find charcoal samples that are both sufficiently large and in unambiguous contexts for conventional dating. Many archaeological contexts remain poorly dated due to the sample requirements of conventional radiocarbon dating.

Even when large pieces of charcoal are found, they are usually wood charcoal, which poses special problems for radiocarbon dating. One problem is that only the outermost layers of wood are living tissue. Previous years' growth rings, which are essentially dead tissue, accumulate over many years in some long-lived trees. If a sample of charcoal comes from wood deep inside a branch or tree trunk, it can yield a date that is many years older than the fire that produced the charcoal. In addition, some types of wood, such as mesquite, are so slow to biodegrade that they remain on the desert floor for decades or even a century or more before being used (Schiffer 1986). This circumstance also causes wood charcoal to yield dates that can be much older than the prehistoric occupation being investigated. Finally, wood charcoal can enter the archaeological record as a product of purely natural events such as a range fire. Because of this, unless a sample of wood charcoal is found in a cultural feature such as a hearth or burned structure, there is no guarantee that it even pertains to human activities.

The AMS dating method circumvents many of these problems because it can routinely date samples as small as 3 milligrams with a precision of ± 50 years and now ± 30 years is common (Smith 1998). This opens a whole new range of radiocarbon dating possibilities. First of all, it is much easier to find small samples of charcoal, and it is now possible to obtain radiocarbon dates from many archaeological contexts that were not previously datable. Second, it is now possible to avoid the "old wood" problem by selecting samples of charcoal produced by short-lived species such as weeds or grasses, and by selecting samples of charcoal from plant tissue that is produced in a single year, such as twigs, seeds or nutshells. Finally, it is possible to select samples of charcoal from plants which do not grow without human intervention in the vicinity of an archaeological site, such as maize. This eliminates the possibility that the radiocarbon date might relate to a natural fire, rather than to a cultural event.

These factors combine to make maize the ideal material for radiocarbon dating. Maize is a plant that is uniquely tied to human beings. There are no wild stands of *Zea mays* because the plant cannot reproduce without human assistance. As a result, maize charcoal always represents human activity, and the possibility that it was introduced through natural processes is negligible. Also, maize is an annual plant, and its ^{14}C content is the product of a single season of photosynthesis, not the many years reflected in charred wood rings from trees. A maize radiocarbon sample dates a precise cultural event lasting from the growth of the maize plant until the cob is picked from the plant, as the radiocarbon clock starts at the moment of an organism's death. Using AMS radiocarbon dating, we can date tiny cupules from corn cobs, and pieces as small as one-third of a maize kernel.

We sometimes obtained radiocarbon dating samples from charcoal found in excavation screens, but more often the samples were recovered through flotation

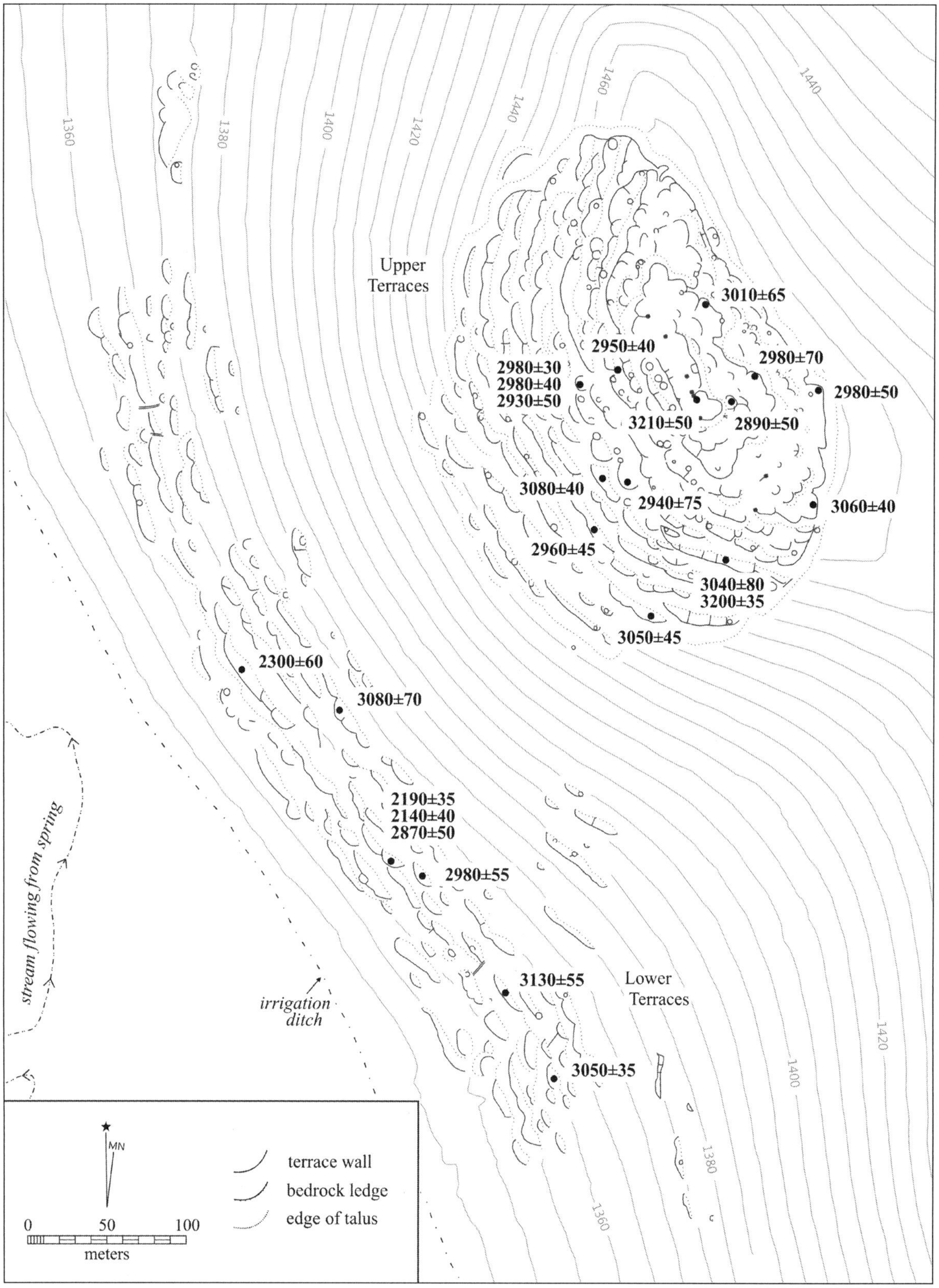

FIGURE 5.1. Map of Cerro Juanaqueña showing the distribution of dated features and their radiocarbon ages.

(Chapter 8). Karen Adams, the project ethnobotanist, identified all potential radiocarbon samples to the lowest possible taxonomic classification, and these identifications guided our selection of samples for actual dating. We were fortunate in that maize was virtually ubiquitous in the features that we excavated, and we could usually find samples of maize charcoal that were suitable for dating. When this was not possible, we tried to select species or plant parts produced in a single year, such as wild gourd seeds or grass stems. When it was necessary to use wood charcoal, we tried to select samples from relatively short-lived species such as ocotillo or saltbush. In only two cases, we were forced to submit wood charcoal from long-lived species such as mesquite. In the case of trying to date a very large circular rock feature at Cerro el Canelo, we dated a long bone from a deer-sized animal. Stafford Research Laboratories, Inc. dated a tiny fish bone from Cerro Juanaqueña to determine if the many fish bones that we found in the heavy fractions of the flotation samples were indeed the same age as the prehistoric occupations.

The geographic location of samples within the site was also an important consideration. We selected specific terraces for excavation based primarily on our estimation of the likelihood that the terrace would yield dateable charcoal and the goal of dating as many parts of the site as possible. In the cases of Cerro Vidal, Cerro los Torres, and Cerro el Canelo, only one or two features were excavated, thus limiting material for radiocarbon dating. We can only hope that the results are representative of the sites as a whole. At Cerro Juanaqueña, we obtained dates from 19 different features (Figure 5.1). We deliberately excavated terraces in different parts of the site in order to explore the possibility that the site grew through accretion or that some areas were occupied at different times.

Our goal was to obtain at least one date for every excavated feature (Chapter 3, Table 3.1). If an initial date from a feature was inconsistent with expectations, we submitted one or two additional samples from the same feature. Individual radiocarbon dating samples usually consisted of a single piece of plant material, but if all maize fragments from a particular context were extremely small, we sometimes combined two or more pieces of charred maize to obtain a sufficient sample.

The radiocarbon dating samples recovered during terrace excavations came from construction fill and midden deposits, as well as from within or under the dense rock accumulations that comprise the berm walls that formed the terraces. As discussed in Chapter 3, there was no evidence for stratigraphically segregated occupation phases within any of the terrace features. While excavating, it was not unusual to see that small particles, such as charred plant material, could descend through the loose, rocky terrace fill; for our purposes the vertical supraposition of the ^{14}C samples was not important. Botanical preservation tended to improve with depth of deposit (Chapter 3 and Chapter 8). However, as it turned out, the terrace features did not yield evidence of long periods of use, either in terms of stratigraphy or the temporal spread within multiple ^{14}C dates from the same feature. We therefore were able to treat each terrace or rock ring as representing a relatively short occupation event.

Results

Table 5.1 lists the results of the 33 AMS radiocarbon dates processed from Early Agricultural period sites in northwestern Chihuahua, and Figure 5.1 shows the spatial distribution of the Cerro Juanaqueña results. The results are remarkably consistent. Dates from different materials show no systematic differences, and all three radiocarbon dating laboratories provided comparable results. In Table 5.1, the raw radiocarbon dating results are listed as "Conventional Radiocarbon Age BP."

Because the amount of ^{14}C in the atmosphere has fluctuated through time, it is necessary to calibrate the radiocarbon age with actual calendar dates using information derived from large datasets of radiocarbon-dated tree-rings of known ages. Fortunately, this is easily done with the computer programs OxCal and CALIB (Bronk Ramsey 2001; Reimer et al. 2004). The result of the calibration is not a single calendar year, but rather a probability distribution which expresses the likelihood that a given radiocarbon age corresponds with a range of calendar years. The ranges of calendar years that correspond to the radiocarbon age of our samples are given in the last two columns of Table 5.1, at one standard deviation and at two standard deviations.

These probability distributions are shown graphically in Figure 5.2, using the OxCal 3.10 calibration results (Bronk Ramsey 2001, 2005; Reimer et al. 2004). The graph accompanying each date represents the probability that the calibrated (cal) calendrical date falls at a particular point. The horizontal X-axis of each graph represents calendar years, while the Y-axis is the probability that the conventional radiocarbon age corresponds with a specific year. The top short line underneath the X-axis represents the one standard deviation probability (68.2%) that the calendar age falls in that range. The bottom line represents the two standard deviation probability (95.4%). This means that there is a 95.4% probability that the radiocarbon age reflects a calendar date within the indicated range. The probability of the date falling at smaller segments can be assessed as well. For example, the probability distribution of NSRL-10861 (2140 ± 40), at two standard deviations, has a 20% probability of falling between 360 cal BC and 280 cal BC

TABLE 5.1. All 33 Radiocarbon Dates in Order by Lab Number.

^{14}C Lab Number[1]	Atomic Accelerator No.[2]	Provenience[3]	Material[4]	Conventional Radiocarbon Age BP[5]	$\delta^{13}C$ ‰[6]	Calibrated Date (1 sigma)[7]	Calibrated Date (2 sigma)[7]
NSRL-3983	CAMS-41687	C.J, T167, U3, L1, 70 cm bs, B42	*Zea mays* cob	2980 ± 50	−12.2	1310–1120 BC	1380–1040 BC
NSRL-3985[8]	CAMS-41688	CJ, T222, U3, L11, 90–100 cm bs, B94	1 *Cucurbita digitata* type seed	3310 ± 60	−25.3	1690–1520 BC	1740–1440 BC
Redate of NSRL-3985	CURL-6918	Same as above	Same as above	2980 ± 30	−24.8	1270–1120 BC	1370–1110 BC
NSRL-3986	CAMS-41689	CJ, T537, U4, L5, 49–59 cm bs, B74	*Zea mays* cob	2890 ± 50	−9.9	1190–990 BC	1260–920 BC
NSRL-3995	CAMS-41690	CJ, T222, U2, L13, 107–117 cm bs, B100	17 *Zea mays* cupules	2930 ± 50	−9.3	1260–1040 BC	1310–970 BC
NSRL-10039	CAMS-45518	CJ, T222, U3, L7, 50–60 cm bs, B79	1 *Cucurbita digitata* type seed	2980 ± 40	−23.6	1300–1120 BC	1380–1040 BC
NSRL-10056	CAMS-45925	CJ, T6, U1, L4, 30–75 cm bs, B93	Ocotillo charcoal	2980 ± 70	−24.3	1370–1050 BC	1400–1000 BC
NSRL-10591	OS-19485	CT T1, U3, L8, 62–72 cm bs, B53	1 *Zea mays* cupule fragment	2920 ± 55	−11.2	1260–1010 BC	1300–930 BC
NSRL-10592	OS-19486	CV, T20, U1, L7, 62–72 cm bd, B31-33	1 *Zea mays* cupule	2100 ± 40	−11.1	170–50 BC	350 BC–AD 10
NSRL-10593	OS-20245	CJ, T387, U2, L8, 90–100 cm bs, B38,43	1 *Zea mays* cupule	3080 ± 70	−11.0	1430–1220 BC	1520–1120 BC
NSRL-10594	OS-20246	CJ, T413, U1, L7, 63–73 cm bs, B47	1 *Zea mays* kernel	2190 ± 35	−10.2	360–170 BC	380–160 BC
NSRL-10595	OS-20247	CJ, T97, U1, L7, 62–71 cm bs, B26	1 *Zea mays* cob fragment	3050 ± 45	−11.3	1390–1220 BC	1420–1120 BC
NSRL-10596	OS-20248	CJ, T297, U2, L7, 73–83 cm bs, B30	1-2 *Zea mays* cupules	2960 ± 45	−10.9	1270–1050 BC	1370–1010 BC
NSRL-10597	OS-20249	CJ, T290, U1, L5, 34–45 cm bs, B38	1-2 *Zea mays* cupules	3080 ± 40	−11.3	1410–1260 BC	1440–1210 BC
NSRL-10598	OS-20250	CJ, T126, U1, L9, 91–101 cm bs, B56	1 *Zea mays* cupule	3040 ± 80	−11 (est.)	1410–1130 BC	1450–1040 BC
NSRL-10599	OS-20251	CJ, T163, U2, L8, 45–65 cm bs, B47	1 *Zea mays* cupule fragment	3060 ± 40	−11.2	1400–1260 BC	1430–1210 BC
NSRL-10600	OS-20252	CJ, T10, U1, L11, 94–104 cm bs, B43	1 *Zea mays* cupule	3010 ± 65	−11.8	1380–1120 BC	1410–1040 BC
NSRL-10712	OS-20425	CV, T20, U1, L5, 42–52 cm bs, B20-23	*Zea mays*	2340 ± 55	−11.0	520–250 BC	800–200 BC
NSRL-10860	OS-21747	CJ, R250, U2, L5, 35–45 cm bs, B41	Ocotillo charcoal	2950 ± 40	−22.7	1260–1050 BC	1300–1010 BC
NSRL-10861	OS-21748	CJ, T413, U1, L3, 23–33 cm bs, B16	*Zea mays*	2140 ± 40	−10.9	350–90 BC	360–40 BC
NSRL-10862	OS-21749	CJ, T413, U1, L9, 83–93 cm bs, B56	*Zea mays*	2870 ± 50	−9.9	1130–940 BC	1220–900 BC
NSRL-11428	OS-25414	CJ, BR1, U6, L3, 28–38 cm bs, B14	1 Mesquite charcoal fragment	3210 ± 50	−20.8	1520–1425 BC	1610–1390 BC
NSRL-11430	OS-25415	CJ, T287, U1, L8, 82–102 cm bs, B62	4 *Zea mays* cupule fragments	2940 ± 75	−11.0	1260–1010 BC	1380–920 BC
NSRL-12480[8]	OS-30608	CC, T94, U2, L4, 41–51 cm bs, B23	Monocot	330 ± 60	−11.6	AD 1490–1640	AD 1440–1670
NSRL-12481	OS-30609	CC, T94, U2, L6, 61–71 cm bs, B29	Ocotillo charcoal	2990 ± 45	−23.8	1370–1120 BC	1380–1040 BC
NSRL-12482	OS-30610	CJ, T415, U1, L2, 20–40 cm bs, B6	1 *Zea mays* kernel fragment	2300 ± 60	−9.8	410–200 BC	550–150 BC
NSRL-12483	OS-30611	CJ, T463, U1, L3, 40–60 cm bs, B3	1 *Atriplex* charcoal fragment	2980 ± 55	−11.3	1370–1120 BC	1390–1020 BC
NSRL-12484	OS-30612	CJ, T487, U4/5, L7, 72–82 cm bs, B94	1 *Zea mays* cob fragment	3130 ± 55	−10.7	1500–1310 BC	1520–1260 BC
NSRL-12485	OS-30613	CJ, T508, U2, L5, 46–56 cm bs, B16	1 *Zea mays* cob fragment	3050 ± 35	−11.0	1390–1220 BC	1420–1120 BC

Beta-142771[8]		CC, Large Circle, U1, L3, 6–27 cm bs, B9	Med/large mammal long bone fragment	630 ± 50	–15.8	AD 1295–1395	AD 1280–1410
Beta-153665		OW, BHT5B, 118 cm bs, B7	Wood charcoal	2490 ± 50	–25.6	770–520 BC	790–410 BC
Beta-172602		CT, T1, U3, L10, 82–92 cm bs, B61	1 *Zea mays* kernel fragment	2960 ± 40	–9.1	1260–1050 BC	1320–1020 BC
Beta-172603		CC, T94, U1, L4, 52–62 cm bs, B11	1 Ocotillo fragment	3000 ± 40	–24.7	1370–1130 BC	1390–1110 BC
SR-6353	CAMS-94860	CJ, T126, U1, L9, 170–180 bs, B57	basioccipital bone Cyprinidae (minnow family)	3200 ± 35	–12 (est)	1515–1430 BC	1530–1400 BC

[1] INSTAAR-NSRL = Institute of Arctic and Alpine Research–Laboratory for AMS Radiocarbon Preparation and Research, University of Colorado at Boulder; Beta = Beta Analytic Inc.; SR = Stafford Research Laboratories, Inc.

[2] CAMS = Center for Accelerator Mass Spectrometry, Lawrence Livermore National Laboratory; OS = National Oceanic Sciences AMS Facility at Woods Hole Oceanographic Institute. CURL = INSTAAR Laboratory for AMS Radiocarbon Preparation and Research, samples run at University of California (UC) Irvine. Beta Analytic does not provide accelerator lab information.

[3] CJ = Cerro Juanaqueña, CT = Cerro los Torres, CV = Cerro Vidal, CC = Cerro el Canelo, OW = Ophelia's Well. T167 = Terrace 167, R250 = Rock Ring 250, BR1 = Burned Rock 1, U3 = Unit 3, L1 = Level 1, 100–110 cm bs = 100–110 cm below surface, B42 = Bag 42.

[4] All plant material charred.

[5] Calculated using a half-life of 5,568 years and corrected for $\delta^{13}C$ isotopic fractionation.

[6] Measured $^{13}C/^{12}C$ ratio relative to the PDB-1 standard.

[7] Atmospheric data from Reimer et al. 2004; OxCal v3.10 Bronk Ramsey (2005).

[8] Date rejected (see text). Excluded from further analysis.

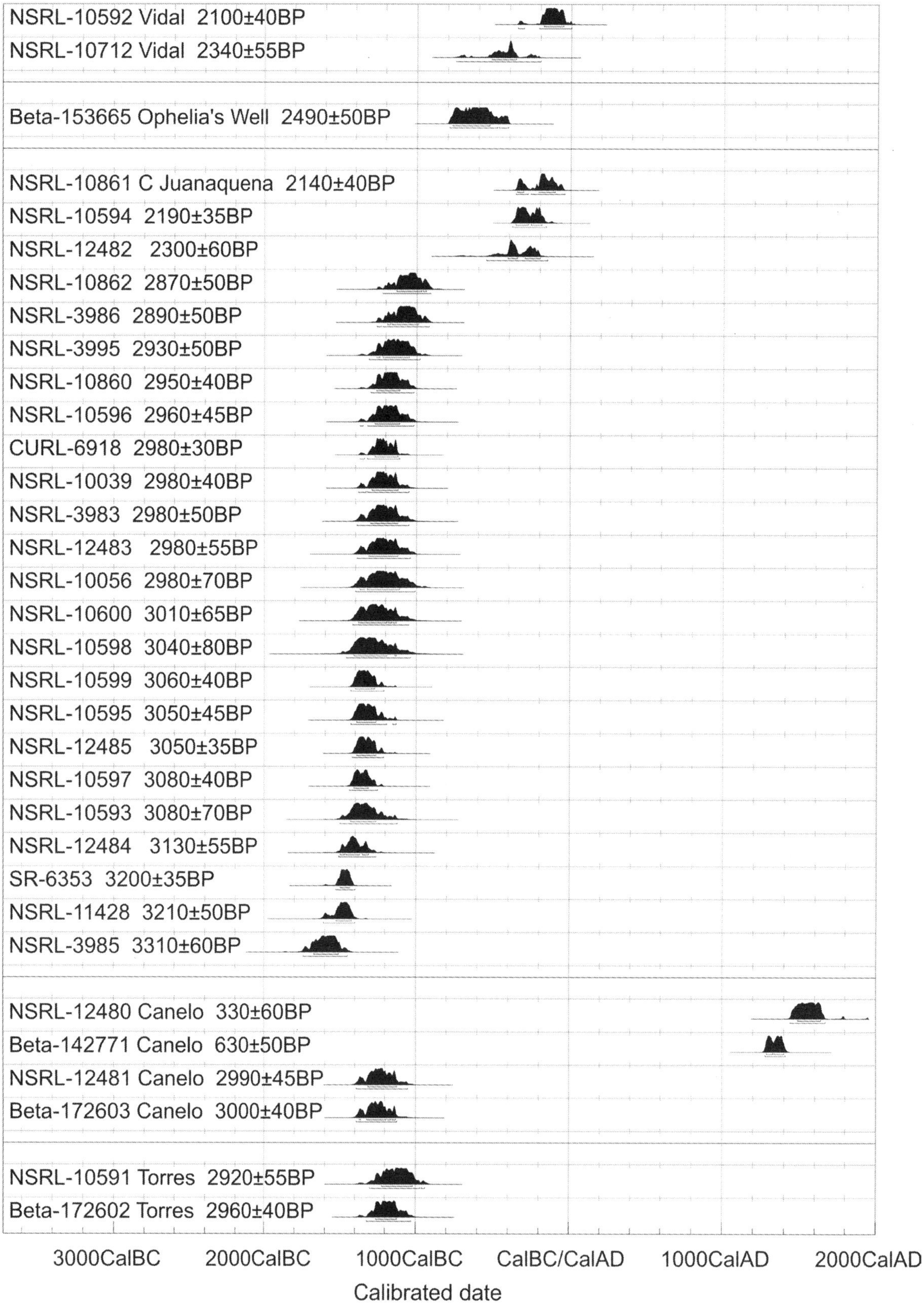

FIGURE 5.2. Oxcal 3.5 tree-ring calibrations of all 33 ^{14}C dates clustered by site and in chronological order.

and is represented by a small spike on the left side of the graph. The same radiocarbon age has another 75.4% probability of falling between 260 cal BC and 40 cal BC as indicated by the higher and wider peak on the right side.

A glance at Figure 5.2 provides a preliminary answer to several of the dating questions posed earlier in this chapter. The oldest occupations are around 1200 cal BC at Cerro Juanaqueña, Cerro el Canelo, and Cerro los Torres. Ophelia's Well, the remnants of a badly damaged open site buried in the floodplain, dates to about 600 cal BC. Hilltop sites were again used at about 200 cal BC, as evidenced by dates from Cerro Vidal and two lower terraces at Cerro Juanaqueña (Figure 5.1). Finally, there are two late prehistoric or early historic dates from Cerro el Canelo. Many of the radiocarbon dates from the first occupation at Cerro Juanaqueña fall within a very brief span, suggesting that occupation of the hilltop site may have taken place during a short period of time. Also, dates from Cerro el Canelo and Cerro los Torres are remarkably similar to those from the early occupation at Cerro Juanaqueña suggesting that at this point in time all three cerros de trincheras were simultaneously occupied. Further, the dates at Cerro Vidal correspond with the second occupation of Cerro Juanaqueña. These patterns were both striking and unexpected. In order to address these issues as precisely as possible we treated the dates to a series of analyses.

Validity of the Radiocarbon Dates

Several sources of error can affect radiocarbon dating, so suites of radiocarbon dates are examined for internal consistency and also compared against expectations from other archaeological evidence. There is an extraordinary level of consistency among the dates and only three dates appear inconsistent.

The first of these three is the oldest date (NSRL-3985) from Cerro Juanaqueña and it comes from a wild gourd seed (*Cucurbita digitata* type). With a radiocarbon age of 3310 ± 60 BP, this sample is quite a bit older than others from Cerro Juanaqueña and is clearly a statistical outlier compared to all other dates. For this reason, we submitted two additional samples from the same terrace. One of the additional samples (NSRL-3995) was from a corn cupule and gave a radiocarbon age of 2930 ± 50 BP, which agreed closely with other dates from the site. The other sample (NSRL-10039) was from another fragment of *Cucurbita digitata*. This particular terrace, Terrace 222, is the only provenience on the entire site yielding gourd remains, and it is likely that this sample represents the same cultural event as the original sample. It gave a radiocarbon age of 2980 ± 40 BP, over 300 radiocarbon years more recent than the original gourd seed date. This new date agrees closely with the age of the maize cupule from the same terrace, as well as other dates from the site. With these results in hand, we contacted the Institute of Arctic and Alpine Research (INSTAAR) at the University of Colorado, and they reviewed the documentation relating to the processing of the original date. They could find no evidence of any processing aberrations that would account for the outlying date. Fortunately, however, they retained a portion of the original seed and processed the sample again to re-date it. The results confirmed our suspicions. The result of the re-dating of NSRL-3985 is 2980 ± 30 (identified as CURL-6918), identical to the second gourd seed date of 2980 ± 40 (NSRL-10039). Therefore, the original date, NSRL-3985, is rejected as a statistical anomaly and the re-dated result (CURL-6918) is used in its place. We also count these two runs on the same sample as only one date in our tally of the number of radiocarbon dates.

Two other radiocarbon dates, both from Cerro el Canelo, are questionable because they appear to be too recent. Sample NSRL-12480 is a charred monocot tissue fragment from Terrace 94 that dated to 330 ± 60 BP. Two different charred ocotillo stem fragments from the same terrace yielded nearly identical dates, 2990 ± 45 BP (NSRL-12481) and 3000 ± 40 BP (Beta 172603). This feature revealed no archaeological evidence of a late or protohistoric occupation. While two Ceramic period arrow points were present on the surface of other parts of Cerro el Canelo, all other evidence from the site is consistent with a Late Archaic occupation. It is most likely that the 330 ± 60 BP date is related to ephemeral historic period use of the site, or to a natural fire event.

The second questionable date from Cerro el Canelo is from a long bone of a deer-size animal. This sample was recovered from a 70 m diameter feature known as the Large Circle (Chapter 4, Figure 4.5), and it yielded a radiocarbon age of 630 ± 50 BP (Beta 142771). We did not find any temporally diagnostic artifacts near the Large Circle, but we strongly suspect that it is contemporaneous with the main occupation of Cerro el Canelo. It is spatially associated with the Early Agricultural terrace features, and it is constructed in an identical manner. Pottery or other archaeological evidence of late prehistoric or protohistoric occupation was not found at this location. Finally, the degree of patination on the rocks that define the Large Circle feature is consistent with that seen on the remainder of the site. Therefore, these two recent dates (NSRL-12480 and Beta-142771) from two widely separated parts of Cerro el Canelo will not be considered further.

Two other radiocarbon dates are considered accurate but imprecise because of the material that comprised the sample. Sample NSRL-11428 (3210 ± 50 BP) is from mesquite charcoal from a burned rock feature (BR 1) near the summit of Cerro Juanaqueña. As the site had only a few burned rock features, it was important to

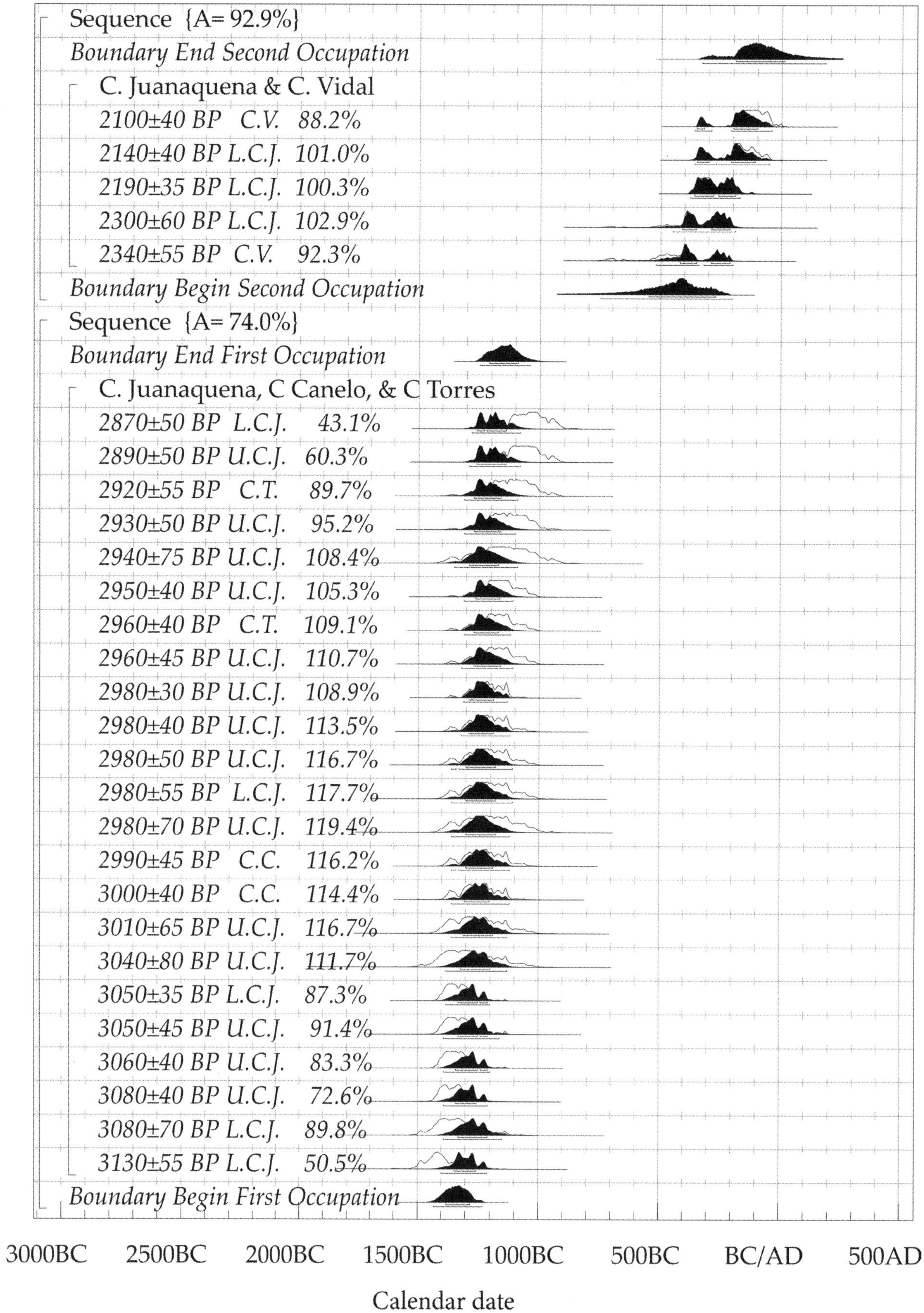

FIGURE 5.3. The OxCal 3.5 posterior probabilities of 28 radiocarbon dates from short-lived species or plant parts. (C.V. = Cerro Vidal; L.C.J. = Cerro Juanaqueña, lower terraces; U.C.J. = Cerro Juanaqueña, upper terraces; C.T. = Cerro los Torres; C.C. = Cerro el Canelo.)

acquire a date from at least one to confirm that these features are associated with the main occupation of the site. Since maize or other short-lived material was not recovered from the feature, we submitted mesquite wood charcoal for dating. The date indicates that the feature is part of the occupation of Cerro Juanaqueña but falls just outside the main cluster of dates; it overlaps many of the Cerro Juanaqueña dates at two standard deviations. Mesquite wood charcoal, as a long-lived species, potentially has lower precision than short-lived specimens so this date is not included in the statistical analyses below which require high precision dates.

Finally, a single date from a fish bone from Cerro Juanaqueña (SR-6353, 3200 ± 35 BP) was acquired to verify if the numerous small fish bones found in terraces across the site were indeed associated with the site and were a part of the human diet. Alternatively, birds of prey could have deposited the fish bone (Chapter 11). Precise radiocarbon dating of bone is difficult because the inorganic (carbonate) phase is prone to irreversible contamination by foreign carbon. The protein (collagen) phase, while subject to contamination, can be purified. Unless removed, carbon contaminants commonly result in bone ages 200 to 1,000 years too young. Stafford Research Laboratories extensively purifies collagen to avoid this issue (Stafford et al. 1991). The fish bone date is 3200 ± 35 BP and clearly demonstrates that fish are contemporaneous with the main occupation of the site and were an element in the prehistoric diet. At two standard deviations, this date overlaps over half of the main cluster of Cerro Juanaqueña dates. However, it will also not be included in the analysis below in order to minimize variability in sample type and optimize precision by including only high-precision, short-lived plant samples in the analyses.

Statistical Analysis

We can now return to the issue of more precisely assessing when these hilltop sites were occupied and evaluating the length of the occupation span. The analysis includes 28 short-lived samples from the four cerros de trincheras and excludes the five samples discussed above and the single date from Ophelia's Well which is not a hilltop site. These 28 dates fall into two discrete clusters and are plotted in Figure 5.3. The first, and largest group contains 23 remarkably clustered dates, including 19 from the upper and lower terraces of Cerro Juanaqueña, two from Cerro el Canelo, and two from Cerro los Torres. These dates clearly indicate that the principal occupation of these sites occurred at about 1200 cal BC. The later cluster contains five dates—three from two lower terraces at Cerro Juanaqueña and two from Cerro Vidal. This cluster indicates a second use of cerros de trincheras at about 200 cal BC.

The OxCal 3.10 radiocarbon dating computer program was used to estimate the beginning and ending dates of the occupations as well as the total span of the occupations (Bronk Ramsey 2001, 2005). The OxCal program has *boundary* and *span* functions based on Bayesian statistics that estimate beginning and ending dates, as well as the span or length of the occupation. We applied these functions to the early cluster of 23 dates from Cerro Juanaqueña, Cerro los Torres, and Cerro el Canelo, as well as the later cluster of five dates from Cerro Juanaqueña and Cerro Vidal.

The conventional approach to estimating beginning and ending dates and span of occupation from a suite of radiocarbon dates consists of simply inspecting the calibrated ranges of the earliest and latest date and then subjectively estimating the beginning and ending of the occupation. In contrast, the Bayesian methods allow statistical combination of the probabilities from all the dates, based on the assumption that all of the dates defined in a group or cluster are part of one uniformly represented occupation with a definite beginning and end. The algorithm also assumes that the sample of dates is a random sample of that occupation (Bronk Ramsey 2001).

The Bayesian statistics are illustrated in Figure 5.3 where each date has two probability graphs. The ones shown only in outline are the *prior* probabilities, which are the same as original calibrated probabilities shown in Figure 5.2. The solid graphs are *posterior* probabilities, new probability distributions that result from combining the probabilities of all the dates in the cluster. For example, the prior probability (original) calibration of 3130 ± 55 BP centers on about 1400 cal BC (1520 to 1260 cal BC at two standard deviations). However, the posterior probability pulls the calibration to the right of its original distribution, closer to the mass of all other dates and now centers it at about 1300 cal BC (1430 to 1210 cal BC). Therefore the OxCal analysis indicates that if the assumptions described above are valid and if all dates in the analysis are considered, then it is more likely that the true calibrated age of 3130 ± 55 BP is a century or so later than the conventional calibration might indicate (Bronk Ramsey 2001).

In Figure 5.3, the percentage value given to the right of the date and site name is the OxCal *Agreement Index* (A). This index reflects the degree to which the final (posterior) distribution overlaps with the original prior tree-ring calibration distribution. If the posterior distribution is the same as the prior distribution, it will have an index of 100%. Note, for example, that the original 2190 ± 35 BP distribution in outline is completely covered by the black posterior distribution and it has an Agreement Index of 100%. The value rises above 100% if the final distribution overlaps with only the very highest or most likely part of the prior distribution (Bronk Ramsey 2001). For example, the 2980 ± 70 BP plot has an

TABLE 5.2. Probabilities of Boundaries and Spans as Shown in Figure 5.3 and Figure 5.4 by Occupation.[1]

Occupation & Sites	OxCal Function	68%	95%
A. First Occupation Cerro Juanaqueña (upper & lower terraces), Cerro los Torres, Cerro el Canelo. 23 dates.	Beginning End Span	**1385**–1285 cal BC 1210–**1090** cal BC 90 to 270 years	**1430**–1230 cal BC 1250–**1040** cal BC 10 to 320 years
B. First Occupation, upper terraces Cerro Juanaqueña (12 upper terraces only). 14 dates.	Beginning End Span	**1380**–1260 cal BC 1230–**1100** cal BC 30 to 220 years	**1430**–1220 cal BC 1260–**1030** cal BC 0 to 300 years
C. Second Occupation Cerro Juanaqueña (2 lower terraces) and Cerro Vidal. 5 dates.	Beginning End Span	**550**–300 cal BC 200 cal BC–cal AD **20** 70 to 340 years	**800**–200 cal BC 340 cal BC–cal AD **150** 20 to 410 years

[1] The beginning and ending dates of each range are in **bold**. Conservatively rounding, the main occupation of Cerro Juanaqueña was underway no later than 1300 cal BC and ended by 1100 cal BC.

agreement index of 119.4%, and the posterior probability centers on only the most likely (highest) part of the original distribution. Note that 13 of 23 dates in the first occupation and three of the five in the second occupation have agreement indices above 100% reflecting the tight clustering of the dates and the short span of these two occupations. The overall Agreement Index for the 23 dates from the main occupation is 74% and from the five dates in the second occupation is 92%.

The validity of an individual date might be questioned if the Agreement Index falls below 60%, since the original probability distribution of the date is notably different from the posterior probability based on the mass of all the dates (Bronk Ramsey 2001). Note that in Figure 5.3, Agreement Indices for only two of the 28 dates fall below 60% and they fall on both the earliest and latest points of the distribution. However, with 28 dates, it is expected, based on statistics alone, that one or two will fall below this level.

The OxCal *boundary function* calculates the probability distribution of beginning and ending dates of the occupation. These can be seen in Figure 5.3 at the top and bottom of each group of dates. The actual numeric ranges are listed in Table 5.2. These calculations indicate that the original occupation of the cerros de trincheras started sometime after 1385 cal BC but before 1285 cal BC and ended sometime between 1210 and 1090 cal BC (68% probability). Note these findings do *not* indicate that the occupation began *at* 1385 cal BC and ended *at* 1090 cal BC, but that these events occurred sometime during these ranges. The beginning and ending points of these ranges are in bold font in Table 5.2 (section A).

Inspection of the beginning and ending posterior probability curves suggests that use of the more restricted one sigma values (68%) is warranted. For example, the one sigma posterior distributions of all of the dates with Agreement Indices over 100% fall between 1300 cal BC and 1150 cal BC. Conservatively (i.e., later) rounding the one sigma, beginning values of 1385 to 1285 cal BC and the abandonment values of 1210 to 1090 cal BC, allows us to say with some confidence that the main occupation of Cerro Juanaqueña was clearly underway no later than 1300 cal BC and ended by 1100 cal BC.

With only two dates each for Cerro el Canelo and Cerro los Torres, the boundary and span OxCal functions would not yield reliable results for these sites alone. However, their high Agreement Indices with the large pool of Cerro Juanaqueña dates indicates that Cerro el Canelo and Cerro los Torres were occupied during the same time that Cerro Juanaqueña was occupied. Also notably, the material culture, the construction, and the organization of the sites are similar. The lack of evidence for occupations from earlier and later time periods is likewise remarkable.

The second occupation of two of the Cerro Juanaqueña lower terraces and Cerro Vidal began sometime between 550 and 300 cal BC and this later occupation ended sometime between 200 cal BC and cal AD 20 (68% probability) (see Figure 5.3 and Table 5.2, section C). The ranges of the distributions with the highest agreement indices fall between 400 and 100 cal BC. By happenstance this range on the tree-ring calibration curve produces bimodal distributions—note the double peaks for each these five dates in Figure 5.3. By conservatively rounding (later in time), we can say that the second occupation was underway no later than 400 cal BC and the sites were abandoned no later than ca. cal AD 1. These data suggest that the second occupation was also a single brief episode analogous to the first occupation.

The OxCal *span function* is a particularly useful way to estimate the length of these occupation events (Bronk Ramsey 2001). It combines probabilities of a suite of dates to infer the length of an occupation, rather than simply subtracting the beginning date from the ending

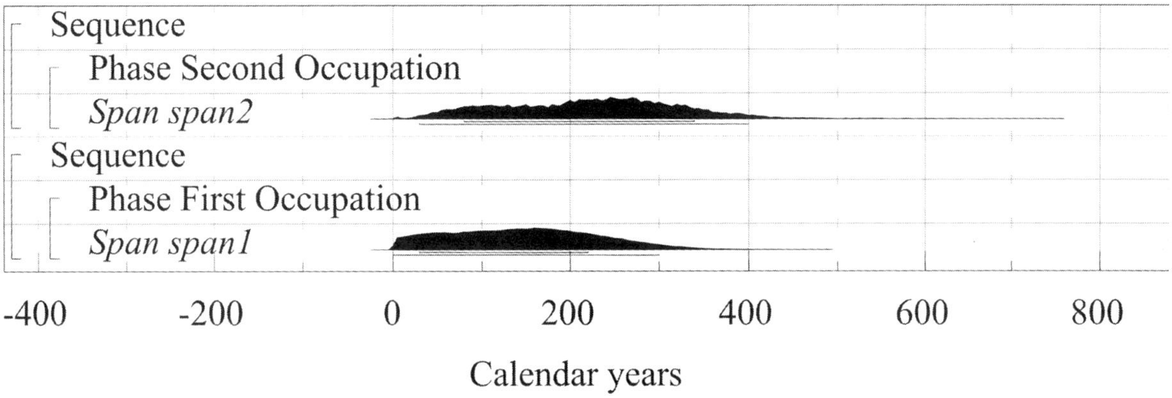

FIGURE 5.4. The OxCal probability span showing estimated duration of first occupation (at Cerro Juanaqueña, Cerro el Canelo, and Cerro los Torres) and estimated duration of second occupation (at Cerro Juanaqueña and Cerro Vidal).

date. Considering all of the dates from the first occupation of the three sites (Cerro Juanaqueña, Cerro el Canelo, and Cerro los Torres), OxCal calculates that the likely span of occupation ranges from as little as 90 years to no more than 270 years (68% probability) (Table 5.2A). The span of the second occupation is similar, 70 to 340 years (Table 5.2, section C and Figure 5.4).

The length of occupation of the upper terraces of Cerro Juanaqueña is of particular interest since this portion of the site represents more construction effort and contains more cultural material, as well as being the most intensively studied portion of the site. This tight cluster of dates has an Agreement Index of 90% with nine of the 14 exceeding 100%. Utilizing only these 14 dates, OxCal estimates, at a 68% probability, that Cerro Juanaqueña's upper terrace occupation endured for as little as 30 years to no more than 220 years (Table 5.2, section B). Figure 5.4 shows this probability distribution. Even though the upper terraces were occupied more intensively, the overall span of occupation is shorter on the upper terraces than on the lower terraces. This is related to the fact that both the earliest and the latest radiocarbon dates from Cerro Juanaqueña come from the lower terraces (see Figure 5.3).

We only saw one case of terrace reoccupation across the two periods of occupation and this occurred on the lower terrace where all other evidence of the second occupation was found. Terrace 413 yielded radiocarbon dates of 2190 ± 35, 2140 ± 40, and 2870 ± 50 indicating the terrace was initially constructed during the first occupation of the site and it was reoccupied during the second occupation. Similar evidence of reoccupations of terraces was not seen elsewhere on the site.

Other archaeological evidence corroborates a short occupation at Cerro Juanaqueña, particularly on the upper terraces. The terraces form a compact and well-organized complex of features. Their coherence and continuity and the overall symmetry of the perimeter berm suggest rapid construction in one or a few building episodes (Chapter 3). These observations are consistent with a short occupation. On the other hand, an occupation period of much less than 200 years appears unlikely as well, particularly given the large number of heavily worn metates. Combining the archaeological evidence, especially an analysis of the wear represented by all of the basin metates, with the analysis of radiocarbon dates suggests that an occupation of the upper terraces of about 200 years is reasonable, and this estimate is discussed further in Chapter 13.

The artifact assemblages from the cerros de trincheras sites discussed here are consistent with the radiocarbon evidence. Ceramics are virtually absent, and diagnostic artifacts such as projectile points and ground stone are overwhelmingly types characteristic of the Late Archaic/Early Agricultural period. Evidence from later time periods either pertains to specialized uses (rock art), or reflects ephemeral use at very specific locations within the site, such as the few isolated ceramics found on the surface.

Conclusion

A systematic radiocarbon dating program defined two restricted occupations of the four cerros de trincheras sites. The initial and by far the largest occupation of Cerro Juanaqueña was underway by 1300 cal BC and the site was abandoned by 1100 cal BC. Cerro el Canelo and Cerro los Torres were also occupied during this same interval. The radiocarbon dates indicate that Indigenous people lived on the upper terraces of Cerro Juanaqueña for only about two centuries before abandoning the site. A restricted reoccupation of at least two of Cerro Juanaqueña's lower terraces, as well as the site of Cerro Vidal, was underway by about 400 cal BC and these locations were abandoned by cal AD 1. This reoccupation lasted between 100 and 300 years.

CHAPTER 6

Overview of Material Culture

Introduction

When Rex Gerald visited Cerro Juanaqueña in 1988 (Gerald 1990), he described dense vegetation that made his search for features and artifacts very difficult. Our own work at the site coincided with a severe drought which affected large areas of northern Mexico in the late 1990s. Thousands of cattle perished as a result of the drought, and people living in the ejidos and other rural communities of Chihuahua suffered great hardship. However, one result of this was that the surviving livestock stripped almost all of the grassy and herbaceous vegetation from large areas of the landscape, including Cerro Juanaqueña. Conditions for finding artifacts on the surface of the site were near ideal.

During our initial visits to Cerro Juanaqueña and other cerros de trincheras, cultural materials were simply sketched in the field and left in place. Later, with permission from the Instituto Nacional de Antropología y Historia (INAH), we began to collect formal artifacts after carefully plotting their location on maps of the sites. Finally, at Cerro Juanaqueña we completed a systematic inspection of the entire surface of the site. As a result of these efforts, we documented almost 900 formal artifacts from the surface of Cerro Juanaqueña alone.

At first, we were most interested in projectile points and in determining whether or not pottery was present. These artifacts were our best initial indications of the age of the site. Projectile points known to have been used during the Late Archaic period were overwhelmingly dominant and pottery was almost completely absent. By 1996 we were able to make a strong case that the site was Late Archaic in age, based on artifacts alone (Roney 1996b).

As radiocarbon dates began to confirm this finding, and as our perception of the artifact assemblages grew, we began to apply this new information to questions of site function. In northwestern Mexico and the southwestern United States, there were many different ideas about how cerros de trincheras might have been used (Chapter 7). Some were thought of as purely agricultural, while others were interpreted as monumental religious structures similar to actual pyramids further south in Mesoamerica. Some were described as fortresses, while others were seen as places of temporary refuge during times of strife.

As we began to grapple with these differing viewpoints, one thing became increasingly clear to us—whatever else was going on, Archaic peoples were actually living on the terraces at Cerro Juanaqueña for long periods of time. The discovery of midden deposits and other features was part of this argument, but a key piece of evidence was the sheer breadth of the artifact assemblage we were finding at Cerro Juanaqueña and other Late Archaic cerros de trincheras. Cores, hammerstones, and bifaces in all stages of completion showed that a full sequence of stone tool manufacture was taking place. Phenomenal numbers of grinding tools demonstrated that plant foods were being processed on a large scale, while unprecedented numbers of projectile points spoke of the importance of hunting (or perhaps warfare—more about that will be discussed later on). Besides artifacts used in everyday life, we found small but significant numbers of unusual artifacts like drills, cruciforms, stone pipes, small mortars, and miniature pestles. By 1998, we argued convincingly that Cerro Juanaqueña, at least, was a hilltop village (Roney and Hard 1998).

Meanwhile, more and more radiocarbon dating results became available, and it became clear that the span of occupation at Cerro Juanaqueña and the other cerros de trincheras in northwestern Chihuahua was very brief, at least in archaeological terms (Roney and Hard 2002). There was a principal occupation between 1300 cal BC and 1100 cal BC, and a much lighter occupation around 300 cal BC. We found no chronometrically dated evidence of occupation after 300 BC. In this context, the artifact assemblages took on increased importance because they offered a window into a particular point in time. Although any individual artifact might pertain to later time periods, we can be certain that the assemblage as a whole has a great deal of chronological integrity. It provides reasonably clear insight into the Late Archaic period.

This chapter provides an overview of the artifact assemblages found at Cerro Juanaqueña and other Late Archaic cerros de trincheras in northwestern Chihuahua. The chapter is based mostly on information collected during systematic surface recording and surface collection at Cerro Juanaqueña, although data from excavated contexts is included where relevant. The chapter also includes a summary of information about artifact assemblages from other cerros de trincheras that we investigated in northwestern Chihuahua.[1] More detailed considerations of specific classes of artifacts, such as ground stone and chipped stone, are given in later chapters.

Cerro Juanaqueña

Table 6.1 tabulates all artifacts recovered from Cerro Juanaqueña, and it is presented here in order to allow the reader to gain some sense of the relative frequency of various classes of artifacts. Most of this chapter is organized to parallel presentation of data in this table. The table distinguishes between artifacts recovered from excavated contexts, and those collected or recorded in surface contexts. This differentiation is significant since some artifacts (such as informal chipped stone tools) were not monitored during surface documentation and others (such as cores and hammerstones) are underrepresented in excavated contexts. The distributions of many classes of artifacts on Cerro Juanaqueña can be found in Appendix 6.2.

Cores and Hammerstones

Cores

Cores are nodules from which one or more flakes have been removed, and aside from debitage, these were easily the most common artifacts found on the surface at Cerro Juanaqueña. In Chapter 14, Bradley J. Vierra analyzes 72 cores from excavated contexts and a sample of 45 others from surface contexts. During the site-wide reconnaissance and mapping phase of the project we excluded tested material (cores with only a single flake removed) because there were so many. Even with this restriction, we tabulated and plotted the locations of over 1,800 cores on the upper and lower terraces. Subtypes were recorded for a sample of 241 of these cores and raw material type was recorded for 1,199 examples (Table 6.2).

The surface recording results corroborate Vierra's more focused analysis. Most of the cores are made from cobbles that occur in river gravels immediately below the site. Rhyolite is the dominant raw material, followed by chalcedony and chert. The most common subtypes are cobble unifaces, multidirectional cores, bidirectional cores, and unidirectional cores. The sample reported by Vierra in Chapter 14 includes fewer multidirectional cores, and it is likely that the specimens he described as flake cores and core fragments were tabulated as multidirectional cores during the comprehensive tabulation described here.

Most of the 1,800 cores on Cerro Juanaqueña were opportunistic, with flakes apparently removed from convenient platforms. However, a small number of bifacial cores were discoidal or ovate in outline, and seem to have resulted from a more deliberate reduction strategy (Figure 6.1). Formalized, bifacial cores like these are sometimes linked to mobility (Andrefsky 1998) and are sometimes thought to be on a trajectory toward large, finished bifacial tools. However, we did not see obvious evidence that these more formalized cores were used as tools, and we did not find large, finished bifacial tools that might represent intended end products for these cores.

Choppers

Although we did not specifically monitor edge damage while documenting surface artifacts, we saw no indication that large numbers of cores at Cerro Juanaqueña were used as choppers. Vierra examined 177 cores and cobble unifaces from the site, and he found only two cores and five cobble unifaces that exhibited damage indicating use as choppers. If this ratio held for the site as a whole, we would expect that approximately 75 cores out of the 1,853 that were mapped would have been used as chopping tools. Most of the cores tabulated in Table 6.1 were primarily sources for flakes.

Hammerstones

Hammerstones were also ubiquitous at Cerro Juanaqueña (Figure 6.2). We tabulated and plotted the locations of over 500 examples and collected more detailed information from over 300 (Table 6.3). Based on the sample analyzed in more detail, relative frequencies of raw material types for hammerstones are comparable to those observed for cores, and like cores, most hammerstones were originally river cobbles available immediately below the site. In the surface sample, most were described as either angular or spheroidal.

Hammerstones probably served a variety of purposes. Angular varieties may have been used for knapping, but we believe that spheroidal hammerstones were sometimes used to sharpen grinding tools. These artifacts exhibit dense impact fractures from repeatedly

TABLE 6.1. Summary of Artifacts Collected and Documented at Cerro Juanaqueña.[1]

	Excavated Contexts		General Surface Collection/Recording	
Cores and Hammerstones				
Cores	72		1853	
Prepared cores		(56)		
Cobble unifaces		(14)		
Split pebbles		(2)		
Hammerstones	3		524	
Informal Tools				
Marginally retouched pieces	23			
Notches	14		2	
Unifaces	7		11	
Denticulates	3			
Perforators	2			
Formal Chipped Stone Tools				
Generalized Bifaces	53		229	
Early stage		(8)		(41)
Middle stage		(7)		(72)
Late stage		(7)		(105)
Undetermined		(31)		(11)
Projectile Points	22		307	
Large		(22)		(293)
Small				(14)
Drills			15	
Cruciforms	2		15	
Ground Stone				
Manos	38		651	
Basin metates	18		690	
Grinding slabs	5		211	
Large formal pestle	1			
Large expedient pestles			19	
Small pestles	1		5	
Small mortars	6		64	
Shaft straightener			1	
Shaped abraders			3	
Unshaped abraders	1		7	
Miscellaneous	1		2	
Other/unknown	14		50	
Bedrock Features				
Grinding slicks			26	
Cupule Locations			4	
Bedrock Mortars			6	
Shaped Stone				
Stone Pipes	4		11	
Finished		(1)		(3)
Unfinished				(5)
Indeterminate		(3)		(3)
Spindle Whorls	1		3	
Bone Tools				
Awls	3		1	
Needle	1			
Weaving tool or awl	1			
Fragments	1		2	
Ornaments				
Pendants	1		1	
Beads				
Bone	4			
Shell	4			
Ceramics				
Prehistoric			5	
Prehistoric/Historic			1	
Historic			116	
Other Historic Artifacts				
Metal bowl			1	
Metal fragment	1			
Rifle cartridge case			1	

[1] Numbers in parentheses provide more detailed typological information for selected artifact categories.

striking a hard surface. Repeated blows result in dense crushing scars focused on the corners and edges of the hammerstones. The accumulation of this wear results in rounding the surface, creating convex surfaces and spheroidal forms. These tools were likely used to peck or roughen or shape manos and metates. This activity was integral to the use of grinding tools because if the grinding surfaces became too smooth they became ineffective. In the 1980s, Robert Hard observed similar tools being used by the Tarahumara to roughen their manos and metates. However, these hypotheses have yet to be evaluated rigorously. Certainly more research is needed to determine the strength and utility of this interpretation.

A small sample of 16 spheroidal hammerstones was collected, some as a grab sample from the surface and others as a result of excavation. More detailed laboratory observations about this small sample are summarized in

TABLE 6.2. Cores from Cerro Juanaqueña ($N = 1853$).[1]

Raw Material	*n*	%	Subtype[2]	*n*	%
Rhyolite	848	71	Cobble Unifaces	81	34
Chalcedony	155	13	Cobble Bifaces	5	2
Chert	158	13	Unidirectional Cores	23	10
Quartzite	10	1	Bidirectional Cores	55	23
Basalt	9	1	Multidirectional Cores	73	30
Local Basalt	7	t	Unknown/Other	4	1
Unknown/Other	12	1			
Total	1199	100	**Total**	241	100

[1] Information about raw materials and subtypes is not available for all 1853 cores.
[2] *Cobble Uniface*—a cobble or pebble with two or more flakes unifacially removed along a single edge margin, usually at one end.
Cobble Biface—a cobble or pebble with two or more flakes removed from two surfaces originating from a single edge, usually at one end.
Unidirectional Core—a core with a single striking platform, so flakes are removed from one direction only.
Bidirectional Core—a core with two platforms so that flakes are removed from two directions.
Multidirectional Core—a core with three or more platforms, so that flakes are removed from three or more directions.

TABLE 6.3. Hammerstones from Cerro Juanaqueña ($N = 524$).[1]

Subtype	*n*	%	Raw Material	*n*	%
Angular	176	54	Rhyolite	239	75
Spheroidal	130	40	Chalcedony	47	15
Irregular	11	3	Chert	19	6
Other/Unknown	10	3	Other	12	4
Total	327	100	**Total**	317	100

[1] Information about subtypes and raw materials is not available for all 524 hammerstones.

Table 6.4. The laboratory analysis divided these 16 specimens into two groups. The first exhibits only battering, while the second has both battering and grinding. Almost all have battering over 50% or more of their surface area, more extensive wear results in rounding the tool. The preferred materials are chalcedony and rhyolite.

The second group is distinctive in that after use as hammerstones, these items were reused as grinding stones. Most have grinding wear on multiple edges and less frequently at the faces. The grinding is typically sufficient to locally obliterate the battering wear. These tools may have been used to shape and smooth other ground stone tools, such as the mortars or some manos. They could have also been used to grind material in the mortars.

Informal Chipped Stone Tools

In Chapter 14, Vierra describes several categories of informal tools, artifacts that are only marginally retouched and are presumably intended for short use lives. These artifacts all lack extensive facial retouch and do not have standardized outlines. For these reasons, they are more difficult to recognize during surface survey and we did not attempt to monitor their occurrence during that phase of the project. Instead, almost all examples were identified during analysis of chipped stone from excavated contexts (Table 6.1) and they are discussed in Chapter 14. The tabulation from excavated contexts is a small sample, but should give some indication of their relative frequencies in the Cerro Juanaqueña assemblage.

Formal Chipped Stone Tools

Generalized Bifaces

Analysis of debitage from Cerro Juanaqueña indicates an emphasis on bifacial reduction, and generalized bifaces were among the most common bifacial artifacts found at this site. Generalized bifaces are artifacts that exhibit retouch scars extending over one-third or more of both of their surfaces, but that do not conform to one of the other artifact categories described below (projectile point, drill/biface, drill, or cruciform). Bifaces were

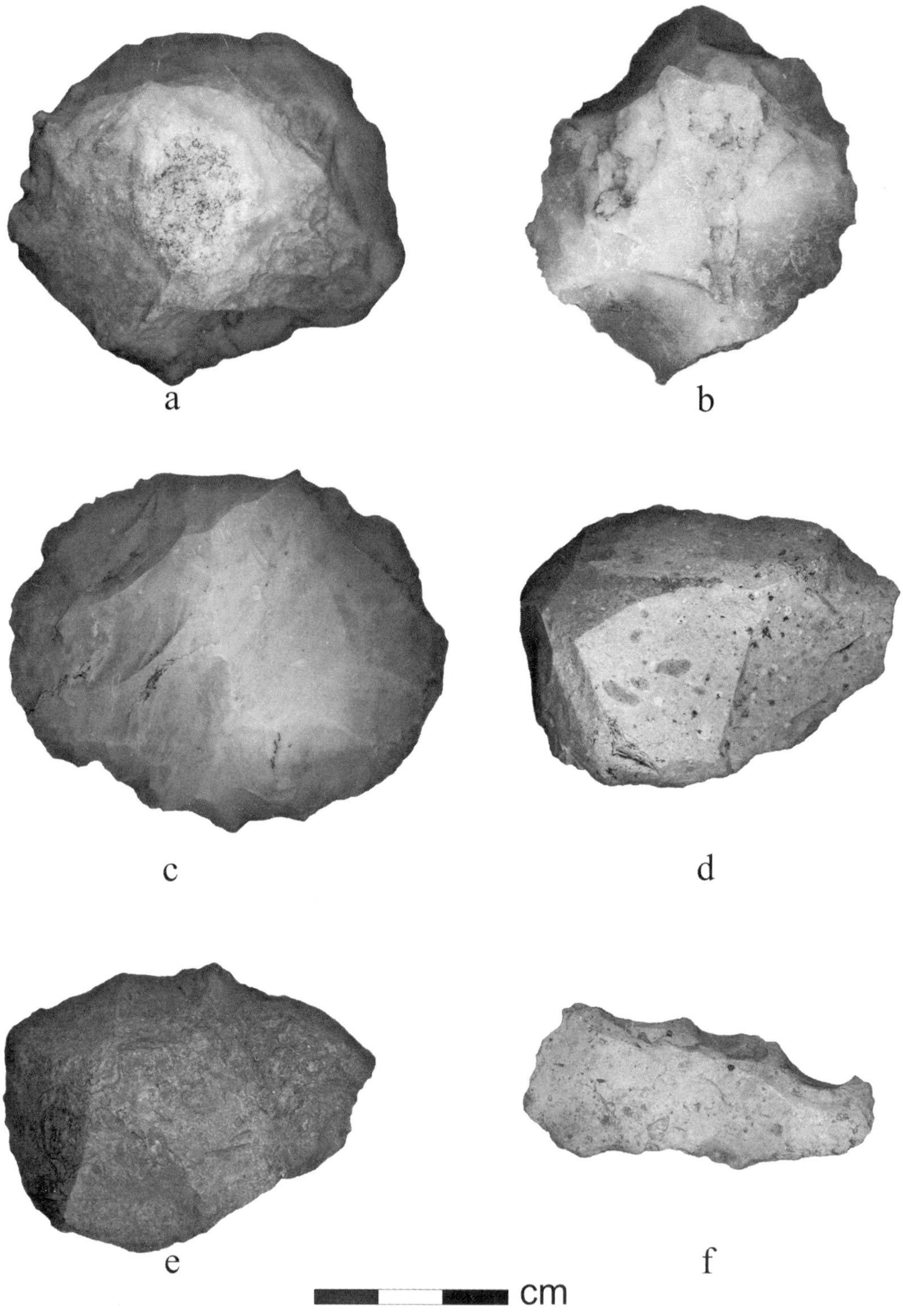

FIGURE 6.1. Bifacial cores. Specimen (*a*) from Cerro el Canelo; specimens (*b*) through (*f*) from Cerro Juanaqueña. Catalogue/bag numbers: (*a*) cc13, (*b*) T487-75, (*c*) sc101, (*d*) sc843, (*e*) sc430, (*f*) sc657.

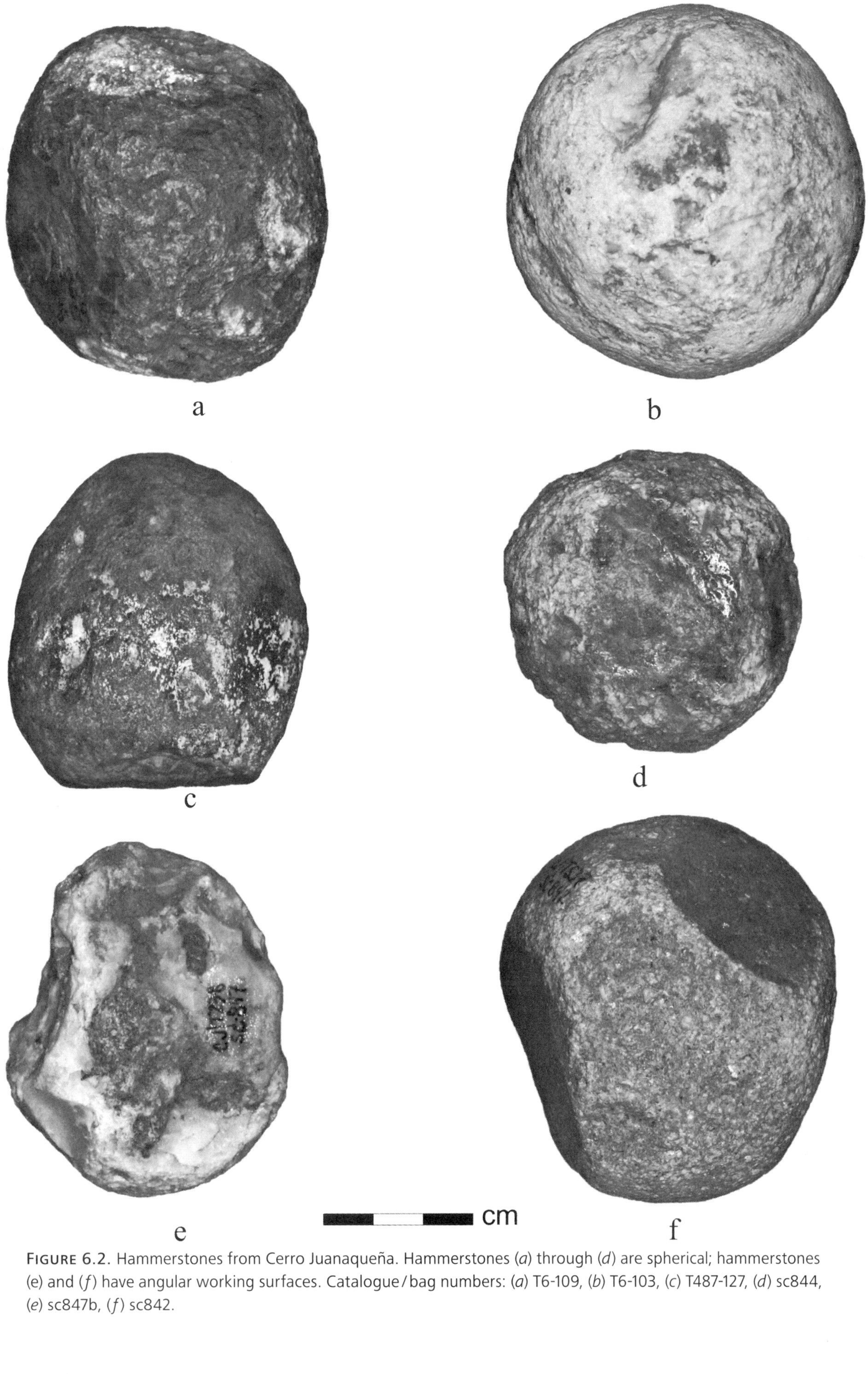

FIGURE 6.2. Hammerstones from Cerro Juanaqueña. Hammerstones (*a*) through (*d*) are spherical; hammerstones (e) and (*f*) have angular working surfaces. Catalogue/bag numbers: (*a*) T6-109, (*b*) T6-103, (*c*) T487-127, (*d*) sc844, (*e*) sc847b, (*f*) sc842.

FIGURE 6.3. Early stage bifaces from Cerro Juanaqueña. Catalogue/bag numbers: (*a*) sc462, (*b*) sc709, (*c*) sc455, (*d*) sc675, (*e*) sc433, (*f*) sc792, (*g*) sc595, (*h*) sc583, (*i*) sc99, (*j*) sc827, (*k*) sc538, (*l*) sc761, (*m*) sc146, (*n*) sc674, (*o*) sc625, (*p*) sc670.

Table 6.4. Hammerstones Analyzed in Laboratory.

	Group 1	Group 2
Total *n*	10	6
Wear	Battering	Battering, grinding
Excavation	4	2
Surface	6	4
Complete	8	6
Fragment	2	0
Materials	Chalcedony, rhyolite	Chalcedony, rhyolite
Shape	Spherical, oval/convex	Spherical, oval/convex
Length cm		
Mean	7.0	8.0
Min–max	5.4–9.5	6.9–9.6
Std dev	1.1	1.0
n	10	6
Width cm		
Mean	6.3	7.0
Min–max	5.8–7.2	6.0–7.6
Std dev	0.6	0.7
n	7	5
Thickness cm		
Mean	5.0	6.0
Min–max	4.1–6.2	5–7.5
Std dev	0.8	1.1
n	8	5
Use	Used to peck ground stone	Used to peck ground stone & for grinding

distinguished from formalized bifacial cores by their smaller size, thinner cross section, and more careful flaking. In Chapter 14, bifaces were classified into three categories intended to reflect different stages of manufacture: early stage, middle stage, and late stage. These are tabulated in Table 6.1 and additional specimens are illustrated in Figures 6.3, 6.4, and 6.5. Note that many of the distal and medial fragments tabulated here as late stage bifaces may well be tips or midsections of finished projectile points.

Several observations can be made concerning the late stage bifaces. Most of these artifacts could be used as knives or cutting tools as they are; and with the addition of notches or hafting modifications, many could have become projectile points. In terms of morphology, most are triangular, with distinct juncture between their lateral edges and their basal edge. Only specimens with straight or convex lateral and basal edges are described here as late stage bifaces. Two un-notched late stage bifaces with concave bases are discussed in the next section as Cortaro points, and two specimens with concave lateral edges and distinctive wear are described below as drills.

The larger late stage bifaces with strongly convex bases are similar in shape to Abasolo points in the Texas projectile point typologies (e.g. Figure 6.5 a and b), and smaller specimens resemble the Catan type. Larger specimens with straight bases resemble the Tortugas type (e.g. Figure 6.5 j). However, none exhibit the distinctive beveled and steeply chipped edges that are often seen on these types (Turner and Hester 1995).

Projectile Points

Projectile points are common on the cerros de trincheras of northwestern Chihuahua. These artifacts are bifaces that show evidence of modification for hafting, and by this definition we documented 329 at Cerro Juanaqueña. In the larger region, projectile points were made in a wide range of shapes and sizes, and this variability is often patterned through both time and space. Projectile points with specific shapes are often limited to particular chronological intervals and geographic areas, although we do not fully understand the reasons for these patterns. Some characteristics such as weight and stem width have clear functional implications and changes in these attributes likely reflect changing technologies. Other characteristics may be stylistic, unrelated to purely functional considerations (McBrinn 2005; Wills 1988). Because of their apparent chronological and spatial patterning, projectile points often have special significance to archaeologists. Analysts group morphologically similar specimens into more or less formally defined types, and plot their distributions through both space and time (Justice 2002; Loendorf and Rice 2004; Sliva 2009, 2015; Turnbow 1997; Turner and Hester 1995).

Implementation of this approach has been very successful in surrounding regions such as the Great Basin and Central Texas, where projectile point morphology can be summarized nicely by a relatively small number of types with predictable temporal and geographic distributions. The resulting typologies are especially useful in the definition of broad chronological sequences, with specific projectile points recognized as diagnostic of particular intervals. Even though we may not know exactly why forms changed through time, the consistent association between specific types and specific chronological intervals makes these taxonomies very useful tools.

Development of this approach has lagged somewhat in the southwestern United States and northwestern Mexico. Although there are several reasons for this, two are especially pertinent here. First, at least during some time periods, flint knappers in this region apparently did not adhere closely to ideal templates when they made projectile points. Morphology seems to have been constrained primarily by functional considerations, and it has proven difficult to summarize the range of

FIGURE 6.4. Middle stage bifaces from Cerro Juanaqueña. Catalogue/bag numbers: (*a*) sc242, (*b*) sc330, (*c*) sc186, (*d*) sc241, (*e*) sc169, (*f*) sc700, (*g*) sc821, (*h*) sc830.

variability by the delineation of few well-defined, contrasting, and mutually exclusive types. Second, relatively few deeply stratified sites yielding large numbers of projectile points have been found. Many of our notions of chronological relationships among Southwestern projectile point forms are derived primarily from two sites, Ventana Cave (Haury 1950) and Bat Cave (Dick 1965). More recent studies have begun to clarify chronological relationships by compiling information about individually dated specimens (Chapin 2017; Justice 2002; Sliva 2005, 2015; Turnbow 1997, 2000).

In this context, projectile points from Cerro Juanaqueña are especially important because they represent a very large collection from a specific place and a limited time period. Elsewhere in this volume we have argued that the major occupation at Cerro Juanaqueña occurred during a brief interval between 1300 and 1100 BC, with a minor reoccupation between 400 BC and AD 1 that appears to have been limited to the lower terraces. Although curated specimens from earlier time periods could be present, it is very unlikely that any substantial occupation on the hill occurred prior to construction of the terraces. Similarly, evidence of Ceramic and Historic period use of the hill is restricted to a very few specific locations and appears to represent transitory and specialized activities. Based on these observations, we believe that the chronological integrity of the Cerro Juanaqueña assemblage is high. There could be a few examples from earlier or later time periods, but the overwhelming majority of artifacts on this site were made and used during an interval between 1300 and 1100 BC, and most of the rest date from an equally brief interval around 300 BC. Therefore, the projectile point assemblage from this site provides important information about the range of projectile point shapes characteristic of this chronological interval in northwestern Chihuahua.

The sample from Cerro Juanaqueña consists of 329 projectile points and projectile point fragments. These include specimens collected in the course of this project, as well as specimens documented early in the project when artifacts were drawn or photographed in the field but not collected. Measurements and observations pertaining to the 329 projectile points are presented

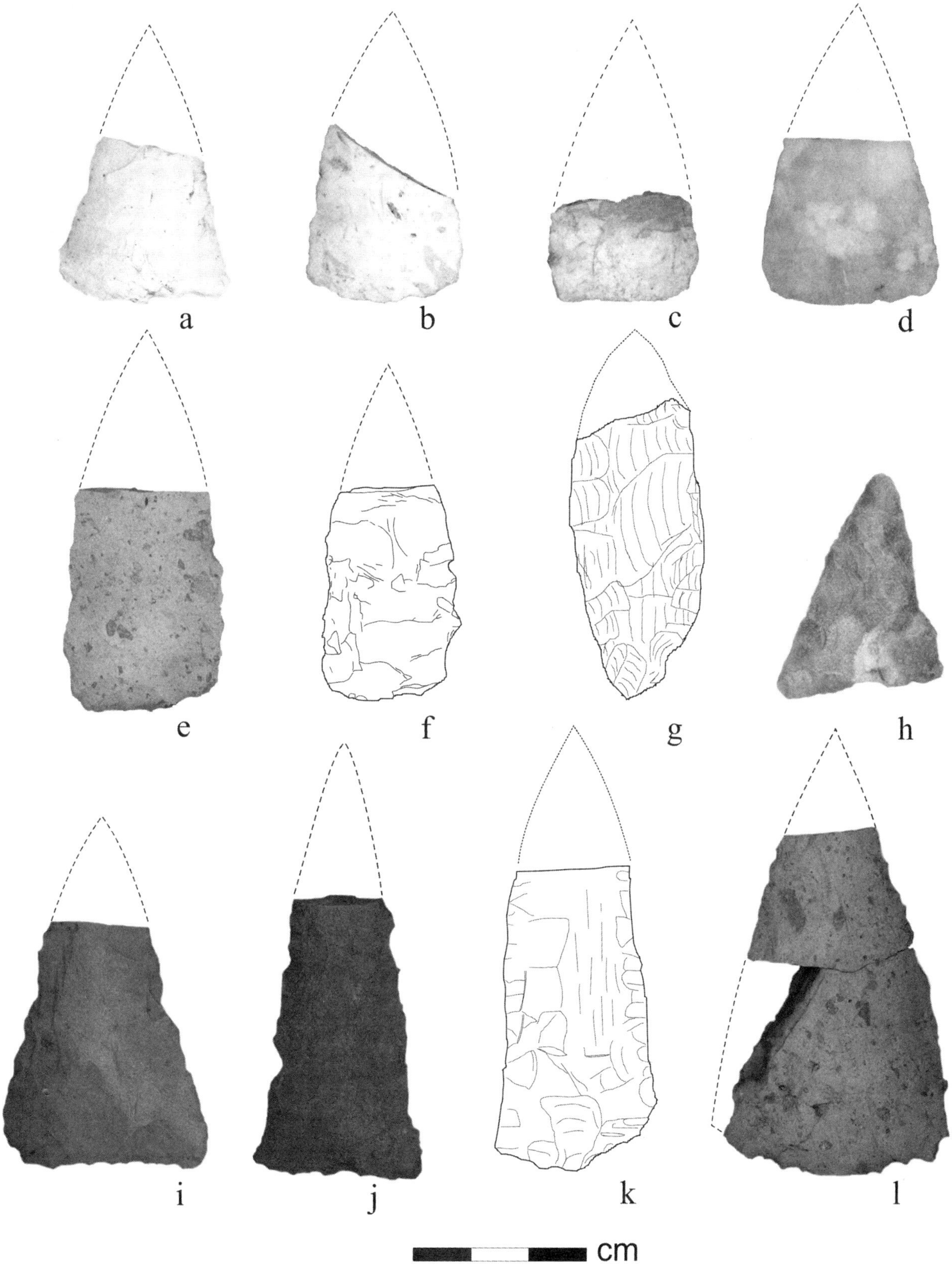

FIGURE 6.5. Late stage bifaces from Cerro Juanaqueña. Catalogue / bag numbers: (*a*) sc173, (*b*) sc246, (*c*) sc243, (*d*) sc296, (*e*) sc172, (*f*) sc186, (*g*) sc838, (*h*) sc265, (*i*) sc161, (*j*) sc124, (*k*) sc11, (*l*) sc135.

TABLE 6.5. Distinctions between Large and Small Projectile Points.[1]

Attribute	Large Points	Small Points
Weight	> 1.5 g	< 1.5 g
Length	> 25 mm	< 25 mm
Thickness	> 3 mm	< 3 mm
Stem Width	> 8 mm	< 8 mm

[1] Based on Turnbow (1997).

in Appendix 6.1 and most are illustrated in Figures 6.6 through 6.16. In this section, we characterize the projectile points in terms of traditional culture-historical types, with the objective of refining their definitions and better understanding their chronological implications.

Different investigators have classified Southwestern projectile points in ways that are not necessarily in agreement. We have relied heavily on the compendium of projectile point information by Justice (2002) for several reasons. First, it is based on systematic consideration of well-controlled collections from a broad geographic area. Second, it makes very good use of the chronological information that is available for various types. Third, Justice also provides relatively detailed discussions of his categories and often contrasts them against each other. Finally, he illustrates a number of examples for each of his types, making it possible to more confidently replicate his classification decisions.

The first broad distinction that we made in analyzing the collection from Cerro Juanaqueña is based on size. Throughout much of North America, projectile point assemblages dating later than about AD 900 are dominated by small specimens which many analysts characterize as arrow points intended for use with the bow and arrow. These forms contrast with larger specimens described as dart points intended for use with an atlatl. Different researchers have used different criteria to distinguish between large and small projectile points. Chris Turnbow (1997) has proposed polythetic criteria based on four attributes: weight, length, thickness, and stem width. These distinctions are summarized in Table 6.5.

We checked all 329 projectile points against these criteria, and found that most specimens could be unambiguously classified as large or small. In only three cases, length and thickness suggested large size while weight and stem width indicated small size. Those specimens are analyzed here with the small points. Using these criteria 315 projectile points were classified as *large* and only 14 as *small* (including the three ambiguous specimens).

Large Projectile Points

A typological classification of the large projectile points is presented in Table 6.6. Of the 315 large projectile points, 53 were broken in ways that precluded further analysis (i.e. notching and stem characteristics had been lost), leaving 262 specimens that were relatively complete. Large side- or corner-notched projectile points with expanding stems and straight or convex bases are one of the hallmarks of the Late Archaic period, and they occurred in large numbers at Cerro Juanaqueña. In this discussion, following Justice (2002), we have classified most of these as either San Pedro or Tularosa Corner-notched.

TABLE 6.6. Typological Classification of Projectile Points from Cerro Juanaqueña.

	Total	% of Classified	Illustration
Large Projectile Points			
San Pedro	149	54	Figure 6.6–6.8
Tularosa Corner-notched	23	8	Figure 6.9
Tularosa Basal-notched	18	7	Figure 6.10
Todsen	12	4	Figure 6.11
Gypsum	8	3	Figure 6.12
Datil	11	4	Figure 6.13
Cortaro	2	1	Figure 6.14
Miscellaneous	39	14	Figure 6.15
Fragments	53		
Small Projectile Points	14	5	Figure 6.16
Total Excluding Fragments	276	100	
Grand Total	329		

San Pedro

Two types discussed in this section are closely related, San Pedro and Tularosa Corner-notched. San Pedro points were first illustrated by Sayles and Antevs (1941) in their discussion of late preceramic times in southern Arizona. Later Haury (1950) provided an expanded discussion of their morphology. Tularosa Corner-notched is a type recently defined by Justice (2002) and is thought to have developed from San Pedro varieties.

These two forms are very similar in overall shape. Justice distinguishes Tularosa Corner-notched and San Pedro specimens on the basis of size, and especially singles out narrow neck width and narrow notch openings as distinguishing characteristics for the Tularosa Corner-notched type. Looking at metric data provided by Justice we found that neck width distributions for these two types overlap (5.5 to 12.6 mm for Tularosa Corner-notched and 10.5 to 25.4 mm for San Pedro). We have chosen to use a neck width measurement of 11.15 mm to distinguish between the two types. Specimens with neck widths of 11.1 mm or less are classified as Tularosa Corner-notched, and those with neck widths of 11.2 mm or greater were classified as San

TABLE 6.7. Varieties of San Pedro Points from Cerro Juanaqueña.

	Convex Base	Straight Base	Total
Corner-notched	66	33	99
Side-notched	47	0	47
Total [1]	113	33	146

[1] Three specimens were identified as San Pedro but were too fragmentary to classify further.

Pedro points. (This value was obtained by calculating the standard deviation of the neck width measurements given by Justice for the two types, and then picking the midpoint between the lower range for San Pedro and the upper range for Tularosa Corner-notched.) Using this criterion we classified 149 specimens as San Pedro Points and 23 as Tularosa Corner-notched.

Justice (2002) describes the San Pedro type as "a corner-notched to expanded stemmed point with a straight to convex basal edge," although occasionally San Pedro points have slightly concave bases and several of the specimens that Justice illustrates would have been described here as side-notched. The notch "is a wide curve with varying degrees of closure depending on the amount of notch depth and width and blade resharpening." Resharpening often resulted in dramatic changes in the shape of the blade. As defined by Justice, the San Pedro type includes forms sometimes referred to as Hueco, Basketmaker II, and En Medio, as well as examples of Types 5, 6, 8, and 15 from Bat Cave (Dick 1965). These points are distributed throughout Arizona and the western half of New Mexico and extend southward into Mexico at least as far as the state of Durango. Dates are usually given as 1500 BC to AD 300, although occasionally 1000 BC is given as the beginning date for this type. The occurrence of large numbers of San Pedro points at Cerro Juanaqueña implies that 1500 BC is probably a more accurate estimate.

In the collection of San Pedro projectile points from Cerro Juanaqueña, it was relatively easy to distinguish varieties with straight bases (Figure 6.6) from those with convex bases (Figure 6.7). With a little more difficulty we could distinguish corner-notched specimens from side-notched specimens (Figure 6.8). These differences are tabulated in Table 6.7. However, just because we can make these morphological distinctions, it does not follow that they are typologically important. We do not know if these distinctions have any functional, temporal, cultural, or spatial significance.

Tularosa Corner-notched

Classification of 23 projectile points as Tularosa Corner-notched (Figure 6.9) raises several issues. Tularosa Corner-notched points are relatively small- to medium-sized corner-notched forms with a narrow neck width, narrow notches, expanding stems, and convex to straight basal edges. As noted above, this type is distinguished from San Pedro by size, and we restricted the designation to specimens with neck widths of 11.15 mm or less. Tularosa Corner-notched points are dated from about 100 BC until AD 700 or 900, and they have a range extending from southwestern Colorado to northern Sonora and Chihuahua, and including much of New Mexico and eastern Arizona (Justice 2002).

This type closely resembles the San Pedro type and is thought to have developed from it by around 100 BC. Thus we would not expect Tularosa Corner-notched points to be well-represented in the Cerro Juanaqueña assemblage. However, after San Pedro, it is the second most common named type, according to our classification. Does this imply that the beginning date for Tularosa Corner-notched points should be revised downwards? This question is difficult because there are no clear criteria allowing a distinction between San Pedro and Tularosa Corner-notched points. We emphasized neck width because it is easily quantified. Had we also considered notch opening width, fewer specimens would have been classified as Tularosa Corner-notched. However, some of the San Pedro points also have notch openings that are well within the range of the Tularosa Corner-notched type. In our classification San Pedro points greatly outnumber Tularosa Corner-notched, and based on notch width, few specimens are unambiguously recognized as Tularosa Corner-notched. These observations reinforce the idea that there is a general trend through time toward smaller and more narrowly notched San Pedro-like points. However, we suspect that this trend is expressed by gradual changes that will be difficult to capture through typological analysis.

Tularosa Basal-notched

Tularosa Basal-notched is the third most common typological category in this analysis, with 18 specimens (Figure 6.10). This type is characterized by a strongly triangular shape, a wide blade, narrow basal notches, and prominent barbs. Justice considered Tularosa Basal-notched and Tularosa Corner-notched types to be closely associated with each other, and to have been used during the same chronological interval, 100 BC to AD 700 or 900. The occurrence of large numbers of these points in the Cerro Juanaqueña assemblage raises the possibility that their span is longer. In fact, they could well range as early as the San Pedro type, beginning around 1500 BC.

This form has sometimes been classified as the Shumla type, which was defined in the lower Pecos area of Texas, with a chronological range of about 1000 BC to 200 BC (Turner and Hester 1995). This chronological range is closer to that implied by the Cerro Juanaqueña data, but the specimens from northwestern Chihuahua

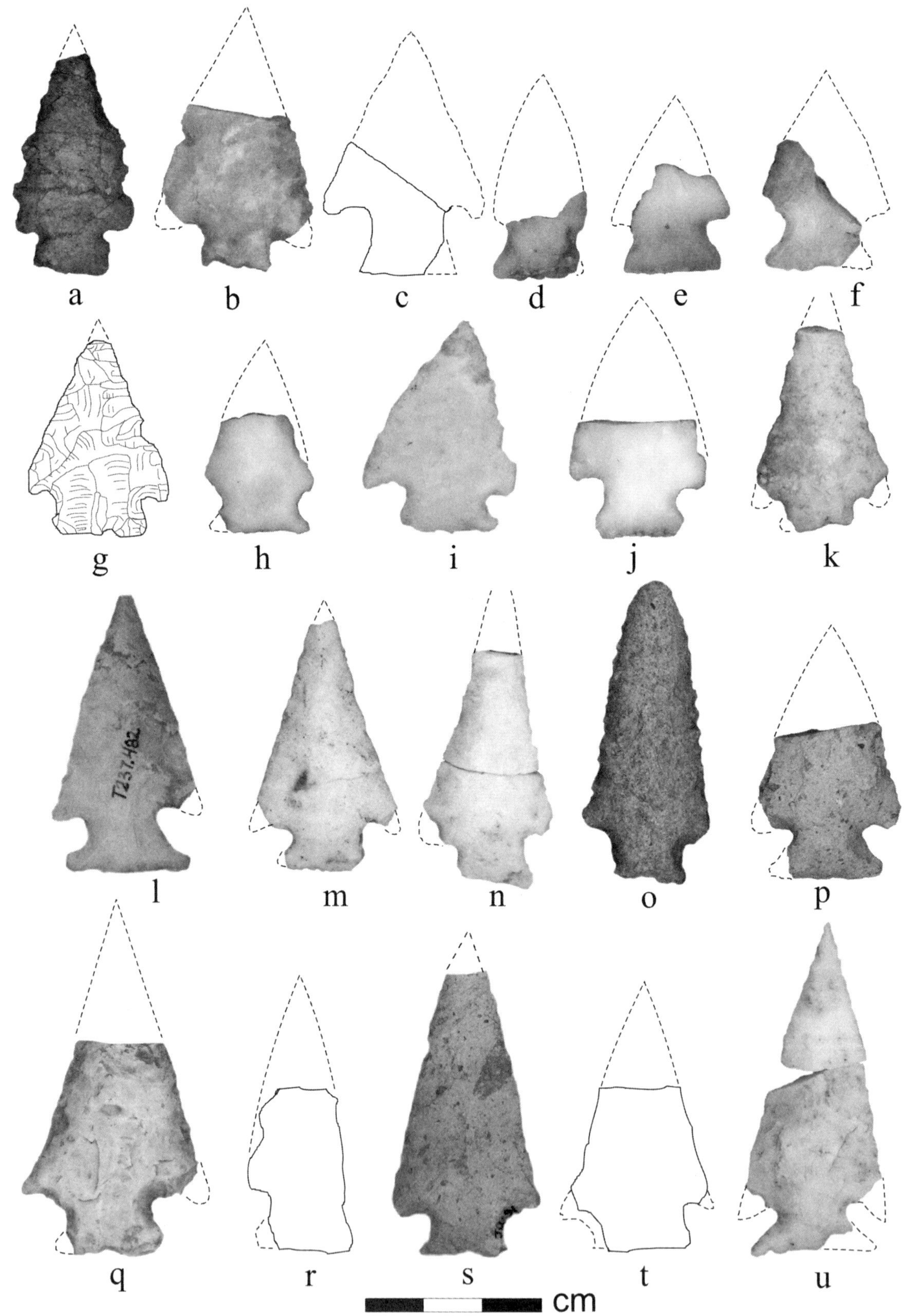

Figure 6.6a. San Pedro Corner-notched points with straight bases from Cerro Juanaqueña. Catalogue/bag numbers: (*a*) sc690, (*b*) sc769, (*c*) sc10.2, (*d*) sc732, (*e*) sc130, (*f*) sc459, (*g*) sc38, (*h*) sc364, (*i*) sc90, (*j*) sc77, (*k*) sc849, (*l*) sc482, (*m*) sc156, (*n*) sc119, (*o*) sc589, (*p*) sc712, (*q*) sc49, (*r*) sc21, (*s*) sc91, (*t*) 7008, (*u*) sc174.

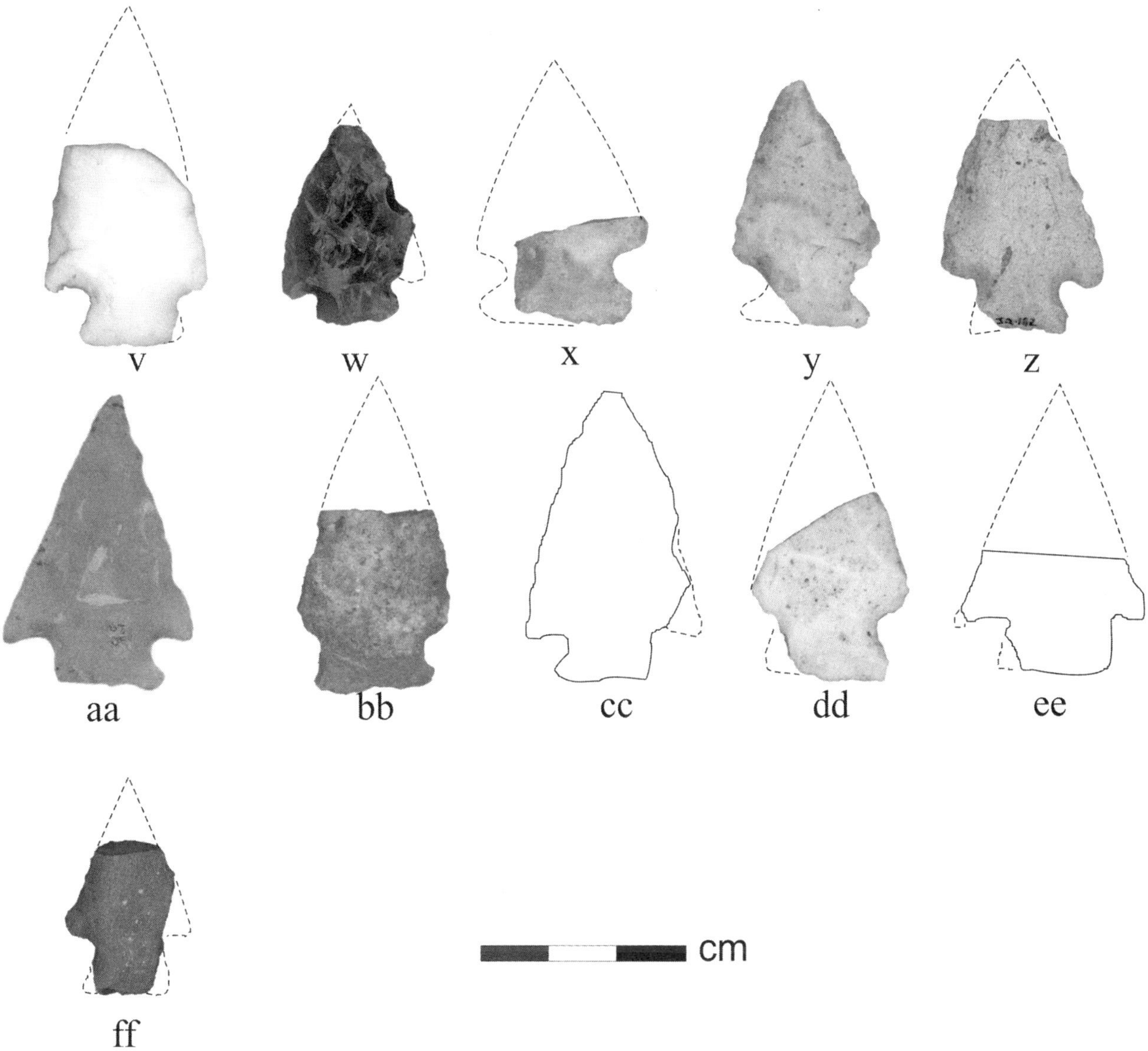

FIGURE 6.6b. San Pedro Corner-notched points with straight bases from Cerro Juanaqueña. Catalogue/bag numbers: (*v*) sc730, (*w*) sc581, (*x*) sc497, (*y*) sc412, (*z*) sc182, (*aa*) sc56, (*bb*) sc63, (*cc*) 7003, (*dd*) sc481, (*ee*) sc18, (*ff*) sc447.

are seldom serrated, and their stems are more commonly expanding with a convex base, rather being rectangular, as is the case for Shumla points.

Some Tularosa Basal-notched points have the massive barbs characteristic of the central Texas Castroville type (Turner and Hester 1995). However, when those specific specimens are considered in the overall range of variation present among the Cerro Juanaqueña projectile points, it is apparent that the Castroville type is not an appropriate classification for these points. In central Texas, Castroville points date between about 800 and 400 BC, which is later than the likely date of the Cerro Juanaqueña basal-notched specimens.

The morphology of this group is very comparable to that recently described from northern New Mexico by Chapin (2017:108–109) as Collier Basal-notched. Chapin reports this type from stratigraphic contexts predating 2000 to 1500 BC, also noting that Elyea (1995) recovered another example from contexts dated by wood charcoal to ca. 2100 BC. Chapin suggested that this form is a late Middle Archaic type in northern New Mexico.

Todsen

We have tentatively classified 12 concave-based specimens as Todsen points (Figure 6.11). This type was defined by MacNeish (1993) and has not yet been widely recognized by other researchers. However, among the currently defined morphological types it seems to be the best fit for specimens from Cerro Juanaqueña. Todsen

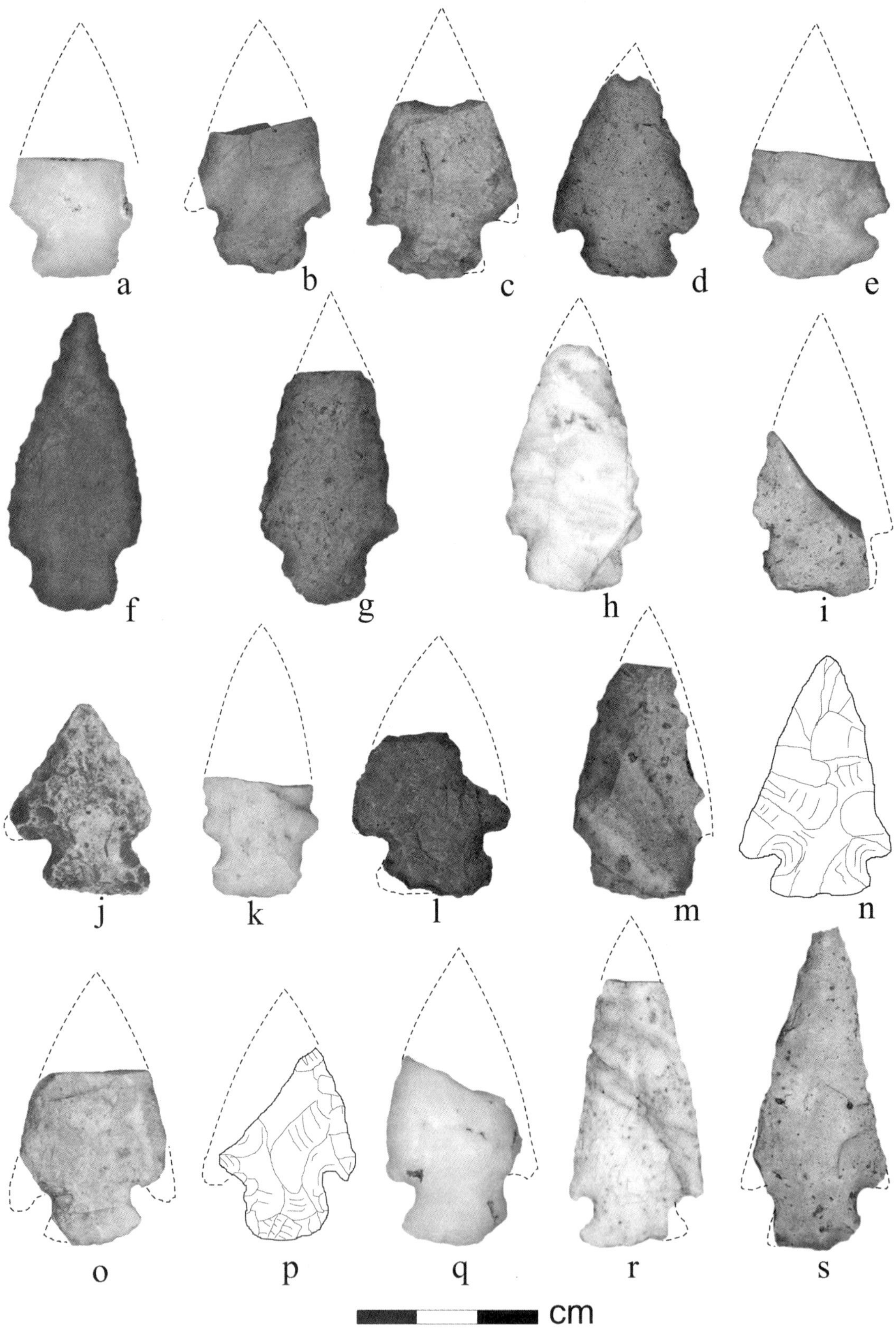

FIGURE 6.7a. San Pedro Corner-notched points with convex bases from Cerro Juanaqueña. Catalogue/bag numbers: (*a*) sc245, (*b*) sc96, (*c*) sc837, (*d*) sc162, (*e*) sc109, (*f*) sc107, (*g*) sc848, (*h*) sc171, (*i*) sc121, (*j*) sc463, (*k*) sc190, (*l*) sc217, (*m*) sc 112, (*n*) sc12, (*o*) sc170, (*p*) sc50, (*q*) sc204, (*r*) sc75, (*s*) sc134.

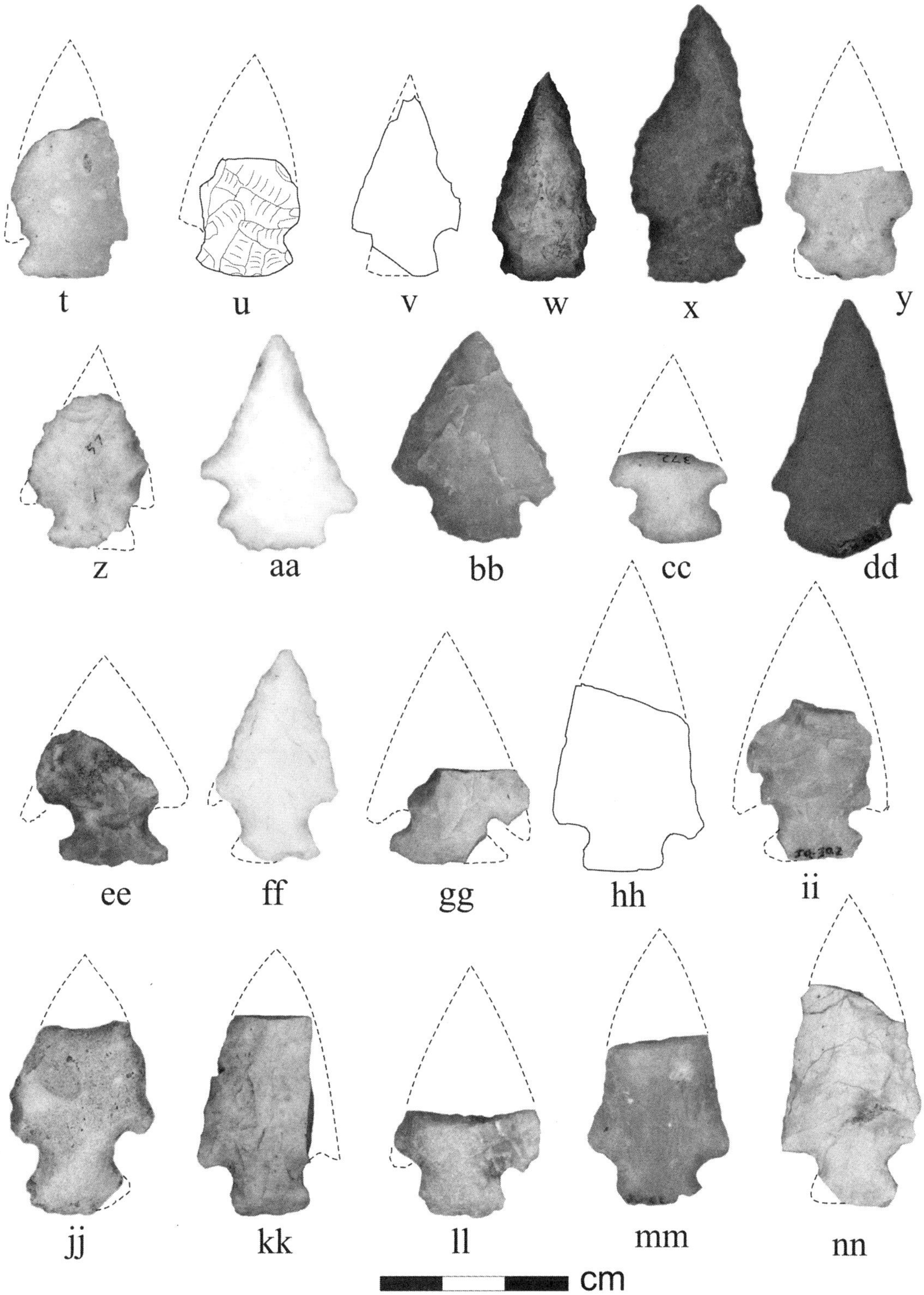

Figure 6.7b. San Pedro Corner-notched points with convex bases from Cerro Juanaqueña. Catalogue/bag numbers: (*t*) sc522, (*u*) sc48, (*v*) 9005, (*w*) sc527, (*x*) sc819, (*y*) sc145, (z) sc57, (*aa*) sc791, (*bb*) 148, (*cc*) sc372, (*dd*) sc100, (*ee*) sc79, (*ff*) sc258, (*gg*) sc69, (*hh*) sc553, (*ii*) sc302, (*jj*) sc89, (*kk*) sc444, (*ll*) sc88, (*mm*) sc68, (*nn*) sc598.

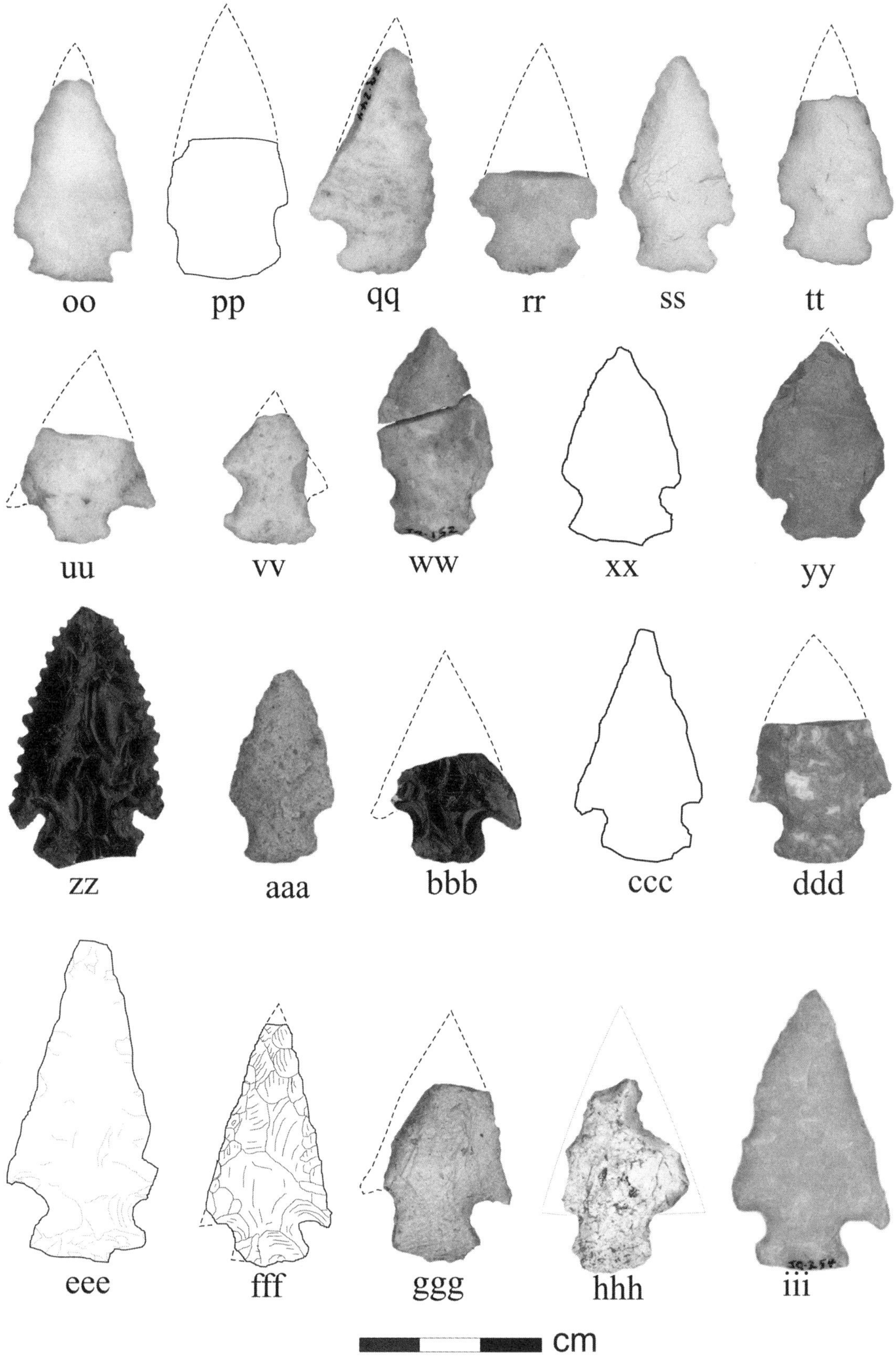

FIGURE 6.7c. San Pedro Corner-notched points with convex bases from Cerro Juanaqueña. Catalogue/bag numbers: (*oo*) sc155, (*pp*) 8001.2, (*qq*) sc244, (*rr*) sc187, (*ss*) sc252, (*tt*) sc549, (*uu*) sc116, (*vv*) sc758, (*ww*) sc152, (*xx*) 7026, (*yy*) sc268, (*zz*) sc804, (*aaa*) sc502.2, (*bbb*) sc502.1, (*ccc*) 7027, (*ddd*) sc351, (*eee*) 7016, (*fff*) sc28, (*ggg*) sc184, (*hhh*) sc376, (*iii*) sc254.

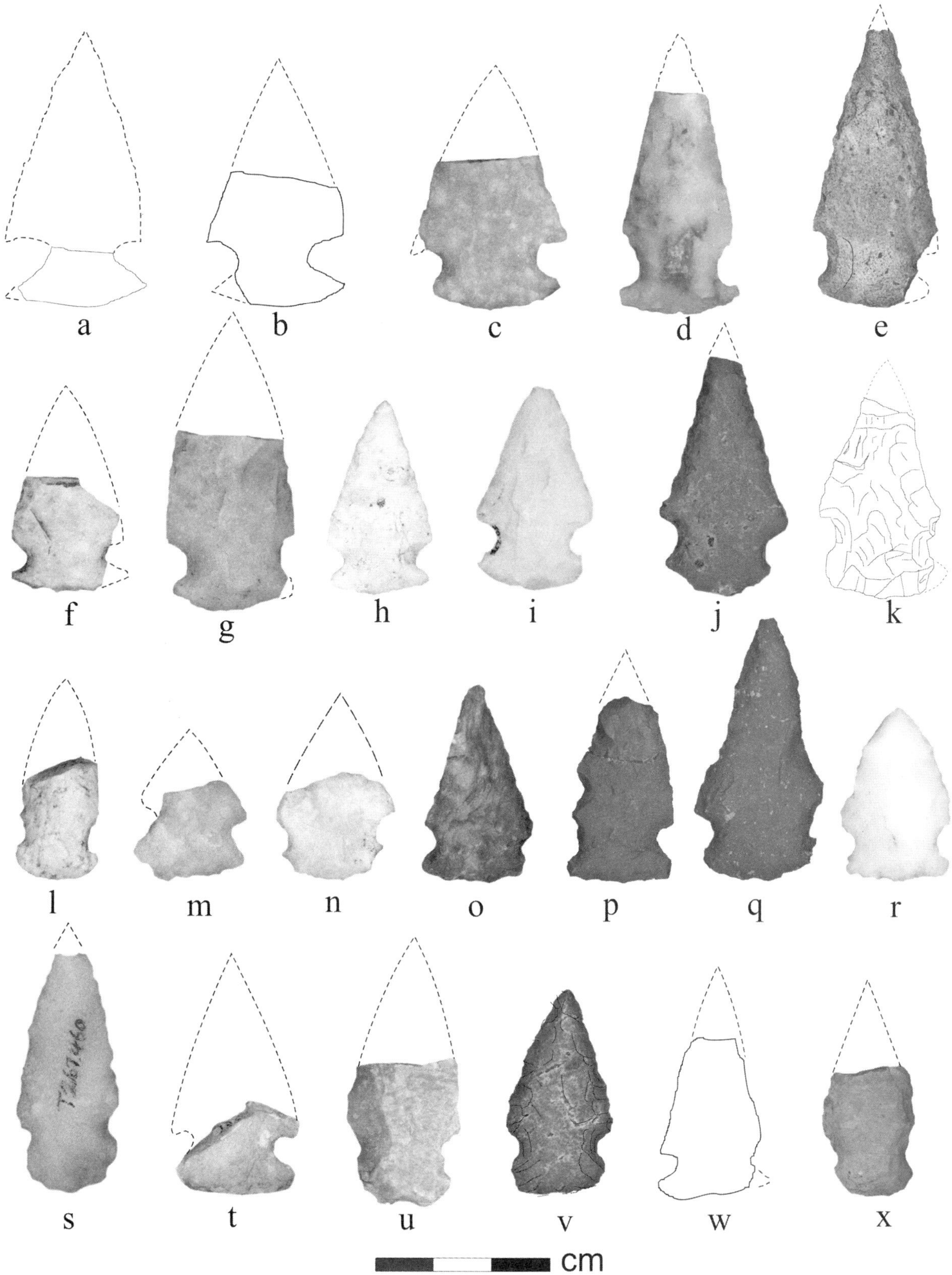

FIGURE 6.8a. San Pedro Side-notched points from Cerro Juanaqueña. Catalogue/bag numbers: (*a*) sc19, (*b*) sc13, (*c*) sc282, (*d*) sc339, (*e*) sc297, (*f*) sc362, (*g*) sc832, (*h*) sc58, (*i*) sc274, (*j*) sc345, (*k*) sc5, (*l*) sc773, (*m*) sc117, (*n*) sc93, (*o*) sc473, (*p*) sc490, (*q*) sc766, (*r*) sc251, (*s*) sc460, (*t*) sc387, (*u*) sc62, (*v*) sc301, (*w*) sc20, (*x*) sc418.

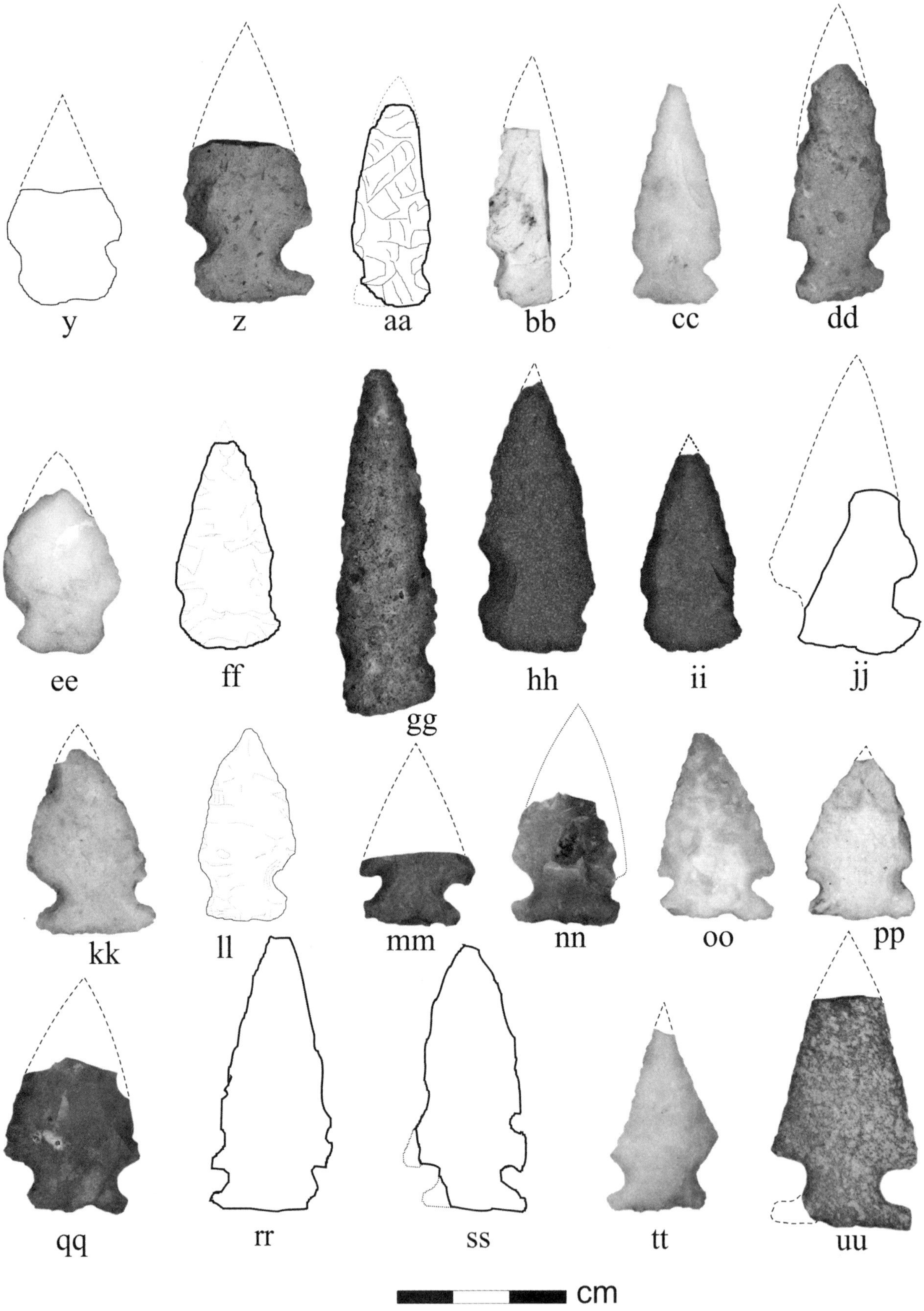

FIGURE 6.8b. San Pedro Side-notched points from Cerro Juanaqueña. Catalogue/bag numbers: (*y*) sc34, (*z*) sc824, (*aa*) sc3, (*bb*) sc229, (*cc*) sc92, (*dd*) sc694, (*ee*) sc294, (*ff*) 7012, (*gg*) sc309, (*hh*) sc373, (*ii*) sc360, (*jj*) 7023, (*kk*) sc260, (*ll*) 7009, (*mm*) sc623, (*nn*) sc785, (*oo*) sc271, (*pp*) sc422, (*qq*) sc371, (*rr*) 8003, (*ss*) 7002, (*tt*) sc336, (*uu*) sc596.

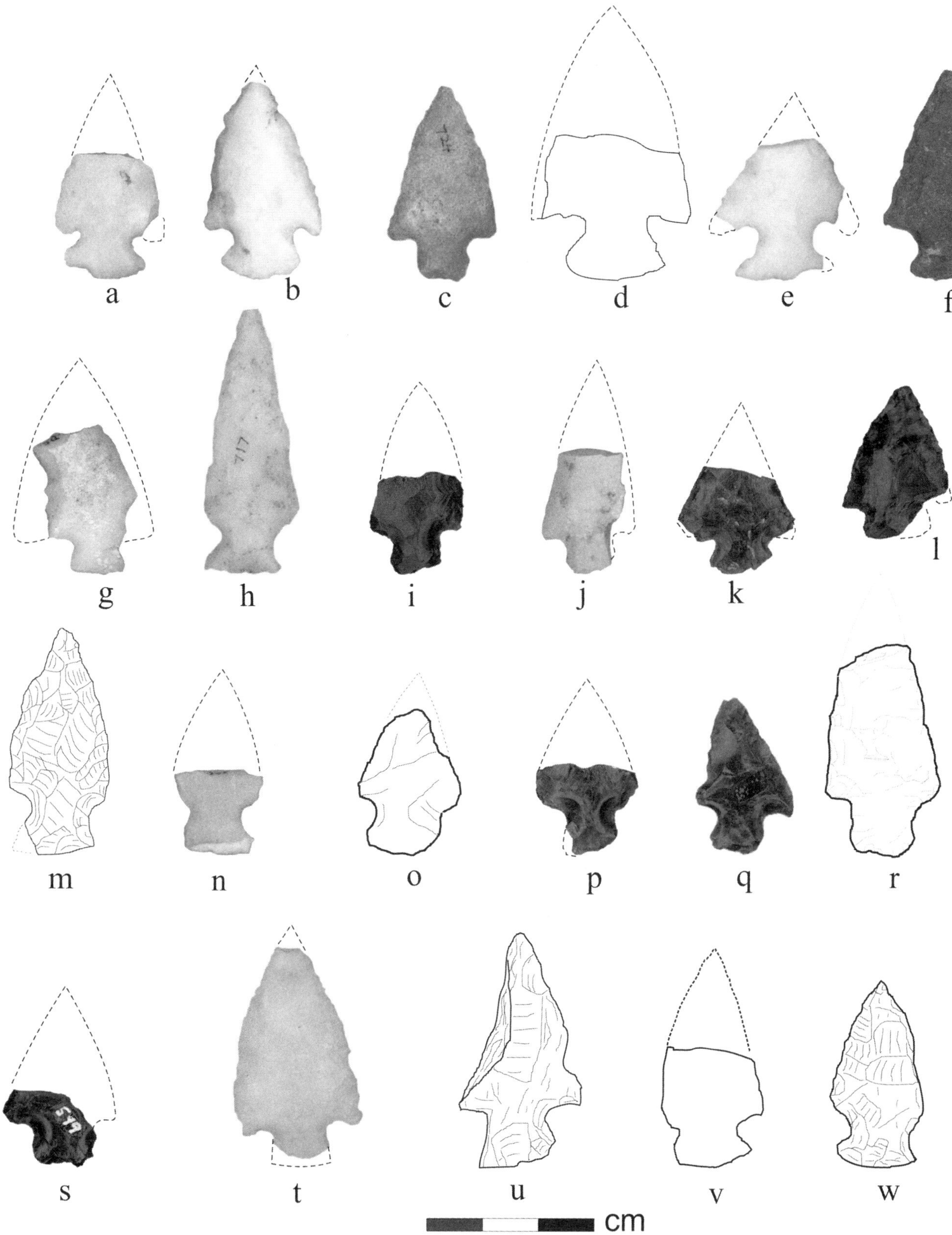

FIGURE 6.9. Tularosa Corner-notched points from Cerro Juanaqueña. Catalogue/bag numbers: (*a*) sc59, (*b*) sc123, (*c*) sc724, (*d*) 7005, (*e*) sc111, (*f*) sc179, (*g*) sc311, (*h*) sc717, (*i*) sc201, (*j*) sc552, (*k*) sc624, (*l*) sc661, (*m*) sc54, (*n*) sc375, (*o*) sc349, (*p*) sc219, (*q*) sc449, (*r*) 7010, (*s*) sc599, (*t*) sc305, (*u*) sc178, (*v*) 7022, (*w*) 7019.

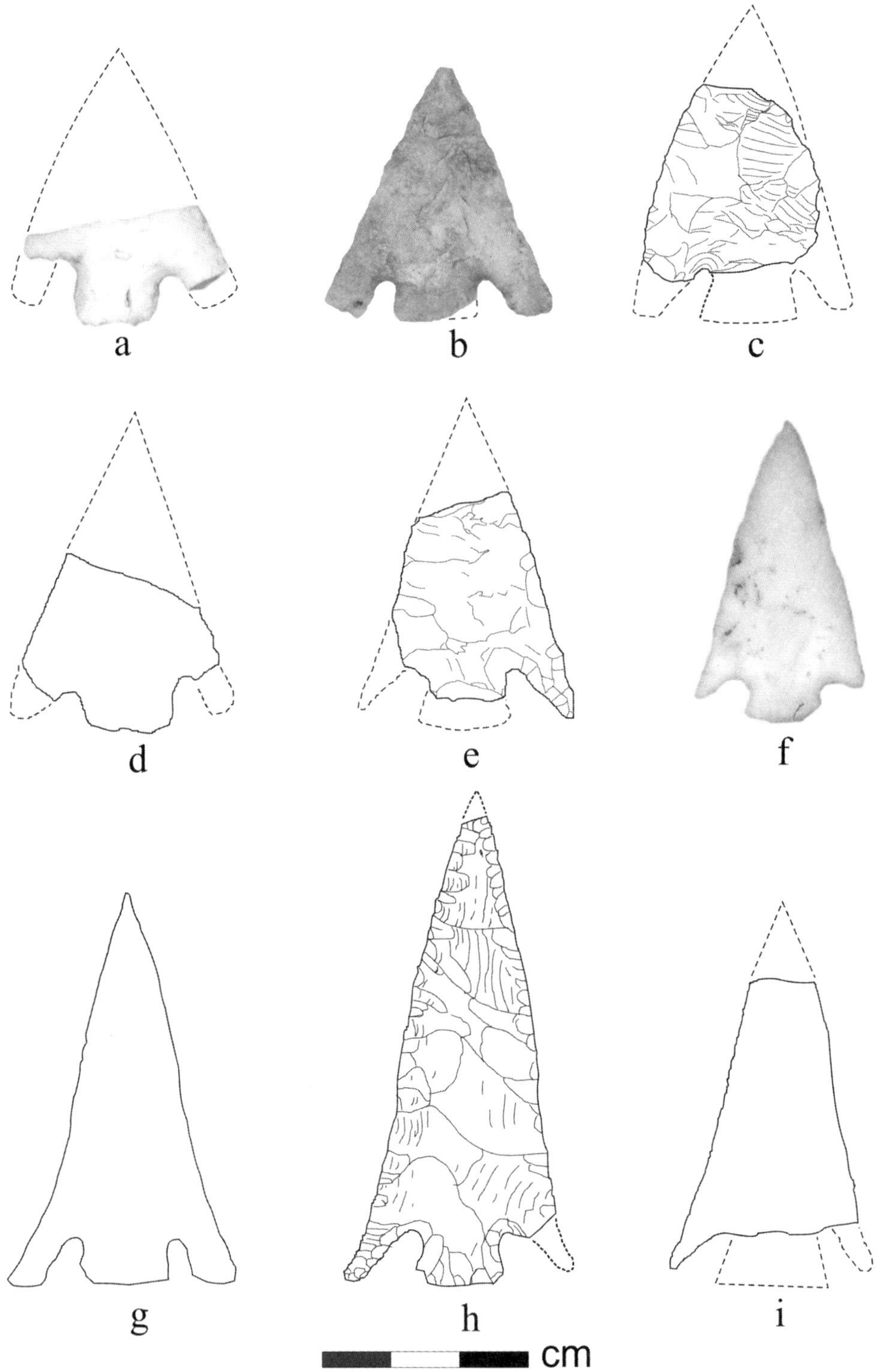

FIGURE 6.10a. Tularosa Basal-notched points from Cerro Juanaqueña. Catalogue/bag numbers: (*a*) sc281, (*b*) 7029, (*c*) 7014, (*d*) sc41, (*e*) 7015, (*f*) sc157, (*g*) 7001, (*h*) sc7, (*i*) sc15.

point blades are triangular with straight or slightly convex edges. The stem is created by either corner-notching ($n = 8$) or side-notching ($n = 4$), and is usually relatively short, expanding sharply to form outward flaring ears. The basal concavity is generally shallow. Among the illustrated examples, three specimens most closely resemble those illustrated by MacNeish (Figure 6.11 a, b, and i). As originally defined, Todsen points also exhibit crude percussion flaking, but this does not seem to be characteristic of the specimens from Cerro Juanaqueña.

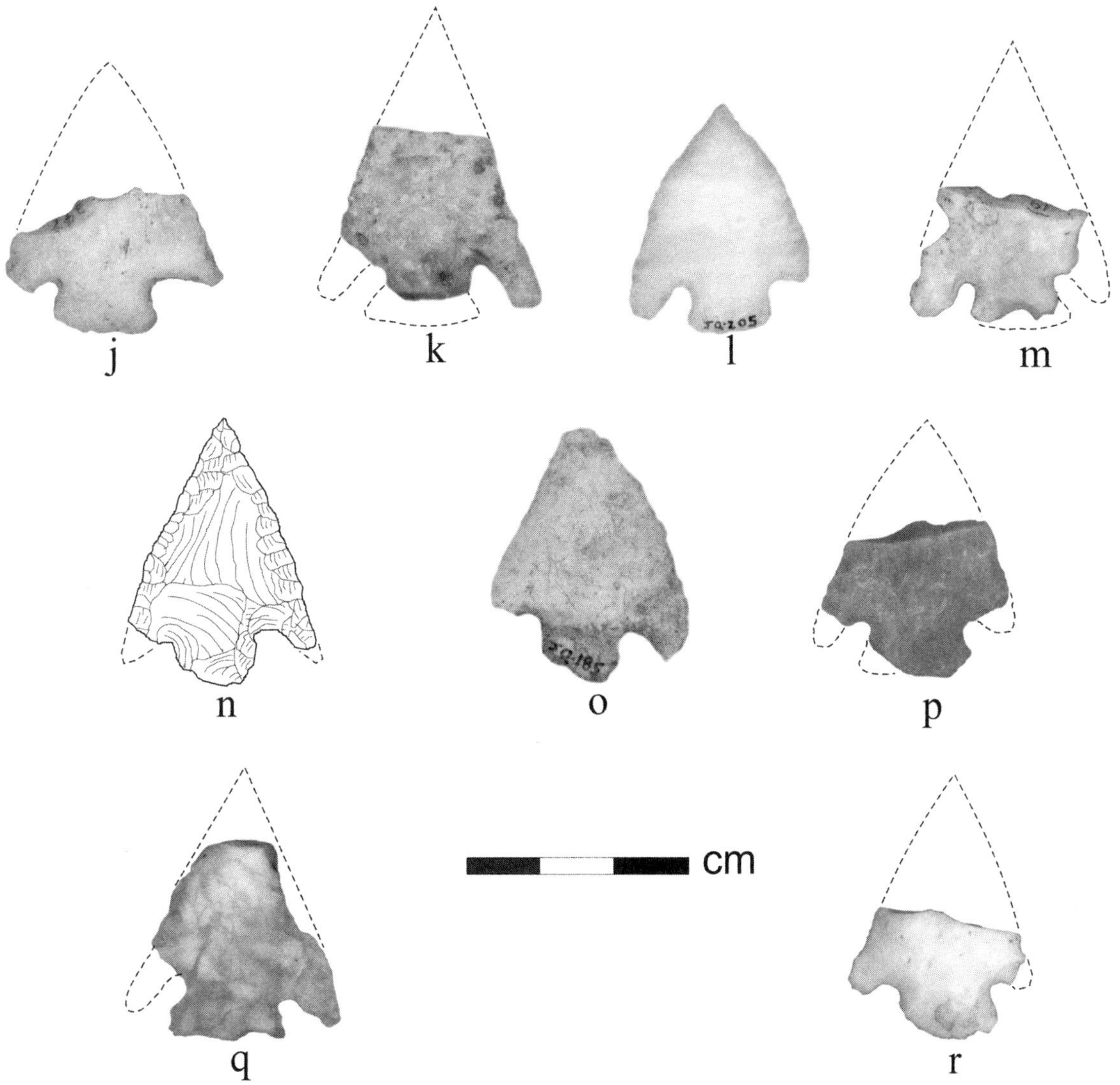

FIGURE 6.10b. Tularosa Basal-notched points from Cerro Juanaqueña. Catalogue/bag numbers: (*j*) sc347, (*k*) sc60, (*l*) sc205, (*m*) sc61, (*n*) sc36, (*o*) sc185, (*p*) sc131, (*q*) sc196, (*r*) sc679.

MacNeish dates this form to Middle Archaic times, 3500 to 2000 BC. The radiocarbon dating evidence from Cerro Juanaqueña (Table 5.2B) indicates that the early occupation of the upper terraces was between 1380 and 1100 cal BC (68% probability) or 1430 to 1030 cal BC (95%), implying that this form may continue somewhat later in northwestern Chihuahua.

Several alternative classifications were considered, but rejected. Bases of the Cerro Juanaqueña specimens have outward flaring ears, which contrast with the downward flaring ears characteristic of Pinto points. The basal concavity is not as deep as that found in Chiricahua or Ventana Side-notched points, and over half of the specimens from Cerro Juanaqueña are corner-notched, rather than side-notched. The shape of the stem is reminiscent of some varieties of the San Jose type (Chapin n.d.), but other San Jose characteristics such as serrated blades, haft grinding, and extreme modification through reworking of broken blades are absent. Significantly, all of the alternative classifications for these 12 specimens also date to Middle Archaic times. Note, however, that Chapin's (2017:107–108) En Medio Eared type from northern New Mexico is similar and may date from the Late Archaic period.

Gypsum

Gypsum points are distinguished by contracting stems. As currently defined (Justice 2002:186–194), other characteristics can be quite variable, with short or long stems, pointed or convex bases, and shoulders ranging from absent to strongly barbed. Eight specimens from Cerro Juanaqueña have contracting stems and are

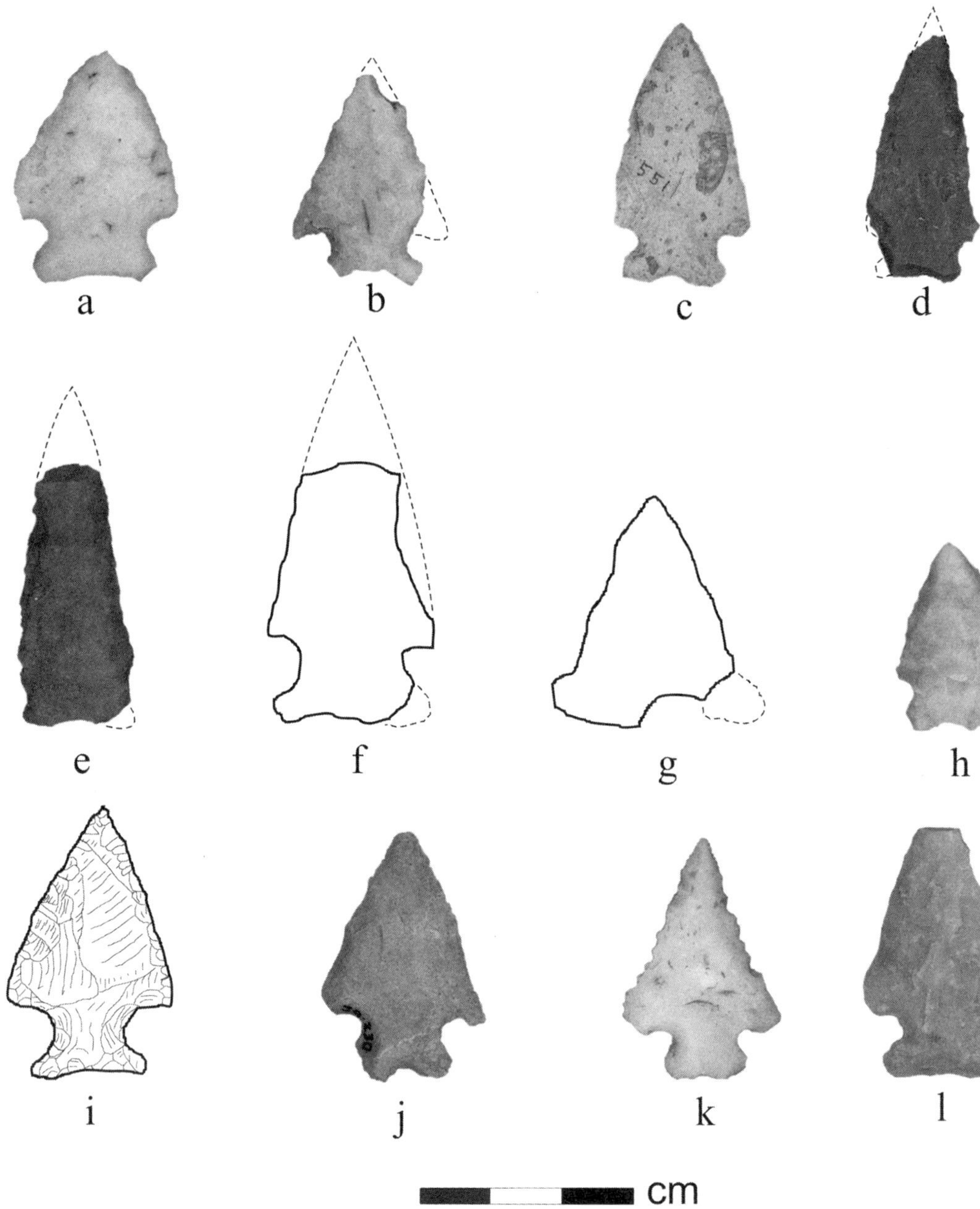

FIGURE 6.11. Todsen points from Cerro Juanaqueña. Catalogue/bag numbers: (*a*) sc491, (*b*) sc357, (*c*) sc551, (*d*) sc755, (*e*) sc266, (*f*) sc6.1, (*g*) sc17, (*h*) sc202, (*i*) sc29, (*j*) sc230, (*k*) sc528, (*l*) sc458.

therefore classified here as Gypsum points (Figure 6.12). Originally the Gypsum type was thought to date as early as 9700 BC, but a more recent consideration of dated contexts for this type suggest that it spans the transition between the Middle and Late Archaic periods in the Southwest, dating between 2000 and 800 BC (Justice 2002).

Datil

We tentatively classify 11 specimens from Cerro Juanaqueña as Datil points, characterized by an essentially rectangular stem, with parallel stem edges and a straight base. The specimens from Cerro Juanaqueña (Figure 6.13) compare favorably with illustrated examples of this type, although two differences are noted. First, most specimens from Cerro Juanaqueña have barbs or strong shoulders. This is within the range of variation of Datil points, but weak shoulders are more characteristic of the type in general. Second, Datil points from elsewhere in the southwestern United States are often heavily reworked, which was not the case for those from Cerro Juanaqueña. Datil points are distributed throughout the

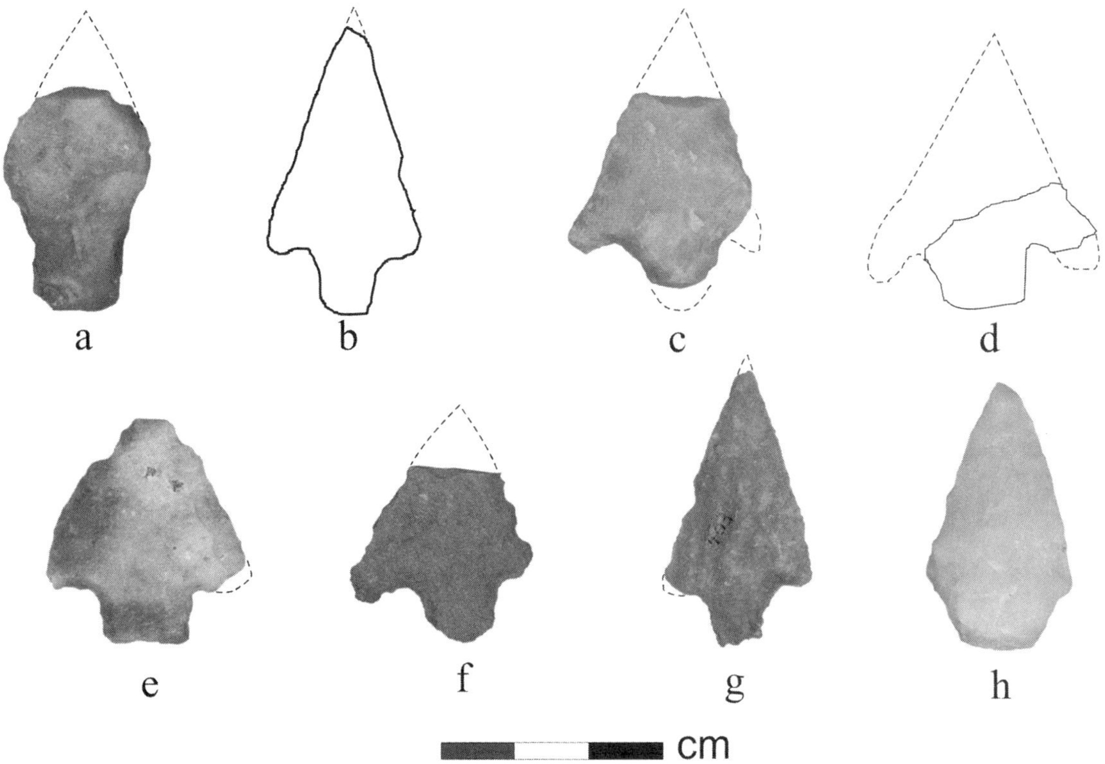

FIGURE 6.12. Gypsum points from Cerro Juanaqueña. Catalogue/bag numbers: (*a*) sc150, (*b*) sc9, (*c*) sc742, (*d*) 7006, (*e*) sc269, (*f*) sc340, (*g*) sc403, (*h*) sc487.

southwestern United States and into Sonora, although they are not common at any single location. Justice (2002:174) estimates their age to be between 1600 BC and AD 300, consistent with their occurrence at Cerro Juanaqueña.

Cortaro

Cortaro points are a projectile point style defined by Roth and Huckell (1992) on the basis of material from southern Arizona. Subsequently, Knight (2003) has reported that this type occurs in significant numbers in the basin and range country of southwestern New Mexico, immediately to the north of Cerro Juanaqueña. Cortaro points are un-notched bifaces with a triangular blade and a concave base. Roth and Huckell (1992) note that many Cortaro points exhibit diamond-shaped cross sections, leading Sliva (2005) to suggest that this was an expedient form, or perhaps not even primarily a projectile point. Instead it may represent a multipurpose hafted biface similar to other bifaces described in the preceding section. This form is thought to have originated during the latter portion of the Middle Archaic, and to have persisted through the Early Agricultural period.

Only two Cortaro points were found at Cerro Juanaqueña (Figure 6.14). Both are thin and relatively well made. Given the fact that Cerro Juanaqueña dates primarily to a point near the transition between Middle and Late Archaic periods, it is surprising that more examples were not found.

Miscellaneous

A total of 39 relatively complete large projectile points are not assigned to any specific type (Figure 6.15). A large majority of these have expanding stems with straight or convex bases ($n = 32$) and are generally consistent with the San Pedro type. However, all of these specimens exhibit morphology that is just a little different from that type. Some (Figure 6.15 t and u) are probably heavily resharpened San Pedro points. Others have stems that are shorter than is normal for the San Pedro type, and still others have shallow or asymmetrical notches. These specimens illustrate our earlier observation that Late Archaic flint knappers did not necessarily adhere closely to a rigid and well-defined mental template when they made projectile points.

Fragments

A total of 53 biface fragments were too incomplete to classify, but retained evidence of notching and were clearly fragments of large projectile points. Most are distal or medial fragments with missing or heavily damaged

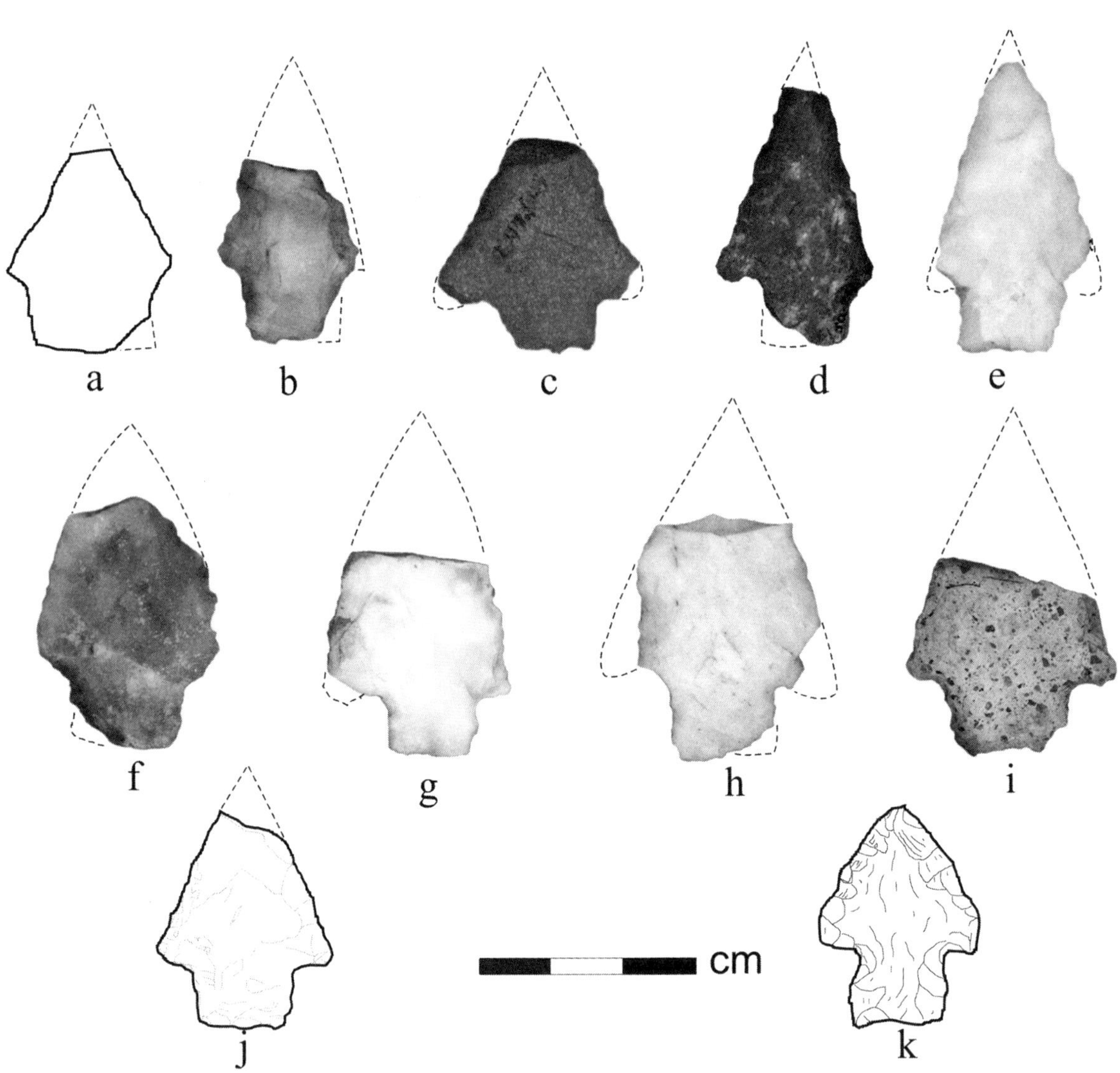

FIGURE 6.13. Datil points from Cerro Juanaqueña. Catalogue/bag numbers: (*a*) sc30.1, (*b*) sc195, (*c*) sc649, (*d*) sc211, (*e*) sc86, (*f*) sc350, (*g*) sc326, (*h*) sc286, (*i*) sc159, (*j*) 7013, (*k*) sc31.

stems. Five specimens were stems, broken at the neck, and all five of these are probably from San Pedro points. In addition, a number of the late stage bifaces discussed in the preceding section appear to be projectile point tips or midsections, but are not considered here because they do not preserve any evidence of hafting.

Small Projectile Points

As discussed above, small projectile points are those that satisfy most of the following criteria: weight less than or equal to 1.5 g, length less than or equal to 25 mm, thickness less than or equal to 3 mm, and/or stem width less than or equal to 8 mm. By this definition 14 projectile points from Cerro Juanaqueña were recognized as small (Figure 6.16). These specimens do not fit any well-defined type. Five or six (Figure 6.16, a through f) resemble one another, with triangular blades, moderate shoulders, and expanding stems with convex bases. One specimen (Figure 6.16 g) was manufactured expediently with minimal retouch on a small blade-like flake. Another (Figure 6.16 h) appears to be a heavily reworked dart point. Three others are idiosyncratic (Figure 6.16 i, j, and k), and three are distal, medial, or lateral fragments (Figure 6.16 l, m, and n).

Based on their small size, most investigators would classify these specimens as arrow points, and presumably they date to the Ceramic period, rather than the Archaic. However, Ceramic period projectile points in this region are not well-documented, and some researchers have recently suggested that experimentation with the bow and arrow may have begun as early as 800 BC in southern Arizona (Sliva 1999a). The 300 BC reoccupation of Cerro Juanaqueña is within this proposed time frame.

FIGURE 6.14. Cortaro points from Cerro Juanaqueña. Catalogue/bag numbers: (*a*) sc189, (*b*) sc222.

Discussion

Overall, the projectile point assemblage from Cerro Juanaqueña is consistent with expectations derived from the broadly recognized projectile point sequence for the southwestern United States. The projectile points can be comfortably classified using the Southwestern typology, and the specific types are about what we expect in this time period. Radiocarbon evidence shows that the assemblage dates primarily to the early part of the Late Archaic period, and it is heavily dominated by San Pedro and related forms such as Tularosa Corner-notched, Tularosa Basal-notched, and other miscellaneous forms with expanding stems and straight or convex bases. However, earlier forms usually associated with the Middle Archaic period are also present in smaller numbers. These include Todsen, Gypsum, Datil, and Cortaro points. Altogether these Middle Archaic forms comprise about 13% of the assemblage (n = 33), while 71% are Late Archaic forms (n = 190). The remaining 16% are not assigned to a specific type, but are generally consistent with Late Archaic projectile point morphology (see Table 6.6).

Elsewhere we argue that it is very unlikely that there is a Middle Archaic occupation at Cerro Juanaqueña. The artifacts found on the constructed terraces were almost certainly deposited after 1300 BC or so, including the Middle Archaic projectile points. It is possible that these are curated specimens, collected elsewhere by the inhabitants of Cerro Juanaqueña. However, it is equally possible that these earlier forms continued to be used for a time after the shift to San Pedro and related types.

Several different typological systems (mostly extrapolated from neighboring regions) are commonly used in southern New Mexico and northern Chihuahua, and chronological inferences can vary, depending upon which system is applied. In addition, variations in projectile point style are sometimes taken as indicators of social distance and boundaries (McBrinn 2005). For these reasons we have compared the typological results from Cerro Juanaqueña with expectations derived from several other typological/chronological schemes.

MacNeish summarized data from five seasons of research at a series of sites near Las Cruces, New Mexico (MacNeish 1993). His phase sequence and the projectile point types associated with each are illustrated in Table 6.8. On the right we have listed the Cerro Juanaqueña equivalent for each type, and the number of specimens identified at Cerro Juanaqueña. Chronologically, the principal occupation at Cerro Juanaqueña falls squarely within the Fresnal phase. Several of the types that MacNeish attributes to this phase do occur at Cerro Juanaqueña, but the vast majority are types attributed to the succeeding Hueco phase.

These observations could imply that some adjustment is needed in definition and dating of the phase sequence in the Las Cruces area. An alternative possibility is that the advent of San Pedro and related forms is closely tied to the development of maize-dependent economies, as some have argued (Berry and Berry 1986; B. Huckell 1995; Matson 1991). If so, this transition could have occurred earlier along the river valleys of northwestern Chihuahua than elsewhere in the region. Perhaps the older, Middle Archaic forms persisted longer in the Las Cruces area for this reason. As we accumulate more chronometric dates associated with projectile points, more macrobotantical remains, and more bone isotope data, a systematic test of these alternative hypotheses should be possible.

Mallouf (1985) has summarized archaeological information from the eastern Trans-Pecos region, centered on the Davis Mountains 350 km east of Cerro Juanaqueña. In this area, Langtry and Val Verde points characterize the interval between 3000 BC and 1750 BC. None of the projectile points from Cerro Juanaqueña are comparable to these two types. However, there are similarities in projectile point forms between 1750 BC and 500 BC. Shumla points in the eastern Trans-Pecos sequence resemble Tularosa Basal-notched points at Cerro Juanaqueña, although their stems tend to be more rectangular (see above). Some of the contracting-stemmed Almagre points are reminiscent of the Gypsum type. The un-stemmed Tortugas points resemble some of the late stage bifaces from Cerro Juanaqueña, but differ in having blade edges that are alternately beveled. This distinctive characteristic was not observed on any of the unstemmed bifaces from Cerro Juanaqueña. Several Trans-Pecos forms, including Lange, Marcos, and

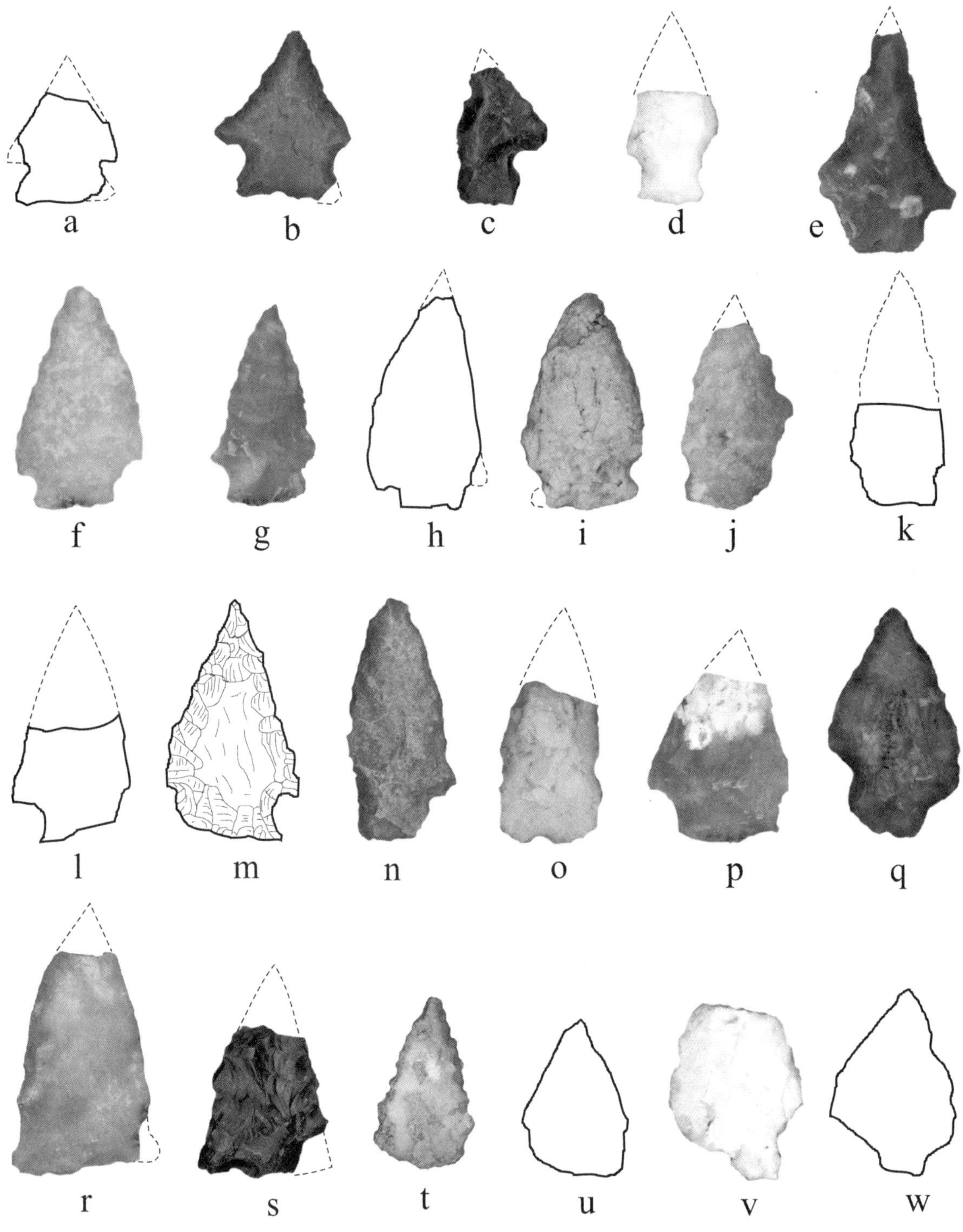

FIGURE 6.15a. Unclassified points from Cerro Juanaqueña. Catalogue/bag numbers: (*a*) 9003, (*b*) sc632, (*c*) sc685, (*d*) sc113, (*e*) sc770, (*f*) sc66, (*g*) sc87.2, (*h*) 9002, (*i*) sc817, (*j*) sc669, (*k*) sc16, (*l*) 7004, (*m*) sc33, (*n*) sc427, (*o*) sc175, (*p*) sc224, (*q*) sc424, (*r*) sc81, (*s*) sc363, (*t*) sc389, (*u*) 8001.1, (*v*) sc379, (*w*) sc35.

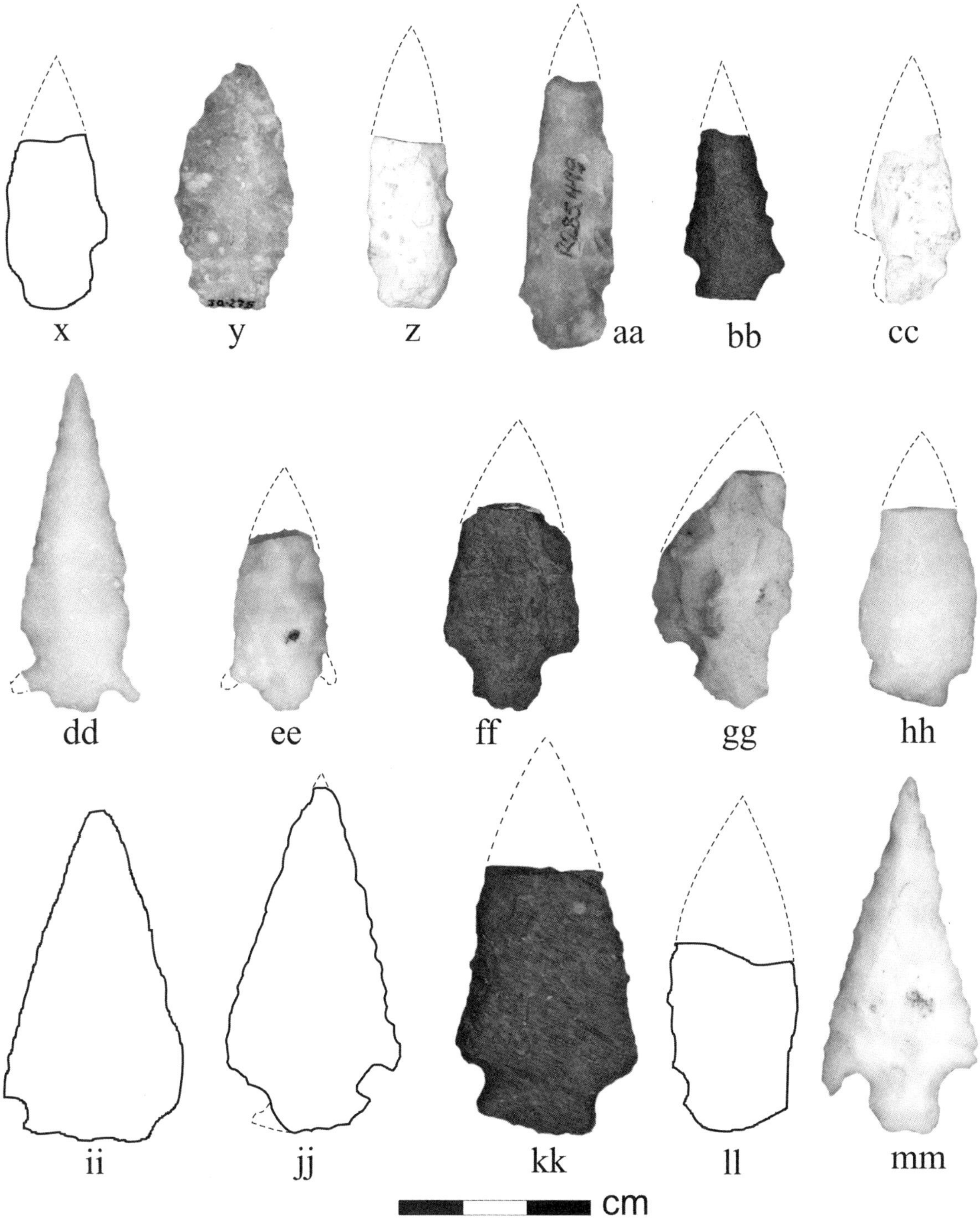

FIGURE 6.15b. Unclassified points from Cerro Juanaqueña. Catalogue/bag numbers: (*x*) sc40, (*y*) sc275, (*z*) sc249, (*aa*) sc479, (*bb*) sc104, (*cc*) sc691, (*dd*) sc149, (*ee*) sc73, (*ff*) sc648, (*gg*) sc83, (*hh*) sc67, (*ii*) sc43, (*jj*) 8002, (*kk*) sc120, (*ll*) sc4, (*mm*) sc85.

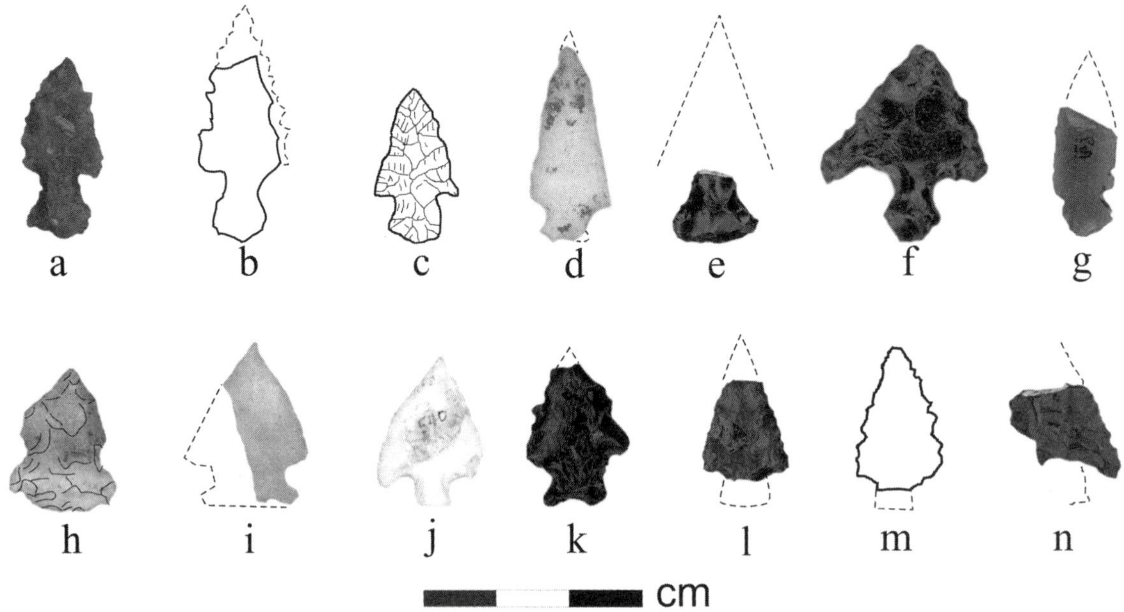

Figure 6.16. Small points from Cerro Juanaqueña. Catalogue/bag numbers: (*a*) sc454, (*b*) 9001, (*c*) sc32, (*d*) sc346, (*e*) sc14, (*f*) sc191, (*g*) sc641, (*h*) sc105, (*i*) sc740, (*j*) sc540, (*k*) sc526, (*l*) sc234, (*m*) sc555, (*n*) sc374.

Table 6.8. Phase and Projectile Point Sequence from Las Cruces, NM Area.[1]

Projectile Point Type	Cerro Juanaqueña Equivalent	Number of Examples
PaleoIndian Pre-6000 BC		
Angostura, Folsom, Clovis, Ayacucho	None	
Gardner Springs Phase 6000 to 4300 BC		
Jay, Bajada	None	
Keystone Phase 4300 to 2600 BC		
Bat Cave, Lerma, Amargosa-Pinto	None	
Pelona, Gypsum-Almagre	Gypsum	8
Todsen	Todsen	12
Fresnal Phase 2600 to 900 BC		
San Jose, Armijo, Chiricahua, La Cueva	None	
Pelona, Augustin	Gypsum	8
Fresnal	Datil	11
Todsen	Todsen	12
Shumla-like	Tularosa BN	18
Hueco Phase 900 BC to AD 200		
Padre Gordo, Pendejo	None	
San Pedro large, Hatch, Hueco	San Pedro	149
San Pedro small, En Medio	Tularosa CN	23

[1] After MacNeish 1993.

Williams points resemble San Pedro points in having expanding stems with straight or convex bases. However, in contrast to the San Pedro points from Cerro Juanaqueña, the Trans-Pecos types have strong shoulders and/or deep corner notches, and are usually well-barbed. Very few specimens from Cerro Juanaqueña resemble the concave-based Marshall type, and none are similar to the distinctive Conejo points.

In the succeeding period, 500 BC to AD 250, Ensor points sometimes resemble side-notched varieties of the San Pedro type, and occasional San Pedro specimens with bulbous stems might be confused with the Palmillas type. Paisano points from the Trans-Pecos area are similar to some points described here as Todsen points, but these are uncommon.

This comparison shows some general similarities in projectile point forms between northwestern Chihuahua and the Trans-Pecos area, but does not support subsuming the two regions under the same typological system. The sequence of projectile point types described for the Trans-Pecos area does not predict the range of types found in the well-dated Cerro Juanaqueña assemblage.

Drills

Twenty-one items were recognized as drills, including 15 from Cerro Juanaqueña, one from Cerro los Torres, and five from Cerro Vidal (see Table 6.9). Figure 6.17 illustrates all examples from Cerro Juanaqueña, as well as two relatively complete specimens from Cerro Vidal. Drills are distinguished from other bifaces on the basis of both morphology and wear. These items have long, narrow blades with thick cross sections and distinctive wear patterns. Most drills were manufactured by careful bifacial reduction, but one example (Figure 6.17 o) is a bi-pointed piece of angular debris, included here because it shows the same massive use wear seen on other drills (described below).

Despite the small number of examples, these artifacts exhibit a range of forms. Two have bulbous bases (Figure 6.17 h and i), likely intended to be grasped between the thumb and forefinger. Another related form, a t-shaped drill, was documented at Cerro Vidal (Figure 6.17 q). Three examples (Figure 6.17 j, k, and l) are of a type described as a "straight drill" (straight edges and no defined stem) by some analysts (Sliva 1997). Three are bi-pointed (Figure 6.17 m, n, and o), including a piece of angular debris used as an expedient drill. One relatively complete biface was assigned to this category because of distinctive wear (Figure 6.17 p). It is wider than most of the other examples, and may be a reused projectile point. Most of these drills have thick lenticular cross sections. However, the working edges of two (Figure 6.17 e and m) were formed by steep retouch on opposite faces, resulting in a parallelogram-shaped cross section. Edges of the expedient drill (Figure 6.17 o) were not modified.

All of the drills show very heavy wear consisting of crushing and polish on blade edges, especially toward the distal end. On over half of the examples the blade edges were actually rounded by use wear, and almost half of the specimens exhibited light polish on blade arrises. A few examples had some crushing on their basal edge and/or facial polish near the base or stem, which could imply that they were hafted. These wear patterns are consistent with use of the drills for manufacture of stone pipes (Sezate 2017).

Cruciforms

Seventeen cruciforms were found at Cerro Juanaqueña, plus one at both Cerro el Canelo and Cerro Vidal respectively (Table 6.10). These are small stone objects, 5 cm or less in maximum dimension, flaked to form a cross or "X" shape (Figure 6.18). These enigmatic artifacts are distributed through southern New Mexico, southern Arizona, western Chihuahua, and Sonora, occasionally extending south as far as Sinaloa and Nayarit. However, their center of distribution seems to be western Chihuahua, and reportedly they are especially common in the vicinity of Villa Ahumada, 150 km east of Cerro Juanaqueña (Phelps 1966).

Alan Ferg (1998) has reviewed published information pertaining to these artifacts. He concludes that they were made during the time span of the San Pedro and Cienega phases in southern Arizona, and that the few examples recovered from Ceramic period contexts are likely curated specimens. The function of these artifacts is still very much a matter of speculation.

Ten of the cruciforms reported here were made of basalt, and four were made of rhyolite. Chert, chalcedony, jasper, and siltstone accounted for only one or two specimens each. All of the cruciforms were presumably initially shaped by percussion flaking, but finishes showed considerable variability. Three examples are described here as fully finished. All surfaces have been ground smooth, so only very faint traces of original flake scars remain. Three specimens were heavily polished. On these examples most flake scars have been obliterated, especially on the body of the artifact. One specimen was characterized by moderate polish. On this example there was heavy polish on the thickest part of body, but elsewhere most arrises were unpolished. Seven of the cruciforms are lightly polished, meaning that under 10x magnification at least traces of polish can be seen on edges or arrises. The remaining four specimens had no evidence of polish, their arrises were more or less pristine. Varying degrees of polish are sometimes thought to reflect varying degrees of completion, with the fully polished examples representing completed

TABLE 6.9. Drills from Cerro Juanaqueña, Cerro los Torres, and Cerro Vidal.

Specimen	Fragment	Base[1]	Cross-section	Observations	Figure
Cerro Juanaqueña					
sc6b	Base	Straight	Thick biconvex	Outline drawing only.	6.17 j
sc151	Tip	—	Thick biconvex	Rounding and heavy polish on edges. Some crushing near tip. Light polish on arrises. Continued in use after very tip was broken.	6.17 f
sc207	Midsection	—	Lenticular	Crushing and rounding on all edges. Polish on most distal portions. Arrises indeterminate. Could not tell if hafted.	6.17 p
sc233	Midsection	—	Parallelogram	Crushing and polish on edges. Light polish on some arrises.	6.17 e
sc291	Whole	Pointed	Irregular	Piece of shatter used as drill. Crushing and polish on edges, heavy polish on tip.	6.17 o
sc300	Midsection	—	Thick biconvex	Edges crushed, rounded, polished. Could not determine arrises.	6.17 c
sc324	Midsection	—	Thick biconvex	Crushing and some polish on edges. Light polish on some arrises.	6.17 b
sc353	Base	Bulbous	Thick biconvex/ diamond	Some crushing and rounding on edges. Heavy crushing on base and light polish on arrises in middle of base imply hafting.	6.17 i
sc390	Base	Straight	Thick biconvex	Edges partly obscured by mineral deposit. Crushing and rounding, heaviest at distal end, but extending along entire edge to base.	6.17 k
sc529	Tip	—	Thick biconvex	Heavy polish on edges and arrises near tip. Edges ground to form flat surfaces perpendicular to faces.	6.17 d
sc586	Base	Straight	Thick biconvex/ diamond	Heavy crushing, rounding, and some polish on distal edges. Less polish on proximal edges. Polish on arrises of both faces throughout. Basal edge shows little damage.	6.17 l
sc715	Base	Pointed	Thick biconvex	Heavy crushing, rounding, and some polish on distal edges. Lighter wear on proximal edges. Heavy crushing on basal edge. Polish on arrises and flake scars toward proximal end.	6.17 n
sc788/789	Midsection	—	Irregular	Two conjoinable fragments. Crushing and rounding on edges. Heavily crazed.	6.17 a
T168-89	Tip	—	Thick biconvex	Heavy crushing, rounding, and polish on edges. Light polish on arrises. Rounding at distal end after very tip was broken.	6.17 g
T175-09	Whole	Pointed	Rectangular	Piece of shatter used as drill. Rounding and heavy polish, along with some hinge fracturing on edges. Very light polish on arrises	6.17 m
Cerro los Torres					
ct06	Tip	—	—	Outline drawing only.	6.31 t
Cerro Vidal					
cv6	Base	T-shaped	—	Outline drawing only.	6.17 q
cv11	Midsection	—	—	Outline drawing only.	6.32 w
cv12	Midsection	—	—	Outline drawing only.	6.32 u
cv15	Midsection	—	—	Outline drawing only.	6.32 v
cv36	Base	Bulbous	Thick biconvex	Crushing on all edges. No rounding or polish. No obvious polish on arrises.	6.17 h

[1] May be pointed, straight, bulbous, or "T"-shaped. "Straight" means that there is no shape or flaring at the base. Instead, tool edges simply extend directly to the basal edge.

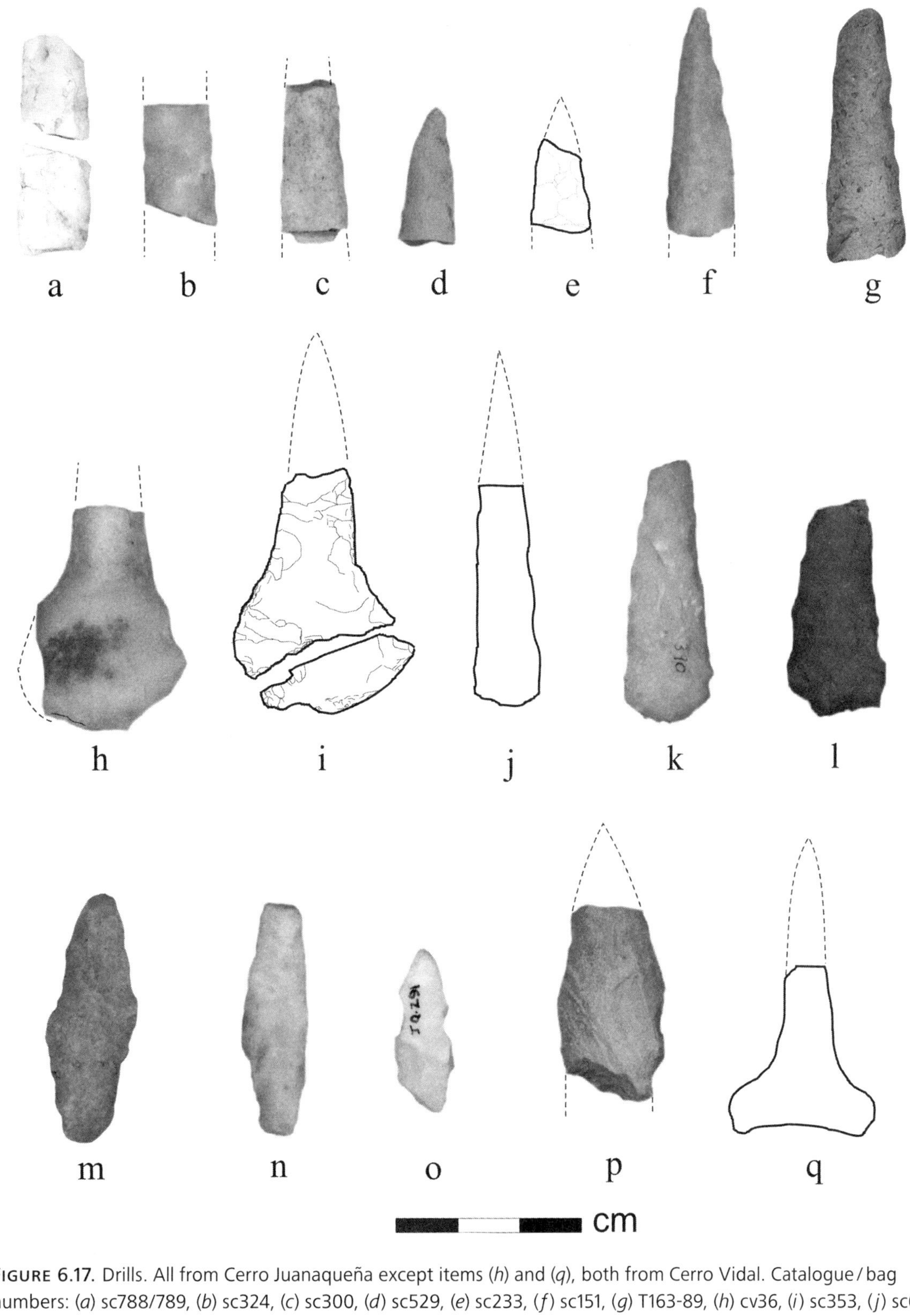

FIGURE 6.17. Drills. All from Cerro Juanaqueña except items (*h*) and (*q*), both from Cerro Vidal. Catalogue / bag numbers: (*a*) sc788/789, (*b*) sc324, (*c*) sc300, (*d*) sc529, (*e*) sc233, (*f*) sc151, (*g*) T163-89, (*h*) cv36, (*i*) sc353, (*j*) sc6b, (*k*) sc390, (*l*) sc586, (*m*) T175-9, (*n*) sc715, (*o*) sc291, (*p*) sc207, (*q*) cv6.

TABLE 6.10. Cruciforms from Cerro Juanaqueña, Cerro los Torres, and Cerro Vidal.

Specimen	Material	Finish	Thickness[1]	Minimum Dimension[1]	Maximum Dimension[1]
Cerro Juanaqueña					
sc338	Basalt	Fully polished	6.7	20.3	42.7
Abajo 6	Chalcedony	Light polish	7.0	16.9	30.3
Abajo 7	Chert	Heavy polish	7.2	18.3	—
sc115	Basalt	Fully polished	7.4	19.8	36.8
sc8[2]	Rhyolite	—	9.0	23.0	39.0
sc94	Chert	Heavy polish	9.6	19.6	—
sc662	Basalt	None	9.7	20.6	—
sc461	Jasper	Light polish	10.0	21.7	—
sc341	Siltstone	Light polish	12.5	24.6	33.0
sc292	Rhyolite	Moderate polish	13.3	23.0	42.9
sc734	Rhyolite	Heavy polish	14.6	23.3	37.7
sc738	Rhyolite	None	15.1	25.3	41.7
sc70	Basalt	Light polish	15.8	28.7	41.4
sc361	Basalt	Light polish	19.0	32.4	46.0
sc688	Basalt	None	19.5	30.2	48.9
T6-28	Basalt	None	16.4	27.6	45.2
R28-22[2]	Basalt?	Fully polished	—	15.0	30.0
Cerro el Canelo					
cc8	Basalt	Light polish	10.0	24.0	—
Cerro Vidal					
cv3[2]	Basalt	Light polish	—	1.0	2.5

[1] In millimeters.
[2] Not collected/exported. Measurements based on drawing only.

specimens. However, the light polish on arrises of seven examples from Cerro Juanaqueña is more or less evenly distributed on the artifact surfaces, and could just as easily be "soft bag wear."

Unifacial Scrapers

Although informal tools sometimes exhibited unifacially modified edges, we did not recognize any formally shaped tools characterized by unifacial retouch.

Ground Stone

Manos

The tabulation of manos in Table 6.1 includes 228 complete specimens and 461 mano fragments. In Chapter 13, we provide detailed analyses of several subsets of this material. The great majority of the manos are oval-shaped, one-hand manos with an average length of about 12 cm. Almost all were made of locally available basalt or rhyolite. A majority are expedient, with no evidence of deliberate shaping, but about a third have shaped, rounded edges. These one-hand manos have one or two grinding surfaces in equal proportions, and the grinding surfaces may be either convex (60%) or flat (40%). Wear was characterized as medium or heavy for 80% of the manos included in the detailed analysis. In Chapter 13, we argue that the unusually low ratio of manos to metates may imply that manos were being routinely removed from Cerro Juanaqueña.

Basin Metates

Any visitor to Cerro Juanaqueña is immediately impressed by the large number of deep basin metates found across the site. We documented over 700 of these impressive artifacts, and estimate that at minimum over 200 complete specimens were originally present (see Chapter 13). All were made from local basalt boulders ranging from 32 to 94 cm in maximum dimension. Perhaps a third of the basin metates evidence some trimming and shaping to achieve a rough rectangular shape, and the upper corners were sometimes rounded by pecking. The basins themselves are worn as deep as 14.5 cm, with an average depth of about 8 cm. About 10% of the basin metates are worn on both sides, and almost one-quarter of these have holes manufactured in them. In the greater Southwest, these deep basin metates occur in Middle Archaic, Late Archaic, and Early Formative

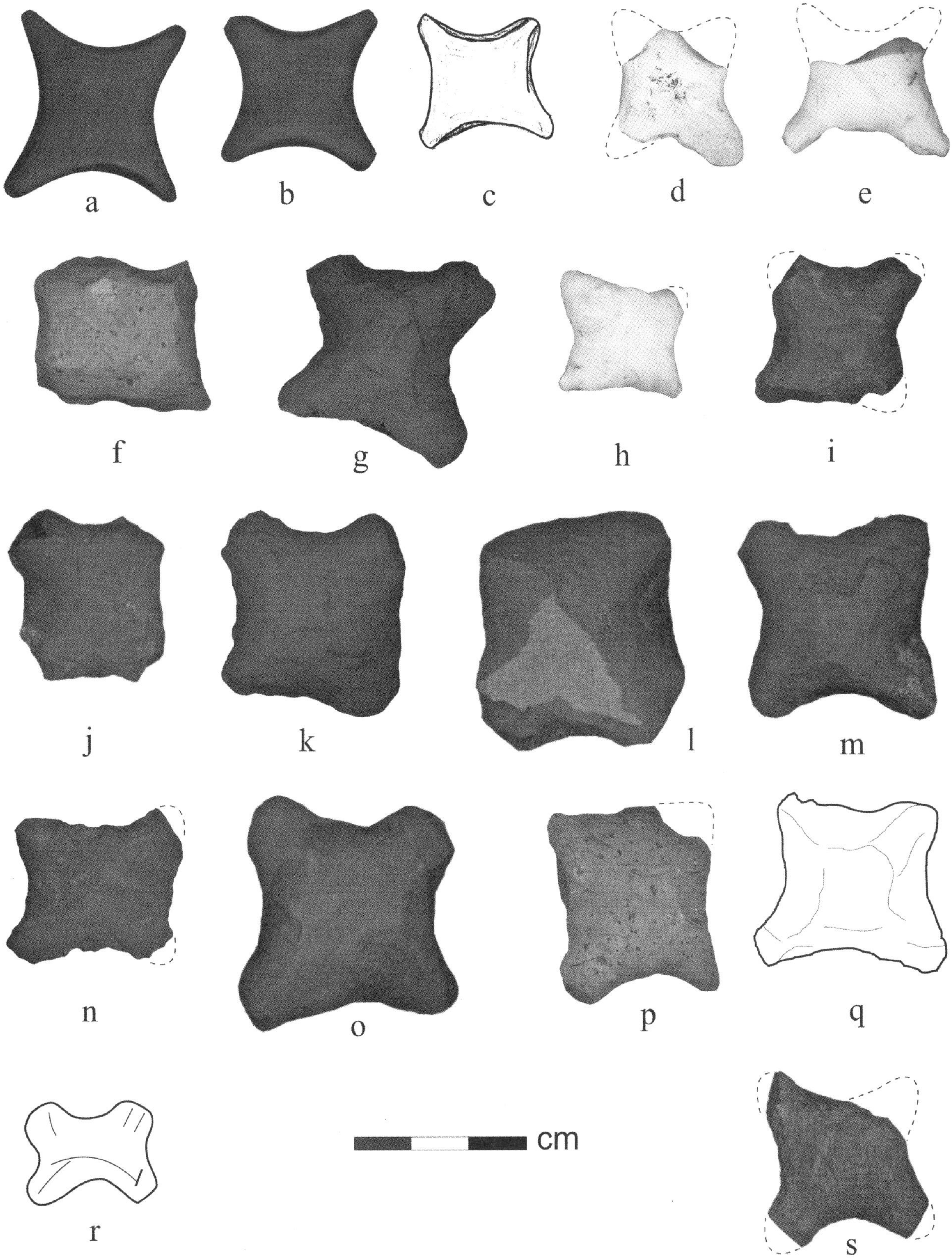

FIGURE 6.18. Cruciforms. Specimens (*a*) through (*q*) from Cerro Juanaqueña; specimen (*r*) from Cerro Vidal; specimen (*s*) from Cerro el Canelo. Catalogue / bag numbers: (*a*) sc338, (*b*) sc115, (*c*) R28-22, (*d*) sc9007, (*e*) sc94, (*f*) sc734, (*g*) sc292, (*h*) sc9006, (*i*) sc461, (*j*) sc341, (*k*) sc70, (*l*) sc361, (*m*) T6-28, (*n*) sc662, (*o*) sc688, (*p*) sc738, (*q*) sc8, (*r*) cv3, (*s*) cc8. Scale approximate.

FIGURE 6.19. Large pestles from Cerro Juanaqueña. (*a*) Large formal pestle (T167-34a and b). (*b*) Large informal pestle (sc322).

TABLE 6.11. Large Expedient Unshaped Pestles from Cerro Juanaqueña Surface Contexts.[1]

	L (cm)	W (cm)	T (cm)	Material	Condition	Notes
T297 sc322	34	15	13	Local basalt	Complete	Possibly used in gyratory crushers and as mano
T251 Ma-71	23.2	14.8	8.3	Local basalt	Complete	Possibly used in rotary querns and as mano
T196 Ma-31	24.4	13.7	8.2	Local basalt	Complete	Possibly used in rotary querns and as mano
T181 Ma-51	20.5	9.0	8.8	Local basalt	Complete	Possibly used in rotary querns and as mano

[1] Ma enumeration used for collections made during ground stone transect sampling effort.

contexts, and they seem to be limited to a handful of sites in southern Arizona, northern Sonora, and northwestern Chihuahua. It is possible that they are related to the substantial use of early varieties of flinty maize.

Close examination of basin metates with holes worn completely through shows multiple flake scars around the perimeter of the holes and use wear on their interior rims, suggesting intentional manufacture. We believe that these metates may have been used with stone or wooden pestles as rotary querns (Hayden 1969) in order to process mesquite pods or edible roots (see discussion in Chapter 13).

Grinding Slabs

In addition to the deep basin metates, we recognized another class of expedient nether stones referred to as grinding slabs. During excavations and systematic surface recording, we found 102 whole and 114 fragmentary examples of this artifact. Grinding slabs have flat working surfaces, rather than well-defined basins. Like basin metates they are made from the local basalt, but on average their exterior dimensions and grinding surfaces are only half as large. In contrast to basin metates, very few grinding slabs were deliberately shaped. A majority of these artifacts exhibited only light to moderate wear, and most were used on only one surface. Grinding experiments indicated that it is very difficult to process seeds of domestic or wild plants on grinding slabs because of their small size and because the seeds are not easily confined on the working surface. In Chapter 13, we conclude that these artifacts were not used for routine food processing.

Large Pestles

Two very different kinds of artifacts in the Cerro Juanaqueña assemblage are referred to as pestles. Large pestles are between 20 cm and 34 cm in overall length, while small pestles are approximately 10 cm in length. In this section we describe the large pestles found on Cerro Juanaqueña.

Only one of the large pestles was a formally manufactured cylindrical pestle. It was recovered in two pieces from the excavations at Terrace 167. It is well made by pecking and smoothing (Figure 6.19 a). There is a lack of any type of wear at the break. The distal end shows substantial grinding, but little evidence of crushing. With a broken length of 27.6 cm and a maximum diameter of 6.9 cm, this pestle is long and slender enough to work in some of the bedrock mortar holes on the site (their diameters are 7, 7.5, 14, and 18.5 cm). It is similar to typical pestles forms found around throughout the Southwest and northwest Mexico.

The remaining large pestles are expedient tools, mostly unshaped, angular in cross section with light to medium grinding wear on one or two faces and significant battering and/or grinding on one or both ends (Figure 6.19 b). We identified 19 of these artifacts on Cerro Juanaqueña, and collected four for more detailed analysis (Table 6.11). These tools have wear indicating that they served as both pestles and manos, and in fact, we originally recorded some as manos. Most exhibit little or no shaping, although at least one example is shaped at both ends. Two of these unshaped pestles have damage and wear to their ends that is consistent with articulating in the pivot holes of deep basin metates or rotary querns. This damage extends over 2 to 4 cm of the end of the artifacts.

We envision that the operator placed one end of the pestle in the hole in the metate, where it pivoted, while the side of the pestle rubbed against the sloping side of the basin metate—in mano fashion—while the proximal end was moved back and forth. This system would provide a mechanical advantage and would have been effective for crushing and grinding hard, large items like mesquite beans. The dry product would likely run out the gaps between the hole and the mortar and would have been caught with a basket or other container placed underneath.

This hypothetical reconstruction accounts for the wear patterns on the ends and sides of these large, unshaped pestles. It is also consistent with the flake scars, size, positioning, worn lips, and oval shape of the holes in the basin metates. However, only 19 of these artifacts were recognized, in contrast to 30 whole basin metates with holes. If expedient large pestles were the primary tool used with these metates, we would expect more examples. One possibility is that these stone artifacts were not the primary pestle used with the basin metate/rotary querns. Instead, they may represent an experimental form that failed to perform adequately. A wooden pestle may have been the regular tool form used in the metate pivot holes. A wooden pestle may have been easier to manufacture and more effective to use for crushing than the stone version. This contrast in activity may also account for the variability in degree of wear on the internal lip of the metate holes, since a wooden pestle should cause slower wear than a stone pestle.

Small Pestles

Small pestles have an overall length no greater than 10.5 cm, and therefore they are very different in scale and function from the large pestles described above. Small pestles are of a size and form that would make them suitable to be held in one hand and used with the small mortars described below. We found six examples at Cerro Juanaqueña (Figure 6.20, Table 6.12). Four of these were carefully manufactured by pecking and grinding to obtain a symmetrical shape with a smooth finish (Figure

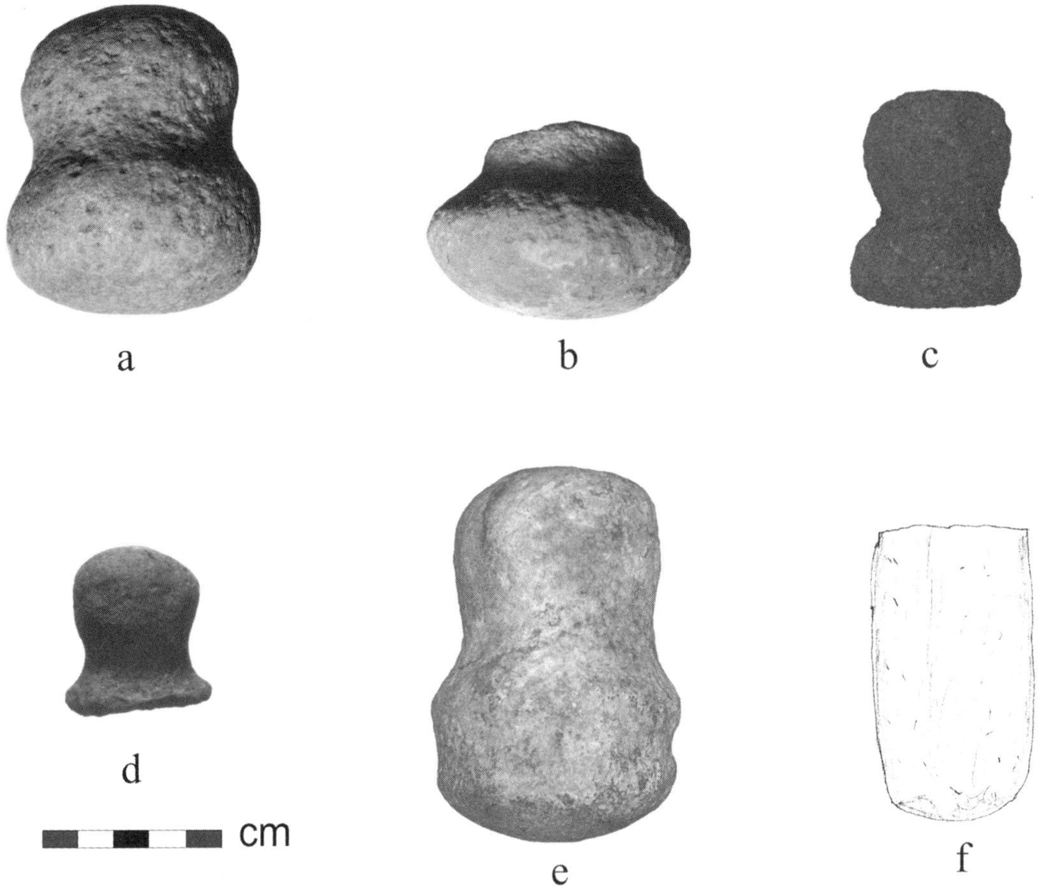

FIGURE 6.20. Small pestles from Cerro Juanaqueña. Catalogue/bag numbers: (*a*) sc237, (*b*) sc164, (*c*) sc539, (*d*) sc181, (*e*) T537-82, (*f*) sc180.

TABLE 6.12. Small Pestles from Cerro Juanaqueña.

Provenience [1]	L (cm)	W (cm)	T (cm)	Material	Complete	Notes	Figure
sc237	8.5	7.0	6.0	Local basalt	Complete	Well-made, heavily used, dumbbell shaped	6.20a
sc164	5.3+	7.6	7.6	Local basalt	Broken	Well-made, heavily used, dumbbell shaped	6.20b
T537-82	10.1	6.6	6.5	Local basalt	Complete	Well-made, grooved, dumbbell shaped	6.20e
sc539	6.1	4.7	4.7	Vesicular basalt	Complete	Not used, rough surfaces, dumbbell shaped, toy?	6.20c
sc181	4.8+	4.5	3.2	Rhyolite	Broken	Small, broken handle, well-made, heavily used, dumbbell shaped [2]	6.20d
sc180	8.2+	4.6	4.0	Nonlocal basalt	Used after broken	Unshaped, small	6.20f

"+" indicates broken in this dimension.
[1] All from surface except T537-82, which is from excavation.
[2] Item sc181 is morphologically similar to a pestle, but wear patterns indicate it is a broken handle to a mortar.

6.20 a, b, d, and e). Their distal ends are heavily ground, and a constriction in their midsections gives them a distinctive "doorknob" or "dumbbell" shape. The proximal ends are rounded and show wear from handling as well as some evidence of grinding wear. The rounded proximal ends and the constrictions increase the comfort of the tool as it is held in the user's hand. The overall size and curvature of the working ends are highly compatible with the size and shape of the stone mortars described below and these pestles were almost certainly used with them.

One of these four complete pestles (Figure 6.20 e) warrants some additional description. It has a second constriction or groove about midway down on the distal end. It also has a slight concavity on the upper portion of the proximal end that may be the result of erosion from crushing hard, small items, such as small pieces of ocher. This pestle has fresher pecking on one side perhaps to slightly modify or roughen that face during its use life.

A second small pestle from Cerro Juanaqueña (Figure 6.20 c) is quite different from the others; it is much smaller and is made from vesicular basalt. There is no evidence that it was ever used. Although it is also well made, it appears unfinished as it is much rougher overall and not as symmetrical as the three examples described above. There is no grinding wear present and all of the surfaces are rough as if it were handled only a little prior to its loss. This item is small enough to have been a child's toy. It would seem that vesicular basalt would have been a poor choice for grinding dry products in a mortar as the material would likely clog the vesicles. However, this vesicular basalt may have been easier to shape and as a toy its functional characteristics may not have been important.

A third example from Cerro Juanaqueña (Figure 6.20 f) is a simple tool made from a small, naturally shaped, cylindrical, water-worn piece of nonlocal basalt. It is broken but it continued to be used after the break as the broken end shows handling wear. Both crushing and abrading are on the narrow (3 by 2 cm) distal end which is likely too small for grinding dry material in the mortars, but this tool may have been effective for pecking and smoothing other stone tools.

The presence of only six small pestles in the assemblage contrasts with the large number of small mortars (see below). This suggests that these pestles were not in widespread use, even assuming complete pestles were carried away at the time the site was abandoned. Small pestles that are functionally equivalent to these are also reported from Early Agricultural sites in the Tucson Basin, where some are thought to have been used with small mortars to process pigments (J. Adams 1998:385–388, 2015c:141; Ferg 1998:589–595). However, those from the Tucson Basin seem stylistically different, and none exhibit the distinctive dumbbell shape found at Cerro Juanaqueña.

Small Mortars

Mortars are small, very carefully made nether stones that are clearly unlike the metates and grinding slabs described above. We found 70 of these at Cerro Juanaqueña, two at Cerro el Canelo, one from Cerro los Torres, and one from an aceramic artifact scatter at the base of Cerro la Virgen. We referred to the mortars as "stone bowls" at the outset of the fieldwork; however, the grinding wear present on virtually all of them shows they were made and used for grinding rather than as containers. Their small, bowl-shaped grinding surface could have held small quantities of pigments or condiments (Figure 6.21). They seem functionally equivalent to the more elaborate trays or mortars reported from Early Agricultural sites in the Tucson Basin and elsewhere in the region (J. Adams 2015c; Lumholtz 1912; Roney et al. 2016; Turnbow 2000).

Small mortars were made from a variety of raw materials, but almost half were manufactured from the local basalt found on Cerro Juanaqueña (Table 6.13). Measurable dimensions of a large sample of these artifacts ($n = 64$) are summarized in Table 6.14. They average about 18 cm in length and 5 cm in thickness (height). All have a single flat or concave working surface, with a mean depth of 2.3 cm, and almost half have a carefully manufactured lip around the outer edge of the mortar.

Almost three quarters are circular or ovoid in overall shape, and over 75% show clear evidence that the exterior sides were shaped. The exterior wall may be a gentle arc that roughly parallels the interior wall, or it may be somewhat blocky so that the exterior side is more vertical than the interior wall. Most have flat exterior bottoms (Table 6.15). The thickness of the bottom of the mortars has a normal distribution with a mean of 3.6 cm and low variability, suggesting uniformity in manufacturing these artifacts.

Interior walls are usually defined by a carefully shaped lip that slopes toward the bottom at an angle typically slightly less than 45 degrees (but ranging from 30 to 60 degrees). The bottoms are almost always concave in profile, usually with evidence of grinding. The grinding surface concavity was likely manufactured to a depth of 1 to 3 cm, based on a few of the mortars that have manufactured concave grinding surfaces but no evidence of grinding. The concave grinding surfaces reach, or come within a centimeter or two of reaching the lip. In virtually all cases the lip is intentionally rounded and rises slightly from the grinding surface to produce a low profile similar to a shallow bowl, tray, or a modern pasta bowl.

FIGURE 6.21a. Small mortars. Specimens (*a*) through (*c*) from Cerro Juanaqueña; specimen (*d*) from Cerro el Canelo. Catalogue / bag numbers: (*a*) sc488, (*b*) sc261, (*c*) T6-65, (*d*) cc32.

FIGURE 6.21b. Small mortars. Specimen (*e*) from Cerro el Canelo; specimens (*f*) through (*h*) from Cerro Juanaqueña. Catalogue / bag numbers: (*e*) cc9a,b, (*f*) sc342, (*g*) sc114, (*h*) sc315.

Three mortars are distinct and warrant separate descriptions. Specimen sc342 from the surface of Cerro Juanaqueña (Figure 6.21 f) was manufactured with a protruding handle. The handle is nearly cylindrical in cross section and is about 4.3 cm long and 4 cm in diameter. It is made from local basalt, is well-shaped, and is slightly constricted near the middle. A broken handle from another mortar is also present in this collection (Figure 6.20d). This handle is rounded and constricts close to where it was attached to the mortar and is similar to the handle on the example illustrated as Figure 6.21 f. Certainly manufacturing mortars with handles was more difficult, perhaps prone to failure.

The second unique item is sc612 (not illustrated), also from the surface at Cerro Juanaqueña. It is different in a number of respects from other mortars as it is made from a geode, a nodule of local basalt with crystals in the center. The natural bowl shape of this geode shows little evidence of shaping or other manufacturing but the interior was clearly used for grinding and the exterior

FIGURE 6.21c. Small mortars from Cerro Juanaqueña. Catalogue/bag numbers: (*i*) sc445, (*j*) sc429, (*k*) sc483, (*l*) sc437.

exhibits substantial wear from handling. It is complete and is 9 cm by 8 cm with a height of 7.7 cm.

The third unique mortar is from the surface of an apparent Late Archaic site at the base of Cerro La Virgen (Figure 6.21 q). We discovered this item while recording the site and collected it, as the mortar would be so easily lost to collectors. It is 28.7 cm long, 15.2 cm wide, and 9.5 cm high forming an elongated shape. It is manufactured from pink rhyolite. The grinding surface is concave and pecked, and it appears not to have been used after it was manufactured as no grinding wear is present. Calcite deposits on the grinding surface also suggests the artifact had been buried for an extended time. The depth of the manufactured concavity is .5 cm and the thickness of the bottom is 4.7 cm. It has a spatulate-shaped handle that is constricted at the neck and is 4 cm long by 5 cm wide by 3 cm thick. It appears the handle was also shaped by pecking as most of its surface is rough, although one side is ground flat. The sides of the mortar are ground smooth and the bottom shows no modification. It appears that this mortar was virtually complete, but unused at the time it was deposited into the archaeological record.

A distinctive attribute of 19 of the mortars (29.7%) is that they have clear evidence that at some point in their use life they were utilized as either manos or metates. This is visible in the form of a relatively flat grinding surface on the underside of the mortar. In some cases it appears that the mortar was made from a fragment of a basin metate, in other cases it was made from a mano, yet in others it is difficult to distinguish. In some examples it is likely that the mortar was used as a mano

FIGURE 6.21d. Small mortars from Cerro Juanaqueña. Catalogue / bag numbers: (*m*) sc365, (*n*) sc139, (*o*) sc108, (*p*) sc317.

FIGURE 6.21e. Small mortar. Specimen collected from aceramic site at base of Cerro la Virgen. Catalogue/bag number: (q) clv.

TABLE 6.13. Material Type of Small Mortars from Cerro Juanaqueña.[1]

Material Type	Count
Local Basalt	31
Nonlocal Basalt	14
Quartzite	1
Rhyolite	11
Volcanic Tuff	1
Unknown	6
Total	64

[1] Six additional specimens were unavailable.

TABLE 6.14. Dimensions of Small Mortars from Cerro Juanaqueña (n = 64).

	Mean (cm)	Standard Deviation	Maximum (cm)	Minimum (cm)	Valid *n*
Length	17.5	4.3	24.0	9.0	8
Width	13.8	3.0	19.2	8.2	11
Overall thickness	5.3	1.3	8.8	2.8	56
Lip thickness	1.5	0.5	2.7	0.8	34
Bottom thickness	3.6	1.1	7.1	1.0	62
Depth Wear	2.3	1.0	3.8	0.2	13

TABLE 6.15. Plan and Cross Section Shapes of Small Mortars from Cerro Juanaqueña.

Cross Section Shape	Plan Shape					
(working surface/bottom)	Circular	Irregular	Ovoid	Rectangular	Unknown	Total
Concave/convex	1	0	4	1	2	8
Concave/irregular	1	0	1	1	0	3
Concave/plano	7	4	22	6	11	50
Plano/plano	0	0	0	0	1	1
unknown	0	1	0	0	1	2
Total	9	5	27	8	15	64

FIGURE 6.22. Shaft straightener from Cerro Juanaqueña.

during or following its use as a mortar. Typically, these mano-metate mortars are more variable in overall appearance than the well-shaped, oval mortars. In many, but not all cases, the bottoms and lips of the mano-metate mortars are not as well-shaped and the exterior walls tend to be blocky and not angle upwards.

The bottoms of the mortars not only revealed evidence that some of the mortars' use lives included use as metates or manos, but also yielded evidence of how the mortars were handled and used. Two additional types of use wear patterns are visible on the bottom of the tools. This wear is visible to the unaided eye but was closely examined with a binocular microscope. One type of wear is visible as rounding of the asperities. Unlike grinding wear, where the asperities are ground to a plane, in this type of wear the asperities are rounded, yet are variable in the degree that they protrude and degree of roundness. This wear is visible on 18 mortars from the surface of Cerro Juanaqueña and tends to occur on the outer sides and bottoms of the tools. This wear appears consistent with the wear from frequent handling and could perhaps also occur if the mortar was resting on a soft surface (perhaps a hide) while it was being used. Another type of wear on the exterior of the mortars is in the form of crushing of the asperities in an uneven manner. This damage tends to be more localized and is seen only on the bottoms of the mortars, visible in six instances among the surface assemblage. This type of abrasion most likely occurred when the mortars were being used while they were on a rough or hard surface, such as a compact floor, wood, or another rock surface.

Since only five of the 64 analyzed mortars were complete, it is clear that most mortar discard events represent broken mortars at the end of their use life. Breakage could occur either by dropping or through use. Even though the mortars only average 5 cm in height, they had a tendency to fracture when the base approaches a thickness of 3 cm. It appears that the mortars were designed in a way that made them portable and easy to handle with working surfaces that were readily accessible. While a deeper design and thicker base may possibly have yielded longer use life and possibly improved product containment, these characteristics were not as important as ease of access and handling. Nevertheless, the mortars had long use-lives as curated tools that were strategically manufactured (J. Adams 2002:21). Whatever was being processed was only needed in small quantities and the tool was likely needed in multiple locations. Unfortunately, none of the mortars have embedded pigment or other substances that reveal what product was ground, as most are from surface proveniences where preservation of such evidence is poor.

Shaft Straightener

The single shaft straightener from Cerro Juanaqueña is made from volcanic tuff and was used as a mano before being made into a shaft straightener (Figure 6.22). It has

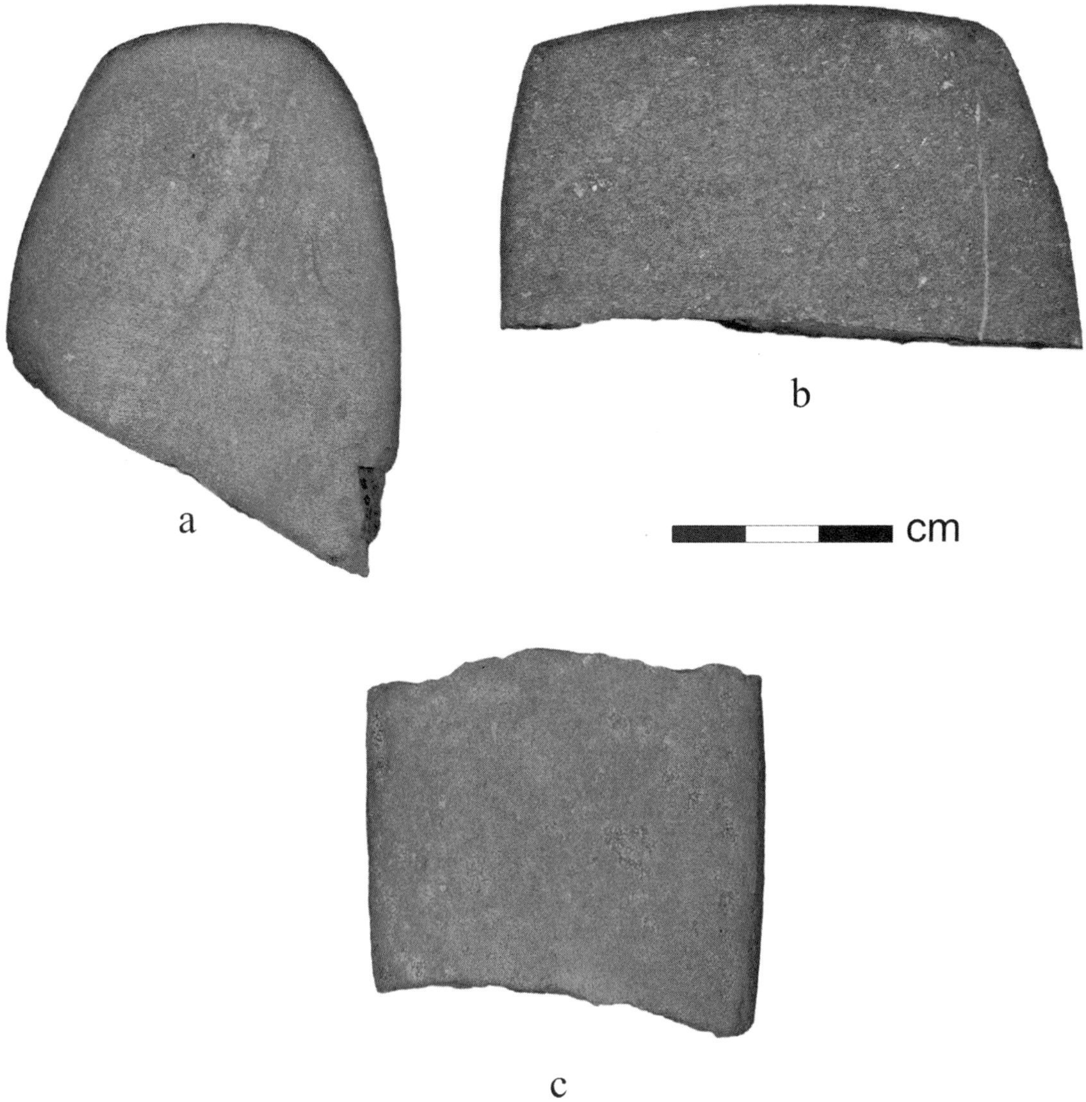

FIGURE 6.23. Shaped abraders from Cerro Juanaqueña. Catalogue/bag numbers: (*a*) sc247, (*b*) sc158, (*c*) sc545.

grinding wear on one face and battering on one end. The groove for straightening shafts is 7.1 cm long, 1.8 cm wide, and .5 cm deep. The interior of the groove is polished throughout. The tool appears to be heat-fractured.

Shaped Abraders

We found fragments of three shaped ground stone artifacts that resemble small palettes in form and manufacture (Figure 6.23). However, all three lack the concave grinding surface which is one of the hallmarks of palettes. These carefully made objects are manufactured from fine-grained stone and are about the size of a person's palm (8 cm by 6 cm or so in size and only about 1.5 cm thick). They are symmetrical, with rectilinear to oval outlines and carefully shaped edges which are rounded or squared. Two have striations on their surfaces. Although we have termed these items "abraders," we do not know their actual function.

Unshaped Abraders

Eight items are fragments of small, unshaped polishing stones from Cerro Juanaqueña; seven were found on the surface and one in excavations (Figure 6.24). These objects differ from manos in that they are smaller, the stone is finer grained, the grinding surfaces tend to be polished, and they are not pecked. All were small enough to fit in the palm of a hand (Table 6.16). Their surfaces range from flat to slightly convex. The absence of shaping and the small degree of polish on them suggests they were expedient tools used to shape or polish hard

FIGURE 6.24. Unshaped abraders from Cerro Juanaqueña. Traces of red ocher were found on item (*b*). Catalogue/bag numbers: (*a*) sc569, (*b*) R250-23. Not illustrated: sc263, sc566, sc568, sc627.

material. Two examples are illustrated in Figure 6.24. One is the only piece of ground stone from Cerro Juanaqueña on which powdered red ocher was found and it is a small, complete, unshaped, rectanguloid artifact (Figure 6.24 b; Table 6.16, specimen R250-23). This item was from excavation and certainly its burial contributed to the preservation of the ocher. The grinding wear was only along the edges of the tool and the margins of the adjacent faces. The faces themselves were not ground. The red ocher was present on one corner of one edge. The general morphology of this tool suggests it could have been used with many of the small mortars from the site.

Miscellaneous Ground Stone

The three spherical handstones have grinding surfaces but their size and shape are inconsistent with that of a mano. These objects range from 6.1 to 7.8 cm in maximum diameter and are made of local basalt or sandstone (Figure 6.25). All three have been shaped, and two of them show medium-to-heavy grinding wear on all surfaces, as well as some crushing wear. These objects would be suitable for use in the stone mortars or other applications where only a small grinding surface is needed.

Other/Unknown Ground Stone

Most items tabulated as Other/Unknown Ground Stone in Table 6.1 are fragments of ground stone tools that could not be confidently classified as either manos or metates.

Bedrock Features

Three classes of bedrock features described in Chapter 3 are also tabulated in Table 6.1. These are grinding slicks, bedrock mortars, and cupules. Because these are not portable artifacts, they are described as site features in Chapter 3. However, some of them may well have functioned in concert with ground stone artifacts described in this section. We recorded 26 grinding slicks, and six bedrock mortars. Although the cupules seem generally distinct from bedrock mortars, there is some overlap in

Table 6.16. Unshaped Abraders from Cerro Juanaqueña.

Abrader Type	Catalogue Number and Provenience	L (cm)	W (cm)	T (cm)	Material	No. of Worked Surfaces	Wear	Plan shape	Shape of Working Surfaces	Comments
Unshaped	R250-23	11.9	6.5	2.8	Shale	4	Medium grinding & polish	Rect	Plano/Plano	Red ocher present on a 1.7 × 1.5 cm area on corner edge of the tool. Grinding and pecking on edges, polish on faces.
Unshaped	sc263	7.1+	6.0	2.5	Unknown Volcanic	2	Polish and light grinding	Unk	Convex/ Convex	End battered, both faces w/ light grinding to edges, too small to be a mano, no pecking, surface more polished than a mano.
Unshaped Polishing stone	sc566 T137	7.8	3.3+	2.2	Volcanic Tuff	2	Polish	Unk	Plano/Plano	Both sides polished. Under micro-scope asperities flattened, possibly used to polish stone.
Unshaped Polishing stone	sc568 T136	5.6+	2.9	2.0	Volcanic Tuff	1	Light Polish	Rect	Plano/Plano	Slight polish on one side, similar to sc566.
Unshaped Polishing stone	sc569 T136	6.7+	3.5+	2.1	Volcanic Tuff	2	Polish	Unk	Plano/Plano	Polish on both sides. Very smooth but light wear.
Unshaped Polishing stone	sc627 R1006	5.3+	3.5	1.9+	Volcanic Tuff	1	Polish	Unk	Plano	Polishing stone, light polish on one side.
Unshaped Polishing stone	sc638 T99	8.7+	7+	2.5+	Sandstone with hematite	1	Medium grinding	Unk	Plano	Unusual small fragment from an unknown tool. Split, only one face observable.
Unshaped Polishing stone	R51-1	4.8+	6.2+	1.0	Unknown	1	Light grinding	Oval	Plano	Unusual thin grinding stone, possibly used as a polishing stone. From surface of R51.

"+" indicates broken in this dimension

FIGURE 6.25. Spherical handstones from Cerro Juanaqueña. Catalogue/bag numbers: (*a*) T6-4, (*b*) sc729.

appearance and dimensions, and some of these might also have been used in conjunction with the ground stone assemblage.

Cupules are small shallow concavities ground into boulders or bedrock. Patination suggested that the cupules can be attributed to the Archaic occupation. Petroglyphs made by pecking or scratching also occur on Cerro Juanaqueña, but most of these are attributed to post-Archaic times. Both cupules and rock art are mentioned here for consistency, but they are considered in more detail in Chapter 3.

Shaped Stone Artifacts

Stone Pipes

Fifteen mostly tubular stone pipes in various stages of completion were found at Cerro Juanaqueña. Their characteristics are summarized in Table 6.17 and six of the most complete examples are illustrated in Figure 6.26. The single complete specimen (Figure 6.26 a, specimen T167-82) was recovered during excavations. It is a superbly manufactured cylindrical tube with a perfectly circular hole bored through the center, leaving walls only 5 mm thick. It is made of soft volcanic rock, and striations running lengthwise along its exterior surface are presumably vestiges of its manufacture. They suggest that final stages of manufacture involved scraping, rather than grinding. One end tapers slightly so that the exterior diameter is only 1.4 cm and the wall thickness is less than 1 mm at its proximal end. The opposite end has a less pronounced taper and is darkened by sooting. A small hole on one side of the pipe does not penetrate to the central axis and appears to be natural rather than deliberately bored. The sediment inside the tube was matrix soil that provided no information about its use. However, sooting of the distal end suggests use as a smoking pipe.

Of the relatively complete examples, four appeared to be finished artifacts, while five were in various stages of manufacture. Degree of completion could not be determined for the remaining six pipes. Most were made of tuff, a soft volcanic stone that is easy to shape by grinding and scraping. Traces of manufacturing marks appear on two examples, indicating that they were being shaped by scraping, rather than grinding. Intense wear on at least some of the chipped stone drills found on Cerro Juanaqueña suggest that they may have been used to drill the bore through the center of these tubular pipes. All but one example were cylindrical tubular pipes with lengths ranging between 6.0 and 14.2 cm. Except for the cloud-blower described below, none had a bowl, and none of the pipe ends preserved evidence of a tenon (a narrowing of the shank designed to fit into a separate tobacco chamber).

The single non-tubular pipe is a "cloud-blower," a piece of scoria 5.5 cm in diameter and almost 4 cm long with bowl-like depressions on each end, connected by a 1.15 cm diameter hole (Figure 6.26 e, specimen sc343). This object is poorly preserved, and it is not impossible that it is a natural object. In any case, it is unlike the other examples in both shape and raw material.

Ferg (1998:596) points out that that Middle Archaic pipes in California and the southern Southwest are primarily short, conical pipes with relatively large, deep mouths, perhaps resembling the scoria specimen (Figure 6.26 e) from Cerro Juanaqueña. However, by the Cienega phase in southern Arizona, pipes were made in

FIGURE 6.26. Stone pipes from Cerro Juanaqueña. Catalogue/bag numbers: (*a*) T167-82, (*b*) sc165, (*c*) sc438, (*d*) sc293, (*e*) sc343, (*f*) sc456. Nine additional specimens not illustrated.

a wide range of sizes and shapes. J. Adams (2002:205–208, 2015c) also provides useful discussions of these artifacts. Stone pipes are now recognized at a number of Early Agricultural period sites in the southern Southwest. Vint and Mills (2016) state that pipes are rare in the southern Southwest after the Early Agricultural period.

Spindle Whorls

Until recently spindle whorls were not known from the southwestern United States and northwestern Mexico prior to AD 500, and stone disk whorls were known only in post-AD 1000 contexts (Teague 1998:47–64). However, J. Adams (2015c:131–144) reports two stone spindle

TABLE 6.17. Stone Pipes from Cerro Juanaqueña.

Specimen	Condition	Stage	Material	Length (cm)	Diameter (cm)	Tube Interior Diameter (cm)	Wall Thickness (cm)	Figure
sc10a[1]	Fragment	Unknown	Tuff	—	—	—	ca. 0.5	—
sc46	Item not collected			—	—	—	—	—
sc165	End	Blank	Tuff	—	ca. 3.20	0.63	1.20	6.26b
sc250	Fragment	Unknown	Tuff	—	ca. 2.50	ca. 1.00	0.84	—
sc293	Half	Finished	Tuff	—	2.65	0.96	0.75	6.26d
sc343	Whole?	Finished?	Scoria	3.72	5.01–5.6	1.15	2.40	6.26e
sc438	Whole	Blank	Tuff	14.20	4.16	1.30	1.41	6.26c
sc456	End	Blank	Tuff	—	3.12–4.19	n.a.	n.a.	6.26f
sc464	Whole	Blank	Tuff	6.00	3.67–4.23	n.a.	n.a.	—
sc608	End	Finished	Tuff	—	ca. 2.80	1.06	0.91	—
sc613	Fragment	Blank	Tuff	—	5.16	1.10	2.17	—
T6-8	Fragment	Unknown	Tuff	—	—	ca. 1.00	0.46	—
T6-32	Fragment	Unknown	Tuff	—	—	ca. 1.00	0.42	—
T6-170	Fragment	Unknown	Tuff?	—	—	—	0.44	—
T167-82	Whole	Finished	Tuff	9.63	2.14	1.14	0.5	6.26a

[1] Not collected. Documentation estimated from field sketch.

whorls and a spindle base from late San Pedro phase contexts at the site of Las Capas in the Tucson Basin. At Cerro Juanaqueña we found four artifacts which we believe were likely spindle whorls (Figure 6.27. Table 6.18). All were fragments preserving less than half of the circumference. This leaves some uncertainty about their overall shape, but all indications are that these were small, symmetrical stone disks with a hole drilled through their centers. Three examples were made of soft, tuff-like stone and were carefully shaped by grinding. These ranged between an estimated 2.8 and 5.0 cm in diameter and were 6 to 7 mm thick, with a 3 mm diameter hole drilled through their center (Table 6.18). Two examples are slightly biconvex in cross section, with a rounded edge and a thickness at the edge that is only slightly less than thickness at the center. The third example was plano-convex in cross section. One face was essentially flat, while the other was convex with a carefully ground 0.6 cm wide bevel around its edge. The fourth example was more fragmentary and not available for close examination, but our scaled photo shows it to be convex on one side with carefully rounded edges and thicker at the perimeter than the center. The central hole, if one was present, has been completely broken away, and if complete the disk would have been about 6.75 cm in diameter, somewhat larger than the better preserved examples. The two spindle whorls reported from the Tucson Basin had incised decorations on their faces, a characteristic not observed on any of the examples from Cerro Juanaqueña.

These artifacts should not be confused with the "perforated disks," "stone rings," or "doughnut stones" discussed by J. Adams (1998) in her description of artifacts from the Tucson Basin. Nor do these specimens seem to correspond with artifacts that B. Huckell (1995) describes as perforated stone rings, stone discoidals, or ground stone disks. In comparison, the specimens from Cerro Juanaqueña are smaller and more delicately made. Ferg (1998) discusses "disks" reported from various early agricultural sites in southern Arizona and Sonora. However, the artifacts he describes were first flaked to shape and were then ground to varying degrees. They also lack a central perforation.

Bone Tools

Bone was well-preserved at Cerro Juanaqueña, and the analyzed faunal assemblage from this site is one of the largest in the Southwest (Chapter 11). Despite this, very few bone tools were found (Figure 6.28).

Awls

The tips of four bone awls were recovered, three from excavations and a fourth from a rodent burrow during surface artifact recording (Figure 6.28). Overall dimensions could not be determined, because the artifacts are all incomplete, but the tips are about 1.7 to 2.0 cm in length and about .5 cm in width. All were manufactured from medium-sized mammal long bone shafts. A fifth small tip (not illustrated) may be an awl or a needle made from an unidentifiable bone, it is similar in form to other four but slightly smaller in diameter and 1.1 cm long. A sixth largely complete bone tool was also found in excavation, but lacks a sharp point, it may be another awl or perhaps a weaving tool (Figure 6.28 e).

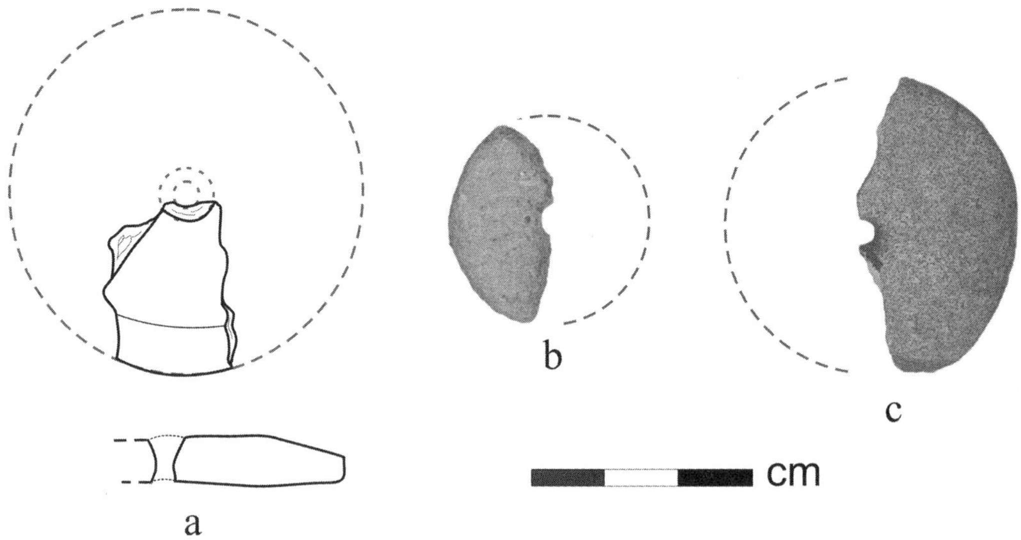

FIGURE 6.27. Spindle whorls from Cerro Juanaqueña. Catalogue/bag numbers: (*a*) sc55, (*b*) sc575, (*c*) sc452. Specimen T175-33 not illustrated.

TABLE 6.18. Stone Spindle Whorls from Cerro Juanaqueña.

Specimen Number	Material	Diameter (cm)	Thickness (mm)	Hole Dia. (mm)	Cross Section
sc55	Tuff	5	6	3–4	Plano-convex, with .6 cm wide bevel
sc452	Tuff	4	7	3	Biconvex
sc575	Tuff?	2.8	7	3	Biconvex
T175-33	Unk	6.75	—	—	Convex, thickened at margin, fragment is 3 × 3.2 cm

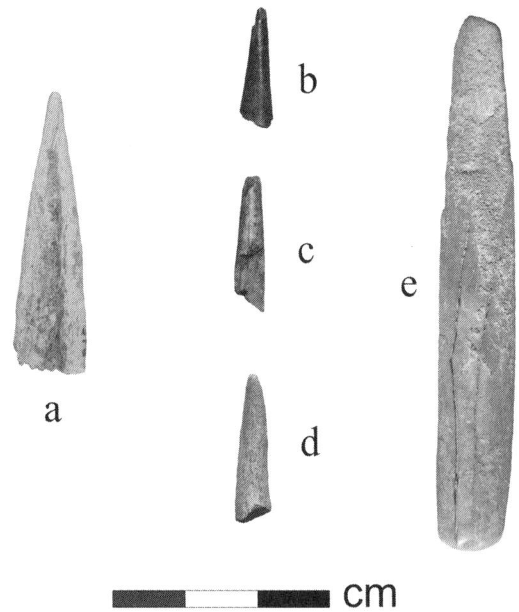

FIGURE 6.28. Bone tools from Cerro Juanaqueña. Specimens (*a*) through (*d*) are awl tips. Catalogue/bag numbers: (*a*) sc378, (*b*) T175-51, (*c*) T163-37, (*d*) T10-31, (*e*) T10-42. Specimen T163-41 not illustrated.

Fragments

Two additional fragments of highly polished large or medium-sized mammal bone were found on the surface (sc193, sc209) and a third was recovered from excavations (Terrace 87). It is clear that these are fragments of a worked bone artifacts, but they are too incomplete to interpret further.

Ornaments

Pendant

One of the few items of jewelry found at Cerro Juanaqueña was a pendant made from a segment of a large mammal long bone shaft (Figure 6.29 a). The item is 5.2 cm long and 1.5 cm wide with a 3 mm diameter hole drilled through at one end. The dorsal surface is highly polished, and both ends have been cut, ground smooth, and polished. On the ventral surface, edges of the bone walls are polished, but the interior bone surface is unfinished (Table 6.19).

A second item tentatively identified as a pendant was found and sketched during one of our early visits to the site (Figure 6.29 b). This object was made of soft, chalky stone and was shaped like a small rod, about 1 cm in di-

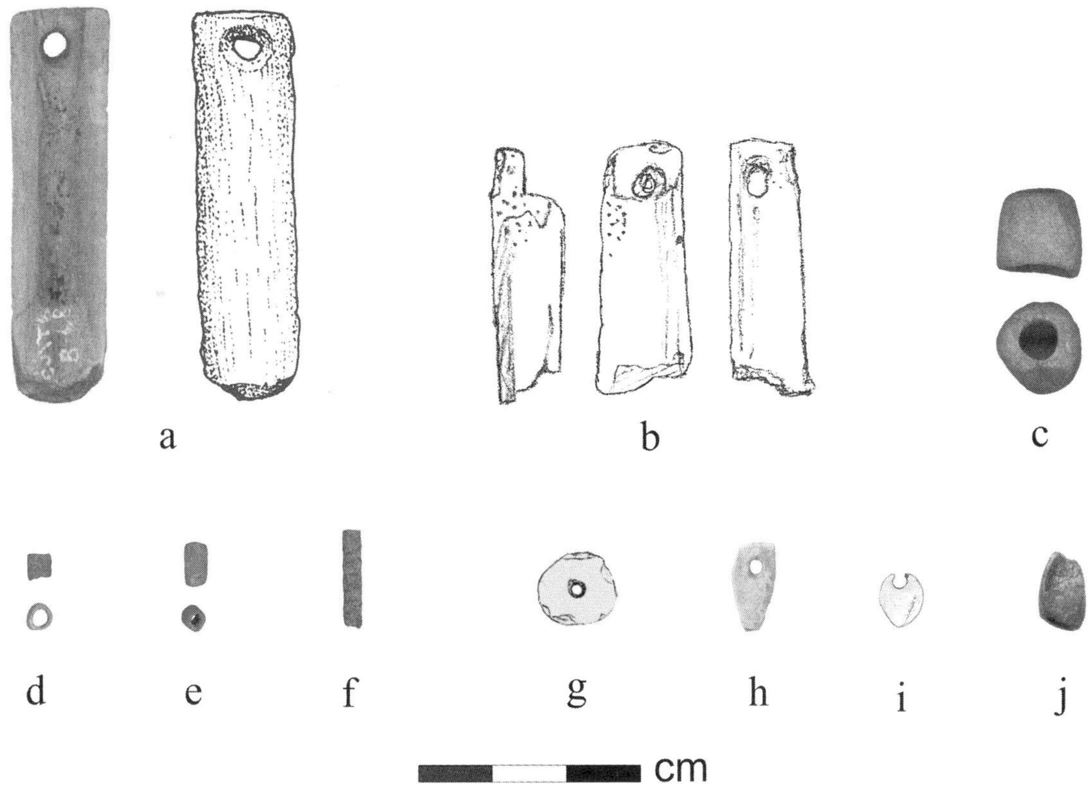

FIGURE 6.29. Ornaments from Cerro Juanaqueña. Catalogue/bag numbers: (*a*) bone pendant T6-68, (*b*) stone pendant (?) sc2, (*c*) bone bead T487-16, (*d*) bone bead T487-125a, (*e*) bone bead T487-125b, (*f*) bone bead T487-130, (*g*) shell bead T6-138, (*h*) shell bead T222-80, (*i*) shell bead T487-168, (*j*) olivella shell bead T487-145.

ameter and 3.5 cm long. However, one end was broken, so its complete original length is unknown. The remaining intact end had been notched and perforated with a small hole. We describe this item here as a pendant, but its design is unusual, and it could just as easily be part of a complex composite artifact.

Beads

Eight beads were found. These items are described in Table 6.19 and are illustrated in Figure 6.29. Four beads are made of shell and four are made of bone. All four bone beads are tubular, cut from long bones. Three of these are small, with maximum diameters of 3.6 mm or less (Figure 6.29 d–f). The fourth is much more robust (Figure 6.29 c). Three of the four shell beads are flat and thin, made of nacreous shell, either round with a hole in the center or oblong in shape with perforations that are slightly off-center (Figure 6.29 g–i). One of these (Figure 6.29 i) is probably made from wing-oyster shell (cf. *Pteria*), a marine shell from the Gulf of California. The remaining bead is a medium-sized (maximum diameter 6.6 mm) spire-lopped *Olivella* shell bead (Figure 6.29 j), also likely from the Gulf of California. Although burned, this bead remains largely intact. Its spire has been ground down to the penultimate suture.

Ceramics

Pottery was rare at Cerro Juanaqueña, and we found pottery at only seven discrete locations on the site (Table 6.20). There were 122 sherds representing 11 vessels: seven dating to historic times, three dating to prehistoric times, and one that could be either historic or prehistoric.

On the lower terraces, seven sherds of a single historic Green Glazed Ware vessel were found on Terrace 486, and an eighth sherd of the same vessel was located 20 m away on a lower terrace (Table 6.20:1). This vessel was probably a medium to large olive jar. Fournier (1997:241) has compared similar material found near El Paso, Texas, at Presidio San Elizario to Guanajuanto Green, dating from the seventeenth to nineteenth centuries.

Another cluster of likely historic period pottery was found a short distance upslope on Terraces 501 and 502. Here a total of 26 sherds were found in an area no more

TABLE 6.19. Ornaments from Cerro Juanaqueña.

Specimen Number	Material	Length (cm)	Width/Dia. (cm)	Thickness (cm)	Hole Dia. (cm)	Figure	
Pendants							
T6-68	Bone	5.2	1.5	0.4	0.3	6.29 a	
sc2[1]	Stone	3.5	1.25	1.0	0.3	6.29 b	
Beads							
T487-125b	Bone	0.61	0.36	—	0.27	6.29 e	Tubular. Both ends slightly tapered and cut smoothly perpendicular to long axis. Some polish on surfaces at end.
T487-125a	Bone	0.30	0.35	—	0.16	6.29 d	Tubular. Ends somewhat irregular and not tapered. However, some polish on surfaces and ends.
T487-16	Bone	1.11	1.13	—	0.47	6.29 c	Tubular. One end cut perpendicular to long axis, the other slightly irregular. Both ends slightly tapered and smoothed.
T487-130	Bone	1.33	0.24	—	0.13	6.29 f	Tubular. One end cut perpendicular to long axis, other end irregular. Both ends somewhat rounded and polished by wear. Exterior surface on all sides also appears polished by wear.
T487-145	Olivella Shell	1.05	0.66	0.60	—	6.29 j	Spire-lopped bead.
T 6-138	Nacreous Shell	—	1.07	0.11	0.20	6.29 g	Discoidal. Flat and thin, with irregular circular outline. Central hole drilled from one side.
T487-168	Pteria (?) Shell	0.71	0.61	0.06	0.16	6.29 i	Flat and elongated with hole drilled off-center, creating a pendant-like object. The hole has worn through to the edge of the artifact.
T222-80	Nacreous Shell	1.20	0.63	0.07	0.21	6.29 h	Flat and elongated, with hole drilled from one side towards wider end.

[1] Measurements estimated from field drawing.

than 5 m in diameter. These included 18 gray/brownware sherds from a small to medium sized jar (Table 6.20:2), and eight sherds from two different red-slipped bowls (Table 6.20:3). The red-slipped bowls are comparable to Carnue and Valle Bajo brownware bowls (M. Marshall, personal communication 2006).

The third cluster of probable historic period ceramics was on the talus in front of Terrace 167, in the upper terrace complex. It included sherds from at least three vessels. One was a large, deep, red-slipped bowl with a flat bottom, represented by 23 sherds (Table 6.20:4). A second large bowl was made of brownware and was represented by about 32 sherds (Table 6.20:5). The third vessel was a large brownware olla, represented by 27 sherds (Table 6.20:6). Within a few meters of these sherds we also recovered a small metal bowl (see below), lending credence to interpretation of these materials as historic in age. The red-slipped bowl seems comparable to Rio de Janos Red, and the brownware jar is similar to Rio de Janos Brown (Gerald 1957).

Prehistoric pottery occurred at four additional locations on the upper terraces, either as isolated sherds or pot drops. On Terrace 301b we found two sherds of a Three Circle Red-on-white bowl (Table 6.20:7). A single sherd of Pilon Red-on-brown was found on Terrace 83 (Table 6.20:10) and two probable brownware bowl sherds were found on Terrace 106 (Table 6.20:8). The final ceramic artifact was a brownware bowl sherd which could be either prehistoric or historic, found on Terrace 277 (Table 6.20:9).

TABLE 6.20. Cerro Juanaqueña Ceramic Descriptions.

1. Green glazeware jar sherds

Paste: Hard and dark brick red in color.
Temper: Fine crushed quartz.
Finish: Exterior is glazed and highly polished, with occasional streaks that have appearance of unintentional incising. Interior glazed, unpolished, with large scraping marks.
Color: Glaze on exterior and interior ranges from brown-green to olive green.
Vessel thickness: 7.0 to 8.0 mm, but up to 23.0 mm at juncture of vessel wall and base.
Rim: Not found.
Number of sherds: 8.
Comment: Wheel thrown.
Provenience: Terraces 486 and 479, lower terrace complex.

2. Gray/brownware jar sherds

Paste: Gray.
Temper: Quartz sand with some soft opaque rock or sherd.
Finish: Light to medium polish on both interior and exterior.
Color: Gray to brown.
Vessel thickness: Mostly 4.0 to 5.0 mm, but up to 6.0 mm.
Rim: Slightly everted near rim edge.
Number of sherds: 18.
Provenience: Terraces 501 and 502, lower terrace complex.

3. Red-slipped bowl sherds, two vessels

Paste: Brown, gray, or orange.
Temper: Quartz sand with some white opaque rock, soft opaque rock, and/or sherd.
Finish: Good polish on exterior and interior surfaces.
Color: Interior slipped red. Exterior unslipped buff, but red slip extends over rim onto exterior for a distance of 1.5 cm below edge, giving bowl an exterior banded appearance.
Vessel thickness: Usually 6.0 mm, but ranging between 4.0 and 7.0 mm.
Rim: Slightly incurved.
Number of sherds: 8.
Comment: Overall appearance similar to Carnue and Valle Bajo brownware bowls.
Provenience: Terraces 501 and 502, lower terrace complex.

4. Red polished deep hemispheric bowl

Paste: Tan-buff with traces of carbon streak in central wall.
Temper: Clear and opaque sand (quartz and felspar?).
Finish: Interior polished with only traces of polish streak visible. Exterior not as well polished as interior, with many polishing streaks.
Color: Interior red slipped. Exterior unslipped, tan to buff in color. Some slip drip occurs over edge of rim on exterior.
Vessel thickness: 4.5 to 9.8 mm.
Rim: Direct and rounded.
Number of sherds: ca. 23.
Comment: Does not resemble Mogollon red polished material such as San Francisco Red. Vessel bottom flat and thick.
Provenience: Terrace 167, upper terrace complex.

5. Brownware deep bowl

Paste: Reddish-tan with light carbon streak.
Temper: Clear and opaque angular sand (quartz and felspar?).
Finish: Interior polished with only traces of polish streak visible. Exterior not as well polished as interior, with many polishing streaks.
Color: Buff, with fire clouds.
Decoration: Broad zone of dark maroon/black mineral paint.
Vessel Thickness: 6.0 to 7.5 mm.

TABLE 6.20. (cont'd.) Cerro Juanaqueña Ceramic Descriptions.

Rim: Direct and rounded. Uneven outline.
Number of sherds: ca. 32.
Comment: Does not resemble late Hispanic pottery wares from Rio Grande corridor.
Provenience: Terrace 167, upper terrace complex.

6. Large brownware olla

Paste: Reddish tan with light carbon streak.
Temper: Mostly light, angular mineral. Some clear quartz-like particles and occasional organic inclusions.
Finish: Well-polished exterior with streaky polish on interior limited to 6 cm from rim.
Color: Buff with fire clouds.
Vessel Thickness: 3.9 to 7.9 mm.
Rim: Direct and rounded. Orifice flares slightly with diameter estimated to be ca. 20 cm.
Number of sherds: ca. 27.
Comment: Does not resemble late Hispanic pottery wares from Rio Grande corridor.
Provenience: Terrace 167, upper terrace complex.

7. Three Circle Red-on-white

Paste: Brownware paste. Grades from orange to white-gray in color.
Temper: Quartz sand.
Finish: Polished.
Color: Brown where not slipped. Slip white and badly eroded.
Paint: Red mineral paint.
Vessel thickness: 8.0 mm.
Rim: Not found.
Number of sherds: 2.
Provenience: Terrace 301b, upper terrace complex.

8. Brownware probable bowl

Paste: Brown with central gray carbon streak.
Temper: Coarse opaque sand.
Finish: Interior and exterior smoothed to very lightly polished.
Color: Brown.
Vessel thickness: 7.0 mm.
Rim: Not found.
Number of sherds: 2.
Provenience: Terrace 106, upper terrace complex.

9. Brownware bowl

Paste: Dark brown with gray interior edge carbon streak.
Temper: Fine opaque and clear sand.
Finish: Light interior and exterior polish.
Color: Brown.
Vessel thickness: 6.0 to 7.0 mm.
Rim: Not found.
Number of sherds:1.
Comment: Does not resemble Jornada Mogollon brownware.
Provenience: Terrace 277, upper terrace complex.

10. Pilon Red-on-brown

One sherd, not described in detail.
Provenience: Terrace 83, upper terrace complex.

Historic Artifacts

Aside from pottery, only three historic artifacts were found on Cerro Juanaqueña. One of these is the metal bowl found on the talus in front of Terrace 167, along with sherds from three ceramic vessels described above. The bowl is round, about 19 cm in diameter and 4.2 cm deep, with a small half-moon cut out from its rim (3.4 cm wide and 1.5 cm deep). Although covered with dents, this artifact was not manufactured by hand hammering. Instead it was stamped from a larger sheet of ferrous metal. The half-moon cut-out on its rim is very regular in appearance, arguing for original production rather than post-production modification. These characteristics suggest a twentieth century age (Clinton McKenzie, personal communication 2017).

A small fragment of metal was recovered during excavations on Terrace 6, in the upper terrace complex, but little can be said about its age or function. The final historic artifact collected on Cerro Juanaqueña was a brass rifle cartridge case manufactured by Remington Arms, Bridgeport, Connecticut, between 1960 and 1989 (headstamp "30-30 WIN R-P").

Artifact Assemblages from Other Cerros de Trincheras in Northwestern Chihuahua

In the course of this project we investigated 13 additional cerros de trincheras in the Río Casas Grandes and Río Santa Maria drainages. These are described in detail in Chapter 4. Cerro Juanaqueña was clearly the most intensively used cerro de trincheras, with a much higher overall density and higher absolute numbers of artifacts. However, several of the other sites also had significant assemblages. Our investigations at each of these other sites were more limited. All of the sites were mapped, and test excavations were completed at Cerro el Canelo, Cerro los Torres, and Cerro Vidal. We did not conduct systematic surface reconnaissance at any of these other sites, but did make general observations as well as collections of specific artifacts. For these reasons the discussions below emphasize particular classes of artifacts, especially those that may be time-sensitive.

Cerro el Canelo

Artifacts collected or documented at Cerro el Canelo are tabulated in Table 6.21, and a selection of formal artifacts from this site are illustrated in Figure 6.30. Although only one core was collected (Figure 6.1 a), a number of cores and hammerstones are present at this site. No attempt was made to monitor the occurrence of informal chipped stone tools, but the range of bifaces is similar to that found at Cerro Juanaqueña (Figure 6.30 q through v), and a single cruciform was collected (Figure 6.30 p). We found 17 projectile points at Cerro el Canelo, including San Pedro (Figure 6.30 a through i), Tularosa Corner-notched (Figure 6.30 j and k), Tularosa Basal-notched (Figure 6.30 l and m), and Datil (Figure 6.30 n) types, as well as one unclassified dart point (Figure 6.30 o) and a small corner-notched projectile point fragment (not illustrated).

TABLE 6.21. Summary of Artifacts Collected or Sketched at Cerro el Canelo.[1]

	Excavated Contexts		General Surface Collection/ Recording	
Cores			1	
Formal Chipped Stone Tools				
Generalized Bifaces	2		16	
Early Stage				(3)
Middle Stage				(6)
Late Stage		(2)		(7)
Projectile Points			17	
San Pedro				(9)
Tularosa Corner-notched				(2)
Tularosa Basal-notched				(2)
Datil				(1)
Miscellaneous				(1)
Fragments				(1)
Small Projectile Points				(1)
Cruciforms			1	
Ground Stone				
Small Mortars			2	
Historic Artifacts				
US Army Knife, 1904			1	

[1] Numbers in parentheses provide more detailed typological information for bifaces and projectile points.

Two small mortars were collected from Cerro el Canelo resembling those described above from Cerro Juanaqueña. The first (Figure 6.21 e) consists of two reconstructed halves of a well-shaped rhyolite oval bowl form. Its overall dimensions are 12.6 long by 11.7 wide by 4.9 cm high, and the bowl's interior diameter is 8.9 cm with a depth of 2.9 cm. The absence of grinding wear suggests it may have been abandoned or broken before it was used. The second (Figure 6.21 d) is a heavily worn basalt oval form that is shaped on all faces. It is 16 cm long by 11.2 cm wide by 6.8 cm high and the grinding surface is 2.5 cm deep.

Oval one-hand manos, basin metates, and grinding slabs are found at this site, although not in the same frequencies seen at Cerro Juanaqueña. The local bedrock at Cerro el Canelo is apparently unsuitable for use as grinding stones, and most, if not all of the ground stone

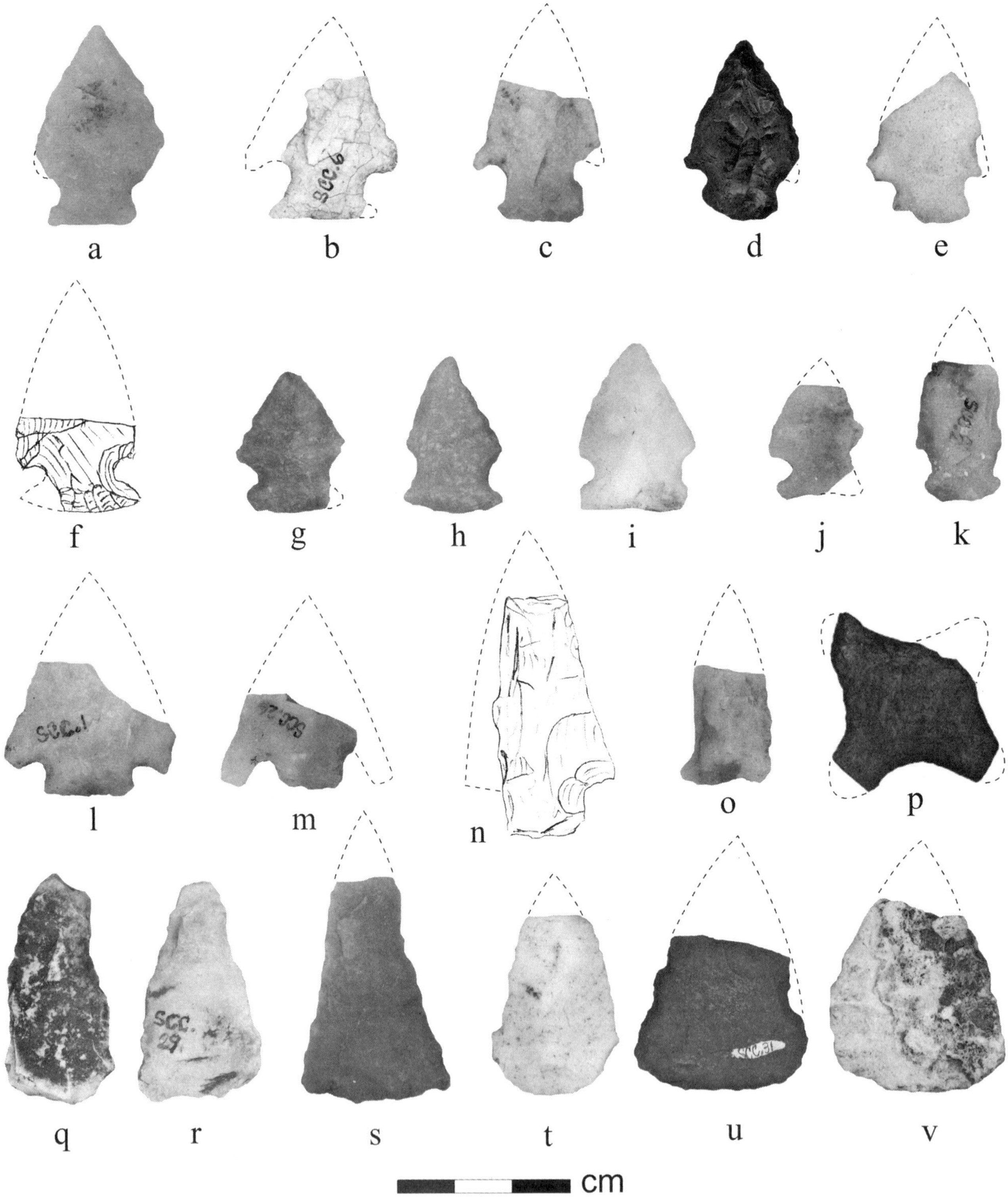

FIGURE 6.30. Selected artifacts from Cerro el Canelo: (*a*) and (*b*) San Pedro Corner-notched, straight base; (*c*) through (*e*), San Pedro Corner-notched, convex base; (*f*) through (*i*), San Pedro Side-notched; (*j*) and (*k*) Tularosa Corner-notched; (*l*) and (*m*) Tularosa Basal-notched; (*n*) Datil; (*o*) miscellaneous; (*p*) cruciform; (*q*) and (*r*) middle stage bifaces; (*s*) through (*v*), late stage bifaces. Catalogue/bag numbers: (*a*) cc14, (*b*) cc6, (*c*) cc12, (*d*) cc24, (*e*) cc20, (*f*) unnumbered specimen, (*g*) cc11, (*h*) cc4, (*i*) cc25, (*j*) cc16, (*k*) cc2, (*l*) cc1, (*m*) cc26, (*n*) unnumbered specimen, (*o*) cc23, (*p*) cc5, (*q*) cc22, (*r*) cc29, (*s*) cc3, (*t*) cc19, (*u*) cc31, (*v*) cc36.

artifacts were made of non-local stone. Cupules similar to those at Cerro Juanaqueña were recorded, and we noted at least two bedrock mortar holes.

Rock art and ceramics seem to be completely absent at Cerro el Canelo. The only historic artifact collected was a U.S. military issue mess knife, 8.5 inches (21.7 cm) long, with a carbon steel blade set in a cast aluminum handle. The inscription "R.I.A. U.S. 1904" is stamped along one side of the handle. This type of knife was only made between 1904 and 1910 at the Rock Island Arsenal, Illinois (Clinton McKenzie, personal communication 2017). We speculate that this is a relic of General John J. Pershing's Mexican Expedition of 1916 and 1917, which passed through this part of Chihuahua in pursuit of Poncho Villa.

Cerro Los Torres

Cerro los Torres had more evidence of Ceramic period use than any of the other cerros de trincheras reported here. We found 12 prehistoric sherds and at least five of the 20 projectile points collected or documented here are small points possibly dating to the Ceramic period. In addition, there are a number of petroglyphs that we attribute to the Ceramic period on stylistic grounds, and several local residents described a whole, red-slipped jar that was found in a cairn at the summit of the hill. Nevertheless, both the features and the overall assemblage at this site are consistent with those observed at other cerros de trincheras in this region, a finding that is corroborated by radiocarbon dating. Almost all of the artifacts potentially attributed to the Ceramic period were found on the summit of the hill.

Artifacts documented at Cerro los Torres are tabulated in Table 6.22 and are illustrated in Figure 6.31. Although none were collected, a number of cores made on fist-sized rhyolite river cobbles were noted. Most of the bifaces found at Cerro los Torres could be confidently identified as projectile points or projectile point fragments, although one drill tip was observed.

One-hand cobble manos and a few portable metates were also present. Most of these grinding tools are made of non-local stone. In addition to manos and metates, we found a single small stone mortar similar to others on Early Agricultural period sites and an unusual, large, shaped, vesicular basalt pestle. The pestle (ct24) is oval in cross section, about 11 cm in diameter, and appears to be the distal fragment of a large pestle. The fragment is unusual as the working end forms a distinct concavity rather than a rounded end suitable for crushing. The interior of the concavity is small with dimensions of 5.3 long by 5.0 cm wide by 1.8 cm deep. The interior and lip of the concavity show substantial crushing with minimal grinding wear. This concavity may be the product of the pestle fragment being used to crush hard, rounded material. For example, crushing small stones using repeated blows would result in the erosion and formation of such a concavity and the minimal grinding wear is consistent with this scenario. Alternatively, this concave surface may have served as a small mortar in which small, hard material was contained and crushed. However, it is smaller than other mortars in the project collections.

TABLE 6.22. Summary of Artifacts Collected or Sketched at Cerro los Torres.[1]

	Excavated Contexts	General Surface Collection/ Recording	
Formal Chipped Stone Tools			
Late Stage Bifaces	2	2	
Large Projectile Points		14	
San Pedro			(4)
Todsen			(2)
Large Unclassified			(2)
Large Fragments			(4)
Cienega Short			(1)
Guadalupe			(1)
Large/Small Fragments		2	
Small Projectile Points		4	
Desert Side-notched			(1)
Small Unclassified			(3)
Drills		1	
Ground Stone			
Small Mortar		1	
Large Pestle		1	
Ceramics			
Prehistoric		12	

[1] Numbers in parentheses provide more detailed typological information for bifaces and projectile points.

Cerro Vidal

Cerro Vidal is the final cerro de trincheras in which we excavated a test unit (Table 6.23). Abundant debitage, as well as cores and hammerstones indicate that lithic reduction was an important activity. Metates and metate fragments were rare, but expedient one-hand manos were relatively common. The 20 projectile points documented here include San Pedro Side-notched (Figure 6.32 a through c), Todsen (Figure 6.32 d), and other large forms (Figure 6.32 e, and g through k). Although we found no ceramics at Cerro Vidal, there were six small projectile points (Figure 6.32 n through s), raising the possibility of some post-Archaic activity at the site. Other chipped stone artifacts include a generalized

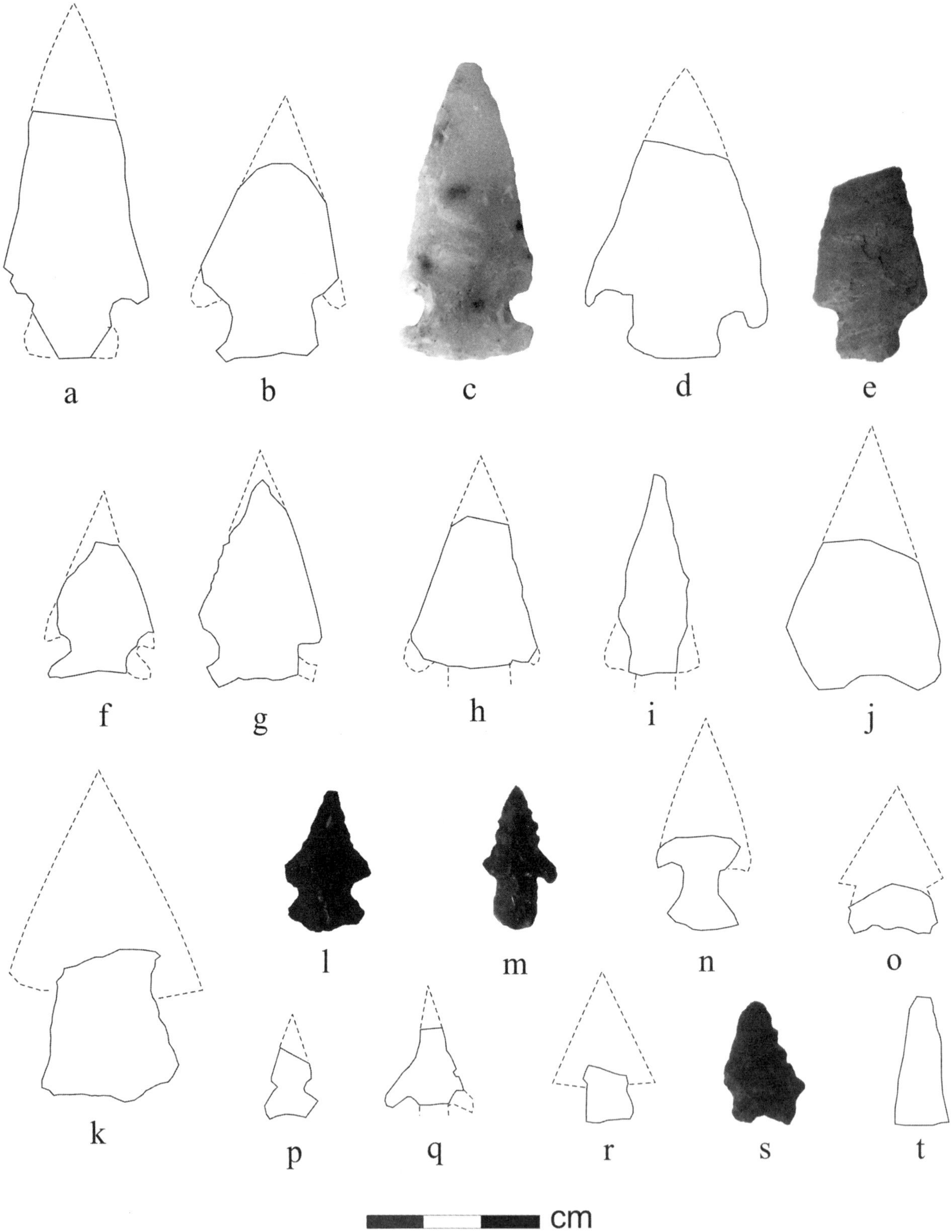

FIGURE 6.31. Artifacts from Cerro los Torres: (*a*) and (*b*) San Pedro Corner-notched; (*c*) San Pedro Side-notched; (*d*) San Pedro Basal-notched; (*e*) unclassified dart point; (*f*) and (*g*) Todsen; (*h*) through (*k*) unclassified dart points; (*l*) Cienega Short (Sliva 2015); (*m*) Guadalupe (Justice 2002: 231–237); (*n*) and (*o*) unclassified projectile points which may be either large or small; (*p*) Desert Side-notched; (*q*) through (*s*) small unclassified projectile points; and (*t*) drill fragment. Catalogue/bag numbers: (*a*) ct1, (*b*) ct2, (*c*) ct17, (*d*) ct27, (*e*) ct25, (*f*) ct7, (*g*) ct13, (*h*) ct14, (*i*) ct3, (*j*) ct12, (*k*) ct5, (*l*) ct20, (*m*) ct16, (*n*) ct4, (*o*) ct9, (*p*) ct8, (*q*) ct15, (*r*) ct10, (*s*) ct19, (*t*) ct6.

TABLE 6.23. Summary of Artifacts Collected or Sketched at Cerro Vidal.[1]

	Excavated Contexts	General Surface Collection/ Recording
Cores	1	
Formal Chipped Stone Tools		
Generalized Bifaces	2	
Early Stage		(2)
Middle Stage		
Late Stage		
Projectile Points	20	
San Pedro		(3)
Todsen		(1)
Tularosa Basal-notched		
Datil		
Miscellaneous		(4)
Fragments		(6)
Small Projectile Points		(6)
Drills	5	
Micro-eccentric	1	

[1] Numbers in parentheses provide more detailed typological information for bifaces and projectile points.

biface (Figure 6.32 f), several early stage bifaces, five drill fragments (Figure 6.32 t through w), a cruciform (Figure 6.32 m), and a micro-eccentric (Figure 6.32 l).

The micro-eccentric is a bifacially retouched obsidian flake 1.6 cm long, 1.1 cm wide, and perhaps 2 mm thick. It is roughly diamond-shaped in outline, with one projection or serration on one side and two on the other. A concentration of small obsidian flakes in the immediate vicinity of the object suggested that it could have been made and discarded in a single episode.

Micro-eccentrics are an unusual category of lithic artifacts described some years ago by John Greer (1969), but seldom reported since then. These items consist of thin, notched flakes, usually of high-grade lithic material, characterized by many narrow notches, sometimes resembling deep serrations. They are small, with lengths usually between 1 and 4 cm and are usually circular or amorphous in outline. Retouch is often restricted to the immediate area of the notches so that most of one or both faces may be free of secondary flake scars. Greer (1969) described 13 specimens from the Four Corners region of the southwestern United States. Most were found on sites with ceramic types implying dates between AD 500 and AD 900. Greer speculated that the objects could have been used as a sort of clothing or body decoration. The occurrence of one in a burial with apparent ceremonial objects suggested "a special use or significance." Sliva (1999b:38) describes an object which could be comparable from the Middle Archaic site of Los Pozos in the Tucson Basin, referring to it as a microdenticulate.

CONCLUSIONS

Systematics

Typological analysis of projectile points, as well as informal consideration of other artifact types, suggests that the assemblages fit comfortably within frameworks developed for the southwestern United States and northwestern Mexico, especially southeastern Arizona and northern Sonora. There is no compelling indication of an extra-regional origin for the material culture reported here.

In terms of systematics, we are comfortable subsuming the early occupation of Cerro Juanaqueña, Cerro el Canelo, and Cerro los Torres into the San Pedro phase, as it is currently defined in southern Arizona (Mabry 2008b; Roth 2016a:108 ff.; Vint 2015b:3). Remains documented on these three sites are slightly earlier in age than the date range usually given for the San Pedro phase, but they include the same range of projectile points and other distinctive elements of material culture such as stone pipes, small carefully shaped stone mortars, small carefully made pestles, and chipped stone cruciforms. More importantly, these sites in northwestern Chihuahua show an emphasis on cultivation of maize, association with riverine floodplains, increasing sedentism, and a growing capacity for collective action, thus paralleling similar San Pedro phase developments in southern Arizona and northern Sonora. In these respects, they contrast with contemporaneous developments in the Jornada Mogollon area to the east and in the Mogollon Rim country to the north. It is clear that the people who lived on Cerro Juanaqueña, Cerro el Canelo, and Cerro los Torres participated in a wider interaction sphere that included La Playa in Sonora and other San Pedro phase settlements in the Tucson Basin.

Our sample of material culture from the later occupation of the cerros de trincheras is much smaller, and only Cerro Vidal appears to have been occupied solely at this time (based on a single radiocarbon date). Nevertheless, we are reasonably comfortable subsuming these remains under the Cienega phase. Several of the smaller projectile point forms found at Cerro Vidal could be Cienega forms. More importantly, continued construction of large-scale cerros de trincheras and an emphasis on floodplain agriculture seem to link developments along the Río Casas Grandes to those on the Río Boquillas in Sonora and the Santa Cruz River in southern Arizona.

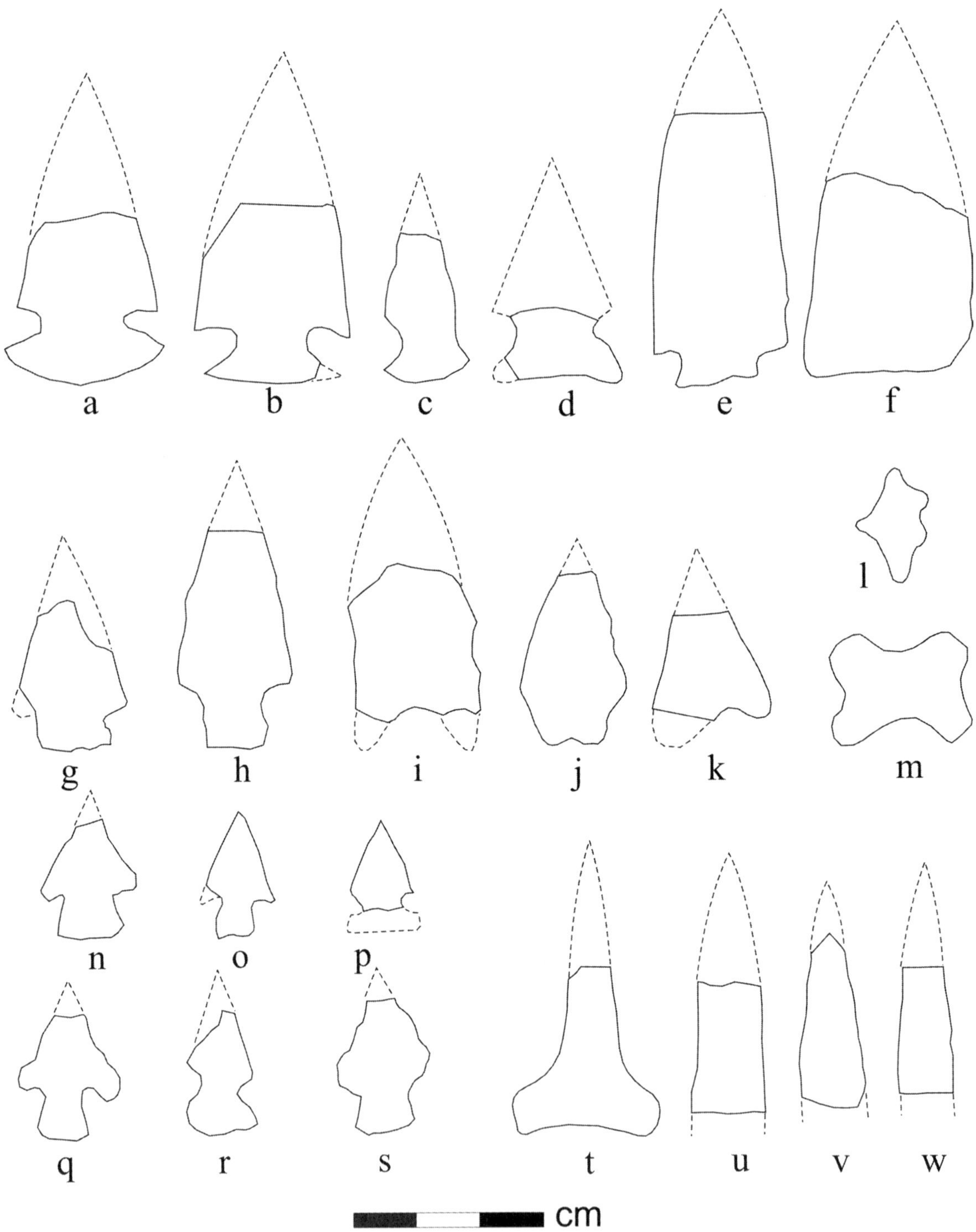

FIGURE 6.32. Artifacts from Cerro Vidal: (*a*), (*b*), and (*c*) San Pedro Side-notched; (*d*) Todsen; (*e*) unclassified large projectile point; (*f*) late stage biface; (*g*) through (*k*) unclassified large projectile points; (*l*) micro-eccentric; (*m*) cruciform; (*n*) through (*s*) small projectile points; and (*t*) through (*w*) drills. Catalogue/bag numbers: (*a*) cv25, (*b*) cv13, (*c*) cv26, (*d*) cv22, (*e*) cv18, (*f*) cv19, (*g*) cv28, (*h*) cv2, (*i*) cv1, (*j*) cv17, (*k*) cv20, (*l*) cv14, (*m*) cv3, (*n*) cv16, (*o*) cv23, (*p*) cv29, (*q*) cv4, (*r*) cv5, (*s*) cv9, (*t*) cv6, (*u*) cv12, (*v*) cv15, (*w*) cv11.

Residential Occupation

Along with information about features, macrobotanical remains, and faunal evidence, the artifacts described in this chapter are also a key element in our argument that Cerro Juanaqueña, Cerro el Canelo, Cerro los Torres, and Cerro Vidal were loci of intensive residential occupation. Cerro Juanaqueña, the most thoroughly investigated of these sites, includes strong evidence of primary lithic reduction, but also exhibits the full range of nonperishable artifacts expected in a residential site of the Late Archaic period.

Social Artifacts

These sites also include a variety of artifacts which are not obviously or directly related to subsistence activities, including stone pipes, small stone mortars and small pestles, cruciforms, and a few ornaments made of exotic shell. To this list, we might also add evidence for the use of red pigments. These items are likely related to social organization and social integration, rather than to subsistence practices. Binford (1962) referred to these as socio-technic or ideo-technic artifacts. Here, in northwestern Chihuahua and adjacent areas of the southwestern United States, they occur in the context of unprecedented population aggregation and ambitious construction projects requiring cooperation among multiple households. Although their numbers are small relative to other artifact types, these items reflect a growing concern with ritual and other social activities that promote group cohesion.

Notes

1. The system of catalogue/bag numbers uses these conventions: "sc," 8000-series, and 9000-series indicate items found on the surface of Cerro Juanaqueña, while T(errace) and R(ing) feature numbers followed by a bag number or 7000-series indicate artifacts from Cerro Juanaqueña excavations. Further details and examples include: T487-75 = Cerro Juanaqueña Terrace 487 excavation, bag no. 75; cc13 = Cerro el Canelo surface collection no. 13; ct1 = Cerro los Torres surface collection no. 1; and cv1 = Cerro Vidal surface collection no. 1.

CHAPTER 7

Warfare and Society

Introduction

The construction of each of the four major cerros de trincheras sites represents a substantial investment of energy. On the basis of a replication experiment, we estimate conservatively that at least 30 person years were required to build the terraces at Cerro Juanaqueña (Chapter 3; Hard et al. 1999), and perhaps one half to one quarter of that total was expended at Cerro el Canelo, Cerro los Torres, and Cerro Vidal. Aside from their initial construction, considerable energetic costs were required to sustain occupation in these inconvenient settings. All fuel, water, food, and most construction materials had to be carried on a daily basis as much as 160 vertical meters from floodplains at the foot of the hills. Moreover, the hilltop settings are exposed to strong winds in the spring; extreme heat in the early summer; wind, rain, and lightning during the monsoon season; and cold winds and inclement weather in the winter. Despite these costs people chose to live in these settings, and they must have had strong reasons to do so. We believe that understanding this phenomenon is an important key to understanding the social and material processes surrounding the initial diffusion of agriculture into this region.

Vignettes

Today the terraces of Cerro Juanaqueña, as well as the other sites, with their flat, arcing surfaces bordered by tumbles of dark basalt cobbles, are generally so well-preserved that they would make fine modern campsites, except for the long climb up the slope. There is sufficient room for a small tent, campfire, an outside eating area, and even a folding chaise lounge. Peering across the slope in almost any direction one sees many dozens of similar terraces, some short and wide, others long and skinny. Many articulate one to another at the same contour level, so a walk some distance around the hills is the easiest walking on the site. The terraces looking downward have a symmetry and appear stair-like. However, looking upslope the view is very different; one cannot see much above the next flowing talus of cobbles that forms the foundation of another terrace. Looking westward across the expansive floodplains of the Río Casas Grandes and Río San Pedro, there is an expanse of fields, some planted, many fallow due to current dry conditions. Several ribbons of cottonwoods thread their way down the flat floodplain demarcating the channels of both rivers as well as irrigation canals that have fallen into disuse since the spring ran dry. The small community of Colonia Oaxaca with its silver water tower and houses, some spaced among fields and others bunched along the dirt roads, lies in the mid-distance. As we considered this landscape during our hot June and July days in the field, our minds' eye flashed with alternative scenes of what Cerro Juanaqueña, the floodplain, and surrounding region would have looked like 3,200 years ago.

Fortified Village

One possibility is that the slope is covered with families and dozens of houses. As the morning sun begins to push at the shadows, fires crackle as they are flamed to life. The slow morning breeze carries the smell of smoke. Children have yet to start the games that will bring them running down the hill with laughter, but a baby cries. As women roughen their metates prior to the morning's grind, the rhythmic taps of hammerstones sound like fingers rapping incessantly on window glass. The high-pitched whish-whish of mano rubbing against metate can only be heard from the closest neighboring terraces as each woman starts grinding maize, amaranth, and whatever seeds the land offers; certainly one woman at every household starts this long task. Children are sent to the spring at the bottom of hill with gourds to fetch water.

The men inspect the distance as if studying a painting, taking in every detail of the landscape. Someone walks across the floodplain over a half kilometer away.

The distance is too great to discern facial features, but the steady stride of a man from the community below soon allows an identification to be made, bringing a moment of reassurance that no strangers can be seen moving among the brush in the distance. A constant worry looms, never far from their mind, that a raiding party may be forming, that enemies are lurking below hoping to catch a lone man or woman moving about to collect firewood, check fields, or hunt jackrabbits. Last month the community successfully repelled nighttime raiders after their stores of maize, confirming constant vigilance was needed. Soon a group of men will move off to an unoccupied terrace further down the slope to repair and refurbish dart points that could be needed soon.

Agricultural Terraces

Or are the terraces covered with maize, squash, or amaranth gardens, or perhaps plantings of agave and prickly pear with scattered small houses among them? The gardens would have benefitted from the warmer slopes above the spring while fall frosts settled in the valley below.

Seasonal Aggregation Camp

Or does Cerro Juanaqueña lie abandoned for most of the year, only to be the scene of a series of bands gathering together to enjoy the harvest of maize, amaranth, and jackrabbits, and one or two deer or pronghorn from the neighboring plains. These bands may spend a month or more trading, talking about distant territories and where pronghorn have been seen lately. Men and women look for partners as other groups of women aggregate around fires preparing the feast for the gathering. Perhaps they bring items from their home territory to trade. Or perhaps they engage in ritual dances that celebrate their identity, supernatural world, and the special meaning of this sacred hilltop.

Display

The mind reaches to conjure up the final image. Maybe a new big man has held a series of feasts and negotiated himself into a position of respect. He and his brothers and cousins now have greater informal influence over their neighbors. Bands from the nearby settlements come to visit and participate in ceremonies. They bring small donations of food to the new leader for the next feast. From a distance Cerro Juanaqueña stands out, as the new terrace walls show as garish white streaks because the freshly moved rocks still retain their rind of carbonate deposits. When a new group arrives, the leader encourages them to construct a new line of terraces for themselves along the southwest quarter of the hill. He suggests that they form a neat line and connect to each other. They will feel a sense of belonging by participating in these ceremonies hosted by the new leader. They hope he will be fair but not bother them too much to help with donations for yet another feast. They look forward to returning to their small settlement down the valley and escaping the busy slopes of Cerro Juanaqueña.

These four imaginary scenes are all distinct possibilities as well as recombined elements from different scenarios as we consider what we know about the subsistence, ground stone, chipped stone tools, paleoenvironmental context, and other basic facts about the site. We can potentially fashion multiple scenarios involving the social and political context. Three general explanations have been offered in the literature for cerros de trincheras: defense, agriculture, and display. We also considered the possibility that Cerro Juanaqueña represented a seasonal aggregation site, held a symbolic role, or offered particular environmental advantages, or a combination of factors. From the outset, the material remains at the site clearly indicated primary use as a residential site rather than a special activity locale. However, special activity use is still possible when the site was not occupied by a residential group. There also is evidence of nondomestic use of the site during the Ceramic and Historic periods that we briefly address. In this chapter, we bring together information from previous chapters, coupled with new analyses, and attempt to scrutinize each of these scenarios in turn.

Background

While these roles are presented as alternatives, it is possible that the site may have served multiple roles. The number of roles that any location plays likely grows as the number of people and duration of its occupation increases. Certainly a village likely serves a greater number of roles than a hamlet. Investigators propose multiple uses for cerros de trincheras, concluding that no single explanation accounts for all sites and that individual cerros de trincheras probably served multiple roles (Downum 2007; Fish et al. 2007a; Fish and Fish 2007; Fontana et al. 1959; Stacy 1974; Wallace et al. 2007). Roles of cerros de trincheras may change through time, and later roles may not be the same as the purpose for which a location was constructed (Hard and Roney 2007; Kowalewski 2007; O'Donovan 2002). Nelson (2007) refers to "placecrafting" as the enhancement of a location's cultural significance through, for example, expanded ritual use and monumental construction. It is also acknowledged that the function of the terraces per se is not necessarily equivalent to the role of the site (O'Donovan 2002). Presence or absence of domestic use crosscuts these alternatives. Key hypotheses relevant to this study are summarized below, while reviews of these and related issues are available elsewhere (Downum

2007; Fish et al. 2007a; Kowalewski 2007; O'Donovan 2002; Stacy 1974; Wallace et al. 2007).

Domestic settlement

Most cerros de trincheras that have been the subject of recent substantial research and excavation clearly include domestic occupations, including Cerro de Trincheras in northern Sonora (McGuire and Villalpando 2007; O'Donovan 2002), and Linda Vista, Cerro Prieto, and Tumamoc Hill in the Tucson Basin (Downum 1986, 2007; Downum et al. 1994; Fish et al. 2011; Fish et al. 2013). At these sites the presence of multiple domestic structures, as well as substantial artifact assemblages representing a range of household activities, are consistent with domestic occupation.

The numerous, large, heavily worn basin metates are the obvious material indicating that Cerro Juanaqueña served as a residential site. Other evidence for residential occupation includes house foundations indicated by rock rings, midden accumulations, and a variety of other artifacts. By "domestic settlement," we mean a location where multiple complete domestic social units reside for significant lengths of time. Indications of the domestic nature of the settlement are discussed throughout this book. We would expect a wide range of activities to occur in a domestic settlement of the scale of Cerro Juanaqueña, including those related to subsistence, as well as social, political, and religious activities among others.

Defense

Defense is often offered to explain the construction and residential occupation of cerros de trincheras. Multiple authors have argued that cerros de trincheras in southern Arizona and northern Sonora served a defensive function (Fontana et al. 1959; Hoover 1941; LeBlanc 1999; McGuire and Villalpando 2015; Sauer and Brand 1931; Wallace and Doelle 2001; Wallace et al. 2007; Wilcox 1979; Wilcox et al. 1979). Tumamoc Hill in the Tucson Basin has long been interpreted as a defensive location (e.g., Wilcox 1979). Wallace argues for Tumamoc Hill's defensive character during both the Cienega (800 BC to AD 50) and Tortolita phases (AD 550 to 650), and he suggests that its unique qualities are likely related to its use by populations from different locations as a short-term locus of aggregation for defensive purposes. However, coauthors Paul and Suzanne Fish emphasize that it served non-defensive roles, including that of ritual, but these purposes did not necessarily exclude defensive roles (Wallace et al. 2007). McGuire and Villalpando (2007, 2015) argue that Cerro de Trincheras played a defensive role, as well as that of a regional center with a ceremonial and administrative precinct. Wallace and Doelle (2001) consider Classic period warfare in the Tucson Basin and argue that the hilltop sites of Linda Vista and Rillito Peak were defensive locations based on difficulty of access and wall constructions. They maintain that Cerro Prieto also was defensive, despite the fact that much of it was constructed on lower slopes, because access is limited by large terraces and compound walls.

Downum (2007) rejects the defensive hypothesis for Linda Vista and Cerro Prieto because of their apparent vulnerability to attacks, including low walls, lower slope constructions, the presence of trails that would provide access to site interiors, and the distance between cerros de trincheras and valley settlements (also see Downum 1986, 1993; Downum et al. 1994).

The defensive refuge hypothesis considers some sites as short-term refuges used in the face of conflict, rather than long-term settlements (Fontana et al. 1959; Johnson 1963; Rice and LeBlanc 2001; Sauer and Brand 1931; Stacy 1974; Wallace and Doelle 2001; Wilcox 1979; Wilcox et al. 1979). Although they all appear to be secondary, historical accounts indicate that Native Americans indeed used hilltop sites as refuges during conflicts with neighboring groups in the protohistoric and early contact periods (e.g., Sauer and Brand 1931; Stacy 1974). However, the defensive refuge hypothesis has been challenged on multiple fronts particularly with respect to cerros de trincheras which have clear domestic components (Downum 1986, 1993, 2007:114; Downum et al. 1994; Fish and Fish 2007; Kowalewski 2007; McGuire and Villalpando 2007). Few recent studies, however, have followed Stacy (1974) and tackled the difficult problem of analyzing archaeological data from cerros de trincheras with wall and terrace complexes but with sparse evidence of substantial domestic use.

Agricultural Terraces

Some terraces at Tumamoc Hill, Cerro Prieto, Linda Vista, Cerro de Trincheras and other locations are thought to have served as agricultural terraces (Downum 2007; Downum et al. 1994; Fish and Fish 2007; Fish et al. 1984; Huntington 1914; Katzer 1993; McGuire and Villalpando 2007; Stacy 1974; Wallace et al. 2007). Evidence for use as agricultural terraces includes their geometry, northerly exposure, elevation above cold air drainages, presence of winter moisture, and the presence of maize pollen (Downum et al. 1994; Fish et al. 1984). Dooryard garden crops in terraces may have included prickly pear, agave, yucca, as well as maize (Fish et al. 1984). Stacy (1974) generated hypotheses considering the soil and other microenvironmental conditions that could have made terraces in the Baboquivari Valley favorable for agriculture.

Wallace points out that the stratigraphy behind terraces on Tumamoc Hill is incompatible with an agricultural function. Terrace fill sequences indicate sedimentation was a product of natural runoff rather than intentional construction. Furthermore many walls are constructed at angles and positions that are inconsistent with an agricultural role (Wallace et al. 2007:72–73). Many of the walls and other constructions at Black Mountain, Tumamoc Hill, Linda Vista, and Rillito Peak are constructed with orientations and positions inconsistent with that of an agricultural function (Wallace and Doelle 2001; Wallace et al. 2007:72–73). Sauer and Brand (1931) reject an agricultural function of Sonoran cerros de trincheras based on the generalized nature of terrace sediments and their lack of potential advantages relative to farming floodplains and flatlands. However, some do agree that the gentle lower slopes of a number of cerros de trincheras were indeed used for agriculture (Wallace and Doelle 2001). Were terraces at Cerro Juanaqueña built as agricultural features? We will return to this question later in this chapter.

Display

It has also been proposed that cerros de trincheras were constructed as a form of display (Haury 1962). The concept of display includes a wide range of behavior. For example, Linda Vista and Cerro Prieto are thought to have been placed in a highly visible context in order to send an aggressive symbolic and political message regarding territorial control and political power (Downum 2007). Some cerros de trincheras are thought to represent a form of monumental architecture, that is a major construction project which in scale and elaboration exceeds the requirements of any practical function that it was intended to perform (Trigger 1990). Monuments may be built of durable materials, and through extraordinary size and design, their builders seek to mark the significance of places. In this way they play a role in the construction of societal memories. As noted earlier, Nelson (2007) terms this investment "placecrafting." O'Donovan (2002, 2004) proposed that domestic terraces at the late prehistoric site Cerro de Trincheras were elaborated to monumental proportions for the purpose of visual display conveying messages regarding ethnic identity, power, and social inequities, while the hilltop location itself communicates the town's importance (McGuire and Villalpando 2007). The placecrafting of hilltop sites is tied to status enhancement that occurs within complex societies, and Cerro de Trincheras represents an early phase of monumentalization (Nelson 2007).

For most investigators, monumental construction is associated with complex social and political organization (Kowalewski 2007; McGuire and Villalpando 2007; Nelson 2007). In our own cross-cultural study, summarized below, monumental architectural elements correlate with complex societies (Hard and Roney 2007). However, monumental construction is not necessarily tied to social stratification or to political hierarchy (Rosenswig and Burger 2012; Sassaman and Randall 2012; Saunders 2012). Did Cerro Juanaqueña have a display role? Certainly, the occurrence of monument building at 1300 cal BC in the Southwestern culture area would be quite unexpected. We will return to the display hypothesis below.

Symbolic/Ideological

Did Cerro Juanaqueña and the other sites serve a particular role in the ideology of its occupants? Ceramic period cerros de trincheras in Arizona, including Cerro Prieto, Linda Vista, Tumamoc Hill, and the cerros de trincheras in the Magdalena Valley, Sonora, are thought to have played important ideological roles and to have been loci of social integration (Fish and Fish 2007; Wallace et al. 2007). Stone corrals are large, unroofed enclosures that occur in distinct summit precincts and are thought to have been a part of ideological themes and ritual practices (Fish and Fish 2007). Rock art is evidence of such practices and is abundant at some cerros de trincheras, but certainly not all (Fish and Fish 2007).

Symbolic roles of cerros de trincheras are not mutually exclusive of other roles including that of defense (Fish and Fish 2007; Wallace et al. 2007). Defense and ritual can be expected to have a close dialectal relationship for multiple reasons: social integration is fundamental to warfare; fortifications have symbolic power; people ritualize war; communities seek spiritual protection in war and try to destroy the spiritual power of enemies; settlements, including fortified settlements, would be expected to have loci of ritual practices; and important places, including ritual spaces, are defended (Keeley et al. 2007; McGuire and Villalpando 2015). However symbolic utility is ultimately predicated on a location's military value (Keeley et al. 2007). Archaeologically it would not be surprising to find evidence of ritual co-occurring with fortified settlements.

Seasonal Aggregation

Were the Río Casas Grandes cerros de trincheras unique places that served as seasonal aggregation locations where multiple groups gathered and shared in ceremonial and/or economic activities such as feasting, obtaining mates, sharing information about resources, and trading? Such events may have occurred on a seasonal basis and are well-known among multiple traditional societies. Roth (2016a) suggests such a scenario for Cerro Juanaqueña, although she acknowledges that the massive labor investment is inconsistent with a seasonal aggregation scenario. However, she ignores other contradictory evidence such as the hundreds of deeply worn metates found widely distributed over the site. We return to this hypothesis and others at the end of this chapter.

Environmental Conditions

Environmental possibilities are not well-developed in the literature. Were Cerro Juanaqueña and the other cerros de trincheras constructed and occupied to take advantage of particular microenvironmental factors? Possibilities could include the view itself; to monitor game; to escape from insects; improved temperature, solar, or wind conditions; or avoidance of floods. These possibilities are evaluated later in this chapter.

Nondomestic, special activity location

This category of use includes multiple activities by people visiting the site when it was not occupied by the residential group, such as plant collecting, hunting, ritual use, lookout, etc. Stacy (1974) considered these issues for cerros de trincheras of the Baboquivari Valley in Arizona's Papagueria. They contained walls and terraces with circular features that were associated with sherds and lithics but tended not to have ground stone, ashy deposits, or obvious hearths representative of everyday living, and she therefore inferred a specialized function (Stacy 1974).

If special activities occur at the same location as residential occupation, it becomes difficult to segregate special activities from the mass of residential material. At Cerro Juanaqueña, a few artifacts from the Ceramic and Historic periods as well as petroglyphs are testimony to the occurrence of other, less visible uses.

Cultural Connections

A concern with cultural origins and affiliations of cerros de trincheras naturally engenders reference to both northern Mexico and Mesoamerica, but the wide geographic and temporal distributions of cerros de trincheras clearly do not reflect historical relationships; cerros de trincheras are not a unitary phenomenon and there is little progress in defining historical continuities (Fish et al. 2007a; Kowalewski 2007). However, some researchers suggest cerros de trincheras may represent a horizon style reflecting broad regional intercommunication (Wallace et al. 2007). Fish and Fish (2007) suggest that the post AD 1200 cerros de trincheras in Tucson and the Río Magdalena valley may be related to concepts shared across northwestern Mexico and the southwestern United States. Nelson (2007) asserts a more or less continuous distribution of cerros de trincheras between northern Mexico and the Loma San Gabriel and Chalchihuites sites in Durango, and that cerros de trincheras are a Mesoamerican-derived architectural form. We report a cluster of cerros de trincheras in southern Chihuahua in the vicinity of Parral with most dating between ca. AD 600 and 900, and one in central Chihuahua, although vast regions remain to be searched (Hard et al. 2015). Our attempts to find Early Agricultural period cerros de trincheras in southern Chihuahua identified the site of La Noria with both ca. AD 1 and AD 650 components (Hard et al. 2015:Table 8.2).

Roth (2016a) thinks that Cerro Juanaqueña is more closely tied to identity formation in northern Mexico. Not only are Early Agricultural period cerros de trincheras found in the Tucson Basin and the Upper Gila River in Arizona (Hard et al. 2016a; Roney et al. 2015b; Roney et al. 2016), but also Cerro Juanaqueña material culture shares many aspects with that of the southern Southwest as well as La Playa, Sonora. Similarities are found in the ground stone tools, cruciforms, stone pipes, and projectile points (Chapters 6, 13, and 14) forming broad continuity across northern Mexico and the southern Southwest, but with regional differences. We return to this issue in the final chapter.

While the Río Casas Grandes cerros de trincheras represent the earliest known cerros de trincheras, it is difficult to trace the diffusion or identity formation of the builders of cerros de trincheras. Although geographic continuity at some scales may be present, the available chronological data suggests clustered appearances of cerros de trincheras in time and space across the borderlands. For example, cerros de trincheras appear at 1300 to 1100 BC (Río Casas Grandes), 800 BC to AD 1 (Río Casas Grandes, Tucson Basin, and Upper Gila River), AD 100 to 700 (Upper Gila River and Tucson Basin), AD 600 to 900 (southern Chihuahua), and AD 1200 (or earlier) to AD 1450 (Río Magdalena, Altar Valley, Baboquivari Valley, and Tucson Basin). Further work will be needed to define these clusters more precisely and to evaluate the possibility that they may be important indicators of other aspects of cultural systems.

Summary

Obviously direct evidence for or against some of these alternative interpretations is readily available while other possibilities are more difficult to assess. A number of these notions are met with skepticism by researchers with different biases. Indeed, our own personal biases tend to favor notions related to infrastructure, economics, and ecology, while other investigators tend to favor symbolic and related approaches. Certainly researcher bias can influence which alternatives and evidence are emphasized. One problem we faced is the lack of understanding of the cultural context of hilltop living. While familiarity with the range of variability and contexts for a variety of kinds of settlements is a staple of anthropology, settlements on top of hills present some fundamental and perhaps unique challenges.

Cross-cultural Study

Why do people live on hills? Should hilltop settlements be considered as typical villages that just to happen to be located on hills? Some segments of our own society prefer hilltops as locations for their homes for status, views, improved climates, elite access, recreation, and personal preferences. Is this a predilection found among nonindustrial societies without motorized or other low cost transportation? Are settlements located on hills for single reasons, multiple reasons, or for no particular reason? Are the conditions that motivate the original siting of a settlement the same as the conditions that encourage continued hilltop living?

In order to answer some of these questions, we turned to the ethnographic record and conducted a cross-cultural survey of the use of hilltops (Hard and Roney 2007). The purpose of this study was not to find direct analogies with cerros de trincheras in the ethnographic literature, but rather to provide a frame of reference through cross-cultural data that would allow us to better contextualize and evaluate the plausibility of alternative explanations that are current in the cerro de trincheras literature summarized above. While cross-cultural studies cannot represent all possible scenarios that may occur in the archaeological record, they can provide a larger context for consideration of archaeological phenomenon and may provide insight into how some alternative explanations of hilltop use can be addressed. Below we summarize the study omitting many details found in the original (Hard and Roney 2007).

Methods

The goals of the cross-cultural study were to systematically examine the world's cultures for examples of societies that established residential communities on hilltops and to understand why they elected to live on hilltops. We used both the microfiche and web versions of the Human Relations Area Files (HRAF and eHRAF), examining a total of 186 cultures around the world. This sample represented more than 50% of all cultures in the HRAF files and provided a statistically representative sample of the world's cultures.

The study was conducted in two phases. The first step was to search for ethnographic groups that established domestic communities on elevated landforms, and the second was to research each case to understand the explanations for this behavior. During the search phase we only selected groups if they established a normal residential unit on an elevated landform for at least part of the year. While nonresidential use of hilltops is also important from an archaeological perspective, it was beyond the scope of this study. However, we were careful to document nonresidential use of hilltops by those groups that also made residential use of hilltops.

During the search phase of the study, we used HRAF's cultural material codes and search capacity to examine text related to settlement patterns and to search for terms related to elevated locations (e.g. hill, bluff, mountain, ridge, terrace, escarpment, mesa). Once we found evidence that a society lived on hilltops, we had to make a series of definitional decisions to evaluate whether or not it should be included in the sample. Cases had to belong to an identifiable ethnic group for a known time and place. A domestic settlement could include social units as small as a single household or as large a city. Nondomestic units such as schools, priests' quarters, or convents were excluded. We also excluded nonresidential hill uses, atypical residential locations, and rockshelters. Field houses and herding camps were considered to be special use sites rather than residential sites because they are typically utilized by a subset of the normal domestic unit. For this reason, they were also excluded.

A society was included if they normally located residential settlements on the top or side of any pronounced elevated landform, regardless of whether the location was considered high or low. However, locations described as small knolls, slight-slopes, low-rises, lower slopes of a valley, or other situations where it appeared that slopes were gentle or elevation changes were slight were not considered pronounced elevated landforms. To be accepted into the dataset we also required that specific settlements be identified on specific elevated landforms, not simply described as living "in the mountains." This criterion eliminated a number of pastoralists. We applied these search criteria to 186 cases in the HRAF and identified 42 ethnographic groups that customarily use elevated landforms for residential sites. With 22% of the world's nonindustrial societies choosing to live on hilltops, we certainly wanted to explore why such residential decisions were made.

We then returned to each of these 42 groups to ask the question, "Why did members of that society elect to live on hills?" Guided by our archaeological hypotheses related to hilltop living, we searched the full-text ethnographies in HRAF and eHRAF using material codes for settlement pattern, land use, architecture, public structures, political organization, religious practices, sacred objects, warfare, agriculture, and monuments. We specifically looked for text that referred to hills and its synonyms. In the discussion below, we use "hill" and "hilltop" as shorthand for elevated locations.

Explanations for Hilltop Living

We were interested in understanding why, in each case, people chose to live on hilltops. To answer this question, we relied on statements by investigators and their informants. Investigators' explanations were critically evaluated with corroborating evidence as needed and, in most cases, accepted. We specifically searched for multiple explanations of single cases, rather that stopping research as soon as one explanation was found. We sought firsthand observations, detailed descriptions, and situations where multiple investigators reported similar explanations. However, we did not attempt to collect data on the specific activities that occurred at these settlements, such as ritual aggregations, agricultural practices, or military activities. After eliminating two cases with insufficient information, we arrived at a set of 40 ethnographic groups with enough detail to infer the ethnographers' explanation of hilltop use (Hard and Roney 2007:Appendix 2.1). We classified the investigators' explanations into five categories. Each explanation was critically evaluated and corroborated with other details and sources as needed. In most cases we accepted the original investigators' explanations. Certainly not all cases had similar levels of information, but cases were accepted when there was sufficient oral history, historic, ethnographic, and/or archaeological information in the HRAF full-text ethnographies to achieve a basic understanding of the hilltop settlements with the material at hand; we examined literature outside of the HRAF files in only a few cases. We expect that further research will corroborate our findings, while at the same time contributing additional explanations for particular cases. The following section explains our classification system of the investigators' explanations of hilltop living.

Limited Options

We classified explanations for hilltop living as having "limited options," to signify cases where there appear to be few reasonable alternatives to settling on hills. In most cases the ethnographers drew this inference, but in a few cases, such as the Inca, we made the inference ourselves based on descriptions of topography, as well as the location and context of other settlements. In extremely steep, mountainous terrain where most settlements are on steep slopes and little bottomland is available, such an inference is justified. This classification was also used where authors clearly indicated that all other potential settlement locations were occupied, as is the case with the Thai Miao. In some other cases, hills are occupied because low lying ground floods or remains damp, such as among the Ganda of Uganda.

Defense

We were particularly careful not to accept unsubstantiated statements about defense or warfare. We required specific discussions of fortifications, raiding, or other direct evidence of the defensive nature of a location before an inference of defense was accepted. In a few instances, an elevated settlement location was attributed to warfare without any corroborating evidence, and those explanations were not accepted. However, we did accept physical evidence of defense at a hilltop location, such as fortifications and city walls, as prima facia evidence that defense was at least one of the reasons for living on hilltop locations.

Agriculture

Explanations for hilltop living were classified as related to agriculture if hill residences provided access to terraced or unterraced hill slope fields and there were statements to that effect. Sometimes both "limited options" and "agriculture" were accepted as explanations if little bottomland was available for living and hill slopes were both farmed and used as residential locations. However, "agriculture," rather than "limited options," was accepted as an explanation in cases where people could have lived on lowlands, but instead elected to live near their hill slope fields, as among the Ifugao of the Philippines.

Display

We also pursued information relating to the display and monumentality hypothesis. However, a concern with display (in the modern anthropological sense) was outside the repertoire of nineteenth and twentieth century observers, which undercut the analysis. While we could not assess whether or not the hills themselves served as elements of display, as aspects of the hypothesis predict, we could determine whether or not monuments were present. We identify monumental architecture as large public buildings (including palaces) that required major labor investments and involved centralized control or planning (McGuire 1998; O'Donovan 1997). Following O'Donovan (1997), we considered the visual impact of monumental construction. We focused on constructions that were elements of towns or other residential centers and excluded monuments that were isolated from the domestic settlement. Therefore, huge mountaintop shrines to Buddha and isolated temples were not considered.

Symbolic/Ideological

This explanatory classification was used when descriptions of spiritual associations, ethnic identification, or other ideological concerns are strongly associated with particular, occupied landforms. Examples occur among the Hopi, the Dogon of Mali, and the Ashanti of Ghana. This classification was added to our analysis after encountering some of these cases.

Environment

This classification was used whenever hill residences were described as providing access to particular resources, other than fields. Such resources could include food, water, wind protection, game lookout, healthful location, insect avoidance, etc. Examples include the Mbuti, Copper Inuit, Pawnee, and Khasi of India.

Results

The ethnographers' explanations for hilltop living are highly patterned (Table 7.1). Two explanations, "defense" and "limited options," account for 85% of the cases (34 of 40 cases) and can be further unpacked. Defense without limited options is implicated in 50% of the cases (20 cases), and limited options with no reference to defense was cited as an explanation for 20% of the cases (eight cases). Fifteen percent (six cases) were related to both defense and limited options. The remaining 15% (six cases) were related to three other conditions—agriculture, microenvironment, and symbolism. The final factor, the presence of monuments, will be addressed separately below. Breaking these results down, excluding monuments, shows that 23 cases had only a single explanation and 17 cases had these multiple explanations. Considering single and multiple explanation separately (excluding monuments) yields similar patterns as the aggregated data. Within the single explanation group, 65% (15 of 23 cases) were defense, 13% (three cases) were limited options, and 22% (five cases) were other conditions.

There are 17 cases with two to five explanations (median = 2). Defense, limited options, and access to agricultural land are each represented 11 times. Combined, defense and limited options account for 94% (16 of 17) of multi-explanation cases. Limited options co-occurs with defense six times, and five of these also include access to agricultural land. Five of 17 cases include limited options without defense, and all five include accessibility to agricultural land. Societies occupy hilltops peacefully to access living space and agricultural land, but frequently people live on hilltops in a defensive posture when there is a shortage of places to live and farm.

Environmental conditions co-occur with other conditions in 10 cases, and eight of these include defense. For the remaining two environmental cases, one includes both limited options and access to agricultural land. There are three multi-explanation cases that include symbolic explanations, and two of these include defense. One exceptional case is the Ashanti who live on hilltops for both environmental and symbolic reasons, rather than for defense or other conditions.

Twenty-eight percent of all cases (11 of 40 cases) included monumental architecture at the hilltop settlements, and all of these cases had complex social organizations. Nine examples had political monuments such as castles, palaces, or city gates, and two examples contained ritual monuments including Maori large carved posts and Siwan mosques. In nine of the 11 monumental examples, defense was a condition of hilltop living, including the Inca, Okinawans, Maori, Hadramaut,

Table 7.1. Ethnographic Sample Showing Reasons Given for Hilltop Settlement.

Cultural Group	Defensive	Limited Options	Agricultural	Environmental	Symbolic/Ideological	Monuments
Dogon	×	×	×	×	×	
Khasi	×	×	×	×		
Inca	×	×	×			×
Okinawa	×	×	×			×
Miao	×	×	×			
Buka	×	×		×		
Hopi	×		×	×	×	
Alor	×			×		
Copper Inuit	×			×		
Garo	×			×		
Pawnee	×			×		
Hadramaut	×					×
Iroquois	×					×
Korea	×					×
Kurds	×					×
Maori	×					×
Siwan	×					×
Slovenes	×					×
Andamans	×					
Eastern Toraja	×					
Iban	×					
Lau Fijian	×					
Luo	×					
Maya	×					
Shluh/Berbers	×					
Tanala	×					
Tarahumara		×	×	×		
Amhara		×				
Aymara		×	×			
Basques		×	×			
Highland Scots		×	×			
Rundi		×	×			
Ganda		×				×
Pomo		×				
Ifugao			×			
Serbs			×			
Akan (Ashanti)				×	×	×
Kuna				×		
Mbuti				×		
Tzeltal				×		
Total 40	26	14	13	13	3	11

Siwans, Koreans, Iroquois, Slovenes, and Kurds. Of the remaining two, one is due to limited options (the Ganda), and the Ashanti mentioned above.

Once hostilities ceased, five of the monumental, defensive locations were abandoned (Iroquois, Siwans, Okinawans, Maori, and Slovenes) and three remained occupied (Kurds, Hadramaut, and Koreans). The Inca continued to live on elevated locations after hostilities ceased, but this was also a case of limited options.

We only had information on the circumstances that prevailed when hilltop locations were initially occupied in two cases— the Hopi and the Icaiche Maya—and both were defensive reactions to conflict. Certainly, most ethnographers did not explore the original conditions of hilltop occupation versus ongoing conditions (i.e. primary versus secondary reasons). However, following the cessation of hostilities, 13 of the 26 defensive-related cases relocated, wholly or in part, to lower elevations. These cases include the Dogon (in part), Alor, Maori, Tanala, Siwans, Eastern Toraja, Garo, Khasi (relocated to slopes), Copper Inuit, Hopi (in part), Iroquois, Lau Fijians, and Maya of Icaiche. Substantial segments of the Dogon and Hopi populations continue to occupy elevated locations due to the access to other resources that their elevated homes provide, their symbolic value, and limited options for moving elsewhere (Dogon). Clearly, at least in a few examples, the reasons for continuing to occupy hilltops are different from the reasons that motivated their original use.

In summary, 65% of cases (26 of 40) are related to defense. A full 85% (34 of 40) can be accounted for by defense or an absence of other places to live. Clearly people without mechanical transport generally do not find the costs of hill living to be worth the benefits unless they are in need of defense or there are no other places to live. Only 15% of cases (six) are related to other conditions. These include three that can be attributed solely to environmental conditions (Mbuti seasonal honey camps, Tzeltal for improved drainage, and Kuna of Panama to avoid floods) and two (Serbs and Ifugao) to gain access to elevated agricultural fields. The Ashanti/Akan are the only case attributed exclusively to combined environmental and symbolic reasons as they had a capital on a low hill that was both imbued with symbolic meaning and elevated them above a permanently wet valley.

One of the most striking findings of this survey is the distinct absence of incidental explanations to establish settlements on hilltops. Given our own frameworks we expected to find a wide variety of explanations such as health, the view, personal preference, ritual locations, recreation, insect avoidance, solar exposure, and others. In fact, we found virtually no explanations of these types.

Information on special functions of hilltops was recorded for these 40 groups who maintained hilltop domestic settlements. In contrast to residential use, the conditions of specialized hilltop use are quite varied. Hilltops are common locations for symbolic and ritual activities including shrines, crosses, statues of Buddha, churches, temples, priests' houses, cemeteries, mausoleums, and tombs and sacred landscape elements including sacred groves and rocks. Economic activities also take advantage of elevated locations and include herding camps, field houses, hunting lookouts, schools, fortifications, defensive refuges, agricultural terraces, roads, garrisons, and markets. Apparently the cost of using hills for special activities is justifiable for many specific tasks, but not for residential living by the family and community.

Conclusions of the Cross-cultural Study

Settlement choices generally represent spatial compromises that facilitate access to environmental and social resources within particular ecological and cultural settings. In many cases living on pronounced landforms increases the cost of accessing and transporting resources. Based on this representative global sample, we suggest hilltop residential living is generally avoided due to the increased transport costs involved in moving resources to the settlement. Despite the prevalent modern use of hills, most nonindustrial people of the world choose to live on elevated landforms only when faced with an absence of available lowland settlement space or when hilltops can contribute to a defensive need. Occasionally people elect hill sites in the absence of these two conditions. Notably there are no examples of people occupying hills solely for incidental reasons such as microclimatic conditions, the view, or improved health. We found no cases of people either initially occupying hills or continuing to live on them only for symbolic or ritual reasons. In our sample, symbolic justification always co-occurs with other conditions. Keeley and his colleagues (2007) note that fortifications frequently had monumental and symbolic qualities, but defensive qualities were prioritized. We also found no cases of seasonal aggregations occupying hilltops.

When hilltops are occupied for defensive purposes, we can expect that most people will abandon their hills when hostilities cease. However, symbolic significance, along with other reasons, may play a role in continuing occupation of a place after the defensive need is passed. In this sample, people do not appear to occupy hills solely for their monumental characteristics, monumental structures, or their symbolic display qualities. Although it is not unusual for monumental constructions to occur as elements of hilltop locations, they are always a product of complex political organizations. Usually the settlements, and sometimes the monuments themselves, have defensive qualities. More frequently than not, defensive locations with monuments also are abandoned once hostilities cease. Occasionally occupation of hill sites with or without monuments continues in peacetime.

We do not suggest that this ethnographic sample is a complete inventory of all possible hilltop residential use either in the past or ethnographically. But we do suggest that all alternative explanations of hilltop residential use do not have equal probability of being correct. Archaeologists can give priority to evaluating whether or not a hilltop site is a residential location and whether or not there are alternative, lower cost locations closer to resources that would have been available for settlement. We expect that hilltop communities would encompass a similar range of activities as lowland settlements including domestic, social, political, and ritual practices. In fact, we found that these particular practices were rarely given as the motivation for hilltop living. In sharp contrast, however, special activity use of hills can be far more variable and may include an array of ritual and economic activities, and the hilltop setting may be integral to that activity.

If a hilltop site is a residential location and if alternative lowland options for settlement are available, then a defensive role should be evaluated with independent archaeological data beyond the hilltop setting itself. We might also expect that, with the cessation of hostilities, hilltop settlements may be abandoned. In a few ethnographic cases, hills continued to be occupied after hostilities ceased and the explanation for remaining at a location may be different from the original reason for settlement. These included cases with monuments and societies with strong symbolic attachments to elevated locations. We can also expect that hilltop settlements may have monumental aspects, but these are most likely to occur in the context not only of political complexity but also defense. This implication is consistent with the statement that hilltop settlements can be the source of powerful offensive and defensive cultural messages (Downum 2007; Kowalewski 2007). The cross-cultural data also clearly indicate that after eliminating the limited options alternative and defense, 15% of cases were related to other conditions. So clearly investigators must continue to evaluate other possibilities as well as alternatives that did not emerge from the ethnological review.

Archaeological Evidence

How do these findings apply to residential occupation on cerros de trincheras in northwestern Mexico? The four sites under consideration here, Cerro Juanaqueña, Cerro los Torres, Cerro Vidal, and Cerro Canelo are residential sites, not special activity locations, given ground stone tools, rock rings, and the density of domestic trash. The limited options explanation for hilltop living explored in the cross-cultural study does not apply to this archaeological setting as it is obvious there are many suitable places to reside other than hilltops in the basin and range province. Thousands of ancient people built residential pueblo and pithouse communities in valley bottom settings along the Río Casas Grandes and its tributaries in the millennia following the abandonment of the cerros de trincheras. Here we review the primary hypotheses that exist in the literature, as discussed in the beginning of this chapter, drawing on multiple lines of archaeological evidence and bringing in implications of the cross-cultural study.

Defense

Were Cerro Juanaqueña, Cerro el Canelo, Cerro los Torres, and Cerro Vidal built and occupied for defensive purposes? In order to evaluate a defensive hypothesis, we consider the specific characteristics of the cerros de trincheras that we investigated most thoroughly. LeBlanc (1999) has developed a series of criteria which he uses to assess the prevalence of raiding and warfare in the prehistoric Southwest, and in this section we will evaluate the Early Agricultural period cerros de trincheras in northwestern Chihuahua against these criteria. The strength of this approach is that LeBlanc's criteria are not post hoc arguments developed to explain selected aspects of the cerros de trincheras. We also include a few lines of evidence from the cross-cultural survey and other investigators' views on the qualities of defensive locations.

The cross-cultural data, consistent with LeBlanc (1999), indicate that defensive sites tend to be constructed on landforms with difficult access, a hallmark of the cerros de trincheras. All four of the sites are on hills that are either isolated or are elements of small isolated ranges. All have steep sides and in most cases these slopes must be climbed before reaching the constructed features (see Chapter 4 for site descriptions). Elevation and difficulty of access are key factors and relative to local alternatives. However, our ethnographic review suggests that defensible postures are characterized by high (ca. 60 to 300 m), isolated locations with difficult access, and that those locations contrast with the elevations of non-defensive hilltop residential sites (Hard and Roney 2007). Cerro los Torres is constructed on an 80 m high hill, and the other three are situated on hills that rise between 120 and 160 m above the surrounding terrain (Table 4.1). Note, however, that some sites, including Cerro Juanaqueña, also have terraces constructed on the lower slopes. Certainly these lower slope terraces have far less defensive potential; perhaps these lower slope terraces were used when perceived threat levels were lower.

These four sites all offer excellent views in 360 degrees as an element of their defensive configuration. Detailed artifact distribution data from Cerro Juanaqueña indicate that residential occupation occurred on all sides of the hill, and it is likely that this is true of the other sites as well. The combination of excellent visibility and steep slopes ensures that the inhabitants of these cerros de trincheras could not be taken unaware. Any potential attacker could be seen from some distance, and attackers could not rush the settlement or even approach it rapidly.

Another of LeBlanc's (1999) criteria is that sites were planned and laid out for defense. In the case of these four cerros de trincheras, perimeter berms are the most obvious evidence for planning and defensive layout. Although individual terraces are the basic constructions on these sites, the terraces are often contiguous, forming linear groups which we have termed macrofeatures. On the four cerros de trincheras discussed here, macrofeatures defining the exterior boundaries of the sites are very large. For example, at Cerro Juanaqueña, the macrofeature that defines the northern, eastern, and southern perimeter of the site consists of 22 adjoined terraces forming a 400 m long alignment. Where this macrofeature runs perpendicular to the hillslope it resembles other macrofeatures composed of individual terraces, but on the northern and southern sides of the site, it runs parallel to the slope and has the form of a large cobble berm, rather than a terrace. This feature is also different from most other macrofeatures in that the cobble berm extends well above terrace surfaces in many places.

A similar feature defines much of the exterior boundary of Cerro el Canelo, and Cerro los Torres is surrounded by a striking perimeter berm. Similarly, Cerro Vidal is enclosed by a perimeter berm on three sides while the fourth side consists of a near vertical slope. These berms did not consist of freestanding masonry walls, which characterize later periods of Southwestern prehistory. However, they would have offered protection to defenders standing behind them, and because of their height and the insecure footing which they offer, they would have impeded attackers. Within the sites a maze of successive macrofeatures composed of adjacent terraces would have provided alternative lines of defense.

Other investigators make observations consistent with the defensive qualities of cerros de trincheras in the Río Casas Grandes. Peregrine (2004) suggests that if an attacker needs to cross three obstacles or more the site is likely defensive. Constructions that are continuous and closed and increase the cost of ascent are likely built for defense (McGuire and Villalpando 2015). In contrast, if constructions reduce the cost of movement on the slope of the hill and are not continuous and closed, then other roles are likely such as symbolizing space or creating privacy (McGuire and Villalpando 2015). Defensive architecture is not designed so much to block attackers, but rather to slow them down, forcing them to scramble, impeding their use of weapons, and making it likely they will not escape alive. The placement of plantings, walls, fences, and houses can also create entanglements that confuse and scatter attackers so they cannot mass at focal points (McGuire and Villalpando 2015; Roscoe 2008). The arrangement of walls and terrace constructions at the Río Casas Grandes cerros de trincheras tend to be continuous and closed. Rock ring dwellings are inside encircling walls that impede movement uphill. The numerous and concentric lines of terraces, perimeter berms, and steep slopes found on the cerros de trincheras themselves create multiple obstacles and form entanglements consistent with defensive fortifications.

LeBlanc argues that larger communities have a significant military advantage over smaller neighbors. Cerro Juanaqueña may have had 200 occupants, and the other three cerros de trincheras may have included as few as 50 or as many as 150 inhabitants. Certainly, in the context of the southern Southwest in the first and second millennia BC, these communities are large.

LeBlanc (1999) also points out that rapid construction is another characteristic of many defensive sites. He argues that when small, dispersed communities are abandoned, large defensive sites are often built immediately thereafter using techniques particularly suited to rapid construction. Several lines of evidence suggest that the cerros de trincheras in northwestern Chihuahua were built rapidly, rather than growing slowly through accretion. First, a tight series of 19 radiocarbon dates from Cerro Juanaqueña, combined with other evidence, suggest that its principal period of occupation lasted only about 200 years, and potentially even less. This leaves little time for accretionary growth. Second, the perimeter berms at the four sites described here suggest that large portions of these sites did not grow through accretionary, uncoordinated construction episodes. Instead the features appear to have been planned to accommodate specific spatial needs and constructed in large-scale, coordinated episodes.

The synchronous occupation of four sites in two episodes indicates this was a regionalized, rather than local phenomenon. During the principal period of use, between 1300 and 1100 cal BC, Cerro Juanaqueña, Cerro los Torres, and Cerro el Canelo were occupied. A second occupation occurred at Cerro Vidal and on a limited number of the lower terraces of Cerro Juanaqueña at 400 cal BC to cal AD 1 (Chapter 5). These data suggest that within northwestern Chihuahua the residential occupation of cerros de trincheras was limited to brief episodes which were synchronous throughout the area. We do not see evidence of residential reoccupation for the remaining two millennia, despite the sometimes dense occupations of the Río Casas Grandes valley.

Several of LeBlanc's expectations are either not met or cannot be addressed using data from northwestern Chihuahua. None of the sites have protected domestic water supplies, a characteristic which is largely precluded by their hilltop location. However, all are adjacent to the floodplain of the Río Casas Grandes, with relatively easy access to water.

LeBlanc also suggests that smaller sites needed more defensive capabilities than larger sites, hypothesizing that smaller sites are more likely to be located in highly defensible settings. We have found no evidence of this phenomenon. Indeed, Cerro Juanaqueña is almost certainly the largest Early Agricultural period site in the region and it is among the most defensible. LeBlanc does state that during intervals of extreme warfare small sites may cease to exist. Instead, everyone resides in some form of larger, fortified community. Given the brief span of occupation at Cerro Juanaqueña and the apparently synchronous use of cerros de trincheras in this area, this scenario may well be the case during two distinct episodes in the Early Agricultural period.

Some aspects of the site at first do not appear to support a defensive hypothesis. The large group of terraces constructed on the lower western slope of Cerro Juanaqueña appear to be particularly vulnerable to attack and similar lower terraces occur on other cerros de trincheras as well. Perhaps these terraces were utilized during intervals of reduced threat as they certainly provide greater access to the floodplain's resources. Wall and terrace heights themselves do not seem high enough to deter attackers, but they do elevate defenders enabling an enhanced view of the slopes below. The concentric layout of walls, terraces, and the loose rubble of terrace taluses creates multiple obstacles to cross and would certainly slow attackers. Low stone walls may also have been supplemented with brush. For example, Apaches often enclosed their camps with fences of cactus and the sharp-pointed yucca plant for protection (Opler 1935:704).

Other criteria that cannot be satisfactorily addressed on the basis of current information pertain to site distributions. During many periods of warfare in the Southwest, settlements tend to occur in distinct clusters separated by a no-man's land which is devoid of settlements (LeBlanc 1999). Sites are clustered in order to provide mutual aid, and the empty zones between clusters are not exploited intensively because they are not secure. The cerros de trincheras in northwestern Chihuahua do form a distinct cluster as LeBlanc predicts, but settlement pattern data from the Early Agricultural period are so incomplete that we are reluctant to stress this argument. Similarly, arguments related to the sequence of site abandonment within clusters and the sequence of cluster abandonment within a region simply cannot be evaluated at this time.

We do know that line-of-site communication between many of the cerros de trincheras in northwestern Chihuahua is possible. Figure 7.1 shows all of the cerros de trincheras that we have documented, as well as the visual linkages among them. The 14 sites form three, or perhaps four, separate groups. Cerro Tascate in the north appears to be isolated,[1] and three sites in the Río Santa María drainage form a separate isolated group. Seven sites in the Río Casas Grandes drainage are definitely intervisible: Cerro Juanaqueña, Cerro la Virgen, Cerro el Canelo, Cerro la Angostura, Cerro la Fundición, Cerro los Torres, and Cerro la Cruz. Further south, Cerro la Boquilla de San Diego, Cerro Vidal, and Cerro la Tinaja also form a definite group of visually interconnected sites. Theoretically these two groups are connected by a line-of-sight between Cerro la Cruz and Cerro la Boquilla de San Diego. However, this distance stretches over 30 km, and in practical terms the sites are probably intervisible only under favorable conditions.

LeBlanc and other authors often consider poorly treated human remains as evidence of warfare. At Cerro Juanaqueña we excavated only 1/10 of 1% of the cultural deposits (52.4 m^3 excavated out of 31,000 m^3), and we did not encounter any formal burials. However, four different proveniences on the site did yield 13 small fragments of human bone, and three of the human bone fragments from two different proveniences were burned (Table 11.17). These bones may reflect poorly treated human remains, although other explanations for them are equally as likely and cannot be addressed with the small excavation sample at hand.

The large number (329) of projectile points from Cerro Juanaqueña (Chapter 6) contrasts dramatically with artifact assemblages from many other Early Agricultural period sites in the Southwest. They accompany a faunal assemblage which is dominated by rabbits, a prey species most often taken by clubs, throwing sticks, and nets rather than by projectiles. These considerations raise the possibility that the artifact assemblage at Cerro Juanaqueña, and perhaps at some of the other cerros de trincheras as well, reflects armament for raiding, warfare, and defense. A spatial analysis shows that projectile point spatial distribution is also different from all other classes of tools, being more evenly distributed across all sectors of the site (Appendix 6.2).

The sum total of observations is difficult to reconcile with the agriculture and display models or other proposed uses of cerros de trincheras. It is likely that

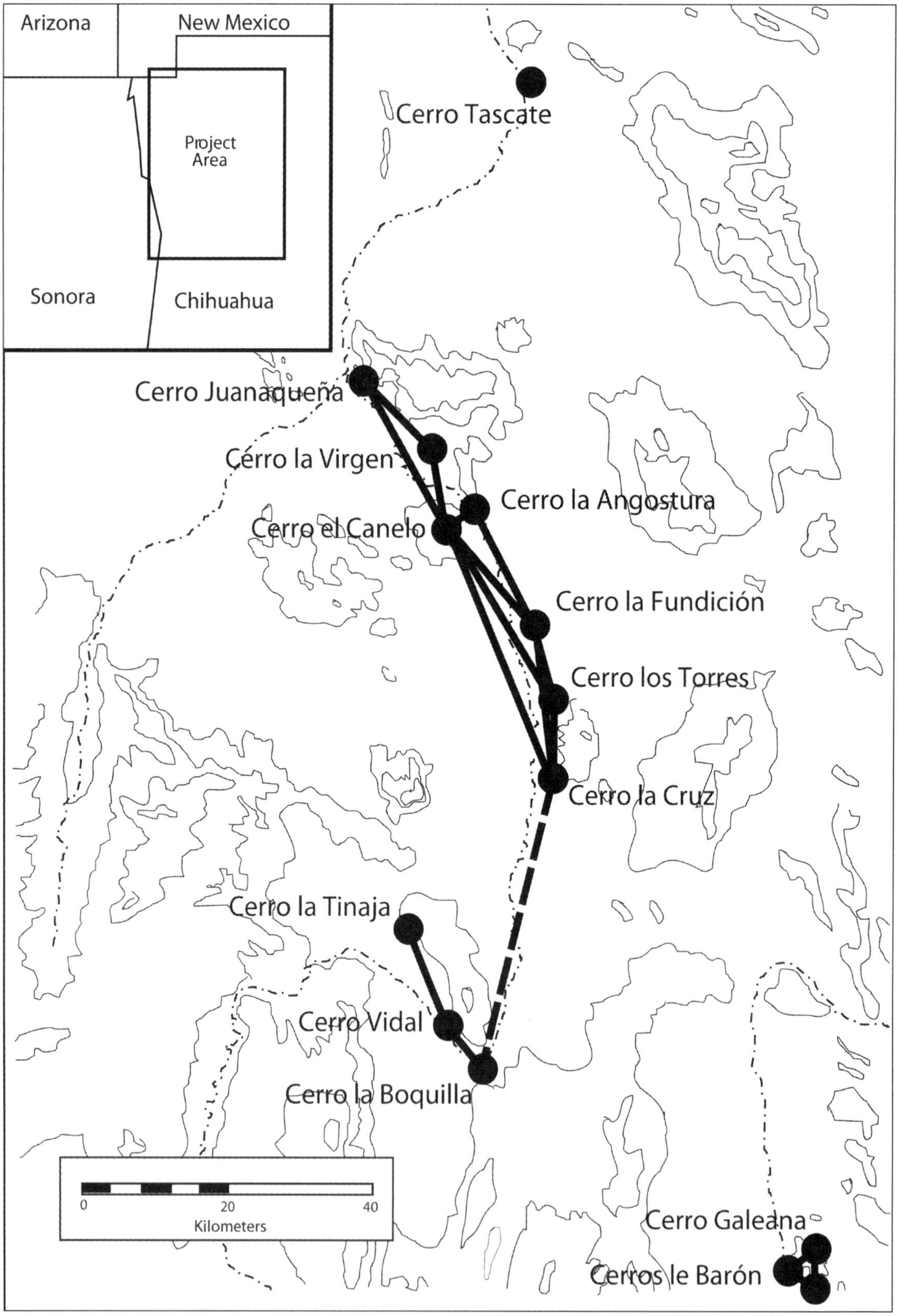

FIGURE 7.1. Map showing all northwestern Chihuahua cerros de trincheras sites and line-of-sight relationships among sites. Solid lines indicate intervisibility. Dashed line indicates theoretical intervisibility, but actual intervisibility is limited by distance.

people were willing to incur the daily inconvenience and costs of hilltop residence only when compelled to do so by periods of particularly intense raiding and warfare.

LeBlanc argues that increasing site size is another line of evidence suggesting warfare, based on the simple principal that larger communities have a significant military advantage over smaller neighbors. We suggest Cerro Juanaqueña may have contained 200 occupants (see Chapter 13). We do not have good information about settlement sizes in the times preceding 1250 BC, but few prehistorians would expect any settlements approaching the scale of Cerro Juanaqueña at this time. Extrapolating this result, the other three cerros de trincheras may have included as few as 50 or as many as 150 inhabitants. While these estimates seem extraordinary, in the Tucson Basin the San Pedro phase irrigation community of Las Capas is modeled to have contained 78 to 234 people (Mabry et al. 2008b:241–245; Vint 2015a:493). Early Agricultural period floodplain community sizes contrast sharply with models of preagricultural times, when small populations of foragers are thought to have spent much of the year dispersed into individual nuclear family units. Response to raiding and warfare is just one of several processes which can lead to population aggregation, but certainly these observations are consistent with other evidence supporting this explanation.

In summary, we argue Cerro Juanaqueña, Cerro el Canelo, Cerro los Torres, and Cerro Vidal were fortified communities. As multiple investigators have pointed out, a defensive use does not exclude the possibility that these cerros de trincheras were occupied for other additional reasons, or that functions other than residential occupation took place on them. Below we explore some of these possibilities.

Agriculture

Were the terraces at Cerro Juanaqueña built for agricultural purposes? The size of the terraces, as well as their geometry, construction, fill, and moisture availability, are inconsistent with an agricultural purpose. Extensive agricultural terrace systems were indeed built in this area particularly around AD 1200, but these late prehistoric features are very different. They were constructed along small drainages and on alluvial slopes where runoff is concentrated and where sediment can be trapped (DiPeso 1974:341; Herold 1965; Howard and Griffiths 1966). In contrast, the terraces that were built on cerros de trincheras show no concern with availability of runoff nor with soil moisture retention. Moreover, they were built in areas with as little as 334 mm of annual precipitation, which is at best marginal for dryland farming. Finally, the small potential farming space formed behind the terraces limits its productivity, particularly when compared to the cost of constructing terraces.

If all of the 3.6 ha of flat surface found behind the 550 terraces at Cerro Juanaqueña were planted in maize, the harvest would only support about six adults for a year. In the Sierra Tarahumara, for example, each person requires the production from 0.6 ha per year (Hard and Merrill 1992). Our experimental terrace and detailed volumetric data provide another way to evaluate the possibility that the terraces served as agricultural features (Chapter 3; Hard et al. 1999). We estimated the labor cost of constructing the potential planting area behind the terraces at about 4000 person days per ha. This calculation is four times or more than the estimated labor cost of constructing modern Mexican agricultural terraces as well as ancient Maya agricultural terraces (Hard et al. 1999; Turner 1983a, 1983b; Wilken 1976: 417–418). In fact, this level of investment is greater than that of the ancient Maya raised fields with estimated construction costs of between 945 and 3020 person days per ha (Turner 1983a). Certainly there is no reason to expect this level of agricultural intensification at this point in time, particularly in comparison to potential for productive farming in the floodplain below without this level of labor investment.

A number of morphological characteristics of the terraces can also be utilized to evaluate their potential use as agricultural features. Agricultural terraces are frequently built with a forward slope that serves to reduce fill requirements and therefore labor costs, but does not impact farming function. The Cerro Juanaqueña terrace surfaces are generally horizontal or have only a slight forward slope. In contrast, horizontal agricultural terraces, particularly in the arid areas of Mesoamerica and South America, are constructed with slight end-to-end gradients to facilitate water movement from one terrace to the one below (Donkin 1979; Spencer and Hale 1961). No water control features appropriate for farming are present on the site. In fact, most terraces on Cerro Juanaqueña have small parapets that impede water run-off to the terraces below and retain colluvium.

Other aspects of terrace construction would seem to minimize the agricultural potential of the terraces. A thin layer of sediment drapes rubble core fill so that the terraces have poor water retention properties, a notable issue given the aridity and high evaporation rate of the region. In comparison, the silty soils of the floodplain have superior water retention properties, as detailed by Lee Nordt in Chapter 12.

In both archaeological and ethnographic contexts in Arizona and other locations, terraces are thought to have been constructed on slopes that place them above cold air inversions of nearby lowlands and therefore extend the growing season (e.g., Downum 1993; Fish et al. 1984; Hack 1942:20). However, cool weather agricultural terraces would still require soil moisture.

In order to evaluate the potential growing season advantage of terrace versus floodplain settings, we placed a series of recording thermometers along the slope of Cerro Juanaqueña and in the floodplain (Sidebar 2.1). On average the growing season is much longer on the terraces, increasing from 177 days in the floodplain to 258 days on the upper terraces. Would the addition of around 80 frost-free days be sufficiently beneficial to warrant the cost of construction? Unlike the Sonoran Desert, the northern Chihuahua Desert has only a summer-dominant rainfall pattern. Given the poor water-holding properties of the terraces, planting dates likely would need to correlate with the arrival of summer rains to provide soil moisture for germination. When recent farmers in the community of Oaxaca take care to preserve the slight soil moisture derived from January and February rains by mulching their fields, they can plant between May 15 and June 4 rather than waiting for the summer rains, but this strategy seems unlikely on the rocky, porous terraces. Nonetheless, the elevated terraces would potentially extend the growing season beyond that of the valley bottom for about a month in the fall; that is from mid- to late October to early to late November. However, given the marginal rainfall regime for dry farming, the poor water retention properties of the terraces, high labor costs, and small size, the construction of these terraces for the purpose of farming is unlikely.

In addition, during the following two millennia apparently no other prehistoric farmers in the area selected this type of terrace as a farming strategy. As discussed in Chapter 2 and Chapter 12, the floodplain provides far superior agricultural potential. Instead, we argue that the terraces were constructed as house and living-area platforms.

Display

Were the cerros de trincheras constructed as a form of monumental display? As discussed earlier, societies that construct monuments tend to have a level of political complexity and population density that is not expected for the Early Agricultural period in northwest Chihuahua. On a more subjective note, the terrace constructions on Cerro Juanaqueña are difficult to see from as little as 5 km away. The stones of freshly constructed terraces would have had a white calcium carbonate encrustation that may have enhanced their visibility. However, the form of the constructions does not appear to enhance visibility. Vertical walls would probably have greater visibility than the rubble berms used. Nor is the perimeter berm particularly visible from below. The small diameter of the rock rings suggests that the house superstructures were of a low profile as well. In fact, the terraces and houses themselves do not meet the definition of monumental architecture as their construction size and scale do appear to be commensurate with their role as living areas and dwellings. We do not see evidence of the scale and visibility of construction nor the social and political context that would be consistent with the display hypothesis.

Symbolic/Ideological

Were Cerro Juanaqueña and related sites constructed as ritual locations? Did their important functions include ideological or symbolic roles (Downum 2007; Fish and Fish 2007)? Certainly, evaluating the symbolic role of a location in the past is difficult at best. Although it is not particularly visible, due to its size and scale Cerro Juanaqueña would have been well-known to the people of the region. We do see evidence that the site was abandoned rapidly. It appears that portable objects with remaining use life were removed, such as small stone mortars and some manos, although the heavy metates were abandoned. These observations suggest that the site was not maintained for reuse on special occasions. The site apparently did not have a long history like that seen at Hopi and some other Southwestern pueblos that relocated to defensive positions in the Historic period but continued to be occupied for centuries following the cessation of hostilities apparently for both symbolic and environmental reasons.

In terms of evidence for ritual use, a few enigmatic items and a feature could take the ritual label in the absence of functional assignment. Artifacts that are either rare or else do not relate clearly to everyday subsistence and manufacturing activities include cruciforms, small mortars, small pestles, shaped abraders, spindle whorls, and stone pipes. In addition, one piece of miscellaneous ground stone retained traces of powdered red oche, another potential indicator of ritual use. At Cerro Juanaqueña, a very small number of abstract petroglyphs and several boulders or outcrops bearing cupules are completely repatinated, implying that they pertain to Early Agricultural period occupation. Finally, the linear features outside the perimeter berm at Cerro Juanaqueña (Chapter 3) could be considered ideological in nature. They consist of two sets of parallel lines that extend eastward for at least 300 m.

One feature on Cerro Juanaqueña, Rock Ring 28, is unusual as it is rectanguloid in shape, although in size it is only slightly larger (3.5 m × 4.5 m) than other rock rings. It is constructed at the very top of the hill, on a naturally level space defined by about a dozen small terraces around its perimeter and by natural breaks in the slope. Excavation revealed no special features or particularly unusual artifact associations, although a single cruciform was recovered. A spatial analysis revealed no apparent clustering of unusual or rare artifacts here or elsewhere on the site (Appendix 6.2). This spatial analysis did reveal a higher proportion of mano fragments and projectile points and low numbers of hammerstones in this summit sector. Although ritual usage cannot be positively eliminated for Rock Ring 28 and vicinity, no compelling evidence supports such usage either.

However, at Cerro el Canelo we mapped a well-defined, outsized rock ring, 70 m in diameter, located in a saddle 650 m northeast of the main site complex. We argue in Chapter 4 that this feature is contemporaneous with the Cerro el Canelo site complex. The feature was constructed by clearing rocks from the surface of the enclosure and piling them to form a low circular berm with an opening to the southwest (Figure 4.5 and Plate 8). This feature likely played a role in social integration and presumably ritual use, and it may be the earliest such public architecture in the southwestern United States or northwestern Mexico. If we are correct, then Cerro el Canelo is consistent with the view that defense and ritual use are not mutually exclusive (Fish and Fish 2007; Wallace et al. 2007). In fact, evidence of ritual activities on sizable village settlements is not surprising. The outsized rock ring also represents a spatially separate precinct for social integrative activities, a pattern seen at later cerros de trincheras (Downum 2007; McGuire and Villalpando 2007). In addition, Medio and Historic period petroglyphs, as well petroglyphs of unknown age, are present on Cerro Juanaqueña, and may be associated with ritual usage. One Historic petroglyph depicts a church steeple with a cross and some of the Medio period rock art shows masked figures and other elements related to Casas Grandes iconography (Chapter 3).

Seasonal Aggregation

Was Cerro Juanaqueña a seasonal aggregation location? Mobile hunters and gatherers are known to periodically participate in aggregations where a number of bands will temporarily camp together for a period of time in order to participate in ceremonies, seek mates, share information, trade, and visit. The six mounds of burned rock that we documented near the summit of Cerro Juanaqueña were likely roasting features (Chapter 3) and might be cited as evidence for seasonal aggregation. Burned rock features are common on palimpsest sites in northwestern Chihuahua, but those at Cerro Juanaqueña are unusual in that they all occur in a limited area near the summit of the hill.

While the seasonal aggregation hypothesis is one that we considered and returned to a number of times, we do feel that it can be eliminated for several reasons. First, the scale of construction at Cerro Juanaqueña and other cerros de trincheras is inconsistent with relatively brief, seasonal occupation. The four cerros de trincheras which we investigated most intensively all involve significant labor investments that do not appear to have been accretionary. The construction of Cerro Juanaqueña alone required approximately 30 person-years of labor. Moreover, the layout of these sites, and especially the presence of massive perimeter berms, seem to imply planning and rapid construction, rather than accretionary growth during separate, unrelated episodes (Chapters 3 and 4).

Second, for Cerro Juanaqueña, at least, there is strong evidence for intensive occupation combined with a brief span of occupation. This is demonstrated most clearly by literally hundreds of deep, basin metates which reflect years of intensive use (Chapter 13). More subjectively, there are large amounts of ashy, charcoal-rich soil, vast quantities of lithic debitage, a phenomenal number of formal artifacts, and an immense quantity of burned animal bone, as well as scattered human remains (Chapters 6, 8, 11, and 14). Such evidence could represent a palimpsest which accumulated gradually over millennia in the course of relatively brief seasonal occupations, but a large suite of radiocarbon dates show that the principal span of occupation was only about 200 years, perhaps less (Chapter 5).

Third, there is direct biological evidence for use of the site in spring, summer, and fall (Chapters 8 and 11). The winter season is not reflected by direct biological indicators, but in Chapter 11, Kari Schmidt points out that lagomorph remains from Cerro Juanaqueña include a very low proportion of juveniles. This observation rules out the possibility that the cerro de trincheras were occupied only during the summer and fall, when juvenile numbers are at their maximum. The high ubiquity of maize at Cerro Juanaqueña may also be an indirect indication of winter-time occupation, assuming that it resulted from reliance on stored food stuffs. Evidence for year-round occupation is inconsistent with hypothesized seasonal use.

Finally, we found no comparable examples of residential use of hilltops in the ethnographic literature. Of course, this observation does not eliminate the possibility, but it does weaken its likelihood.

Environment

Were Cerro Juanaqueña and the other cerros de trincheras constructed and occupied to take advantage of particular microenvironmental factors, such as improved temperature or solar conditions, or avoidance of floods? The cerros de trincheras were not occupied to escape from seasonal flooding as there are abundant settings either on ancient alluvial terraces or adjacent to the floodplain that do not flood. Thousands of ancient people found such places to locate their settlements along the Río Casas Grandes for over two millennia without resorting to living on hilltops. Lying above the cold air that settles in the valley bottom, the slopes of Cerro Juanaqueña do have warmer daily minimum temperatures than the floodplain, which might attract occupation. However, Cerro Juanaqueña and other cerros de trincheras also have a number of negative attributes including extremely high winds in the spring, extremely high temperatures in the summer, exposure to violent storms during the monsoon, and lack of firewood in the winter. Whatever environmental benefits may accrue by hilltop living, they seem unlikely to outweigh the costs of transporting wood, water, and food up the hill.

Special Activities

Did Cerro Juanaqueña serve as a place that was visited to conduct special activities in addition to its residential role? Certainly the overwhelming domestic evidence would likely mask such evidence if it occurred during the Early Agricultural period. Several categories of archaeological remains could be interpreted as evidence of special activities, when taken individually. These include cupules, burned rock features, linear features, and several types of unusual artifacts. However, when considered in their larger, overall context, it seems most likely that these remains are elements of a relatively fixed and intensively occupied residential complex. In contrast, a number of Ceramic and Historic period artifacts and petroglyphs show that the site was visited from time to time over the millennia following its abandonment.

Artifacts attributed to the Ceramic period included six sherds of prehistoric pottery and possibly 14 small projectile points (see Chapter 6). These items were scattered widely across the upper terraces, essentially as isolated artifacts. There were no particular concentrations that might be interpreted as a discrete activity area.

Historic artifacts, on the other hand, occurred in three distinct clusters. At Terrace 167, in the upper terrace complex, there was a cluster of about 80 sherds from at least three different vessels, and nearby was a small metal bowl. On the lower terraces, a small concentration of 26 Historic period sherds representing three different vessels was found on Terraces 501 and 502. A small rectangular basalt mano was also discovered in this cluster. These finds suggest this area may have been a campsite or perhaps a seasonal residential location used by a small group, such as a family, to care for fields. The third concentration of Historic artifacts was a scatter of eight green-glazed sherds found on Terraces 479 and 486. These artifacts represent a single medium to large jar.

We also recorded petroglyphs at 26 loci widely scattered on boulders on the highest part of the upper terrace complex of Cerro Juanaqueña. These are described in Chapter 3 and listed in Table 3.6. Many represent Medio period motifs as well as Historic period elements including representations of churches and a man on a horse. These artifacts and petroglyphs represent reuse of the site for multiple special activities consistent with the pattern we identified in the cross-cultural study; that is, hilltop locations are visited for a wide range of purposes.

Why Warfare during the Early Agricultural Period?

Growing recognition suggests that raiding and warfare are not limited to societies with higher levels of political organization. Publications argue persuasively that anthropologists have underestimated both the occurrence and importance of conflict in small-scale societies (Ember and Ember 2014; Keeley 1995; Lambert 2002; Otterbein 1999). Reconsideration of the significance of warfare in the prehistoric Southwest emerged in the 1990s and remains a key topic (Cole 2012; Haas 1990; Haas and Creamer 1996; LeBlanc 1999; Lekson 2002; McGuire and Villalpando 2015; Rice and LeBlanc 2001; Schaafsma 2000; Wilcox and Haas 1994). Anthropologists have considered raiding and warfare from a wide range of perspectives, but for the purposes of this discussion our immediate concern is with the contexts that lead to these kinds of behavior. That is, what general social, economic, and technological conditions are commonly associated with raiding and warfare? Keeley (1996:117) poses just this question, and he summarizes empirical studies that have addressed the issue. Surprisingly, he found that raiding and warfare are not strongly correlated with either high population densities or with higher levels of social complexity, as we might expect.

Keeley does specify four conditions associated with increased bellicosity: 1) rapid population growth, which increases economic demands and also stresses the capacity of social institutions; 2) development or introduction of new technology in food production, transportation, and weaponry; 3) frontier zones, such as zones of contact between stratified versus egalitarian societies, between pastoral nomads and village farmers, or between farmers and foragers; 4) expectable but unpredictable disasters such as droughts, floods, and insect infestations.

Current evidence suggests that 3,200 years ago people living along the riverine oases of northwestern Chihuahua were using agriculture in new ways. This powerful technology opened a whole range of subsistence possibilities and at the same time imposed new constraints. People in different settings responded to agriculture in different ways. In northwestern Chihuahua we do see a dramatic population increase, a strong emphasis on settlements in floodplain settings, increased sedentism, and population aggregation around particularly well-watered points along floodplains. Using agricultural technology people created new niches and moved quickly to exploit increasing population growth at these locations. These conditions must have required major changes in social institutions and, in our view, would have led to the kind of instability that Keeley describes. Instability would have increased, with competition for these particular locations and their products. Fluctuations in agricultural production, which can result from a variety of causes, would have accentuated these conditions, perhaps leading to the expectable but unpredictable disasters that Keeley found to be associated with increased raiding and warfare. Proximity to foraging people who continued a very different lifestyle in the basin and range province to the east implies that the early agricultural sites in northwestern Chihuahua were also a frontier zone, which may also have increased the intensity of conflict in the region. In short, all of the conditions that Keeley identifies were present in northwestern Chihuahua 3,200 years ago, leading to a strong theoretical expectation of violent conflict during this period.

Early Agricultural period warfare was not limited to the Río Casas Grandes during the Early Agricultural period, but was present at multiple well-watered locations along main stem drainages in the southern basin and range. Evidence for raiding and warfare is based on both the occurrence of cerros de trincheras and on osteological data representing the San Pedro and Cienega phases in at least three other valleys—the Santa Cruz River in the Tucson Basin, La Playa on the Río Boquillas in northern Sonora, and along the Upper Gila River in southeastern Arizona.

Violent trauma from blunt force or projectiles was present in 17.8% ($n = 28$) of San Pedro phase and 7.9% ($n = 38$) Cienega phase burials from the Tucson Basin (Watson and Phelps 2016:Table 1). At La Playa, evidence of violence was present in 41.7% ($n = 12$) of San Pedro phase burials, 17.9% ($n = 28$) of Cienega phase burials, and 15.2% ($n = 66$) of undated burials (Watson and Phelps 2016:Table 1). In sum, violent trauma was present in 16.3% ($n = 172$) of Early Agricultural individuals in their study. Watson and Phelps argue the violence was related to warfare among communities of the same society as part of revenge cycles, however, it is difficult to determine the identities of warring communities.

Cienega phase constructions on Tumamoc Hill indicate a defensive posture (Wallace et al. 2007; Wilcox 1979). Along the Upper Gila River in southeastern Arizona, Round Mountain is another Cienega phase cerro de trincheras. The site has multiple defensive characteristics including difficult access, continuous and concentric wall constructions, and walls constructed in locations to impede access along least-cost pathways (Hard et al. 2016a; Roney et al. 2014; Whisenhunt et al. 2016). The Río Casas Grandes cerros de trincheras are the local expression of rapid cultural change across the southern basin and range resulting from increasing dependence upon agricultural production and the competition it brings for particularly well-watered locations.

Conclusions

We have argued here that the Early Agricultural period cerros de trincheras in northwestern Chihuahua were most likely occupied during periods of especially intense raiding and warfare resulting from fundamental changes associated with increased reliance upon agriculture. The conditions that make violent conflict more likely were all present 3,200 years ago. Worldwide ethnographic data suggest that hilltop residential settlements are strongly associated with raiding and warfare. However, such analogical models require archaeological data independent of hill location to evaluate along with alternative hypotheses.

We could find little support for the hypothesis that site occupations were due to the environmental advantages of hilltop living. We rejected the ideas that Cerro Juanaqueña was constructed for agricultural purposes, display, or seasonal aggregation. Roth's (2016a) suggestion that Cerro Juanaqueña was a seasonal aggregation location is inconsistent with much of the information presented in this book, including evidence for intensive occupation (particularly the hundreds of heavily worn metates), radiocarbon data, site layout, the occurrence of multiple other cerros de trincheras in this vicinity, and indications of multi-season occupation, as well as the massive labor investment. However, we did find evidence of social integrative activities with the discovery of a 70 m diameter rock circle at Cerro el Canelo, which may be one of the earliest communal structures in the region. If we are correct in our temporal assignment, this finding is consistent with the Fishes' (2007) expectations that ritual and defense are not mutually exclusive.

However, several aspects of the sites are inconsistent with defense. The lower slope terraces at Cerro Juanaqueña would certainly appear to be vulnerable, and perhaps they were used when threats were minimal. A number of scholars have suggested that cerros de trincheras would be susceptible to siege tactics given the absence of water supplies on hilltops. However, sieges were generally not in use in societies without complex social structures (Keeley 1996; McGuire and Villalpando 2015).

The sites also exhibit many characteristics associated with a defensive pose. These aspects include the hills' height, difficulty of access, defensive site layouts including perimeter berms and 360 degree orientations, large site sizes, rapid construction, and site intervisibility. Several other more circumstantial lines of evidence reinforce these conclusions, including burned and widely scattered human remains and unusually large numbers of projectile points. The evidence that cerros de trincheras in this region were abandoned and not reoccupied is consistent with ethnographic examples of abandonment following disruption of hostilities. More importantly several sites were occupied synchronously during two brief, discrete episodes.

Based on the fortified sites in the Río Casas Grandes and osteological evidence for violence in burials in the Santa Cruz River basin and at La Playa, there is evidence of warfare by 1300 to 1200 BC in the river valley communities of northern Sonora, southern Arizona, and northwestern Chihuahua. By the Cienega phase, Tumamoc Hill in the Tucson Basin and Round Mountain on the Upper Gila River in southeastern Arizona were built as fortified communities (Fish et al. 2013; Wallace et al. 2007; Wilcox 1979). Evidence for conflict also occurred during the Cienega phase at La Playa and along the Río Casas Grandes. In all four cases, these settlements were involved with farming at highly desirable, well-watered points along main stem drainages where resources were a source of violent competition.

Loss of floodplain productivity may have ended the Early Agricultural period cerros de trincheras occupations. Two erosional events (Chapter 2, 12, and 15) correlate with the first and second abandonments of Cerro Juanaqueña. The destruction of alluvial terraces diminished riverine agricultural potential, and this setting was no longer worthy of defense.

Notes

1. We recently became aware of two additional cerros de trincheras within 15 km of Cerro Tascate. See Chapter 4, endnote 3.

CHAPTER 8

Plant Use

Karen R. Adams and J. Kevin Hanselka

Introduction

The plant record from four cerros de trincheras sites in northern Chihuahua is exceptionally informative for open archaeological sites dating back 3,000 years. Multiyear excavations at Cerro Juanaqueña, and limited excavations at the sites of Cerro Vidal, Cerro el Canelo, and Cerro los Torres are reported here, providing insight into food reliance, seasonality of resource harvesting, and the nature of habitats visited. A few ethnobotanical samples are also reported from a fifth site, Ophelia's Well, a badly damaged subsurface site found while searching for buried deposits in the Río Casas Grandes floodplain (see Chapter 4). This chapter summarizes the extraction, identification, and interpretation of the plant remains collected from the project (K. Adams 1998, 1999, 2002; Hanselka 2000; Hard and Roney 1999a). We divided our ethnobotanical work in the following way: Adams examined and identified the specimens and wrote the interpretations, and Hanselka floated many of the samples and reviewed the ethnographic and archaeological literature on the plants.

The vegetation in the area surrounding these sites is composed primarily of plants occupying the Plains and Great Basin Grassland and Semidesert Grassland floras (Chapter 2; Brown 1982c, 1982d). To better understand available resources in the immediate area of Cerro Juanaqueña, a collection of modern plants was obtained in September 1998 by K. Adams (1998) and are listed in Table 8.1. This modern list supports archaeological plant identifications and also provides insight into additional useful resources likely available in prehistory. Nomenclature in text and tables of this chapter primarily conforms to *Arizona Flora* (Kearney and Peebles 1960) and other published taxonomic resources available in the late 1990s.

The ancient plant record has the potential of contributing to many of the project's research themes, in addition to the obvious theme of ancient subsistence. These include: non-subsistence plant use; the age and time depth of site occupation; seasonality and short-term duration of site occupation; the residential character of occupations; paleoenvironment; site formation processes; and the adoption of agriculture in northwestern Chihuahua.

Subsistence

The nature of subsistence at cerros de trincheras in the region was initially unknown. The possibilities included: full reliance on wild plant and animal resources; full dependence on maize agriculture; or some combination of agricultural efforts and wild resource use. Any of these strategies may also have included, or heavily relied upon, the farming of indigenous cultigens. Learning the basis of the economy was critical to understanding many aspects of the site, such as residential character, seasons of residency, and general degree of sedentism. Knowing the wild plants sought for food would also reflect local and regional vegetation at the time of occupation. Because the archaeobotanical record can be incomplete, the modern list of plants (Table 8.1) provides a broader perspective on resources available for meeting a wide range of human needs both currently and in prehistory.

Plants Sought for Non-Subsistence Reasons

Examination of charred wood fragments and other non-reproductive parts offers insights into some of the trees and shrubs sought by ancient groups for fuelwood, construction elements, and other material culture needs. This information complements the subsistence record and provides further information for reconstructing the nature of regional vegetation at the time of occupation.

Table 8.1. Modern Plants Observed and Collected in September 1998 at Cerro Juanaqueña and in the Janos Region (UTM Zone 12R, 3424600N, 774700E. Elevation: 1,480 meters.)

Plant Name	Field No	LST	HS	UST	AA	RB	JTE	LPT	MRF	MIF	JTW	OT
Agave aff. *palmeri* Engelm	37-98		×									×
Agave sp. (possibly *parryi*; need flowers)	35-98		×									
Albutilon malacum S. Watson	15-98	×										
Allionia incarnata L.	03-98	×										
Allionia sp.							×					
Aloysia wrightii		×	×	×								
Amaranthus palmeri Wats. (male)	57-98								×	×		
Amaranthus palmeri Wats. (male)	27-98			×								
Amaranthus palmeri Wats. (male)	43-98						×					
Ambrosia spp.							×					
Ambrosia trifida							×					
Anemopsis californicus (Nutt.) Hook & Arn.	53-98							×				
Apodanthera undulata Gray	31-98			×								
Aristida ternipes Cav. Var. *ternipes*	18-98		×									
Aristolochea aff. *Wrightii* Seem	22-98		×									
Asclepias aff. *Brachystephana* Torr.	32-98			×								
Atriplex canescens												×
Atriplex wrightii Wats.	46-98						×		×		×	×
Baccharus sarothroides						×						
Bebbia juncea (Benth.) Greene	38-98		×									
Berula erecta (Huds.) Cov.	54-98							×				
Bouteloua curtipendula			×	×								
Bouteloua Rothrockii Vasey	17-98		×									
Brickellia			×	×								
Cephalanthus occidentalus L.	45-98						×		×			
Cevallia sinuata Lag.	19-98		×	×								
Chenopdodium sp. (need seeds)	61-98						×		×			
Chenopodium incanum (Wats.) Heller	04-98	×	×	×								
Chloracantha spinosa (Benth). Nason	60-98						×		×			×
Chloris virgata Swartz	58-98			×			×					
Croton lindheimerianus Scheele	06-98	×	×	×								
Cucurbita digitata Gray	51-98		×				×					
Cucurbita foetidissima							×		×			
Cucurbita sp. (need fruit; possibly *palmata*)	50-98						×					
Cyperus odoratus L.	52-98							×				
Dalea wrightii Gray	14-98	×										
Dasylirion wheeleri Wats.	40-98		×									
Datura							×	×				
Digitaria californica (Bentham) Henrard	41-98		×									
Echinochloa colonum (L.) Link	44-98						×					
Ephedra				×								
Eragrostis cilianensis (All) Vign. Lutati ex Janchen	62-98									×		
Eragrostis intermedia Hitchc.	33-98		×									
Erigeron												×
Eriogonum spp.			×	×								
Erioneuron pulchellum (H.B.K.) Tateoka	11-98	×		×								

Table 8.1. (cont'd.) Modern Plants Observed and Collected in September 1998 at Cerro Juanaqueña

Plant Name	Field No	LST	HS	UST	AA	RB	JTE	LPT	MRF	MIF	JTW	OT
Ferocactus		×	×	×								
Fouquieria			×									
Fraxinus								×				
Gutierrezia				×								
Helianthus annuus							×		×			
Hibiscus denudatus Benth.	16-98	×		×								
Hibiscus sp.										×		
Ipomoea spp.							×		×	×		
Iva ambrosiaefolia (Gray) Gray	10-98	×	×	×								
Jatropha macrorhiza Benth.	25-98			×								
Juglans						×						
Kallstroemia grandiflora Torr.	47-98						×					
Kallstroemia grandiflora Torr.	02-98	×	×									
Lycium ?		×										
Macroptilium atropurpureum (D.C.) Urban	24-98		×									
Manihot davisiae Croizat	21-98		×									
Matricaria			×									
Mentzelia multiflora (Nutt.) Gray	29-98			×			×		×			
Muhlenbergia porteri Scribner ex. Beal	34b-98		×									
Nastursium								×				
Nicotiana obtusifolia Mart. & Gal.	26-98			×								
Opuntia (cholla)		×	×	×							×	×
Opuntia (prickly pear)		×	×								×	
Opuntia leptocaulis			×									
Physalis hederaefolia Gray	23-98		×									
Pleuraphis mutica Buckley	13-98	×										
Polygonum								×	×			
Populus						×						
Portulaca halimoides L.	59-98						×					
Proboscidea			×				×					
Prosopis glandulosa Torr.	09-98	×		×			×				×	×
Robinia pseudoacacia L.	63-98									×		
Salix spp.						×						
Salsola pestifer							×					
Scirpus pungens Vahl.	56-98							×				
Senna bauhinoides Gray	42-98				×							
Senna wislenzenii Gray	20-98		×	×								
Setaria macrostachya H.B.K.	01-98	×										
Solanum eleagnifolium				×			×		×			×
Solanum sp.							×					
Sorghum halapense							×		×			
Sphaeralcea angustifolia (Cavanilles) G. Don	49-98						×					
Sporobolus airoides (Torrey) Torrey	48-98						×					×
Sporobolus wrightii Munro ex. Scribner	30-98			×								
Stipa comata			×	×								
Talinum aurantiacum Engelm.	05-98	×		×								
Talinum paniculatum (Jacq.) Gaertn.	07-98	×	×	×								
Trianthema portulacastrum L.	08-98	×					×					×

Table 8.1. (cont'd.) Modern Plants Observed and Collected in September 1998 at Cerro Juanaqueña

Plant Name	Field No	LST	HS	UST	AA	RB	JTE	LPT	MRF	MIF	JTW	OT
Tridens muticus (Torrey) Nash. var. *muticus*	12-98	×										
Trixis californica Kellogg	28-98			×								
Typha								×				
Unknown shrub	36-98		×									
Verbesina												×
Wislizenia refracta Engelm.	55-98							×				
Xanthium							×	×				
Yucca (narrow leaf)			×									×
Zizyphus obtusifolia (Torr. & Gray) Gray	34a-98	×										

Explanation of abbreviations:
Field No: Adams plant collection number for herbarium specimens. Plants are curated at the University of Arizona Herbarium (Tucson, Arizona). Grasses were verified by Dr. John and Charlotte Reeder. Remaining plants were verified by Paul Jurgens.
LST: Cerro Juanaqueña, lower site constructed terraces.
HS: Cerro Juanaqueña, unmodified hill slopes.
UST: Cerro Juanaqueña, upper site terraces.
AA: Cerro Juanaqueña, alluvial apron of site.
RB: Modern river bottom, Río Casas Grandes.
JTE: Jorge Terrace on east side of Río Casas Grandes (see Figure 12.5).
LPT: La Palotada, a warm springs area.
MRF: Modern rain-fed agricultural field.
MIF: Modern irrigated agricultural field.
JTW: Jorge Terrace on west side of Río Casas Grandes (see Figure 12.5).
OT: Oaxaca Terrace (see Figure 11.02).

Age of the Occupation

Cerros de trincheras sites in southern Arizona and northern Sonora (see Chapter 1) have generally dated late within the prehistoric occupation of the southwestern U.S. and northwestern Mexico (Downum et al. 1994; Fish et al. 2007b; McGuire and Villalpando 2015). Knowing the time period when the Río Casas Grandes cerros de trincheras were both constructed and occupied placed early research emphasis on recovery of short-lived organic remains suitable for AMS radiocarbon dating. Products of agricultural efforts, such as maize kernels or cob fragments, received highest priority, as these items are both inherently cultural and can be directly dated using AMS methods (see Chapter 5).

Seasonality and Sedentism

The question of whether Cerro Juanaqueña and the other early cerros de trincheras represented seasonal aggregations of individuals, or full-year occupations, became critical to understanding the nature of the Late Archaic occupation of the region. As reviewed in Chapter 1, existing models of Early Agricultural period adaptations incorporate a range of mobility options (Diehl 2005c; Gregory and Diehl 2002; B. Huckell 1995; Mabry, 2008; Vint 2015a; Whittlesey 2016; Wills and Huckell 1994). The subsistence record can offer insight into both seasonality of occupation and degree of sedentism. Individual taxa/plant parts often have defined seasons of availability, though these can vary in length and predictability (K. Adams and Bohrer 1998). Taken as a whole, the entire plant record helps define the minimum number of seasons people gathered plants in the region, and hence were likely occupying the cerros de trincheras.

If the Cerro Juanaqueña and other cerros de trincheras occupants were primarily farmers, they would have been tied to their agricultural fields for portions of the agricultural cycle. Tasks related to field preparation, planting, tending, and harvesting span many months of the growing season from early spring through late fall. Agriculturalists often reside in the vicinity of their farming fields for a major portion, if not all, of the calendar year.

Paleoenvironment

Evidence from packrat middens in the American Southwest indicates that modern Sonoran and Chihuahuan Desert vegetation assemblages have been largely in place for the past 4000–4500 years (McAuliffe and Van Devender 1998; Van Devender 1990). Cerro Juanaqueña was first occupied at an ideal time when effective moisture remained high but flash flooding had diminished, enhancing conditions for an aggrading floodplain on the Río Casas Grandes (Chapter 2).

Plant records from archaeological contexts can shed some light on ancient environments. However, plants gathered by ancient people reflect what is culturally preferred, as well as what is available in the environment where they live. Plants can also represent travels to distant locales for special plant needs. Because people alter environments intentionally to promote preferred plants (K. Adams 2004; K. Adams and Fish 2006) and as a result of long-term occupation of a landscape (K. Adams and Bowyer 2002), ancient plant records are more likely to reflect the plants present in ancient vegetation assemblages, rather than the relative proportions of the plants themselves.

Site Formation Processes

Aspects of the ethnobotanical research provided information concerning site formation processes. Postoccupational disturbance was evident via heavy seed rain from plants growing on sites. Increased preservation of organic remains and the vertical downward movement of some seed/fruit size-classes were reflected in the contents of strata with increasing site depth.

Methods

Definitions

Plant materials in archaeological sites are collectively termed "archaeobotanical" specimens and come in two sizes. The larger of these are *macrofossils*, or easily recognized and collected plant remains, including agricultural products, charred wood fragments, and larger seeds and fruit. Macrofossils are generally retrieved in the field during excavation and screening, and constitute a subjective sample dependent on the judgment of the excavator. *Microfossils* are tiny or microscopic plant parts such as seeds, fruit and other reproductive parts, small fragments of charred wood, and pollen grains. Microfossils, those plant specimens too small to be seen during excavation, are routinely acquired in bulk sediment samples designated as either "flotation samples" or "pollen samples." Because microfossil samples are systematically collected and processed, they can potentially provide a less subjective look at prehistoric plant use at a site than macrofossils can. This chapter reports on the recovery of macrofossils from excavation screens and microfossils via flotation. A number of attempts by palynologists to recover pollen from Cerro Juanaqueña deposits, from ground stone tools, and from backhoe trenches in the Río Casas Grandes alluvial terraces are discussed in Chapter 10.

Recovery

Numerous factors affect the preservation of archaeological plant materials, including traditional plant processing techniques and local environmental conditions (Miksicek 1987). Since plant preservation is often poor in open, well-drained deposits such as those found on the cerros de trincheras, it was necessary to use several archaeobotanical recovery methods over the multiyear project. Larger plant macrofossils, such as charred wood and maize cob fragments, were collected both when encountered in the excavation units and as excavated sediments were passed through one-eighth-inch screens. A total of 285 containers of macrofossil samples were collected from all sites. In the artifact catalog, these were identified as "^{14}C" samples when they appeared to be charred wood and "ethnobotanical" samples when they appeared to be seeds or maize. Ultimately, in the lab some of these turned out not to be charred plant material at all.

Seeds and other small items were obtained through water processing of flotation samples. Since its popularization by Struever (1968), water separation has revolutionized research into prehistoric plant subsistence by allowing for the recovery of tiny items that are easily missed during excavation. In its simplest form, flotation involves adding archaeological sediments to water or some other liquid. Heavy material in the sediment sample sinks, while lightweight organic material (buoyant because air is trapped inside plant cells) floats into the water column, where it can be scooped up or otherwise collected.

Excavators routinely collected sediment samples for flotation from discrete levels in every excavation unit, and retained them in plastic bags. In order to enhance recovery, samples were relatively large in volume, with a mean of 6.5 liters per sample, and a median and mode of 4 liters. Whenever possible, flotation samples were collected from 50 cm or more below the modern surface, for reasons that are discussed in greater detail below. Overall, a total of 401 flotation samples were collected from all sites. These samples were processed in the field laboratory in Janos.

Flotation Procedures

Two flotation methods were employed over the course of the project. In 1997, samples were processed using a simple apparatus consisting of a full bucket of water suspended over a large funnel. Sediment samples were carefully poured into the bucket, and the mixture was gently agitated. Buoyant "light fraction" materials within the sediment sample floated to the surface, overflowed out of the bucket, and were collected in a mesh bag situated at the funnel mouth. Because there was some question regarding the effectiveness of this method, in 1998 the project adopted a variation of the decanting method of field flotation developed by Bohrer and Adams (1977). Using this procedure, lab technicians slowly added each sediment sample to a bucket of water and gently agitated the mixture, allowing lightweight material to be released from the dirt matrix and into the water column. The water and the floating matter within it were then carefully hand-poured through a 0.5-mm geological sieve, while heavier materials were retained in the bottom of the bucket. The sieve collected any floating items not small enough to slip through. The light fraction was then carefully removed from the sieve and placed in a mesh bag to dry, and the heavy fraction in the bucket was emptied onto a window screen, rinsed free of sediment, and also collected in a separate mesh bag. Micro-flakes and bone were removed from the sieve, bagged separately, and ultimately combined with the artifactual and faunal collections for their respective analyses. Heavy fractions were later screened and carefully sorted through, particularly in search of small pieces of bone (see Chapter 11).

Selection of Flotation Samples for Analysis

Far too many flotation samples were collected and processed than could be analyzed, so 194 light fractions out of the 401 floated samples were selected for examination. Light fractions were chosen and placed within four priority levels based on the following criteria: Priority level one samples were those taken from 50 cm or more below the modern surface where better plant preservation occurred, and had highly visible charred material within the sample itself or in its excavation provenience. Priority level two samples were those taken from deep deposits but lacking in notable associated charred materials. Priority level three samples were those from proveniences with visible charred remains, but from relatively shallow contexts. Priority level four samples were from excavated features not otherwise represented in the above samples, so that all feature types were included in the analysis.

Analysis Procedures

Macrofossils

Up to 20 pieces of charred wood per sample were snapped in half to expose a fresh transverse (cross) section for viewing tree ring traits (e.g., nature of rings, vessels, rays, background patterning) under a Zeiss binocular microscope at magnifications of 8 to 50×. Each piece was identified to the most specific taxonomic level possible, generally to genus, but at times to the broader family level. A collection of modern charred wood specimens representing many of the tree and shrub species of the Sonoran and Chihuahuan deserts provided comparative materials for this effort.

All charred non-wood specimens were pulled from macrofossil samples that had been thinly spread out so that all specimens were visible. All such items were examined under the microscope, and identified, if possible. These segregated materials were then bagged and labeled separately.

TABLE 8.2. Total Flotation and Macrofossil Samples Analyzed from Each Site.

Site	Flotation Samples	Macrofossil/ Charred Wood
Cerro Juanaqueña	157 (53)	167
Cerro Vidal	5 (3)	
Cerro el Canelo	21 (13)	19
Cerro los Torres	10 (3)	
Ophelia's Well	1	17
Totals	194 (72)	203

Note: Numbers in parentheses indicate a subset of sterile flotation samples from each site that contained no identifiable materials, or charred materials too small to identify with confidence.

Microfossils

The buoyant light fraction recovered from each flotation sample was subdivided into a series of graded particle sizes (>4.0 mm, 2.0–4.0 mm, 1.0–2.0 mm, 0.5–1.0 mm, <0.5 mm) for ease of analysis, and all materials larger than 0.5 mm in size were examined under a microscope. Up to 20 pieces of charred wood were also pulled for examination, as described above. Reproductive parts were identified with the aid of a collection of southwestern U.S. and northern Mexican plant reproductive parts, plus comparison to specimens in the University of Arizona Herbarium (ARIZ) and to reference texts (Martin and Barkley 1961). Unidentifiable items were described. Gayle Fritz at Washington University in St. Louis examined the amaranth (*Amaranthus*), chenopod (*Chenopodium*), and indeterminate cheno-am seeds, and searched samples of the 0.5–1.0 mm and < 0.5 mm fractions for additional seeds of these taxa. Her analysis of these materials is described in Chapter 9.

Results

This section summarizes the results of examination of 194 flotation samples and 203 separate bags of macrofossils, including charred wood (Table 8.2). The majority of the samples represent Cerro Juanaqueña, with the remainder from the other sites. Thirty-seven percent ($n = 72$) of all flotation samples were sterile, either containing no charred materials, or charred materials much too small to identify with confidence. These sterile samples are noted in Table 8.2, and are included in all calculations of presence (ubiquity) in the tables within this chapter. A number of tables in this chapter were prepared based solely on plant remains recovered in flotation samples, which are considered superior by virtue of their systematic collection in the field, and because a large number of flotation samples are available for in-

TABLE 8.3. Presence of Uncharred Plant Parts and other Items in Flotation Samples.

	Cerro Juanaqueña *n* = 157	Cerro Vidal *n* = 5	Cerro el Canelo *n* = 21	Cerro los Torres *n* = 10	Ophelia's Well *n* = 1
Plant Parts					
Boerhaavia type seed	1			1	
Cactaceae type seed	27		14		
Celtis type seed	2		6		
Cheno-am seed	154	5	15	9	1
Compositae type achene	14		5		
Echinocereus type seed	2				
Erodium type seed			2		
Euphorbia type seed	16	1	5	5	
Poaceae type caryopsis	2				
Kallstroemia type seed	14		2	2	
Leguminosae type seed	3				
Liliaceae type seed			3		
Opuntia (p. pear) type seed	1				
Paniceae type caryopsis	63	5	14	8	
Portulaca type seed	15	3	4	3	
Salvia type seed			1		
Solanaceae type seed	1				
Sphaeralcea type seed	4	1	2	2	
Trianthema type seed	128	5	14	9	
Non-Plant Parts					
Bug body part	154	5	21	10	
Rodent pellet	150	5	21	10	
Snail shell	112	5	8	10	1

Note: *n* = total number of flotation samples examined for each site. Count equals number of flotation samples that contained uncharred plant parts.

terpretation. The results of Fritz's cheno-am analyses are integrated into the appropriate tables and summaries in this chapter, and the details are reported in Chapter 9. Notations of plant remains recovered in macrofossil/charred wood samples are included, whenever appropriate.

Many taxonomic names are followed by the word "type" in the tables of this chapter, which conveys that the prehistoric specimen(s) closely resemble the taxon/taxa named, but may also compare well to other related (or occasionally unrelated) taxa. This conservative approach is used because of the similarity in appearance of parts of southwestern plants, especially when they are burned and degraded, and to the incomplete nature of most modern comparative collections. The specific criteria of identification of all parts, including charred wood specimens, can be found in Appendix 8.1, and the entire database of raw data comprises Appendix 8.2.

Both charred and uncharred plant remains were preserved in the cerros de trincheras samples. Charring is generally considered to be the result of prehistoric cultural activities, and uncharred plant remains are often introduced into cultural deposits through noncultural processes such as rodent activity and wind (Minnis 1981:147). Charred materials reveal the most about past subsistence and wood use; however, uncharred items offer their own set of insights into site formation processes and local flora.

Uncharred Specimens

For this project, uncharred specimens were routinely noted in the light fraction components of flotation samples (Table 8.3). Cheno-am seeds (primarily *Chenopodium*), carpetweed (*Trianthema* type) seeds, bug body parts, and rodent pellets all occurred in over 80% of the Cerro Juanaqueña samples and were predominant in other sites as well. A general type of grass grain (Paniceae type) and snail shells were encountered in many samples examined. Other taxa, including *Portulaca, Euphorbia*, Cactaceae, Poaceae, and Solanaceae, occurred sporadically to rarely.

These uncharred materials reveal something about the natural formation processes operating at these sites, say something about the composition of the surrounding flora, and provide perspective on certain available natural resources. The repeated occurrence of bug body parts and rodent pellets suggests that insects and rodents have continuously disturbed the site sediments. Mixing of stratigraphic layers and homogenization of contents might result from this process. Some of the uncharred taxa recovered (*Trianthema, Portulaca, Euphorbia*, Solanaceae) are plants known to thrive in disturbed habitats, so this record reflects such activities as domestic animal grazing, and rodent and insect burrowing.

The presence of uncharred items in flotation samples establishes them as members of the modern local flora. This record reveals that *Chenopodium* and *Amaranthus* are common components of the area vegetation. These and most of the other uncharred seed and grain resources have been known ethnographically as foods in the Southwest. This can be taken further to suggest they may have also been potentially available food resources in prehistory.

This uncharred record also holds a cautionary note. Any seed or grain type that is constantly available to enter site sediment, such as the chenopod, amaranth, and carpetweed seeds, may have been similarly available in prehistory, and their seeds could have become inadvertently charred in ways unrelated to actual human use. Charred *Trianthema* seeds were recovered in Cerro Juanaqueña deposits, potentially documenting human use or inadvertent natural charring. However, the ethnographic and prehistoric uses of chenopod and amaranth plants and seeds as resources are so extensive (K. Adams 1988) that it is less likely their presence is due to anything other than intentional human use, if only as a resource to provide steam during roasting events of other food products.

Charred Specimens

A minimum of 22 separate edible taxa represented by charred reproductive parts (Table 8.4) have been identified from all sites. These include: possibly domesticated *Amaranthus, Zea mays, Astragalus nuttalliana, Atriplex, Chenopodium, Cucurbita digitata, Eragrostis intermedia, Euphorbia, Ferocactus*, Poaceae, *Helianthus, Juglans*, two different Leguminosae, *Physalis, Portulaca, Prosopis, Rhus aromatica, Salvia, Scirpus, Sphaeralcea*, and *Trianthema*.

Thirteen taxa represented by charred wood or vegetative parts (Table 8.5) have also been identified from flotation and macrofossil samples from the cerros de trincheras. These include: *Atriplex*, Compositae, *Fouquieria, Fraxinus*, Poaceae, *Juglans, Juniperus, Larrea*, Leguminosae, Monocotyledon, *Pinus, Populus/Salix*, and *Prosopis*. The charred items are assumed to relate to some prehistoric activity associated with fire. The wide variety of plant parts recovered will be discussed in detail below.

Plant Preservation Issues

Early in the project it became clear that site formation processes had resulted in substantial mixing of deposits. Due to the nature of terrace construction, there has been substantial downward filtering of particles and cultural materials within the archaeological deposits. After initial construction, sediments and cultural remains alike traveled downward through the voids between the cobbles and boulders that form the terrace walls and fill (see Chapter 3). This process necessitated continued maintenance of the terrace living surfaces by the prehistoric occupants, at least in the early years of terrace occupation. Root, rodent, and insect activity has further exacerbated the disturbance problem by churning up the deposits over thousands of years. Cultural materials are, therefore, intermixed throughout all levels, even down into the pre-occupation layers in the colluvium on which the terraces are built. Fortunately, since AMS radiocarbon dates seem to indicate that these sites were occupied for a relatively short length of time, the adverse effects of deposit disturbance on cultural reconstructions are relatively low.

The results from all field seasons seem to suggest that the depth from which a flotation sample was collected heavily influenced the amount of plant remains contained within that sample. In 1997, the field strategy emphasized a mix of vertical and horizontal excavation units. Although plant remain returns from this season were generally low, the most productive samples with the most abundant plant records were from deeper deposits within the terrace fill. Cognizant of this, in 1998 deep excavations were implemented to target recovery of ethnobotanical material as well as to understand terrace construction. Deep trenches were necessary to bisect selected terrace features and resulted in the collection of flotation samples from deeper contexts that were substantially more productive than those collected in 1997. The primary goal of the 1999 field season was to horizontally expose and define possible prehistoric occupation surfaces within the terraces, leading once again to shallow block excavations, and to a correspondingly reduced level of plant part recovery. In 2000, excavation focused on the band of terraces located on the lower western slope of the site which happened to have

TABLE 8.4. Presence of Charred Reproductive Plant Parts in F (Flotation) and M (Macrofossil) Samples, Including Results of Cheno-am Analysis.

Taxon	Common Name	Part	Cerro Juanaqueña	Cerro el Canelo	Cerro los Torres	Cerro Vidal	Ophelia's Well
Domesticates							
Amaranthus	amaranth	seed	F*	F*	F*	F*	
Zea mays	maize, corn	cob fragment	F,M				
Zea mays	maize, corn	cob segment	F,M				
Zea mays	maize, corn	cupule	F,M		F	F	F
Zea mays	maize, corn	embryo	M				
Zea mays	maize, corn	kernel, kernel fragment	F,M		F		
Wild Resources							
Reproductive parts							
Astragalus nuttalliana type	milkvetch	seed	F				
Atriplex	saltbush	seed	F*		F*		
Cheno-am	cheno-am	seed	F	F	F	F	F
Chenopodium	chenopod	seed	F*	F*	F*	F*	
Cucurbita digitata type	cucurbit	seed interior	M				
Eragrostis intermedia type	lovegrass	caryopsis	F		F		
Euphorbia type	spurge	seed			F		
Ferocactus type	barrel cactus	seed	F				
Poaceae type	grass	caryopsis	F		F*		
Helianthus type	sunflower	achene		F			
Juglans type	walnut	nutshell fragment	M				
Leguminosae type	legume	cotyledon fragment	F				
Leguminosae type	legume	seed	F		F		
Physalis type	groundcherry	seed			F		
Physalis/Solanum	groundcherry/ nightshade	seed	F*				
Portulaca type	portulaca	seed	F*	F			
Prosopis type	mesquite	cotyledon fragment	F				
Rhus aromatica type	lemonade berry	seed	F				
Salvia type	chia	seed	F				
Scirpus type	bullrush	achene	F				
Sphaeralcea type	globemallow	seed	F				
Trianthema type	carpetweed, horse purslane	seed, seed fragment	F				
Unknown		seed, seed fragment	F	F	F*	F*	F

F* = Recovered in cheno-am analysis of flotation samples, see Chapter 9.

deep deposits. In short, these findings appear to indicate that on these terrace features, plant preservation and the chance for plant preservation generally increases with the depth from which a sample is taken.

This argument is illustrated in Figure 8.1, where the x-axis represents arbitrary 10 cm excavation levels below surface, and the y-axis represents the total number of charred reproductive plant parts found in a level for each excavation unit. This chart includes only those reproductive parts found in flotation samples and only those samples that had at least one charred reproductive part present. It includes seeds, maize kernels/kernel fragments, and cupules (the structural element on a maize cob that once held two kernels). It does not include charred wood specimens or maize cob fragments, as these items tend to preserve better

TABLE 8.5. Presence of Charred Wood and Charred Vegetative Plant Taxa/Parts in F (Flotation) and M (Macrofossil) Samples.

Taxon	Common Name	Part	Cerro Juanaqueña	Cerro el Canelo	Cerro los Torres	Cerro Vidal	Ophelia's Well
Atriplex type	saltbush	wood	F, M	M	F		
Compositae type	composite	wood	F, M				
Fouquieria type	ocotillo	wood	F,M	M			
Fraxinus type	ash	wood	F	M			
Poaceae type	grass	stem fragment	F,M		F		
Juglans type	walnut	wood	F,M				
Juniperus type	juniper	wood	F,M				
Larrea type	creosote bush	wood	F				
Leguminosae type	legume	wood	F,M				
Monocotyledon type	monocot	stem fragment	F				
Monocotyledon type	monocot	tissue fragment	F,M	F,M			
Pinus type	pine	wood	F				
Populus/Salix type	cottonwood/willow	wood	F,M				
Prosopis type	mesquite	wood	F,M	F,M			
Unknown		wood, twig fragment	F,M	F,M			
Unknown		tissue fragment	F	F			F

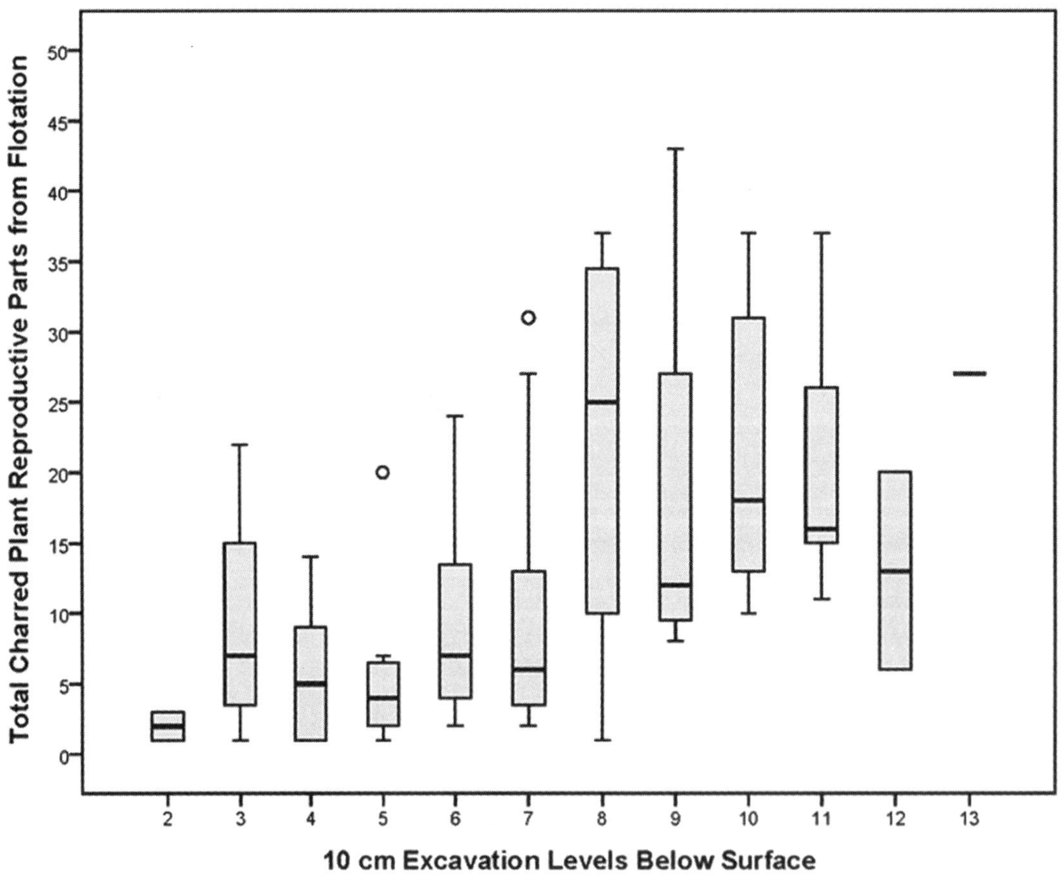

FIGURE 8.1. Boxplot of total charred reproductive plant parts for all levels.

than smaller, more fragile remains, and are more likely related to tinder/fuel usage. Nor did it include the results of the cheno-am analysis. The boxplot shows that the number of plant parts recovered from flotation samples per level is highly variable but there is an increase in the median number of plant reproductive parts recovered per level after level 7.

Two terrace features at Cerro Juanaqueña (Terrace 222 and Terrace 487) were particularly well-represented by food remains. Preservation most likely played a key role here. For example, both terraces contained deep deposits, some of which were likely rock and sediment used to construct the terraces, or midden material discarded as part of everyday life. Terrace 487 also had a large storage pit (Subfeature pit B, Chapter 3) which could have filled with and protected discarded plant debris following its abandonment. The locations of Terrace 222 and Terrace 487 are shown on Figure 3.2, and additional details can be found in Table 3.1, as well as the annual field reports (Hard and Roney 1998b, 1999a; Appendices 3.2 and 3.3).

Flotation samples from 17 Cerro Juanaqueña terrace features were examined to determine the geographic breadth of maize and other food use by households across the site (Table 8.6). Only 17 of the 29 excavated features (see Table 3.1) produced sufficient ethnobotanical remains to be useful for comparative analysis. Each terrace included in Table 8.6 is represented by an average of eight flotation samples. In a few cases, the macrofossil record has been included to complete this view. Sufficient remains were not recovered from 12 of the 29 features to include them in Table 8.6, among them: one excavated terrace (T273), three features that were only tested with 50 cm × 50 cm units (T463, T465, R307A), seven excavated rock-ring features (R1A, R28, R51, R234, R239, R250, R286), and the burned-rock feature (BR1). If identifiable plant remains were found in these features they were from macrofossil or charred wood samples, not from flotation samples. The rock-ring features yielded particularly poor results as the features were small and deposits were uniformly shallow. The data can be found in Appendix 8.2.

Subsistence: Domesticates

Amaranthus seed

Gayle Fritz analyzed a sample of cheno-am seeds which potentially represent both *Chenopodium* and *Amaranthus* seeds from two separate plant families (see below) and her detailed results and discussion are described in Chapter 9. She successfully identified possibly domesticated *Amaranthus* seeds as well as wild *Chenopodium* seeds. It is reasonable that the cheno-am seeds contain additional examples of possibly domesticated, thin-seed coat amaranth. A wild species of amaranth, *Amaranthus palmeri*, currently grows on upper terraces at Cerro Juanaqueña, and is profuse in both irrigated and rainfed modern fields.

Potentially domesticated amaranth seeds were commonly found on all four sites, consistently ranking high in frequency of occurrence in the flotation samples (Tables 8.4 and 8.7). They were present in 13% of all flotation samples examined from Cerro Juanaqueña, ranking only below maize and cheno-ams, and slightly above chenopods (Table 8.8).

Zea mays cob fragment, cob segment, cupule, embryo, kernel, and kernel fragment

Domesticated maize (*Zea mays*) parts were present in three of the four cerros de trincheras sites as well as Ophelia's Well (Table 8.4). Most of the project's maize radiocarbon dates place the domesticate at 3,200 to 3,000 calendar years old, a period in which maize is widespread in the Southwest, but is generally thought to have played a minor role in the diet, except in the Tucson Basin and southeast Arizona (Chapter 1; K. Adams 1994; Hard and Roney 2004; B. Huckell 1995; Mabry 2002; Mabry, ed. 1998, 2008; Mabry et al. 1997; Smiley 1994; Vint 2015). Maize dates of ca. 4000 years old are known from west-central New Mexico, southwestern New Mexico, the Tucson Basin, northeastern Arizona, and east central Arizona (Huber 2005; L. Huckell 2006; Laumbach 2013; Mabry, ed. 2008; Merrill et al. 2009; Shackley 2005a; Smiley 1994). The variety of maize parts preserved in flotation samples and as macrofossils provided the bulk of specimens submitted for radiocarbon dating. Their presence represents food use of the kernels, and suggests burning the leftover cobs as fuel or tinder.

The maize at Cerro Juanaqueña was fairly well-preserved. Sixteen cob segments were complete around their circumference for at least a portion of the cob, permitting accurate assessment of row number and a few additional observations (Table 8.9). Cobs had from 8 to 12 rows with a mode of 12 (second mode = 10, median = 11.0, mean = 10.6). All were quite small and round in cross section. Only a single, round, flint kernel was complete enough to be measured at 0.5 cm in length, 0.6 cm in width, and 0.6 cm in thickness. Other kernel fragments appear to have a mix of flint and flour endosperm.

The morphology of burned maize in archaeological sites derives from at least three sources. Genetic make-up is clearly responsible for the standard morphological traits of each maize variety, as well as their range of expression. However, environmental parameters such as moisture and temperature (K. Adams et al. 1999; Doebley 1990), along with the length of charring (K. Adams 1994:275–276), can alter morphology of some maize traits in significant ways, complicating efforts to determine racial classification when the maize is both fragmented and charred.

TABLE 8.6. Presence of Charred Reproductive Parts within Flotation Samples in 17 Cerro Juanaqueñna Terraces, Including Results from Cheno-am Analysis.

Taxon	Part(s)	T222	T487	T10	T163	T97	T387	T126	T413	T6	T537	T167	T175	T287	T290	T297	T508	T415	Total No. Flotation Samples
	Terrace Number =	T222	T487	T10	T163	T97	T387	T126	T413	T6	T537	T167	T175	T287	T290	T297	T508	T415	Total No. Flotation Samples
	No. of Flotation Samples =	8	27	6	7	2	15	5	15	14	8	4	5	5	2	8	5	0	136
Taxon	Part(s)																		Total Terraces with each Taxon/Part
Domesticates																			
Zea mays	cob fragment, cob segment, cupule, embryo, kernel, kernel fragment	×	×	×	×	×	×	×	×		*	*		*	×	×	×	*	15
Amaranthus	seed	×**	×**	×**	×**		×**		×**	×**	×**								8
Wild Resources																			
Cheno-am	seed	×	×	×	×	×	×	×	×	×	×					×	×		12
Chenopodium	seed	×**	×**		×**		×**		×**	×**	×**					×**			8
Unknown	seed, seed fragment	×	×**	×**	×**		×**						×						6
Poaceae type	caryopsis	×	×**		×**		×**												4
Scirpus type	achene	×		×			×												3
Salvia type	seed	×			×														2
Trianthema type	seed, seed fragment		×	×															2
Portulaca	seed		×**	×**			×**												3
Astragalus nuttalliana type	seed	×																	1
Cucurbita digitata type	seed interior	*																	1
Eragrostis intermedia type	caryopsis	×																	1
Ferocactus type	seed					×													1
Juglans type	nutshell fragment		*																1
Leguminosae type	cotyledon fragment				×														1
Leguminosae type	seed		×																1
Atriplex	seed	×**																	1
Physalis/Solanum	seed		×**																1
Prosopis type	cotyledon fragment		×																1
Rhus aromatica type	seed		×																1
Sphaeralcea type	seed	×																	1
Total taxa within each terrace		13	13	7	8	3	8	2	4	3	4	1	1	1	1	3	2	1	74

* Recovered only as a macrofossil specimen in this terrace.
** Recovered during cheno-am analysis (see Chapter 9).

Table 8.7. Relative Presence of Charred Reproductive Parts in all Flotation Samples (Including Sterile Samples) for all Sites Other Than Cerro Juanaqueña.

		Cerro el Canelo (*N* = 21)				Cerro los Torres (*N* = 10)				Cerro Vidal (*N* = 5)				Ophelia's Well (*N* = 1)	
Taxon	Part(s)	No. of samples with taxon	No. samples with taxon in cheno-am analysis	Total	%	No. of samples with taxon	No. samples with taxon in cheno-am analysis	Total	%	No. of samples with taxon	No. samples with taxon in cheno-am analysis	Total	%	No. of samples with taxon	%
Domesticates															
Zea mays	cob fragment, cob segment, cupule, kernel, and kernel fragment					5		5	50.0%	2		2	40.0%	1	100.0%
Amaranth sp	seed		1	1	4.8%		5	5	50.0%		1	1	20.0%		
Wild Resources															
Cheno-am	seed	5	Present	7	33.3%	7	Present	9	90.0%	1	Present	1	20.0%	1	100.0%
Chenopodium sp.	seed		4	4	19.0%		6	6	60.0%		1	1	20.0%		
Atriplex type	seed						1	1	10.0%						
Poaceae	caryopsis						2	2	20.0%						
Eragrostis intermedia type	caryopsis					3		3	30.0%						
Euphorbia type	seed					1		1	10.0%						
Helianthus type	achene	1		1	4.8%										
Leguminosae type	seed					1		1	10.0%						
Physalis type	seed					5		5	50.0%						
Portulaca type	seed	1		1	4.8%									1	100.0%
Unknown	seed	1	2	3	14.3%		5	5	50.0%		1	1	20.0%		

TABLE 8.8. Cerro Juanaqueña Ubiquity or Presence in Flotation Samples in Rank Order of Charred Reproductive Parts in all 157 Flotation Samples (Including Sterile Samples).

		Flotation Samples (*N* = 157)				
Taxon	Part(s)	No. of samples with taxon	No. samples with taxon in cheno-am analysis	Total	%	Rank Order
Domesticates						
Zea mays	cob fragment, cob segment, cupule, kernel, kernel fragment	80*		80	51.0%	1
Amaranthus sp.	seed		21	21	13.4%	3
Wild Resources						
Cheno-am	seed	41	present	40	25.5%	2
Chenopodium sp.	seed		16	16	10.2%	4
Unknown	seed, seed fragment	5	present	15	9.6%	5
Poaceae type	caryopsis	3	8	11	7.0%	6
Trianthema type	seed, seed fragment	6		6	3.8%	7
Portulaca	seed		4	4	2.5%	8
Salvia type	seed	4		4	2.5%	8
Scirpus type	achene	4		4	2.5%	8
Eragrostis intermedia type	caryopsis	3		3	1.9%	9
Astragalus nuttalliana type	seed	2		2	1.3%	10
Leguminosae type	seed	2		2	1.3%	10
Atriplex	seed		1	1	0.6%	11
Ferocactus type	seed	1		1	0.6%	11
Leguminosae type	cotyledon fragment	1		1	0.6%	11
Prosopis type	cotyledon fragment	1		1	0.6%	11
Rhus aromatica type	seed	1		1	0.6%	11
Physalis/Solanum	seed		1	1	0.6%	11
Sphaeralcea type	seed	1		1	0.6%	11
Cucurbita digitata type	seed interior	*				
Juglans type	nutshell fragment	*				

* Also or only in macrofossil samples, see Table 8.4.

With these factors in mind, the traits of the Cerro Juanaqueña maize do not overlap those of Chapalote maize, although it is generally consistent with a popcorn, having small, rounded kernels and flinty endosperm (with some lesser portion of the endosperm floury). Chapalote, considered an early and long-persistent type in the American Southwest, is described as a 12 to 14 rowed flint or popcorn type with isodiametric kernels marked with striations (K. Adams 1994). Nor does the Cerro Juanaqueña maize fit the general descriptions of other common Southwestern maize types (K. Adams 1994, 2015). Other archaeologists have published descriptions of archaeological maize in the region, for example from a series of caves in Cave Valley along the Río Piedras Verdes southwest of Nuevo Casas Grandes (Mangelsdorf et al. 1956). At Swallow Cave, maize is preserved in the undated earliest pre-ceramic level, 76.2–91.4 cm below levels more confidently dated to AD 900 and later. The earliest intact charred cob, labeled a pre-Chapalote type, had 12 rows and was tapered at both ends, and in shape was quite similar to the earliest cobs from Bat Cave in New Mexico. Archaeologists speculated that rapid change in maize types in northwest Mexico was initiated with the post-AD 725 introduction of more than one type of new germplasm, spurring development of new types still common in the region (Mangelsdorf et al. 1956). Maize from AD 1060–1205 Cuarenta Casas, located to the west of Cave Valley, consisted of two highly variable types: (1) an eight-rowed type with relatively shallow cupules, and (2) a type with a greater number of rows, and long deep cupules (Benz and Alvarez del

TABLE 8.9. Maize (*Zea mays*) Cob Segment Measurements from Cerro Juanaqueña.

	Cob		Cross	Cupule						
No. of rows	Length (mm)	Width (mm)	Section Shape	Length (mm)	Width (mm)	Depth (mm)	Terrace	Unit	Level	Bag
12	7	6	round	2	2	1.5	T387	1	11	59
12	12	5	round	2	2	1	T387	2	8	40
08	4	4	round	2	4	1	T387	2	8	40
12	6	4	round	2	2	1	T097	1	7	26
10	10	4	round	2	2	0.8	T387	2	5	23
12	6	5	round	1.5	2.5	1.5	T413	2	7	53
10	10	6	round	2	2	2	T487	6	6	117
10	5	5	round	2	2	1	T487	6	6	117
10	11	6	round	2	2	1	T508	2	5	16
08	6	3	round	2	2	1	T487	4,5	7	94
12	9	6	round	1.5	1.5	1	T487	2	2	9
12	12	7	round	2	2	1	T487	2	9	32
12	7	6	round	2	2		T537	4	5	74
12	13	7	round	1.8	2.3		T167	3	1	42
10	7	4	round	1.5	1.5	1	T163	2	8	40, 48, 50, 52
08	7	4	round	2	2	1.5	T297	1	10	47, 48, 49

Castillo n.d.). As with Mangelsdorf and Lister (1956), the researchers at Cuarenta Casas suggested that the inability to correlate the Cuarenta Casas maize types with extant northwestern Mexican races (Apachito, Azul, Blando, Cristalino de Chihuahua) reflected relatively recent development of these modern varieties via hybridization and introgression of preceramic period indigenous maize with late-arriving eight-rowed maize and teosinte types from the southern Sierra Madre Occidental. Charred maize cobs from Casas Grandes Medio period deposits (Cutler and Blake 1974) and from Waterfall Cave (Cutler 1960) were very similar to those types identified from Cave Valley caves (Mangelsdorf and Lister 1956).

However, the Cerro Juanaqueña maize does share some characteristics with other Early Agricultural period maize. Recent studies (Diehl 2005b; L. Huckell 2005, 2006) of the morphology of Early Agricultural period maize (ca. 2200 BC to ca. AD 50) show that cobs were relatively tiny. Only McEuen Cave yielded uncharred cobs, which are more reliable for metric data than charred cobs. The oldest of these McEuen Cave cobs varied in length but they seldom exceeded 10 cm and had a mean rachis diameter of 5.7 mm (L. Huckell 2006). The San Pedro phase maize has a modal row number of 10 (median = 11.6) that varied from 8 to 14 rows, consistent with the Cerro Juanaqueña maize (Diehl 2005b; L. Huckell 2006). Cienega phase maize has greater variability but reflects a trend of increasing row numbers; ranging from 8 to 20 rows with a mode of 12 rows (L. Huckell 2006). Changes in morphology reflect increasing productivity, a trend that is underway by the Early Agricultural period and appears to continue through time (Diehl 2005b; L. Huckell 2006). After AD 1000, maize cobs are notably larger (K. Adams 1994; Diehl 2005b).

At Cerro Juanaqueña, charred maize reproductive parts were recovered more often than any other taxon (Table 8.8). Examples of one or more maize parts present in nearly 51% of all flotation samples examined give maize a rank order of number one among foods recovered. This finding suggests a fairly constant level of maize use in prehistory. Although the number of flotation samples examined from the remaining sites is extremely low, maize, amaranth, chenopod, or cheno-ams ranked first or second at nearly every other trinchera site, based on relative presence in flotation samples (Table 8.7). In addition to its economic importance, two factors may contribute to the high ranking of maize: (1) the fact that cobs are used as a fuel or tinder once the kernels are removed increases chances of preservation by burning; and (2) the cob, cupule, and kernels have a high likelihood of preservation, due to their durable nature (Hard et al. 1996).

Maize parts were recovered in 15 of the 17 terraces examined, confirming the broad reliance on maize by occupants of Cerro Juanaqueña (Table 8.6). Although ground stone tools, which could have been used for both maize and wild seed processing, are more prevalent on upper terraces, maize was recovered on all five lower, and on 10 of 12 upper terraces, suggesting that all households were involved in all aspects of food preparation and use.

Subsistence: Wild Plants

At least 18 wild plant resources were gathered by the occupants of Cerro Juanaqueña for food (Table 8.8) and these include: *Astragalus nuttalliana*, *Atriplex*, *Chenopodium*, *Cucurbita digitata* (macrofossil sample only), *Eragrostis intermedia*, *Ferocactus*, Poaceae, *Juglans* (macrofossil sample only), two different Leguminosae, *Physalis*, *Portulaca*, *Prosopis*, *Rhus aromatica*, *Salvia*, *Scirpus*, *Sphaeralcea*, and *Trianthema*.

However, with the exception of cheno-ams, *Chenopodium*, and "unknown" seeds, no single wild taxon was present in more than 7% of all flotation samples examined (Table 8.8), nor recovered in more than four of the 17 Cerro Juanaqueña terrace features studied for plant remains (Table 8.6). This high diversity of wild resources contrasts with their relatively low ubiquity. Other than the cheno-ams, no single species predominated across the site. With the exception of deep and well-preserved terraces Terrace 222 and Terrace 487 (Table 8.6), the diversity within most terraces was low. Given the large number of samples examined, sampling does not seem an issue. The flotation samples examined from the final year of excavation produced no new taxa, suggesting that the majority of preserved taxa had already been recovered. Aside from the cheno-ams, it is possible that no single wild resource was substantially more important than any of the others. This finding might suggest wide diet breadth, where return rates are not substantially different among some of the resources gathered. Assuming the two well-preserved terraces more accurately reflect prehistoric diet, their high wild plant diversity also implies broad diet breadth.

The wild resources includes weeds of disturbed places (*Trianthema*, *Salvia*, *Sphaeralcea*, *Chenopodium*, *Portulaca*), grasses (*Eragrostis intermedia*, Poaceae), plants that prefer damp habitats (*Scirpus*, *Prosopis*), and established perennials (*Ferocactus*, *Rhus aromatica*, *Atriplex*). The macrofossil record adds two likely food products, *Cucurbita digitata* and *Juglans*, bringing the total number of domesticated and wild foods in Cerro Juanaqueña deposits to at least 20, including two different legumes (Leguminosae) and excluding unknowns and cheno-ams (Table 8.8).

The remaining four sites reveal limited use of wild foods in prehistory (Table 8.7). Despite relatively small numbers of samples examined, Cerro los Torres and Cerro el Canelo preserved two additional plant taxa/parts (*Euphorbia* seed and *Helianthus* achene) not recovered in Cerro Juanaqueña (Table 8.4). Although differential preservation and recovery cannot be eliminated, perhaps resource preferences are being expressed. Or resource availability may have differed, despite general similarity in environmental settings. The following section provides details on the wild plants recovered, including comments on where the plants might be found in the area. Additional information on presence of these taxa in other archaeological sites, with an emphasis on the Early Agricultural period and examples of ethnographic uses for these plants, can be found in Appendix 8.3.

Astragalus nuttalliana type seeds

Three charred, well-preserved milkvetch (*Astragalus nuttalliana*) seeds were identified from a Cerro Juanaqueña terrace. Milkvetches (family Leguminosae) tend to be herbaceous perennials, but annual species do exist (Wooton and Standley 1915:356). *Astragalus nuttalliana* is one of the most common species of annual milkvetch in Arizona (Kearney and Peebles 1960:468). *A. nuttalliana* can grow in a wide variety of hostile environments, but it flourishes under enhanced moisture conditions (Bohrer 1987:111–112). It commonly occurs on plains, mesas, and slopes (Kearney and Peebles 1960:468). *A. nuttalliana* flowers from February to May, and sometimes into late summer (Kearney and Peebles 1960:468). A persistent pod with enclosed seeds would extend the gathering period of this seed type by weeks to months (Barneby 1964:52).

Cheno-am seeds

Charred seeds identified as "cheno-ams" were recovered in a number of samples from all sites, and in 12 of 17 individual terraces at Cerro Juanaqueña (Table 8.6). The term "cheno-am" is used when referring to seeds that may belong to either of two different plant genera, *Chenopodium* (family Chenopodiaceae) and *Amaranthus* (family Amaranthaceae). Since the seeds of these two plants are quite similar, it can be difficult to tell which of these taxa is represented by a charred and/or highly degraded specimen. When such a seed is encountered, the "cheno-am" label is employed. Intact specimens of both seed types do, however, have diagnostic characteristics that can be used to distinguish between them (Fritz 2007). Seeds designated as cheno-ams were submitted to Gayle Fritz for more precise identification, and these results are presented in Chapter 9 along with more details about *Amaranthus* and *Chenopodium*. Fritz identified a possibly domesticated variety of *Amaranthus* as well as non-domesticated members of the genus *Chenopodium* in this sample.

Chenopodium and *Amaranthus* are similar in growth habit and environmental requirements, though *Chenopodium* often sprouts earlier in the growing season, while *Amaranthus* usually waits for summer rains. Flowering and seed production typically occur in the summer and early fall (K. Adams 1988:168; Kearney and Peebles 1960). *Chenopodium* is an annual herb, with alternate, often fleshy leaves, and small green flowers; amaranth is also an herbaceous, weedy annual, with ovate leaves and small green flowers. Both plants produce large amounts of tiny, compressed seeds, and both commonly occupy disturbed places, such as cultivated fields, river bottoms, and roadsides (Kearney and Peebles 1960; Wooton and Standley 1915). Currently, *Chenopodium incanum* and wild *Amaranthus palmeri* are common weeds on the Cerro Juanaqueña terraces and unmodified hillslopes. *A. palmeri* also occurs as a weed in modern irrigated and non-irrigated maize fields in the area.

Columnar-Celled Seed Coat Fragments (CCSCs)

During examination of samples for cheno-am seeds and possibly domesticated *Amaranthus*, Fritz and her students recognized small, irregularly fractured seed coat fragments characterized by a relatively smooth exterior surface and a thick cross section showing vertically oriented, tightly arranged, elongated cells. These have previously been called "columnar-celled seed coat fragments," or CCSCs (L. Huckell 2002:669). Since these were not systematically recovered during examination of the flotation samples, they are not listed in the tables or appendices of this chapter; however, a brief discussion is included here.

Although a number of plant families are being considered as potential candidates for this item (L. Huckell 1995:94, 2002:669), one highly likely candidate is the squash family, including both *Cucurbita* and *Apodanthera*, whose seeds display distinctive CCSCs as an exterior coat. CCSCs are common components in midden deposits at late preceramic, early ceramic, and Hohokam sites in the southern Southwest. More than 1,000 were reported from Donaldson and Los Ojitos (L. Huckell 1995:94), and seed coat fragments with a similar internal structure were recovered from Late Archaic/Early Agricultural period contexts at the Boatyard site (AA U:3:286/1352) in the Tonto Basin of central Arizona (L. Huckell 2002:674). Columnar-celled seed coat fragments have also been found at Stone Pipe, Santa Cruz Bend, Square Hearth, and Houghton Road (L. Huckell 1998). While absent from Cienega phase components at East Bank in southeastern Arizona, CCSCs did occur in Hohokam period contexts at that site (Diehl 1999a:85). Their high ubiquity in feature fill indicates that these items probably represent an important prehistoric food resource (L. Huckell 1995:94).

Cucurbita digitata type seeds

Two *Cucurbita digitata* type (family Cucurbitaceae) seeds, along with two other specimens possibly representing the seeds of wild cucurbits, were recovered from one terrace on Cerro Juanaqueña. The genus *Cucurbita* is confirmed to be among the first domesticated plants in the New World (Smith 1997), predating the domestication of maize (*Zea mays*). *Cucurbita pepo* was independently domesticated in both the southeastern U.S. and in Mexico (Robinson and Decker-Walters 1997). There is currently no evidence that the Cerro Juanaqueña remains represent anything other than wild species. *Cucurbita digitata*, *C. foetidissima*, and *C.* aff. *palmata* currently grow on the first river terrace above the Río Casas Grandes near Cerro Juanaqueña, and *C. digitata* grows on the lower cultural terraces on the site.

Wild members of the genus *Cucurbita* are coarse, perennial, herbaceous vines sprouting from thick roots. The plants are common on dry plains and mesas, and in alluvial soils. *C. digitata* has large (up to 10 cm in diameter), globe-shaped, pale yellow fruits with longitudinal stripes; the mature fruits of *C. foetidissima* are similar, except they are dark green with pale green stripes. In Arizona, *Cucurbita digitata* flowers from June to October, *C. foetidissima* from May to August, and *C.* aff. *palmata* from April to September (Kearney and Peebles 1960:822). The fruits typically ripen two to three months after flowering (K. Adams 1988:190), and each fruit can hold as many as 300 edible seeds (DeVeaux and Schultz Jr. 1985:454–455).

Eragrostis intermedia type caryopses (grains)

Fifty-two charred lovegrass (family Poaceae) grains were recovered from a terrace on Cerro Juanaqueña, and four from a terrace on Cerro los Torres. Plains lovegrass (*Eragrostis intermedia* Hitchc.) is a warm-season tufted perennial with culms (stems) up to 90 cm in height (Powell 1994). This grass is presently a component of the Plains and Great Basin Grassland and Semidesert Grassland floras (Brown 1982c, 1982d), and currently grows on Cerro Juanaqueña (Table 8.1). It is known to occupy rocky hills, canyons, and plains (Kearney and Peebles 1960:887; Powell 1994:132). Plains lovegrass flowers from July to September, and grain maturity proceeds rapidly thereafter. Ripe grains are available as early as July, and can be harvested through November (K. Adams 1988:229; Diehl 1997, 1999b).

Euphorbia type seed

One *Euphorbia* type seed was recovered from a Cerro los Torres terrace. Spurge plants (family Euphorbiaceae) are mostly weedy, annual or perennial herbs, but can vary in morphology from trees to shrubs, or even succulent, cactus-like species. They favor grasslands, pastures, roadsides, and waste places (Muenscher 1939). Some species in Arizona may flower and fruit from February into December (L. Huckell 1998). Although Brown (1982d) does not list *Euphorbia* as present in the Semidesert Grassland, and our recent survey failed to note any spurge plants near Cerro Juanaqueña, their current presence in the region has been documented by the frequent recovery of uncharred spurge seeds in flotation samples. Indeed five species of Euphorbia (*E. albomarginata* Torr & A. Gray, *E. cinerascens* Engelm., *E. hyssopifolia* L., *E. prostrata* Aiton., *E. radians* Benth.) were identified in the Janos Biosphere Reserve that includes most of the western portion of the municipio of Janos (Comisión Nacional de Áreas Naturales Protegidas 2006; Melgoza 2005). The use of spurge plants as food is not commonly assumed; however when recovered with other subsistence resources, it may be that prehistoric groups learned how to render their toxic components harmless (K. Adams 1980:41).

Ferocactus type seed

The recovery of a single charred *Ferocactus* (family Cactaceae) seed from a Cerro Juanaqueña terrace may only hint of the level of use of this valuable food in prehistory. Barrel cacti are often large plants (one species, *F. wislizeni*, can grow over 3 m in height), globular to cylindrical in shape, with thick, prominent ribs and well-developed hooked or straight spines. They are a common component in the Semidesert Grassland community (Brown 1982d) and grow today on the slopes of Cerro Juanaqueña and likely on the other archaeological sites as well. Flowering occurs in spring and summer, and the fruit typically ripens in the late summer and fall; ripe fruit can be retained on the plants for months (Kearney and Peebles 1960). Such "on the plant" storage reduces the need for human storage efforts.

Poaceae type caryopses (grains)

Besides lovegrass, caryopses of possibly two other grasses were recovered from four terraces on Cerro Juanaqueña and from one on Cerro los Torres. Cerro Juanaqueña and the other early cerros de trincheras are situated in and adjacent to two grassland communities that are well represented by a diversity of grass taxa (Chapter 2, Brown 1982c, 1982d). In a September 1998 visit to the area (Table 8.1), a wide variety of grasses were collected and identified (*Aristida ternipes*, *Bouteloua curtipendula*, *B. rothrockii*, *Chloris virgata*, *Digitaria californica*, *Echinochloa colonum*, *Eragrostis cilianensis*, *E. intermedia*, *Erioneuron pulchellum*, *Muhlenbergia porteri*, *Pleuraphis mutica*, *Setaria macrostachya*, *Sporobolus airoides*, *S. wrightii*, *Stipa comata*, *Tridens muticus*). These grasses and others have had a long history of importance to Southwestern groups (Doebley 1984).

Helianthus type achene

A single charred sunflower (*Helianthus annuus*, family Compositae) achene (fruit) was preserved within a terrace on Cerro el Canelo, hinting at the use of this food resource in prehistory. The common sunflower is a 0.5 to 2.5-m-tall annual, herbaceous plant that occupies roadsides and other disturbed situations (Correll and Johnston 1970; Kearney and Peebles 1960:903). Sunflowers currently grow in the Cerro Juanaqueña region on river terraces and in modern rainfed agricultural fields, where they produce ripe achenes in the late summer and early fall. The achenes were an important food to many prehistoric groups.

Common sunflower was domesticated in prehistoric times, indicated by archaeological recovery of achenes that are larger than the upper size limits of wild populations. It has long been widely accepted that the center of this event occurred in eastern North America (Smith 2006). However, this finding has been challenged in recent years by the discovery of large sunflower specimens directly AMS dated to about 4500 BP at the San Andres site on the southern Gulf Coast of Mexico (Lentz et al. 2001). The two specimens in question include a "...carbonized sunflower seed and a partially carbonized achene" (Lentz et al. 2001:370). Additional arguments for a Mexican origin were garnered from archaeological, ethnohistoric, ethnographic, and linguistic sources (Lentz et al. 2008). However, many scholars still overwhelmingly favor an eastern North American origin (Brown 2008; Harter et al. 2004; Heiser 2001, 2008a, 2008b; Reiseberg and Burke 2008; Smith 2006, 2008). It is possible that sunflowers were domesticated more than once in prehistory.

Regardless of the regional origin of domesticated sunflowers, nothing indicates that the Cerro el Canelo *Helianthus annuus* achene represents anything but a wild plant. While Heiser questions the presence of domesticated sunflower in prehistoric Mexico, he does acknowledge that "...the wild sunflower may have grown in northern most Mexico in early times" (Heiser 2008a:9). The Cerro el Canelo specimen confirms this, and that it was used by Late Archaic populations in Chihuahua.

Juglans type nutshell

A single piece of carbonized walnut (*Juglans*) (family Juglandaceae) nutshell was identified from a Cerro Juanaqueña terrace. Members of this genus vary from shrubs to very tall trees (Kearney and Peebles 1960:214). *Juglans* trees produce fruit in the form of a large (2.5 to 4.0 cm) globose nut enclosed within a dry husk. The nutshells of southwestern species are thick and dense, with shallow fissures on the external surface (Kearney and Peebles 1960; Powell 1998). Walnut trees favor canyon bottoms, streams, and other watercourses. In southern Arizona, *J. major* starts producing fruit by June, nut maturity is reached in September, and the trees are mostly devoid of nuts by the end of November (K. Adams 1988:269).

In addition to this single shell fragment, charred fragments of *Juglans* wood also occurred in a number of the terraces on Cerro Juanaqueña. This will be discussed in more detail below.

Leguminosae type seed

Two charred legume seeds were recovered from a terrace on Cerro Juanaqueña, a charred cotyledon fragment was recovered from a different terrace on Cerro Juanaqueña, and a single seed came from a Cerro los Torres terrace. Plants in the legume family (Leguminosae) are widespread in arid areas and include many plants used as human food, such as beans and peas. The family also contains poisonous members, including locoweeds (*Astragalus*) and other species useful as fish poisons (Kearney and Peebles 1960:395). The three charred legume seeds recovered are unlike *Astragalus nuttalliana*, which was found in other contexts on Cerro Juanaqueña. Therefore, another one or possibly two types of legume became burned on a terrace feature, possibly unintentionally. Legumes currently known in the area include mesquite (*Prosopis*), *Dalea*, *Macroptilium*, *Robinia*, and two species of *Senna*. The unknown specimens are likely not any of these.

Physalis type seed

Five charred *Physalis* seeds (family Solanaceae) were recovered in three flotation samples from Cerro los Torres and a *Physalis/Solanum* seed from Cerro Juanaqueña was found during the cheno-am analysis (see Chapter 9). Members of the genus *Physalis*, commonly known as groundcherry or husk tomato, are low growing annual or perennial herbs that favor disturbed areas such as agricultural fields or roadsides (Powell 1998:350). All species produce small, edible berries resembling miniature tomatoes encased in a papery husk (K. Adams 1980: 53; L. Huckell 1995:88). In Chihuahua, the plants may be most productive following the late summer rains. Although this genus is not listed as present in the grasslands of northern Chihuahua (Brown 1982d), Table 8.1 lists *Physalis hederaefolia* Gray as locally collected and *Solanum elaeagnifolium* Cav. and *S. rostratum* were identified in the municipio of Janos (Comisión Nacional de Áreas Naturales Protegidas 2006; Melgoza 2005).

Portulaca type seed

A single charred *Portulaca* seed (family Portulacaceae) was recovered from a Cerro el Canelo terrace and five from three Cerro Juanaqueña terraces were found during the cheno-am analysis. Purslane plants are small annual or perennial, somewhat succulent herbs that prefer disturbed, dry soils in sunny habitats (Kearney and Peebles 1960:290). The plants can often be found covering the ground on plains and mesas after summer rains. One species, *P. halimoides*, has recently been seen growing on a river terrace above the Río Casas Grandes, and uncarbonized seeds were common in the flotation samples. Many species of *Portulaca* are available for use in the late summer/early fall (K. Adams 1988:416; Kearney and Peebles 1960:290–291).

Prosopis type cotyledon fragment

A single charred cotyledon (seed half) fragment, with the distinctive markings of *Prosopis* (mesquite), was recovered from a Cerro Juanaqueña terrace. Mesquite plants (family Leguminosae) are thorny shrubs or trees (Kearney and Peebles 1960:401–402). The fruit is a coiled, slightly curved, or straight pod with embedded and partitioned seeds (Powell 1998:177). Although mesquite commonly occurs along watercourses or in other areas where the water table is relatively high, it has been reported that the roots can occasionally reach a depth of 60 feet (Kearney and Peebles 1960:402). They often form dense stands on pasture and rangeland, and may be scattered as a smaller shrub on lower mountain slopes and grasslands (Kearney and Peebles 1960). *Prosopis glandulosa* and *P. juliflora* are common components in the grasslands of northern Chihuahua today (Brown 1982c, 1982d), and *P. glandulosa* was observed growing on the slopes and terraces on Cerro Juanaqueña, and on the river terraces above the Río Casas Grandes. In southern Arizona, mesquite trees typically produce fruit in May and June; fruit maturity follows in late June and July (Cornejo et al. 1982).

Rhus aromatica type seed

Two charred *Rhus aromatica* seeds were recovered from a terrace on Cerro Juanaqueña. Sumac, or lemonade berry (family Anacardiaceae), plants are shrubs that occur in dry or rich soils, on mesas, slopes, and dry cliffs (Kearney and Peebles 1960:322). Although not recorded in our plant survey (Table 8.1), *R. aromatica* var. *flabelliformis* Shinners was identified in the region (Comisión Nacional de Áreas Naturales Protegidas 2006; Melgoza 2005). A number of *Rhus* species are also present in the grasslands of northwestern Chihuahua (Brown 1982c, 1982d), including *R. choriophylla* and *R. virens* (Comisión Nacional de Áreas Naturales Protegidas 2006). The plants produce fruit in the summertime, but the fruit of one species in southeastern Arizona (*Rhus trilobata*) has been observed clinging to the plants between October and January (K. Adams 1988:458–459; Kearney and Peebles 1960:522–524).

Salvia type seeds

Six charred seeds of a seed type resembling chia (*Salvia columbariae*) were recovered in two terraces at Cerro Juanaqueña. Chia, an annual member of the mint family

(Labiatae), prefers sandy soils in disturbed habitats, such as arroyo bottoms. It is a cool-season plant, blooming and setting seed in the springtime, from March to May (Kearney and Peebles 1960:741). Although this plant was not seen in the September 1998 visit to the area, by late summer all visible evidence of the spring-active plant had usually disappeared from the landscape, except for seeds in the soil.

Scirpus type achene

Four charred bulrush (*Scirpus* type) achenes preserved in three Cerro Juanaqueña terraces. Bulrushes belong to the sedge family (Cyperaceae). They may be annuals or perennials, sometimes small and grass-like, or sometimes over a meter tall (Kearney and Peebles 1960:151; Wooton and Standley 1915:114). They thrive in moist habitats, favoring marshes, stagnant ponds, ditches, and stream banks (Kearney and Peebles 1960:151). Presently, *Scirpus pungens* grows at La Palotada, a warm springs habitat in the Janos area. Flowering and fruiting occurs in the summer, and mature achenes would be available in August (K. Adams 1988).

Sphaeralcea type seed

Archaeologists recovered one globemallow (*Sphaeralcea* type) seed from a terrace on Cerro Juanaqueña. Globemallows (family Malvaceae) are perennial herbs that often form dense patches in disturbed soils (L. Huckell 1998:72; Wooton and Standley 1915:420). *Sphaeralcea angustifolia* presently grows in the Janos area on a river terrace above the Río Casas Grandes. Globemallow may produce fruit and seeds both in the spring and once again in the late summer, in response to the summer rainy season (Kearney and Peebles 1960:541).

Trianthema type seed

Nine charred *Trianthema* (family Aizoaceae) seeds were preserved in one Cerro Juanaqueña terrace, and a seed and seed fragment were recovered from a separate terrace on this site. Carpetweed, or horse purslane, plants are annual herbs with long prostrate stems and bright green, fleshy leaves. The plants prefer disturbed locations and are common weeds on irrigated lands (Kearney and Peebles 1960:281). *Trianthema portulacastrum* was present in the Janos area in September 1998, growing on lower site terraces on Cerro Juanaqueña and on river terraces in the floodplain. Uncharred, likely modern carpetweed seeds preserved frequently within the archaeological deposits, reflecting its current abundance in the vicinity. Horse purslane plants tend to come up quickly and so thickly that they literally "carpet" the landscape following the onset of the summer rains. The fleshy leaves can be harvested from late spring to late fall, and large quantities of seeds are produced in the summer and early fall (Curtin 1984; L. Huckell 1998:64).

TABLE 8.10. Rank Order, Based on Relative Presence, of Charred Wood Types and other Vegetative Parts in 91 Flotation Samples (Including Sterile Samples) for Cerro Juanaqueña.

		Flotation Samples (*N* = 91)		
Taxon	Part	Number	%	Rank order
Prosopis type	wood	33*	36.3%	1
Fouquieria type	wood	12*	13.2%	2
Atriplex type	wood	11*	12.0%	3
Unknown	wood	9	9.9%	4
Poaceae type	stem fragment	6*	6.6%	5
Juglans type	wood	4*	4.4%	6
Larrea type	wood	4	4.4%	6
Leguminosae type	wood	4*	4.4%	6
Populus/Salix type	wood	4*	4.4%	6
Monocotyledon type	tissue fragment	3*	3.3%	
Pinus type	wood	2	2.2%	
Compositae type	wood	1*	1.1%	
Fraxinus type	wood	1	1.1%	
Juniperus type	wood	1*	1.1%	
Monocotyledon type	stem fragment	1	1.1%	
Unknown	tissue fragment	1	1.1%	

* Also recovered in macrofossil samples.
Note: Wood charcoal was not examined from 66 flotation samples from 1998 and 2000.

Charred Wood Types

Wood identification was attempted within only a portion ($n = 91$) of the total analyzed flotation samples ($n = 157$), with samples from 1997 and 1999 excavations being targeted for charred wood analysis. Charred wood fragments were also examined within macrofossil samples from all years. The record of charred wood at Cerro Juanaqueña is extensive (Table 8.10). Mesquite (*Prosopis*) ranks highest in recovery. Ocotillo (*Fouquieria*) and saltbush (*Atriplex*) were also present. It is possible the ocotillo may represent the burned framework of temporary brush dwellings or the interior support posts of some wattle and daub structures. In addition, the extensive presence of maize cob parts suggests use of the leftover cobs as a fuel or tinder. Woody materials from at least nine additional taxa were rarely recovered.

As with the food plants, flotation samples from the same 17 terrace features were examined to determine the breadth of wood use across the site (Table 8.11). This analysis confirms that mesquite, saltbush, and ocotillo were the most broadly used, along with maize cobs. As with the food record, Terrace 487 and Terrace 222 each preserved the highest number of charred wood types, some present only as macrofossils. Again, enhanced preservation conditions are likely an important factor, although variation in plant usage among occupants of different terraces cannot be ruled out.

Charred fragments of mesquite (*Prosopis*) and ocotillo (*Fouquieria*) were the only wood types identified from the Cerro Juanaqueña rock rings, which may have served as the bases of wattle and daub structures used for dwellings or possibly other functions such as storage (see Chapter 3). Rock ring size, presence of metates on floors ($n = 2$), and presence of a hearth on one floor supports the argument that the rock rings served for domestic use. If this is the case, perhaps the use of ocotillo as part of the above ground infrastructure mirrors historic uses reported for the Seri. Mesquite could have provided both construction elements and fuelwood.

Only two of the other sites preserved charred wood fragments (Table 8.12). The limited data suggest that Cerro el Canelo occupants chose ash (*Fraxinus*) and other unknown wood(s). At Cerro los Torres, the pattern of wood use is based on very few samples.

Atriplex type wood

Charred saltbush (family Chenopodiaceae) wood fragments were found in four terraces on Cerro Juanaqueña, and in one terrace each on Cerro los Torres and Cerro el Canelo. Saltbush plants exist in both annual and perennial forms, and are mostly shrubs or herbs covered with white, meal-like particles. The fruit is small and bladder like (utricular), with a thin pericarp and one free seed (Kearney and Peebles 1960:254; Powell 1998:111). *Atriplex* species are very salt-tolerant plants, preferring relatively compact soils, often in open flats and disturbed areas (Powell 1998:111). Four-wing saltbush (*A. canescens*) and *A. wrightii* were observed growing on a river terrace above the Río Casas Grandes near Cerro Juanaqueña in September 1998 (Table 8.1). Most *Atriplex* species flower and set seed in summer and early fall (Kearney and Peebles 1960:255).

Compositae type wood

Charred wood fragments from an unidentified member of the sunflower family (Compositae) were recovered from three terrace features on Cerro Juanaqueña. Generally members of the sunflower family are shrubs or herbaceous perennials or annuals (Kearney and Peebles 1960). Numerous Compositae have been reported in the grasslands of northern Chihuahua, including species of *Aster, Bahia, Brickellia, Cirsium, Ericameria, Flourensia, Gutierrezia, Isocoma, Parthenium, Ratibida, Senecio, Viguiera*, and *Zinnia* (Brown 1982d). The stems of most Compositae species do not produce much heat or lasting coals when burned, so their use exclusively as a fuelwood is doubtful.

Dicotyledon type wood, stem fragment

The large circle feature at Cerro el Canelo produced charred wood fragments and a stem fragment representing one or more unidentified dicotyledon. Dicots are one of two major groups of angiosperms (flowering plants), characterized by embryos with two cotyledons, or seed leaves, and flower parts that occur in multiples of four or five (Powell 1998:13). The archaeological specimens could represent any of hundreds of dicot species present in the local environment.

Fouquieria type wood

Charred ocotillo (family Fouquieriaceae) wood preserved in seven terraces and one rock ring (R250) on Cerro Juanaqueña, and in one terrace on Cerro el Canelo. Ocotillo plants are shrubs with little or no trunk and thorny, wand-like stems that can grow up to 9 m in length (Powell 1998:273–274). Brown (1982d) lists *F. splendens* as a dominant component of the scrub-shrub component in present-day northwestern Chihuahua. The plant is very common on dry mesas and slopes (Kearney and Peebles 1960:640; Wooton and Standley 1915:428), and it presently grows on the unmodified slopes of Cerro Juanaqueña and other hill slopes in the Janos area. Ocotillo produces flowers and fruit from April until June, sometimes later, depending on rains (Kearney and Peebles 1960:640).

Fraxinus type wood

Abundant charred ash (family Oleaceae) wood fragments were recovered from the large stone circle feature near the base of Cerro el Canelo. Some of these fragments are only partially charred, raising the possibility that they are of recent origin. Charred *Fraxinus* wood was also recovered from Terrace 487 at Cerro Juanaqueña. These plants are usually trees, but occasionally occur as large shrubs (Kearney and Peebles 1960:641; Powell 1998:315). While some species of ash (i.e., *Fraxinus greggii* and *F. cuspidata*) typically occupy steep, rocky slopes, others (i.e., *Fraxinus velutina* and *F. lowellii*) prefer more mesic (damp) habitats along streams and creeks (Kearney and Peebles 1960:642; Powell 1998:315–318). In September 1998, *Fraxinus* trees were observed growing adjacent to the La Palotada warm springs habitat near Janos (Table 8.1). Ash trees

Table 8.11. Presence of Charred Wood and Other Charred Vegetative Parts from 17 Terraces at Cerro Juanaqueña in Flotation Samples from 1997 and 1999 Excavations.

		Terrace Number																	Total Terraces with each
Taxon	Part(s)	T487	T222	T387	T413	T163	T10	T6	T537	T175	T287	T508	T297	T97	T126	T415	T290	T167	Taxon/Part(s)
Prosopis type	wood	×	×	×	×	×	×	×	×	*	*	*	×	*	*	*			15
Atriplex type	wood	×	×	*	×	*	×	*											7
Fouquieria type	wood	×	×	×		×	×	*					×						7
Poaceae type	stem fragment	*	×	×	*		×												5
Juglans type	wood	×		×	×					*		*							5
Populus/Salix type	wood	×		×	*							*							4
Compositae type	wood	*	×								*								3
Leguminosae type	wood		×					×	×										3
Unknown	wood, twig fragment					×				×	×								3
Juniperus type	wood				*	×													2
Larrea type	wood		×						×										2
Monocotyledon type	tissue fragment	×		*															2
Pinus type	wood	×			×														2
Fraxinus type	wood	×																	1
Monocotyledon type	stem fragment		×																1
Unknown	tissue fragment	×																	1
Total taxa within each terrace		11	8	7	7	5	4	4	3	3	3	3	2	1	1	1	0	0	63

* Recovered only as a macrofossil specimen in the terrace.

TABLE 8.12. Rank Order, Based on Relative Presence, of Charred Wood and Other Vegetative Parts in Flotation Samples (Including Sterile Samples) from Sites Other than Cerro Juanaqueña.

		Cerro el Canelo (*n* = 13)			Cerro los Torres (*n* = 2)		
Taxon	Part(s)	No.*	%	Rank	No.*	%	Rank
Atriplex type	wood				1	50.0%	2
Fraxinus type	wood	2**	15.4%	1			
Poaceae type	stem fragment				2	100.0%	1
Monocotyledon type	tissue fragment	1**	7.7%	2			
Prosopis type	wood	1**	7.7%	2			
Unknown	wood, twig fragment	2	15.4%	1			
Unknown	tissue fragment	2	15.4%	1			
Fouquieria type	wood	**					

* No. is the number of flotation samples containing the taxon/part, excluding flotation samples not examined from 1998 and 2000 for charred wood. Nothing was recovered from Ophelia's Well.
** Also or only in macrofossil samples.

usually produce fruit in early summer (K. Adams 1988:241; Kearney and Peebles 1960:642; Powell 1998: 315–318).

Poaceae type stem fragment

Grass (Poaceae) stem fragments preserved in flotation samples from three Cerro Juanaqueña terraces, and one on Cerro los Torres. These specimens are harder to identify to genus than the grass caryopses (grains) previously discussed. Although these remains possibly represent waste material from processing grass grains for meals, grass stems were likely also gathered for material culture needs.

Juglans type wood

Charred fragments of walnut (family Juglandaceae) wood were recovered in five Cerro Juanaqueña terraces. As noted above, one *Juglans* type nutshell fragment was also recovered from a terrace there. Walnut trees presently grow along the Río Casas Grandes in the valley and would have provided a ready source of fuelwood for the inhabitants of Cerro Juanaqueña. For more information about the ecology of walnut, see the discussion of the *Juglans* nutshell fragment above.

Juniperus type wood

Four fragments of charred juniper type (family Cupressaceae) wood were recovered from two terraces on Cerro Juanaqueña. Members of this genus are evergreen trees or shrubs with scale-like or awl-shaped leaves, short trunks with reddish wood, and shredded or furrowed bark. They have small, berrylike cones with one to several seeds (Kearney and Peebles 1960:58; Powell 1998:24). Juniper is common at moderate elevations in rocky, well-drained, and/or poor soils, commonly in association with oak (*Quercus* sp.) and piñon (*Pinus edulis*) (Kearney and Peebles 1960:58; Powell 1998:24). Throughout New Mexico, the trees frequently occur on foothills and high plains (Wooton and Standley 1915). Juniper is listed as a member of the Semidesert Grassland flora (Brown 1982d); although our vegetation survey failed to recognize it in the immediate vicinity of Cerro Juanaqueña, *Juniperus monosperma* (Engelm.) Sarg. and *J. deppeana* Steud. was recorded in the region (Comisión Nacional de Áreas Naturales Protegidas 2006; Melgoza 2005). K. Adams (1988:283) reports that in southeastern Arizona, alligator juniper (*J. deppeana*) produces fruit starting in April, but these ripen as late as the following September. The ripe fruit may remain attached to the plants for up to nine months, offering on-the-plant storage.

Although juniper is not common in the area at present, it does occur in shrub form in association with grasslands (Melgoza 2005). It is readily found at higher elevations within 40 km or less to the west. A limited record of charred *Juniperus* wood fragments at Cerro Juanaqueña suggests that juniper wood was possibly more accessible in the past either as living trees or shrubs or carried as driftwood by the Río Casas Grandes. Otherwise, people must have occasionally traveled some distance to acquire this resource.

Larrea type wood

Larrea type (family Zygophyllaceae) charred wood fragments were recovered in flotation samples from two terraces at Cerro Juanaqueña. Creosote plants are shrubby evergreens, up to 3 m tall, with glossy green leaflets (Kearney and Peebles 1960). This plant is now a common member of the local grassland flora (Brown 1982d). The shrubs particularly favor alluvial, hardpan soils. Although flowering occurs sporadically throughout the year, it happens most profusely in the spring (Kearney and Peebles 1960; Powell 1998; Wooton and Standley 1915).

Leguminosae type wood

A limited amount of charred wood representing the family Leguminosae was recovered from three terraces on Cerro Juanaqueña. These specimens could represent wood from legume trees such as *Mimosa*, *Acacia* or *Prosopis*, all members of the modern flora (Brown 1982d), but they could not be reliably identified beyond the family level. Mesquite (*Prosopis*) was extensively used as a fuelwood at Cerro Juanaqueña and is well represented at Cerro el Canelo. Other local legume species likely were also sought for this purpose.

Monocotyledon type stem fragment, tissue fragment

Limited numbers of unidentified monocot stem and tissue fragments were recovered from three Cerro Juanaqueña terraces, and from one on Cerro el Canelo. Monocots are a major group of angiosperms, characterized by an embryo with one seed leaf, and flower parts present in multiples of three (Powell 1998:13). Monocots include some of the largest and most familiar groups of plants, such as agaves, palms, and grasses (Powell 1994, 1998:42–60). The stem and tissue fragments recovered from archaeological contexts could derive from succulents such as *Agave* or *Yucca*, or from Poaceae stems, including *Zea mays*. Presently, *Agave* aff. *palmeri*, *Agave* aff. *parryi*, and a species of narrow-leaf *Yucca* all grow on unmodified hillslopes of Cerro Juanaqueña and Cerro el Canelo. *Yucca elata* Engelm. was also recorded in the region (Comisión Nacional de Áreas Naturales Protegidas 2006; Melgoza 2005).

Pinus type wood

Charred pine (family Pinaceae) wood was recognized in two separate terraces at Cerro Juanaqueña. Members of this genus are evergreen trees, with thin, scaly bark, and needlelike leaves. The seeds are encased in woody, scaly cones. Upon maturity, the scales spread and separate, releasing the seeds (Kearney and Peebles 1960:50; Powell 1998:33). Most *Pinus* species commonly occupy slopes, mesas, and canyons at higher elevations (Powell 1998:34–38). The nearest pine-oak woodlands are 60 to 70 km to the southwest, west, and west-northwest of Cerro Juanaqueña, among the higher peaks of the Sierra Madre at elevations greater than 2300 m. Pine species in these areas include *Pinus ponderosa* var. *arizonica*, *P. cembroides*, and *P. engelmannii*. These pine-oak woodlands also contain *Juniperus deppeana*, *J. monosperma*, and a variety of oaks. In addition to these pine and juniper species, *Pinus edulis* is also reported in the region (Comisión Nacional de Áreas Naturales Protegidas 2006). Generally, pine nuts are ripe and available for collection in the mid-fall (K. Adams 1988:386).

The presence of charred pine wood at an archaeological site in the Río Casas Grandes floodplain suggests the possible prehistoric use of mountain resources. Pines and junipers could have been closer to Cerro Juanaqueña in sheltered canyons in the past, particularly if the environment was wetter and potentially cooler than today. Alternatively, limited amounts of pine and juniper wood may have found their way to the area as driftwood carried by the waters of the Río Casas Grandes.

Populus/Salix type wood

Cottonwood (*Populus*) and/or willow (*Salix*) type (family Salicaceae) charred wood fragments were preserved in four terraces at Cerro Juanaqueña. Cottonwood is a large tree, between 15 and 30 m in height, with rough, light-colored bark, and deltoid-, ovate-, or lanceolate-bladed leaves (Kearney and Peebles 1960:207; Powell 1998:67). Willow trees grow up to 12 m high and have linear, lanceolate, elliptic, or oblong blades either with toothed or untoothed margins (Kearney and Peebles 1960:209; Powell 1998:61). Both trees are present in the Semidesert Grassland community today. They are usually confined to more mesic locations such as canyons and arroyos in the hills and adjacent to waterways. Modern specimens are currently found in the Río Casas Grandes river bottom below Cerro Juanaqueña. Both genera produce fruits in late spring and early summer (K. Adams 1988; Powell 1998:67).

Prosopis type wood

Mesquite (family Leguminosae) was the most ubiquitous charred wood type recovered in the Cerro Juanaqueña flotation samples (Table 8.10), and preserved in nearly all the terrace features (Table 8.11). It was also recovered at Cerro el Canelo (Table 8.12). In a vegetation survey of the area, *Prosopis glandulosa* was seen growing on the lower and upper site terrace features at Cerro Juanaqueña and on the alluvial terraces in the valley bottom below. However, although clearly a major component of this community, in none of these locations was the species dense enough to presently be considered a mesquite *bosque*.

Discussion

Age of the Occupation

Agricultural debris in the form of maize cob and kernel fragments, and other short-lived plant remains, provided the bulk of materials submitted for radiocarbon dating of the cerros de trincheras. Archaeobotanical analysis of plant parts provided the means to obtain a tight array of AMS radiocarbon dates (Chapter 5) and to demonstrate that these cerros de trincheras were occupied substantially earlier than other known trincheras sites in the southern Southwest.

Subsistence

The archaeobotanical evidence in these cerros de trincheras indicates that maize agriculture was an important economic pursuit. This is evident through recovery of durable maize parts preserved by charring in all four of the cerros de trincheras sites, including in 51% of flotation samples and 15 of 17 terraces at Cerro Juanaqueña. However, if it can be assumed that the plant record in the two best-preserved terraces at Cerro Juanaqueña (Terrace 222 and Terrace 487) provides an approximation of the range of recovery, a slightly different picture emerges. In these locations, maize parts were preserved in 25% (two of eight) of flotation samples from Terrace 222, and in 85% (23 of 27) of flotation samples from Terrace 487. This range of maize recovery may be due to sampling variability, or possibly differential use of the terraces in prehistory.

Possibly domesticated *Amaranthus* was also present. Garden weeds such as *Chenopodium* appear to also have played a repeating role in subsistence. The common occurrence of both maize and plants of disturbed habitats together reflects use of cultigens and their associated useful field weeds. The record of additional wild plants sought as food differs to some extent among the sites, and factors related to archaeological sampling and/or preservation cannot be ruled out. However, the presence of other wild plants is extremely low compared to that of maize; for example, the most frequently recovered species, after cheno-ams, is *Trianthema* type and it occurred in only six flotation samples of 157 examined (3.8%) at Cerro Juanaqueña (Table 8.8). Excluding the "unknown" seeds, only 29 flotation samples (18.5%) from Cerro Juanaqueña contained other wild edible, reproductive specimens.

The findings at Cerro Juanaqueña can be examined in light of economic models that question the importance of maize in the Late Archaic period (Gregory and Diehl 2002) (Chapter 1). A number of Early Agricultural sites ($n = 5$) in southern Arizona are considered to have a very diverse subsistence economy based on: (a) high maize ubiquity, above 80% in some cases (usually counted by its presence in features excavated); (b) correspondingly high ubiquities of several wild resources; and (c) high mean diversity (20.6) of wild plant resources (Gregory and Diehl 2002:217). Six later Ceramic period sites presumably reflect more entrenched agricultural economies, with similar high maize ubiquity but a statistically significant lower mean diversity of wild resources (10.8). Gregory and Diehl (2002:213) point out that high maize ubiquity, in and of itself, should not be considered the sole indicator of agricultural dependence nor of sedentism. Archaeobotanical results of well-sampled Cerro Juanaqueña reveal: (a) maize ubiquity of 51% for the site as a whole; (b) extremely low ubiquity of all wild resources except weedy cheno-ams (25.5%, Table 8.8); and (c) a high diversity of total plant food taxa ($n = 20$). This Cerro Juanaqueña patterning suggests that diet breadth was broad, and that both agricultural products and wild plant foods played important roles in subsistence, similar to many other Early Agricultural period sites in the Tucson Basin.

The most likely location for successful farming would have been the floodplains located at the base of each cerro de trincheras. These large expanses of alluvial sediment were periodically susceptible to summer monsoonal flooding of rivers, but would have provided fairly good farming locations, particularly considering their locations at stream reach boundaries (Nials et al. 2011), although drought and low nitrogen levels also may have provided obstacles to farming (Chapters 2 and 12). Although floodplains can at times receive more water than plants need or can sustain, the risk of occasional crop loss could have been offset by frequent successes.

Plants Sought for Non-Subsistence Reasons

Identification of nonreproductive plant parts has revealed that common fuelwood and construction elements utilized in the past were mesquite, saltbush, and ocotillo. Ocotillo may have provided framework for temporary brush dwellings and interior supports of wattle and daub structures, or have been used to construct fencing as in the historic period. Leftover maize cobs routinely provided tinder or fuel, and people occasionally carried in parts of many additional woody species, likely to meet a wide variety of material culture needs.

Seasonality and Sedentism

The record of plant reproductive parts contains some information on seasonality of site occupation. Assuming agricultural fields were relatively close, some individuals minimally occupied the cerros de trincheras during the agricultural growing season. Tasks would have included field preparation in the springtime through harvesting and drying of crops in the fall. Most people likely resided in the vicinity of their farming fields for a major portion of the year.

The wild plant record confirms this multi-season interpretation. Chia seeds are a springtime resource, and lovegrass grains a late summer/fall ripening grain. Many wild plants were available following summer rains, such as amaranths, wild cucurbits, carpetweed, portulaca, sunflowers, and groundcherry. Other plants retain useful parts on the plant for some time following ripening, such as the persistent pods of the cool season plant *Astragalus nuttalliana*.

Together, the agricultural and wild plant resources suggest multiple season occupation from spring through fall. The plant record remains mute regarding winter occupation, true of most prehistoric plant records (K. Adams and Bohrer 1998). However, winter occupation of the cerros de trincheras is a reasonable possibility, especially if people felt obligated to remain to watch over and have access to stored crops.

Paleoenvironment

Although plants gathered by ancient people reflect what is culturally preferred, they also reveal resources available in the local environment. People living at Cerro Juanaqueña harvested resources from a number of habitats. Numerous foods came from agricultural fields and the disturbed edges of these fields. People sought mesquite, ash, and cottonwood/willow wood, chia seeds, bulrush achenes, and walnut nuts and wood from damper locations such as the edges of the nearby river or arroyo bottoms. The drier hillslopes offered barrel cactus seeds and lemonade berry fruit. Stands of native grasses likely covered river terraces and uplands in prehistory, as they do today. Most of these plants sought in prehistory grow in the local area today (Table 8.1), although the relative proportions may well have differed. Historic activities on the landscape, including domestic animal grazing, have contributed to local extirpation of taxa, including altered fire regimes associated with grassland reduction, shrub expansion, and an increase in weeds from other continents. The modern and ancient environments are discussed more fully in Chapters 2 and 12.

The ancient plant record suggests people may have traveled to acquire a few plants for special needs. Limited recovery of charred juniper and pine wood fragments from Cerro Juanaqueña suggests access to woodlands in the past. Although juniper occurs today in shrub form associated with grasslands 30 km southwest of Cerro Juanaqueña, both taxa grow in pine-oak woodlands some 60 to 70 km distant from the study area. It is unknown whether they grew nearer to Cerro Juanaqueña in the past, possibly in sheltered canyons where relative available moisture and cold air drainage provided appropriate growing conditions. An alternative explanation might be that people opportunistically gathered small pieces of pine and juniper wood carried as driftwood by the nearby Río Casas Grandes.

Site Formation Processes

The individual terraces at Cerro Juanaqueña differ notably in their records of charred plant materials. Those terraces with deeper deposits had some of the best preserved records. The plant record also revealed continual deposition of uncharred items into site stratigraphy, including vertical movement of smaller items downward. These insights supported arguments that many terraces were locations of residential occupation, where continuous deposit of debris of daily living occurred.

Early Agriculture

Early farming in the southwestern United States and northwestern Mexico seems clearly established at Cerro Juanaqueña by 1300 cal BC, and likely at the other Early Agricultural period cerros de trincheras in the region. Maize provided an important component of the local economy, along with a possibly domesticated amaranth and numerous wild plant resources. Based on the plant record and evidence presented elsewhere in this volume, these early farmers in northwest Chihuahua successfully supported a substantial population as they continued to harvest wild plant resources.

CHAPTER 9

Analysis of Amaranth and Chenopod Seeds

Gayle J. Fritz[1]

Introduction

Chenopod (*Chenopodium* spp.) and amaranth (*Amaranthus* spp.) seeds from the American Southwest are routinely classified as "cheno-ams" because seeds from these two genera are morphologically similar and have been used in much the same way by native peoples in the region. In eastern North America, however, fewer species are native, and seeds differ more in size and shape, allowing archaeobotanists to discriminate more easily between the two genera. The need to do so is heightened by the fact that *Chenopodium berlandieri* ssp. *jonesianum* was one of several native seed-bearing plants domesticated in the Eastern Woodlands (Fritz and Smith 1988:287; Smith 1984, 1985). Although domesticated amaranths were grown by a number of indigenous farmers in what is now the southwestern United States and northwestern Mexico both before and after contact (Nabhan 1979b), the practice of lumping charred archaeological specimens into a "cheno-am" category makes it difficult to determine when cultivation of amaranth as a grain crop began north of Mesoamerica (Fritz 2007).

Early in the cerros de trincheras project, Karen Adams's analysis (Chapter 8) of the light fractions from flotation samples showed that seeds in the "cheno-am" category were both ubiquitous and problematic. The first year's assemblage (1997) introduced the possibility that small-seeded annuals such as amaranth and chenopod might have been major constituents of the plants being processed in the many basin-shaped metates found on the artificial terraces. My close examination of these seeds revealed that many of the cheno-ams could in fact be identified to genus, and that those classifiable as amaranths (*Amaranthus* sp.) had traces of very thin seed coats (testae) making them resemble domesticated rather than wild species. Those classifiable as chenopods (*Chenopodium* sp.), on the other hand, had relatively thick seed coats with surface sculpturing resembling wild or weedy species rather than known crops.

Subsequent seasons altered assessments because the small grains were not as ubiquitous as maize, which was abundant enough to have served as a staple food source. Still, we wanted to know whether or not a crop such as amaranth was grown along with maize in the ancient fields along the Río Casas Grandes. Therefore, specimens identified as "cheno-ams" were sent to me in St. Louis for further examination after project paleoethnobotanist Karen Adams completed her work with them. Key samples were targeted for re-sorting of the 0.5–1.0 mm portion of the light fractions, since the abundance of modern, uncharred (but still black or dark red) chenopod and amaranth seeds in these samples made sorting exceedingly difficult, and therefore reexamination was advisable. In addition, fine fractions with particles larger than 0.425 mm and smaller than 0.5 mm or 0.7 mm that had not been previously examined were scanned. These consisted of particles that fell below the cut-off point for initial examination. Although cheno-am specimens from these sites exceed 0.5 mm in diameter, with some as large as 1.0 mm, many can fall through a 0.5 mm geological sieve if oriented diagonally to the mesh. In summary, the specialized analysis of cheno-ams consisted of the following steps and methods:

1. Identification to genus, whenever possible, of seeds originally identified as "cheno-ams." These are listed in Table 9.1.
2. Re-scanning of particles 0.5–1.0 mm of selected samples from which "cheno-ams" had already been pulled. These samples are from the same proveniences as the Table 9.1 samples. The detailed results of this analysis are listed in Table 9.2.
3. Scanning of the fine component of the light fraction not previously examined. This usually consisted of particles larger than 0.425 mm and smaller than 0.5 mm or 0.7 mm. (The normal cut-off point for examination by project paleoethnobotanist, Karen Adams, was 0.5 mm, but some samples were cut off at 0.7 mm.) These are listed in Table 9.3, and the results of all three sample types are summarized in Table 9.4.

Table 9.1. Reexamination of Cheno-am Seeds.

Site Name	Terrace	Unit	Level	Bag	Year	Original Id Cheno-ams	Amaranth	Chenopod	Cheno-ams	Atriplex	Uncharred ch. or am.	Unidentified Seed	Unidentifiable Seed Fragment	Total Examined
C. Juanaqueña	6	1	4	93	1997	3	2		1					
C. Juanaqueña	6	1	4	108	1997	3	1	1	1					
C. Juanaqueña	6	1	5	120	1997	1			1					
C. Juanaqueña	10	1	9	38	1998	4	4							
C. Juanaqueña	10	1	11	43	1998	2	1		1					
C. Juanaqueña	97	1	7	26	1998	2			1		1			
C. Juanaqueña	126	1	10	60	1998	1			1					
C. Juanaqueña	163	2	7	40	1998	2			2					
C. Juanaqueña	163	1	7	43	1998	7			5		2			
C. Juanaqueña	163	2	8	48	1998	3			3					
C. Juanaqueña	222	3	8	85	1997	4			4					
C. Juanaqueña	222	3	11	95	1997	11	1	3	7					
C. Juanaqueña	222	2	13	100	1997	6		5	1					
C. Juanaqueña	222	3	6	1002	1997	6		2	3	1				
C. Juanaqueña	387	2	4	20	1998	3					3			
C. Juanaqueña	387	2	5	23	1998	5	1	1	3					
C. Juanaqueña	387	2	6	29	1998	1							1	
C. Juanaqueña	387	2	7	34	1998	2	2							
C. Juanaqueña	387	2	8	38	1998	1		1						
C. Juanaqueña	387	1	8	41	1998	2			2					
C. Juanaqueña	387	1	11	57	1998	2	1	1						
C. Juanaqueña	487	2	8	27	2000	1	1							
C. Juanaqueña	487	2	10	33	2000	2		1	1					
C. Juanaqueña	537	4	6	73	1997	2	1		1					
Subtotal	24 samples						15	15	38	1	6	0	1	76
C. el Canelo	94	2	4	22	2000	3		2				1		
C. el Canelo	94	2	6	28	2000	5	1	2	1			1		
C. el Canelo	94	2	9	36	2000	10		8	2					
Subtotal	3 samples						1	12	3	0	0	2	0	18
C. los Torres	1	1	7	27	1998	2		2						
C. los Torres	1	1	8	32	1998	4		3	1					
C. los Torres	1	3	8	53	1998	2	1	1						
C. los Torres	1	3	9	55	1998	5		2	3					
C. los Torres	1	3	10	59	1998	5/6*	2	1	2					
C. los Torres	1	3	11	62	1998	6	2	1			1	1	1	
Subtotal	6 samples						5	10	6	0	1	1	1	24
Cerro Vidal	20	1	5	20	1998	2	1	1						
Subtotal	1 sample						1	1	0	0	0	0	0	2
Total	34 samples					120/121	22	38	47	1	7	3	2	120
Percent							18%	32%	39%	1%	6%	3%	2%	100%

*Only five seeds in a capsule that should have contained six.

TABLE 9.2. Seeds from Re-examination of Fine Particle Fraction that Previously Yielded "Cheno-ams."

Site Name	Terrace	Unit	Level	Bag	Year	Amaranth	Chenopod	Cheno-am	Grass Seed	*Portulaca* sp.	*Solanum/ Physalis*	Unidentified Seed	Unidentifiable Seed Fragment	Total
C. Juanaqueña	6	1	4	93	1997									
C. Juanaqueña	6	1	4	108	1997									
C. Juanaqueña	6	1	5	120	1997									
C. Juanaqueña	10	1	9	38	1998			2						
C. Juanaqueña	10	1	11	43	1998			2						
C. Juanaqueña	97	1	7	26	1998									
C. Juanaqueña	126	1	10	60	1998									
C. Juanaqueña	163	2	7	40	1998			1				1		
C. Juanaqueña	163	1	7	43	1998			9						
C. Juanaqueña	163	2	8	48	1998	3	2	9	1				9	
C. Juanaqueña	222	3	8	85	1997							1		
C. Juanaqueña	222	3	11	95	1997			1	3			1		
C. Juanaqueña	222	2	13	100	1997	5		2				1		
C. Juanaqueña	222	3	6	1002	1997							1		
C. Juanaqueña	387	2	4	20	1998									
C. Juanaqueña	387	2	5	23	1998	2		3				1		
C. Juanaqueña	387	2	6	29	1998	1		4	1	1			2	
C. Juanaqueña	387	2	7	34	1998	2		8						
C. Juanaqueña	387	2	8	38	1998	2	2	5	7	1		1	1	
C. Juanaqueña	387	1	8	41	1998			2						
C. Juanaqueña	387	1	11	57	1998									
C. Juanaqueña	487	2	8	27	2000	1		2				1	2	
C. Juanaqueña	487	2	10	33	2000	4	1	2	3	2	1		2	
C. Juanaqueña	537	4	6	73	1997		1							
Subtotal	24 samples					20	6	52	15	4	1	8	16	122
C. el Canelo	94	2	4	22	2000		1	2	1					
C. el Canelo	94	2	6	28	2000		18	6						
C. el Canelo	94	2	9	36	2000		17							
Subtotal	3 samples					0	36	8	1	0	0	0	0	45
C. los Torres	1	1	7	27	1998			2					2	
C. los Torres	1	1	8	32	1998	2	2	2	2				6	
Cerro los Torres	1	3	8	53	1998		3	1						
Cerro los Torres	1	3	9	55	1998		1	4	1					
Cerro los Torres	1	3	10	59	1998									
Cerro los Torres	1	3	11	62	1998	1								
Subtotal	6 samples					3	6	9	3	0	0	0	8	29
C. Vidal	20	1	5	20	1998									
Subtotal	1 sample					0	0	0	0	0	0	0	0	0
Total	34 samples					23	48	69	19	4	1	8	24	196
Percent						12%	24%	35%	10%	2%	1%	4%	12%	100%

TABLE 9.3. Seeds from Re-examination of Fine Fraction Samples Not Previously Examined.

Site Name	Terrace or Ring	Unit	Level	Bag	Year	Amaranth	Chenopod	Cheno-am	Grass Seed	*Portulaca* sp.	Unidentifiable Seed Frag.	Total
Cerro Juanaqueña	T10	2	9	36	1998					1		
Cerro Juanaqueña	T10	1	12	45	1998							
Cerro Juanaqueña	T126	2	9	36	1998							
Cerro Juanaqueña	T126	1	9	56	1998							
Cerro Juanaqueña	T222	2	6	43	1997							
Cerro Juanaqueña	T297	2	9	40	1998							
Cerro Juanaqueña	T297	1	10	47	1998		1	1				
Cerro Juanaqueña	T297	1	12	54	1998			4				
Cerro Juanaqueña	T387	1	4	18	1998							
Cerro Juanaqueña	T387	1	5	25	1998			1	1			
Cerro Juanaqueña	T387	1	6	27	1998			5				
Cerro Juanaqueña	T387	1	10	51	1998							
Cerro Juanaqueña	T387	2	10	53	1998			4			2	
Cerro Juanaqueña	T413	2	3	25	1998							
Cerro Juanaqueña	T413	2	4	29	1998							
Cerro Juanaqueña	T413	1	5	35	1998							
Cerro Juanaqueña	T413	2	5	37	1998							
Cerro Juanaqueña	T413	2	6	43	1998							
Cerro Juanaqueña	T413	2	7	46	1998							
Cerro Juanaqueña	T413	1	8	50	1998	1		2				
Cerro Juanaqueña	T413	1	9	55	1998		1					
Cerro Juanaqueña	T413	1	10	58	1998			1				
Cerro Juanaqueña	T487	2	4	14	2000							
Cerro Juanaqueña	T487	2	9	29	2000							
Cerro Juanaqueña	T487	3	5	46	2000	1		1				
Cerro Juanaqueña	T487	3	6	50	2000		1	2				
Cerro Juanaqueña	T487	3	7	52	2000							
Cerro Juanaqueña	T487	3	8	56	2000				1			
Cerro Juanaqueña	T487	3	9	59	2000							
Cerro Juanaqueña	T487	4	2	67	2000							
Cerro Juanaqueña	T487	4,5	3	78	2000							
Cerro Juanaqueña	T487	4,5	4	81	2000							
Cerro Juanaqueña	T487	4,5	5	85	2000			1				

Cerro Juanaqueña	T487	4,5	6	89	2000							
Cerro Juanaqueña	T487	4,5	7	92	2000							
Cerro Juanaqueña	T487	2	11	95	2000							
Cerro Juanaqueña	T487	2	12	98	2000							
Cerro Juanaqueña	T487	6	9	124	2000	1						
Cerro Juanaqueña	T508	2	4	11	2000							
Cerro Juanaqueña	T508	2	5	14	2000							
Cerro Juanaqueña	T508	2	6	18	2000			2				
Cerro Juanaqueña	T508	2	7	21	2000							
Cerro Juanaqueña	T508	2	8	24	2000			1				
Cerro Juanaqueña	R234a	1	na	20	1998				1			
Subtotal	44 samples					3	3	25	3	1	2	37
Cerro el Canelo	94	1	5	13	2000							
Cerro el Canelo	94	2	5	25	2000							
Cerro el Canelo	94	2	7	31	2000		4	1				
Cerro el Canelo	94	2	8	34	2000							
Subtotal	4 samples					0	4	1	0	0	0	5
Cerro los Torres	1	2	7	25	1998							
Cerro los Torres	1	1	9	37	1998	2		1			3	
Cerro los Torres	1	3	5	45	1998							
Cerro los Torres	1	3	7	49	1998			4			1	
Subtotal	4 samples					2	0	5	0	0	4	11
Cerro Vidal	20	2	6	26	1998							
Cerro Vidal	20	1	6	27	1998							
Cerro Vidal	20	1	7	31	1998			2			1	
Subtotal	3 samples					0	0	2	0	0	1	3
Total	55 samples					5	7	33	3	1	7	56
Percent						9%	13%	59%	5%	2%	13%	100%

TABLE 9.4. Summary of Seed Counts by Feature from all Three Sample Types.

Site	Fea.	From Table	Sample Type	Amaranth	Chenopod	Cheno-am	Atriplex	Grass Seed	*Portulaca* sp.	*Solanum/ Physalis*	Unidentified Seed	Unidentifiable Seed Fragment	Total
Cerro Juanaqueña	T6	1	Seeds Originally Id'd as "Cheno-ams"	3	1	3							
		2	Re-examined, prev yielded Cheno-ams										
		3	*										
Cerro Juanaqueña	T10	1	Seeds Originally Id'd as "Cheno-ams"	5	0	1							
		2	Re-examined, prev yielded Cheno-ams			4							
		3	Previously Unexamined Fine Fraction						1				
Cerro Juanaqueña	T97	1	Seeds Originally Id'd as "Cheno-ams"			1							
		2	Re-examined, prev yielded Cheno-ams										
		3	*										
Cerro Juanaqueña	T126	1	Seeds Originally Id'd as "Cheno-ams"			1							
		2	Re-examined, prev yielded Cheno-ams										
		3	Previously Unexamined Fine Fraction										
Cerro Juanaqueña	T163	1	Seeds Originally Id'd as "Cheno-ams"			10							
		2	Re-examined, prev yielded Cheno-ams	3	2	19		1			1	9	
		3	*										
Cerro Juanaqueña	T222	1	Seeds Originally Id'd as "Cheno-ams"	1	10	15	1						
		2	Re-examined, prev yielded Cheno-ams	5		3		3			4		
		3	Previously Unexamined Fine Fraction										
Cerro Juanaqueña	T297	1	*										
		2	*										
		3	Previously Unexamined Fine Fraction		1	5							
Cerro Juanaqueña	T387	1	Seeds Originally Id'd as "Cheno-ams"	4	3	5						1	
		2	Re-examined, prev yielded Cheno-ams	7	2	22		8	2		2	3	
		3	Previously Unexamined Fine Fraction			10		1				2	
Cerro Juanaqueña	T413	1	*										
		2	*										
		3	Previously Unexamined Fine Fraction	1	1	3							

Cerro Juanaqueña	T487	1	Seeds Originally Id'd as "Cheno-ams"	1	1	1							
		2	Re-examined, prev yielded Cheno-ams	5	1	4		3	2	1	1	4	
		3	Previously Unexamined Fine Fraction	2	1	4		1					
Cerro Juanaqueña	T508	1											
		2											
		3	Previously Unexamined Fine Fraction			3							
Cerro Juanaqueña	T537	1	Seeds Originally Id'd as "Cheno-ams"	1	0	1							
		2	Re-examined, prev yielded Cheno-ams		1								
		3											
Cerro Juanaqueña	R234a	1	*										
		2	*										
		3	Previously Unexamined Fine Fraction					1					
			Subtotal	38	24	115	1	18	5	1	8	19	229
Cerro el Canelo	T94	1	Seeds Originally Id'd as "Cheno-ams"	1	12	3					2		
		2	Re-examined, prev yielded Cheno-ams		36	8		1					
		3	Previously Unexamined Fine Fraction		4	1							
			Subtotal	1	52	12		1			2		68
Cerro los Torres	T1	1	Seeds Originally Id'd as "Cheno-ams"	5	10	6					1	1	
		2	Re-examined, prev yielded Cheno-ams	3	6	9		3				8	
		3	Previously Unexamined Fine Fraction	2		5						4	
			Subtotal	10	16	20		3			1	13	63
Cerro Vidal	T20	1	Seeds Originally Id'd as "Cheno-ams"	1	1								
		2	Re-examined, prev yielded Cheno-ams										
		3	Previously Unexamined Fine Fraction			2						1	
			Subtotal	1	1	2						1	5
			Total	50	93	149	1	22	5	1	11	33	365

* Samples of that type were not analyzed.

4. Selection of well-preserved amaranth seeds for seed coat measurement using Scanning Electron Microscopy, in order to compare testa thicknesses with those of wild, weedy, and domesticated amaranths.

This degree of effort is necessary in order to understand the economic role of amaranth and other small grains in early agricultural systems of the American Southwest and Northern Mexico; however, it remains uncertain if a domesticated amaranth was grown in northern Chihuahua or elsewhere in the Southwest at this early point in time.

Overview of Modern Cultigen Amaranths and Their Archaeological Counterparts in the Southwest

The Grain Amaranths

There are three important modern species of domesticated amaranth (also called the grain amaranths). *Amaranthus caudatus* is a South American grain still grown in traditional Andean communities of Ecuador, Peru, Bolivia, and Argentina (National Research Council 1984:27). *Amaranthus cruentus* was and is economically most important in central and southern Mexico and Guatemala, but it spread into the Pueblo region of Arizona where it was used historically as a red dye (Nabhan 1979b; Sauer 1993). In Mexico, millions of people still enjoy the seeds as a sweet snack called *alegrías*. *Amaranthus hypochondriacus* was widely grown from southern Mexico to Arizona and made its way in pre-Contact times as far east as the Ozark highlands of Arkansas and Missouri (Fritz 1984, 2007). This species became established in Africa, South Asia, and Mongolia soon enough after trans-Atlantic European exploration that, until recently, botanists speculated that it might have been in those locations in more ancient times. Now all three species of grain amaranths are held to be entirely of New World origins (Sauer 1950, 1967, 1993). In Mesoamerica, amaranth was at times a highly valued crop. The provinces sent 20,000 tons of amaranth per year to the Aztec emperor Montezuma, where it played a major role as both a ritual and food item (National Research Council 1984:1).

Domesticated amaranth species have bushier and more compact inflorescences than their free-growing counterparts, resulting in higher seed yield. When grown for grain, their seeds are white or yellow rather than black, but seed diameter does not necessarily increase (although some modern *A. cruentus* seeds are several tenths of millimeters larger than wild counterparts). Pale-seeded amaranth populations tend to retain an average diameter of approximately 1 mm, but a seedhead can be as long as 50 cm with as many as 50,000 seeds and weigh more than 1 kg (Sauer 1993; Tucker 1986:9). The crops bear plumper, non-shattering seeds whose enlarged embryos often give a distinct equatorial rim. Although this rim is not exactly truncate (i.e., the sides are not vertical, and the cross section is not strictly rectangular), dried archaeological specimens sometimes have upper and lower rim ridges on the cotyledon side (the side containing embryonic seed leaves opposite the root or radicle), enabling them to stand freely on edge as if they did have truncate margins (Figure 9.1). Cultigen amaranth seeds are pale rather than black due to reduced seed coat thickness. Seeds with thinner seed coats apparently germinate more readily and may be preferred for consumption (McClung de Tapia et al. 1996:22). Thicker seed coats of wild amaranths delay germination until optimal conditions are present; however, higher levels of tannins may make them less favored for consumption (McClung de Tapia et al. 1996: 22). These tough seeds have long viability and can germinate after 40 years (National Research Council 1984: 17). Pale-seeded populations do not exist in the wild (Robertson 1981:287).

Ethnography of Amaranth in the Southwest

Ethnobotanists have known for centuries that both *Amaranthus hypochondriacus* and *A. cruentus* were grown historically in Arizona, New Mexico, Sonora, and Chihuahua. Hopi, Zuni, and Rio Grande Pueblo groups grew an amaranth for red face paint and for mixing the crushed red inflorescences with corn meal to enhance the meal's color. This species is in some cases known to be *A. cruentus* and in other cases assumed to be that species (Sauer 1993). John Wesley Powell acquired pale *A. hypochondriacus* seeds from Southern Paiutes in northern Arizona in the 1870s, and Isabela Kelly observed weedy and domesticate mixes used by Southern Paiutes as late as the 1930s (Sauer 1993:12). On linguistic grounds, Amadeo Rea (1997:289) thinks that an early twentieth-century reference to a cultivated seed plant called "ki'ak:" by the Pima refers to *A. hypochondriacus*.

As summarized by Gary Nabhan (1979b), early twentieth-century ethnobotanists observed amaranth cultivated for use of its pale seeds for food in Sonora, Chihuahua, and northern Sinaloa by the Yaquis, Pima Bajo, Warihio, Tarahumara, and Mayo. A few Warihio and Tarahumara families still grow grain amaranth (Robert Bye, Instituto de Biología, UNAM, personal communication, 2017; Suzanne Nelson, Native Seeds/SEARCH, personal communication 2002).

Archaeological Amaranth

Domesticated amaranth may date as early as 4000 BC in Mexico, based on pale seeds from Tehuacán Valley paleofeces. *Amaranthus cruentus*, represented by both flower parts and pale seeds, came from Abejas phase

FIGURE 9.1. Photo of pale, uncharred *Amaranthus hypochondriacus* seeds from Holman Shelter, Arkansas, dated to 920 ± 109 RCYBP (uncal, SMU 1632 [Fritz 1984]). University of Arkansas Museum catalog number 32-22-4. Note upper and lower rim ridges, characteristic of cultigen amaranth. Seeds range from 0.7 to 1.1 mm in diameter.

deposits (3500 to 2300 BC) (Sauer 1969). I strongly believe that any rockshelter specimen needs to be directly AMS-dated to test for secure context.

Evidence available since the 1960s demonstrates that *A. hypochondriacus* was grown in Arizona before European contact. Desiccated pale seeds and flower parts came from Salado contexts (ca. AD 1400) in Tonto National Monument (Bohrer 1962). A single pale amaranth seed that might be older was mixed with thousands of desiccated cotton seeds stored in pots in the Pinaleño Mountains near Tucson (Haury and Huckell 1993). A concentration of charred amaranth seeds from the Las Canopas site, a late Sedentary period (AD 1020 to 1150) Hohokam site in the Phoenix Basin, is argued to represent a mixture of originally pale (prior to carbonization) and dark but very thin-coated cultigen grains (Fritz et al. 2009). These seeds were found inside a ceramic bowl on the floor of a Hohokam house, where ancient farmers had intentionally mixed them with maize kernels, tepary beans, and cotton seeds.

Archaeobotanist Charles Miksicek (1987) pioneered research into morphological signs of amaranth domestication in the Southwest using charred specimens. Miksicek suggested that seeds from five Hohokam sites—all postdating AD 1000—were *A. hypochondriacus*. Scott Kwiatkowski (1989) attributed seeds from a sixth site to the same species. Miksicek's criteria were diameter size and margin configuration. Reliance on seed size is problematic, however, because wild and cultigen amaranth seeds overlap significantly in diameter, and other researchers (McClung de Tapia et al. 1996; Sauer 1950, 1993) have found no appreciable size increase between domesticates and non-domesticates. Both size and shape can be distorted during charring, since amaranth seeds puff up and pop. Although none of the past candidates for cultigen status were illustrated, Miksicek's descriptions of them as having truncate margins with distinct ridges holds true for uncharred pale seeds and for charred ones that are sufficiently well-preserved to indicate that they were pale prior to carbonization, such as many of the seeds from Las Canopas, mentioned earlier. Truncate margins are not characteristic of wild amaranth. Therefore, I believe that he is right about the presence and possible widespread distribution of cultigen amaranth in the late prehistoric Sonoran Desert Hohokam culture.

While clearly cultigen amaranth was grown north of Mesoamerica after AD 1000, its earlier history in the desert borderlands region is less securely documented. Based on high ubiquities and frequencies from various Middle and Late Archaic sites, Jonathan Mabry (2005a) includes amaranth, chenopod, and other native plants (tansy mustard, beeweed, Indian rice grass, and dropseed grasses) in an "early Southwestern crop complex" that developed as early as the Middle Archaic period and persisted as a component of early maize farming traditions in this region. According to Mabry's model, these species may have been formally cultivated and/or

TABLE 9.5. Summary of Seed Counts by Sites for all Samples.

	Amaranth	Percent Amaranth	Chenopod	Percent Chenopod	Cheno-ams	Percent Cheno-ams	Total	Total Percent
Cerro Juanaqueña	38	21%	24	14%	115	65%	177	100%
Cerro el Canelo	1	2%	52	80%	12	18%	65	100%
Cerro los Torres	10	22%	16	35%	20	43%	46	100%
Cerro Vidal	1	25%	1	25%	2	50%	4	100%
Total	50		93		149		292	

domesticated, or their natural abundances augmented through anthropogenic niche construction and management practices. This scenario is relevant to the evidence presented in this chapter for possibly domesticated amaranth from Cerro Juanaqueña and nearby trincheras sites. Furthermore, a report of possible Late Archaic (ca. 1500 BC) *A. cruentus*, based on desiccated utricle parts (papery tissue making up the amaranth fruit coat), comes from Fresnal Shelter in southwestern New Mexico (Tagg 1996). I have examined these specimens, and none of the associated seeds are pale.

RESULTS

Analysis for this study focused on three types of samples received from project ethnobotanist Karen Adams who processed the flotation samples: 1) seeds previously identified as "cheno-ams" (Table 9.1); 2) re-examination of the fine particle fraction (0.5–1.0 mm) of selected samples from which "cheno-ams" had already been removed (Table 9.2); and 3) scanning of the fine component of the light fraction not previously examined (0.425–0.5/0.7 mm) (Table 9.3). All three sample types are summarized in Tables 9.4 and 9.5.

Seeds Previously Identified as "Cheno-ams"

In Chapter 8, Adams and Hanselka indicate that cheno-ams were found in 41 flotation samples from Cerro Juanaqueña (Table 8.8) and they were found in 14 samples from the other sites (Table 8.7). From these, a subset consisting of 120 seeds classified as cheno-ams from 34 flotation samples were submitted to me for examination. Table 9.1 lists the detailed results of this part of the analysis.

I identified 22 of these 120 seeds as charred *Amaranthus* sp. (Table 9.1) and 38 as charred *Chenopodium* sp. Figure 9.2 illustrates examples of *Amaranthus* sp. and *Chenopodium* sp. seeds with SEM images. Forty-seven remained in the cheno-am category, meaning that absence of seed coat or generally fragmentary or otherwise poor condition made identification to the genus level impossible. One charred specimen could be identified as *Atriplex* sp. on the basis of well-preserved, adhering pericarp. Seven seeds in the original cheno-am category were found to be uncharred—therefore modern—after close inspection and, in a few cases, the breaking open of the seed to reveal the fresh perisperm within. Finally, I removed five of the original cheno-am seeds from that category due to the absence of anatomical features (beak, encircling embryo, or groove left after loss of embryo) associated with chenopods and amaranths. None of these could be classified beyond "unidentified seed" ($n = 3$) or "unidentifiable seed fragment" ($n = 2$).

Re-examination of Fine Particle Fraction from Samples Previously Yielding "Cheno-ams"

Targeted scanning of small-sized particles from the 34 submitted samples that had yielded amaranth and chenopod resulted in recognition of 23 new charred amaranth, 48 new charred chenopod, and 69 new charred cheno-am specimens, more than doubling the previous number (Table 9.2). Newly pulled specimens follow the pattern observed earlier: the chenopod seeds seem morphologically wild or weedy, with relatively thick, patterned seed coats and rounded margins, whereas the amaranth seeds have traces of very thin, smooth seed coats and hints of the less-rounded margins and embryo ridges that characterize domesticated cultivars. All of the amaranth seeds, however, are in less than perfect condition, and none has a seed coat totally encircling the embryo and beak. I tried to be conservative in labeling any specimen to genus, hence the dominance of new "cheno-ams," most of which I believe are more likely to be amaranths than chenopods. Although I know of no other taxon that might be represented by the better-preserved specimens, the flaws in their condition introduce an element of uncertainty.

In addition to amaranth, chenopod, and cheno-am seeds, I identified 19 new tiny grass seeds (Poaceae family), 4 charred purslane (*Portulaca* sp.), 1 nightshade (*Solanum* sp. or *Physalis* sp.), 8 unidentified seeds, and 24 unidentifiable seed fragments from the previously unsorted or resorted portions of samples that had been found to contain cheno-ams (Table 9.2). The high ratio of chenopods and amaranths to other small-sized seeds reinforces the observations by Adams and Hanselka that these taxa were the most abundant of the potential food sources in this size group.

Examination of Previously Unsorted Fine Particle Fractions

For the third component of the analysis, I scanned the very fine fractions (0.425–0.5/0.7 mm) of 55 samples that had not previously been sorted. This very fine portion tends to yield very few seeds so was not examined during Adams's analysis. Adams found cheno-ams in the other portions of the light fractions from 10 of these samples. Most of the 55 samples had been labeled either "high" or "medium" priority by the excavators on the basis of richness of biological remains, although few if any approached the yield of the best samples from which cheno-ams were already known. Eighteen of the newly resorted fractions were found to contain one or both of the relevant taxa (Table 9.3), including 5 amaranth, 7 chenopod, and 33 cheno-am seeds. Other taxa found in these samples include three grass seeds and one purslane (*Portulaca* sp.), along with seven unidentifiable seed fragments. Again, this strengthens the observation that chenopod and amaranth are the dominant small grains in the assemblage overall.

Summary of Samples

A total of 41 samples out of 89 fine sediment fractions examined yielded either amaranth, chenopod, or cheno-am seeds, indicating a high ubiquity level. Table 9.4 summarizes all three components of the analysis and total counts: 50 amaranth, 93 chenopod, and 149 cheno-ams. The new assemblage, including all three sample types, is large enough to display apparent intra- and inter-site distributional patterns (Tables 9.4 and 9.5). Amaranth was somewhat concentrated at Cerro Juanaqueña's Terrace 387, with 11 seeds identified to that genus and 37 additional cheno-ams, many of which are probably amaranth. Another concentration of cheno-ams came from Terrace 163 at Cerro Juanaqueña, but these are in especially poor shape and few could be assigned to genus (Table 9.4).

Terrace 94 at Cerro el Canelo had the densest concentration of *Chenopodium* (total = 52), contributing more than half the count (93) of that genus (Table 9.4).

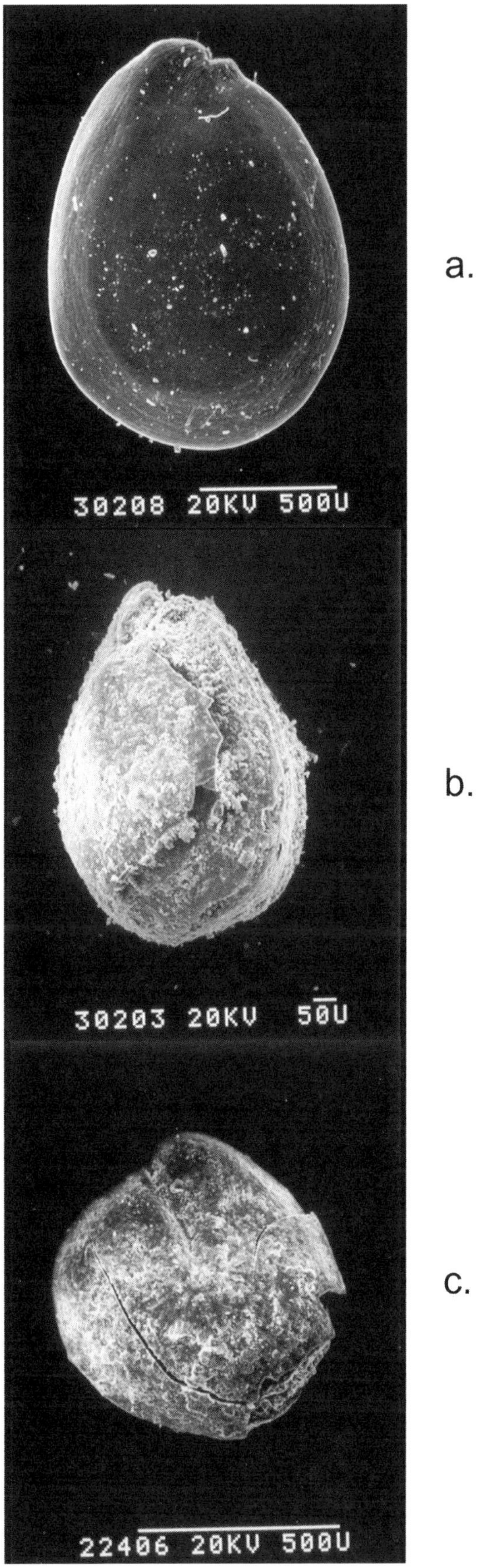

FIGURE 9.2. Scanning electron micrograph images of (*a*) modern, black-seeded cultigen *A. cruentus* (scale bar = 500 microns); (*b*) charred *Amaranthus* sp. from Cerro Juanaqueña T. 10 Unit 1 Level 9 (scale bar = 50 microns); and (*c*) charred *Chenopodium* sp. from Cerro Juanaqueña T. 387 Unit 2 Level 8 (scale bar = 500 microns). Note that the chenopod specimen has a wider beak and, unlike amaranth, a distinct sulcus reaching from the margin to center of seed.

TABLE 9.6. Testa Thickness.

Status	Species	Color	Domesticated	Mean testa thickness µm	Standard Deviation µm	Min µm	Max µm	No.	Source	Notes
1. Grain Crop	*A. hypochondriacus* Azteca	pale	Y	9.66	2.48	2	18	30	McClung de Tapia et al. 1996:26	Sample is a normal distribution, side wall measurement listed, mean of end is 10.53 µm.
2. Grain Crop	*A. cruentus* Mexicano	pale	Y	10.7	2.21	4	20	52	McClung de Tapia et al. 1996:26	Sample is normally distributed, side wall measurement listed, mean of end is 10.9 µm.
3. Grain Crop	*A. hypochondriacus* Rio San Lorenzo	pale	Y	5.8		3.5	8.8	4	This study	From Native Seeds/SEARCH
4. Grain Crop	*A. hypochondriacus*	pale	Y	6.5	2.72	3.8	8.8	3	K. Moses, personal communication 1998	From Native Seeds/SEARCH, SEM measurements by Kim Moses at University of Texas at San Antonio, each seed measured in multiple locations and mean value used for each seed. Min and max represents the thinnest and thickest locations among the three seeds.
5. Grain Crop	*A. cruentus*	pale	Y	5				1	This study	From U.S. Arboretum, Washington DC
6. Archaeological	*Amaranthus* sp.	unknown	Y ?	8.9	1.996	5	13	14	This study	Charred archaeological sample, C. Juanaqueña & C. los Torres.
7. Ornamental	*A. hypochondriacus* Mixteco	dark	Y	25.54	4.51	16	34	61	McClung de Tapia et al. 1996:26	Sample is normally distributed, side wall measurement is listed, mean of end is 22.7 µm.
8. Ornamental	*A. cruentus* (ornamental)	dark	Y	31				1	This study	Botanical Garden, Bonn Germany, modern ornamental, seed coat of the single seed ranged 30–32 µm.
9. Wild/weedy	*A. hybridus* L.	dark	N	31.7				1	McClung de Tapia et al. 1996:23	
10. Wild/weedy	*A. retroflexus* var. *salicifolius*	dark	N	20.8				1	McClung de Tapia et al. 1996:23	
11. Wild/weedy	*A. powellii* S, Wats	dark	N	24.3				1	McClung de Tapia et al. 1996:23	
12. Wild/weedy	*A. retroflexus*	dark	N	22				1	This study	Hortus Botanicas, Szeged, Hungary, range of seed wall 21–23 µm.
13. Wild/Weedy	*A. palmeri*	dark	N	18.5				1	This study	Río Casas Grandes floodplain, range of seed wall 17– 20 µm.

Table 9.5 summarizes all four sites and shows that Cerro el Canelo's pattern is strikingly different from that of Cerro Juanaqueña and Cerro los Torres. There is no apparent reason that readily accounts for Cerro el Canelo's high frequency of chenopod and virtual lack of amaranth. At Cerro Juanaqueña and Cerro los Torres, the proportion of amaranth is nearly identical at 21% and 22%, respectively. Cerro los Torres, Cerro Juanaqueña, and Cerro el Canelo are all contemporaneous, dating to ca. 1200 BC, while Cerro Vidal was occupied later at ca. 200 BC (Chapter 5). Unfortunately, the sample size at Cerro Vidal prevents any temporal comparisons. Two of the lower terraces at Cerro Juanaqueña also represent the later occupation. Only Terrace 13 is represented in this analysis, and recovery here is also too slight to make any diachronic comparisons.

Measurement of Amaranth Seed Coats by Scanning Electron Microscopy

Modern Amaranth Seed Coat Thicknesses

Unlike many cultigens, the seeds of domesticated amaranths cannot be distinguished from their wild counterparts on the basis of size (McClung de Tapia et al. 1996: 20–21; Sauer 1993). The clearest morphological difference is that domesticated grain amaranths have seeds that are pale in color due to reduction in seed coat thickness, whereas wild and weedy amaranths are reddish to black due to the presence of much thicker seed coats. An analogous contrast in seed coat thickness between crop and wild or weedy plants holds for cultigen vs. wild *Chenopodium berlandieri*, although many thin-coated chenopod seeds remain black whereas others are pale (Fritz and Smith 1988; Smith 1985).

Emily McClung de Tapia and her colleagues (1996) in Mexico City conducted baseline seed coat measurements of modern *Amaranthus hypochondriacus* and *A. cruentus*, the two cultigen species of amaranth known ethnographically from Mexico and the southwestern United States, and compared them to three wild/weedy amaranths. To summarize their results (Table 9.6), pale-seeded *A. hypochondriacus* of the 'Azteca' cultivar ($n = 30$) had seed coats between 2 and 18 microns in thickness, with an average of approximately 10 microns. Pale *A. cruentus*, cultivar 'Mexicano' ($n = 52$), seed coats measured between 4 and 20 microns, with an average of approximately 11 microns (Table 9.6). Seed coats of a dark-seeded variety (cultivar 'Mixteco') of *A. hypochrondriacus* ($n = 61$), used primarily as an ornamental and dye, ranged between 16 and 34 microns in thickness (average = 26 microns). Their study also reported only one seed each of three wild or weedy species. These black-seeded amaranths had seed coats as thin as 20.8 microns (for *A. retroflexus*) and as thick as 31.7 microns (for *A. hybridus*), with an intermediate value for *A. powellii* (Table 9.6).

In order to increase the database for modern reference specimens, I measured the seed coats of four pale-seeded *A. hypochondriacus* (C-7, "Rio San Lorenzo," purchased from Native Seeds/SEARCH), one pale-seeded *A. cruentus* (from the U.S. Arboretum in Washington D.C.), one dark-seeded ornamental *A. cruentus* (grown at the Botanical Garden in Bonn, Germany), one wild or weedy *A. retroflexus* (grown at the Hortus Botanicus, Szeged, Hungary), and one weedy *A. palmeri* collected from the Río Casas Grandes floodplain below the site of Cerro Juanaqueña (Table 9.6). The pale grain amaranths measured an average of 5.8 (for *A. hypochondriacus*) and 5.0 (for the single *A. cruentus*) microns thick. The single wild *A. palmeri* seed coat varied from 17–23 microns (median = 18.5), while the single wild *A. retroflexus* was 21–23 microns (median = 22) (Table 9.6). The thickest seed coat, measuring between 30 and 32 microns, came from an ornamental *A. cruentus* with large, dark red seeds. This cultivar is probably a result of modern horticulture.

Student Kim Moses at The University of Texas at San Antonio, as a class project, also measured three pale *A. hypochondriacus* seeds (also from Native Seeds/SEARCH) using the same SEM techniques. She took multiple measurements of each seed coat and calculated the mean, minimum, and maximum for each. Her values are very similar to mine for this species (Table 9.6).

Archaeological Amaranth Seed Coat Thicknesses

The measurements of 11 charred amaranth seeds from Cerro Juanaqueña and three from Cerro los Torres were measured using a Hitachi S-450 Scanning Electron Microscope in the Biology Department at Washington University in St. Louis. The instrument was operated by Microscopy Technician Michael Veith. These 14 specimens with visible seed coat fragments were mounted on 12-mm-diameter aluminum stubs to which carbon adhesive tabs were affixed. Stubs were sputter-coated with gold prior to placement in the electron microscope. Examining each specimen in turn, the stub was rotated until a cross section of the broken seed coat was positioned at right angles to the viewing plane. Images were captured at 1000x to 2000x magnification on Polaroid Positive/Negative film (Figure 9.3). A scale in microns at lower right of each image was used to calculate seed coat thickness. The preserved seed coat on an individual specimen might vary in thickness by a few microns, but most are relatively uniform. Maximum seed coat thickness was recorded unless variation was as much

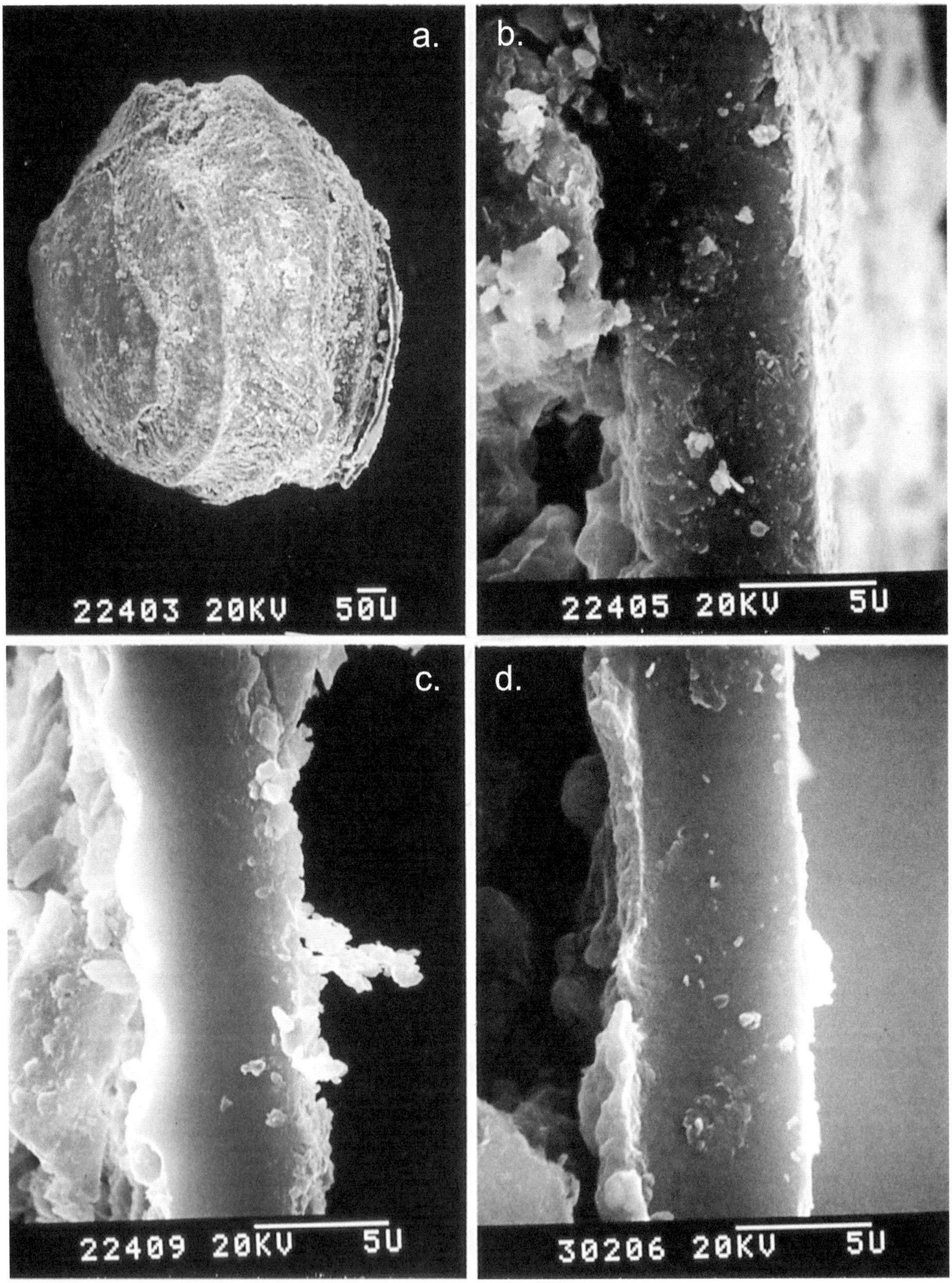

FIGURE 9.3. (*a*) SEM image of amaranth seed from Cerro los Torres, Terrace 1 Unit 3 Level 10, showing broken testa (seed coat) on both sides, with embryo exposed. Scale bar = 50 microns. (*b*) Cross section of amaranth seed coat from Cerro los Torres T. 1 U. 3 L.8. (*c*) Amaranth seed coat from Cerro Juanaqueña T. 387 U. 2 L. 7. (*d*) Amaranth seed coat from Cerro Juanaqueña T. 222 U. 2 L. 13. Scale bars of (*b*), (*c*), and (*d*) = 5 microns. (See Table 9.7 for seed coat thicknesses as measured under the SEM.)

TABLE 9.7. Seed Coat Thickness of Archaeological Amaranth Seeds from the Cerros de Trincheras Project.

Site	Feature	Unit	Level	Thickness microns (µm)	Average microns (µm)
Cerro Juanaqueña	T 10	1	4	10	
Cerro Juanaqueña	T 6	1	5	7–10	8.5
Cerro Juanaqueña	T 10	1	9	10	
Cerro Juanaqueña	T 163	2	8	13	
Cerro Juanaqueña	T 222	2	13 (1 of 2)	8	
Cerro Juanaqueña	T 222	2	13 (2 of 2)	7	
Cerro Juanaqueña	T 387	1	11	9	
Cerro Juanaqueña	T 387	2	7 (1 of 2)	5	
Cerro Juanaqueña	T 387	2	7 (2 of 2)	7	
Cerro Juanaqueña	T 487	2	8	11	
Cerro Juanaqueña	T 487	3	5	9	
Cerro los Torres	T 1	3	8	7	
Cerro los Torres	T 1	3	10	8–10	9
Cerro los Torres	T 1	3	11	9–11	10
			Mean	8.8	
			Std. Deviation	1.996	

as 2 microns, in which case median thickness was used. The thinnest seed coat is 5 microns thick; the thickest is 13, and the average thickness is 8.8 microns (Table 9.7).

Figure 9.4 plots all testa values from Table 9.6. When available, error bars show the standard deviation. The archaeological sample, with a mean testa thickness of 9 microns, is intermediate among the sample of modern pale-seeded grain crops that average 5–11 microns (2 minimum to 20 maximum). This is well below the mean values of 18–32 (16 minimum to 34 maximum) microns of dark-colored populations measured either by McClung de Tapia et al. (1996) in Mexico City or by me in St. Louis. Although alternative interpretations are discussed below, I believe it is likely that these amaranth grains represent domesticated or at least cultivated populations, harvested from agricultural fields along with maize or from other anthropogenic landscape zones where selection favored rapid germination of thin-coated seeds.

Discussion

Plant domestication is a gradual evolutionary process that can be difficult to track archaeologically and is best studied by combining multiple sources of evidence (Langlie et al. 2014). Amaranth seeds were relatively ubiquitous in the assemblages from the cerros de trincheras sites reported in this chapter, but this case study is constrained by less than ideal preservation and by relatively small sample sizes. Alternative scenarios to full domestication must be considered, the first being that a very thin-coated wild or weedy amaranth is represented. If the thin amaranth seed coats from Late Archaic period cerros de trincheras reflect non-domesticated populations, I argue that they probably evolved as weedy components of the agroecology and are therefore "artifacts" along with crops (cf. Anderson 1952). However, a weed amaranth with seed coats this thin has not been documented, and such a thin seed coat is likely to be maladaptive outside of agricultural contexts. Second, I have been asked if these archaeological seed coats might be thinner than modern wild seeds because testa thicknesses shrank during carbonization. Studies by Smith (1985) and others (Bruno 2006) indicate that chenopod diameters shrink an average of 5% when charred, but that seed coat thicknesses do not change, with the single possible exception of *Chenopodium ambrosioides*. If seed coat shrinkage accounts for the difference in testa thickness, charring needs to have reduced seed coat thickness by about one-third around the entire seed on 14 different specimens. Although systematic, controlled experimental studies have yet to be conducted to verify that carbonization has minimal impact on amaranth seed coat thickness, McClung de Tapia and her colleagues (1996:26) did report that the testa and embryo of *A. hypochondriacus* showed no apparent change following experimental charring. They apparently did not report post-carbonization measurements.

Third, the charred amaranth (and possibly also chenopods) might not be contemporaneous with

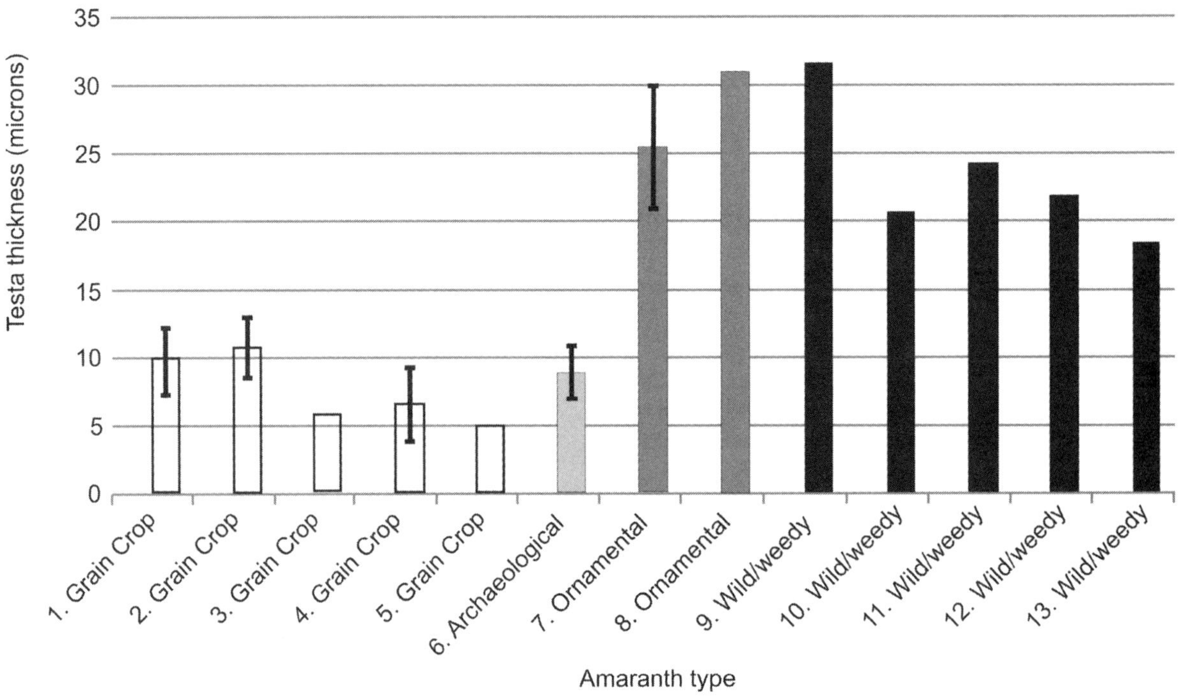

FIGURE 9.4. Graph of mean amaranth testa thicknesses by taxon of various samples in Table 9.6; error bars are one standard deviation.

ancient contexts, but rather filtered down into lower levels after more recent grass fires; that is, they are contaminants from later times. This would necessitate an explanation for amaranth seed coat reduction as a byproduct of agriculture or perhaps historic cattle grazing, as mentioned above. The archaeological team and I carefully considered AMS dating the charred amaranth seeds, but we did not do so as it would have required the destruction of more than half of the amaranth seeds. As Chapter 5 describes, 23 AMS dates from Cerro Juanaqueña were processed on multiple species and no intrusive dates were obtained, nor were modern or recent artifacts found in lower excavation levels. The probability of historic charring of an unknown, historic, very thin-testa species that burned as a result of grass fires and then worked its way into the archaeological deposits on two different sites seems low but is not impossible. Fourth, some taxon other than amaranth might be represented with the imperfect condition of these specimens masking their true identity. This possibility seems unlikely because of commonalities between vegetation of the Río Casas Grandes landscape and that of nearby southern New Mexico, where far more botanical studies have been conducted, but to my knowledge, no other candidates are known. The modern weed, *Amaranthus palmeri*, however, is probably much more common today than it was in the past, and it is not impossible that it rose to dominance at the expense of another taxon with similar seed morphology whose identity is now unrecognized.

For these reasons, I cannot rule out other scenarios to account for the presence of thin-testa amaranth seeds, but believe that cultigen amaranth may have been a part of early agricultural systems at Cerro Juanaqueña and other early cerros de trincheras. The most compelling evidence will likely come from directly dated, desiccated seeds from dry rockshelter sites or from other unusual contexts at an open site, such as a storage facility where large numbers of thin-testa seeds are preserved in extremely good condition.

Conclusions

Maize was probably not the only crop grown by early agriculturalists in what is now northwest Mexico and the southwestern United States. Directly dated seed and rind fragments of *Cucurbita pepo* document the presence of squash/pumpkin by 2000 cal BC at McEuen Cave, and bottle gourd (*Lagenaria siceraria*)—although not directly dated—occurs in preceramic levels of sites such as Tularosa Cave and Cordova Cave (Cutler 1952; Kaplan 1963; Merrill et al. 2009; Shackley 2005a). Archaeologists believe that grain amaranths were domesticated in Mexico as far back as 4000 BC, and we know they were grown in the Southwest during late precontact and historic times. Pale-seeded *A. hypochondriacus* is still a minor crop planted by traditional Tarahumara and Warijio farmers.

The relatively high ubiquity of amaranth seeds with seed coat thicknesses well within the cultigen range indicates that grain amaranth may have been part of the early agricultural system in the Río Casas Grandes valley (Fritz 2007). Cultigen amaranth could have made a difference by adding diversity and, given its drought-resistant properties, reducing risk. Production of thin-coated grains on amaranth plants with large, dense inflorescences would have provided more palatable and more easily harvested and processed food than if wild seeds alone were utilized. Based on ethnographic evidence, we can infer that cultigen amaranth was sometimes added to ground maize, resulting in a nutritionally superior (lysine-rich) meal or flour than if maize had been used alone (National Research Council 1984). Because of the potential difference a crop such as amaranth might have made to early Southwestern farmers, the challenging work of recovering and carefully examining "cheno-am" seeds should be continued and intensified.

Notes

1. Acknowledgments: I greatly appreciate having had the opportunity to participate in this project, including the adventure of working in the field with Robert Hard, John Roney, Karen Adams, Kevin Hanselka, and all the other intrepid team members at Cerro Juanaqueña and the nearby trincheras sites. Michael Veith, microscopist for the Biology Department at Washington University in St. Louis, deserves my special thanks for his exceptional technical skills and artistry with the scanning electron microscope. I am grateful to Robert Hard for drafting Figure 9.4.

CHAPTER 10

Pollen, Phytolith, Starch, and Stable Isotope Analyses

Introduction

This chapter presents additional analytical efforts to obtain data about human interactions with the ancient environment and includes palynological and stable isotope analyses. We submitted pollen, phytolith, and starch samples from a variety of contexts to explore their potential for obtaining information about prehistoric plants and their uses. Human and jackrabbit bone fragments were analyzed for stable isotopes with the goal of gaining data about ancient subsistence practices and past environments.

Pollen, Phytolith, and Starch Analyses

Three different small studies were conducted to evaluate the potential of pollen, phytolith, and starch proxies to enhance our understanding of Cerro Juanaqueña and the ancient environment. Here we integrate and summarize the three studies and include the original analytical reports in appendices. The studies were conducted by Suzanne Fish of the Arizona State Museum (Appendix 10.1; Hard and Roney 1999b), Lisa Lavold from Vaughn Bryan's Palynology Laboratory at Texas A&M University (Appendix 10.2), and Linda Scott Cummings of Paleo Research Institute (Appendix 10.3). Researchers examined a total of 21 pollen samples including sediment from floodplain backhoe trenches, sediment samples from Cerro Juanaqueña features, and excavated ground stone from the site. The field team collected all sediment samples with implements cleaned between each sample and placed the sediment in paper coin envelopes.

Background

Suzanne Fish's study (Appendix 10.1) examined 10 pollen samples, including five from dated buried floodplain deposits, three from pollen washes from ground stone tools from Cerro Juanaqueña, one sample from a midden deposit behind Terrace 6, and a tenth from the surface of a modern fallow agricultural field in the Río Casas Grandes floodplain. These pollen washes were conducted in the field lab and the extract sent to her. She reported results in terms of percentages of 200 pollen grain counts and found pollen in nine of the 10 samples. Lavold (Appendix 10.2) examined six samples for pollen and phytoliths from dated floodplain backhoe trench deposits. She found the pollen to be badly degraded and in low densities. She concluded that further analysis would not be productive and did not identify any pollen, so her findings are not discussed further. Linda Scott Cummings (Appendix 10.3) examined five samples, including two from backhoe trenches, two mano pollen washes, and one sediment sample from the storage pit at Terrace 487. Cummings conducted the pollen washes in her lab and extracted and searched for pollen, phytoliths, and starch in all five samples. She reported percentages in chart form on the basis of 100 or 200 grain counts with separate counts for pollen and phytoliths. She found pollen and phytoliths in all five samples but did not recover any starch. Less than ideal pollen preservation, low pollen densities, and budgetary considerations inhibited further analyses beyond these three initial studies. Further details of the methods, samples, and results can be found in the electronic appendices.

Summaries of Fish's and Cummings' results are provided here, beginning with the pollen samples from backhoe trenches (Table 10.1), followed by the pollen from five Cerro Juanaqueña ground stone tool washes and two archaeological sediment samples (Table 10.2). In order to summarize the data across analysts, we use ordinal categories of pollen frequencies.

These analyses were in part driven by an interest in developing paleoenvironmental sequences from dated alluvial deposits. Therefore a total of 14 pollen/phytolith samples were processed by the three labs from the floodplain surface and buried geomorphological units dated by Lee Nordt (Chapter 12). One sample was from a fallow modern agricultural field on the floodplain surface. The dated geomorphological samples that were processed include: one sample from Unit G (cal AD 1400), two from Unit F-Viejo paleosol (cal AD 600), three from the Unit E-Janos paleosol (appx. cal AD 1), three from the Unit D-Trincheras paleosol (approximately 1400 cal BC), one from the Unit D-Antonio paleosol (3500 cal BC), one from the Unit C Petra paleosol (4300 cal BC), and two samples from the Unit B-Guiermo peat layer (8000 cal BC). A total of seven of these samples analyzed by Fish and Cummings produced results (Table 10.1). For comparisons, this discussion relies on the inventory on modern plants presented in Chapter 8.

Results and Discussion

The sample collected from a modern fallow agricultural field in the floodplain reflected modern conditions, with medium levels of aster family (high spine Asteraceae and low spine Asteraceae) and cheno-ams (*Chenopodium* and *Amaranthus*) and low amounts of grass family (Poaceae). These taxa are consistent with modern vegetation in an agricultural field. Rare levels of nightshade family (Solanaceae) and willow (*Salix*) pollen were present which is expectable along the floodplain since willow is a riverine tree found in the floodplain today and nightshade taxa are found in disturbed habitats. Rare to low levels of pine (*Pinus*), oak (*Quercus*), and juniper (*Juniperus*) are present, although these taxa are not found nearby. Instead they were likely transported by wind or water from higher elevations.

The youngest buried sample from the backhoe trenches came from Unit G which dates to cal AD 1285 to 1420. This sample contains rare to medium levels of *Artemisia* (e.g., sagebrush), aster, grass, four o'clock flower (*Boerhavia*), willow, buckwheat (*Eriogonum*) and cheno-am pollen, all of which are consistent with the floodplain setting. *Tidestromia* (e.g., honeysweet) and buttonbush (*Cephalanthus*) are also rare in this sample. Rare cattail (*Typha*) pollen suggests the presence of a continually wet habitat. Today aster family, various grasses, willow, and buttonbush grow in the floodplain below Cerro Juanaqueña but not cattail. Cattail is found however at the La Palotada hot spring, about 20 km west of Cerro Juanaqueña, in the presence of standing water (Chapter 8). Taxa from high elevations include medium levels of pine and low to rare levels of juniper, oak, and alder (*Alnus*) that were likely wind- or water-deposited.

The sample from the Viejo paleosol (cal AD 420 to 650) contains rare to medium frequencies of *Artemisia*, aster, grass, four o'clock flower, buckwheat, willow as well as a high frequency of cheno-am pollen. In addition, rare levels of creosote bush (*Larrea*), *Ephedra*, hackberry (*Celtis*), and *Kallstroemia* (e.g. poppy) are present. Liguliflorae, a subfamily of the sunflower family that contains many lettuce plants, is also present. Rare levels of intrusive pine, oak, and juniper appear as well. Arizona poppy is present in the floodplain today. The willow suggests higher levels of moisture while the other taxa represent desert grassland and disturbance plants.

The Janos paleosol sample (cal AD 20 to 330) contains the similar suite of taxa that are typical of these floodplain samples: *Artemesia*, aster family, high levels of cheno-am, grass, four o'clock flower, *Ephedra*, and willow. In addition upland pine, oak, juniper, and alder were wind or water deposited from upland elevations. Other rare taxa in this sample includes globemallow (*Sphaeralcea*), primrose (Onagraceae), and mesquite (*Prosopis*), all of which are expectable in this desert grassland floodplain habitat (Kearney and Peebles, 1969). Rare *Umbelliferae* (carrot or parsley family) pollen is present and typically these plants prefer marshy or damp soil. The only Umbelliferae in the modern plant inventory is water parsnip (*Berula erecta*). Globemallow and, of course, mesquite are present in the modern floodplain.

There are two samples from the Trincheras paleosol (1620 to 1130 cal BC) located 2 m below the modern surface, which would have been the surface used and farmed by the occupants of Cerro Juanaqueña. The taxa present in these two samples are similar, although one sample has more taxa represented than the other and the pollen frequencies are somewhat different. The core set of taxa present in these two samples include: *Artemesia*, aster, cheno-am, grass, four o'clock, *Ephedra*, globemallow, mesquite, and willow. Liguliflorae, mint family (Labiatae), and sedge (Cyperaceae) are also present. Intrusive pine, oak, juniper, and alder are present including one sample with a high frequency of pine. Although we hoped to find *Zea mays* pollen, none was found. However the backhoe trenches only sampled a tiny fraction of this buried Trincheras paleosol. The only sedge in the modern plant inventory (Chapter 8) is fragrant flatsedge (*Cyperus odoratus* L.) and it was found in the hot spring setting not the floodplain. The rare presence of cattail pollen, sedge, and trilete and monolete fern spores suggests higher levels of soil moisture than found today. Cattail requires constantly damp conditions while sedge and fern thrive under conditions of high soil moisture.

Cummings examined two backhoe trench samples and found phytoliths in both. All phytoliths found represent grasses, suggesting that no other phytolith-bearing plants other than grasses were present (Appendix 10.3). Cummings separated phytoliths into chloridoids, festucoids, and panicoid or bilobate forms. Representing other grass cells are buliforms, trichomes, and elongate forms. While 97% of native US grasses can be subdivided

TABLE 10.1. Backhoe Trench and Modern Pollen Results.

Analyst	Fish	Fish	Fish	Fish	Cummings	Cummings	Fish
Analyst Sample No.	2	3	4	1	BHT 2	F.2-00	10
Sample type	BHT	BHT	BHT	BHT	BHT	BHT	Modern Floodplain
Field Provenience	BHT6-99 C5b 241–251 cm	BHT10 Ab2 169–178 cm	BHT12 Ab 111–112 cm	BHT2 AbK 201–216 cm	BHT 2, Akb4 182–202 cm	BHT F.2-00 O1b5 100 cm	Old Field
Alluvial Unit/Paleosol	G	F-Viejo	E-Janos	D-Trincheras	D-Trincheras	B- Guiermo (Peat)	
Chapter 12 Backhoe Trench reference	BHT-y	BHT-z	BHT-bb	BHT-h	BHT-h		
^{14}C date BP, Chapter 12	590 ± 70	1520 ± 60	1860 ± 60	3140 ± 100	3140 ± 100	8950 ± 70 (date from BHT-2)	
Calendar Years (two sigma calibrated)	AD 1285–1420	AD 420–650	AD 20–330	1620–1130 BC	1620–1130 BC	8075–7920 BC	Modern
Intercept (cal)	AD 1400	AD 560	AD 160	1410 BC	1410 BC	8000 BC	

Scientific name	Common name	Frequency [1]						
Artemisia	Sagebrush	rare	rare	rare	rare	rare		
High spine Compositae	Aster family, e.g. snakeweed, sunflower	rare	rare	rare	rare	medium*	low*	medium
Low spine Compositae	Aster family, e.g. ragweed	medium	medium	medium	medium	rare	rare	medium
Cheno-am	Chenopod and amaranth	rare	high	high	medium	low	rare	medium
Poaceae	Grass family	medium	rare	rare	medium	low	low*	low
Boerhavia-type	Four o'clock, spiderling	rare	rare	rare	+		rare	+
Sphaeralcea	Globemallow			rare	rare			
Kallstroemia	Poppy		+	+				
Onagraceae	Primrose			rare				
Eriogonum	Buckwheat	rare	rare					
Euphorbia-type	Spurge							
Cylindropuntia	Cholla							
Larrea	Creosote bush		rare					
Prosopis	Mesquite			rare	+		rare	
cf. Leguminosae	Legume							rare

Ephedra	e.g. Mormon tea		rare	rare	rare	rare		
Simmondsia	Jojoba							
Alternanthera	Joyweed							
Cleome-type	e.g. Beeweed, spiderflower							
Rhamnaceae	Buckthorn family							
Pinus	Pine	medium	rare	rare	rare	high	high	low
Quercus	Oak	low	rare	rare	low	rare		rare
Juniperus	Juniper	+	rare	rare	rare	rare	rare	rare
Salix	Willow	rare	rare	rare	rare			+
Alnus	Alder	rare		rare	+			
Zea	Maize							
Typha	Cattail	rare				rare*		
Solanaceae	Nightshade family							rare
Labiatae	Mint family				rare			
Tidestromia	Honeysweet	rare						
Cephalanthus	Buttonbush	rare						
Celtis	Hackberry		rare					
Umbelliferae	Carrot or parsley family			rare				
Liguliflorae	Lettuce, e.g. dandelion, chicory		rare			rare	rare	
Cyperaceae	Sedge family				rare	rare		
Platyopuntia	Prickly pear							
Cereus-type	Cholla, hedgehog, others							
Agave	Agave							
Asteraceae (tissue fragment)	Aster family					rare		
Monolete smooth	Fern spore					rare		
Trilete smooth	Fern spore					rare	rare	
Sporormiella	Dung fungus					rare		
Indeterminate		low	rare	rare	low			rare
No of taxa		16	17	19	17	16	10	12

[1] Ordinal classification of percentages. 0.5–5% = "rare," 5.1–10% = "low," 10.1– 40% = "medium," and greater than 40% = "high." Percentages based on 200 grain standard sum.
"+" indicates a pollen type observed only in scanning material after tabulation of the 200 grain standard.
"*" indicates a pollen type occurring in aggregates of six or more grains.

Table 10.2. Cerro Juanaqueña Pollen Results.

Analyst		Fish	Fish	Fish	Cummings	Cummings	Cummings	Fish
Analyst Sample No.		6	7	8	23	29	88	9
Sample type		Ground stone	Ground stone	Ground stone	Ground stone	Ground stone	Archaeological sediment	Archaeological sediment
Field Provenience		T6 Bag 65 U1/L3	T6 Bag 90 U2/L9	T6 Bag 130 U7/L2	T10 Bag 23 U1/L6/ 142–152 cm bd	T297 Bag 29 U2/L6 160–170 cm bd	T487 Bag 88 U4-U5/L6 156–166 cm bd	T167 Bag 71 U1-U2 115 cm bd
Context		Mortar	Metate	Mano	Mano	Mano	Storage pit B sediment	Terrace midden sediment
Scientific name	**Common name**	**Frequency[1]**						
Artemisia	Sagebrush	rare	rare	rare	rare	rare	rare	rare
High spine Compositae	Aster family, e.g. snakeweed, sunflower	rare	low	rare	low	low	low	low
Low spine Compositae	Aster family, e.g. ragweed	medium	medium	medium	rare	medium	medium	medium
Cheno-am	Chenopod and amaranth	high*	high*	low	high*	high	high*	high*
Poaceae	Grass family	rare	rare	rare	rare	low	rare	rare
Boerhavia-type	Four o'clock, spiderling	low	rare	+				rare*
Sphaeralcea	Globemallow	+	rare					rare
Kallstroemia	Poppy	+	+	+				rare
Onagraceae	Primrose		+					
Eriogonum	Buckwheat	rare	+		rare		rare	rare
Euphorbia-type	Spurge		rare	+				
Cylindropuntia	Cholla		+	+				
Larrea	Creosote bush		rare	rare				
Prosopis	Mesquite	rare						
cf. Leguminosae	Legume							
Ephedra	e.g. Mormon tea	rare		rare				rare
Simmondsia	Jojoba					rare		
Alternanthera	Joyweed						rare	

Cleome-type	e.g. Beeweed, spiderflower						rare	
Rhamnaceae	Buckthorn family						rare	
Pinus	Pine	rare	rare	rare		rare	low	low
Quercus	Oak	rare	rare	rare			rare	+
Juniperus	Juniper	rare	+		rare	rare	rare	
Salix	Willow			rare				
Alnus	Alder	rare						
Zea	Maize			rare			rare	
Typha	Cattail							
Solanaceae	Nightshade family							
Labiatae	Mint family							
Tidestromia	Honeysweet							
Cephalanthus	Buttonbush							
Celtis	Hackberry							
Umbelliferae	Carrot or parsley family	rare						
Liguliflorae	Lettuce, e.g. dandelion, chicory			rare		rare		
Cyperaceae	Sedge family						rare	
Platyopuntia	Prickly pear			+				
Cereus-type	Cholla, hedgehog, others			+				rare
Agave	Agave							rare
Indeterminate		low	low	rare				rare
No of taxa		17	17	19	7	9	14	15

[1] Ordinal classification of percentages. 0.5–5% = "rare," 5.1–10% = "low," 10.1–40% = "medium," and greater than 40% = "high." Percentages based on 200 grain standard sum.
"+" indicates a pollen type observed only in scanning material after tabulation of the 200 grain standard.
"*" indicates a pollen type occurring in aggregates of six or more grains.

among three subfamilies Pooideae, Chloridoideae, and Panicoideae, there is not always a consistent match between phytolith forms and subfamily identification (Appendix 10.3). Although festucoid phytoliths are typical of the subfamily Pooideae grasses, small amounts of festucoid phytoliths are produced by nearly all grasses. Pooideae grasses tend to be cool season C_3 grasses and include the large genus of *Festuca* with *F. arizonica* as an example. Chloridoid phytoliths are found typically in the subfamily Chloridoideae and usually represent C_4 short grasses such as *Muhlenbergia*, *Bouteloua*, and *Sporobolus* among others (McNamee 2012). However, some Chloridoideae grasses produce both bilobate (panicoid) and festucoid phytoliths (Appendix 10.3). The grasses in the Panicoideae subfamily are generally warm season C_4 genera and include *Digitaria*, *Panicum*, and *Setaria*. Buliform phytoliths forms are produced by grasses in response to wet conditions, such as floodplains and other places that may have moist soil conditions.

The Trincheras paleosol phytolith sample yielded a high frequency of chloridoids, low frequency of festucoids, and rare to low levels of panicoids. Buliforms were present in low levels. None are suggestive of the presence of *Zea mays*. The chloridoids and panicoids typically represent C_4 grasses while the festucoids typically represent C_3 grasses. The cheno-am pollen from the Trincheras paleosol represents both C_4 amaranth and C_3 chenopod. The grass pollen likely represents C_4 grasses, and much of the remaining taxa identified in the pollen and phytolith studies are C_3 plants.

Nordt (Figure 12.6, Table 12.4) indicates the Trincheras paleosol supported about 60% C_4 plants, a value that does not contradict the phytolith and pollen data. The stable isotope, pollen, and phytolith data are mutually supportive and indicate a floodplain during the occupation of Cerro Juanaqueña that supported a mixed desert grassland vegetative community, as well as disturbance plants, habitats with damp soils, and locations with continual moisture.

The final backhoe trench pollen sample is from Unit B, the Guiermo peat deposit (8075 to 7920 cal BC). It contained the suite of floodplain taxa at rare to low levels including aster, cheno-am, grass, four o'clock, mesquite, and Liguliflorae. A high level of pine and rare juniper pollen again imply water or wind deposition. The phytoliths from this sample are similar to those in the Trincheras paleosol. However the C_3 festucoid frequency is low, but more common than in the Trincheras paleosol. The C_4 chloridoids, with a medium frequency, remain the most common phytolith taxa present.

In summary the six productive backhoe trench pollen samples indicate that the floodplain supported a diversity of species and habitats, including disturbed habitats, desert grasslands, and desert shrubs. The pollen samples also indicate higher levels of moisture than at present. Although mesquite is normally produced at low frequencies, these pollen spectra suggest that the floodplain was not a mesquite bosque during the periods represented by these samples. All productive backhoe trench samples yielded at least one or two taxa, at rare frequencies, suggesting that mesic habitats were present in the floodplain. However, the Trincheras paleosol, which was contemporaneous with the primary occupation of Cerro Juanaqueña, yielded the most taxa that reflect moist conditions, including sedge, fern, cattail, and willow.

Suzanne Fish analyzed three pollen washes from ground stone tools and Linda Scott Cummings analyzed two. These five samples came from one metate, three manos, and one mortar. Pollen was found on all samples. Cummings also searched for phytoliths and starch, finding the former but not the latter. The pollen results were similar to those from the modern agricultural field with the presence of aster, cheno-am, grass, pine, oak, and juniper. However cheno-ams were present in high levels on four of the five ground stone tools, and three of these included aggregates of cheno-am pollen. Large clusters of pollen are typically not efficiently transported by wind and may indicate immediate introduction of the pollen from the plant or possibly the processing of cheno-ams that would include immature floral parts. As discussed in Chapter 8, chenopod and amaranth are well-known economic plants, and in Chapter 9, Gayle Fritz discusses the recovery of possibly domesticated amaranth and non-domesticated chenopod seeds from the site. *Zea mays* pollen was recovered from one mano, which is consistent with the evidence of the use of maize on the site and the arguments presented throughout this volume of the economic importance of maize.

Investigators also found spurge (*Euphorbia* type), cholla (*Cylindropuntia*), prickly pear (*Opunita*), and hedgehog or related cacti (*Cereus* type) pollen at rare densities on the ground stone tools. Relatively scarce cactus pollen is seldom recovered in samples from natural vegetation and likely reflects economic resource residues (Appendix 10.1).

Two sediment samples were analyzed from good archaeological contexts from Cerro Juanaqueña, one analyzed by Fish and the other by Cummings. One sample was from a midden deposit from Terrace 167 and the other was from a storage pit in Terrace 487. Most of the taxa found in these two samples likely represent background pollen rain, as a similar suite of types are present as found on the ground stone tools and modern agricultural field. However high levels of cheno-ams are present, including aggregates of cheno-ams in both

TABLE 10.3. Human Stable Isotope Samples from Cerro Juanaqueña.

Feature	Unit	Level	Bag	Depth below datum cm	Count Frags	Element	Comments	Bone wt g	U of Calgary Lab #	$\delta^{13}C$ collagen ‰	$\delta^{15}N$ collagen ‰	Corrected C/N	Percent collagen yield	Notes
T10	1	2	5	102–112	1	Cranium	Cf. w/human	0.9	123			none	0.47	Yield low
T10	2	7		152–162	3	Cranium	Reconstructable fragments	5.7	124			none	0.93	Yield low
R239	1	3	10	149–156	2	Cranium	Weathered	0.7	125	−12.7	8.7	3.8	0.19	C/N high, yield low
T287	6	99	28	130–139	6	3 Femoral head & 3 cancellous fragments	Cancellous fragments are indeterminant & burned	2.5	126	−17.9	8.6	3.5	0.35	Yield low

contexts. *Zea mays* pollen was found in the storage pit on Terrace 487 (Chapter 3), consistent with the recovery of high frequencies of charred maize from this feature and our expectations that maize was stored at the site.

Summary

Despite small samples and limited preservation, the pollen, phytolith, and starch studies did produce useful results. At the time Cerro Juanaqueña was occupied the floodplain was not a mesquite bosque but supported a diversity of species and habitats including disturbed habitats, desert grasslands, and desert shrub habitats. Most importantly the area had higher levels of moisture than at present, including standing water, indicated by the presence of sedge, fern, cattail, and willow. The ground stone samples yielded cheno-am and maize pollen consistent with other evidence that these plants were being extensively utilized and processed. Cactus pollen was also recovered from ground stone and likely reflects use of these plants. The Terrace 487 storage pit yielded maize pollen consistent with our expectations that maize was stored at the site.

Stable Isotope Analyses of Human Bone

Stable isotope studies of human bone have the potential of producing critical data regarding ancient diet and ecology, and this brief section reports on our work on four samples (Coltrain et al. 2007; Hard and Katzenberg 2011; Katzenberg 2008; Lee-Thorp 2008). Chapter 11 and Table 11.17 describe the eleven small human bone fragments found in excavations at Cerro Juanaqueña. In an effort to glean additional paleodietary information, four of the largest human bone samples (0.7 to 5.7 g) were selected for stable isotope analysis. Three of the samples were cranial fragments, while the fourth sample consisted of femoral head fragments and partially charred cancellous bone fragments. All samples were analyzed by Anne Katzenberg's stable isotope laboratory at the University of Calgary.

Collagen was isolated following the procedures described by Sealy (1986). The collagen was analyzed on a Thermo Finnigan MAT Delta^{PLUS}XL mass spectrometer linked to a Carlo Erba gas analyzer in the Isotope Science Laboratory, University of Calgary. The data include atomic carbon nitrogen ratios (C/N), collagen yields, and $\delta^{13}C$ and $\delta^{15}N$ values. The results, presented in Table 10.3, indicate poor preservation. All four samples produced low amounts of collagen, ranging from 0.19% to 0.93%. Normal sample collagen yield is expected to be 15% to 30%, and sample yields below 1% merit caution. Two samples did not produce results for nitrogen isotopes, which is common when yields are so low, and therefore did not yield C/N ratios. C/N values are used as quality control and should fall between 2.9 and 3.6 (DeNiro 1985). The two other samples produced C/N ratios of 3.5 and 3.8 which are high but acceptable in light of the low yields

The $\delta^{15}N$ values 8.7‰ and 8.6‰ are similar and fall within the range of Southwestern Basketmaker II and III individuals (Coltrain et al. 2007). However the two $\delta^{13}C$ values of −12.7‰ and −17.9‰ are disparate. This range of 5.2‰ is substantially greater than that found in most samples of a single human population and suggests that one or both of these $\delta^{13}C$ values are problematic. The former value indicates diets with substantial contribution in C_4 resources, such as maize, amaranth, and C_4 grasses, consistent with subsistence and other data from Cerro Juanaqueña. However, the latter $\delta^{13}C$ value of −17.9‰ reflects little C_4 input and suggests a diet largely of C_3 resources, which is inconsistent with all other evidence from the site. Given the limitations of these data, including the small number of samples, low collagen yield, elevated C/N values, large $\delta^{13}C$ range, and burning present on one sample, it is difficult to place much confidence in these results.

Stable Isotope Analysis of Leporids

Introduction

Isotopic composition of vertebrate teeth, bone, otoliths, and eggshells has proven to be a useful paleoecological proxy for ancient environments. The use of vertebrate isotopic studies has been recently extended to the stable carbon isotope analysis of ancient leporid (rabbits and hares) bone by Raymond Mauldin, his colleagues, and others (e.g., Mauldin 1993a; Kemp et al. 2013; Munoz et al. 2011; Somerville et al. 2016) as they may provide additional data for reconstructing environmental conditions. Because leporids preferentially feed within a limited area, stable carbon isotope analysis of preserved bones may reflect the extent of C_4 versus C_3 vegetation present during the mammal's short life span and from within their small home range. Leporid bones are ubiquitous in archaeological sites and can be readily dated either through associated archaeological material or directly dated by AMS radiocarbon methods. Leporid bones offer a potential source of paleoecological data that may reflect ancient vegetation patterns at a far finer scale than most other proxies (Kemp et al. 2013; Munoz et al. 2011).

Background

Examination of stable isotopes in leporid remains in order to assess environmental conditions includes previous work by Mauldin (1993a) who compared modern leporid $\delta^{13}C$ values from upland and lowland settings in southern New Mexico. Leonard Kemp and his colleagues (2013) examined $\delta^{13}C$ data from series of ancient and modern black-tailed jackrabbit (*Lepus sp.*) and cottontail (*Sylvilagus sp*) from locations in southern New Mexico and west Texas. Smith and his colleagues (Smith et al. 2014) presented modern cottontail and jackrabbit $\delta^{13}C$ results from southern New Mexico and west Texas and compared them with those collected from central Texas. Ugan and Coltrain (2011) evaluated stable isotope data from modern and prehistoric black-tailed jackrabbits from the Eastern Great Basin to evaluate the relationships between $\delta^{15}N$ values and soils, temperature, precipitation, vegetation, and topography. Somerville et al. (2016, 2017) found correlations between leporid stable isotope values and ancient human modifications to the landscape in central Mexico. Somerville and colleagues (2018) also examined a large sample of leporids from multiple ecoregions across Mexico and the United States to examine the relationships between a range of environmental variables and stable isotope values.

In this pilot study, we sampled bone from 12 ancient and two modern *Lepus sp.* (jackrabbit) individuals from the Río Casas Grandes study area. The only jackrabbit species occurring in the area around Janos, Chihuahua, today is *Lepus californicus* spp, the common black-tailed jackrabbit that inhabits the grasslands, plains, and deserts of the western U.S and Mexico. Two other jackrabbit species are recorded in western Chihuahua. *Lepus callotis* Wagler, the white-sided jack rabbit, is known from somewhat higher elevations within 75 km to the northwest, west and south of Janos (Anderson 1972:267). *Lepus alleni* (antelope jackrabbit) is considered rare in Chihuahua and has only been identified about 400 km south of Janos (Anderson 1972:265), but is common in south central Arizona and into Sonora (Burt and Grossenheider 1976:206–207; Cockrum 1960:67–68). The archaeological fauna are classified only as *Lepus* sp. since most archaeological jackrabbit bone cannot be positively identified beyond the genus level, and the other species are present in adjacent regions. However, the identifications from the cerros de trincheras compared favorably with black-tailed jackrabbit in both size and form, and it is likely that the majority of the jackrabbit remains are *L. californicus* (Chapter 11). Jackrabbits were the most commonly recovered taxa at Cerro Juanaqueña.

Jackrabbit Feeding and Foraging Behavior

The assumption that leporid bone stable isotope values may be a proxy for ancient localized vegetation conditions is derived from the feeding and foraging behavior of these small herbivores (Kemp et al. 2013; Munoz et al. 2011). Jackrabbits are opportunistic generalists that utilize grass, forbs, shrubs, and succulents but show a strong preference for green grass and forbs whenever they are present. In the southern Southwest and northwest Mexico most grasses and some forbs use the C_4 photosynthetic pathway, while most forbs and shrubs are C_3 plants. Jackrabbits also feed on desert succulents, particularly prickly pear, which uses the CAM (Crassulacean Acid Metabolism) pathway.

Feeding studies of jackrabbits consistently show that grasses and forbs are a fundamental part of their diet. In the northern Chihuahua Desert near Las Cruces, New Mexico, overall diet across multiple years and study sites consisted of 22% grasses, 32% forbs, and 47% shrubs (Daniel et al. 1993:298). Jackrabbits readily switch among grass, forbs, shrubs, and succulents as the available forage fluctuates across seasons and years (Wansi et al. 1992:300). Jackrabbits first consume green grass and forbs during the growing season but change to shrubs as green herbage declines in the non-growing season or in times of drought, and there is an inverse relationship between forbs and grasses relative to shrubs (Vorhies and Taylor 1933).

Jackrabbits generally do not feed on plants in proportion to presence in the environment, but preferentially select forbs and grasses. In summer, grass consumption is highest; in winter, shrub consumption is highest; and forb consumption is similar across spring, summer, and fall (Daniel et al. 1993:299). Although jackrabbits feed on a large number of species, only a few species made up the vast majority of their diet (Wansi et al. 1992:300). During drought shrubs replace grasses and forbs (Daniel et al. 1993:299; Wansi et al. 1992). In New Mexico, the favored grass species are dropseed, black grama, three-awn, and fluffgrass. All of these are C_4 species, with the exception of fluffgrass which is a non-native C_3 species. The major forbs consumed are leatherweed croton, dwarf dalea, silverleaf nightshade, wooly paperflower, scarlet globemallow, spectacle pod, and snakeweed—all C_3 plants. The major shrubs consumed were honey mesquite, broom snakeweed, soaptree yucca and prickly pear cactus. All but the cactus are C_3 species, and prickly pear is a CAM plant (Daniel et al. 1993:304; Wansi et al. 1992:298).

Prickly pear varies from only a trace to as much as 11% of the diet (Daniel et al. 1993:298). Succulents are selected when little annual or new growth of perennial plants is available, from November to February and particularly in dry years (Hoffman et al. 1993:313). Prickly pear is also used in some years in June, depending upon the availability of other forage, rainfall, and temperature. Many studies have reported that *Opuntia* cactus is a minor dietary item (less than 5%) and indicated that cacti and succulents are water sources during dry periods (Hoagland 1992:283). Vorhies and Taylor (1933:529) report that jackrabbits feed on cactus during the drier times of the year. In the extremely arid conditions with the absence of grasses and forbs, cactus can become a major dietary item as it did on the dry Carmen Island off of Baja California (Hoagland 1992:283).

Apparently, jackrabbits do not prefer maize plants as there are few such reports. In the prairie of northeast Colorado, introduced and cultivated species dominated jackrabbit diets throughout the year and tended to replace native shrub consumption during the winter (Flinders and Hansen 1972). Maize was a minor component of the diet and was far outweighed by the use of other plants such as alfalfa, western wheatgrass, winter wheat, and summer cypress.

Studies in the Chihuahuan desert of southern New Mexico, as well as northern Mexico and southern Arizona, have yielded similar results, reflecting jackrabbits' preference for C_3 forbs and C_4 grasses during the growing season and use of C_3 shrubs when grasses or forbs are not available during the nongrowing season, drought, or intervals of poor range conditions (Dabo 1980; Fatehi 1986; Marin et al. 2003; Nelson et al. 1997; Vorhies and Taylor 1933). These researchers consistently find that jackrabbits prefer C_4 grasses and C_3 forbs when they are available and consume C_3 shrubs when they are not. Studies from the Great Basin, Plains, and Plateau confirm a similar pattern of jackrabbit feeding, characterizing this species as a generalist that prefers grasses and forbs when they are available, but readily browsing a variety of shrubs when they are not (Currie and Goodwin 1966:307; Flinders and Crawford 1977:219; Hayden 1966:44–45; Johnson and Anderson 1984:81). Prickly pear, a CAM plant, is consumed when little other new plant growth is available.

Dietary intake is also a product of foraging area. Home ranges are the areas used by animals on a day-to-day basis and do not include seasonal migrations. In the Jornada Experimental Range near Las Cruces black-tailed jackrabbits move between shrubby cover and grassy feeding areas, a distance that could be 2 to 3 km (1 to 2 miles). Home ranges may be several miles in diameter on the Santa Rita Experimental Range in southern Arizona (Vorhies and Taylor 1933:485–486). A study in the shrub-steppe vegetation in northern Utah found that home ranges varied from 1 to 3 km^2 (Smith 1990:251). Other studies in California, Idaho, and Kansas have reported smaller home ranges of < 0.2 km^2 (Smith 1990:253). Generally leporid home ranges are small and do not exceed 500 ha (Munoz et al. 2011).

The isotopic signature of bone reflects a lifetime average diet of the individual, including the gestation period. Adult hares have an average lifespan of one to three years (French et al. 1965; Munoz et al. 2011), therefore presumably the collagen isotopic value would reflect the average diet over the turnover period. Since jackrabbit diet shifts considerably between C_4 grasses during the summer and C_3 shrubs during the winter,

TABLE 10.4. Jackrabbit (*Lepus sp.*) Stable Isotope Samples from the Río Casas Grandes Region.

Site	Prov/Fea	Unit	Bag	Level	Depth below datum cm	Count	Sample Weight (g)	Element(s)	U of Calgary Lab #	$\delta^{13}C$ collagen ‰	$\delta^{15}N$ collagen ‰	Corrected C/N	Collagen yield	Note
Cerro Juanaqueña	T163	4	5	2	109–117	7	1.77	Mandible, metapodials	CCT 01F	−12.8	8.9	3.3	1.47	
Cerro Juanaqueña	T537	5		1	110–180	2	2.96	Innominate	CCT 02F	−14.6	7.0	3.5	0.16	
Cerro Juanaqueña	T487	4.5	86	5	146–156	1	1.35	Calcaneus	CCT 03F	−13.8	5.5	4.1	0.16	C/N out of range, not used
Cerro Juanaqueña	T6	2	69	6	146–156	1	2.11	Tibia	CCT 04F	−14.7	7.2	5.5	0.10	C/N out of range, not used
Cerro Juanaqueña	T297	1	26	6	160–170	1	3.13	Innominate	CCT 05F	−12.6	6.2	3.6	0.26	
Cerro Juanaqueña	T126	2	37	9	210–220	1	1.7	Femur	CCT 06F			none	0.06	not used
Sitio Antonio	Backdirt					1	1.87	Innominate	CCT 07F	−14.4	6.5	3.2	2.10	
Sitio Antonio	Backdirt					3	0.75	Scapula, thoracic vertebra	CCT 08F	−11.9	5.5	3.2	8.65	
Sitio Janos	Backdirt					1	2.19	Humerus	CCT 09F	−12.5	5.7	3.2	4.88	
Sitio Janos	Backdirt					1	2.44	Innominate	CCT 10F	−11.4	7.2	3.2	6.97	
Sitio Janos	Backdirt					1	2.14	Femur	CCT 11F	−17.3	6.6	3.2	6.09	
Sitio Janos	Backdirt					1	1.38	Humerus	CCT 12F			none	0.09	not used
Modern *L. californicus*	40 km W of CJ					1			TEXM01F	−19.0	4.7	3.3	0.43	
Modern *L. californicus*	Base of CJ					1			TEXMO2F	−19.7	7.0	3.6	6.90	

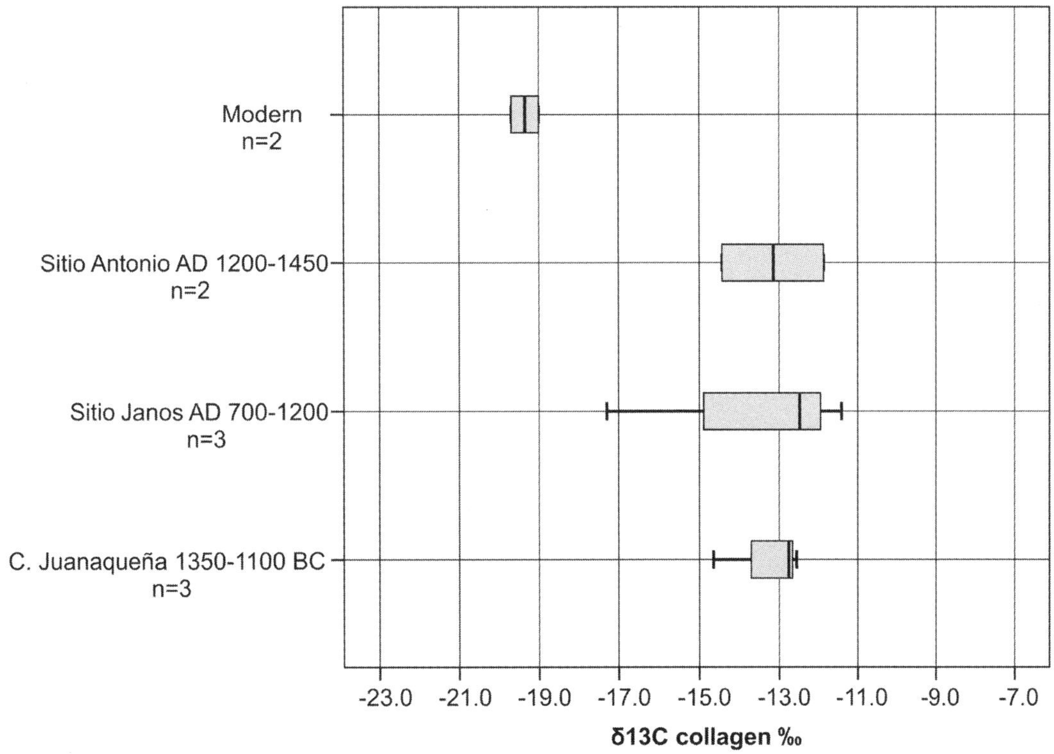

FIGURE 10.1. Boxplot of $\delta^{13}C$ values from prehistoric and modern jackrabbit collagen from the Río Casas Grandes region. Note small sample sizes.

the isotopic values of immature rabbits only reflect their diet during the portion of year they lived. Therefore immature jackrabbits would be expected to give more variable results than adults.

Methods

For this pilot study, six jackrabbit bones from Cerro Juanaqueña were submitted to Anne Katzenberg at the University of Calgary for stable isotope analysis. Comparative samples were also recovered during minimal examination of looters' backdirt at two later, ceramic period sites located along the floodplain (Appendix 4.1). We submitted four samples from the Sitio Janos site that is located on the western outskirts of the town of Janos, 9 km southwest of Cerro Juanaqueña. Based on surface ceramics, the site dates to the Viejo period (AD 700–1000/1200). Sitio Antonio is a Medio period pueblo (AD 1200 to 1450) (Dean and Ravesloot 1993; Whalen and Minnis 2001) located about 1 km west of Cerro Juanaqueña and two samples from this site were included. In addition, two modern black-tailed jackrabbits were collected for comparison, one near the base of Cerro Juanaqueña and the second 40 km west of Cerro Juanaqueña.

The jackrabbit bone samples were prepared at Anne Katzenberg's laboratory and analyzed in the Isotope Science Laboratory at the University of Calgary using the same methods as the human bone analysis described in the earlier section of this chapter. Fourteen bone samples were processed as listed in Table 10.4, six from Cerro Juanaqueña, two from Sitio Antonio, four from Sitio Janos, and two modern black-tailed jackrabbits. Four of the prehistoric samples produced invalid results, two with elevated C/N ratios and two that did not yield $\delta^{15}N$. These four samples are not utilized in the remaining analyses or discussion, leaving eight prehistoric and two modern samples with valid results.

Results and Discussion

The results are plotted in Figure 10.1. The medians of all three prehistoric sites fall between $\delta^{13}C$ −13.1 to −12.8‰ indicating that C_4/CAM resources were a substantial part of the diet. In contrast, the median of the two modern samples is −19.3‰ indicating C_3 plants played a larger role in the diet than in the prehistoric samples.

While this pilot study involved only a small number of samples the results are of interest. They are comparable to results that Kemp et al. (2013) obtained for *Lepus sp.* samples from the central Tularosa Basin in southern New Mexico. Their study included eight modern jackrabbits from the Tularosa Basin and 28 prehistoric samples. The prehistoric samples were from three

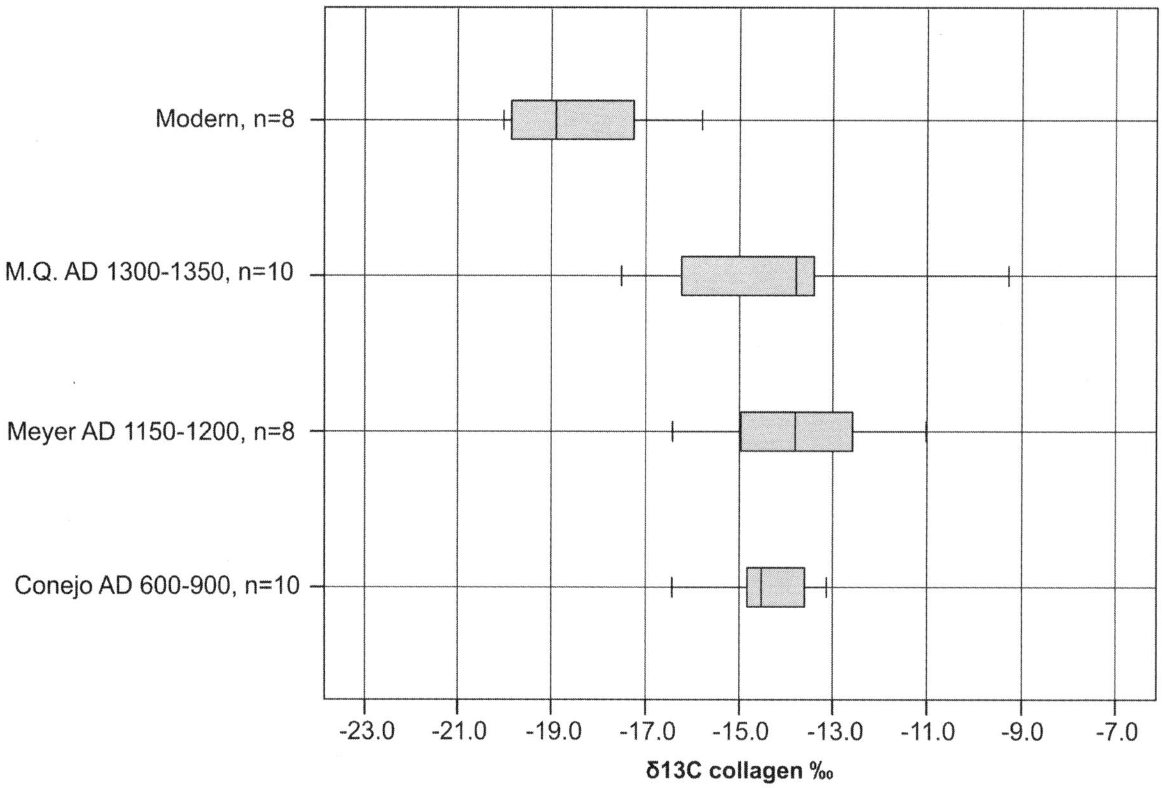

FIGURE 10.2. Boxplot of $\delta^{13}C$ values from prehistoric and modern jackrabbit collagen from the southern Tularosa Basin, New Mexico (Kemp et al. 2013:Figure 2). M.Q. = Madera Quemada site.

sites located on the alluvial fans of the Tularosa Basin and represented three discrete time periods. The medians of the three sites fall between $\delta^{13}C$ −14.4‰ to −13.8‰ and are plotted in Figure 10.2. The median of the modern samples is −18.5‰.

Even after considering the depletion in atmospheric carbon values due to burning of fossil fuels, the Tularosa Basin samples reflect a substantial decline in $\delta^{13}C$ values from prehistoric to modern times (Kemp et al. 2013). This decline reflects a decreased intake in C_4/CAM plants relative to C_3 resources. The well-documented historic period shrub expansion and corresponding reduction of grasslands (Chapter 2) is likely responsible for this shift (Ceballos González et al. 2010; Dick-Peddie 1993; Hruska et al. 2017; Van Auken 2000; York and Dick-Peddie 1969). However Kemp and his colleagues (2013) note that the prehistoric sites were on alluvial fans with more grasses while the modern samples are from the central part of the shrubbier basin. The Río Casas Grandes samples reflect a similar sharp decline from prehistoric to modern that is consistent with patterns of historic shrub invasion.

Considered together, the prehistoric Río Casas Grandes $\delta^{13}C$ values trend somewhat higher (medians $\delta^{13}C$ = −13.1 to −12.8‰) than the prehistoric Tularosa Basin samples (medians $\delta^{13}C$ = −14.4‰ to −13.8‰). This difference is likely related to the prevalence of C_4 grasslands in the Río Casas Grandes compared with that of the Tularosa Basin. The Río Casas Grandes floodplain and surrounding area is at an elevation of 1350 to 1400 m and is described as Semidesert Grassland supported by 33.4 cm average annual rainfall (Chapter 2). The central Tularosa Basin has an elevation of 1200 to 1300 m and is largely Chihuahuan Desertscrub with Semidesert Grasslands on the bordering alluvial fans (Brown and Lowe 1980). Average annual precipitation is comparatively lower—26.9 cm at Chaparral, New Mexico.

A third comparison is consistent with this preliminary pattern of increasing dominance of C_4 grasslands correlating with increasing leporid $\delta^{13}C$ values. Two prehistoric *Lepus* sp. samples from two Viejo period sites in central Chihuahua, 200 km south of Cerro Juanaqueña, yielded $\delta^{13}C$ values of −10.0‰ and −9.0‰, indicating that leporids were largely feeding on a C_4/CAM plant diet (Webster 2001:96). The elevation of the area is 1880 m, higher than Cerro Juanaqueña, and is mapped as Plains Grassland with a rainfall of 40 to 50 cm (Brown and Lowe 1980; Webster 2001).

As described in Chapter 2 and 12 and earlier in

this chapter, the buried floodplain Trincheras paleosol (contemporaneous with the occupation of Cerro Juanaqueña) supported a mixed desert grassland community based on $\delta^{13}C$ soil stable isotope values and the pollen and phytolith results. Although they could have been harvested from any floodplain or upland location, the stable isotope values of the Cerro Juanaqueña jackrabbits are consistent with a paleoenvironmental reconstruction that the landscape 3000 years ago was dominated by C_4 grasslands, with grasses more prevalent than the mixed shrub landscape present today.

Future work with leporid stable isotopes as paleoenvironmental indicators requires larger samples, consideration of other species including cottontail, modern comparative samples, consideration of leporid feeding patterns, and in particular the role of CAM plants as well as C_4 and C_3 resources (Kemp et al. 2013; Munoz et al. 2011).

CHAPTER 11

Faunal Remains

Kari M. Schmidt

Introduction

The animal bones recovered from Cerro Juanaqueña and the three other cerros de trincheras, Cerro el Canelo, Cerro los Torres, and Cerro Vidal, are important for understanding these Early Agricultural period sites. The role of faunal exploitation at these sites is critical to understanding human interactions with the environment and is particularly relevant to subsistence and mobility. This chapter addresses the methods and results of the faunal analysis and places them in ecological context. This faunal material has also been discussed in my dissertation (Schmidt 2008) and the project's annual reports (Hard and Roney 1998b, 1999a; Roney and Hard 2000; Schmidt 2000; Schmidt and Nisengard 2001) (see Appendices 3.2 to 3.5). I address formation processes, variability in taxonomic abundance within and between sites, evidence for intrusive remains, and implications for mobility patterns.

Methods

Taxonomy

During excavations at Cerro Juanaqueña and the other cerros de trincheras, animal bones were consistently found throughout the deposits and were removed from the screens as well as from the flotation samples (Chapter 3). The animal bones recovered from the sites were assigned to the lowest taxonomic level possible. Genus and species identifications remained the goal throughout the analysis, but assignment to that level was not always possible. Table 11.1 lists the scientific taxonomy and common names to the lowest taxonomic level reached during the identification process. It provides the class, order, family, genus, and species names for all identified taxa, as throughout the remainder of the text I primarily use common names.

Identifications were made using the comparative collections at the Oklahoma Archeological Survey in Norman, Oklahoma, the Museum of Southwestern Biology affiliated with the University of New Mexico, and the Zooarchaeology Laboratory at the University of Arizona. Additionally, identifications were augmented by osteological manuals when necessary (e.g., Cohen and Serjeantson 1996; Gilbert 1993; Lawrence 1951; Olsen 1964, 1968).

The analysis followed standard zooarchaeological procedures (Grayson 1984; Klein and Cruz-Uribe 1984). For each bone, I recorded the following attributes: lowest taxonomic identification, element (e.g., tibia, ulna), portion of element (e.g., proximal, distal, or complete), side (if able to be determined), age (presence or absence of epiphyseal fusion lines, epiphyses, etc.), fusion (fused or unfused), presence and degree of burning, natural taphonomic factors (e.g., root-etching, weathering), break patterns (if other than natural), and pathologies. In addition, other surface modifications (e.g., rodent or carnivore gnawing), if present, were recorded. These items were coded on data sheets and entered into a computer spreadsheet.

Certain specific identifications were not made because of the presence of multiple species in the site area. For example, two species of jackrabbit are found in northwestern Chihuahua (*Lepus californicus* and *L. callotis*), black-tailed and white-sided jackrabbit respectively (Anderson 1972; Desmond 2004). Jackrabbit identifications, therefore, were made to *Lepus* spp. For the most part, the identifications from the cerros de trincheras compared favorably with black-tailed jackrabbit in both size and form, and it is likely that the majority of the jackrabbit remains are *L. californicus*.

There are also two species of cottontail rabbits in the region, the desert cottontail (*Sylvilagus audubonii*) and the eastern cottontail (*Sylvilagus floridanus*) (Anderson 1972; Findley et al. 1975). Consequently, as with the jackrabbits, cottontail identifications were made to the generic level, *Sylvilagus* spp., and not to the specific level. However, it is probable that the majority of the cottontail remains recovered in this assemblage are desert cottontail (*S. audubonii*).

Similar difficulties occurred with specific identifications in the order Artiodactyla. Pronghorn (*Antilocapra americana*), mule deer (*Odocoileus hemionus*), and white-tailed deer (*Odocoileus virginianus*) are all found in northwestern Chihuahua and adjacent regions of New Mexico and Texas (Anderson 1972; Findley et al. 1975; Schmidly 1994). Distinctions between pronghorn and deer can be made based on certain elements (Lawrence 1951), but these distinctions were not always possible given the elements represented in the assemblages (e.g., phalanges). Additionally, distinctions between mule deer (*O. hemionus*) and white-tailed deer (*O. virginianus*), both present in the region (Anderson 1972; Ceballos González and List 2011; Findley et al. 1975), were difficult to make. Deer identifications compare favorably with mule deer, and most specimens in this assemblage likely belong to *O. hemionus* given their dominance in the region. Nevertheless, the category *Odocoileus* spp. was used to encompass both species.

Specific identifications in the canid family (dogs, wolves, and coyotes) were also complicated since skeletal distinctions between the domestic dog (*Canis familiaris*), coyote (*Canis latrans*), and wolf (*Canis lupus*) are subtle. Dogs and coyotes are common while the wolf has been largely extirpated from its former range in northwestern Chihuahua and southwestern New Mexico in the recent past (Anderson 1972; Findley et al. 1975). Coyotes are generally smaller than wolves, but there is some overlap in size between the smallest North American wolf subspecies, the Mexican wolf (*Canis lupus baileyi*), and large coyotes. In general, coyote bones are less robust than those attributed to the wolf, and this attribute was the primary distinguishing element used in the few examples recovered in this assemblage. Coyote and dog remains are also similar, again differing primarily in robustness and size, but remains in this assemblage did not compare favorably with domestic dog specimens.

In addition to the methods discussed in the preceding paragraphs, certain analytical techniques were used to quantify both the unidentifiable and identifiable remains. These techniques include the use of the number of identified specimens (NISP), the minimum number of individuals (MNI), and the calculation of lagomorph and artiodactyl indices.

Number of Identified Specimens (NISP) and Minimum Number of Individuals (MNI)

The NISP is the number of bones in an assemblage that can be assigned to a particular taxon (Klein and Cruz-Uribe 1984). Determining NISP values for an assemblage is fairly simple: bones are identified to the lowest possible taxonomic level (e.g., jackrabbit) and the numbers of identifiable specimens within that taxon are tabulated (Grayson 1984).

Despite the importance of quantifying taxa in faunal assemblages, problems are inherent with using the NISP method (Grayson 1984). First, the use of NISP assumes that all bones are equally affected by breakage. As a result, this method cannot control for interdependence, as one bone has the potential to be broken into many fragments, and creates problems for using statistical measures. Problems of interdependence means that every element is not independent of every other element, since multiple elements can be from each individual animal and individual bones are broken into multiple fragments. An extreme case, for example, are the 238 venomous snake vertebrae that were found at Terrace 126. Do these represent 238 snakes or one snake? Because of this problem, NISP is considered a maximum value. The issue of interdependence also impacts the utility of NISP values as ratio values (Grayson 1984; Lyman 2008). How should the ratio of 238 snake bones be compared with the 99 jackrabbit bones from the same feature? Were venomous snakes more than twice as important as jackrabbits? Probably not, and as discussed below, MNI values are used to potentially eliminate the problem of interdependence. In addition, NISP measures do not control for the fact that all species do not have the same number of elements. The measure does not account for differential preservation of bones, which likely patterns with density and porousness, and NISP is affected by differing recovery techniques (Cannon 1999; Shaffer and Sanchez 1994). Finally, NISP values are affected by butchering and transport decisions. Small animals killed away from a site are more likely to be carried back whole, while only the meatiest elements of larger animals are returned to the site.

The MNI of an assemblage signifies how many animals represent a particular taxon. White (1953:397) described the MNI calculation as, "separating the most abundant element of the species found into right and left components and using the greater number as the unit of calculation." An assemblage with 10 left jackrabbit tibiae and 8 right jackrabbit tibiae has an MNI of 10 jackrabbits. This quantification method has the potential to alleviate some of the shortcomings of NISP, especially in terms of interdependence and fragmentation, but it is not problem-free. MNI calculations are affected by the relative degree that the analyst chooses to aggregate the data (e.g., pit structures, terraces, excavation units, levels, site as a whole, etc.), and also in terms of how to compute the MNI (e.g., which element will be used for calculation).

In this analysis, both NISP and MNI calculations were used in order to take advantage of the best of both methods without ignoring their weaknesses. Aggregating

Table 11.1. Scientific and Common Names of Identified Animals

Class	Order	Family	Genus species	Identified animal
Actinopterygii (Ray-finned fishes)				Indeterminate fish
	Cypriniformes (Carps, minnows, and suckers)	Cyprinidae (Cyprinids)		Minnow
	Cypriniformes	Catostomidae (Catostomids)		Sucker
Reptilia (Reptiles)	Testudines (Turtles)			Indeterminate turtle
	Testudines	Emydidae (Pond turtles)	*Terrapene ornata*	Western box turtle
	Squamata (Lizards and snakes)	Viperidae (Vipers)		Venomous snake
	Squamata	Colubridae (Colubrids)		Nonvenomous snake
	Sauria (Lizards)			Indeterminate lizard
Amphibia (Amphibians)	Anura (Frogs and toads)	Scaphiopodidae (Spadefoot toads)		Spadefoot toad
Aves (Birds)				Indeterminate bird
	Galliformes (Quails, pheasants, peacocks)	Odontophoridae (Quails)	*Callipepla* sp.	Quail
	Passeriformes (Passerines)			Perching bird
	Anseriformes (Waterfowl and allies)	Anatidae (Swans, geese, and ducks)	*Anas* sp.	Duck
Mammalia (Mammals)	Rodentia (Rodents)			Indeterminate rodent
	Rodentia	Sciuridae (Squirrels and allies)		Indeterminate squirrel
	Rodentia	Sciuridae	*Tamias* sp.	Chipmunk
	Rodentia	Heteromyidae (Kangaroo rats and pocket mice)	*Perognathus* sp.	Pocket mouse
	Rodentia	Heteromyidae	*Dipodomys* sp.	Kangaroo rat
	Rodentia	Cricetidae (Native rats and mice)	*Peromyscus* sp.	Deer mouse
	Rodentia	Cricetidae	*Neotoma* sp.	Wood rat
	Rodentia	Cricetidae	*Sigmodon* sp.	Cotton rat
	Rodentia	Geomyidae (Pocket gophers)		Pocket gopher
	Lagomorpha (Lagomorphs)	Leporidae (Leporids; rabbits and hares)		Indeterminate rabbit
	Lagomorpha	Leporidae	*Lepus* sp.	Jackrabbit
	Lagomorpha	Leporidae	*Sylvilagus* sp.	Cottontail
	Carnivora (Carnivores)	Canidae (Canids; dogs and allies)	*Canis* sp. (Dogs, coyotes, wolves)	Indeterminate canid
	Carnivora	Canidae	*Canis latrans*	Coyote
	Carnivora	Mustelidae (Weasels, skunks, allies)	*Taxidea taxus*	Badger
	Artiodactyla (Artiodactyls, odd-toed ungulates)			Indeterminate artiodactyls
	Artiodactyla	Cervidae (Elk and deer)	*Odocoileus* sp.	Indeterminate deer
	Artiodactyla	Antilocapridae (Pronghorn)	*Antilocapra americana*	Pronghorn
	Artiodactyla	Bovidae (Cows, sheep, allies)	*Ovis canadensis*	cf. Bighorn sheep
	Artiodactyla	Bovidae	*Bison bison*	cf. Bison

MNI values at the site level (MNI_{site}) were calculated using the most abundant element in the assemblage, which was the mandible. This approach eliminates all problems of interdependence and represents only the minimum number of animals in a taxon needed to account for the assemblage (Lyman 2008). This approach is the most common one used in the Southwest literature and is therefore reported throughout this study.

But MNI has the problematic effect of substantially reducing the data, and these reductions are not proportional across an assemblage. MNI reduces large NISP values to a far greater proportion than small NISP values. Because of this, MNI values for individual feature assemblages were calculated using the most abundant element in that sub-assemblage and these values were summed. This sum of feature MNI values for the site is indicated as $MNI_{feature}$. The element used in a feature to calculate MNI may vary from feature to feature. If a taxon is found in more than one feature, $MNI_{feature}$ values are greater than MNI_{site} values. $MNI_{feature}$ values eliminate the problem of interdependence within a feature, but do not eliminate it among features. However, the probability that bone fragments from a same animal were recovered from two different features is unlikely as only a small proportion of each feature was excavated (see Table 3.1) and the excavated features were widely spaced across the site. $MNI_{feature}$ counts provide a better assessment of the relative importance of different animals though the problem of aggregation still exits. In addition, $MNI_{feature}$ is not necessarily a superior estimate to those of NISP (Grayson 1984; Lyman 2008). Here I view NISP as a maximum value, $MNI_{feature}$ as an intermediate value, and MNI_{site} as a minimum value for any particular taxon. $MNI_{feature}$ values are reported as well as MNI_{site} values for the excavated data, while MNI_{site} values are only used for the flotation data. Most of the analyses rely on NISP values following the prescriptions of Lyman (2008:77) and Grayson (1984).

Fauna of Northern Chihuahua

All of the cerros de trincheras discussed in this chapter are in the Semidesert Grassland biotic community as discussed in Chapter 2 (Brown 1982d). While the Semidesert Grassland is transitional in the sense of being geographically positioned between the Chihuahuan Desertscrub community to the east and the Plains Grassland community to the west, it is a distinct biotic community with a diverse array of animal species (Brown 1982d). Many of these species have adapted to scarce and unpredictable precipitation, high variability in temperatures, and minimal vegetative protective cover.

A high mammal species diversity is found in the desert grasslands, particularly due to the high diversity of rodent species found there (Ceballos González et al. 2010; Parmenter and Devender 1995:198). Some of the major mammals that are well-represented in the Semidesert Grasslands, including those of northwestern Chihuahua, are black-tailed jackrabbit (*Lepus californicus*), desert cottontail (*Sylvilagus audubonii*), spotted ground squirrel (*Spermophilus spilosoma*), hispid pocket mouse (*Perognathus hispidus*), Ord's, banner-tailed, and Merriam's kangaroo rats (*Dipodomys ordii, D. spectabilis, D. merriami*), white-footed mice (*Peromyscus leucopus*), cotton rats (*Sigmodon hispidus, S. fluviventer*), southern plains and white-throated woodrats (*Neotoma micropus, N. albigula*), badger (*Taxidea taxus*), coyote (*Canis latrans*), mule deer (*Odocoileus hemionus*), and pronghorn (*Antilocapra americana*) (Anderson 1972; Ceballos González et al. 2010; Parmenter and Devender 1995). Grasslands in the area are the home to one of the largest remaining black-tailed prairie-dog (*Cynomys ludovicianus*) communities in North America (List et al. 2010). Additionally, where dense ocotillo and/or thornscrub elements are present, species such as collared peccary (*Pecari tajacu*) and white-tailed deer (*Odocoileus virginianus*) occur (Brown 1994; Ceballos González and List 2011).

Pronghorn and mule deer are presently found close to the site while bighorn sheep are not. However bighorn sheep were found in the high desert mountain ranges throughout Chihuahua (Anderson 1972:394). They have been reestablished in the mountains of southwestern New Mexico, about 65 km north of Cerro Juanaqueña, and in the late nineteenth century, they were in the mountains about 130 km east southeast of the site (Anderson 1972). In the Late Archaic period, they would have occupied the semi-arid ranges in the vicinity of Cerro Juanaqueña. Bison were identified in the Historic period around Paquimé some 50 km to the south and have been recently reintroduced into the Janos Biosphere Reserve (DiPeso et al. 1974:Vol 8:243; List et al. 2010).

The variety of birds found in desert grasslands is also great (Ceballos González et al. 2010; Parmenter and Devender 1995), and a list of the most prominent species includes Swainson's Hawk (*Buteo swainsoni*), prairie falcon (*Falco mexicanus*), kestrel (*Falco sparverius*), mourning dove (*Zenaida macroura*), scaled quail (*Callipepla squamata*), roadrunner (*Geococcyx californianus*), burrowing owl (*Athene cunicularia*), common poor-will (*Phalaenoptilus nuttallii*), ladder-backed woodpecker (*Picoides scalaris*), western kingbird (*Tyrannus verticalis*), ash-throated flycatcher (*Myiarchus cinerascens*), common raven (*Corvus corax*), and cactus wren (*Campylorhynchus brunneicapillus*). Gambel's quail (*Callipepla gambelii*), which prefers shrub habitat, is also present. A number of endemic reptiles and amphibians also are in the semidesert grassland (Parmenter and Devender 1995) including the western yellow box turtle (*Terrapene ornata luteola*), desert-grassland hognose snake (*Heterodon nasicus kennerlyi*), western hooknose snake (*Gyalopion canum*), desert-grassland whiptail (*Aspidoscelis uniparens*), southwestern earless lizard (*Holbrookia texana scitula*), and western green toad (*Anaxyrus debilis insidior*).

Modern ranching practices have significantly affected the landscape of northern Chihuahua. In the last 150 years, the number of non-native animals has increased, and native animals have been either extirpated or have decreased dramatically in numbers (e.g., wolf, pronghorn, and elk). The modern environment of northern Mexico likely resembles that found prehistorically, but differs in degree and density of fauna (Chapter 2). While the animal taxa remain similar, the relative abundance of species has changed. Over the last 100 to 150 years, dramatic shrub invasion and other changes in the composition of the biotic communities of the region have occurred, although the essential environmental conditions in the region today are similar to those experienced by prehistoric inhabitants. Given this information, it is possible to examine the subsistence behaviors associated with the prehistoric inhabitants at the Chihuahuan cerros de trincheras, and to see which of these animals and biotic communities they exploited.

Results

Most work was focused on excavations at Cerro Juanaqueña with smaller excavations conducted at three other Early Agricultural period cerros de trincheras (Chapter 4). In addition, small samples of backdirt were screened at two Ceramic period sites in the area in order to provide comparative data. Finally, the remnants of a buried, badly damaged Late Archaic period site (Ophelia's Well) were exposed. Faunal remains were recovered at each of these sites, and these analyses are presented in the following pages.

Cerro Juanaqueña

From four seasons at Cerro Juanaqueña, 33,173 bones were analyzed; 54% (17,799) were from excavation screens while the balance was from flotation contexts (Table 11.2). Thirty features were excavated, 24 of which yielded bone. Bone was recovered from 17 terraces, six rock rings, and a burned-rock feature. Five of the six features that lacked bone were only tested with 30 cm × 30 cm units (Chapter 3), and the sixth was a rock ring with shallow deposits. The six features that lacked bone are listed in Table 11.2 but are not included elsewhere in this chapter. Of the bone recovered from excavation, 14.4% (n = 2558 out of 17,799) were identified to at least the level of class.

Appendix 11.1 presents the Cerro Juanaqueña taxa identifications by feature for the excavation data, and Appendix 11.2 provides the flotation data. These appendices also present both NISP and MNI values for the individual features. In general, preservation of the bones was good, and bones were often recovered in large fragments. Weathering on the bones was present, although the frequency and severity was generally low, suggesting the faunal material was quickly buried. The bones typically show minimal evidence of root-etching, rodent-gnawing, and carnivore-digestion as only a few instances of each were noted. Nearly a quarter of the bones (24.1%) contain evidence for burning, and the significance of this finding is discussed below.

The median excavated bone sample from each feature is 397.5 with wide variability [Interquartile Range (IQR) = 1178, n = 24]. Samples from the 13 terraces on the upper part of the site were much larger (median = 1197, IQR = 1036) than samples from the four lower terraces (median = 151, IQR = 717), the six upper rock rings (median = 125, IQR = 210), and the single burned rock feature (135) (Figure 11.1).

The volume excavated per feature (Table 3.1) is not particularly large (median = 1.85 m^3, minimum = 0.1 m^3, maximum = 6.6 m^3, IQR = 1.2 m^3, n = 24). The excavated density of bone recovered from the features trended high but was variable (median = 254.8 fragments of bone/m^3 of excavated sediment, IQR = 579.8/m^3, n = 24). The rock rings had lower density (median = 160.8, IQR = 463.3/m^3, n = 6) than the terraces (median = 309.1/m^3, IQR = 672.9/m^3, n = 17), while the burned rock feature was similar to the rock rings (192.9/m^3). The upper terraces had much higher bone density (median = 520.4/m^3, IQR = 564.3/m^3, n = 13) than the lower terraces (median = 66.6/m^3, IQR = 108.0/m^3, n = 4). Figure 11.1 shows that at Cerro Juanaqueña there is no significant relationship between feature volume excavated and total remains recovered (r^2 = .03, p = .383) due to the variable bone density in the features. Figure 11.1 also reflects the higher bone recovery from the upper terraces relative to the rock rings, burned rock feature, and lower terraces.

All features yielding bone except one date to the primary occupation of the site, 1300 to 1100 cal BC. A single lower terrace, Terrace 413, yielded dates representing both this primary occupation and a reoccupation that occurred between 400 BC and AD 1 (Chapter 5). Since this terrace yielded an NISP of only eight unremarkable bones, these data are included throughout the analysis and are briefly discussed towards the end of the chapter.

Table 11.3 summarizes only the excavated samples from the terraces and rock rings at Cerro Juanaqueña. Data compared in Tables 11.3 and 11.4 includes the following for each feature: the number of taxa identified, the lagomorph and artiodactyl indices for each feature,

TABLE 11.2. Total Number of Bones Recovered from Individual Features at Cerro Juanaqueña.

Feature	Zone	Excavation/Test	Bones recovered in excavation	Bones recovered in flotation	Total
T6	Upper	Excav	3888	2343	6231
T10	Upper	Excav	1364	2510	3874
T97	Upper	Excav	202	215	417
T126	Upper	Excav	1566	1331	2897
T163	Upper	Excav	1488	3827	5315
T167	Upper	Excav	350	36	386
T175	Upper	Excav	1811	2072	3883
T195	Upper	Test	0	0	0
T222	Upper	Excav	1197	116	1313
T273	Upper	Excav	537	0	537
T287	Upper	Excav	680	432	1112
T290	Upper	Excav	591	148	739
T297	Upper	Excav	445	742	1187
T387	Lower	Excav	195	91	286
T413[1]	Lower	Excav	107	34	141
T415[2]	Lower	Test	0	0	0
T463	Lower	Test	0	0	0
T465	Lower	Test	0	0	0
T487	Lower	Excav	960	641	1601
T508	Lower	Excav	33	40	73
T537	Upper	Excav	1383	538	1921
BR1	Upper	Excav	135	143	278
R1A	Upper	Excav	296	0	296
R28	Upper	Excav	0	0	0
R51	Upper	Test	64	0	64
R234A	Upper	Excav	9	0	9
R239	Upper	Excav	172	95	267
R250	Upper	Excav	78	0	78
R286	Upper	Excav	248	20	268
R307A	Upper	Test	0	0	0
Total			17799	15374	33173

[1] One early (ca. 1300–1100 BC) and two later (400 BC–AD 1) radiocarbon dates came from this feature.
[2] One late (400 BC–AD 1) radiocarbon date but no bone.

and the percentage of bones in the excavated feature assemblage that are burned. The number of taxa count includes indeterminate class identifications in all cases (indeterminate artiodactyl, indeterminate rabbit, etc.). It does not include unidentified remains.

Burning was present in 24.1% of all excavated bones (17,799) in the Cerro Juanaqueña assemblage. This high proportion suggests that processes in addition to cooking were involved. A closer examination of the dominant taxon, leporids, enabled a closer analysis. The proportion of burned leporids is 24.5%(475/1937) and mirrors the overall proportion of burned bone. However, this average proportion masks important variability among features. Figure 11.2 plots proportion of burned leporids by total NISP for each feature. Features with less than 25 NISP are to the left of the reference line. The proportion of burned leporid elements in these features varies from 0 to 80% and include rock rings, upper terraces, and lower terraces (Chapter 3). The proportion of burned leporids in the remainder of the features is bimodally distributed and defined by the two ellipses. One group has 10% to 20% burned bone and a second group has 30% to 40% burned bone. The former includes one lower terrace and five upper terraces. The latter group consists of four upper terraces and one burned rock feature. Presumably the high rate of burned leporid bones in the burned rock feature is related to its function. However, there was no such obvious evidence of burning in the remaining four upper terraces. It is likely that high rates of burned bone is related to processes in

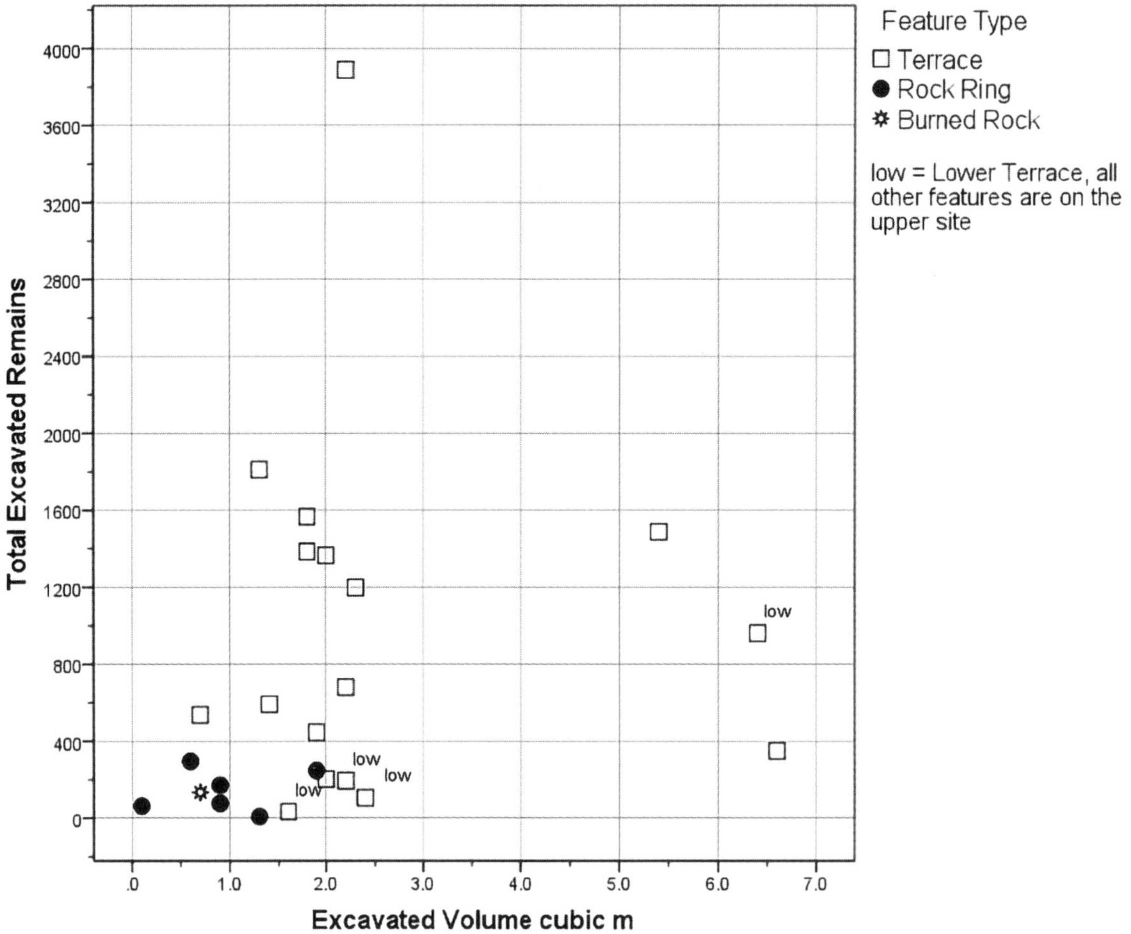

FIGURE 11.1. Total number of excavated faunal elements by excavated volume of features. Total number of bones is unrelated to excavated volume. Note that rock rings and lower terraces (low) produced less bone than upper terraces.

addition to cooking, including burning trash. If burning trash is playing a role, we might expect features with the highest bone density to have the highest rates of burning. While no correlation was found between proportion of burned bone and bone density among features, two features (Terrace 6 and Terrace 175) have both the highest bone density (over 1300 bones per cubic meter) and among the highest rates of burned leporid bone, 35–39%, suggesting that trash burning is playing a role and that burning is unrelated to post-occupational processes such as natural fires. More detailed analyses may be required to further advance our understanding of the high rates of burning.

Other Sites

Appendix 11.3 provides the detailed data from the excavations at the three other cerros de trincheras (Cerro el Canelo, Cerro Torres, and Cerro Vidal), as well as Ophelia's Well, and the two Ceramic period sites, Sitio Janos and Sitio Antonio (Chapter 4 and Appendix 4.1). Sitio Janos is on the western outskirts of Janos, 9 km southwest of Cerro Juanaqueña, and dates to the Viejo period (AD 700–1000/1200). Sitio Antonio is a Medio period pueblo (AD 1200 to 1450) located 1 km west of Cerro Juanaqueña along the Río Casas Grandes. Summaries of the faunal data from these sites are found in Tables 11.5 and 11.6. Table 11.5 presents excavation and flotation data including sample sizes, excavation volume, and NISP. Table 11.6 includes the number of taxa and the lagomorph and artiodactyl indices. Descriptions of the other cerros de trincheras are found in Chapter 4.

The strategy at the other three cerros de trincheras was to excavate one or two terrace features to obtain radiocarbon and other samples using the same excavation strategy as that used at Cerro Juanaqueña. Faunal remains were collected from the excavation screens as well as from processed flotation samples. Two terraces were excavated at Cerro el Canelo and one each at Cerro Torres and Cerro Vidal. Radiocarbon dates show that Cerro el Canelo and Cerro Torres are contemporaneous

TABLE 11.3. Excavated Assemblages from Terraces and Rock Rings at Cerro Juanaqueña.

Feature	Location	Number of taxa	NISP	NISP Lagomorph index	NISP Artiodactyl index
T6	Upper	14	478	0.09	0.02
T10	Upper	15	288	0.15	0.06
T97	Upper	7	23	0.90	0.00
T126	Upper	10	376	0.15	0.06
T163	Upper	13	163	0.18	0.10
T167	Upper	3	25	0.11	0.28
T175	Upper	8	98	0.22	0.18
T222	Upper	12	279	0.13	0.03
T273	Upper	7	49	0.14	0.12
T287	Upper	8	71	0.28	0.17
T290	Upper	6	52	0.24	0.05
T297	Upper	7	102	0.21	0.09
T387	Lower	5	34	0.29	0.15
T413	Lower	4	8	0.50	0.00
T487	Lower	9	154	0.10	0.04
T508	Lower	2	5	0.0	0.00
T537	Upper	11	260	0.08	0.06
BR1	Upper	2	30	0.03	0.0
R1A	Upper	6	24	0.10	0.09
R51	Upper	1	2	0.0	1.00
R234A	Upper	1	8	0.0	0.00
R239	Upper	3	6	0.20	0.17
R250	Upper	4	8	0.0	0.17
R286	Upper	4	15	0.25	0.00
Total			2558		

with Cerro Juanaqueña (Chapter 5). Data in Table 11.5 show that the excavated faunal samples from Cerro el Canelo and Cerro Torres contained less than 350 pieces from each of the three terraces (excluding tiny fragments recovered through flotation); this sample size overlaps with the lower producing terraces and rock rings from Cerro Juanaqueña. Cerro Vidal is contemporaneous with the late occupation (cal 400 BC to AD 1) at two of the lower terraces at Cerro Juanaqueña (Terraces 413 and 415). Unfortunately, like these late terraces at Cerro Juanaqueña (see Table 11.2), Cerro Vidal only produced a tiny sample of fauna. A total of 681 bones were recovered from excavation screens from the other three cerros de trincheras, as well as 1399 from flotation. The three sites produced a median sample of 168.5 excavated bones (IQR = 287, $n = 4$), a sample size more similar to those from the Cerro Juanaqueña rock rings and lower terraces than the Cerro Juanaqueña upper terraces. Data from the three other cerros de trincheras show that 39.4% of the bones were identifiable (268/681), but preservation conditions varied across the four terraces, with a range of 5% to 52.2% identifiable (Table 11.5).

Ophelia's Well is slightly earlier than Cerro Vidal (cal 790 BC to 410 BC), but consisted of only the remnants of a badly damaged buried site in the floodplain and only a few pieces of fauna were non-systematically recovered during profiling operations. Screening of the looter's back dirt at Sitio Janos yielded a sample size of 234, of which 12% were identifiable. The backdirt at Sitio Antonio produced 537 bones, 17.1% of which were identifiable (Table 11.5). Bone preservation from these Ceramic period sites was similar to that of Cerro Juanaqueña, where the proportion of identifiable bone was 14.4%.

Taxa

Table 11.7 presents a summary of the taxa identified at Cerro Juanaqueña. Leporids overwhelm the assemblage with 76% (1937) of the 2558 identified faunal remains consisting of jackrabbits (*Lepus* spp.), cottontail rabbits (*Sylvilagus* spp.), and indeterminate rabbits (Leporidae). Venomous and non-venomous snakes make up 11.2% (286) and a surprising 5.5% (140) are from the large family of rodents including kangaroo rat (*Dipodomys* sp.),

Table 11.4. Comparison of Excavated Burned and Unburned Rodent and Rabbit Remains from Features at Cerro Juanaqueña.

	# Burned Rodent	# Unburned Rodent	# Total Rodent	% Burned Rodent	# Burned Rabbit	# Unburned Rabbit	# Total Rabbit	% Burned Rabbit
T6	6	11	17	17	156	286	442	35
T10	0	24	24	0	21	216	237	9
T97	0	8	8	0	0	10	10	0
T126	1	3	4	25	11	105	116	9
T163	0	3	3	0	20	118	138	14
T167	0	0	0	0	11	7	18	61
T175	0	3	3	0	30	46	76	39
T222	3	15	18	17	88	159	247	36
T273	0	4	4	0	0	38	38	0
T287	0	12	12	0	6	43	49	12
T290	0	7	7	0	0	42	42	0
T297	0	4	4	0	12	76	88	14
T387	0	13	13	0	2	15	17	12
T413	0	1	1	0	0	6	6	0
T487	0	15	15	0	16	75	91	18
T508	0	0	0	0	0	4	4	0
T537	1	4	5	25	71	167	238	30
BR1	0	0	0	0	11	19	30	37
R1A	0	1	1	0	7	13	20	35
R51	0	0	0	0	0	0	0	0
R234A	0	0	0	0	4	4	8	50
R239	0	0	0	0	2	3	5	40
R250	0	1	1	0	4	1	5	80
R286	0	0	0	0	3	9	12	25
Total	11	129	140	7.9	475	1462	1937	24.5

cotton rat (*Sigmodon* sp.), pocket mouse, deer mouse, and squirrels; many of the snakes and rodents appear to be intrusive. An additional 5.1% (131) of the identified sample consists of artiodactyl remains, including bighorn sheep (cf. *Ovis canadensis*), pronghorn (*Antilocapra americana*), indeterminate deer (*Odocoileus* sp.), mule deer (*Odocoileus hemionus*), a single possible bison (cf. *Bison bison*), and indeterminate artiodactyl remains. The remaining 2.5% (64) of the identified specimens in the assemblage include small fish, lizards, terrestrial turtles (*Terrapene* sp.), indeterminate turtles, spadefoot toads (*Scaphiopus* sp.), indeterminate canids, coyote (*Canis latrans*), and badger (*Taxidea taxus*).

Leporids made up 90% (240/268) of the NISP excavated fauna from the three other cerros de trincheras, an even higher frequency than that of Cerro Juanaqueña; more faunal details are in Appendix 11.3. Following leporids and artiodactyls, the balance of the assemblage was made up of rodents, a turtle, a nonvenomous snake, and a coyote. Thus the overall rank order from the three cerros de trincheras of jackrabbits, cottontails, artiodactyls, rodents, followed by taxa that occur in low numbers is similar to that from Cerro Juanaqueña.

The Viejo period site, Sitio Janos, also contained a high proportion of leporids, 96.4% (27/28), but Sitio Antonio, the Medio period site, contained only 32.6% leporids (30/92) as that collection included 54% indeterminate rodents. The absence of artiodactyls from the Ceramic period sites is notable, although perching birds and turtles are present.

Analysis

Aggregation for MNI

MNI calculations are strongly dependent on the level of aggregation. $MNI_{feature}$ assumes that elements of the same animal were not recovered from more than one feature and is therefore not being double counted. As discussed previously for Cerro Juanaqueña, this assumption has strong support. MNI_{site} eliminates this possibility by aggregating data at the site level, which substantially reduces the data from across the site and yields a minimum MNI estimate. Table 11.8 shows the

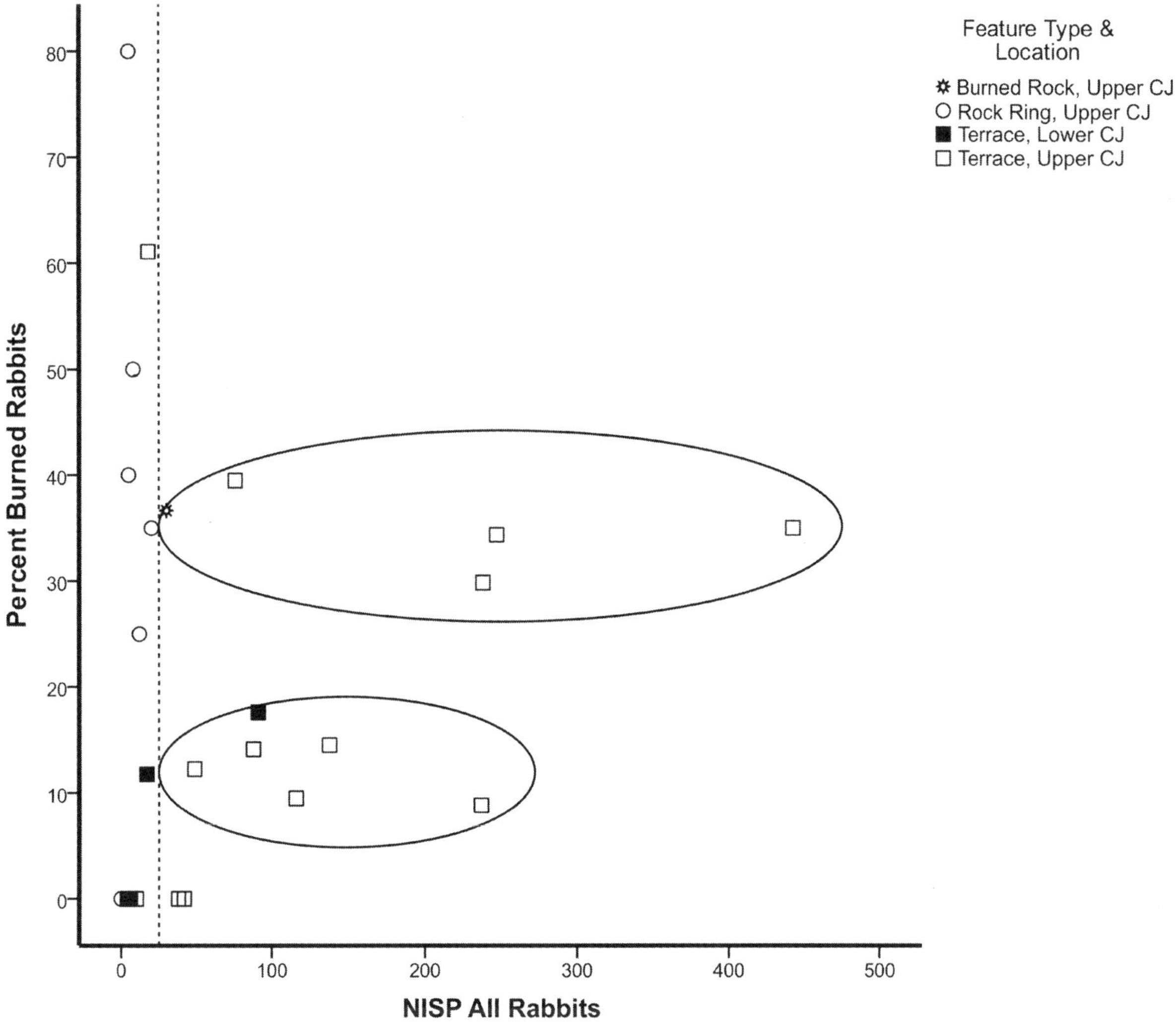

FIGURE 11.2 Percent burned rabbit bones and NISP of all rabbits. Features with a NISP < 25 (left of the vertical reference line) are highly variable in percent of burned bone. Remaining features form two modes as defined by ellipses.

$MNI_{feature}$ calculation for the three key taxa—jackrabbit, cottontail, and artiodactyls. The MNI for each taxon in each feature is listed and then summed across features to arrive at the $MNI_{features}$ value for the site. The element used varies among features and is also listed in the table.

Table 11.7 lists MNI_{site}, $MNI_{feature}$, NISP for all taxa, and the NISP percentage and allows comparison of these approaches. At Cerro Juanaqueña, MNI_{site} for each artiodactyl taxon (indeterminate artiodactyl, indeterminate deer, pronghorn, bighorn sheep, and bison) is 1, resulting in the artiodactyl MNI_{site} being 5. In contrast the artiodactyl $MNI_{feature}$ is 36, and likely provides a more realistic view of the relative importance of these animals.

Grayson (1984) and Lyman (2008) have shown that log MNI and log NISP are closely correlated. Figure 11.3 shows that for the Cerro Juanaqueña data, $\log_{10}$ NISP and $\log_{10}$ $MNI_{feature}$ are significantly and strongly correlated ($r^2 = .907$, $p < .0001$, $n = 30$, intrusive venomous snakes removed). The slope of the line is .578, which is similar to what Lyman (2008:52) and Grayson (1984) found across multiple data sets. A similar relationship (not shown) occurs when $\log_{10}$ NISP is plotted against $\log_{10}$ MNI_{site}, but the relationship is not as strong ($r^2 = .49$, $p < .0001$, $n = 30$). The substantial number of taxa that have an MNI_{site} of 1, yet have an elevated NISP, likely reduces the strength of the correlation (Table 11.7). The slope of the regression line between $\log_{10}$ NISP and $\log_{10}$ MNI_{site} (not shown) is .32. Both the slope and the strength of the correlation (r^2) are well below those seen across multiple assemblages (Lyman 2008:54–55). These observations suggest that in the case of the Cerro Juanaqueña assemblage, $MNI_{feature}$ and NISP are closely related and are better representatives

Table 11.5. Faunal Samples Recovered from Excavation and Flotation from Other Sites.

Site	Location	Volume excavated m^3	Bones recovered in excavation	NISP Excavation	Bones recovered in flotation	NISP Flotation	Total recovered
Cerro el Canelo	T41	0.51	80	4	637	5	717
Cerro el Canelo	T94	1.38	335	175	544	9	879
Cerro el Canelo	Total	1.89	415	179	1181	14	1596
Cerro los Torres	T1	1.76	257	87	218	10	475
Cerro Vidal	T20	1.32	9	2	0	2	11
Ophelia's Well	Profiling/ testing		3	1	7	0	10
Sitio Antonio	Screened looter's backdirt		537	92	33	5	570
Sitio Janos	Screened looter's backdirt		234	28	NA	NA	234

Table 11.6. Excavated Data from Other Sites Including Number of Taxa, NISP, Lagomorph Index, and Artiodactyl Index.

Site	Location	Excavation Number of taxa	NISP	NISP Lagomorph index	NISP Artiodactyl index
Cerro el Canelo	T41	4	4	0.5	0.33
Cerro el Canelo	T94	6	175	0.062	0.041
Cerro los Torres	T1	7	87	0.27	0.026
Cerro Vidal	T20	2	2	0.5	0
Sitio Antonio	Screened backdirt	6	92	0.233	0
Sitio Janos	Screened backdirt	2	28	0	0

of taxonomic abundance than MNI_{site}. Figure 11.3 shows that jackrabbits by far dominate the assemblage at Cerro Juanaqueña, with cottontails a distant second.

Leporids and the Lagomorph Index

Originally developed by Bayham and Hatch (1985), the lagomorph index is calculated by dividing the total cottontail count by the sum of all lagomorphs [S/(S+L); where S = *Sylvilagus* and L = *Lepus*]. Only elements identified to at least the genus level are included in the calculation. The index provides a measure of the relative frequencies of cottontails/jackrabbits in an assemblage. The index compares the NISP of cottontail rabbits to the sum of all rabbits and hares and values less than 0.5 indicate jackrabbits occur more frequently in an assemblage. Zooarchaeologists have used this index to make inferences about taxonomic abundance in an assemblage, and it allows relative comparisons among assemblages (Dean 2007; Dean and Beaver 2017; Lee and Speth 2004; Quirt-Booth 1996; Schmidt 1998; Szuter 1991; Szuter and Bayham 1989). Use of the lagomorph index assumes that potential biases, such as interdependence and differential preservation, affect the NISP of both taxa in a similar fashion.

All data indicate that leporids were key subsistence items, with jackrabbits by far the most important relative to cottontails. Table 11.9 presents the lagomorph and artiodactyl indices calculated with NISP of excavated screen bone from Cerro Juanaqueña.

The lagomorph index of 0.134 derived from NISP values (260/(1656+260+21)) indicates only 13.4% of identifiable lagomorph bones are cottontail; clearly the exploitation of jackrabbits was far more important than cottontails. Inferring relative leporid taxonomic abundance in an environment, known as the biocenosis, from faunal assemblages is a common practice but involves several issues and assumptions (Lyman 2008:69–75).

A number of factors can affect the relative proportion of jackrabbits to cottontails in the environment. Cottontails are generally found in areas with dense vegetative cover including wood and shrub areas, riparian settings, and other closed vegetation where they can hide from predators. Conversely, jackrabbits prefer open spaces where they can flee from predators (Brown and Krausman 2003; Legler 1970; Szuter and Bayham 1996). Today the upland vegetation in the vicinity of Cerro Juanaqueña is Semidesert Grassland and is a mosaic of grasslands and shrublands (Chapter 2). Jack-

TABLE 11.7. Excavated Faunal Remains from Cerro Juanaqueña by MNI_{site}, $MNI_{feature}$, NISP, and Percent NISP.

Identification	Abbrev	MNI_{site}	$MNI_{feature}$	NISP	% of Total NISP
Indeterminate small fish	InFis	1	2	6	0.24
Western box turtle	BTurt	1	6	12	0.47
Indeterminate turtles	InTur	1	2	5	0.20
Nonvenomous snakes	NvSnk	1	7	48	1.88
Venomous snakes	VnSnk	1	1	238	9.33
Indeterminate lizards	InLiz	1	2	4	0.16
Spadefoot toads	Toad	1	2	3	0.12
Indeterminate birds	InBrd	1	4	10	0.39
Quail	Quail	1	2	3	0.12
Perching birds	PBird	1	8	9	0.35
Duck	Duck	1	1	1	0.04
Indeterminate rodents	InRod	1	6	28	1.10
Indeterminate squirrels	InSqr	1	1	5	0.20
Pocket mouse	PoMse	9 (mandible)	12	73	2.86
Deer mouse	DrMse	3 (mandible)	3	8	0.31
Cotton rat	CoRat	5 (femur)	5	20	0.78
Kangaroo rat	KaRat	1	1	1	0.04
Wood rat	WoRat	1	1	1	0.04
Pocket gopher	PoGof	1	1	3	0.12
Chipmunk	Cpmnk	1	1	1	0.04
Jackrabbit	JaRab	25 (mandible)	57	1656	64.63
Cottontail	Cottl	5 (mandible)	24	260	10.20
Indeterminate rabbit	InRab	1	6	21	0.82
Indeterminate canid	InCan	1	2	3	0.12
Coyote	Coyot	1	4	6	0.24
Badger	Badgr	1	2	2	0.08
Indeterminate artiodactyls	InArt	1	15	77	3.02
Indeterminate deer (cf. Mule deer)	InDer	1	8	24	0.94
Pronghorn	Pghrn	1	10	24	0.94
cf. Bighorn sheep	BShep	1	2	5	0.20
cf. Bison	Bison	1	1	1	0.04
Total Identified Remains (14.4%)		—		2558	100
Total Unidentified Remains (85.6%)		—		15241	—
Total Faunal Remains		—		17799	—

rabbits were 10 times more abundant in shrublands than grasslands in the Janos Biosphere Reserve while cottontails showed no difference between the two habitats (Ceballos González et al. 2010). In the past, considering likely wetter conditions, the absence of cattle grazing and fire suppression, grasslands should have been favored relative to shrublands. In addition to the uplands, the nearby floodplains and alluvial terraces were also heavily exploited. Today the floodplains are dominated by fallow and cultivated fields and uncultivated shrublands and grasslands. Paleoecological data (Chapters 2, 10, and 12) indicate that 3000 years ago the floodplains and alluvial terraces supported C_4 grasslands as well as C_3 species, the latter which includes both trees and shrubs. Stable isotope data presented in Figure 10.1 indicate that Cerro Juanaqueña jackrabbits were feeding on a diet dominated by C_4 resources. In the past, as today, both upland and floodplain habitats favored jackrabbits relative to cottontails.

Anthropogenic modifications to landscapes may alter leporid habitats. If an area is intensively or extensively occupied, habitats can become more open and favorable to jackrabbits (Bayham and Hatch 1985; Szuter and Bayham 1989, 1996). In the vicinity of Cerro Juanaqueña, disturbance would have included agricultural fields planted in the floodplains, harvesting firewood

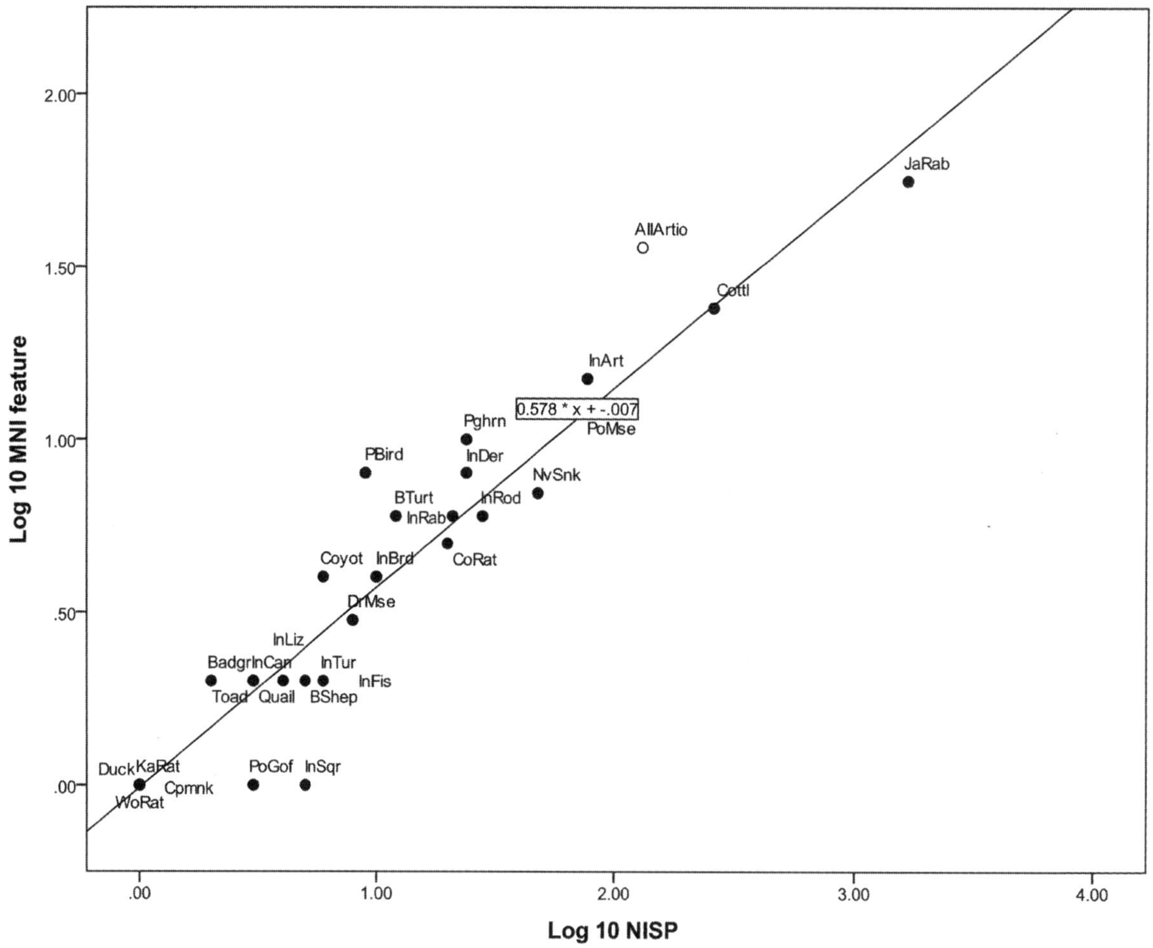

FIGURE 11.3. Taxa plotted by $\log_{10}$ NISP and $\log_{10}$ $MNI_{feature}$. There is a strong and significant linear relationship between $\log_{10}$ NISP and $\log_{10}$ $MNI_{feature}$, suggesting that $MNI_{feature}$ is a good indicator of taxonomic abundance. Abbreviations are listed in Table 11.7. "AllArtio" refers to all artiodactyls, although the correlation coefficient and regression equation are calculated without this value since each artiodactyl taxa is separately included. Venomous snakes are excluded as they are clearly intrusive. Note the dominance of jackrabbits with cottontails a distant second.

and construction wood from the uplands and floodplains, and perhaps burning the landscape. For example, the ethnobotanical results show that mesquite wood was commonly used (Chapter 8). Two centuries of intensive habitat use in the vicinity of Cerro Juanaqueña may have further increased jackrabbit relative to cottontail populations in the anthropogenic environment.

Studies of modern leporid populations in open, upland, Chihuahuan semidesert grassland and desert scrubland communities indicate that *S. audubonii* typically makes up only 0 to 20% of the leporid populations, while *L. californicus* makes up the balance (Lightfoot et al. 2010; Nelson et al. 1997). Open shrublands would have favored jackrabbits relative to cottontails (Ceballos González et al. 2010). Overall the faunal data suggest cottontails and jackrabbits were being taken in proportions that are consistent with expected distributions in their environment.

Artiodactyl Index

Artiodactyls are a highly ranked dietary contributor at Cerro Juanaqueña, and this pattern is seen at most other Early Agricultural period sites in the southern Southwest (Schmidt 2008). The relative importance of artiodactyls is implicated in understanding subsistence and is closely tied to questions of mobility. Table 11.7 shows that following indeterminate artiodactyls, deer and pronghorn occur at the same rate, with bighorn sheep in fourth place, and a single example of possible bison. This relative ranking among artiodactyls remains virtually unchanged whether MNI_{site}, $MNI_{feature}$, or NISP is used. The preferred habitat of both mule deer and pronghorn are accessible from Cerro Juanaqueña.

The artiodactyl index compares the relative contributions of lagomorphs and artiodactyls (Bayham 1982). Table 11.9 presents the Cerro Juanaqueña data used in calculating this index. The artiodactyl index,

TABLE 11.8. Jackrabbit, Cottontail, and Artiodactyl $MNI_{feature}$ Calculations.

	Jackrabbit MNI	Cottontail MNI	Artiodactyl MNI
BR1	3 (calcaneus)	1	0
R1A	2 (calcaneus)	1	2
R51	0	0	1
R234A	1	0	0
R239	1	1	1
R250	1	0	1
R286	1	1	0
T6	6 (mandible)	2 (calcaneus)	2
T10	5 (radius)	1	4
T97	1	1	0
T126	3 (mandible)	1	1
T163	2 (mandible)	1	3
T167	1	1	1
T175	2 (calcaneus)	1	3
T222	4 (calcaneus)	2 (mandible)	3
T273	2 (mandible)	1	2
T287	2 (mandible)	1	4
T290	1	1	1
T297	4 (mandible)	2 (mandible)	2
T387	2 (ulna)	1	1
T413	1	1	0
T487	6 (calcaneus)	1	1
T508	1	—	0
T537	5 (ulna)	2 (femur)	3
$MNI_{feature}$	57	24	36

Note: MNI is aggregated at the feature level and $MNI_{feature}$ is the sum of these values.

using NISP, is .063 (131/(131+1937)) indicating artiodactyl elements are only 6.3% of all artiodactyl and leporid elements, suggesting artiodactyl use is low relative to lagomorphs.

Another approach to evaluating the relative importance of artiodactyls is through rank order distributions (Grayson 1984; Lyman 2008). Notice in Table 11.10, using NISP, jackrabbits are first, followed by cottontails, with artiodactyls last. But if taxa are rank-ordered by $MNI_{feature}$ or MNI_{site}, then artiodactyls rank ahead of or equal to cottontails. Jackrabbits were clearly the most important prey, but this shift in ranking suggests that artiodactyls may have been more important than the artiodactyl index of .06 suggests.

TABLE 11.10. Comparison of Rank Order Distribution of Jackrabbit, Cottontail, and Artiodactyls Using NISP, $MNI_{feature}$, and MNI_{site}.

	NISP	$MNI_{feature}$	MNI_{site}
Lepus sp.	1656	57	25
Sylvilagus sp.	260	24	5
Artiodactyl	131	36	5

Artiodactyl indices throughout the Southwest vary due to site location, population, and site type (Szuter and Bayham 1989). There is a strong trend related to elevation; sites in upland areas (above 800 m) typically have NISP indices above 0.35, while lowland sites (below 800 m) typically have NISP artiodactyl indices below 0.10 (Szuter 1991; Szuter and Bayham 1996). Table 11.11 shows a similar trend for Early Agricultural period desert borderland sites; eight of the 10 sites have NISP artiodactyl indices less than 0.15. The low Cerro Juanaqueña values are consistent with other lowland Early Agricultural sites. The highest artiodactyl index in Table 11.11 is from McEuen Cave, a Middle to Late Archaic site in the Gila Mountains. McEuen Cave is located in a transitional zone between three distinct biomes (woodland, desert grassland, and riparian), which made access to a number of distinct artiodactyl species likely. These trends are clearly connected to the relative potential of different settings to support both leporids and artiodactyls (but see Dean 2003; Szuter 1991). It is difficult to assess if overhunting in particular environments has also impacted the relative importance of artiodactyls without systematic diachronic data (Cannon 1999; Huckleberry et al. 2001; Szuter 1991).

Intrusive Remains

Venomous and nonvenomous snakes (286/2558) are 11.2% and rodents (140/2558) are 5.5% of the Cerro Juanaqueña NISP assemblage ranking them second and third behind leporids and followed by artiodactyls (Table 11.7). Typically, it is difficult to determine if rodents, snakes, and a few other small animal remains found in archaeological deposits are from subsistence activities or are the result of natural deaths. Such small animals were well known as subsistence items among many Native American groups (e.g., Shaffer 1992) but many also burrow and die in archaeological deposits, and discriminating their origin remains a challenging

TABLE 11.9. Cerro Juanaqueña Lagomorph and Artiodactyl Indices Using NISP.

	Indeterminate Rabbit	Sylvilagus sp.	Lepus sp.	Lagomorph Index	All Lagomorph	Artiodactyl	Artiodactyl Index
NISP	21	260	1656	0.134	1937	131	0.06

TABLE 11.11. NISP Artiodactyl Indices at Sites in Northern Mexico and the Southern Southwest.

Site	Age	Lagomorph NISP	Artiodactyl NISP	NISP Artiodactyl Index
Cerro Juanaqueña (northwest Chihuahua)	San Pedro phase	1929	131	0.06
Los Pozos[1] (Tucson, Arizona)	San Pedro phase	2033	262	0.11
Valley Farms[2] (Tucson)	San Pedro phase	69	5	0.07
Wetlands[2] (Tucson)	San Pedro phase	234	16	0.06
Las Capas[2] (Tucson)	San Pedro/Cienega phase	1347	55	0.04
Stone Pipe Site[3] (Tucson)	Cienega phase	491	23	0.04
Santa Cruz Bend Site[3] (Tucson)	Cienega phase	1330	265	0.15
Square Hearth Site[3] (Tucson)	Cienega phase/Early Ceramic	33	2	0.06
Turquoise Ridge[4] (Hueco Bolson, New Mexico)	Formative period	1490	0	0.0
McEuen Cave[5] (east-central Arizona)	Middle to Late Archaic period	388	91	0.19

[1] Wöcherl 2001; [2] Schmidt 2008; [3] Wöcherl 1997; [4] Whalen 1994; [5] Schmidt 2001.

TABLE 11.12. Proportion of Complete Bones of Selected Taxa.

Taxa	No. Complete	NISP	% Complete
Jackrabbit	400	1656	24.2
Cottontail	90	260	34.6
Indeterminate bird	3	10	30.0
Passerines	2	9	22.2
Pocket mice	43	73	58.9
Deer mice	5	8	62.5
Cotton rat	12	20	60.0
Indeterminate rodent	18	28	64.3
Nonvenomous snakes	41	46	89.1

issue. Reliable classification of any single bone as intrusive or not is difficult at best, but an examination of trends can address the question of whether or not snakes and rodents should be highly ranked as economic taxa at Cerro Juanaqueña. For this analysis, I focus on the excavated assemblage, but the issue remains for the flotation samples as well given the difficulties of discerning if small, unburned bones of burrowing species derive from cultural or post-depositional processes. The distribution, general appearance, and treatment of potentially intrusive remains is used to address this issue.

The 238 venomous snake bones appear to be a straightforward case of intrusive remains. They were all found in a single feature, Terrace 126, none were burned, and virtually all appear to be complete. They were, however, spread horizontally among two adjacent one-by-one meter units and distributed vertically over a meter. Presumably one or more snakes died in a burrow and the bones moved downward through the profile. The vertical movement of artifacts through the terrace fill is discussed in Chapter 3. Similarly, 40 of the 48 nonvenomous snake bones are from Terrace 487; none of the 40 are burned, and virtually all are complete. These also appear to be clear cases of intrusive remains. They too were spread among multiple adjacent excavation units and vertically 80 cm through the profile. Clearly snakes were not as economically important as their NISP values would indicate, although three nonvenomous snake bones are burned, all from other features.

Rodent remains are more difficult to assess as they are distributed across the site. In a few cases, rodent remains had a fresh appearance, such as being yellowish to brown in color, or were found in articulated segments making their recent origin clear. However, completeness and burning were more common attributes whose trends can be examined for clues as to the degree rodents were utilized. Rodent bones were frequently complete, consistent with an intrusive origin. Table 11.12 indicates 58.9% to 64.3% of rodent bones were complete. In contrast, only 24.2% to 34.6% of leporid bones were complete and bird bones were similar to leporids with 22.2% to 30.0% complete.

Earlier I argued that the high rate of burning throughout the assemblage was not only due to cooking. Instead, burning typically occurred after the bones were deposited but during the site occupation, suggesting trash deposits were being burned. If a high proportion of burned bones, such as found among the leporids, is consistent with indigenous use and deposition, then a low proportion of burned bones for a particular taxon suggests a contribution from post-occupational origins. Only 7.9% (11/140) of the excavated sample of rodent bones are burned versus 24.5% (475/1937) of the rabbit bones (see Table 11.4). A chi-square test for a two-by-two table indicates this difference is unlikely to be due to chance (chi-square =20.2, $df = 1$, $p < .001$), although there is a substantial difference between the two sample

TABLE 11.13. Faunal Remains from Cerro Juanaqueña Flotation Samples by MNI_{site}, NISP, Percent of Total, and Percentage of NISP Burned.

Identification	Minimum Number of Individuals (MNI_{site})	NISP	Percentage of total NISP	NISP burned (Percentage of NISP burned)
Indeterminate small fish	1	88	26.8	22 (25%)
Nonvenomous snake	1	18	5.5	0 (0)
Indeterminate lizard	1	2	0.6	0 (0)
Indeterminate amphibian	1	4	1.2	1 (25)
Indeterminate bird	1	1	0.3	0 (0)
Indeterminate rodent	1	183	55.8	12 (6.6)
Jackrabbit	1	4	1.2	0 (0)
Cottontail	1	6	1.8	0 (0)
Indeterminate rabbit	1	19	5.8	4 (21.1)
Indeterminate artiodactyl	1	3	0.9	0 (0)
Total Identified Remains (2.1%)	—	328	100	39 (11.9)
Total Unidentified Remains (97.1%)	—	15046	—	2757 (18.3)
Total Faunal Remains	—	15374	—	2796 (18.2)

sizes. The fact that 92.1% (129/140) of rodent remains are not burned suggests that many rodents in the assemblage were not related to subsistence activities at Cerro Juanaqueña and were introduced to the site after its abandonment. However, the 11 burned rodent bones from excavation suggests that rodents were indeed consumed at Cerro Juanaqueña. The burned rodent remains include three pocket gophers, one cotton rat, two ground squirrels, a single pocket mouse, and four indeterminate rodents. If I assume that 25% of the utilized assemblage is burned, then this proportion can be extrapolated to the rodents and can provide a procedure to estimate the NISP of rodents that should be related to the occupation of the site. Therefore an NISP of 44 (11 burned/0.25) rodents are expected to be in the assemblage that would have been contributed by the occupants, suggesting that approximately 70% of the rodents are intrusive ((140 total NISP rodents—44)/140). In a rank order of all taxa, rodents would then rank behind both leporids and artiodactyls, but remain a top-ranked taxon. Snakes, however, would not be highly ranked and were utilized at a low frequency.

Flotation Samples

In addition to the faunal remains recovered from the screens during excavation, the processing of collected flotation samples from Cerro Juanaqueña produced 15,374 bones (Table 11.13). Keep in mind that only a small sample of fill was collected for flotation samples and the heavy fraction was sieved through window screen to recover the bone. In contrast, the excavated sample represents 100% of all fill screened through ⅛-inch mesh. As the two data sets had different sampling protocols and represent different size ranges, they are difficult to combine, but they do provide complementary data sets.

Table 11.13 shows the taxa identified in flotation samples from Cerro Juanaqueña, the MNI identified in the flotation remains, the NISP of each taxon, and the percentage of the identified assemblage that each taxon comprised (see Appendix 11.2 for feature level detail). The team processed 160 heavy fractions from flotation samples that produced 15,374 bones, 328 of which were identifiable. From these identifiable elements, 55.8% were indeterminate rodents, 26.8% were small fish, and 5.8% were leporids. Only 2.1% (n = 328) of the flotation sample fauna were identified to at least the level of class versus 14.4% of the excavated assemblage. The tiny fragments recovered from flotation samples retain fewer distinctive features that allow identification, accounting for the low proportion of leporids. Small bone fragments from very small animals may be more identifiable than small fragments from larger taxa. Leporids and artiodactyls dominate in excavated contexts, while fish and rodents are relatively more abundant in the same contexts from flotation samples. In fact, rodents are identified at a slightly higher rate in flotation versus excavated assemblages. From the excavated assemblage, 0.79% of all bones are identified as rodents (140 NISP all rodents/17,799 total bones) compared with 1.19% of the flotation sample (183 NISP rodents/15,374 total). However, as estimated above, a large proportion of small rodents are intrusive and not a product of subsistence activities as discussed later in this chapter.

Application of the procedure discussed above allows assessment of the degree that rodents are overrepresented by intrusive individuals in the flotation samples. Note that 25% of the fish are burned, consistent with the burned proportion in the excavated sample and unlike rodents and snakes, and even unidentified remains, it is unlikely that fish died naturally in burrows! Since 12 of the 183 indeterminate rodents from flotation samples are burned, extrapolation suggests that 48 rodents (12/.25) could be attributable to cultural activities, with the balance more likely intrusive.

Fish Remains

At Cerro Juanaqueña, a total of six fish bones were found in the screens from two terraces during excavation (Terrace 6 and Terrace 126), but fish bones were recovered with much greater frequency in the heavy fraction of the flotation samples. Fish remains were recovered in flotation samples associated with both the upper and lower terraces, and at both terraces and rock rings. The flotation samples yielded a total NISP of 88 fish bones from Cerro Juanaqueña, five from Cerro los Torres (one from excavation screens), and three from Cerro el Canelo, but not Cerro Vidal where only a tiny sample of bone was recovered (Table 11.13 and Appendix 11.3). The widespread occurrence of fish was a surprising and important contribution to understanding subsistence activities. This discovery set in motion a number of lines of inquiry to understand this unexpected development.

Gobalet (2002), using comparative materials at the Department of Biology, California State University, Bakersfield (CSUB), identified the fish remains to the most specific level possible and his report is found in Appendix 11.4. At Cerro Juanaqueña, only catostomids (suckers) and cyprinids (minnow family including chubs, redheads, and dace) were identified to family, and most would have come from species about 100 mm in length. Because Smith and Miller (1986) reported only two catostomids from the Río Casas Grandes, the remains are definitely either *Catostomus plebius* (Rio Grande sucker) or the undescribed *Catostomus* sp. Smith and Miller (1986) report four species of minnow from the Río Casas Grandes, but the preservation of the bone impeded more precise identification. Three of these four possible species are under 110 mm in length, and the fourth is under 200 mm (Appendix 11.4).

Because of their unexpected presence, we considered the possibility that the fish remains at Cerro Juanaqueña were introduced into the assemblage by nonhuman predators (i.e., raccoons or birds of prey). This contention was refuted on three counts: 1) Twenty-five percent of all the fish remains recovered in the flotation assemblage were burned indicating deposition during the occupation, and this proportion is consistent with the burned proportion of other nonintrusive taxa. 2) The fish remains were in generally good condition and there was a lack of evidence for digestion by larger carnivores. 3) Finally, the largest piece of fish bone was submitted for AMS radiocarbon dating, and the results corresponded exactly (3200 ± 35 radiocarbon years BP, SR-6353; Chapter 5) with the brief interval of human occupation at Cerro Juanaqueña. Based on these factors, the inhabitants of Cerro Juanaqueña extensively utilized the small fish resources in the Río Casas Grandes and the Río San Pedro for a portion of their dietary needs.

Given the differences between the high frequency of fish bones in the flotation samples and their scarcity in excavation samples, estimates were derived to integrate the two data sets by asking how many fish would have been recovered if 100% of the fill was processed as flotation samples? In order to compare the recovery rate of bone from flotation heavy fractions with that from the ⅛-inch screens, we first calculated the flotation sample volumes in proportion to the excavation unit-level volumes from which the flotation sample was drawn. We then extrapolated the number of fish bones that would have potentially been recovered if entire levels had been examined as a flotation heavy fraction sample rather than with ⅛-inch screen. For example, if a single fish bone was recovered in the heavy fraction of a 4-liter (0.004 m^3) flotation sample of sediment, and the entire level consisted of 0.1m^3 of sediment, then we would anticipate recovering 25 fish bones if the entire level had been examined as a heavy fraction [((0.1 m^3/.004 m^3) × 1 bone) = 25 bones]. This was only calculated for levels from which a fish taxon had been recovered from a flotation sample. Only the flotation sample volumes and estimated excavation volume for each particular sample and level from which a fish bone or other taxa was found were used. This procedure is conservative as it only extrapolates to the remainder of the excavation level that was not floated

Using this extrapolation procedure, 734 NISP fish bones would have been recovered from Cerro Juanaqueña. Considering combined extrapolated and screen-recovered values, fish (734 extrapolated heavy fraction + 6 screen = 740 NISP) would have been recovered. Since these small fish were a highly-ranked resource, presumably technologies were used that allowed them to be harvested in large numbers.

Approaches described in the ethnographic record of the American Southwest, northern Mexico, and Texas include bow and arrow, spears, nets, poisons, weirs, and basketry traps and scoops (Alvarez de Williams 1983: 304; Arnon and Hill 1979; Campbell 1983; Crumrine and Crumrine 1967; Davenport 1924:294; Fontana 1983; Griffen 1983; Pennington 1983; Stewart 1957, 1983; Wallace 1955). According to Pennington, the Tarahumara of western Chihuahua

> sought fish everywhere and they [were] caught with hook and line, within funnel-shaped traps

of vegetable materials, behind converging stone walls closed by weirs, and with seines... [as well as by] crushing and tossing into slowly moving or quiet water any one of at least 29 species of plants used as a piscicide agent [Pennington 1983:281].

In central Coahuila, "fishing, participated in by both sexes, was done by shooting the fish with bow and arrow at night by torchlight, the use of nets, and underwater capture by hand from holes or pockets along steep stream banks" (Campbell 1983:353). The Mohave, who inhabited the Colorado River Valley along the California, Nevada, and western Arizona borders, consumed fish as their principal source of meat (Stewart 1983:59). The variety of techniques they used are described:

> Fish were caught in the Colorado River and in the lagoons and sloughs left by its spring flooding. One of the most common techniques used a seine or drag net with poles attached to each end and sticks spaced vertically across the net with stones serving as weights. The net was worked by 2 men at opposite ends who dragged it through the water. A large basketry scoop of willow lashed with bast would have a long handle attached across the center and was used by one man.... Fish were also caught in sieves, dip nets, and weirs, and were shot with bow and arrow.... Angling was not so common, but fishhooks were made from cactus spines that had been moistened, heated, and bent [Stewart 1983:61].

Fishing in the arid areas of the American Southwest and northwest Mexico was clearly carried out where and when the opportunity existed along rivers and in other bodies of water. Fish may be an underappreciated source of protein for prehistoric residents of these areas, as demonstrated by the ethnographic discussion above, by the recovered fish remains from the cerros de trincheras, and based on the extrapolation analysis for Cerro Juanaqueña.

Rodent Remains

Addressing the indeterminate rodents recovered from flotation requires both consideration of their potential as intrusives as well as using the extrapolation procedure. Above I suggested that approximately 48 (12/0.25) unidentified rodents found in flotation may not be intrusive. To use the extrapolation procedure on this value, an average proportion of flotation to excavated fill is needed for those levels for which flotation samples were processed and from which unidentified rodents were recovered. For those levels about 14% of the fill was processed as flotation samples. Extrapolating (48 * (1/0.14)), there would have been an estimated 342 utilized, unidentified rodents recovered if all sediment had been floated. If we combine 342 with the estimate of 44 rodents from the excavation screens that were likely utilized, this yields a total of 386. While this exercise yields only an approximation, it again shows that rodents were a highly ranked taxon, despite the dogged problem with intrusive remains.

The only other taxon with sufficient flotation data to warrant attention and potentially alter the rank ordering based on the screen data is jackrabbits, and they remain the top-ranked taxon (1671 screen + 20 extrapolated = 1691 NISP). Given these rough attempts at integrating the flotation and excavation samples, and accounting for intrusive remains, jackrabbits are followed by fish and then rodents, and the remaining rankings would be dependent on the excavated data.

Taxonomic Diversity

Taxonomic diversity, or the number of taxa in an assemblage, is a measure of the range of species being exploited by ancient hunters. Considering all faunal data, that is the 24 features at Cerro Juanaqueña, the four features from the three other cerros de trincheras, and the two ceramic period sites, 31 taxa are represented, including unidentified classes (unidentified artiodactyls, etc.). The mean taxa count per feature is 6.33 (minimum = 1, maximum = 15, standard deviation = 3.89, $n = 30$). Taxonomic diversity is frequently a function of assemblage size, and this relationship has been observed across a wide range of archaeological phenomenon (Meltzer et al. 1992; Rhode 1988). This relationship is illustrated in Figure 11.4 and shows that among all sites and features there is a direct, positive, and highly significant relationship between $\log_{10}$ NISP and $\log_{10}$ number of taxa ($r^2 = .758$, $p < .001$, $n = 30$).

Cerro Juanaqueña Features

One question that the project asked was: are there potential functional differences between and among the terraces and rock rings? In looking at the individual excavated terrace assemblages, clearly similarities exceed differences among the terraces. For example, the animals found in each are similar from terrace to terrace, given variability in sample sizes.

Similarities as well as differences exist between the rock rings and terraces. Figure 11.1 and Table 11.3 show that the rock ring faunal samples tend to be substantially smaller than the terrace samples, but there are no notable differences in the animals present. Table 11.14 compares the taxa present in the terraces and rock rings for both excavated and flotation data. No unique taxa are identified in the rock rings, and the assemblage seems

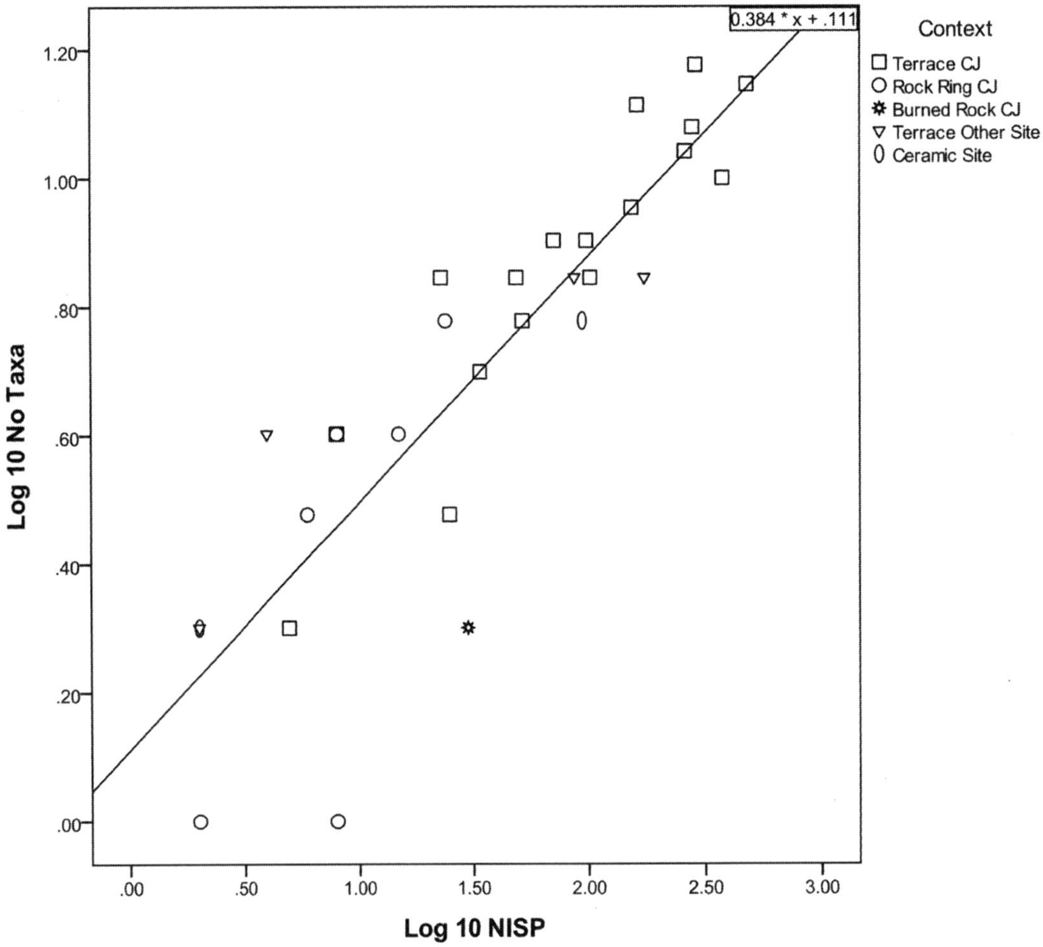

FIGURE 11.4. Feature types plotted by Log_{10} NISP and log_{10} number of taxa. There is a positive, direct correlation between log_{10} NISP and log_{10} number of taxa among all sites and features. Taxa diversity increases with sample size. Some symbols overlap.

to be a subset of taxa that were identified in the terrace assemblages, which is expected given the smaller sample sizes. For example, at Rock Ring 1A, cottontails, jackrabbits, unidentified artiodactyls, deer, indeterminate rodents, and nonvenomous snakes are all present in the feature assemblage. At Rock Ring 250, cottontails, jackrabbits, kangaroo rats, box turtles, and pronghorn remains are identified.

A comparison of the relationship between NISP and taxonomic diversity does show a difference, however. Figure 11.5 shows that the slope of the regression line between log_{10} NISP and log_{10} taxa diversity is steeper for Cerro Juanaqueña rock rings than for the site's terraces. Rock rings have more taxa than expected given their sample size. This pattern is surprising and attempts have been made to determine if these differences are due to sampling. Since the rock rings tend to have smaller samples, comparisons are made (not shown) using only the terraces with small NISPs, but the pattern persists. Removing the outliers and repeating the analysis also has no material effect on the pattern (not shown). The pattern is not due to apparent sampling differences. Further data are needed to explore the meaning behind this pattern.

Upper Terraces versus Lower Terraces

The four terraces excavated on the lower part of the site follow the earlier trends. As Figure 11.5 shows, the taxonomic diversity found in the lower terraces is no different than that in the upper terraces, given the assemblage size, since all lower terrace points fall close to the regression line. Table 11.15 provides a taxonomic breakdown between upper and lower terraces and includes both excavated and flotation data. It shows that all but one taxon (see below) present in the smaller lower terrace assemblages are found in the upper terrace assemblages. Both upper and lower terraces had similar

TABLE 11.14. Comparisons Between Terraces and Rock Rings at Cerro Juanaqueña.

	Terraces (excavation)	Rock rings (excavation)	Terraces (flotation)	Rock rings (flotation)
Small fish	×		×	×
Western box turtle	×	×		
Indeterminate turtles	×			
Venomous snakes	×			
Non-venomous snakes	×	×	×	
Indeterminate amphibians			×	
Indeterminate lizards	×		×	
Spadefoot toads	×			
Indeterminate birds	×	×	×	
Quail	×			
Perching birds	×			
Duck				
Indeterminate rodents	×	×	×	×
Indeterminate squirrels	×			
Chipmunk	×			
Kangaroo rat		×		
Wood rat	×			
Pocket gopher	×			
Jackrabbit	×	×	×	
Cottontail	×	×	×	
Indeterminate rabbit	×		×	
Indeterminate canid	×			
Coyote	×			
Badger	×			
Indeterminate artiodactyls	×	×	×	
Indeterminate deer	×	×		
Pronghorn	×	×		
cf. Bighorn sheep	×			
cf. Bison	×			

Note: × = present.

ranking of taxonomic frequencies; leporid percentage at both upper and lower terraces is approximately 80%, followed by rodents, artiodactyls, and finally by a general mix of all other taxa.

Terrace 413 is unique in that it is a lower terrace and had both early and late occupation radiocarbon dates. It only yielded eight identifiable pieces including three cottontail remains, three jackrabbit remains, one pocket mouse, and the only duck bone from the entire site.

Seasonality

Animal bones can be useful for inferring seasonal activities as they relate to the organization of mobility and how long or often a site was in use (Chapter 8; K. Adams and Bohrer 1998; B. Huckell 1995; Monks 1981; Muñiz 1998; Rafferty 1985; Roth 1989). Although plant taxa representing spring, summer, and fall availability are found in the macrobotanical assemblages from the cerros de trincheras (Chapter 8), seasonal data from faunal remains are somewhat muted. A few of the identified taxa can be useful for inferring possible season of occupation, but are not reliable signatures of seasonal occupation at the cerros de trincheras. The following paragraphs discuss the possible seasonal indicators from the identified faunal remains (but cf. Russo 1998).

The presence of a single duck ulna in the Terrace 413 assemblage may suggest an early winter or late spring occupation at Cerro Juanaqueña, but the sample size of one does not lend strong support to this contention. Further, this is the only excavated mixed early and late occupational context we identified. Annual migratory bird routes include the Río Casas Grandes valley, but an early winter occupation cannot be identified to the exclusion of a late spring occupation; both are possible.

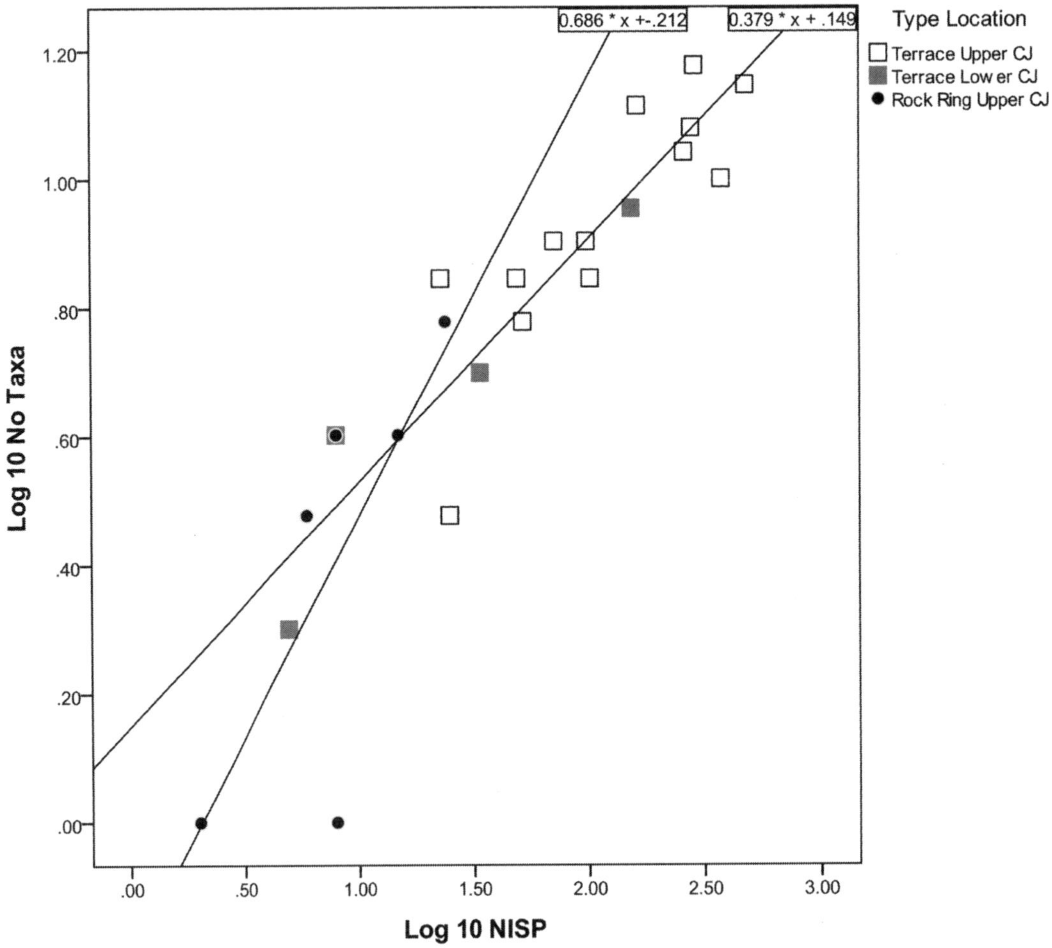

FIGURE 11.5. Upper and lower terraces and rock rings plotted by $\log_{10}$ NISP and $\log_{10}$ taxa diversity. The slope of the regression line between $\log_{10}$ NISP and $\log_{10}$ number of taxa for Cerro Juanaqueña rock rings is steeper (slope = 0.686) than for all Cerro Juanaqueña terraces (slope = 0.379). Two symbols overlap. Rock rings have more taxa than expected given their sample size.

Further, although archaeological studies examining season of hunting have primarily focused on animals that give birth only once annually and have a limited birth season, particularly artiodactyls (Driver and Maxwell 2013; Hillson 2005; Zeder 2001), the sample of artiodactyls from cerros de trincheras is quite low compared to leporids. Rabbits, which make up between 75% and 90% of the cerros de trincheras assemblages, have been used in a few studies to estimate the season of hunting activities represented by comparing the frequency of immature animals with an expected frequency in a natural population (Whalen 1994b; Young 1996). Wildlife studies predict that immature jackrabbits and cottontails comprise less than 10% of the population between December and April or May, and make up more than 50% of the population from June to October, congruent with the growing season when the most forage is available and the young of the year emerge (Bronson and Tiemeier 1958; Davis et al. 1975; Hard 1983; Medlin 1974; Whalen 1994a). However, these trends can vary based on local plant production seasonality and long breeding seasons. Archaeological assemblages that contain low proportions of both immature jackrabbits and cottontails may *possibly* reflect intensive procurement of rabbits in the winter and early spring, before the young of the year emerge. Conversely, assemblages with high proportions of young may reflect a summer-to-fall hunting season. Assemblages with moderate levels of immature animals could be the result of year-round hunting or hunting during the late fall and winter.

At the cerros de trincheras, immature leporids make up a very small percentage of the identified remains. Can this low frequency be used to narrow the range of seasonal possibilities? This pattern suggests that

TABLE 11.15. Comparisons Between Upper and Lower Terraces at Cerro Juanaqueña.

	Upper Terraces (excavation)	Lower Terraces (excavation)	Upper Terraces (flotation)	Lower Terraces (flotation)
Small fish	×		×	×
Western box turtle	×			
Indeterminate turtles	×			
Venomous snakes	×			
Non-venomous snakes	×	×	×	×
Indeterminate amphibians			×	
Indeterminate lizards	×	×	×	
Spadefoot toads	×	×		
Indeterminate birds	×		×	
Quail	×			
Perching birds	×	×		
Duck		×		
Indeterminate rodents	×	×	×	×
Indeterminate squirrels	×			
Cotton rat	×			
Deer mice	×	×		
Pocket mice	×	×		
Chipmunk	×			
Kangaroo rat	×			
Wood rat	×			
Pocket gopher	×			
Ground squirrels	×			
Jackrabbit	×	×	×	
Cottontail	×	×	×	
Indeterminate rabbit	×		×	×
Indeterminate canid	×			
Coyote	×			
Badger	×			
Indeterminate artiodactyl	×	×	×	
Indeterminate deer	×			
Pronghorn	×	×		
cf. Bighorn sheep	×			
cf. Bison	×			

Note: × = present.

summer-to-fall only occupation is not a possibility. The other possibilities that would generate a very low proportion of immature individuals (e.g., winter–spring only, year-round, other multi-season occupations) are difficult to evaluate. However, these data do allow us to say with some confidence that the occupation included a large proportion of the nongrowing season. This finding is significant when placed in context with the plant seasonality data, which, of course, emphasized occupation during the growing season extending through spring, summer, and fall. The leporid age pattern indicates that the occupation included a large proportion of the nongrowing season. The two data sets, together along with other indicators, discussed elsewhere, suggest that the site was occupied for much of the year, perhaps on a sedentary or near-sedentary basis.

Mobility

The faunal remains suggest that the organization of mobility at cerros de trincheras tended toward increased residential stability as agricultural production and population increased. If so, butchering patterns should reflect this trend because people must travel further from a residential location to obtain game as

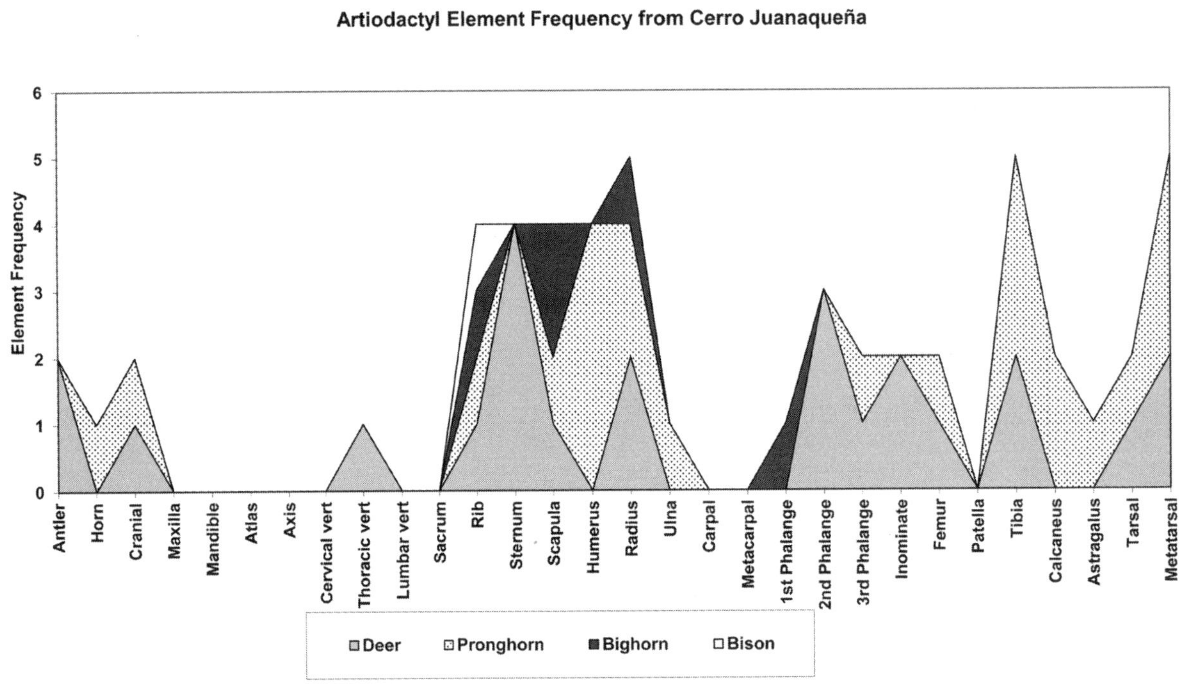

FIGURE 11.6. Artiodactyl element frequencies from Cerro Juanaqueña. Both high meat and low meat elements are well represented suggesting that artiodactyls are not being transported from long distances. Counts are additive, for example, one deer cranial fragment and one pronghorn cranial fragment were found at Cerro Juanaqueña.

populations increase and expand (Kent 1993). As people travel greater distances to procure resources, transport decisions become important, especially for large game, and certain portions of the carcass are left at processing locations. Axial elements (e.g., ribs and vertebrae), as well as meatier portions of long bones (e.g., humeri and femora), are predicted in elevated numbers at residential locales, as meat is likely to be returned for distribution (Bunn 1993; Kent 1993). Conversely, elements with comparatively less meat (e.g., cranial and distal limb bones) are not expected to be found as frequently in assemblages from locales used primarily as residential sites, as these elements will be left where the animal was killed (Binford 1980). This change would not necessarily reflect a change in prehistoric diets per se, but a change in the butchering patterns associated with food procurement (Kent 1993).

The majority of bones in the artiodactyl skeleton were represented in the Cerro Juanaqueña assemblage (Figure 11.6). However, vertebrae were scarce, perhaps due to their low bone density making them more susceptible to taphonomic damage. Low meat elements (e.g., cranial, distal limb bones, phalanges) were apparently not being systematically abandoned at distant field processing locations but instead were being returned to the site along with high meat elements. The presence of both high and low meat elements is consistent with the transportation of complete artiodactyls back to the site for processing and consumption. This pattern is found across deer, pronghorn, and bighorn sheep, ignoring the single bison element. Perhaps animals were being killed relatively close by so that transport costs of low meat element portions was not a significant factor. This pattern may suggest that hunters were not routinely traveling 50 to 70 km to the west in search of deer populations in the massive ranges of the Sierra Madre Occidental but instead were taking animals closer to home.

Other Items

Worked Bone

Prey species were important for their dietary value, but their remains were sometimes used for other purposes as well. Several worked bone specimens were recovered from Cerro Juanaqueña. These include five awl tips and one largely complete possible awl, all manufactured from medium-sized mammal long bone shafts (Figure 6.28), a bone pendant, and four bone beads (Figure 6.29). Three additional pieces of highly polished segments of an unidentified medium-sized mammal long bone fragments were also recovered.

Shell

A small amount of shell, consisting of 68 specimens, was found at Cerro Juanaqueña and has been identified by Arthur Vokes of the Arizona State Museum (personal communication, 2013). The results of Vokes' inspection

TABLE 11.16. Shell from Cerro Juanaqueña.

Feature	Analyst[1]	Identification[2]	No. of occurrences[3]	No. of specimens[4]	Artifact[5]
R1a	AV	cf. Anodonta	1	1	
T 6	Field	Shell	1	1	Bead
T 6	AV	Sonorella	1	1	
T 97	AV	Anodonta	2	2	
T 97	AV	cf. Anodonta	2	2	
T126	AV	cf. Sonorella	2	2	
T126	AV	Sonorella	4	10	
T 222	Field	Shell	1	1	Bead
T 290	AV	Anodonta	3	3	
T 290	AV	cf. Anodonta	1	1	
T 290	AV	Sonorella	2	2	
T 413	AV	cf. Fossaria or Bulimulus	1	1	
T 487	Field	Nacreous Shell	2	2	Bead
T 487	Field	Gastropod	4	5	
T 487	AV	Freshwater/terrestrial snail	1	1	
T 487	AV	Succinea	4	4	
T 487	AV	cf. Succinea	2	2	
T 487	AV	Anodonta californiensis	2	2	
T 487	AV	Anodonta	11	15	
T 487	AV	cf. Anodonta	3	3	
T 487	AV	Olivella cf. dama[6]	1	1	Bead
T 487	AV	Melongena patula[6]	1	1	
T 487	AV	cf. Pteria[6]	1	1	Pendant
T 508	AV	Anodonta	2	2	
T 537	AV	cf. Anodonta	1	1	
T537	AV	Bivalve[6]	1	1	
		Total	57	68	

[1] **Analyst.** Shell was identified by Arthur Vokes, Arizona State Museum. However, for various reasons not all pieces of shell were sent for analysis. In these cases, shell was recognized only by field identification.
[2] **Identification.** The abbreviation "cf." indicates lack of identifying characteristics prevents confirmation of identification but is similar to the named taxon.
[3] **No. of Occurrences.** This entry tabulates the number of distinct proveniences (normally excavation units and levels) within the terrace which yielded specimens of the identified class. This gives a rough indication of frequency, but note that any given occurrence could include more than one shell fragment, and that fragments of a single shell could be distributed across several different occurrences.
[4] **No. of Specimens.** This columns lists the number of distinct individuals that were found in a provenience following Vokes' observations in cases when two or more fragments are likely from the same individual. In a few cases of field observations, multiple tiny fragments from the same provenience are assumed to be from the same individual.
[5] **Artifact.** This column identifies the five actual manufactured shell artifacts that were found.
[6] Indicates a marine species, presumably from the Gulf of California. All others are freshwater or terrestrial taxa and are known from the region.

of these remains are summarized in Table 11.16. Four specimens of marine shell were recognized, including an *Olivella* (cf. *dama*) shell bead, a wing-oyster (cf. *Pteria*) bead pendant, an unworked fragment of crown conch (*Melongena patula*), and an unidentified marine bivalve shell. Olivella, conch, and wing oyster are all found in the Gulf of California, and their presence at Cerro Juanaqueña suggests that by 1300 cal BC social networks reached at least that far.

The most common shell, comprising 47% of the specimens, was a freshwater mussel, *Anodonta* sp. (including those identified as cf. *Anodonta*). Two of these were identified as *Anodonta californiensis*, California floater, and it is likely that this is the principal, perhaps

TABLE 11.17. Human Bones Recovered at Cerro Juanaqueña.

Feature	Element (all are fragments)	Unit	Level	Bag	Depth	Presence of burning	Count	Notes	Isotope sample
T10	Unidentified cranial	1	2	5	102–112	—	1		Yes
T10	Unidentified phalanx	1	10	40	182–192	—	1		
T10	Unidentified phalanx	1	11	42	192–202	—	1	Possible fetus	
T10	Unidentified tooth	2	7		152–162	—	1	3 Reconstructable fragments	
T10	Unidentified cranial	2	7			—	1	3 Reconstructable fragments	Yes
R239	Unidentified parietal	1	3	10	149–156	—	2		Yes
T163	Unidentified cranial	7	3	59	113–123	×	1		
T287	Molar	22	3	48	130–139	—	1		
T287	Femoral head	6	99	28	130–139	—	3		Yes. Femoral head & 3 cancellous fragments are one sample
T287	Cancellous (spongy) bone	6	99	28	130–139	×	2		
T287	Cancellous	6	99	28	130–139	×	1		

the only species represented. Vokes identified most of the *Anodonta* specimens as cf. *californiensis* species. *Anodonta* shell was found in six of the thirty features excavated, but is usually represented by only a few fragments. Nevertheless, a single terrace, Terrace 487 did yield 20 specimens. In this discussion "specimen" refers to one or more fragments from the same unit and level that are thought to belong to the same animal.

Anodonta shell is thin, but can range as high as 12.7 cm in length (Arizona Game and Fish Department 2012). The shell is nacreous (shiny) and was occasionally used for ornaments in both prehistoric and historic times. However, according to Virden-Lange (2015), it becomes especially fragile and brittle, and must be worked when the shell is still green. *Anodonta* shell has been reported from several Ceramic period sites in southern New Mexico (cf. Anyon and LeBlanc 1984:306; Wallace, 1998:316). None of the specimens from Cerro Juanaqueña were clearly worked, and these mussels likely were exploited as a minor food source, as was probably true in the Arizona archaeological record as well (Virden-Lange 2015:230).

The remaining shell were all from air-breathing snails. Most common was *Sonorella* sp., a talus snail, which accounted for 22% of the specimens recovered from Cerro Juanaqueña. As its common name implies, these snails live in talus deposits much like the artificial terraces on the cerros de trincheras. Most, perhaps all of the *Sonorella* shells recovered are likely intrusive. In addition to *Sonorella*, six specimens were identified as *Succinea* sp., the amber snail. Amber snails require a moist habitat, as may have been found on the floodplain below Cerro Juanaqueña.

Human Bone from Cerro Juanaqueña

Fifteen fragments of human bone were recovered from Cerro Juanaqueña. No human remains were identified at Cerro el Canelo, Cerro Vidal, or Cerro los Torres. Table 11.17 lists the human material identified at Cerro Juanaqueña. These isolated fragments suggest that inhumations are likely present at the site. We expect that further excavations would likely encounter human remains as their presence is consistent with the long-term residential character of the site. Four samples of sufficient size were submitted to Anne Katzenberg of the University of Calgary for stable isotope analysis and those results are presented in Chapter 10.

Conclusions

Faunal remains recovered from the cerros de trincheras provide a unique opportunity to better understand exploitation of animal resources at a time when the role of agriculture was rapidly changing in northwestern Chihuahua. These data have been used to explore the characteristics of the assemblage, the relative importance of various taxa, and settlement strategies. Specifically, the analysis of the faunal remains from Cerro Juanaqueña, plus small assemblages from three other cerros de trincheras and two Ceramic period sites in the Río Casas Grandes valley, provides a unique database with which to evaluate settlement and subsistence questions for an area and time period that has not seen intensive archaeological investigations.

Appropriate characterization of taxonomic abundance is critical to the interpretation of the faunal assemblage. It has been argued that NISP is the best measure of taxonomic abundance since $\log_{10}$ $MNI_{feature}$ and $\log_{10}$ NISP are closely correlated, while MNI_{site} appears to overly reduce the data. Since NISP is the primary measure of taxonomic abundance, efforts were made to understand its variability. NISP does not correlate with excavated feature volume since the density of animal bone in individual features varies widely. Finally, taxonomic richness is closely correlated with NISP demonstrating that consideration of the number of taxa must include a concern with sample size. The high diversity of taxa found at Cerro Juanaqueña is a product of sample size, which is closely related to the intensity of use of the site as a residential location by a substantial number of people for an extended period of time.

The abundant faunal remains at Cerro Juanaqueña are well-preserved and yielded important insights into subsistence practices. While the samples from Cerro el Canelo, Cerro Vidal, and Cerro los Torres are much smaller than that recovered from Cerro Juanaqueña, similarities do appear. With the exception of Cerro Vidal (where only two specimens were identified), the percentage of leporids in the excavated assemblages falls between 85% and 91%. The consistent abundance of leporid remains at all of these sites suggests a heavy reliance on these species and also indicate that jackrabbits were heavily exploited relative to cottontails. While the more intensive exploitation of jackrabbits relative to cottontails may be a reflection of ecological differences in the area, the low number of artiodactyls relative to lagomorphs (the artiodactyl index) indicates a pattern that is not distinct from other coeval sites throughout the southern Southwest (Thiel 1998; Waters 1998; Wöcherl 2001).

If the potential intrusives are set aside momentarily, the rank ordering of taxa by NISP or $MNI_{feature}$ yields similar results, with jackrabbits and cottontails followed in roughly similar order by the most important artiodactyls, deer, and pronghorn (Table 11.7). The $MNI_{feature}$ versus NISP rank ordering is somewhat different among the taxa that occur at lower frequency.

Addressing the problem of potentially intrusive taxa required utilizing the proportion of burned bones as an estimator of the proportion of nonintrusive versus intrusive rodents present. Integrating the flotation data with the excavated data, particularly with regard to the relative importance of fish and rodents, involved extrapolation from the flotation data to estimate the frequency

Table 11.18. Synthesis of Excavation NISP, $MNI_{feature}$, Intrusive, and Flotation Data to Form Three Rank-Ordered Groups.

Identification	Group
Leporids	A
Fish	A
All rodents	A
Deer and pronghorn	A
Birds	B
Turtles	B
Canids	B
Bighorn sheep	C
Indeterminate lizards	C
Spadefoot toads	C
Badger	C
Duck	C
Bison	C
Snakes	C

Note: Some taxa are aggregated.

of fish and rodent remains as if all excavated sediment had been processed by flotation. In order to synthesize these estimates with the excavated data, I defined three rank-ordered groups (A, B, and C) that involve collapsing some taxa. I referred to both the NISP and $MNI_{feature}$ data to develop these rankings (Table 11.18). Group A consists of those taxa that clearly dominate. In rank order they are: leporids, small fish, all rodents, and the two key artiodactyl taxa, deer and pronghorn. Jackrabbits and cottontails were combined and are clearly the most important taxa. Small fish are next and land in the top group based on their high rate of recovery from flotation and the extrapolation procedure. As an aggregate, rodent remains are in the top group given both the flotation and intrusive analyses. Deer and pronghorn form a natural grouping and are distant from bighorn sheep and bison.

The B group consists of the birds, turtles, and canids. Their NISP and $MNI_{feature}$ values place them clearly ahead of the third group and well behind the top group. The aggregate "birds" consists of the terrestrial birds (passeriformes, indeterminate birds, and quail). The turtles are the Western box turtle and indeterminate turtles, and the canids are the coyote and indeterminate canids. The C group includes the taxa that occasionally occur, such as bighorn sheep, indeterminate lizards, badger, duck, bison, and all snakes. Their NISP and $MNI_{feature}$ values are all low, and it is difficult to give much significance to rank ordering within this group. For example, the NISP of bighorn sheep is 5, but the $MNI_{feature}$ is 2, shifting its rank order. Snakes fall into the last group despite their elevated NISP, based on their intrusive status.

The primary focus on a small group of taxa, especially leporids and artiodactyls, followed by a large number of taxa that are used to a far lesser extent, is typical of many Southwestern assemblages. The importance of fish is not typical of Southwestern assemblages, a finding that only emerged with the detailed analysis of the flotation heavy fraction. The significance of the use of rodents came into focus only by considering both the flotation evidence and an estimate of their intrusive proportion. Clearly the occupants of the Río Casas Grandes valley exploited a wide range of taxa from a variety of biotic communities. Based on the faunal remains recorded and identified from these sites, similar procurement strategies were used by the inhabitants of all the cerros de trincheras in northwestern Chihuahua, and this primary pursuit of leporids and artiodactyls appears to continue, based on the very small samples, into the Ceramic period. Similar habitats were exploited and similar subsistence activities were undertaken through time in northern Chihuahua.

While seasonality data from the assemblage are limited, the low frequency of young leporids is consistent with the seasonal indicators from the floral data. Occupations that continued through most of the year including fall, winter, spring, and summer would best account for both sets of data, although periodic absences from the site cannot be eliminated. Surprisingly, however Roth (2016a) suggested that conclusions derived from my 2008 dissertation indicated that Cerro Juanaqueña was only occupied on a seasonal basis. She stated:

> Schmidt's (2008) analysis of the faunal remains indicates that these *trincheras* sites in northern Chihuahua, although large and spectacular, were only occupied seasonally. Schmidt argued that groups used the terraces for agriculture during the summer but moved in other seasons to hunt and gather resources. [Roth 2016a:126]

In fact, while conclusions derived from faunal analyses were somewhat muted in terms of a seasonal signature and few conclusions were drawn (Schmidt 2008: 347–348), botanical analyses showed that residential stability and dependence on cultigens were firmly in place at Cerro Juanaqueña by circa 1300 cal. BC. And because successful subsistence studies must incorporate both faunal and floral remains, I actually concluded (Schmidt 2008:354) that the high diversity of animal taxa and the wide representation of skeletal elements in the faunal assemblage corroborated the signatures identified in the botanical assemblages by demonstrating that the occupation of Cerro Juanaqueña was intensive and year-round.

The absence of some taxa from this large assemblage is notable. For example, taxa that should have thrived in better watered, wooded communities, such as turkey, white-tailed deer, beaver, and bear, are not present in the assemblage. This deficiency suggests the inhabitants were not exploiting the Sierra Madre Occidental, with its extensive woodlands and forests, located only 50 to 75 km to the west. In addition, the artiodactyl butchering patterns suggest that both high and low meat elements were being transported back to Cerro Juanaqueña, an observation that suggests that long distance transport of carcasses was not a significant strategy. Both of these factors suggest that large and small mammal hunting was done close to the site, with only occasional forays farther afield to procure additional protein sources. In addition, the enhancement of the local landscape through farming on the alluvial terraces of the floodplains may have provided an attractive habitat for cottontail rabbits and other small animals who enjoy tender young plants and the protective cover they provide.

Merrill and López González (2007) discuss the absence of prairie dogs from the Cerro Juanaqueña faunal assemblage. Their absence is surprising because large prairie dog colonies are found in the vicinity of the site today (Ceballos González et al. 2010; List et al. 2010). The earliest record of their presence in northwestern Chihuahua comes from Spanish missionaries in the late 1700s. Merrill and López González (2007) suggest that prairie dogs may have spread northward in the historic period with the introduction of cattle and horses.

The single charred bison rib in the Cerro Juanaqueña assemblage is tantalizing. In contrast, bison were common in the Paquime assemblage, some 2500 years later (DiPeso et al. 1974:Vol. 8; Merrill and López González 2007). If bison were nearby and plentiful during the occupation of Cerro Juanaqueña, they should be prominent in the assemblage given their high economic return. If people were routinely traveling long distances to transport bison meat, high meat elements, such as ribs and vertebrae, would be expected to be more common in the assemblage. While trade is a possibility, the virtual absence of exotic trade items at Cerro Juanaqueña is notable and is limited to a few pieces of marine shell. Bison were likely rare in the region and their acquisition was infrequent at best, despite the fact that while the site was occupied the region was experiencing higher effective moisture (Chapter 2).

Consideration of faunal assemblages from Early Agricultural period sites in the Tucson Basin (Ezzo and William 1998; Gregory 2001; Mabry, 1998), where early investments in farming as well as foraging also occurred, show similarities in assemblages. The assemblages are also similar to those from later sites throughout the southern Southwest, suggesting a relatively stable pattern of animal procurement through time. The percentage of lagomorphs and artiodactyls in the Early Agricultural period assemblages varies between 75% and 92%, and 4% and 18% respectively (Schmidt 2008). Lagomorphs are most abundant, followed by artiodactyls and then all other taxa, a pattern that is surprisingly consistent from site to site. These similarities suggest that the earliest farmers in the southern Southwest and northern Mexico participated in similar subsistence strategies despite being geographically disparate and inhabiting distinct niches, a pattern that continues into later periods (Assad Hunter 2001; Bayham and Hatch 1985; Cannon and Etnier 1997; Cruz Antillón et al. 2004; Dean 2003; DiPeso et al. 1974:Vol 8; Gallaga Murrieta et al. 2016; Gillespie 1989, 1994; Glass 1984; Hodgetts 1996; James 1989; McCarthy et al. 2013; Schmidt 1998, 1999, 2005, 2008; Szuter and Bayham 1989; Waters 2000; Waters et al. 2015; Whalen 1994b; Whalen and Minnis 2009). The causes of these similarities are likely related to environmental and cultural factors, including biogeographical patterns and human ecological dynamics.

The high ubiquity of maize from analyzed flotation samples, as well as a wide range of other macrobotanical remains including cheno-ams and grasses, offers further support for a diverse subsistence economy (Chapter 8). Given the distribution and diversity of both faunal and botanical remains from the cerros de trincheras, the inhabitants were participating in a mixed foraging and farming economy, and they continued to use wild resources while making substantial use of cultigens for their daily sustenance.

CHAPTER 12

Late Quaternary Landscape Evolution

Lee C. Nordt[1]

Introduction

Episodes of valley erosion, deposition, and soil formation condition the distribution and preservation of the archaeological record in alluvial settings. Coupled with late Quaternary climate and vegetation records, geomorphic investigations also provide insight to settlement patterns and subsistence strategies. In contrast to arid land studies in much of the Southwest, the relationship among climate, vegetation, and fluvial behavior is poorly understood in the Semidesert Grassland bordering the Chihuahuan Desert (Nordt 2003). Yet numerous important archaeological sites occur in the region, which is also recognized as an important setting related to the spread of agriculture into the Southwest (Hard and Roney 1998a; Merrill et al. 2009). The purpose of this chapter is: 1) to reconstruct the late Quaternary fluvial history of the Río Casas Grandes and Río San Pedro near Cerro Juanaqueña and correlate events with the other regions in the southwestern United States and northwestern Mexico; 2) to assess the response of streams in northern Mexico to shifts in climate and ecosystem stability; and 3) to discuss the influence of the alluvial stratigraphic framework on the preservation of the archaeological record and agricultural potentials.

Setting

The study area occurs in the Semidesert Grassland situated between the Chihuahuan Desertscrub biotic community to the east and north, as well as Plains Grasslands and Madrean Evergreen Woodland biotic communities to the west and south (Chapter 2; Brown 1982a, 1982b, 1982c, 1982d). Bedrock in the study area consists of rhyolites and basalts deposited in the Tertiary and early Pleistocene (Moran-Zenteno 1994) (Figure 12.1). The northwest trending basin and range configuration is structurally controlled by tectonic faulting. Late Tertiary conglomerates rich in caliche form bajadas immediately above the lower alluvial plains. Most streams eventually empty into bolsons (Moran-Zenteno 1994).

The Río Casas Grandes flows approximately 200 km across bedrock in the upper basin before entering a broad alluvial valley and successive bedrock constrictions (Figure 12.1). The drainage basin above the Cerro Juanaqueña study area is approximately 8,000 km^2 and ranges in elevation from 2,700 m in the headwaters to 1,350 m in the valley floor near the project area. The Río Casas Grandes eventually flows into Laguna de Guzmán in the Chihuahuan Desertscrub community near the U.S.-Mexico border. The Río San Pedro has a drainage area of approximately 1,600 km^2, which begins in the highlands to the south near an elevation of 2,500 m and joins the Río Casas Grandes in the study area. A severe valley constriction occurs a few kilometers downstream from the confluence of the two rivers (Figures 12.1 and 12.2). Today, both river channels are ephemeral and carry water only after intense rainfall events.

Methods

Geomorphic investigations were conducted during each field season. Because of the absence of alluvial cutbank exposures, 69 backhoe trenches were excavated in the Cerro Juanaqueña study area for temporal and spatial reconstruction of the late Quaternary alluvial stratigraphic record and for archaeological prospecting. Excavations were conducted on private and community lands that were actively being grazed, sometimes limiting the number and location of exposures desired. Trenches were excavated to depths of 4 to 5 m and typically 1 m wide and 3 to 8 m long. Thirty-four key backhoe trenches were aligned to permit the construction of six valley cross sections (Table 12.1, Figure 12.2, BHT-1–5, a–f, g–k, l–p, q–u, v–bb, 25). Soil-stratigraphic descriptions were written following standards and procedures of the Soil Survey Division Staff (1993) and Folk (1980) (Table 12.2). The alluvial sediments were subdivided into strati-

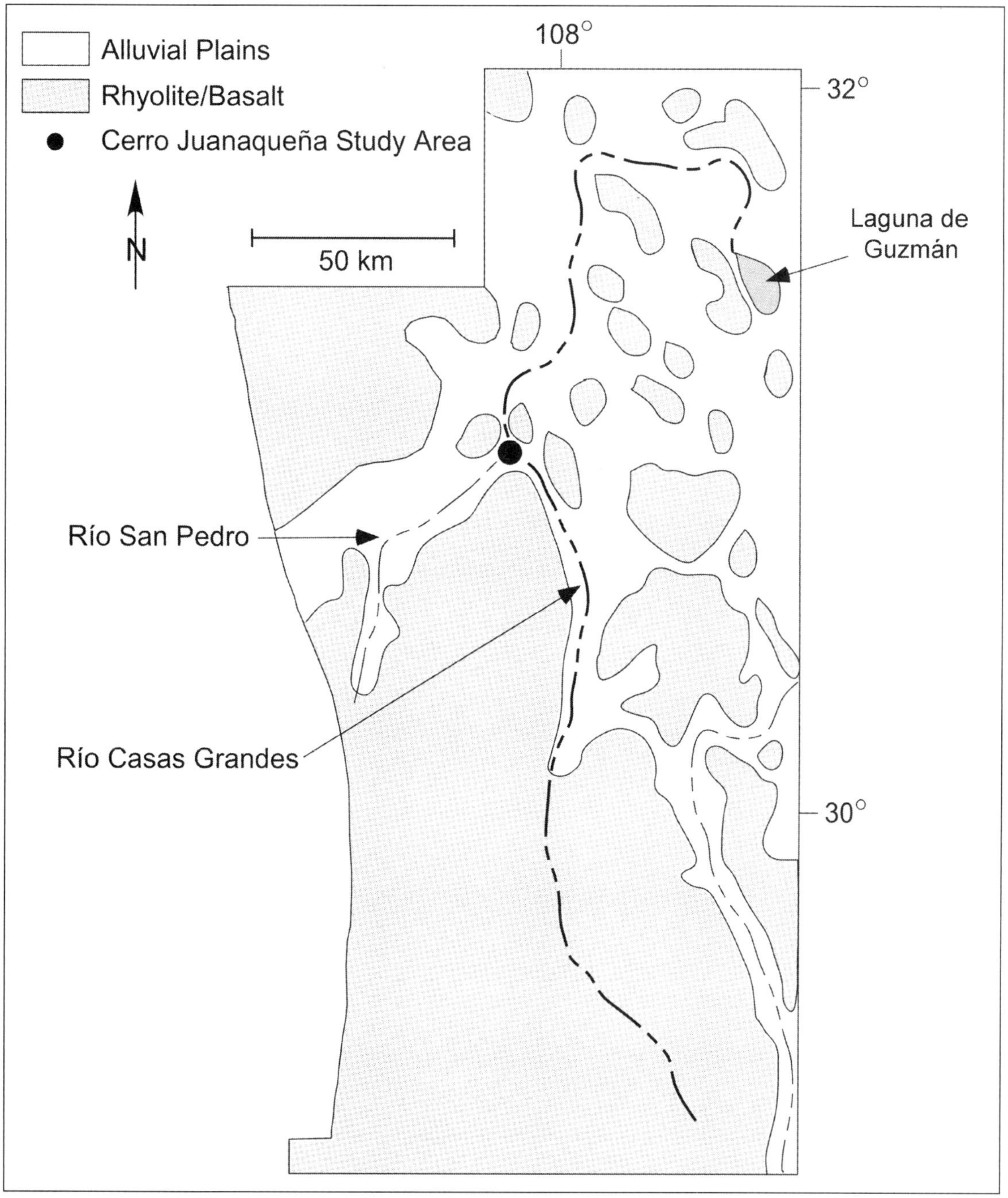

FIGURE 12.1. Regional geology and geomorphology of the Río Casas Grandes and Río San Pedro valleys in the vicinity of the Cerro Juanaqueña study area.

graphic units based on bounding unconformities where sediment-time was missing from either erosion or landscape stability and soil formation (Waters 1992).

The alluvial chronology was based on radiocarbon dating of charred plant material, sediment humates, and soil humates using AMS and radiometric analyses (Table 12.3), and the stratigraphic context of time-diagnostic prehistoric artifacts. Charcoal ages were interpreted as the time of alluvial deposition of the associated sediments. Sediment humate ages obtained from organic rich muds were also assumed to approximate the time of alluvial deposition. Sediment humate ages from paleosol (buried soil) surface horizons were interpreted as the maximum time of burial of the stratigraphic unit containing the paleosol (Matthews 1985). All radiocarbon dates are conventional ages that were corrected for variations in $^{13}C/^{12}C$ content, and unless otherwise stated, reported in radiocarbon years BP (Beta

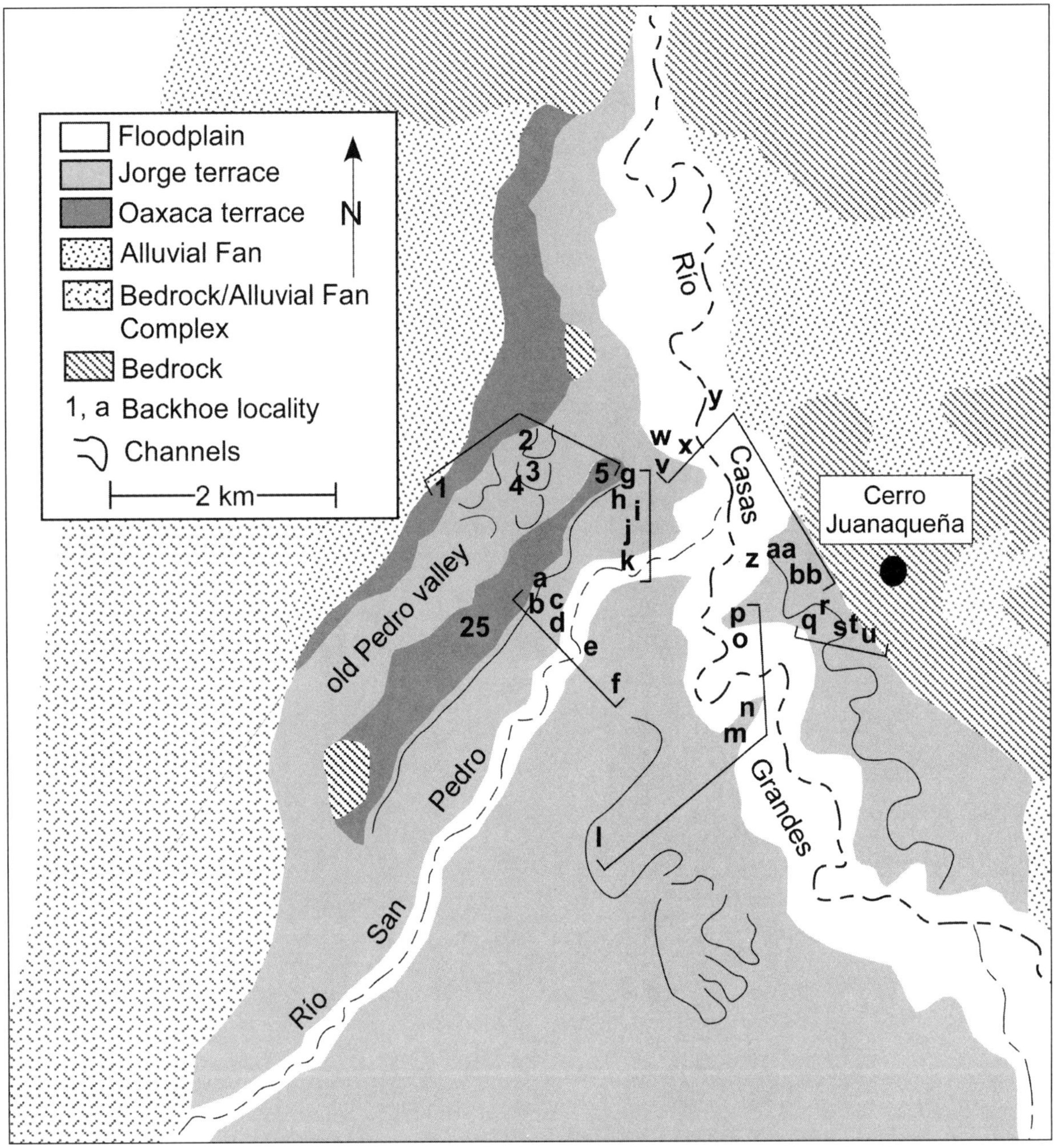

FIGURE 12.2. Geomorphic map of the Cerro Juanaqueña study area at the confluence of the Río Casas Grandes and Río San Pedro. Backhoe localities are represented by numbers and letters and grouped to form six valley transects for reconstruction of the late Quaternary alluvial sequence (modified from Nordt 2003).

Analytic, Inc.). For consistency with other chapters, the two-sigma tree-ring calibrated dates in AD/BC format are indicated in parentheses for key radiocarbon ages. Maximums and minimums of calibrated ages are used as needed to express radiocarbon age ranges tied to particular dates, and median calibrated dates are used in other cases. Table 12.3 provides details of tree-ring calibrations, $\delta^{13}C$ values, and other data for the radiocarbon dates. Ancient ecological patterns were investigated using stable carbon isotope values of paleosol and detrital organic carbon and are reported in $\delta^{13}C$ against the Vienna Pee Dee Belemnite standard (VPDB). Findings were measured by the Texas A&M University Rangeland Ecology Laboratory.

LANDFORMS

The oldest three landforms mapped in the study area—Bedrock, Bedrock-Alluvial Fan Complex, and Alluvial Fan—are discussed in detail in Nordt (2003). The three late Quaternary alluvial landforms relevant to this chap-

TABLE 12.1. Backhoe Trenches Identified in Chapter 12.

Chapter Ref.	Field No.
BHT-1	BHT-26
BHT-2	BHT-22
BHT-3	BHT-G99
BHT-4	BHT-F2-99
BHT-5	BHT-1
BHT-a	BHT-6-00
BHT-b	BHT-10
BHT-c	BHT-P7
BHT-d	BHT-18
BHT-e	BHT-24
BHT-f	BHT-Pr7-99
BHT-g	BHT-5a
BHT-h	BHT-2
BHT-i	BHT-Pr3
BHT-j	BHT-I
BHT-k	BHT-E
BHT-l	BHT-4
BHT-m	BHT-5Pr99
BHT-n	BHT-1Pr99
BHT-o	BHT-A
BHT-p	BHT-C
BHT-q	BHT-B99
BHT-r	BHT-17
BHT-s	BHT-15
BHT-t	BHT-16
BHT-u	BHT-Noria
BHT-v	BHT-H2
BHT-w	BHT-H1
BHT-x	BHT-4
BHT-y	BHT-6-99
BHT-z	BHT-10
BHT-aa	BHT-9
BHT-bb	BHT-12
BHT-25	BHT-25

ter are the Oaxaca terrace, Jorge terrace, and floodplain (Figure 12.2). The late Pleistocene Oaxaca terrace is preserved mainly in the Río San Pedro valley. The Holocene Jorge terrace is topographically inset by 1 m of relief below the Oaxaca terrace and has visible channel scars and avulsed channels. The Jorge terrace also includes the relict surface of the old Río San Pedro (old San Pedro valley) because the upper part of its post-abandonment alluvial history is similar to other areas of the Jorge terrace. The modern floodplain is wider along the Río Casas Grandes than the Río San Pedro. The floodplains are also inset below the Jorge terrace by about 1 m of relief.

TABLE 12.2. Soil Horizon Nomenclature.

Horizon	Characteristics
A	surface mineral horizon
Ap	surface mineral horizon altered by land use disturbance
Ak	surface mineral horizon with calcium carbonate enrichment
Ab	buried soil
O	surface organic horizon
Bt	subsurface horizon with clay enrichment
Bk	subsurface horizon with calcium carbonate enrichment
Bw	weakly developed subsurface horizon
Bg	subsurface horizon with gray colors due to anaerobic conditions
Bkg	subsurface horizon with gray colors due to anaerobic conditions and calcium carbonate enrichment
C	unweathered soil parent material

ALLUVIAL STRATIGRAPHY

Seven unconformably-bound alluvial stratigraphic units were identified in the Río San Pedro and the Río Casas Grandes valleys, and from oldest to youngest, they are labeled Unit A through Unit G (Figures 12.3, 12.4, and 12.5). From oldest to youngest, paleosols are designated as: Guiermo[2] (GI), Petra (PT), Antonio (AN), Trincheras (TR), Janos (JA), and Viejo (VI). The surface soil of the Oaxaca terrace and modern floodplains are not labeled. Backhoe trench excavations never encountered bedrock, but local residents indicate that bedrock occurs at a depth of 6 to 7 m based on water well excavations.

Unit A

The oldest fluvial deposit identified in the project area is Unit A, which forms the constructional surface of the late Pleistocene Oaxaca terrace (Figure 12.2; Figure 12.3A, BHT-1 and BHT-5; Figure 12.5). Unit A is characterized by a fining upward sequence from basal channel gravels (≤ 5 cm diameter) to fine and medium sands and to upper flood basin silts and clays (Figure 12.3A, BHT-1), typical of a mixed load meandering stream. The upper facies has weathered to a thick A-Bt-Bk soil profile with clay enrichment in the reddish brown Bt horizon and 15% to 25% pedogenic carbonate nodules in the yellowish brown Bk horizon. The surface of Unit A in BHT-5 (Figure 12.3A), however, has been truncated where it consists of a 1-m-thick wedge of channel gravels (≤ 5 cm diameter) overlying brown and bedded sands, silts, and clays. The Bk horizon in the upper part of Unit A in this locality also has 15% to 25% pedogenic

TABLE 12.3. Radiocarbon Ages and Calendar Years for Samples Recovered from Backhoe Trenches.

Sample No.[1]	Alluvial Unit, Paleosol	Backhoe Trench	Depth (cm)	^{14}C age[2] (yr BP)	$\delta^{13}C$‰	AMS/ Radiometric	Calendar Years[3] (one sigma)	Calendar Years[3] (two sigma)	Intercept	Material	Interpretation
Beta -123405	G	r	91–101	210±70	−28.2	AMS	AD 1650–1690 AD 1735–1815 AD 1925–1950	AD 1515–1585 AD 1625–1950	AD 1670 AD 1950	detrital charcoal	time of deposition
Beta -136510	G	k	350	500±40	−23.9	AMS	AD 1410–1435	AD 1400–1450	AD1425	dispersed charcoal	time of deposition
Beta -110619	G	y	241–251	590±70	−25.0[4]	Radiometric	AD 1300–420	AD 1285–1445	AD 1400	organic rich clay	time of deposition
Beta -136509	G	w	213	760±40	−28.2	AMS	AD 1250–1285	AD 1210–1295	AD 1270	dispersed charcoal	time of deposition
Beta -136511	F	j	130	1130±40	−11.7	AMS	AD 880–980	AD 795–1000	AD 900	dispersed charcoal	time of deposition
Beta -153658	F-Viejo	f	88–98	1380±80	−16.4	Radiometric	AD 620–690	AD 540–790	AD 660	soil humate	maximum for burial
Beta -110617	F-Viejo	x	87–97	1400±60	−25.0[4]	Radiometric	AD 620–675	AD 560–720 AD 735–760	AD 650	organic rich clay	time of deposition
Beta -110620	F-Viejo	z	168–178	1520±60	−25.0[4]	Radiometric	AD 465–475 AD 515–620	AD 420–650	AD 560	organic rich clay	time of deposition
Beta -136506	F	q	269–271	1690±60	−18.2	Radiometric	AD 260–20	AD 230–465 AD 480–520	AD 385	organic rich clay	time of deposition
Beta -110621	E-Janos	bb	111–121	1860±60	−25.0[4]	Radiometric	AD 80–225[5]	AD 20–330	AD 160	soil humate	maximum for burial
Beta -123413	E-Janos	aa	114–124	1910±60	−19.3	Radiometric	AD 55–160	20 BC–AD 245	AD 100	soil humate	maximum for burial
Beta -110616	E-Janos	t	71–81	1980±80	−25.0[4]	Radiometric	AD 45–110	175 BC–AD 225	AD 45	soil humate	maximum for burial
Beta -123406	E-Janos	d	188–198	2210±80	−25.6	AMS	380–165 BC	400–40 BC	345 BC 310 BC 210 BC	feature charcoal	time of deposition
Beta -153661	E-Janos	u	162–172	2240±40	−19.5	AMS	380–350 BC 310–210 BC	390–190 BC	360 BC	soil humate	maximum for burial
Beta -153660	E-Janos	f	156–166	2250±40	−17.5	AMS	380–350 BC 300–220 BC	390–200 BC	370 BC	soil humate	maximum for burial
Beta -123412	E-Janos	h	80–90	2290±50	−17.0	AMS	395–360 BC 280–250 BC	405–200 BC	380 BC	soil humate	maximum for burial
Beta -153663	E	a	152–170	2380±40	−20.5	AMS	420–400 BC	740–710 BC 530–390 BC	410 BC	feature charcoal	time of deposition

Beta -153665	E	g	103–119	2490±50	–25.6	AMS	780–520 BC	790–410 BC	760 BC 640 BC 560 BC	feature charcoal	time of deposition
Beta -153662	E	g a	264	2500±40	–23.5	AMS	780–740 BC 710–530 BC	790–420 BC	760 BC 620 BC 590 BC	feature charcoal	time of deposition
Beta -112286	D-Trincheras	h	201–211	3140±100	–25.0[4]	Radiometric	1505–1285 BC	1620–1130 BC	1410 BC	soil humate	maximum for burial
Beta -123409	D	h	282–292	3330±50	–20.5	AMS	1670–1525 BC	1730–1500 BC	1615 BC	organic rich clay	time of deposition
Beta -123411	D	c	300–310	3630±80	–22.2	AMS	2120–2080 BC 2050–1890 BC	2195–1755 BC	1965 BC	organic rich clay	time of deposition
Beta -123410	D-Antonio	2	146–156	4540±80	–21.5	Radiometric	3360–3090 BC	3505–3425 BC 3390–2930 BC	3335 BC	soil humate	maximum for burial
Beta -136507	D-Antonio	b	350–360	4710±40	–14.8	AMS	3620–3590 BC 3525–3500 BC 3445–3380 BC	3635–3370 BC	3515 BC	soil humate	maximum for burial
Beta -136508	D	n	300–310	4790±40	–16.7	AMS	3640–3620 BC 3590–3525 BC	3650–3515 BC	3635 BC	organic rich clay	time of deposition
Beta -136505	C-Petra	f	183–193	5510±40	–15.9	AMS	4360–4335 BC	4440–4320 BC	4345 BC	soil humate	maximum for burial
Beta -153659	C	f	246–256	6900±50	–18.8	AMS	5810–5720 BC	5870–5700 BC	5750 BC	organic rich clay	time of deposition
Beta -123408	B-Guiermo	2	215–225	8950±70	–25.0	Radiometric	8040–7960 BC	8075–7920 BC	8000 BC	peat	maximum for burial
Beta -112285	A	5	227–237	14410±60	–20.3	AMS	15630–15415 BC[5]	15710–15275 BC	15515 BC	organic rich clay	time of deposition
Beta -110618	Viejo	L	232–265	470±50	–25.0[4]	Radiometric	AD 1420–1455	AD 1405–1495	AD 1435	wood	intrusive tree root

[1] Beta Analytic, Inc.
[2] Conventional radiocarbon age reported in radiocarbon years before present (yr BP) and corrected for variations in $\delta^{13}C$.
[3] Calibrated to calendar years from Stuiver and Reimer (1993).
[4] $\delta^{13}C$ value estimated.
[5] Calibrated to calendar years from Reimer et al. (2013) using OxCal v4.3.2 (Bronk Ramsey 2017).

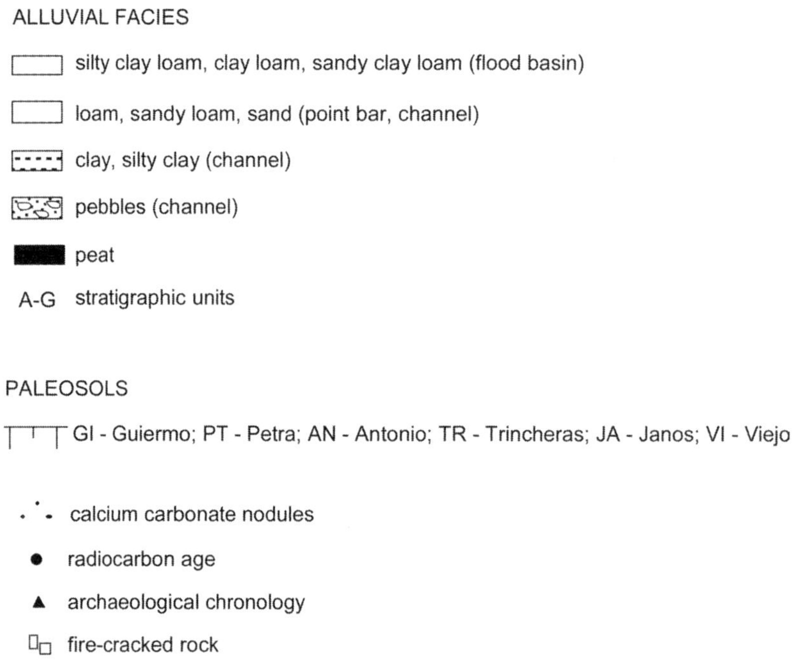

FIGURE 12.3. Legend.

carbonate nodules. A sediment humate radiocarbon age indicates that deposition of Unit A was ongoing approximately 14,410 ± 60 radiocarbon years BP (15,710–15,275 cal BC) (Figure 12.3A; Table 12.3).

Unit B

Unit B is buried in the old Pedro valley as revealed in BHT-2, 3, and 4 (Figures 12.2, 12.3A, 12.5). The Unit B surface is undulating with thick gravelly (≤ 6 cm diameter) and sandy point bars laterally adjacent to deep channels, together suggesting deposition by a meandering bedload stream. The channel facies grades up into a flood basin veneer containing the Guiermo paleosol (GI) with a weakly developed sandy clay loam and noncalcareous A-Bg profile and iron mottles (redox features) (Figure 12.3A, BHT-3). The channels of Unit B are filled with peat, forming an organic phase of the Guiermo paleosol with an O-Bg profile. The peat soil probably formed within about 500 years given commonly accepted organic accumulation rates (see Retallack 2001). A radiocarbon age of 8950 ± 70 radiocarbon years BP from BHT-2 (Figure 12.3A; Table 12.3) indicates that between about 14,000 and 9,000 radiocarbon years ago (15,050 cal BC and 8250 cal BC) the Río San Pedro downcut within the old Pedro valley, abandoning the Oaxaca floodplain and then partially filling the new valley with alluvium. Shortly after 8,900 radiocarbon years ago, the Río San Pedro avulsed to the modern valley, which accompanied by a high water table and reduced sedimentation, permitted the formation of the organic-rich Guiermo paleosol phase. Unit B has not been identified in the Río Casas Grandes valley, presumably because of middle and late Holocene channel erosion.

Unit C

Unit C is buried beneath the Jorge terrace in both the Río San Pedro and Río Casas Grandes valleys (Figures 12.2, 12.3B, 12.4A, and 12.5). The lower channel facies of Unit C consists of gravels (≤ 4 cm diameter) in a brown to olive brown loam matrix grading up into a silt and clay flood basin facies, indicating deposition by a mixed to bedload meandering stream (Figure 12.3B, BHT-f; Figure 12.4A, BHT-l). The Petra paleosol (PT) formed in the upper part of Unit C sediments and has an A-Bk profile with few pedogenic carbonate nodules in the Bk horizon. Channel gravels exposed in BHT-l appear to be associated with surface meander scars on the Jorge terrace that can be traced to the vicinity of Unit C in BHT-f (see Figure 12.2). Coupled with soils having similar morphological properties (e.g., Petra paleosol), deposits in BHT-l are tentatively correlated with Unit C. The larger sizes of the channel scars in this area indicate formation by the Río Casas Grandes rather than the Río San Pedro. A radiocarbon age from organic rich sediments in the upper part of the channel gravel facies in BHT-f (Figure 12.3B; Table 12.3) indicates that deposition of Unit C began sometime between approximately 8900 and 6900 ± 50 radiocarbon years BP (8070–5700 cal BC). A soil humate age of 5510 ± 40 radiocarbon years BP (4440–4320 cal BC) from the surface horizon of the Petra

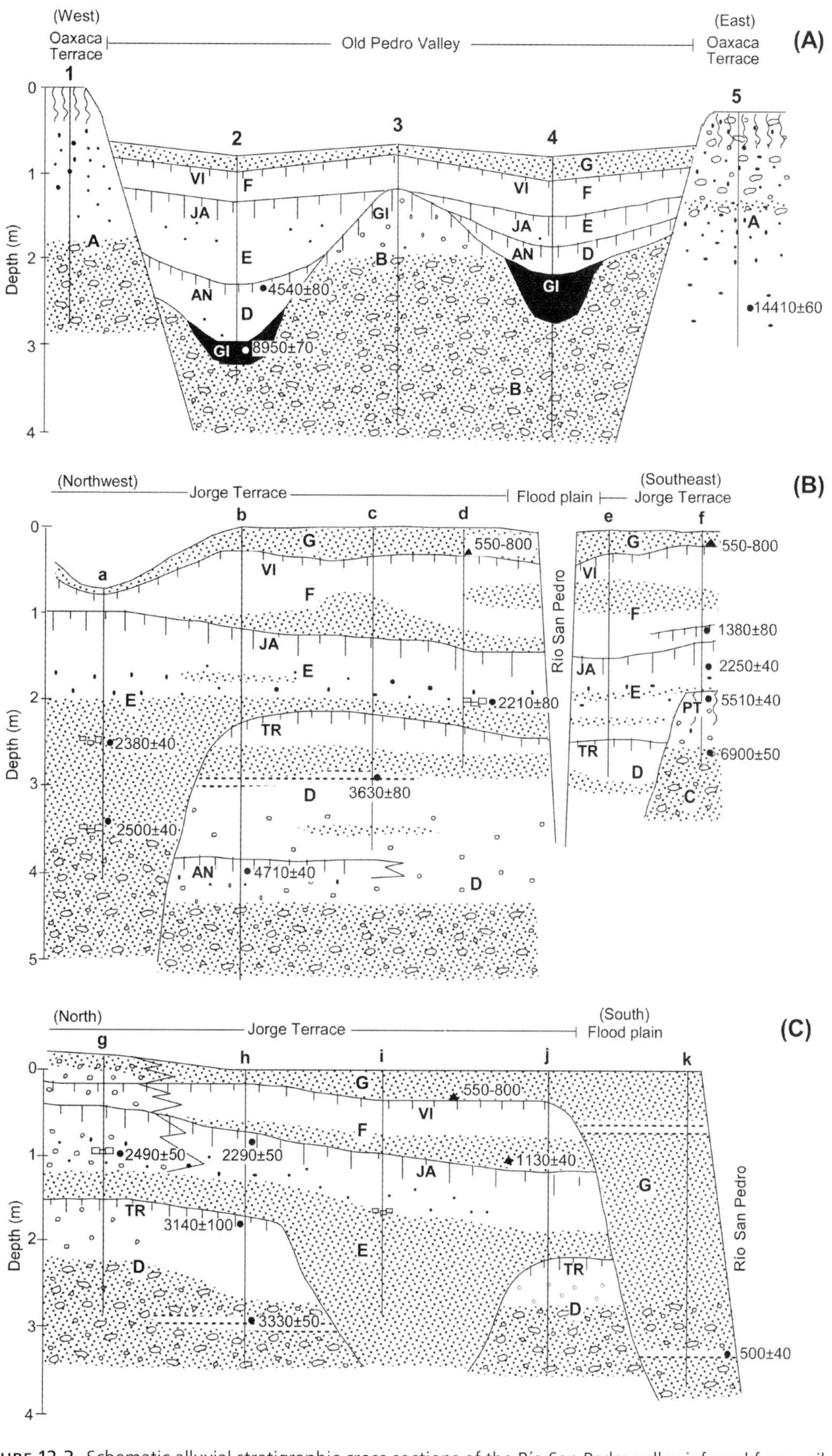

FIGURE 12.3. Schematic alluvial stratigraphic cross sections of the Río San Pedro valley inferred from soil-stratigraphic columns constructed from backhoe trench localities (Cross Section 3A = Backhoe Trenches [BHT] 1–5; Cross Section 3B = BHT a–f; and Cross Section 3C = BHT g–k. Figure 12.2 locates BHTs).

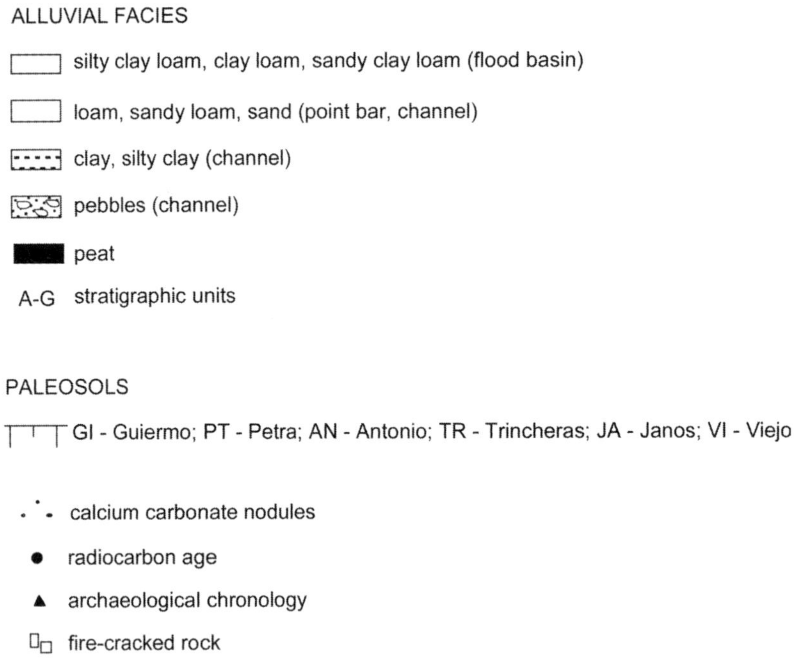

FIGURE 12.4. Legend.

paleosol in BHT-f indicates that deposition of Unit C had terminated by this time and that pedogenesis was ongoing (Figure 12.3B; Table 12.3).

Unit D

Unit D is inset to Unit C beneath the Jorge terrace in both river valleys (Figure 12.2; Figure 12.3A, B, C; Figure 12.4A, B, C; Figure 12.5). Basal channel gravels (≤ 6 cm diameter) are contained in a brown to pale brown sandy matrix. Stratified and brown very fine sands, silts, and clays form the overlying flood basin facies. This sedimentological assemblage points to deposition by a mixed load meandering stream. Two brief episodes of pedogenesis occurred during deposition of Unit D. In the modern Río San Pedro valley, the lower Antonio paleosol (AN) has a grayish brown A-Bkg profile that is somewhat poorly drained with a few groundwater carbonate nodules (Figure 12.3B, BHT-b). This paleosol appears to be localized, having formed in a channel swale. The Antonio paleosol also formed in the old Pedro valley where Unit D fills abandoned channels of Unit B from slack water deposition (Figure 12.3A, BHT-2 and 4). The Trincheras paleosol (TR), with a weakly developed A-Bw profile, formed in the uppermost flood basin facies of Unit D and is laterally traceable beneath much of the Jorge terrace. The Trincheras paleosol was not identified in the old Pedro valley. A sediment humate radiocarbon age from the top of the channel gravel facies of unit D is 4790 ± 40 radiocarbon years BP (Figure 12.4A, BHT-n; Table 12.3). Ages from the Antonio paleosol beneath the Jorge terrace (Figure 12.3B, BHT-b) and from the old Pedro valley (Figure 12.3A, BHT-2) are 4710 ± 40 and 4540 ± 80 radiocarbon years BP, respectively (Table 12.3). Ages from overlying point bar deposits in Unit D are 3630 ± 80 (Figure 12.3B, BHT-c) and 3330 ± 50 radiocarbon years BP (Figure 12.3C, BHT-h; Tablc 12.3). An age of 3140 ± 100 radiocarbon years BP was determined from the surface horizon of the Trincheras paleosol at the top of Unit D (Figure 12.3C, BHT-h), indicating that pedogenesis was ongoing about this time. These ages, when combined with those from Unit C, indicate that channel downcutting prior to deposition of Unit D occurred sometime between 5500 and 4700 radiocarbon years BP (4340–3420 cal BC) and that brief periods of pedogenesis occurred near 4500 radiocarbon years BP (3220 cal BC) and 3100 radiocarbon years BP (1390 cal BC).

Unit E

Unit E is inset to and buries Unit D beneath the Jorge terrace in both valleys (Figures 12.2, 12.3, 12.4, and 12.5). Channel deposits commonly consist of gravels up to 5 cm in diameter in a medium sand matrix. Channel deposits grade up into stratified brown to pale brown sands, silts and clays, with occasional pebble stringers. This sedimentological assemblage illustrates deposition by a sandy meander belt stream. The flood basin facies is pedogenically altered to A-Bk profiles forming the Janos paleosol (JA), typically with 1% to 8% pedogenic carbonate nodules in the Bk horizon. Unit E and the Janos

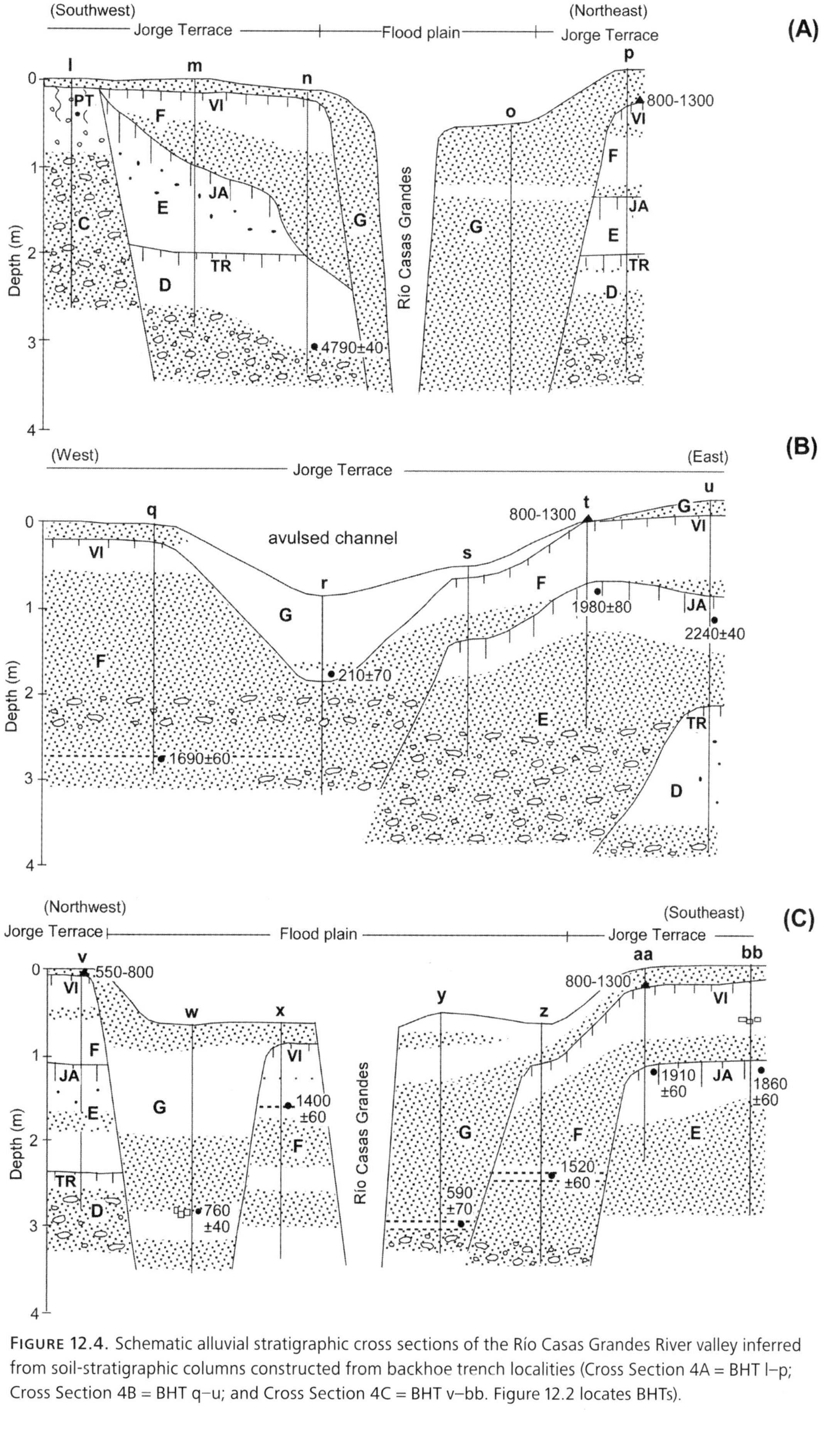

FIGURE 12.4. Schematic alluvial stratigraphic cross sections of the Río Casas Grandes River valley inferred from soil-stratigraphic columns constructed from backhoe trench localities (Cross Section 4A = BHT l–p; Cross Section 4B = BHT q–u; and Cross Section 4C = BHT v–bb. Figure 12.2 locates BHTs).

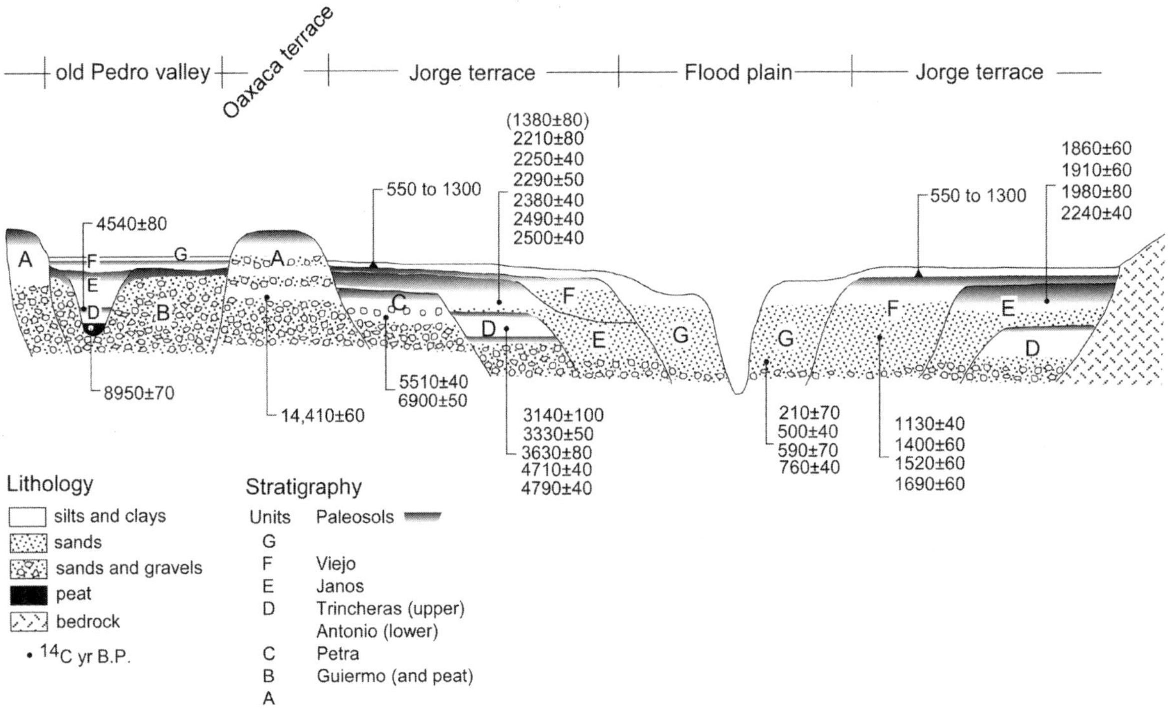

FIGURE 12.5. Composite alluvial stratigraphic cross section of the Río San Pedro and Río Casas Grandes valleys.

paleosol also bury Unit D slack water deposits in the old Pedro valley (Figure 12.3A, BHT-2 and 4). Radiocarbon ages from charcoal in upper channel deposits of Unit E are 2490 ± 50, 2380 ± 40, and 2500 ± 40 radiocarbon years BP (Figure 12.3B and C, BHT-a and g; Table 12.3). Radiocarbon ages from the Janos paleosol are 2290 ± 50, 2250 ± 40, and 2210 ± 80 radiocarbon years BP (Figure 12.3B and C, BHT-h, f, and d), and 2240 ± 40, 1980 ± 80, 1910 ± 60, and 1860 ± 60 radiocarbon years BP (Figure 12.4B and C, BHT-u, t, aa, and bb; Table 12.3). The 2210 ± 80 charcoal sample from the base of the A horizon in BHT-d (Figure 12.3B) demonstrates that deposition was slowing and pedogenesis beginning at about this time. Based on radiocarbon ages from both Unit E and from the overlying Unit F, the Janos paleosol formed during the centuries of pedogenesis between about 2300 and 1800 radiocarbon years BP (390 cal BC–cal AD 230). Radiocarbon ages from Unit D and Unit E show that channel downcutting prior to the deposition of Unit E occurred following the formation of the Trincheras paleosol and before the formation of the Janos paleosol between 3100 and 2500 radiocarbon years BP (approximately 1130 cal BC-790 cal BC, based on the end and beginning of the respective two sigma calibrated dates).

Unit F

Unit F is inset to and buries Unit E beneath the Jorge terrace in both river valleys (Figures 12.2, 12.3, 12.4, and 12.5). At this time the Río Casas Grandes flowed through the avulsed channel visible at the surface of the Jorge terrace on the east side of the valley (Figure 12.2, Figure 12.4B, BHT-q and r). This channel and associated meander belt is filled with thick point bar sands and occasional gravel stringers. Truncated remnants of Unit F channel deposits also occur within the modern floodplain of the Río Casas Grandes (Figure 12.4C, BHT-x and z). Channel margin deposits of Unit F were identified in the Río San Pedro valley (Figure 12.3C, BHT-h, i, and j). Upper deposits of Unit F form a veneer of stratified and brown flood basin silts and clays weathered into weakly developed A-Bw and A-C profiles designated as the Viejo (VI) paleosol. Slack water deposits of Unit F were also flooded into the old Pedro valley above Unit E (Figure 12.3A, BHT-2, 3, and 4). Channel and upper point bar deposits of Unit F date to 1690 ± 60, 1520 ± 60, and 1400 ± 60 radiocarbon years BP from organic rich muds (Figure 12.4B and C, BHT-q, z, and x), and to 1130 ± 40 radiocarbon years BP from charcoal (Figure 12.3C, BHT-j). The surface of Unit F and the Viejo paleosol are temporally constrained by time-diagnostic artifacts in BHT-f, p, and v ranging in age from 1300 to 500 radiocarbon years BP (Figure 12.3B and C; Figure 12.4A–C). In sum, these ages indicate that deposition of Unit F ensued after a brief period of channel erosion between 1800 and 1600 years ago (approximately cal AD 230–500) and terminated no later than 760 ± 40 radiocarbon years BP, based on radiocarbon ages from Unit G (cal AD 1210–1295).

Unit G

Unit G fills the modern floodplain valleys of both rivers (Figures 12.2, 12.3, 12.4, and 12.5). The channel facies is comprised of thick sands, sometimes cross bedded, with thin gravel lenses (≤ 3 cm diameter). The flood basin facies is confined mainly to the modern floodplain valleys, which is inset about 1 m below the Jorge terrace. The surface soil is a very weakly developed, calcareous, brown to dark brown silty clay loam (not designated). Unit G, where deposited across the Jorge terrace, appears to be represented largely by eolian reworking of older surface deposits and consists of calcareous and brown silt loams and very fine sandy loams. Radiocarbon ages from the channel facies of Unit G are 760 ± 40 and 590 ± 70 radiocarbon years BP (Figure 12.4C, BHT-w and y), and 500 ± 40 radiocarbon years BP (Figure 12.3C, BHT-k) from charcoal and organic rich muds (Table 12.3). An age of 210 ± 70 radiocarbon years BP (Figure 12.4B, BHT-r; Table 12.3) was obtained from dispersed charcoal at the base of Unit G in the avulsed channel on the east side of the Río Casas Grandes. The modern rivers have sandy, mixed load meander belts with common channel oxbows.

Stable Carbon Isotopes and C_4 Soil Biomass Production

C_4 plants are typically warm season grasses yielding $\delta^{13}C$ values of approximately −12‰ (Nordt et al. 2008; Von Fisher et al. 2008). These plants respond positively to growing season temperature and secondarily to growing season rainfall. C_3 plants constitute most other plants, including cool season grasses, trees, forbs, and desert shrubs that yield $\delta^{13}C$ values of approximately −26‰ (Nordt et al. 2008; Von Fisher et al. 2008). A mixing line can be established between −12‰ for a pure C_4 plant community and −26‰ for a pure C_3 plant community to estimate relative contributions from a mix of C_4 and C_3 plants. For example, a $\delta^{13}C$ value of −19‰ indicates equal contributions to the soil organic matter pool from both C_4 and C_3 plants. Relative temperature estimates have been made in Texas from $\delta^{13}C$ values of late Quaternary buried paleosols (Nordt et al. 1994; Nordt et al. 2002). However, a distinction must be made between isotopic values emplaced during soil formation in A and B horizons (pedogenic) versus those emplaced during deposition (detrital) (Nordt 2001). For the former, it is assumed that the $\delta^{13}C$ values represent the plant community growing on the soil at the time of soil formation. For $\delta^{13}C$ values emplaced during deposition, if the drainage basin is small, then the values may be representative of the local ecosystem similar to those derived from pedogenesis. If the drainage basin is large, then it becomes more difficult interpreting the source of the isotopic signals. For this investigation, pedogenic and detrital isotopic values are distinguished in the soil isotopic columns in Figure 12.6 and Table 12.4.

Only one $\delta^{13}C$ value was derived from the late Pleistocene deposit (Unit A), which dated to 14,410 ± 60 radiocarbon years BP (15,710–15,275 cal BC). This value of −21.2‰ is detrital, having come from deep in BHT-25 from an upper point bar sand (Figure 12.6A; see also Figure 12.2). This finding indicates that the source area of the Río San Pedro and Río Casas Grandes during the latest Pleistocene consisted of about 35% to 40% C_4 warm season grasses.

The next group of isotopic data comes from BHT-f where between approximately 6900 ± 50 and 5510 ± 40 radiocarbon years BP (5870–4320 cal BC) values trend from −19.5 to −17.2‰ within the Petra paleosol (Figure 12.6B). C_4 plant production was increasing during this interval from a community with approximately equal proportions of C_4 and C_3 plants to one dominated by C_4 plants. Data between approximately 5500 and 4500 radiocarbon years BP (4340–3220 cal BC) were derived from BHT-b (Figure 12.6C). This profile formed in a poorly drained setting and is deeply buried in the modern floodplain of the Río San Pedro. The lower three isotopic values of −24.6, −20.1, and −19.0‰ were possibly influenced by riparian vegetation. The values from the surface of the Antonio paleosol, however, appear to represent at least moderately well-drained conditions supporting a predominantly C_4 grassland community near, and shortly after, 4790 ± 40 radiocarbon years BP (3650–3515 cal BC) (similar to the Petra soil).

Between approximately 3330 ± 50 and 2250 ± 40 radiocarbon years BP (1730–200 cal BC) (BHT-h and f), detrital isotopic values typically show a stronger C_3 signature whereas the Trincheras and Janos paleosols both reveal a C_4 grass-dominated plant community (Figure 12.6B and D). Between approximately 2290 ± 50 and 800 radiocarbon years BP (BHT-h and f) (405 cal BC–cal AD 1240) the isotopic record varies, but suggests that C_4 grasslands continued to dominate (Figure 12.6B and D).

By observing isotopic values of the uppermost buried paleosols in the soil-stratigraphic profiles, the vegetation on the prehistoric landscape during the Contact period can be estimated (see the uppermost $\delta^{13}C$ values of the Viejo paleosols in BHT-f and BHT-h and the surface value from the buried paleosol in BHT-25 in Table 12.4). At this time it appears that the landscape consisted of between 65% and 85% C_4 grasses (approximately −14 to −17‰), consistent with the presence of a C_4 short grass prairie rather than desert shrubs. Surface detrital values from BHT-25 and BHT-h, in association with detrital values from sediments in the modern floodplain BHT-y, indicate that modern eolian and fluvial sediments contain C_4 and C_3 plant contributions ranging from 40% to 60% (approximately −17 to

Table 12.4. Stable Carbon Isotopes of Paleosol and Detrital Organic Matter Recovered from Backhoe Trenches.

Location	Alluvial Unit, Paleosol	Horizon	Depth (cm)	^{14}C age (yr BP)	$\delta^{13}C$ (‰)	Organic Source: Pedogenic	Organic Source: Detrital
BHT-25	G	C1	0–7	modern	−19.2		yes
	G	C2	7–12	modern	−15.4		yes
	A	Ab	12–30	Contact Period	−14.5	yes	
	A	Bt	30–44	unknown	−14.0	yes	
	A	C	275+	14,000	−21.2		yes
BHT-f	F-Viejo	Ab1	12–21	800–1600	−15.2	yes	
	F-Viejo	Bw	21–33	800–1600	−13.0	yes	
	F-Viejo	Ab	33–42	800–1600	−17.2	yes	
	F-Viejo	Bw	42–49	800–1600	−15.7	yes	
	F-Viejo	Bk	49–56	800–1600	−19.2	yes	
	F-Viejo	Ab	56–76	800–1600	−17.5	yes	
	F-Viejo	BC	76–88	800–1600	−15.3		yes
	E-Janos	Ab2	88–107	800–1600	−15.6	yes	
	E-Janos	Bk	107–123	800–1600	−15.6	yes	
	E-Janos	Akb	123–138	2000–3000	−16.2	yes	
	E-Janos	Bw	138–166	2000–3000	−16.7	yes	
	E-Janos	BC	166–183	2000–3000	−16.4		yes
	C-Petra	Ab3	183–217	4500–5500	−17.2	yes	
	C-Petra	Bw	217–246	6000–7000	−18.1	yes	
	C-Petra	BC	246–258	6000–7000	−19.5		yes
BHT-b	D-Antonio	Ab	390–402	4500–5500	−14.5	yes	
	D-Antonio	Bw1	402–419	4500–5500	−13.2	yes	
	D-Antonio	Bw2	419–440	4500–5500	−14.8	yes	
	D-Antonio	Bkg1	440–454	4500–5500	−15.6	yes	
	DAntonio	Bkg2	454–485	4500–5500	−19.0	yes	
	D-Antonio	Ckg	485–504	4500–5500	−20.1		yes
	D-Antonio	C	504+	4500–5500	−24.6		yes
BHT-h	F	Ap	0–15	800–1600	−17.0		yes
	F-Viejo	Ab1	15–24	800–1600	−16.9	yes	
	F-Viejo	Bk	24–35	800–1600	−15.6	yes	
	F-Viejo	Akb	35–53	800–1600	−14.7	yes	
	F-Viejo	BC	53–73	800–1600	−14.7		yes
	E-Janos	Ab2	73–96	2000–3000	−17.0	yes	
	E-Janos	Bk1	96–119	2000–3000	−17.5	yes	
	E-Janos	Bk2	119–159	2000–3000	−18.6	yes	
	E-Janos	Bk3	159–182	2000–3000	−19.1		yes
	D-Trincheras	Ab3	182–202	3000–4000	−18.0	yes	
	D-Trincheras	BC	202–242	3000–4000	−17.4		yes
	D-Trincheras	C1	242–281	3000–4000	−19.5		yes
	D-Trincheras	C2	281–300	3000–4000	−20.5		yes
BHT-y	G	Ap	0–15	<600	−20.7		yes
	G	C1	15–31	<600	−20.6		yes
	G	C2	31–47	<600	−20.1		yes
	G	C3	241–251	<600	−20.4		yes

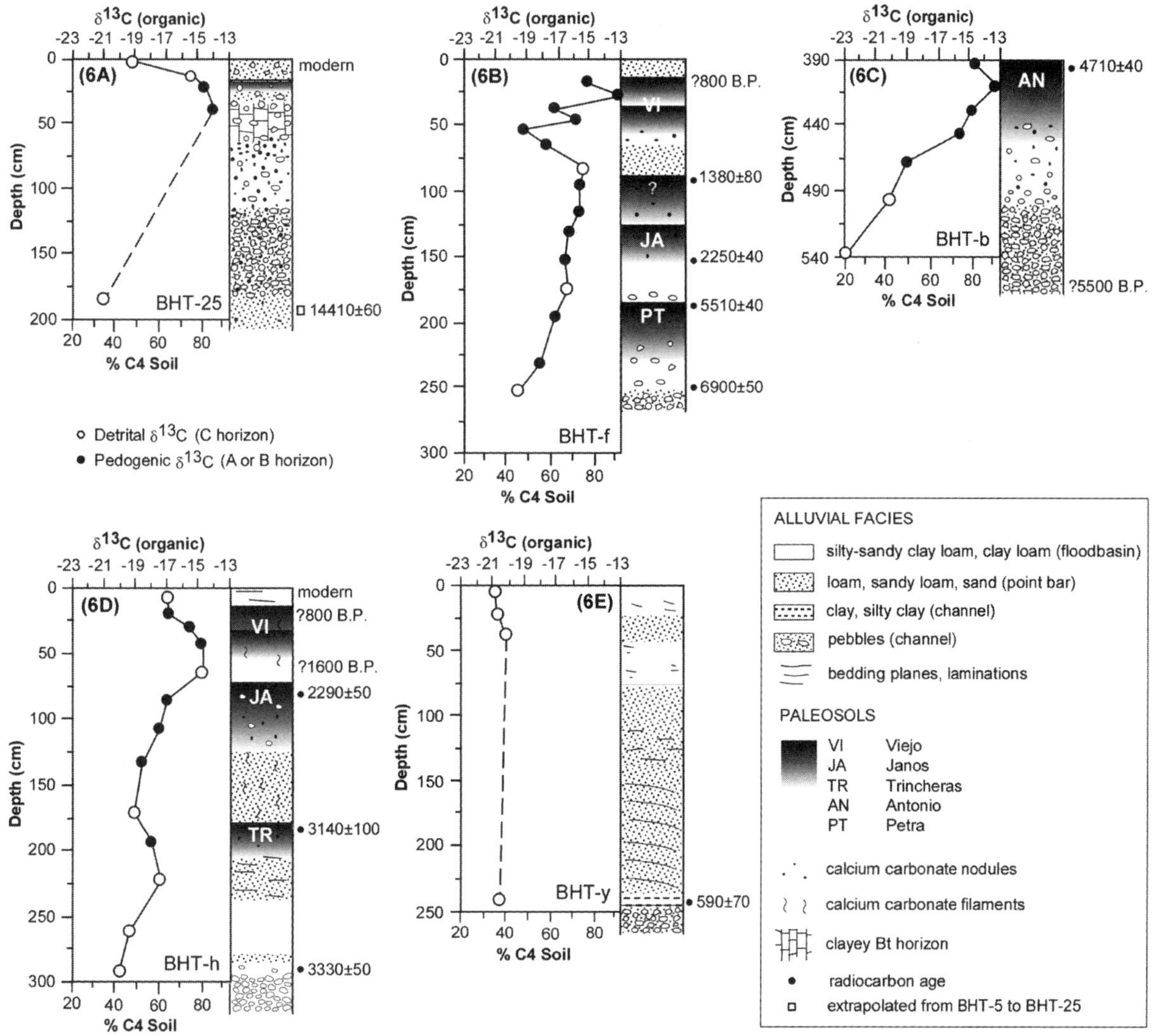

FIGURE 12.6. Stable carbon isotopic distributions of selected soil-stratigraphic columns from the Río San Pedro and Río Casas Grandes valleys.

−21‰). In the floodplain, the C_3 component could consist of either local C_3 desert shrubs or C_3 tree signatures transported from the highlands. Eolian deposits on the terraces probably yield C_3 signatures derived from local desert shrubs.

In sum, it appears that C_4 grasslands have dominated the floodplains and terrace of the Cerro Juanaqueña study during the last 7,000 years (5900 cal BC). Minor perturbations occur but probably represent natural isotopic variability expected across any landscape. Detrital isotopic values are typically more greatly influenced by C_3 plants, but interpreting their source areas is problematic. It seems logical that modern detrital floodplain deposits would have a greater C_3 component from trees in the highlands of the drainage basins, and that eolian deposits draping terraces would have a greater C_3 contribution from local desert shrubs. In any case, during the last 7,000 years detrital isotopic values indicate that C_4 plants constituted at least 50% of the plant communities in both the highlands and floodplains with pedogenic values demonstrating that C_4 plants dominated the floodplains and terraces.

Alluvial Stratigraphy and Climate

The Río Casas Grandes and Río San Pedro experienced numerous periods of channel erosion, deposition, and soil formation during the late Quaternary (Figure 12.7). Erosional episodes occurred sometime between 14,000 and 9000 radiocarbon years BP (15,050 and 8250 cal BC) (E1, E2), near 5000 radiocarbon years BP (3780 cal BC) (E3) and 3100 radiocarbon years BP (1250 cal BC) (E4), and between 1860 ± 60 and 1690 ± 60 radiocarbon years BP (cal AD 20–520) (E5) and 1130 ± 40 and 760 ± 40 radiocarbon years BP (cal AD 795–1295) (E6). Erosional

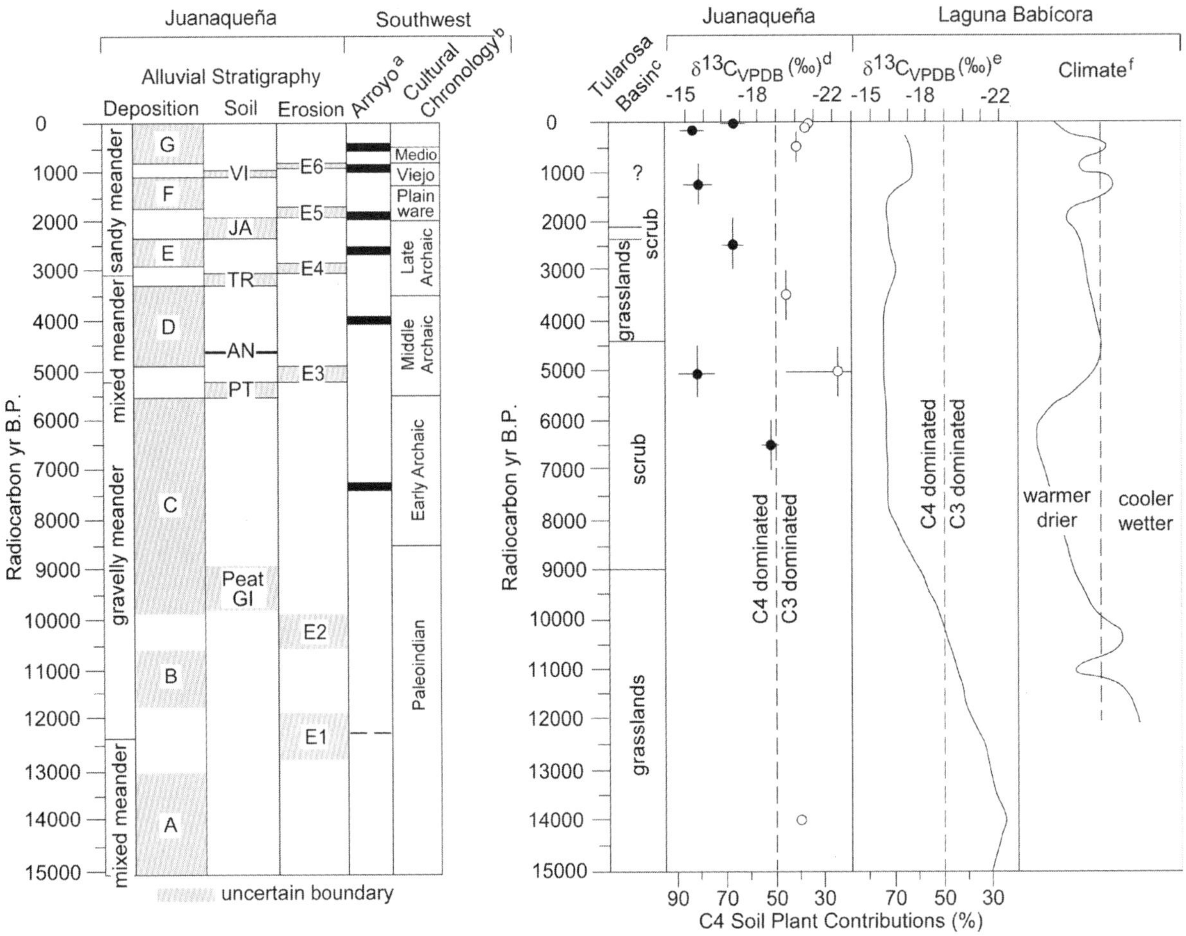

FIGURE 12.7. Summary of regional paleoecology and paleoclimate interpretations in comparison to periods of alluvial deposition, channel erosion, and soil formation in the Cerro Juanaqueña study area and to periods of arroyo formation and high magnitude flooding in the North American Southwest. Modified from Nordt (2003). (*a*) Arroyo formation from Waters and Haynes (2001) in Arizona. (*b*) Prehistoric cultural chronology for the region from Huckell (1996a) and Whalen and Minnis (2001). (*c*) Based on paleosol stable carbon isotopes from southern New Mexico (Buck and Monger 1999). (*d*) Stable carbon isotopes from alluvial paleosols in the Cerro Juanaqueña study area. Vertical lines through data points indicate a time range of averaged isotopic values with horizontal lines representing the standard deviation. (*e*) Stable carbon isotope curve taken from Laguna Babícora sediments (Huang et al. 2001) after conversion of n-alkane $\delta^{13}C$ values to bulk sediment humate $\delta^{13}C$ values using –19‰ as the C_4 end member. (*f*) Lake level fluctuations and sediment geochemistry of Laguna Babícora (modified from Ortega-Ramírez et al. 1998).

events episodically lowered the floodplain surfaces between approximately 14,000 and 3000 radiocarbon years BP (cal 15,050–1250 cal BC) (Units B, C, and D), after which substantial vertical floodplain building began (Units E and F). A last episode of floodplain lowering occurred prior to deposition of Unit G culminating in construction of the modern floodplain. Brief periods of soil formation occurred near 5500 (PT), 4500 (AN), 3100 (TR), and 800 (VI) radiocarbon years BP, (4340, 3220, 1250 cal BC and cal AD 1240) with an extended interval of pedogenesis between 2300 and 1800 radiocarbon years BP (JA) (390 cal BC–cal AD 230). All paleochannels and associated facies indicate deposition by meandering, mixed to bedload streams that in at least two cases were associated with channel avulsions (Units B and F). In addition, meander cutoffs are evident at the surface of the old Pedro valley (Unit B), the Jorge terrace (Units C and F), and the modern floodplain (Unit G).

This complex sequence of alluvial erosion, deposition, and soil formation occurred in desert grasslands where floodplain construction proceeded by meander belt deposition of channel gravels and overbank silts and clays (Figure 12.7). This finding suggests the persistence of reduced overland flow in the region throughout the Holocene, perhaps governed by the presence of grasslands. In contrast, alluvial sequences of large streams in the deserts of the North American Southwest developed with shifts between narrow and deep channels and wide and shallow channels, with arroyos forming by vertical cuts and fills (B. Huckell 1996b; Huckleberry 1995; Waters and Ravesloot 2000). Further, in conjunction with a relatively abrupt shift from gravelly bedload to fine-grained flow regimes sometime between 8000 and 5000 radiocarbon years BP (6920–3780 cal BC) in the American Southwest, streams in desert grasslands bordering the Chihuahuan Desert record do not record this sedimentological shift. This phenomenon was undoubtedly aided by the stabilizing effect of established C_4 grasslands and bedrock valley constrictions in the Cerro Juanaqueña area.

Of the six channel erosion events that occurred in the Río Casas Grandes and Río San Pedro valleys of the Cerro Juanaqueña study area, three correlate with the timing of widespread arroyo formation in the North American Southwest (Figure 12.7, E1, E5, E6). However, incongruence in these records for one episode of erosion (E2) can be attributed to channel avulsion in the Cerro Juanaqueña area. It may be that E3 and E4 correlate with erosional events in the North American Southwest, but that the timing is slightly different or that the radiocarbon chronologies do not offer sufficient resolution for an adequate correlation. In the latest Pleistocene, a shift to warmer and drier climates probably initiated channel downcutting by lowering water tables and increasing flashy, overland flow during brief but intense thunderstorm activity (Bull 1997; Waters and Haynes 2001). Erosion events during the Holocene appear to have occurred just after a period of paleosol formation during similar warm and dry intervals according to data from Laguna Babícora and in part from ecosystem shifts in the Tularosa Basin (Figure 2.3). A shift to cooler and wetter climates at the end of formation of the Petra, Janos, and Viejo soils could have initiated channel downcutting by exposing an increasingly unstable landscape with reduced vegetative cover to a sudden increase in rainfall and overland flow (Bull 1997; Waters and Haynes 2001). The coupled formation of the Trincheras paleosol and the E4 event all occurred within a warming trend. Deposition then followed during cooler and wetter intervals. In the late Holocene, increased periodicity of high magnitude floods (Ely 1997; Ely et al. 1993) and arroyo formation in the North American Southwest are attributed to enhanced development of El Niño cycles causing more frequent winter rains in association with subtropical cyclones. Fluvial hydraulic relations also indicate that with the spreading of upland vegetation during wetter climates, reduction in delivery of sediments to channels leads to incision (Schumm and Brackenridge 1987).

The five intervals of Holocene paleosol formation in the Cerro Juanaqueña study area correlate with the general timing of pedogenesis in the North American Southwest (Waters and Haynes 2001) and exceptionally well with intervals in the San Juan Basin of northwest New Mexico. In fluvial and eolian environments of the latter, periods of soil formation were identified near 8000, 4500, 3100, and 2100 radiocarbon years BP, (6920, 3220, 1390, and 130 cal BC) and sometime between 1300 and 600 radiocarbon years BP (cal AD 690–1350) (Smith and McFaul 1997). This evidence lends further credence to regional correlation of key alluvial events related to climate. Holocene paleosols in the Cerro Juanaqueña area all generally formed under dominance of C_4 grasslands (Figure 12.7). Zonal air circulation patterns in the region during the late Pleistocene led to gentle, winter rains as the westerlies shifted to the south in response to continental glaciation. Even though solar insolation had increased dramatically by the early Holocene, a shift to meridional air flow and increased summer monsoon rains (Bull 1997; COHMAP Members 1988; Ely 1997) apparently delayed the onset of aridity in the American Southwest. This finding is consistent with the emergence and persistence of C_4 grasslands in the Tularosa Basin, Laguna Babícora, and the Cerro Juanaqueña area shortly after 10,000 radiocarbon years BP (9520 cal BC). Although global circulation models suggest the formation of a middle Holocene Altithermal in response to reduction in solar insolation and monsoonal rainfall in the American Southwest (COHMAP Members 1988), there is no evidence of this transformation in the Cerro Juanaqueña area and Laguna Babícora in terms of C_4 plant production. However, the presence of scrub vegetation in the Tularosa Basin and lower lake levels at Laguna Babícora point to a warm and dry interval similar to the Altithermal.

Geoarchaeological Interpretations

Created by multiple episodes of channel erosion, deposition, and soil formation, the alluvial stratigraphic framework influences prehistoric preservation potentials, settlement patterns, and subsistence strategies in the valleys of the Río Casas Grandes and Río San Pedro. This section first discusses the general distribution of potential surface and buried sites from a diachronic, or evolving landscape, viewpoint (Figures 12.8 and 12.9), and then more specifically addresses the landscape setting and farming potential as reconstructed from a synchronic landscape standpoint 3000 radiocarbon years BP (Figure 12.10). Preservation potentials are emphasized because very few of the 69 excavated backhoe trenches yielded archaeological components due to the enormous area being prospected.

Preservation Potentials

Preservation of Paleoindian sites will be restricted to one of two areas in the Río Casas Grandes and Río San Pedro valleys: buried within Unit B below depths of 50 cm in the old Pedro valley, or on the surface of

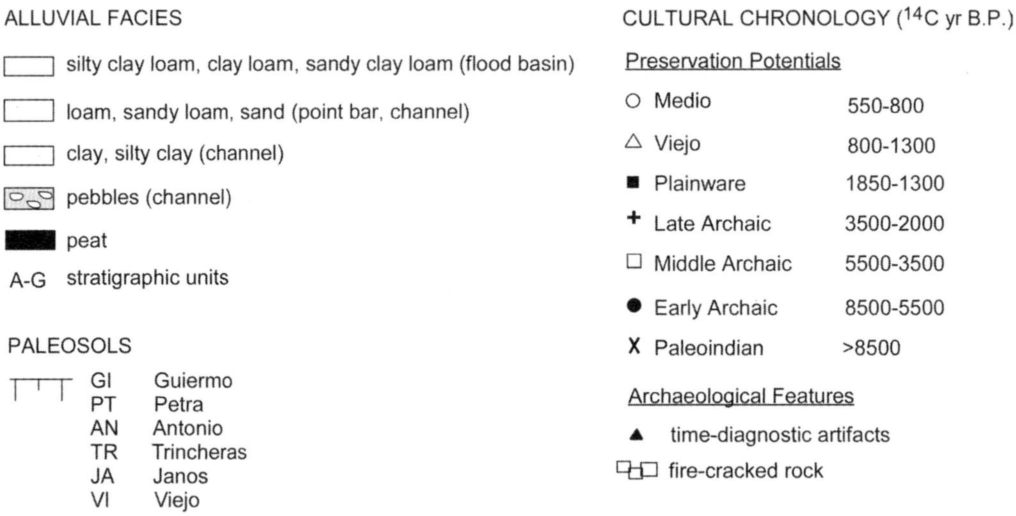

FIGURE 12.8. Legend.

the adjacent Oaxaca terrace (Figures 12.2 and 12.8A). Preservation in primary contexts will be low within Unit B because of excessive stream power that resulted in deposition of a gravelly bed load valley fill. However, features dating to the latter part of the Paleoindian period will be better preserved because of association with the Guiermo soil formed in the veneer of overbank deposits in the upper of Unit B. The climate across the Pleistocene–Holocene was becoming more arid, but still generally mesic, generating a base flow stream rich in water, plant, and wildlife resources. Thus, the Oaxaca terrace bordering the Río San Pedro would have been an ideal location for prehistoric occupation. Sediments dating to Paleoindian times have yet to be discovered along the Río Casas Grandes, however. These deposits may be better preserved below valley constrictions where lateral channel meandering is mitigated because of higher channel gradients.

Early Archaic sites may be preserved in three locations: first, buried in Unit C and the Petra soil beneath the Jorge terrace (Figures 12.2, 12.8B, and 12.9A); second, on the Guiermo soil in the old Pedro valley buried to depths of 50 to 100 cm (Figure 12.8A); and third, on the surface of the Oaxaca terrace mixed with previously deposited Paleoindian components (Figure 12.8A). Preservation in primary contexts will be relatively high within sediments of the Petra soil of Unit C and on the Guiermo soil because burial occurred by subsequent low-energy flooding. Features associated within sediments of the Petra soil will be discrete and not mixed with components from other cultural periods as with the surface of the Guiermo and Oaxaca terrace soils. Unit C deposits have not been identified in the old Pedro valley because after the Río San Pedro shifted to its modern valley around 8900 radiocarbon years BP slack water deposition did not penetrate into the valley until later in the Holocene. The exact distribution of Unit C deposits is not known. However, based on limited exposures, the rivers appear to have been mixed to bedload streams meandering within context of a warming trend and spreading warm season grasslands. Assessing settlement patterns and subsistence strategies at this time is difficult.

Middle Archaic components will be buried in the middle to lower part of Unit D at depths of more than 2 m below the Jorge terrace surface (Figures 12.2, 12.8, and 12.9), on the Guiermo soil mixed with Early Archaic components in the old Pedro valley (Figure 12.8A), on the Oaxaca terrace mixed with both Paleoindian and Early Archaic components (Figure 12.8A), and perhaps partly on the surface of the Petra soil (Figures 12.8B and 12.9A). A dominance of fine-grained sediments indicates that slow, overbank deposition buried and preserved Middle Archaic features and should be found in discrete, primary contexts in the middle part of Unit D, particularly in association with the Antonio soil. Unit D is widely distributed in both the Río Casas Grandes and Río San Pedro valleys and characterized by broad and low-lying floodplains relative to later floodplains. The climate at this time was such that stabilization of warm season grasslands persisted. Channel base flow may have been higher at this time.

Late Archaic sites will be confined to: Unit E beneath the Jorge terrace and in the old Pedro valley, either deeply buried within channel/point bar deposits or within sediments that eventually weathered into the Janos paleosol (Figures 12.2, 12.8, and 12.9); on the Trincheras soil of Unit D (Figures 12.8 and 12.9); on the Petra soil of Unit C (Figures 12.8B and 12.9A); or on the surface of the Oaxaca terrace (Figure 12.8A).

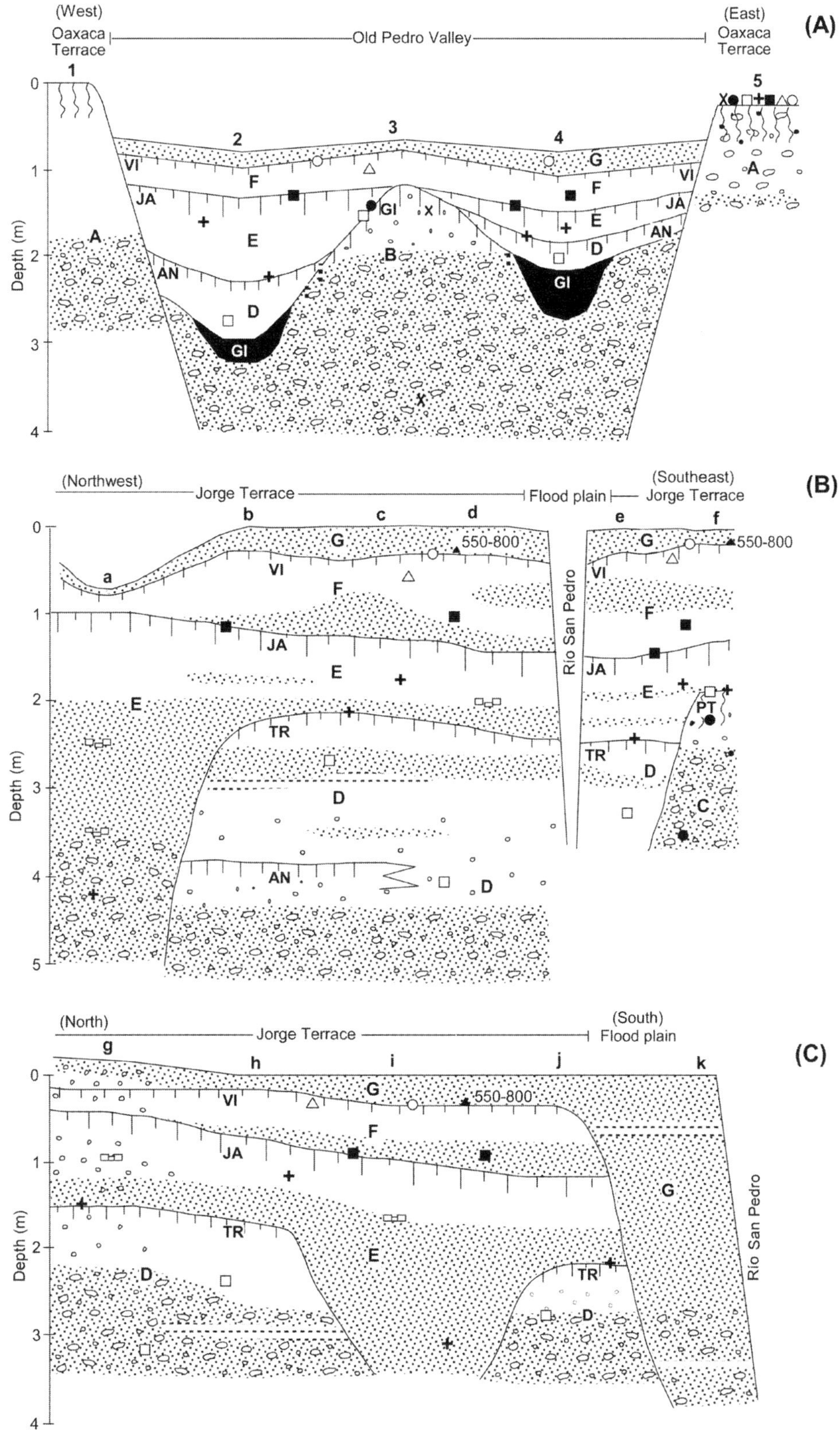

FIGURE 12.8. Schematic alluvial stratigraphic cross section of the Río San Pedro valley illustrating prehistoric preservation potentials by cultural period with symbols.

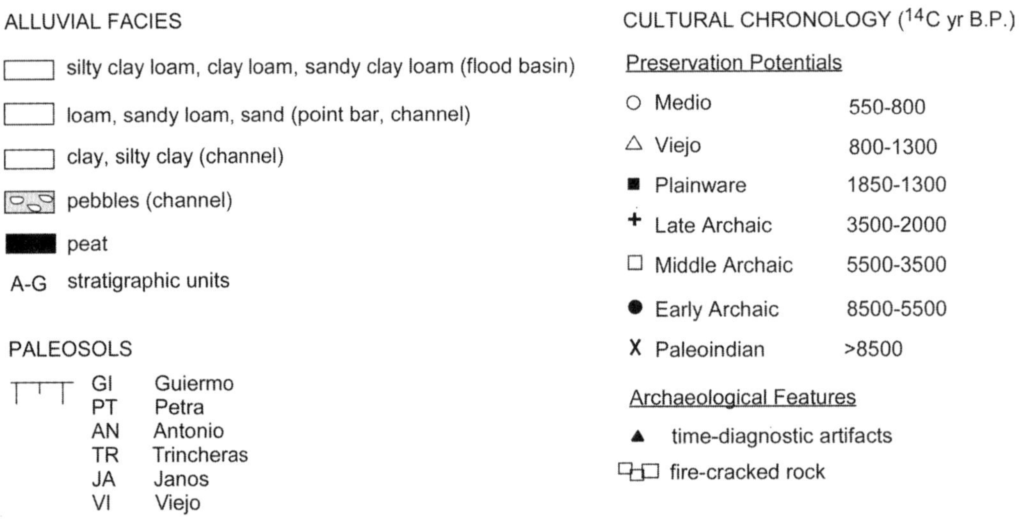

FIGURE 12.9. Legend.

Contextual integrity is indicated, as discrete Late Archaic components would be greatest on the Trincheras soil or in the upper part of Unit E. The Unit E sedimentological assemblage suggests the presence of a mixed load sandy meander belt with low base flow, consistent with a warming and drying climatic trend. In Unit E, the oldest buried features were discovered during the geomorphic investigation. Two occupation zones were identified within a Unit E channel of the Río San Pedro dating from approximately 2380 to 2490 radiocarbon years BP (Figure 12.3B). A date of 2490 radiocarbon years BP (Figure 12.3C) was retrieved from charcoal from the Ophelia's Well site outside of the Unit E paleochannel on the margin of the Oaxaca terrace (BHT-g). A Late Archaic burned rock feature was dated to 2290 radiocarbon years BP (Figure 12.3C) just below the Janos paleosol near the modern Río San Pedro channel. Clearly from these features and the alluvial stratigraphic framework, the floodplain was being actively utilized during the Late Archaic period.

Plainware period (AD 100–650) components will most likely be preserved: on the Janos paleosol within a meter of the surface of the Jorge terrace; in the old Pedro valley (Figures 12.2, 12.8, and 12.9), distributed throughout the lower part of Unit F (Figures 12.8 and 12.9); on the Petra soil in some localities (Figure 12.9A); or on the Oaxaca terrace (Figure 12.8A). Although subjected to disturbance prior to burial, a relatively long-term stable surface was responsible for development of the Janos soil and potentially 500 years of accumulation of Plainware components. Sediments that buried the Janos paleosol were deposited as relatively low-energy, fine-grained overbank deposits, which would have led to optimal preservation. Another location for buried Plainware sites would be along and perhaps within the abandoned Río Casas Grandes paleo-channel on the Jorge terrace just below Cerro Juanaqueña (see Figures 12.2 and 12.9B). Unit F deposits in the Río San Pedro valley appear only to be of overbank origin as the Río San Pedro channel at this time period has not been located. Unit F was deposited from a mixed load, sandy meander belt, during periodic fluctuations in wet/dry cycles.

Viejo period (AD 650–1200) components will be restricted to the upper part of Unit F, particularly in association with the Viejo soil within 50 cm of the surface of the Jorge terrace (Figures 12.2, 12.8, and 12.9). In fact, numerous Viejo period sites (solid triangles on Figure 12.9) have been recorded at the surface because of erosion near the shallow escarpment separating the Jorge terrace and floodplain (Figure 12.9A, B, and C). Viejo period features could also be discovered buried at shallow depths in the old Pedro valley in Unit F (Figure 12.8A), on the Petra soil (Figure 12.9A), and on the surface of the Oaxaca terrace (Figure 12.8A) as a palimpsest with other features. Features at depth in Unit F have greater potential for preservation as discrete components than those associated with the surface of the Viejo soil. Although Unit F was deposited during relatively mesic conditions, the climate had become warmer and drier by the time the Viejo soil formed.

Medio period (AD 1200–1450) features will be: confined to the lower half of the Unit G sediments in the floodplains of both river valleys at depths greater than 2 m (Figures 12.2, 12.8, and 12.9); at depths of less than 50 cm in slack water deposits of the old Pedro valley (Figure 12.8A); on the Viejo soil in all localities at depths of < 50 cm (Figures 12.8 and 12.9); and on the surface of the Oaxaca terrace with prehistoric features from all other time periods (Figure 12.8A). The modern channels of Unit G are mixed load sandy meander belts.

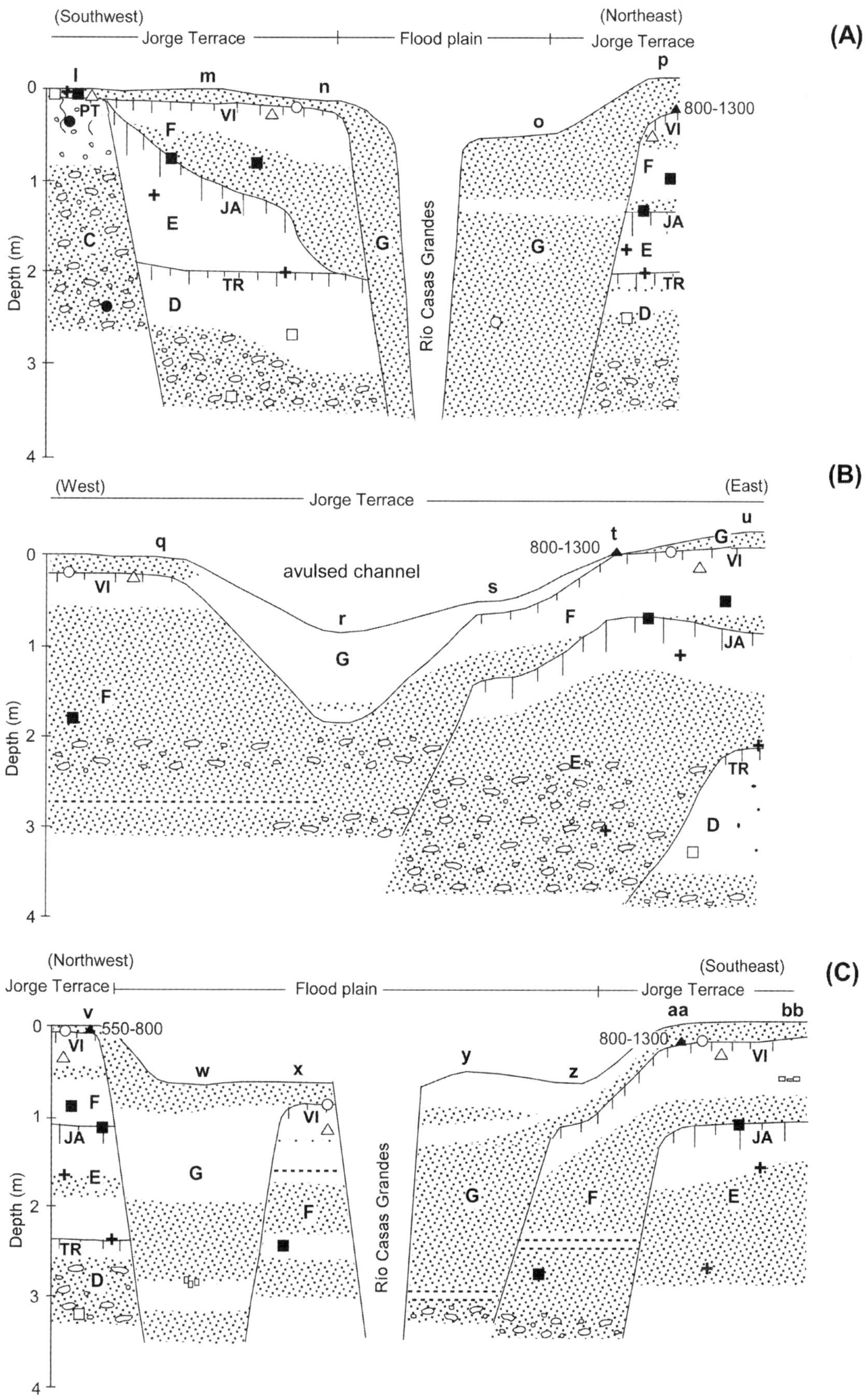

FIGURE 12.9. Schematic alluvial stratigraphic cross section of the Río Casas Grandes valley illustrating prehistoric preservation potentials by cultural period with symbols.

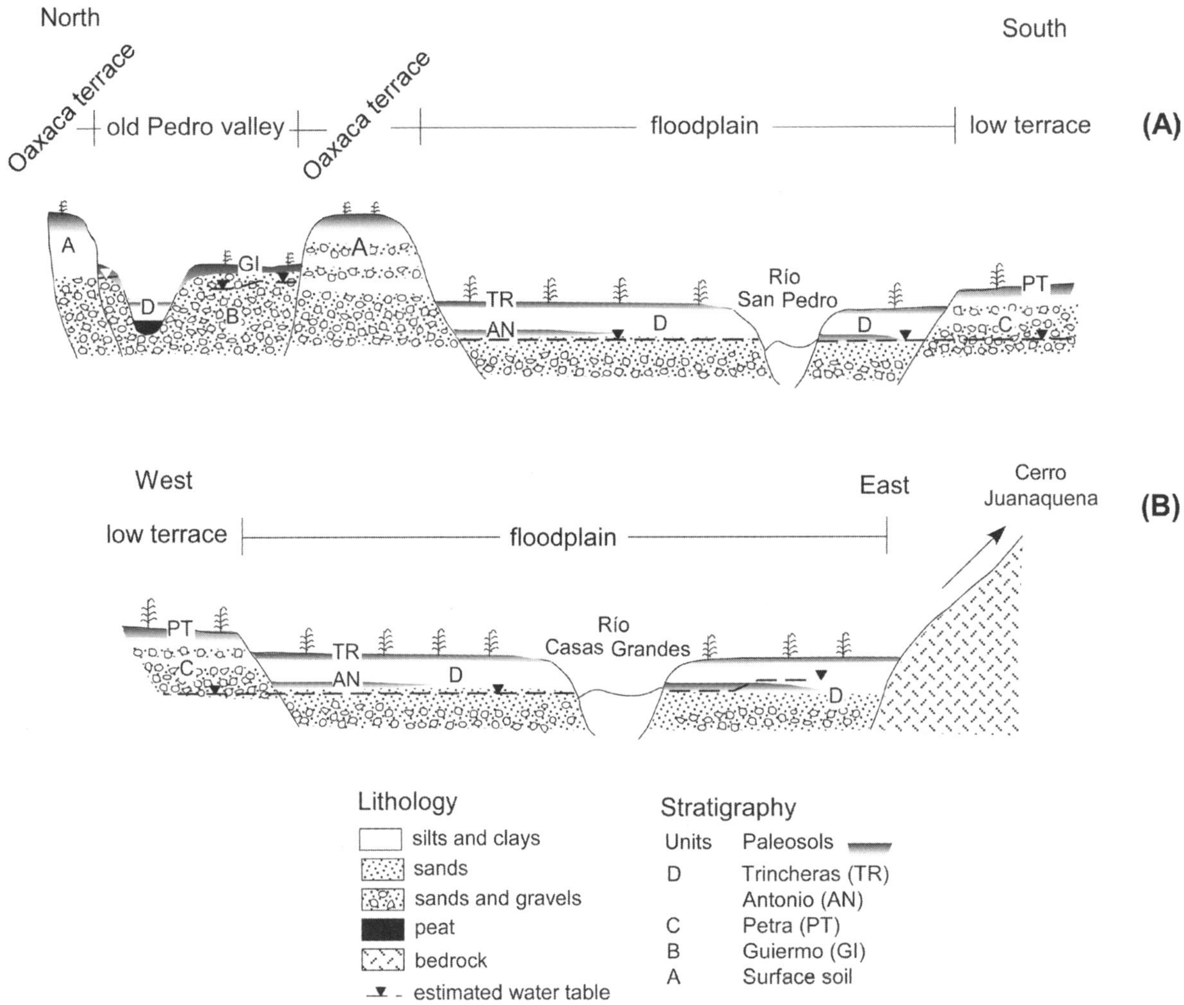

FIGURE 12.10. Schematic alluvial stratigraphic cross section of the Río San Pedro (A) and Río Casas Grandes (B) valleys 3000 ^{14}C yr BP. Height of maize stalks indicates relative potential productivity per landscape setting.

Numerous Medio period sites were located on the Viejo soils indicating continued intensive occupation of both river valleys.

Based on the stratigraphic record and interpreted environments of deposition, Units D, E, and F have the greatest potential for buried features in primary contexts among the sediments of the Río Casas Grandes and Río San Pedro valleys. Because of rapid deposition, short-term encampments probably ensued in the floodplain, with long-term establishments more likely on the Oaxaca terrace or the Petra soil surface prior to burial. Five-hundred years of prehistoric components could be mixed on the Janos soil, with at least several thousand years of components on the Petra soil in some areas, but less than a few centuries of artifacts on the Trincheras, Antonio, and Viejo soils. Sites dating to older than 3000 radiocarbon years BP will not be preserved below the Jorge terrace at the base of Cerro Juanaqueña because of channel erosion associated with deposition of Unit F. Sites older than this are more likely preserved along the Río San Pedro and on the west side of the Río Casas Grandes.

Reconstruction of the 3000 radiocarbon years BP Landscape

During occupation of Cerro Juanaqueña approximately 3000 radiocarbon years BP (1250 cal BC), the alluvial setting of the Río Casas Grandes and Río San Pedro valleys looked much different than today. The Río San Pedro valley included the old Pedro valley and a floodplain approximately 1 km wide (Unit D) between the Oaxaca terrace (Unit A) and the low terrace containing the Petra soil (Unit C) (Figure 12.10A). The Río Casas Grandes floodplain was relatively broader (Unit D), extending from the base of Cerro Juanaqueña to the west approximately 2 km before abutting Unit C and the Petra soil (Figure 12.10B). As viewed in Figures 12.2 and 12.10 for 3000 radiocarbon years BP, the Trincheras floodplain comprised approximately 14 to 16 km^2 of land area, the low terrace and Petra soil approximately 2 km^2, the

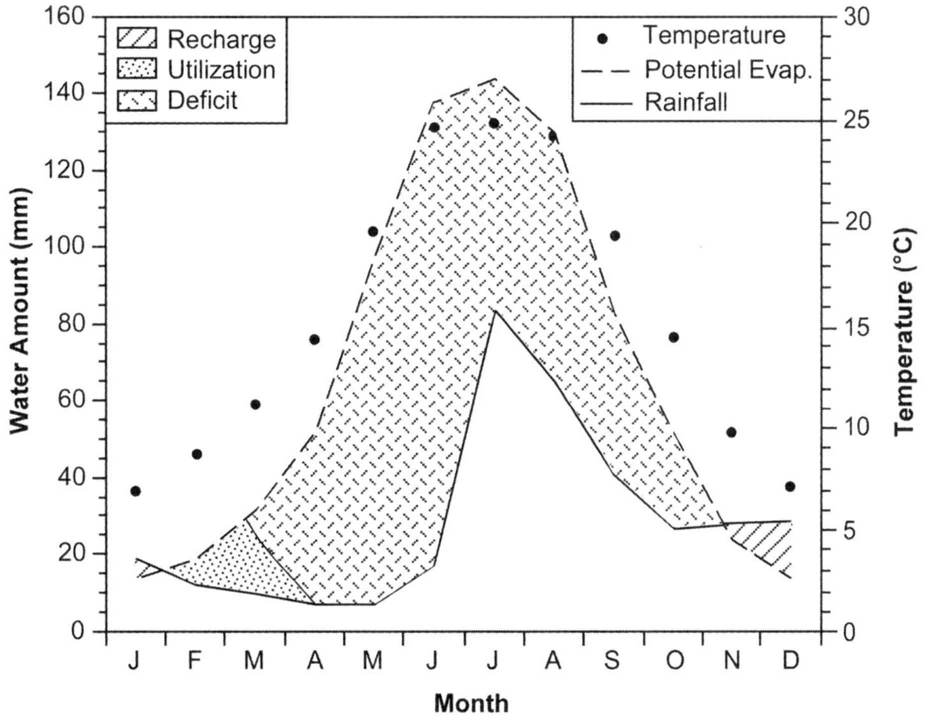

FIGURE 12.11. Thornthwaite soil water model using modern monthly climate data for Janos, Chihuahua (1961–1988).

Oaxaca terrace approximately 3 km², and the old Pedro valley approximately 2 to 3 km². The Río Casas Grandes and Río San Pedro rivers at this time were characterized by greater base flow and lower flood magnitudes, but had greater flood frequency than today. The slowly meandering stream contained well-sorted and fine-to-medium channel gravels with thin sandy point bar deposits grading up into fine-grained overbank deposits capped by the weakly developed Trincheras soil (Unit D). The floodplain and old Pedro valley were covered with warm season C_4 grasses and with clumps of trees in abandoned meander loops and along the active channel. Warm season C_4 grasses also dominated the low terrace and Oaxaca terrace.

Although no agricultural features have been located in the alluvial valleys dating to this time period, there is evidence that a mixed farming and foraging strategy was employed by the occupants (Chapters 8 and 15). Due to the valley constriction 3 km downstream from the confluence of the Río Casas Grandes and Río San Pedro, coupled with a climate somewhat more humid than today and soils conducive to agricultural production, farming potentials were probably greater 3000 radiocarbon years BP relative to earlier parts of the Holocene (Chapters 2 and 15).

Water Resources

The availability of water is the most limiting factor to agricultural production in arid to semiarid climates. It is unknown if irrigation was practiced along the Río Casas Grandes and Río San Pedro at 3000 radiocarbon years BP as the relatively small sample of backhoe trenches failed to locate canals. Modern farmers irrigate, but as late as the 1980s, they preferred to grow maize utilizing only rainfall and would irrigate only if summer rainfall was inadequate (Chapter 2). Although conditions are somewhat more arid today, variable rainfall makes it likely that water management was required for successful agricultural production during Late Archaic times as well. Using modern mean monthly temperatures and precipitation, the Thornthwaite soil water model (Dunne and Leopold 1972) demonstrates that soil water recharge occurs for three months during the winter, followed by three months of soil moisture utilization (Figure 12.11). The next six months, during the late spring, summer, and early fall, exhibit soil moisture deficits from high evaporative demand, consistent with the need for irrigation agriculture in the region today. Within this context, rainfall, flood recession, and water table agricultural potentials are discussed from a water procurement standpoint 3000 radiocarbon years BP (Table 12.5).

Rainfall Agriculture

The Thornthwaite water balance model and associated monthly precipitation and temperature data provide important baseline information for estimating soil-water-plant relations (Dunne and Leopold 1972) (Figure 12.11; Table 12.5). Most modern maize production occurs in climates with mean midsummer temperatures of 19°C

TABLE 12.5. Estimated Farming Potentials Based on Water Sources and Soil Properties for Landscapes Exposed at 3000 ^{14}C yr BP.

	Water Sources			Soil Properties											
										Fertility					
Landform	Rain fall[1,2]	Flood Water[1]	Water Table[1]	Paleosol	Horizons	Depth (cm)	Texture[3]	Reaction to HCl	Water Capacity[4]	N	P	K	Ca	Salinity	Farm Potential
Floodplain	mod	high	mod	Trincheras	A	0–20	Sicl	strong	high	low	low	mod	high	mod	Moderate
(Unit D)					Bw	20–59	Sil	strong							
Low Terrace	mod	low	low	Petra	A	0–63	Cl	strong	high	mod	mod	mod	mod	low	Moderate
(Unit C)					Bk	63–77	Cl	strong							
Old Pedro	mod	low	high	Guiermo	A	0–23	Cl	very weak	mod	low	low	mod	mod	high	Low
(Unit B)					Bg	23–48	Fsl	very weak							
High Terrace	mod	low	low	Surface	A	0–13	Fsl	none	mod	low	low	low	low	low	Low
(Unit A)					Bt	13–44	Scl	strong							
					Bk	44+	Scl	strong							

[1] mod = moderate.
[2] rainfall amount relative to modern, more arid climate.
[3] Sicl = silty clay loam; Sil = ilt loam; Cl = clay loam; Fsl = fine sandy loam; Scl = sandy clay loam.
[4] Estimated using texture parameters of Rawls (1983).

or greater with a frost-free growing season of at least 120 days (Shaw 1988). In the Cerro Juanaqueña area, the mean growing temperature at the town of Janos for the rainy season is 23°C (June to September), which is sufficiently high for maize production.

Modern maize requires approximately 400 to 600 mm of water for a 120-day growing season, although native landraces may have required as little as 150 to 250 mm for a growing season of 60 to 90 days (Muenchrath and Salvador 1995; Shaw 1988). Because of rainfall limitations, one must assume the lower limit of water requirement for native varieties for the Late Archaic people to have had any chance of practicing substantial agriculture in the Cerro Juanaqueña area. Even if the monthly average rainfall of 80 mm for July fell all at once (see Figure 12.11), this would have provided only enough water for 16 days of a modern maize growing season using evapotranspiration rates of 5 mm per day typical for arid to semiarid climates for modern maize (Muenchrath and Salvador 1995; Shaw 1988). Even with enhanced water use efficiency for native maize strains, soil water would have lasted no more than a few weeks.

In addition to rainfall, soil water-holding capacity is important for maize production, and can be estimated from standard field textures (see Rawls 1983). The Trincheras and Petra soils in the upper 1 m have high water-holding capacities (150 and 200 mm) because of loamy textures associated with only a few coarse fragments. The Guiermo and Oaxaca soils will hold only 70 to 120 mm of water because they are shallow, contain coarser textures, and have abundant coarse fragments of gravel or carbonate. No doubt the prehistoric peoples would have been vulnerable to the unpredictable timing of rainfall events, even on high water-holding soils such as the Trincheras and Petra. With respect to rainfall agriculture, low water-holding capacities of the Guiermo and Oaxaca soils would have been no more problematic than the other soils because typical monthly rainfall amounts would still have been less than their water-holding capacities. Thus, all landscapes are rated as moderate for rainfall farming potential (Table 12.5).

Flood Recession Agriculture

From a water resource standpoint, the Trincheras floodplain could have been adapted to flood recession farming during the summer rainy season (Table 12.5; Figure 12.10). Fine textures and minimal development of the Trincheras soil without intervening sand or gravel layers suggest that flooding was frequent but of low magnitude and nondestructive. A flood event covering the floodplain with 150 to 200 mm of water would have filled all of the plant-available soil pores, providing enough water for 30 to 40 days of the growing season using evapotranspiration rates for modern maize, and perhaps longer for native strains. One significant flood event prior to planting in July would have supplied the water needs for maize production for a considerable part of the growing season. However, a flood event of only 75 to 100 mm of water inputs prior to planting would have required an additional 75 to 100 mm of water later in the growing season from yet another flood. Either would be possible given the climate and stream hydrological conditions around 3000 radiocarbon years BP, but large flood events also most likely would have been destructive to the crops.

Based on the absence of Unit D flood deposits on the Petra soil on the adjacent low terrace, flood recession agriculture was unlikely to have been possible in the Unit C area. Unit D deposits were entering the old Pedro valley 3000 radiocarbon years BP, but filling swampy channels abandoned during the early Holocene. There is also no evidence that flood water ever covered the Oaxaca terrace during the Holocene. Thus, potential for flood recession agriculture would have been high in the Unit D floodplains, but low on all terrace soils (Table 12.5).

Water Table Agriculture

Water table levels at 3000 radiocarbon years BP can only be estimated by documenting the distribution of soil iron and manganese redoximorphic features indicative of anaerobic conditions (Table 12.5; Figure 12.10). However, it is not always possible to differentiate between water table conditions in the past from alterations created by more recent, or modern, water tables. Based on observations from backhoe trenches, the modern water table appears to be highest (50 to 150 cm below surface) beneath the Jorge terrace on the east side of the Río Casas Grandes and beneath the old Pedro valley. Water tables appear to be lowest (200 to over 300 cm below surface) beneath the Jorge terrace on the west side of the Río Casas Grandes and along the Río San Pedro, and at intermediate levels (150 to 200 cm below surface) in the modern floodplains. Beneath the Jorge terrace adjacent to Cerro Juanaqueña, high water table levels are generated by springs emanating from the base of the bedrock hill. In the old Pedro valley, elevated water tables are sustained from subsurface flow emanating from adjacent alluvial fans, accompanied by buried channels plugged with fine-grained sediments retaining percolating waters from rainfall events.

Assuming that modern water table levels have not influenced the distribution of redoximorphic features formed in the past, the Trincheras floodplain soil indeed experienced an elevated water table. However, it appears to have fluctuated and was not permanently at shallow depths. If so, then water table farming would have been possible in the Trincheras floodplain soil beneath Cerro Juanaqueña as long as the capillary fringe persisted at the lower depth of the primary root system. In the old Pedro valley, redoximorphic features are even more abundant in the Guiermo soil suggesting elevated water table levels for extended periods. Again, the timing and longevity of the water level would have been crucial to the production of mesophytic plants such as maize. There is no indication of high water tables within the Petra soils of the adjacent low terrace or of the Oaxaca soils. Water table levels were probably at depths of more than 2 m, out of reach of the primary root system of maize.

Soil Fertility

Nitrogen, phosphorous, potassium, and calcium are the most important macronutrients for mesophytic crops such as maize (Blackmer 2000; Olson and Sander 1988) (Table 12.5). Micronutrients in arid to semiarid soils are typically in sufficient abundance given the slow, but continuous weathering of primary minerals releasing iron and manganese (Helmke 2000; Nordt 2003). Soil nitrogen and phosphorous levels are released from organic matter decomposition and potassium from the weathering of clays (Blackmer 2000). Phosphorous can also be obtained from the mineral apatite, whereas calcium is typically derived from the weathering of calcium carbonate. Salinity is another consideration for optimal crop production. High levels commonly occur in subenvironments where saline waters enter the soil from seeps, from elevated water tables, or from irrigation waters.

Floodplain Trincheras Soil

The Trincheras floodplain soils at 3000 radiocarbon years BP had calcareous profiles with few to common iron and manganese redoximorphic features (Table 12.5; Figure 12.10). Light Munsell colors of the surface horizon were typically 7.5YR 4/2 and 3.5/2 indicating low levels of organic matter, and thus nitrogen and phosphorous. Potassium levels would have been moderate given silty clay loam and silt loam textures. Calcium levels were high because of the presence of detrital alluvial calcium carbonate (positive reaction to HCl). Salinity may have been a moderate problem due to evapotranspiration from a periodically elevated water table.

Low Terrace Petra Soil

The Petra soil formed on a low terrace adjacent to the broad Trincheras floodplain 3000 radiocarbon years BP. Because of greater landscape stability and soil development, organic carbon inputs from grasslands were greater, producing darker surface horizon colors ranging from 10YR 3/2 to 2/2 (Table 12.5). Higher organic matter levels would have generated moderate levels of nitrogen and phosphorous. Potassium and calcium levels would have also been high because of clay loam textures and high calcium carbonate content (positive reaction to HCl). However, salinity levels would have been low because of low water tables.

Old Pedro Valley Guiermo Soil

The old Pedro valley contained an array of landscapes and soils at 3000 radiocarbon years BP. Excluding the wetland soils within the abandoned early Holocene channels of the Río San Pedro, the weakly developed Guiermo soil developed as a seasonal wetland with a combination of warm season grasses and hydrophytic plants. This ecosystem produces surface horizon colors of 10YR 3.5/2 and 4/2, indicating low organic matter content and consequently low nitrogen and phosphorous levels. Potassium concentrations would have been moderate because of clay loam over fine sandy loam textures. Calcium levels were moderate because of a weak reaction to HCl and lower calcium carbonate content. Because of persistently high water tables, high salinity and water logging would have posed problems for mesophytic plant growth.

Oaxaca Terrace Soil

The Oaxaca terrace soils have sandy loam surface horizons over clay loam subsurface horizons. Light surface horizon colors (10YR 4/3 and 5/3) point to low levels of organic matter and thus nitrogen and phosphorous. Even potassium would have been low because of low clay content in the surface horizon, and calcium levels were indeed low because the upper profile had been decalcified from leaching. Salinity would not have been a problem.

Summary

Overall, the Trincheras and Petra soils at 3000 radiocarbon years BP would have had moderate potential for maize agriculture under the climate and soil conditions of the time. During the summer months, the Trincheras soil, with high water-holding capacity, received water from rainfall, flood recession, and perhaps a water table. However, nitrogen would have been in low supply and limiting to optimal plant growth without external inputs. The Petra soils would have had higher inherent nitrogen contents but were entirely dependent on rainfall as a water source because water was not available from flood recession or a water table. The Guiermo and Oaxaca soils would have had low potential for agricultural production. Coupled with lower water-holding capacities, the Guiermo soils were susceptible to a persistently high water table, salinity, and anaerobic conditions, and productivity of the Oaxaca soils was limited by very low nitrogen contents and droughtiness.

Conclusions

There is considerable evidence of human occupation in the Cerro Juanaqueña study area in the uplands, on the surfaces of ancient floodplains, or buried within the alluvial valleys. In addition to the potential discovery of buried artifacts dating back until at least the Middle Archaic, the potential for future discoveries spanning all of prehistory in the alluvial valleys is considerable because of: 1) the presence of large floodplains such as the Río San Pedro and Río Casas Grandes; 2) a preponderance of relatively low energy fluvial systems with mixed and suspended low rivers depositing thick sedimentological packages behind bedrock valley constrictions; 3) numerous buried soils or landscapes, indicating surface exposures ranging from a few centuries to thousands of years that became sealed in discrete, primary contexts or as palimpsests; 4) water availability and fertile soils available for agricultural production; and 5) presence of warm season grasses confirming that the current grassland ecosystem persisted throughout the Holocene.

It is unclear whether the alluvial histories reconstructed in the Cerro Juanaqueña study area are typical or atypical for the region. Relief and climates change rapidly across short distances, conditioned by the presence of bedrock valley constrictions. There is evidence, however, for at least some coherence in cut and fill episodes and periods of soil formation in the study area (Nordt 2003) compared to north-central Mexico (Butzer et al. 2008) and to the North American Southwest. Future studies should begin to systematically correlate major fluvial events within and across drainage basins to provide a better understanding of baseline conditions for settlement patterns, subsistence strategies, and preservation potentials for understanding prehistoric migrations and activities in southwestern North America and northwest Mexico.

Notes

1. I am indebted to Robert Hard and John Roney for their generous support of this project with funding from National Science Foundation grant number SBR-97086210. The Instituto Nacional de Anthropología e Historia graciously permitted geological excavations in the project area. Jose Zapata was instrumental in providing much needed logistical support. Parts of this manuscript were generously reviewed by Peter Allen, Stacy Atchley, Robert Hard, Gary Huckleberry, John Roney, and Michael Waters.
2. The "Guiermo" paleosol was named for Guillermo, a resident of the community of Colonia Oaxaca.

CHAPTER 13

Manos, Metates, and Grinding Slabs

Introduction

After first-time visitors catch their breath following their struggle up Cerro Juanaqueña they quickly notice the numerous, massive basin metates that are scattered about the surface. Although the metates were made of the basalt from the hill, their number and distinct shapes make them easy to spot among the jumble of stones. These numerous large, heavily worn basin metates have been important throughout the project because they indicate that substantial food processing and domestic activity occurred at the site. Basin metates are typical of the Archaic period and are not common in later periods. These metates are remarkable due to their massive size and extensive wear, as well as their accompanying manos.

This chapter presents detailed descriptions and analyses of the manos, grinding slabs, and basin metates found during the project. All other ground stone items such as mortars, pestles, abraders, handstones, and other miscellaneous artifacts are described in Chapter 6. The large sample of manos, metates, and small grinding slabs at Cerro Juanaqueña provide particularly important data sets that directly address our major research themes. When we began our work, we did not know if Cerro Juanaqueña was a seasonally reoccupied aggregation site used by bands of hunters and gatherers, or a permanent settlement of agriculturalists, or some other combination of agricultural reliance and sedentism. The ground stone tools described in this chapter became directly relevant to understanding the role of maize, the degree of sedentism, and population size. In order to address these three issues, we first had to understand the basic characteristics of these tools. This chapter addresses variability in form and function as well as the use life of these artifacts, including manufacture, use, and discard. It also refers to grinding experiments that we conducted in order to provide insights into the use of the ground stone tools.

Background

Manos and Metates in the Archaic Period

Studies of ground stone have been fundamental to understanding Archaic period adaptations in the southern Southwest, and manos and metates tend to dominate these assemblages. Large mano and metate assemblages in Archaic period sites in southern Arizona allowed early investigators to make inferences regarding the fundamental importance of plant gathering and processing in the economy (Haury 1950; Sayles 1983; Sayles and Antevs 1941). Archaic period sites yield slab metates, basin metates, and small (one-hand[1]) manos that are typically made from accessible raw materials. These artifacts exhibit minimal investment in manufacture and have long been associated with wild seed processing (J. Adams 2005; Dick 1965; Haury 1950; Sayles 1983; Sayles and Antevs 1941). These forms contrast with formal trough and slab metates utilized with large, two-hand manos that became widespread across the Southwest and northwest Mexico following the early Ceramic period and are associated with intensive maize processing (Haury 1950; Martin and Rinaldo 1947; Sayles 1983; Sayles and Antevs 1941). The changing morphology of Ceramic period ground stone tools includes increasing mano grinding surface area that is likely related to efficiency gains due to the demands of maize processing. This interpretation of increasing mano grinding surface area is referred to here as the *size efficiency hypothesis* (Diehl 1996; Hard 1990; Hard et al. 1996; Lancaster 1983, 1984; Mauldin 1993b).

Multiple investigators have observed that mano and metate forms did not change with the Middle to Late Archaic period introduction of maize, leading to the conclusion that wild seed grinding technology using basin and slab metates and small manos was suitable for processing maize at that time (J. Adams 2005; Dick 1965; Wills 1988). This view is consistent with data prior to the 1990s when it is was thought that maize contributed only a minor proportion of the preceramic diet (Haury 1983; Minnis 1985). However, with overwhelming evidence that maize played a substantial role in preceramic economies in southern Arizona, the apparent persistence of pre-maize grinding technology presents a conundrum, since one might expect changes in grinding technology related to the demands of processing substantial quantities of maize. In fact, early research in southeastern Arizona did document a series of pronounced changes in Late Archaic period ground stone tools. Convex manos, deep basin metates, and hopper-mortars[2] became common, and one-hand manos increased in size, when compared with earlier periods (Sayles 1983:71–79, Sayles and Antevs 1941:24; cf. Wills 1988:14–17). These changes continued into the early Ceramic period and were thought to be related to increased maize processing (Sayles 1983:78).

In contrast to Sayles's observations, Archaic period assemblages from other key locations, including Ventana Cave and Bat Cave, did not show changes in mano and metate morphology during the Late Archaic period (Dick 1965; Haury 1950). Morris (1990) suggested that the larger one-hand manos and deep basin metates that Sayles observed from the Late Archaic period may have been related to agricultural production. He noted that larger Late Archaic period one-hand manos are present along drainages in southeastern Arizona, locations where Late Archaic period farming was likely, but not at Ventana Cave and Picacho Reservoir in southern Arizona, settings unrelated to agricultural production (Morris 1990). Morris's hypothesis that changes in Late Archaic period manos and metates are related to maize production has not been pursued, but his argument is consistent with the ground stone size efficiency hypothesis.

In contrast, Jenny Adams saw few changes in manos and metates related to maize production. She studied multiple ground stone assemblages from Early Agricultural period sites (Middle and Late Archaic periods) in the Tucson Basin, where abundant maize production occurred (J. Adams 1998, 1999, 2001, 2005). In these assemblages the dominant food processing tools were one-hand manos, basin metates, and flat/concave metates. Metates were scarce due to recycling or scavenging, and deep basin metates rarely occurred. Adams suggested that flat/concave metates would have been the likely maize grinding tools, and she documented increasing variability in mano size. She suggested that need to produce maize flour would have been minimal until the Ceramic period and that alternative maize preparation techniques were used during the Early Agricultural period. Maize may have been processed as green corn, which requires little processing effort (J. Adams 1999, 2015a, 2015b). Her studies emphasized a life history and behavioral approach to the study of ground stone, arguing that tool design, planning, intensity, and duration of use drive ground stone morphology, and that overall trends in grinding surface size are not meaningful due to multiple intervening variables.

Problem Orientation

Given the unique assemblage of manos and metates at Cerro Juanaqueña and our interest in the role of agricultural production in the economy, we examined the relationship between mano and metate morphology and maize and wild food processing. Our focus was collecting a range of dimensional and morphological details that would allow us to describe and gain insight into the formation of the mano and metate assemblage including manufacture, use, and discard. As described below, the vast majority of the ground stone data derives from a systematic inventory of the surface artifacts.

The ground stone surface assemblage from Cerro Juanaqueña can be almost totally attributed to the San Pedro phase. The principal occupation of the site, based upon combined radiocarbon and diagnostic artifact data presented in Chapters 1, 5, and 6, falls comfortably in the San Pedro phase and dates between 1300 and 1100 cal BC. Minor reoccupation of the site occurred during the Cienega phase, between 400 cal BC and cal AD 1, based on radiocarbon dates from two lower terraces. It is unlikely that there was any substantial occupation of the site prior to the San Pedro phase. Similarly, evidence of Ceramic and Historic period use of the hill is restricted to a very few specific locations and appears to represent transitory and specialized activities (Chapter 6). Ceramic period metate and mano forms are virtually nonexistent on the site.[3] The general nature of the surface and excavated ground stone assemblage is consistent with that of the Late Archaic period. For these reasons, we assume that the assemblage has substantial chronological integrity and that it can be largely attributed to the primary occupation of the site, 1300 to 1100 cal BC.

Ground Stone Data Collection Procedures

During each of the four seasons of fieldwork we collected data on the ground stone assemblage. The data collection strategies evolved as our knowledge of the site and assemblage grew. Clearly from the outset, far more ground stone tools were visible on the surface of Cerro Juanaqueña than would be recovered from excavation, therefore the surface ground stone became an important component of the project. However, transport, storage, and curation costs made it impractical to collect numerous manos and metates—many of the metates weighed well over 50 kg. Instead, we used systematic in-field data collection strategies. In 1997 and 1998 we documented all complete ground stone artifacts encountered along two transects across the site, taking detailed notes about each tool. We used this information to design a more efficient sampling and in-field recording strategy which was implemented across the entire site in 1999 and 2000 to inventory all ground stone and other tools. Further details on the transect sample and inventory data collection procedures are in the next sections. Collection of ground stone artifacts at Cerro Juanaqueña was limited to the items found in the course of our excavations and to a small number of typical artifacts intended to supplement the detailed in-field measurements and drawings made by the archaeological team. At the other excavated sites, surface assemblages were only described informally. No ground stone artifacts were recovered during excavations at Cerro el Canelo and Cerro los Torres. However, a small number of items were recovered during excavations at Cerro Vidal. All collected ground stone artifacts were curated and more detailed observations were made in the laboratory. The raw data are reported in Appendix 13.1.

Transect Sample

During the two field seasons prior to making the 100% surface inventory described below, we needed a sample of ground stone data from Cerro Juanaqueña in order to begin to understand the variability present in the surface assemblage. Our sampling design consisted of systematic survey along two east–west transects across the site. One was along the 600 N and the other along the 680 N grid lines. We analyzed and described 100% of all complete ground stone tools on all terrace features intersected by these transect lines. We also recorded nether stone[4] fragments during this effort, but not mano fragments. In addition, ground stone on the surface of terraces selected for excavation received the same treatment as the transect-recorded ground stone. Kari Schmidt and Robert Hard made all of these ground stone observations and many tools were drawn to scale. This sample provided valuable insights into ground stone assemblage and were reported in our early publications (e.g. Hard and Roney 1999a, 1999b). These observations are referred to as the *transect sample* in this chapter. The transect sample included some detailed attributes that were not later recorded during the total inventory described below. The ground stone attributes are described in Sidebar 13.1.

Total Surface Inventory

A large portion of data in this chapter is derived from the 100% inventory of all surface tools, an effort described in more detail in Chapter 6. During the 1999 and 2000 field seasons, team members systematically walked over the entire site, terrace by terrace, mapping and inventorying all formal tools.

During this effort there was insufficient time to measure all of the many ground stone tools on the site, so we devised a systematic sampling strategy. One hundred percent of all ground stone tools (including those previously documented during the transect sample) were recorded and point-located. These were divided into two subsamples. We made comprehensive observations on 50% of all complete manos and nether stones, naming this subsample the *detailed sample*. For the other 50%, we simply recorded ground stone type (e.g., basin metate, grinding slab, etc.), calling this group the *typology-only sample*. The team recorded the point locations of all tools on the site map. In practical terms, if the first complete nether stone encountered was recorded in detail, the next whole nether stone located was only listed by type. This was readily accomplished since alternating lines on the coding sheet were marked for detailed or typology-only recording. The same procedure was used for whole manos using separate coding sheets. The detailed attributes recorded are described in Sidebar 13.1.

Fragments of manos and nether stones were recorded in a parallel fashion but only every fifth one encountered (20%) was placed in the detailed sample. The other 80% were placed in the typology-only sample since fragments yielded far less useful information than complete tools.

This systematic sampling strategy is unbiased, since sample assignment was based strictly on the sequence of encountering the artifacts. This strategy avoids the standard sampling problem of first having to identify all items in a population prior to drawing a random sample. The result was a data set which included location and typological information for all manos and nether stones, as well as detailed measurements and observations for a statistically reliable sample of these artifacts. Only a small number of representative artifacts were collected. Most of the data reported in this chapter were derived through this total surface inventory.

Excavated Sample and Surface Collections

We collected, brought to the lab, described, and curated all excavated ground stone. Virtually all came from Cerro Juanaqueña; the excavations at the three other sites, produced only 10 items from Cerro Vidal. Many items were drawn and photographed, and detailed observations were made using coding sheets. Some of the excavated ground stone from excellent contexts was subjected to pollen washes. We occasionally used a binocular microscope in the lab to examine these collections. A number of ground stone tools were collected from the surface prior to the inventory. Those items were added back into the surface inventory during data analysis. Fifty percent were randomly assigned to the detailed sample and the other 50% were added to the typology-only sample so as not to bias the data.

Sidebar 13.1. Ground Stone Attributes

The attributes we observed on the ground stone are explained below. These attributes were applied consistently to the manos, grinding slabs, basin metates, mortars, pestles, and abrading stones, that are discussed in this chapter and in Chapter 6. The observations of the attributes were consistently applied across the years of fieldwork and analysis, during both in-field analysis and during lab analysis. However, more attributes were recorded during the transect sample than during the surface inventory.

Complete versus fragment. If all three dimensions were sufficiently complete to measure a ground stone tool, it was considered complete. If not, it was a treated as a fragment. For example, if all three dimensions of a metate with a corner missing are measurable then the tool is considered complete.

Dimensions. Length, width, thickness, and other dimensions were maximums taken to nearest tenth of a centimeter. If any dimension was broken, an "x" was written before the measurement. All summaries of dimensional data were derived only from complete dimensions unless otherwise noted. For most ground stone tools, length was the longest dimension and width was measured perpendicular to length. Thickness was the vertical dimension of an item in its natural orientation. For grinding slabs and basin metates, length was defined as parallel to the reciprocating direction of mano travel and was usually the longest dimension. The grinding surface length and width on both basin metates and grinding slabs were measured as maximum dimensions of visible wear across a horizontal plane, rather than sloping grinding surfaces. Depth of wear on metates, grinding slabs, and mortars was measured from the lowest interior point perpendicular to the top of the tool.

Plan view. The terms circular, oval, irregular, triangular, cylindrical, rectangular, unknown, and other were used to describe plan view categories.

Number of ground surfaces. The number of ground surfaces were counted and indicated as one or two, etc.

End grinding. The presence of grinding on the ends of a mano was indicated as yes or no.

Material. Raw material was identified using major categories based on hand samples, and was described using terms such as basalt, rhyolite, granite, unknown, etc. "Local basalt" indicates that the material was similar in color, texture, and inclusions to the native Cerro Juanaqueña bedrock. The Cerro Juanaqueña basalt is medium grain, dark, with occasional, small vesicles and largely without inclusions (aphanitic). Further descriptions of the local basalt are found in Chapters 2, 3, and 14.

Depositional context. This observation referred to the position of an artifact on the surface with respect to features. If a tool was on the surface of a feature, the depositional context was identified as feature surface, talus slope, or wall as appropriate. Otherwise, the depositional context was classified as not associated with a feature, or other.

Wear. Grinding surface wear was classified as heavy if the shape of the grinding surface was significantly altered. If only the asperities were worn flat, then the artifact was classified as exhibiting medium wear. Light wear was indicated if only the tops of the asperities showed wear.

Notes. Team members made use of the notes column to systematically record the presence of a hole in the bottom of the metate and its dimensions. Other attributes that could be noted here include unusual wear patterns, striations, shaping, additional grinding surfaces, and so forth. If collections or drawings were made, they were indicated in this column as well.

Pecking. The presence and absence of pecking marks—presumably for resharpening or roughening the grinding faces of manos, metates, grinding slabs, and mortars—was indicated by multiple, small impact fractures of consistent size and depth. This attribute included only pecking on the grinding surfaces. During the in-field analysis this attribute was only systematically recorded during the 1997–1998 transect sample but not during the 1999–2000 surface inventory. We also consistently recorded it for artifacts analyzed in the laboratory, such as the mortars.

Striations. The presence and direction of striations was indicated by the presence of multiple, tiny grooves on the grinding faces. Striations were only systematically recorded during the transect sample and in the laboratory.

Shaping. Shaping was indicated by pecking, flake scars, or polishing. On manos and metates, it was reliably visible only on the edges where evidence of shaping had not been obliterated by use wear. Pecking on the edges of a tool was considered shaping, whereas pecking on the grinding surface was classified as "pecking" and was considered evidence of re-sharpening. Shaping was consistently observed in the laboratory and as part of the in-field transect sample study, but not during the surface inventory.

TABLE 13.1. All Surface and Excavated Manos, Grinding Slabs, and Basin Metates from Cerro Juanaqueña.

Artifact Type and Condition	Surface	Excavation
Manos		
Complete	215	13
Fragments	436	25
Grinding Slabs		
Complete	100	2
Fragments	111	3
Basin Metates		
Complete	110	4
Fragments	580	14
Other/Unknown Nether Stones		
Complete	5	
Fragments	45	8
Unknown Manos or Metates		
Fragments		6
Total	1602	75

In this chapter, data from the transect sample, surface inventory, and excavated contexts are reported separately. Unless indicated, the data from the 100% surface inventory are the focus of this chapter.

Ground Stone Descriptions and Analyses

Table 13.1 lists all manos, grinding slabs, and basin metates from the Cerro Juanaqueña surface inventory and the excavations. The sheer size of the surface assemblage is unusual, particularly for sites of this time period. There are 651 fragmentary and whole manos. Virtually all are small, and fall into the traditional one-hand mano classification. There are 690 fragmentary and whole basin metates and most of these are heavily worn. This assemblage is an obvious indication that the site was the scene of intensive food processing and domestic activity by a sizable population, and the assemblage is rich source of data. Only 38 manos and 18 basin metates were recovered from excavations. There were 211 grinding slabs from the surface and 5 from excavation, a notably lower frequency than the basin metates and manos.

The relatively small proportion of grinding tools from excavation relative to the surface inventory is consistent with other categories of tools including hammerstones, cores, projectile points, and mortars. The excellent surface visibility, high site integrity, shallow deposits, low rate of deposition, low impact of artifact collecting, and the relatively small proportion of the site that was excavated all contribute to this disparity. Since the excavations involved substantially less than 1% of the site, it is easy to imagine that thousands more ground stone tools remain buried.

Manos

Analysis of the manos focuses on three issues. The first is variability in mano form and the relationship between mano form and their use in mano-metate sets. The second is the life cycle of the manos, including raw material acquisition, manufacture, maintenance, use and discard, all of which have implications for understanding the grinding technology at Cerro Juanaqueña. The third examines mano size and its relationship with the role of maize in the subsistence base. Before pursuing these topics, we review sampling results and describe the mano assemblage.

Archaeological Contexts of Manos

Surface Manos

Recall that during the surface inventory all ground stone tools were located and identified by type. In addition, detailed measurements and observations were made on a 50% systematic sample of the complete manos, while the other 50% fell in the typology-only sample. Inspection of Table 13.2 indicates this field sampling approach was effective. There are 215 complete manos on the site, and detailed observations were made for 52% of these (112 of 215). Both in the aggregate and subdivided by type, the detailed treatment sample varies from 46% to 56%, close to the expected 50% distribution.

TABLE 13.2. Whole Mano Types in Surface Inventory.

Mano Type	Number in Detailed Sample	Number in Typology-only Sample	Total	Percentage of All Manos	Percentage in Detailed Sample
Flat	42	33	75	34.9%	56.0%
Convex	64	63	127	59.1%	50.4%
Other/Unknown	6	7	13	6.0%	46.2%
Total	112	103	215	100.0%	52.1%

Table 13.3. Surface Mano Fragments by Mano Type and Sample.

Mano Type	Number in Detailed Sample	Number in Typology-only Sample	Total	Percentage in Detailed Sample
Flat	37	112	149	24.8%
Convex	38	135	173	22.0%
Other/Unknown	21	93	114	18.4%
Total	96	340	436	22.0%

Table 13.4. Excavated Whole and Fragmentary Mano Types.

Mano Type	Complete	Fragment	Total
Flat	7	6	13
Convex	4	5	9
Unknown/Other	2	14	16
Total	13	25	38

Table 13.5. Excavated Manos by Feature Number.[1]

Feature	Count	Percent
R 1.1	3	7.9%
R 234.1	3	7.9%
R 286	1	2.6%
T 6	9	23.7%
T 167	7	18.4%
T 175	4	10.5%
T 222	2	5.3%
T 273	1	2.6%
T 290	1	2.6%
T 297	1	2.6%
T 413	1	2.6%
T 487	3	7.9%
T 537	2	5.3%
Total	38	100.0%

[1] Includes only features that contained manos.

Table 13.3 summarizes the mano fragment data. There were 436 mano fragments on the surface of the site. We recorded only location and typological data for four out of five mano fragments, while location and detailed information were only recorded for every fifth one. As expected, about 20% of each type was included in the detailed sample (18.4% to 24.8%), the balance being in the typology-only sample.

Excavated Manos

Table 13.4 summarizes the 13 complete manos and 25 fragmentary manos recovered from excavations. Ground stone collected from the surface of an excavated unit became part of the appropriate surface collection, not the excavation collection. Table 13.5 shows that a disproportionate number of manos occurred at Terraces 6, 167, and 175. These differences appear unrelated to excavated volumes at each feature. More than 50% of excavated manos came from these three features. However, these features only account for 19% of the total volume of sediment excavated from 30 features on Cerro Juanaqueña (10.1 m^3 of 52.6 m^3 total excavated volume, see Table 3.1). Thirty-one complete and fragmentary manos were from terrace excavations and seven were from rock ring excavations. Given that far larger volumes of sediment were excavated from terraces than from rock rings, this is not surprising.

Other Contexts

Recall that a minor Cienega phase reoccupation was identified at two lower terraces on Cerro Juanaqueña. One of these, Terrace 413 yielded radiocarbon dates representing both San Pedro and Cienega phase occupations (see Chapter 5). Unfortunately, only a single, unremarkable mano fragment, classified as "unknown," was recovered from the excavation and no ground stone was on the surface of Terrace 413. Terrace 415 also yielded a single Cienega phase radiocarbon date but no manos from either the surface or from excavation.

No ground stone was present in the excavations at Cerro Torres or Cerro de Canelo. However, 10 manos were recovered from Terrace 20 on Cerro Vidal, which also dates to the Cienega phase. This small assemblage is unremarkable, consisting of nine mano fragments and one complete mano. One item was from the surface while the remainder was from excavation. Five are made of rhyolite, three from basalt, and two from an unknown material. The complete mano is 13.7 by 8.9 by 7.4 cm in size, was made from basalt, and exhibits only light wear. It is oval in plan and convex in cross section. This limited assemblage is consistent with that from Cerro Juanaqueña, except that it contains a much higher proportion of manos made from rhyolite. Overall, our impression is that the ground stone seen at Cerro Vidal, as well as that which we informally inspected on the surface at Cerro Torres and Cerro de Canelo, are similar to that of Cerro Juanaqueña, although the density of ground stone tools on Cerro Juanaqueña was by far the highest (Chapter 6).

Mano Form

Manos are classified largely based on their cross sections. Traditionally used to subdivide manos into one-hand versus two-hand types, length is not particularly useful for classifying this assemblage, since virtually all of the manos are of the one-hand variety. The shape of a mano's cross section indicates the manner in which it wore against a metate and may be a useful clue as to how manos were used. We classified each mano as convex or flat based on the shape of its grinding surface. We also collected detailed dimensional data and made other observations as described in Sidebar 13.1.

Jenny Adams (2014) outlines wear patterns related to a circular stroke in a typical basin metate versus that of a reciprocal stroke that may be used in basin, trough, or slab metates. A mano used with a circular stroke in a basin metate shows use wear on the proximal[5] edge, the distal edge, and both ends. Wear facets may also become apparent on edges and ends (Adams 2002). Multidirectional or circular striations form on the mano while it is moved along the metate with a circular stroke. In contrast, a reciprocating motion will produce rounding on the proximal and distal edges. Pressure on the mano is lessened on the leading edge and increased in the trailing edge when grinding in a forward motion (J. Adams 2002; Haury 1950; Morris 1990). Through repeated use, the reciprocating mano should develop a distinctly convex working surface, and the convexity should be most pronounced in the proximal-distal cross section. Linear striations form parallel to the mano's reciprocal movement. Reciprocal motion would tend to favor a mano form that is longer than it is wide.

Convex Manos

Convex manos typically have convexity in both width (proximal-distal) and length dimensions (Figure 13.1). The degree of convexity ranges from slight to almost spheroidal. It is clear that the convex grinding surfaces articulate with the concave grinding surfaces of the basin metates; the metates appear to be concave to a similar degree in both dimensions. We suspect that convex manos were used primarily with a reciprocal motion.

Flat Manos

Flat manos have grinding surfaces that are distinctly planar (Figure 13.2). Apparently, their flat working surfaces tend not to fit in the basin metates once their grinding surfaces became steeply worn. They may have been used in a circular or reciprocal motion. In most cases, discriminating in the field between flat and convex forms was straightforward. A review of the field data revealed only a few cases of ambiguity in classification, and any errors introduced appear to be both rare and random and should have no effect on the outcome of the analysis.

TABLE 13.6. Complete Mano Size by Type from Detail Sample of Surface Inventory.

Mano Type	Length (cm)	Width (cm)	Thickness (cm)	Area (cm^2)
Large Convex *n* = 43				
Mean	12.9	10.4	4.9	136.7
Minimum	10.0	7.0	1.5	77.0
Maximum	17.5	17.5	9.5	297.5
Standard Deviation	1.80	1.85	1.73	41.64
Small Convex *n* = 21				
Mean	8.8	7.6	4.5	66.6
Minimum	7.0	6.0	2.5	42.0
Maximum	9.5	9.0	7.5	85.5
Standard Deviation	0.75	0.90	1.16	11.50
Flat *n* = 42				
Mean	12.4	10.6	4.3	133.4
Minimum	7.0	6.0	2.5	49.0
Maximum	17.5	13.5	7.5	201.5
Standard Deviation	1.90	1.75	1.02	33.82
Other/Unknown *n* = 5[1]				
Mean	8.8	7.6	2.7	75.0
Minimum	5.5	4.5	2.5	30.3
Maximum	13.0	13.0	3.0	169.0
Standard Deviation	3.09	3.45	0.25	59.6
Total *n* = 111				
Mean	11.8	9.8	4.5	119.4
Minimum	5.5	4.5	1.5	30.3
Maximum	17.5	17.5	9.5	297.5
Standard Deviation	2.42	2.14	1.42	45.30

[1] There are six items in the Other/Unknown detailed sample, but one is missing data. A total of 112 items are in the detailed sample.

Unknown or Other manos

Unknown/other manos may have an identifiable form but do not fit the convex or flat categories, or they are unidentifiable (Figure 13.2b, Item d).

Mano Assemblage Analysis

There are 215 whole manos and 436 fragments on the surface of Cerro Juanaqueña but only 13 complete and 25 fragments from excavation. At first glance the Cerro Juanaqueña manos appear to be typical one-hand manos, but in a number of significant ways they differ. Histograms (not shown) indicate that all three dimensions for the combined surface inventory sample are normally distributed. Their mean values are: 11.8 cm (length) by 9.8 cm (width) by 4.5 cm (thickness) (Table 13.6). The mean dimensions of the 13 complete excavated manos (Table 13.4) are: 11.7 cm (s = 3.9) by 9.3 cm (s = 2.4) by 5.1 cm (thickness) (s = 1.7). The surface sample and the excavated sample are similar.

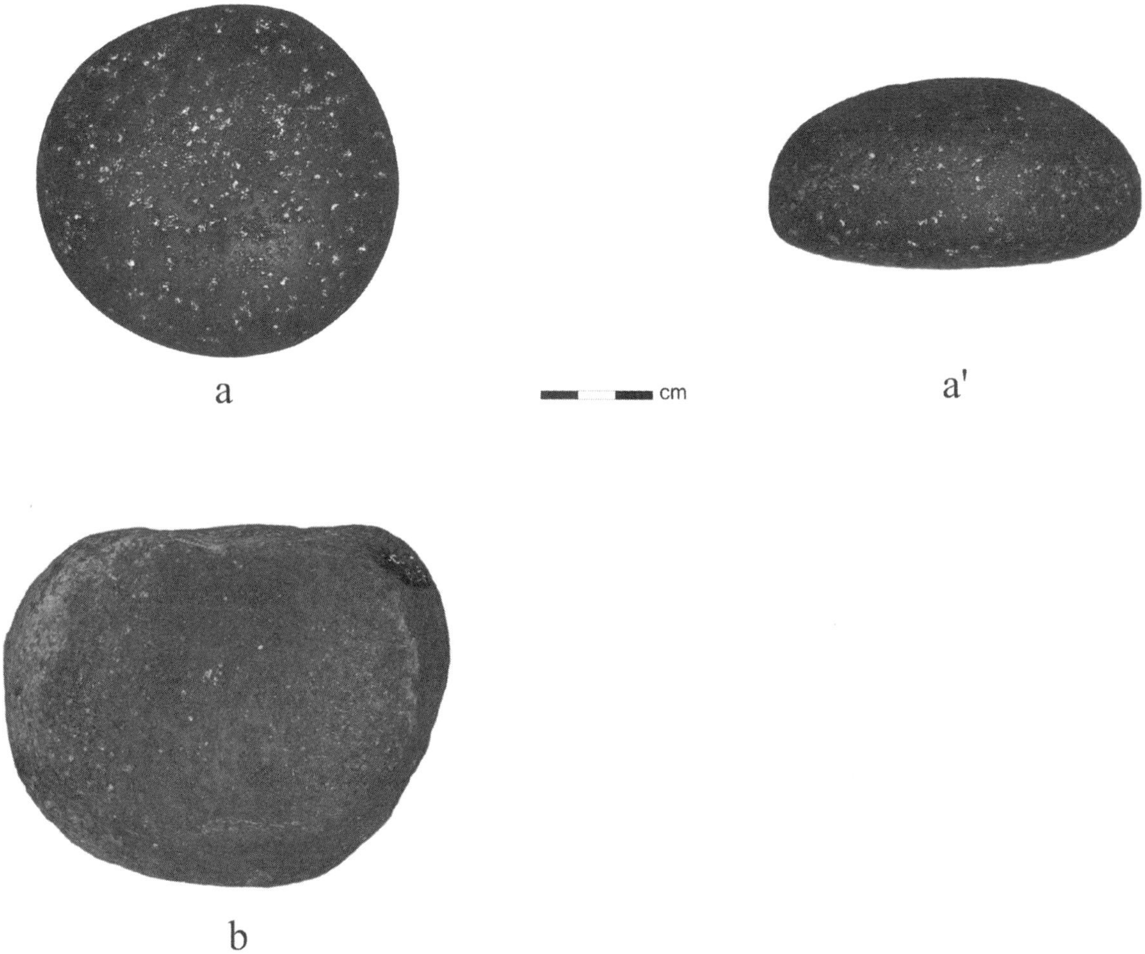

FIGURE 13.1a. Convex manos from Cerro Juanaqueña. Catalogue/bag numbers: (*a*) R51-8, (*b*) R1a-27.

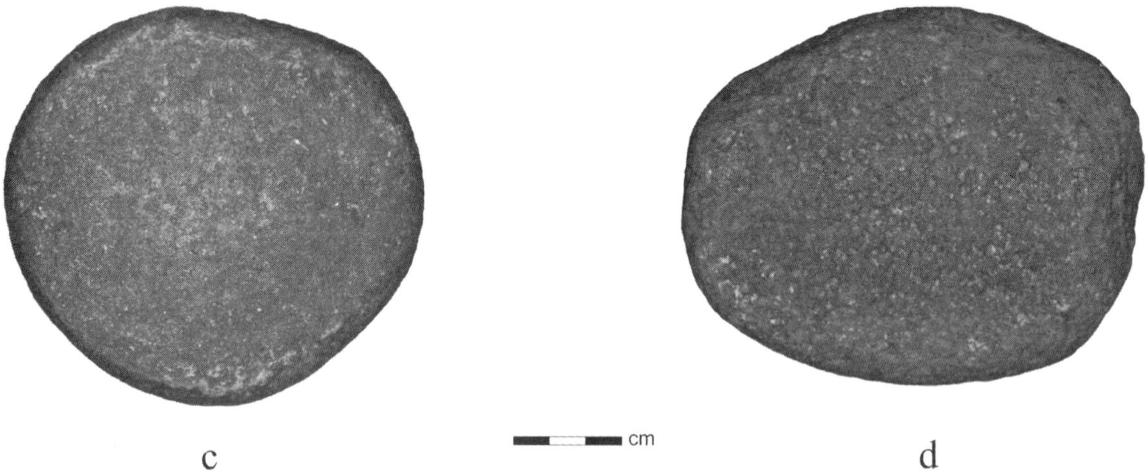

FIGURE 13.1b. Convex manos from Cerro Juanaqueña. Catalogue/bag numbers: (*c*) R234-2, (*d*) T287-61.

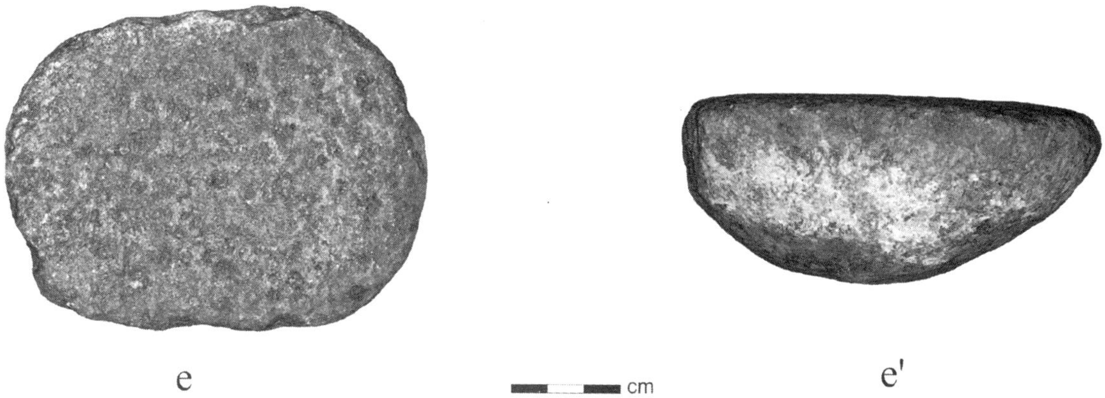

FIGURE 13.1c. Flat/convex mano from Cerro Juanaqueña. Catalogue/bag number: T290-16.

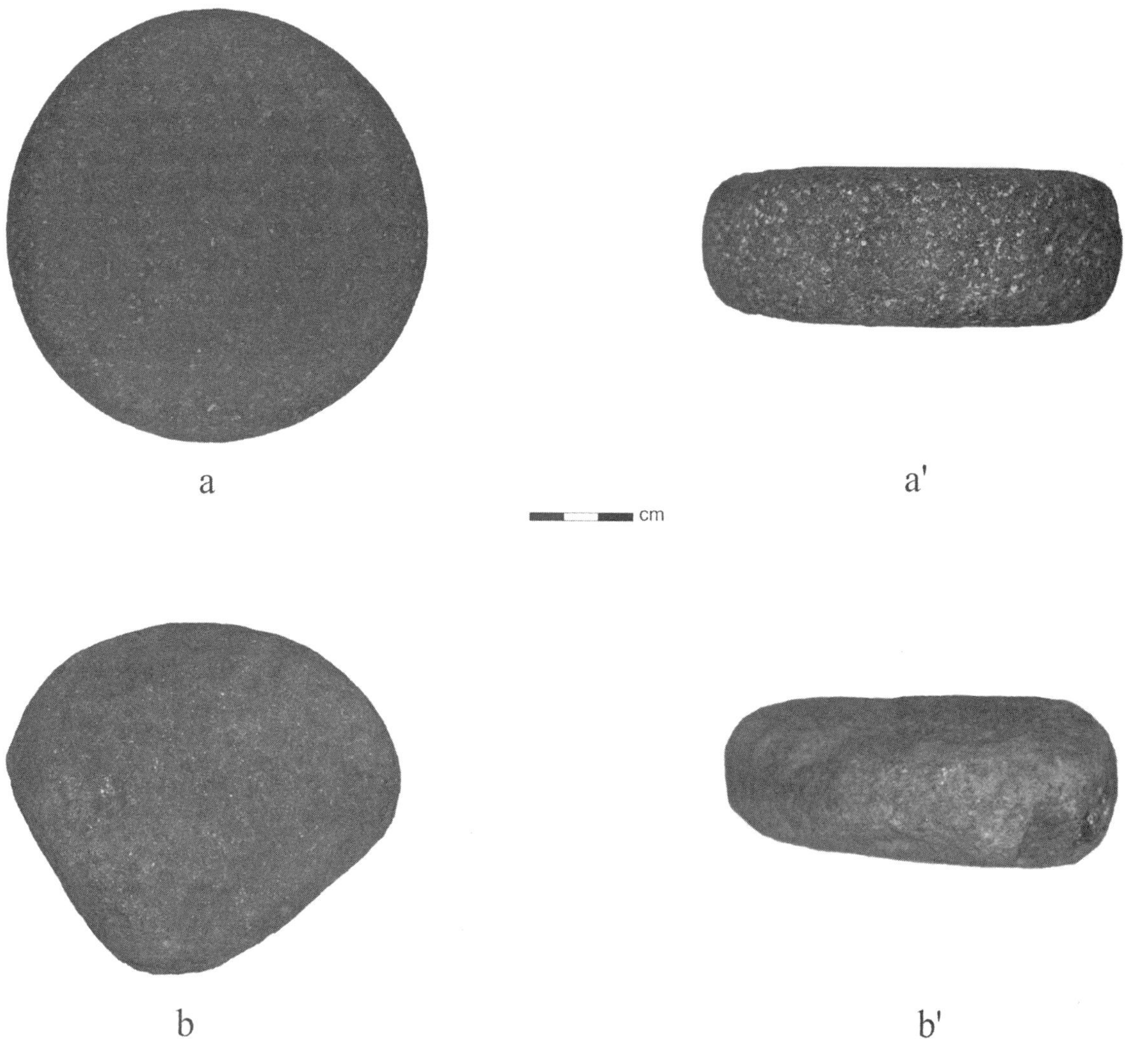

FIGURE 13.2a. Flat manos from Cerro Juanaqueña. Catalogue/bag numbers: (*a*) sc319, (*b*) T6-130.

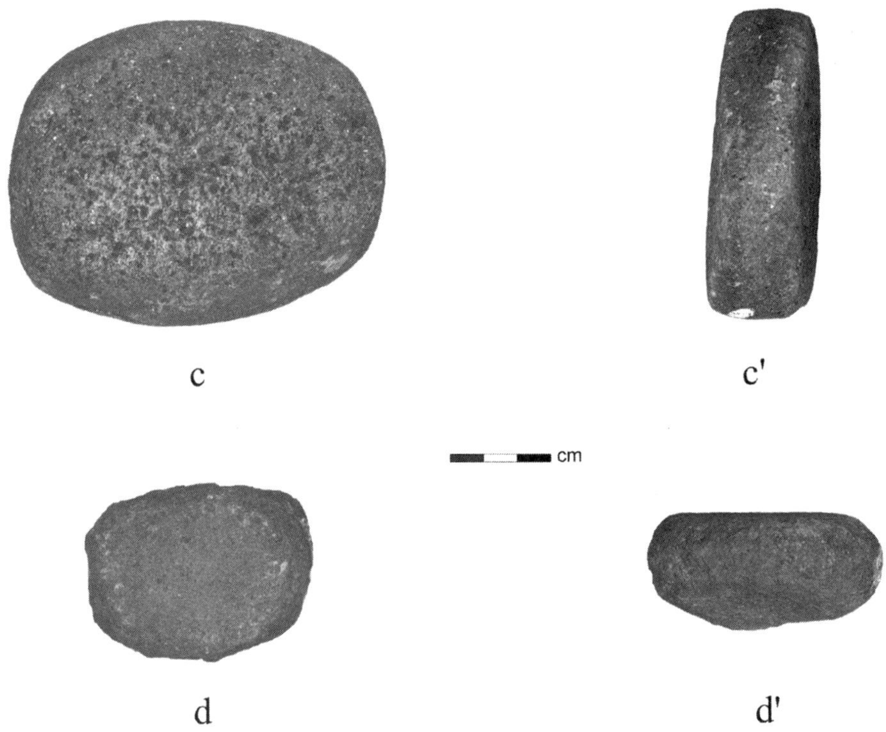

FIGURE 13.2b. Flat and "other" manos from Cerro Juanaqueña. Item c is flat, Item d is classified as "other." Catalogue/bag numbers: (*c*) T6-2, (*d*) T175-41.

The surface manos have grinding areas[6] that average 119.4 cm^2 (Table 13.6). These are larger than typical Archaic period manos in the Southwest. For example, Southern Jornada Archaic manos and those from Utah's Desha Complex are 75.0 and 90.0 cm^2. Below we offer further comparisons and suggest that the large size of manos from Cerro Juanaqueña is related to maize processing via the efficiency size hypothesis.

Within this large assemblage, there are both flat and convex forms and we have attempted to understand the significance of this variability. The other/unknown category does not bear analysis as it is a small sample of variable tools, a few of them quite small in size. Histograms of all the flat mano dimensions are normally distributed, but a histogram of the convex mano lengths has a bimodal distribution (Figure 13.3). For this reason, the convex manos were split into two subgroups, those greater than or equal to 10.0 cm and those less than 10.0 cm. The flat mano group is normally distributed; only three flat manos are less than 10.0 cm in length and these are not split out.

Table 13.6 also lists the average dimensions of the three mano types: small convex, large convex, and flat manos. Comparisons between the large convex manos and flat manos in all three dimensions show that they are similar, while the small convex manos are distinct.

Table 13.7 portrays the search for patterns while considering additional variables that may cross-cut the three mano types. For example, cross-tabulations between the three mano types and five plan view shapes (circular, irregular, oval, rectangular, unknown) show that flat circular, small convex irregular, and large convex oval forms are the most common. Considering manos with one versus two grinding surfaces, there is a roughly even division. The local basalt from the Cerro Juanaqueña hill is by far the most common material utilized and rhyolite is a distant second choice. Medium wear is the most common followed by heavy and then light wear. Grinding is present on mano ends in about half of the assemblage. Manos are typically found on terrace surfaces and talus slopes. They are rarely seen as construction elements in berms or elsewhere. The scarcity of manos and metates used as construction elements is consistent with our understanding that the site was rapidly built rather than growing through accretion. If wall constructions were built by accretion, one would expect that ground stone would be recycled into construction elements.

Do the cross tabulations depicted in Table 13.7 represent a greater degree of clustering of manos with particular suites of attributes than would be expected by chance alone? That is, are there manos with a particular combination of attributes that are unusually common and that may have served a particular function? In order

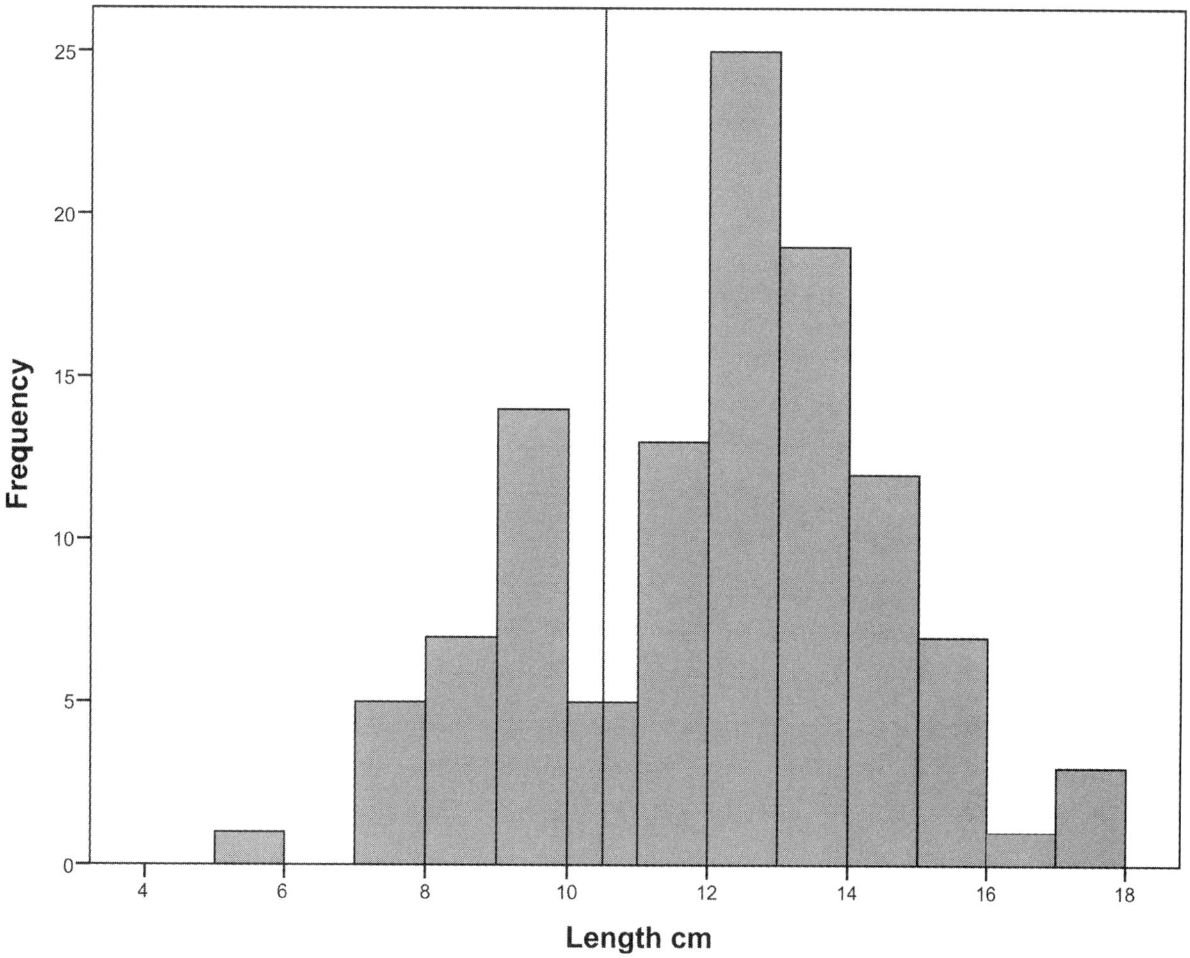

FIGURE 13.3. Histogram of convex mano length showing a bimodal distribution. Convex manos are subdivided into those equal to or greater than 10 cm and those less than 10 cm. Vertical reference line is at 10 cm.

to address this question, the results of a chi-square analysis are depicted in Table 13.8. Ten manos were removed to eliminate problems with low expected cell counts. A comparison of Tables 13.7 and 13.8 shows, for example, that items with unknown plan, three ground surfaces, and granite material are not present in Table 13.8.

The chi-square values in Table 13.8 show no significant difference among the three mano types (large convex, small convex, and flat) with respect to number of ground surfaces, wear, grinding on ends, and depositional context. The three mano groups appear to have substantial similarities. In contrast, the analysis of mano plan shows that the differences among types are significantly different than expected by chance alone (chi-square = 22.34, $df = 6$, $p = .001$). Adjusted standardized residuals (not shown) indicate that flat circular manos, irregular small convex manos, oval large convex manos, and rectangular large convex forms all occur in greater frequencies than expected. These cells are in bold. Material distribution is somewhat significant (chi-square = 5.159, $df = 2$, $p = .076$). Other variability in the mano assemblage appears to be the product of multiple interacting variables, and it is difficult to attach much importance to them. For example, the flat mano form and the large convex form are similar in size and most qualitative aspects. Most attributes do not cluster more than chance alone might predict, as shown in Table 13.8. Generally, we do not see sufficient dissimilarity among the forms to suggest that one or another has highly specialized roles. Occasionally combinations of these forms are found in a single tool. The detailed transect data show that seven out of 87 manos in this sample have both a flat and a convex grinding surface documenting that these two forms can occur as one tool. Even so, these analyses do indicate that manos with flat cross sections tend to be circular in plan and large convex manos tend to be oval in plan, and that small convex manos are commonly irregular in plan.

Further exploration of these data reveals trends and yields inferences about how these various manos with

TABLE 13.7. Mano Type and Qualitative Characteristics in Surface Detailed Sample.[1]

	Mano Type							
	Large Convex		Small Convex		Flat		Total	
	Count	Row *n*%	Count	Row *n*%	Count	Row *n*%	Count	Row *n*%
Plan								
Circular	8	24.2	5	15.2	20	60.6	33	100%
Irregular	4	21.1	10	52.6	5	26.3	19	100%
Oval	23	56.1	6	14.6	12	29.3	41	100%
Rectangular	5	55.6			4	44.4	9	100%
Unknown	3	75.0			1	25.0	4	100%
Total	43	40.6	21	19.8	42	39.6	106	100%
Number of Ground Surfaces								
One	21	42.9	11	22.4	17	34.7	49	100%
Two	21	38.2	9	16.4	25	45.5	55	100%
Three	1	100.0					1	100%
Total	43	41.0	20	19.0	42	40.0	105	100%
Material								
Local basalt	39	43.3	15	16.7	36	40.0	90	100%
Rhyolite	3	20.0	6	40.0	6	40.0	15	100%
Granite	1	100.0					1	100%
Total	43	40.6	21	19.8	42	39.6	106	100%
Wear								
Heavy	17	43.6	6	15.4	16	41.0	39	100%
Medium	18	39.1	10	21.7	18	39.1	46	100%
Light	8	40.0	5	25.0	7	35.0	20	100%
Unknown					1	100.0	1	100%
Total	43	40.6	21	19.8	42	39.6	106	100%
Grinding on End								
No	24	41.4	13	22.4	21	36.2	58	100%
Yes	19	39.6	8	16.7	21	43.8	48	100%
Total	43	40.6	21	19.8	42	39.6	106	100%
Depositional Context								
Feature surface	18	40.0	8	17.8	19	42.2	45	100%
Talus slope	23	39.7	13	22.4	22	37.9	58	100%
Unknown	1	50.0			1	50.0	2	100%
Wall	1	100.0					1	100%
Total	43	40.6%	21	19.8%	42	39.6%	106	100%

[1] Unknown/Other type is excluded.

overlapping characteristics may have been used. The small convex form was defined based on size; it has a higher than expected frequency of irregular plan form and use of rhyolite, but otherwise is similar to the other two forms. The irregular small convex form tends to be more spheroidal in overall appearance, and this shape matches basin metates that are deeply worn. As the grinding trough becomes narrower with greater depth, a smaller, more spheroidal mano form is a better fit.

The oval large convex shape is the most common mano form, as it is an efficient mate for the basin metates. Our experimental use of an oval, large convex mano on a deep basin metate demonstrates that this mano form articulates well with the concave shape of the basin metate. It is effective in reciprocal grinding in a basin metate, although it could be used in a rotary fashion as well.

Manos with flat cross sections likely have advantages in metates that are not deeply worn, where both a reciprocal and circular motion may be more readily

Table 13.8. Chi-square (χ^2) Analysis of Mano Characteristics.[1]

	Mano Type							
	Large Convex		Small Convex		Flat		Total	
	Count	Row *n* %	Count	Row *n* %	Count	Row *n* %	Count	Row *n* %
Plan								
Circular	8	26.7%	4	13.3%	**18**	**60.0%**	30	100.0%
Irregular	4	21.1%	**10**	**52.6%**	5	26.3%	19	100.0%
Oval	**21**	**53.8%**	6	15.4%	12	30.8%	39	100.0%
Rectangular	**4**	**50.0%**			4	50.0%	8	100.0%
Total	37	38.5%	20	20.8%	39	40.6%	96	100.0%
		$\chi^2 = 22.34$, $df = 6$, $p = 0.001$						
No Ground Surfaces								
One	20	42.6%	11	23.4%	16	34.0%	47	100.0%
Two	17	34.7%	9	18.4%	23	46.9%	49	100.0%
Total	37	38.5%	20	20.8%	39	40.6%	96	100.0%
		$\chi^2 = 1.659$, $df = 2$, $p = 0.436$						
Material								
Local basalt	34	41.5%	**14**	**17.1%**	34	41.5%	82	100.0%
Rhyolite	3	21.4%	**6**	**42.9%**	5	35.7%	14	100.0%
Total	37	38.5%	20	20.8%	39	40.6%	96	100.0%
		$\chi^2 = 5.159$ $df = 2$, $p = 0.076$						
Wear								
Heavy	14	42.4%	5	15.2%	14	42.4%	33	100.0%
Medium	16	36.4%	10	22.7%	18	40.9%	44	100.0%
Light	7	36.8%	5	26.3%	7	36.8%	19	100.0%
Total	37	38.5%	20	20.8%	39	40.6%	96	100.0%
		$\chi^2 = 1.153$ $df = 4$, $p = 0.886$						
Grinding on End								
No	21	38.2%	13	23.6%	21	38.2%	55	100.0%
Yes	16	39.0%	7	17.1%	18	43.9%	41	100.0%
Total	37	38.5%	20	20.8%	39	40.6%	96	100.0%
		$\chi^2 = 0.679$ $df = 2$, $p = 0.712$						
Depositional Context								
Feature surf	16	40.0%	7	17.5%	17	42.5%	40	100.0%
Talus slope	21	37.5%	13	23.2%	22	39.3%	56	100.0%
Total	37	38.5%	20	20.8%	39	40.6%	96	100.0%
		$\chi^2 = 0.463$ $df = 2$, $p = 0.793$						

[1] The cells in bold have significantly higher or lower counts than expected based on adjusted standardized residuals (not shown).

used. A subset of the flat circular manos stood out in that they are carefully shaped into a symmetrical circular plan using the local basalt. Pecking is present around the edges indicating a greater level of effort to craft them than most other manos. They have heavy wear, typically on both surfaces, and the edges tend to exhibit wear as well. These "discoid" manos are not common. Only seven out of 106 manos in the detailed surface sample fit this description. It is not clear if the discoid manos are simply a formal variant of the flat forms or a tool with a special role. We used one in our experimental trials, and it was particularly effective for the initial crushing of maize in a basin metate but not a good fit for grinding in these metates.

We suspect that flat, small convex and large convex forms were part of a household's routine inventory of grinding tools used to crush and grind seeds and maize kernels. The obtuse corners on the edges of the flat, circular forms are particularly effective at crushing. Both the flat and convex forms may be used against the sidewall of the deeply worn basin metate, rather than only along the bottom of the metate. But flat manos tend not to fit in the bottoms of the deeply worn basin metates. Small convex forms are likely most effective in the most deeply worn metates where the grinding basin becomes most constricted as it approaches the apex of a "V" shape.

Perhaps as a basin metate went through its use life, starting as a shallow basin and growing deeper with accumulated wear, the concave surfaces of the grinding basin changed. Evolving grinding basin shape would have required shifts in mano morphology. Although given that maize and seeds can be effectively ground on the interior sides of metates and that larger mano forms would have been more efficient than smaller forms, we think it is unlikely that a grinder would exclusively use one mano form. Instead we see these three forms as components of a grinding kit that a grinder could shift among as conditions warrant.

Mano Manufacture, Use, and Discard

Most of the manos were manufactured from the regolith basalt that covers the hill, referred to here as "local basalt." Manufacture, for the most part, involved minimal shaping, and during use little effort was expended in re-sharpening the vesicular basalt or managing wear. Most manos that were abandoned do not appear to have been exhausted. However, the moderate level of wear that they do exhibit indicates that they were likely used for years before being abandoned.

Eighty-five percent (90 of 106, Table 13.7) of the Cerro Juanaqueña manos were apparently manufactured from the local basalt. Rhyolite makes up all but one of the remaining 15.1% of manos (Table 13.7), and rhyolite is widely available off of the hill, particularly in nearby channel gravels. The occupants selected small cobbles that approximated a mano shape, as the majority of the tools are not formally shaped. Evidence of shaping is visible on 37% of the transect sample of complete manos (32 of 87), indicating that shaping was sometimes performed but was not the dominant strategy. There does not appear to have been any substantial effort to manage the wear of these manos (J. Adams 2005). Re-sharpening by pecking is clearly visible on only 43% of the manos in the transect sample (37 of 87), however the somewhat vesicular basalt does not require substantial re-sharpening. Based on the overall shape of the manos, they were used in a reciprocal fashion in the elongated basin metates, but rotary motion was not excluded. Striations are only visible on 10% (9 of 87) of the manos in this sample, probably due to the hardness of the basalt. However, for seven of the nine examples the striations are parallel to the width of the mano. These striations indicate that the manos were reciprocated back and forth, rather than being used in a circular motion.

Most manos are not heavily worn, as 62% (66 of 106) have only medium or light wear. The manos are evenly divided between those that only have wear on one side (46.7%, 49 of 105, Table 13.7) versus those with wear on two or more sides (53.3%). These patterns are in sharp contrast to the heavily worn and maintained basin metates discussed in the next section. Most manos were not exhausted when they entered the archaeological record. Instead, they were abandoned or lost. Clearly, the manos had far shorter use lives than the basin metates. This pattern of somewhat expedient manos also contrasts with that seen on many long-term Ceramic period agricultural sites, where manos show substantial wear management, tend to be heavily worn on both faces, and are often shaped extensively. Instead, manos on Cerro Juanaqueña were largely unshaped tools, made from the small cobbles that are ubiquitous on the surface of Cerro Juanaqueña with little or no modification prior to use.

The contrast between the extremely curated basin metates and the somewhat expedient manos leads us to explore the frequency of surface manos relative to metates. While we have a count of all whole manos on the surface ($n = 215$), we do not know how many complete manos the fragments represent. We can estimate a minimum number of complete manos represented by the fragments by first calculating the average grinding surface area of each mano fragment. This exercise yields a result of 59 cm^2 ($n = 92$, $s = 32.86$). The mean grinding surface area of the complete manos is 119.4 cm^2. Therefore, it takes 2.0 mano fragments to be equivalent to the area of one complete mano (119.4 cm^2 versus 59.0 cm^2). Since there are 436 mano fragments on the site (Table 13.1) these fragments are equivalent to 218 complete manos. Summing 218 estimated manos with 215 actual complete manos indicates that there is at minimum the equivalent of 433 manos on the surface at this site (Hard et al. 1996). In a later section, we use a similar calculation to consider the complete and fragmentary basin metates, estimating that there are a minimum of 202 basin metates on the surface of Cerro Juanaqueña. There are only about twice as many manos on the site as metates (433 manos versus 202 metates). This number is far below our expectations, given the long use life of metates. Later we also estimate that the deep metates have approximately a 25-year use life. Mano use life should be far shorter with higher rates of loss, breakage, and wear relative to metates. Further, households should have multiple manos for each metate. If all manos and metates were being abandoned on the site, we would expect a far higher ratio of manos to metates. However, these data suggest there are far fewer manos relative to metates than expected on the site. Manos likely were being removed from the site while metates generally were not. We suspect that much of the removal of manos was at the hands of the inhabitants, given the high frequencies of other artifacts on the site and the relatively minor evidence of later use of the site.

One possibility is that the short use life and relatively low frequency of manos relative to metates is related to mobility. If families were arriving and departing the site, this movement would result in a high rate of removal, loss, and abandonment of manos. Perhaps when a family departed the site they took manos with them, and therefore manos, heavily worn or not, were not accumulating on the site. When families arrived without manos they quickly manufactured one using the local basalt. In contrast, the massive metates are permanent and valued fixtures at the site. While we argue that the site was used through much of the year, some abandonment and return is expected and the mano assemblage may be an indicator of that.

Mano Grinding Surface Area in the Southwest

Previous studies suggest that mano grinding surface area and its derivatives are related to the role of maize in the economy, and we refer to this as the size efficiency hypothesis (see Diehl 1996, Hard 1990; Hard et al. 1996; Lancaster 1983, 1984; Mauldin 1993b, 1995). A scaling law from the physics of comminution shows that grinding capacity (amount ground in a given time) increases as a cubic function of grinding surface length (Hard et al. 1996). As originally presented, multiple variables affect the morphology of manos and metates but the scaling law suggests that size efficiency relationships are of fundamental importance (Hard et al. 1996). This size efficiency hypothesis suggests that in the aggregate manos and metates will reflect meaningful patterns related to efficiency despite substantial variability in tool function, design, raw material, processing recipes, and morphology. The goal of this approach is not to assign a particular function to a particular tool, but rather to use aggregated data to evaluate general trends. Although the size efficiency hypothesis has a strong theoretical and empirical basis, the relationship is not expected to be linear because multiple variables affect ground stone morphology (Hard et al. 1996). For example, use of maize races with a floury endosperm after ca. AD 500 had an important effect on ground stone morphology (J. Adams 1999; Diehl 1996).

Mabry (2005a) and Diehl (1997) point out the need for efficient maize processing technology. If maize processing costs are too high, then the net return rate of maize drops below that of most other wild plants, making it less likely that maize will be adopted, given the assumptions of human behavioral ecology. Other aspects of the efficiency arguments are developed elsewhere (Hard et al. 1996).

Jenny Adams (1998:416, 1999, 2001:118) has challenged this view suggesting that the scheme does not consider many intervening variables and that dimensional data alone do not capture functional variability. Hard and his colleagues (1996) initially stated that multiple variables affect mano and metate morphology. The principal position is that the central tendency of mano area of an aggregated assemblage is related to agricultural dependence, despite many other variables (Hard et al. 1996). The argument does not attempt to assign function to any particular grinding tool nor does it suggest one-hand manos with seed processing and two-hand manos with maize processing as J. Adams contends (1998:416). In fact Adams's own aggregated mano length data from southern Arizona (1999:Figure 9) shows that mano length is smaller in preagricultural sites than in Early Agricultural period sites (Preagricultural, $n = 86$, median = 10.6 cm; Early Agricultural, $n = 105$, median = 12.2 cm). These observations are consistent with the size efficiency hypothesis and with Morris's (1990) observations.

For a second comparison we integrated both the Cerro Juanaqueña mano area and maize ubiquity data into existing comparisons of multiple Southwestern regions (Hard et al. 1996) focusing on preceramic contexts. Area is preferred to length as it tends to be a more sensitive metric. Cerro Juanaqueña mean mano area of the surface inventory is 119.4 cm^2 (Table 13.6) and maize ubiquity is 51% (Chapter 8). Figure 13.4 is a scattergram of maize ubiquity and mano area for Cerro Juanaqueña and four other preceramic Southwestern assemblages (Table 13.9). Note that contexts with little or no maize to the left of the vertical reference line, including the Southern Jornada Archaic period and Southern Utah Desha Complex Archaic, have manos that are substantially smaller than locations where maize is a major element of the subsistence base, including Cerro Juanaqueña, Black Mesa Basketmaker II, and Cedar Mesa Basketmaker II. Presumably, the occupants at all of these locations were using hard, flinty maize as they all date prior to the appearance of flour maize. The overall trends in these two independent data sets suggest that the central tendency of aggregated mano size does correlate with the presence of substantial maize in the subsistence base. These increases are likely related to the efficiency demands of processing hard, flinty maize. There is no question that each assemblage has many unique characteristics, but the overall trends are of interest as well (cf. J. Adams 1998). We encourage future work on dating the appearance of flour maize, further grinding experiments, and particularly systematic

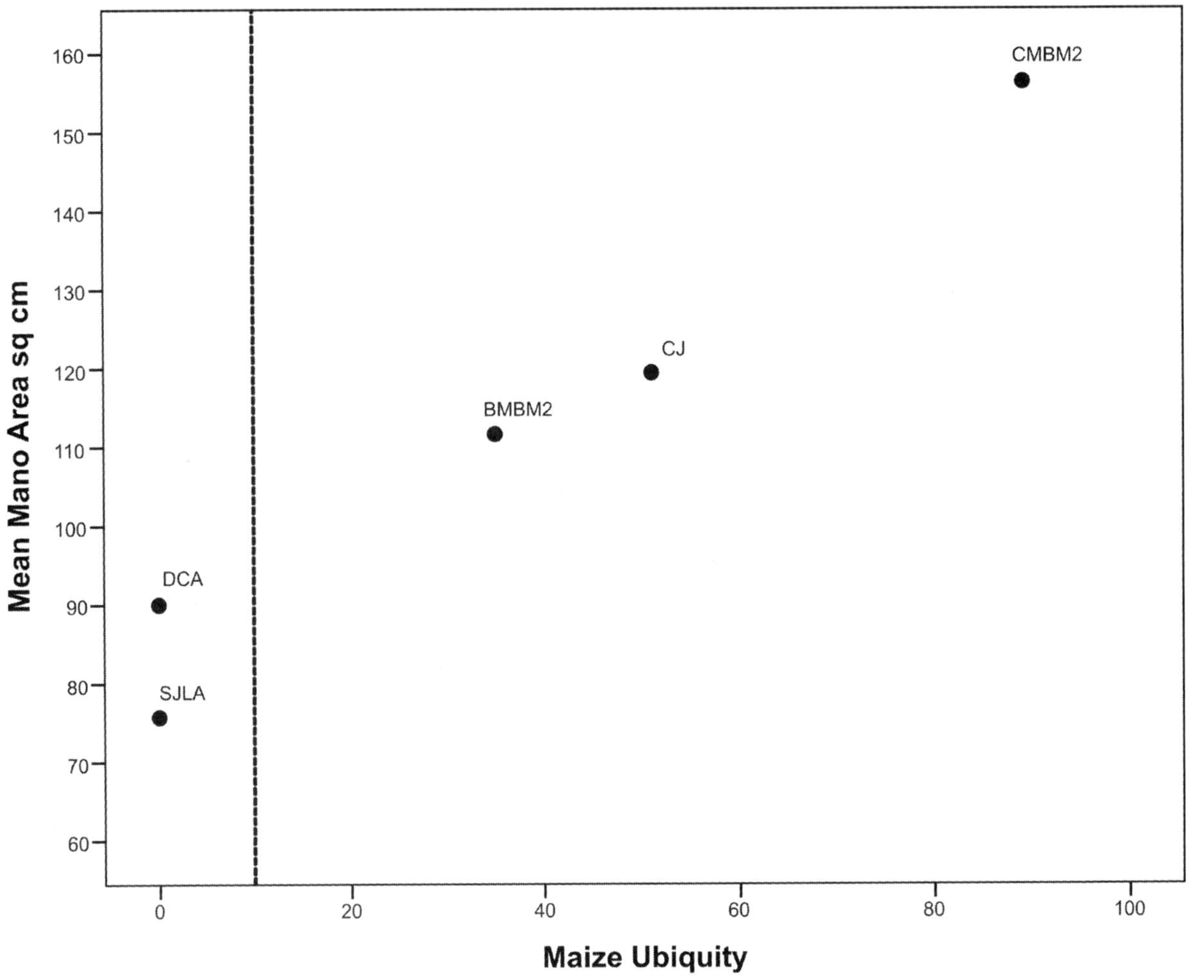

FIGURE 13.4. Scattergram of maize ubiquity and mano area from Cerro Juanaqueña and four other Southwestern Preceramic assemblages. BMBM2 = Black Mesa Basketmaker II; CJ = Cerro Juanaqueña; CMBM2 = Cedar Mesa Basketmaker II; DCA = Southern Utah Desha Complex Archaic; SJLA = Southern Jornada Archaic period. Details are in Table 13.9 and Hard et al. (1996). Vertical line is at maize ubiquity of 10% for reference.

presentation and analysis of dimensional data in order to advance our understanding of how the adoption of maize affected ground stone technology.

Nether Stones: Grinding Slabs and Basin Metates

The basin metates were easy to identify during the surface inventory due to their distinctive elongated, deeply worn grinding surfaces and large size. In all but a few cases, the grinding slabs were easily distinguished from the basin metates, since they have flat grinding surfaces and are far smaller. The basin metates were the primary lower stone for food processing. The grinding slabs are more difficult to understand in that their size and wear, among other characteristics, are inconsistent with routine processing of staple foods. We first present an analysis of the grinding slabs and follow with a discussion of the basin metates.

TABLE 13.9. Mean Mano Area and Maize Ubiquity for Cerro Juanaqueña and Four Other Preceramic Assemblages across the Southwest.[1]

Cultural Component	Abbrev	Mean Mano Area cm²	Maize Ubiquity %
Black Mesa BMII	BMBM2	111.6	35
Desha Complex Archaic	DCA	90.0	0
Cedar Mesa BMII	CMBM2	156.3	89
Cerro Juanaqueña	CJ	119.4	51
Southern Jornada Archaic	SJLA	75.7	0

[1] Details of data are in Hard et al. 1996.

FIGURE 13.5a. Grinding slabs from Cerro Juanaqueña. Catalogue/bag numbers and dimensions: (*a*) T175-12 is 31 cm long by 24 cm wide by 10 cm thick, and the grinding surface is 19 cm long by 12 cm wide. (*b*) T6-99 is 23.4 cm long by 17.4 cm wide by 7.3 cm thick, and the grinding surface is 18.7 cm long by 12.0 cm wide.

Grinding Slabs

The grinding slabs are small stones with a flat grinding surface made from the local basalt (Figure 13.5). They typically have only light to moderate wear and little or no evidence of shaping. They are far too small to be classified as slab metates and were almost certainly not involved with routine vegetal food processing.[7] They are similar to tools identified by others as abrading slabs, grinding slabs, nether stones, lapstones, proto-palettes, paint grinding slabs, and small metate-like grinding stones (J. Adams 1998:385, 394–396; Martin and Rinaldo 1947:332–333; Sayles 1983; Sayles and Antevs 1941:Plate IX). The term "grinding slab" is the best fit for these small, flat nether stones as it carries minimal functional implications.

These data were gathered as described earlier from both the surface inventory and the transect samples. During the in-field data collection, classification of a

FIGURE 13.5b. Grinding slabs from Cerro Juanaqueña. Catalogue/bag numbers and dimensions: (*c*) T175-66 is 22.1 cm long by 17.9 cm wide by 7.2 cm thick, and the grinding surface is 16.5 cm long by 10.8 cm wide. This grinding slab was used for grinding experiments. (*d*) T487-171 is 23.5 cm long by 12.5 cm wide by 8.1 cm thick, and the grinding surface is 14.9 cm long by 7.4 cm wide.

TABLE 13.10. Dimensions of Complete Basin Metates and Grinding Slabs from Detailed Surface Sample.[1]

	Basin Metates			Grinding Slabs		
	Length	Width	Thickness	Length	Width	Thickness
Mean	64.7	44.5	18.6	26.3	19.1	7.8
Maximum	94.0	67.0	50.0	54.0	34.0	16.0
Minimum	32.0	17.0	9.5	14.0	10.5	2.5
Std Deviation	15.3	10.2	8.3	7.6	5.3	3.2
Sample *n*	49	48	47	47	47	48

[1] The sample *n*'s vary slightly among dimensions as there were cases in which tools were partially buried so all dimensions could not be measured. All dimensions approximate a normal distribution.

TABLE 13.11. Basin Metate and Slab Grinding Surface Area Dimensions (cm).

	Basin Metate			Grinding slab		
	Grinding surface length	Grinding surface width	Depth of wear[1]	Grinding surface length	Grinding surface width	Depth of wear
Mean	42.5	24.6	7.97	19.6	12.9	0.06
Maximum	78.0	54.0	14.5	46.0	22.0	1.0
Minimum	15.0	8.0	0.0	5.0	5.0	0.0
Std Deviation	12.1	8.7	3.98	7.2	4.5	0.20
Sample *n*	*n* = 49	*n* = 48	*n* = 48	*n* = 47	*n* = 47	*n* = 48

[1] Depth of wear is bimodally distributed, see text.

tool as a grinding slab or as basin metate was rarely ambiguous, although it did occur. A tool with a flat grinding surface and without a concavity was classified as a grinding slab. The classification of a tool with a slight concavity raises potential ambiguity as to whether it should be classified as a basin metate or a grinding slab (J. Adams 1998:363). During the analysis it became clear that grinding slabs are about half the overall size of basin metates. Reviewing the length, width, thickness, and depth of wear (degree of concavity), data suggests that misclassifications were rare. We recognized five cases of potential ambiguity on the field forms based on patterning of the overall dimensions and grinding surface shapes. The field crews classified three items as basin metates that likely should have been identified as grinding slabs and two grinding slabs that should have been classified as basin metates. These five artifacts were reclassified during analysis. In any case, these issues do not affect the outcome of the analysis.

We recorded 100 complete grinding slabs on the surface of Cerro Juanaqueña during the surface inventory and another 111 grinding slab fragments. Only two complete slabs and three fragments were found during excavation (Table 13.1). Out of 100 grinding slabs, 48 were measured and 52 were listed by type only based on the 50% systematic sampling design.

The overall size of grinding slabs is small, averaging only 26.3 cm long by 19.1 cm wide by 7.8 cm thick (Table 13.10). The grinding slabs are too small to accommodate the movement of a typical 10.0 cm long mano while grinding a product. During our informal grinding experiments we found that maize or seeds are readily pushed off the slab's surface. The grinding slabs likely were used with the unshaped abraders or handstones discussed in Chapter 6, rather than manos. A substantial proportion of the potential working area on most slab metates was not used. Table 13.11 lists grinding surface dimensions. The utilized grinding surface area of the slabs is small (mean = 19.6 cm long by 12.9 cm wide), making it even less likely that manos were routinely used with grinding slabs.

The grinding slabs are expedient tools made from the Cerro Juanaqueña basalt. Among the 48 whole grinding slabs with detailed data, 45 are made from the local basalt, one is rhyolite, and two are basalt that is apparently not from Cerro Juanaqueña. The grinding slabs had relatively brief use lives. Half (24 of 48) of the grinding slabs have only light wear, 27% (13 of 48) have medium wear, and only 23% (11 of 48) have heavy wear. Note also the relatively low number of grinding slab fragments. There are only 111 fragments relative to 100 complete grinding slabs. These tools received so little use and maintenance that a far lower percentage were discarded due to breakage than other ground stone tools.

During the transect survey we recorded further details regarding manufacture and maintenance of 25 grinding slabs. None of these grinding slabs showed evidence of either manufacture or maintenance. None showed evidence of shaping or pecking on the edges or grinding surface. Finally, all 25 were only used on one side. These characteristics are consistent with expedient tools that were manufactured with little planning, received little maintenance, and were readily discarded when no longer needed.

In summary, the grinding slabs were not used for routine food processing. They are expedient tools which overlap with tools identified in the literature as abrading slabs, grinding slabs, nether stones, lapstones, proto-palettes, paint grinding slabs, and small metate-like grinding stones. Small flat stones were selected for immediate grinding tasks and were readily abandoned. Unlike the basin metates, the slabs are highly portable. One possibility is that the grinding slabs were used to grind ocher or some other material used in small quantities, although no residue could be identified on the grinding slabs despite searching. One excavated handstone discussed in Chapter 6 did have red ocher residue. Multiple Early Agricultural period ground stone tools including tools similar to grinding slabs, as well as formally shaped small stone mortars which were used to grind red ocher in the Tucson Basin (J. Adams 2005).

Basin Metates

The deeply worn basin metates are the most striking aspect of the Cerro Juanaqueña ground stone assemblage. Given their sheer size and quantity, the in-field data collection methods were critical to examining these metates. This approach to the large number of tools on the surface provided unique opportunities to address issues of form and function as well as subsistence, sedentism, and population size at Cerro Juanaqueña.

Of the 110 complete basin metates on the surface of Cerro Juanaqueña, 49 were measured in the detailed sample and 61 fell in the typology-only sample. Three of these 61 were collected because they had unusual characteristics, and for this reason we added them back into the typology-only sample, rather than the detailed sample. Including them in the typology-only data set prevented them from biasing the results of the detailed sample. The detailed sample included 44.5% of the 110 basin metates, close to the expected 50%. Another 580 basin metate fragments were also counted, yielding a total of 690 basin metates visible on the site's surface (Table 13.1). This quantity is striking by any standard and is testimony to the intensity of food processing and domestic activities that occurred at Cerro Juanaqueña. Excavations yielded only three complete and 14 basin metate fragments. The following discussions will primarily rely on the surface inventory data, although we will return to a few of the detailed observations recorded during the transect sample.

These tools are massive. With average dimensions of 64.7 cm by 44.5 cm by 18.6 cm thick (Table 13.10), they are larger than most Southwestern metates and they are very heavy. We weighed two of them: one was 40.8 kg (sc735, Terrace 320), the other 64.4 kg (sc384, Terrace 10). Lifting one of these metates is difficult. A vaquero who assisted us tied a single metate between two pack horses to get it down the steep hill! Examples of these metates are illustrated in Figure 13.6.

The basin metates are heavily worn, are made from the local basalt, and had long use lives. All but one of 49 basin metates for which we have detailed data were made from the local Cerro Juanaqueña basalt and the single exception was made from rhyolite. Table 13.11 lists the grinding surface length, width and depth. Note that the grinding surface dimensions average 42.5 cm by 24.6 cm, forming a distinctly elongate grinding surface. The overall dimensions of the metates are 15 to 20 cm longer and wider than the grinding surface. The users could have substantially trimmed the metates to reduce the overall size and mass but did not, indicating that these tools were designed as site furniture, not as portable tools.

The outlines of the basin metates approximate an uneven, rectangular solid. Some of the local stones chosen for metate manufacture had one or more naturally straight sides. Some of the basin metates received minimal trimming and shaping to achieve this rough rectangular form. Large flake scars and sometimes pecking were present on the edges of 32.1% (9 of 28) of whole basin metates recorded during the detailed transect surface sample. This shaping produced roughly vertical sides. At times one side exhibits some trimming while another side is irregular with natural protrusions. The upper corners of the basin metates show occasional rounding by pecking as well. Clearly, the effort in manufacture was limited to that necessary to obtain a rough rectangular shape. Absent is any attempt to obtain symmetry or smooth the exterior surfaces, as seen in the Cerro Juanaqueña small mortars (Chapter 6) or in metates from some late Ceramic period contexts elsewhere in the Southwest and northwest Mexico.

We also suspect that the initial production of a metate included a slight concave grinding surface. Our maize grinding experiments (see Appendix 13.2) clearly showed that even a small concavity on the surface helps to control the product while it is being crushed and ground. A shallow basin could be pecked out during the

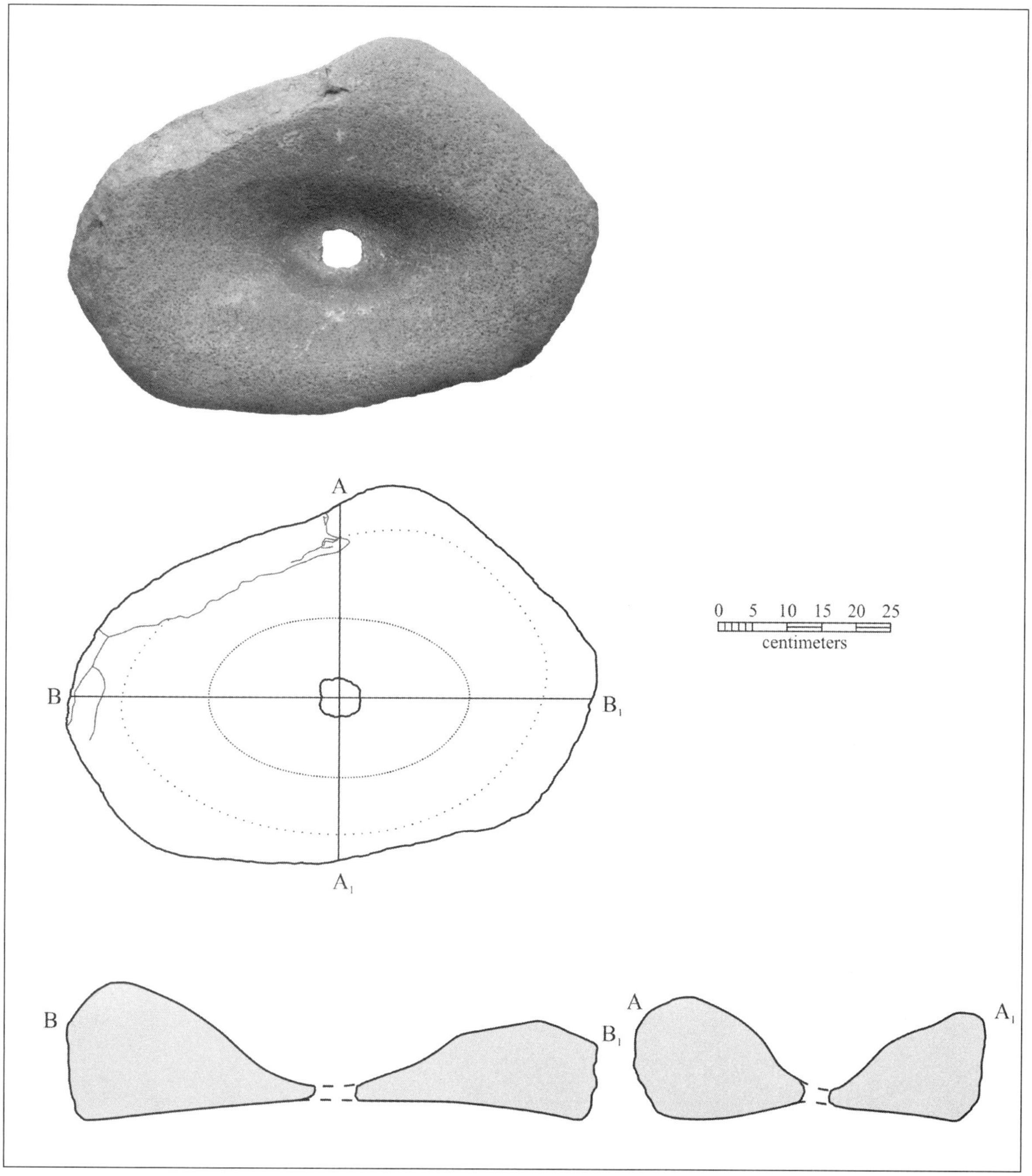

FIGURE 13.6a. Basin metate from Cerro Juanaqueña. Note opening from reuse as a rotary quern. Catalogue/bag number: (*a*) T175-31.

manufacturing process, or a stone could be selected for a metate that already had a natural concavity. Below we use the depth of metate wear to estimate total grinding effort, and in that analysis we assume metates start their use life with 1 cm of wear.

The regolith basalt is slightly vesicular, apparently reducing the need for pecking or re-sharpening. Pecking is visible on 28.6% (8 of 28 from transect sample) of whole basin metates. One hundred percent of basin metates have heavy wear, but this is true by definition since basin metates have a concavity produced by wear. Basin metates typically only show use on a single side. Only three out of 28 (10.7%) whole basin metates show grinding on the second side.

Basin Metate Wear Patterns

The shape of the concave grinding surfaces helps to understand the manner in which they wear. The wear

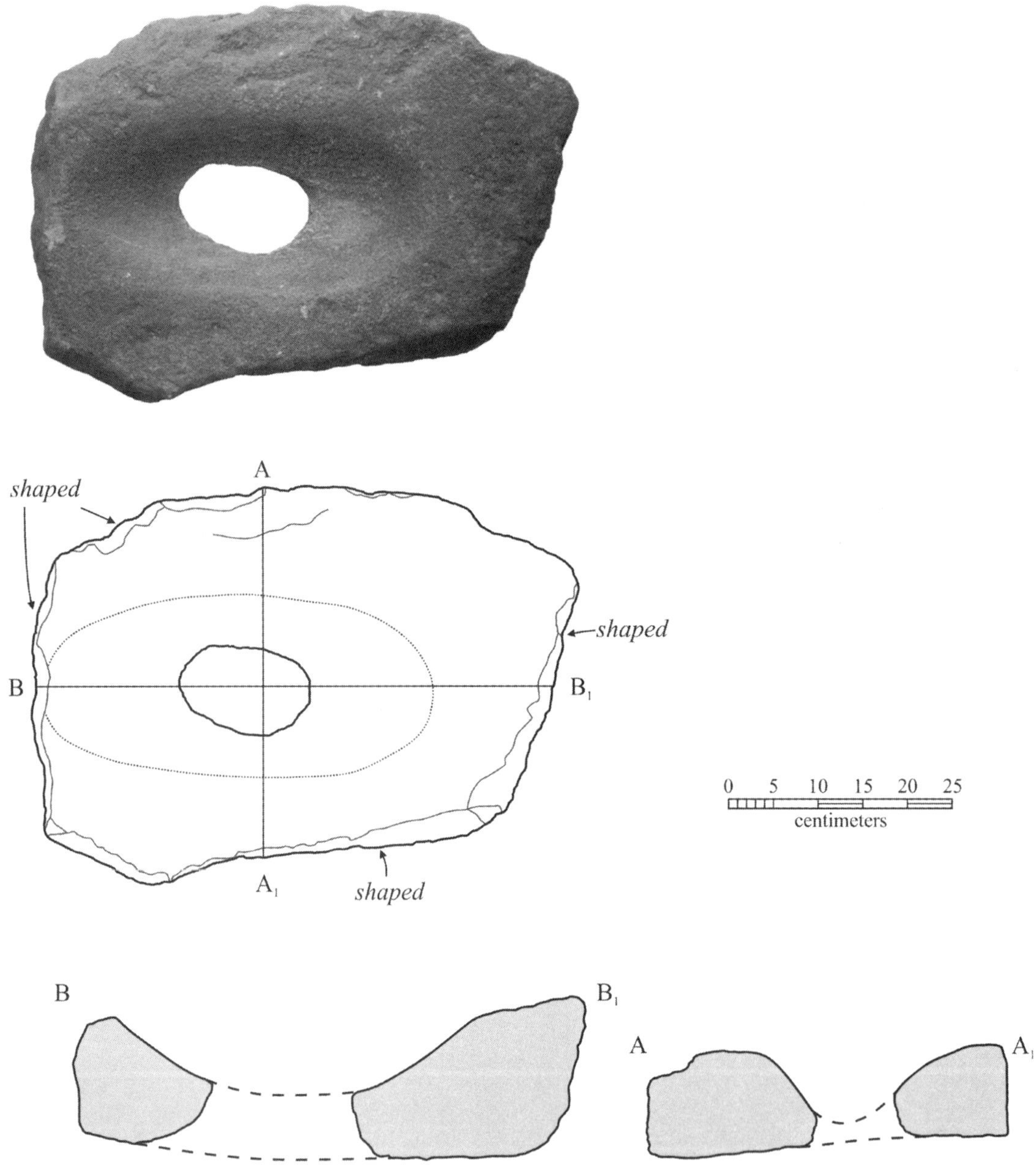

FIGURE 13.6b. Basin metate from Cerro Juanaqueña. Note opening from reuse as a rotary quern. Catalogue/bag number: (*b*) sc735.

patterns of the basin metates are illustrated in Figure 13.6 and show that metates wear towards a "V" shape when viewed in a lateral (across width) cross section and a shape of a shallow bathtub when viewed in a longitudinal cross section. The grinding surfaces are curvilinear in both lateral and longitudinal directions.

The grinding surfaces of the Cerro Juanaqueña basin metates are distinctly longer than they are wide. The mean ratio of grinding surface length to width is 1.78 and is normally distributed (minimum = 1.0, maximum = 2.75, $s = 0.366$, $n = 52$). As discussed later, this elongated grinding surface is due to primarily reciprocal, rather than rotary grinding. Grinding surface length and width are correlated and this relationship is highly unlikely to be due to chance ($r^2 = .651$, $p < .001$, $n = 52$). The elongated shape of the grinding surface is a fundamental characteristic of this basin metate assemblage. The Cerro Juanaqueña metates may represent a tran-

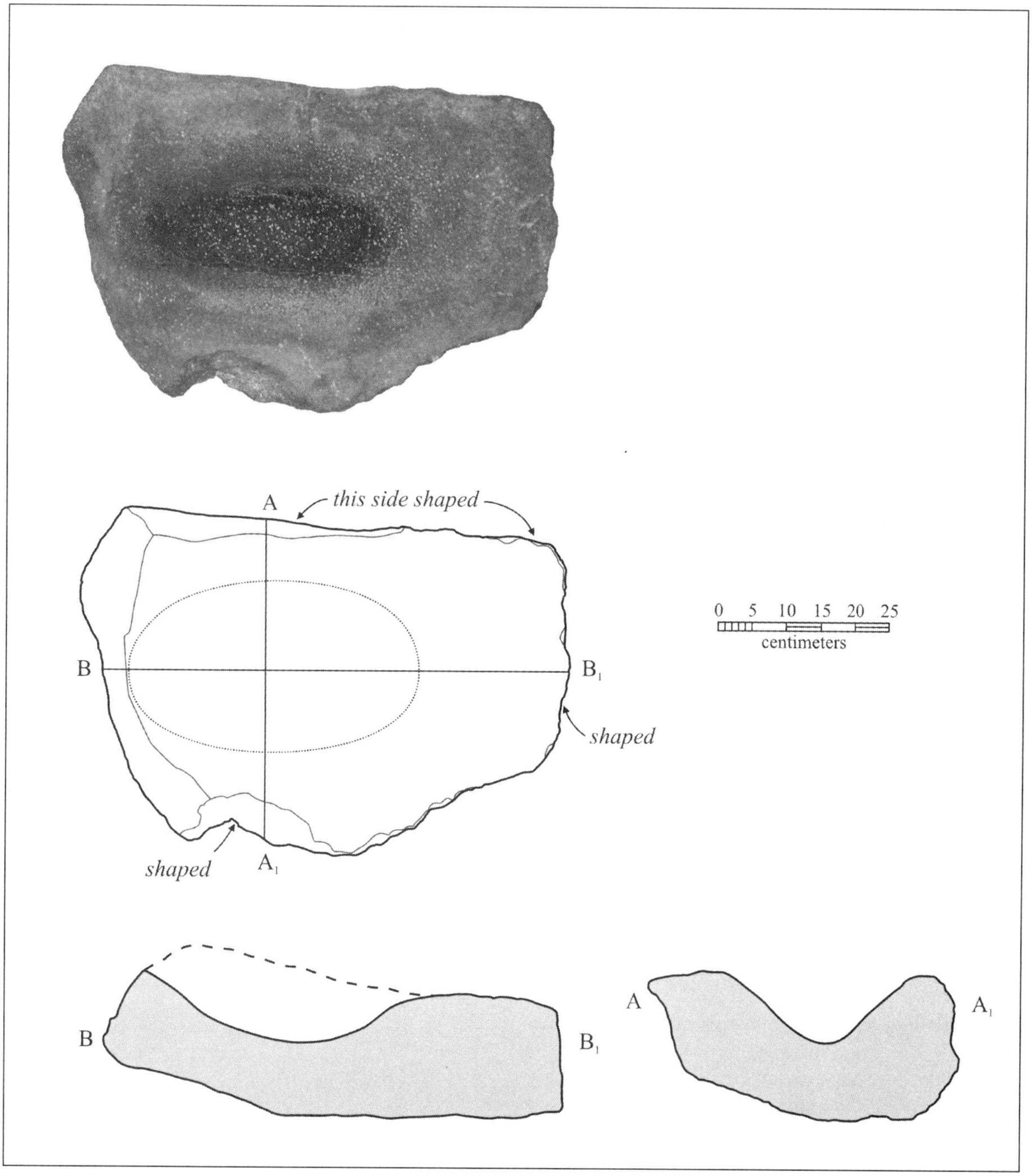

FIGURE 13.6c. Basin metate from Cerro Juanaqueña. Catalogue/bag number: (*c*) sc384.

sitional technology between traditional Archaic period basin metates that were used with both rotary and reciprocal grinding and elongated trough and slab metates of the Ceramic period that were used with reciprocal grinding motions.

Another striking aspect of the basin metate assemblage is the depth of wear. For most metates the grinding surface concavity is more than 7.0 cm deep. However, a histogram (Figure 13.7) of metate wear depth is bimodally distributed, with a clear break between 4.0 and 7.0 cm. Table 13.12 subdivides depth of wear into shallow and deep. The shallow group consists of all basin metates with 4.0 cm or less of wear. The shallow wear group has a mean depth of 2.1 cm ($n = 13$) and the deep group is 10.2 cm ($n = 35$). This bimodal distribution is not well understood. In fact, we cannot rule out sampling error. Metate depth of wear in the transect sample is not bimodally distributed, but it is asymmetrical with a skew towards the small depths ($n = 28$, mean = 8.5 cm, $s = 4.05$, median = 9.4).

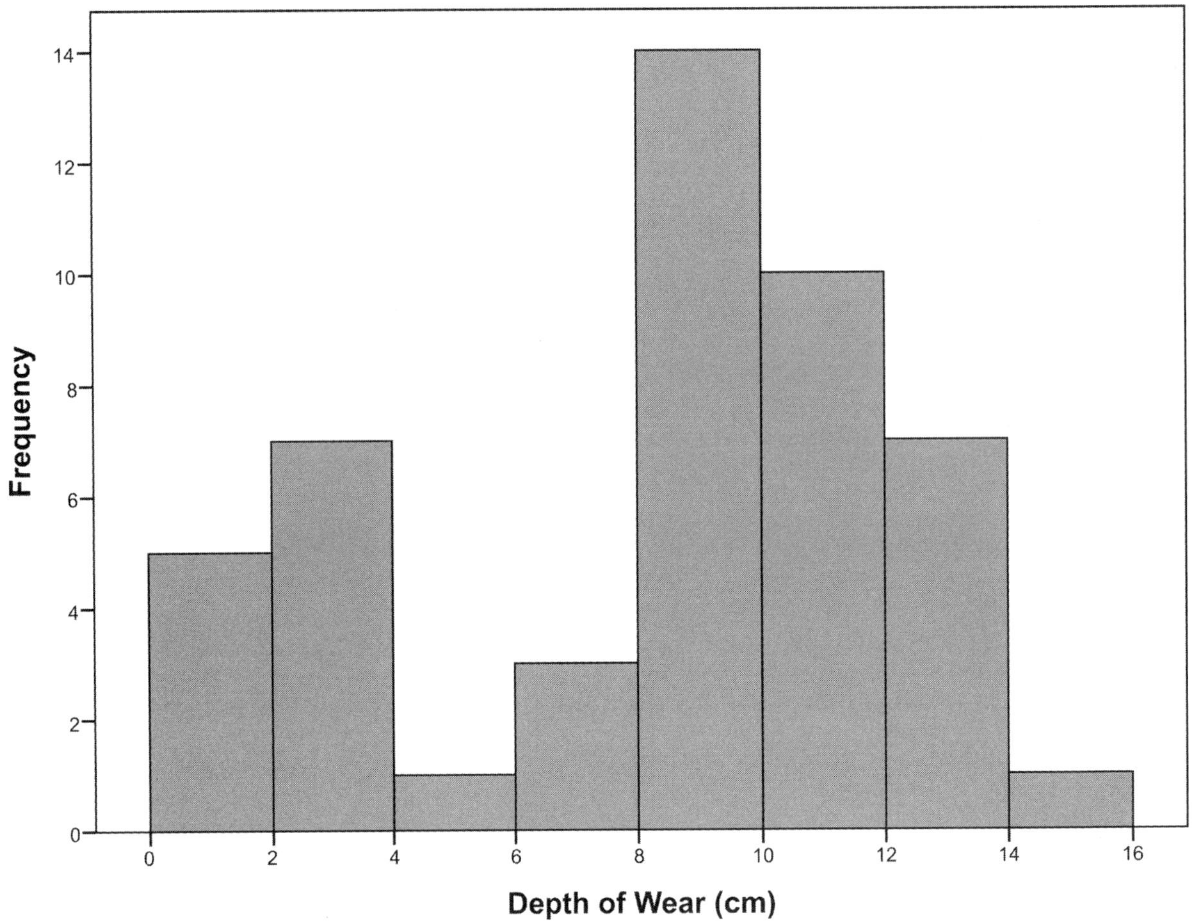

FIGURE 13.7. Histogram of depth of wear for all basin metates in detailed sample (n = 48). Because of the bimodal distribution, the sample was split into two groups for further analysis. The shallow wear group has wear less than or equal to 4.0 cm.

We examined other potential relationships to understand the wear depth patterns of basin metates, but few were revealing. The overall lengths and widths of many of the shallow basin metates are in the same range as those of the deep basin metates. There is no relationship between overall thickness and depth of wear. A comparison with the grinding slabs shows that only about half of the shallow basin metates overlap in size (overall length and width) with the largest grinding slabs. We could find little useful patterning in spatial distributions on the site. Both deep and shallow metates apparently occur in virtually all sectors of the site. Both deep and shallow wear patterns also occur in similar proportions on feature surfaces versus talus slopes. No evidence suggests one form is older than the other. Recall that two lower terraces had later Cienega phase radiocarbon dates (Terraces 413 and 415). No metates were found in these excavations, and only a single grinding slab fragment was on the surface of Terrace 415.

TABLE 13.12. Depth of Wear for the Deep and Shallow Basin Metate Groups.

	Deep Basin Metates	Shallow Basin Metates
n	35	13
Mean cm	10.2	2.1
Median cm	10.0	2.0
Minimum cm	7.0	0.0
Maximum cm	14.5	4.0
Standard Deviation cm	1.76	1.30

Depth of wear is strongly patterned with grinding surface area. Figure 13.8 plots grinding surface area (grinding surface length multiplied by width) and wear depth, showing two distinct swarms. Grinding surface lengths of less than about 40 cm and widths less than 20 cm strongly cluster with shallow depths. A vertical reference line at 800 cm^2 on Figure 13.8 indicates this pattern.

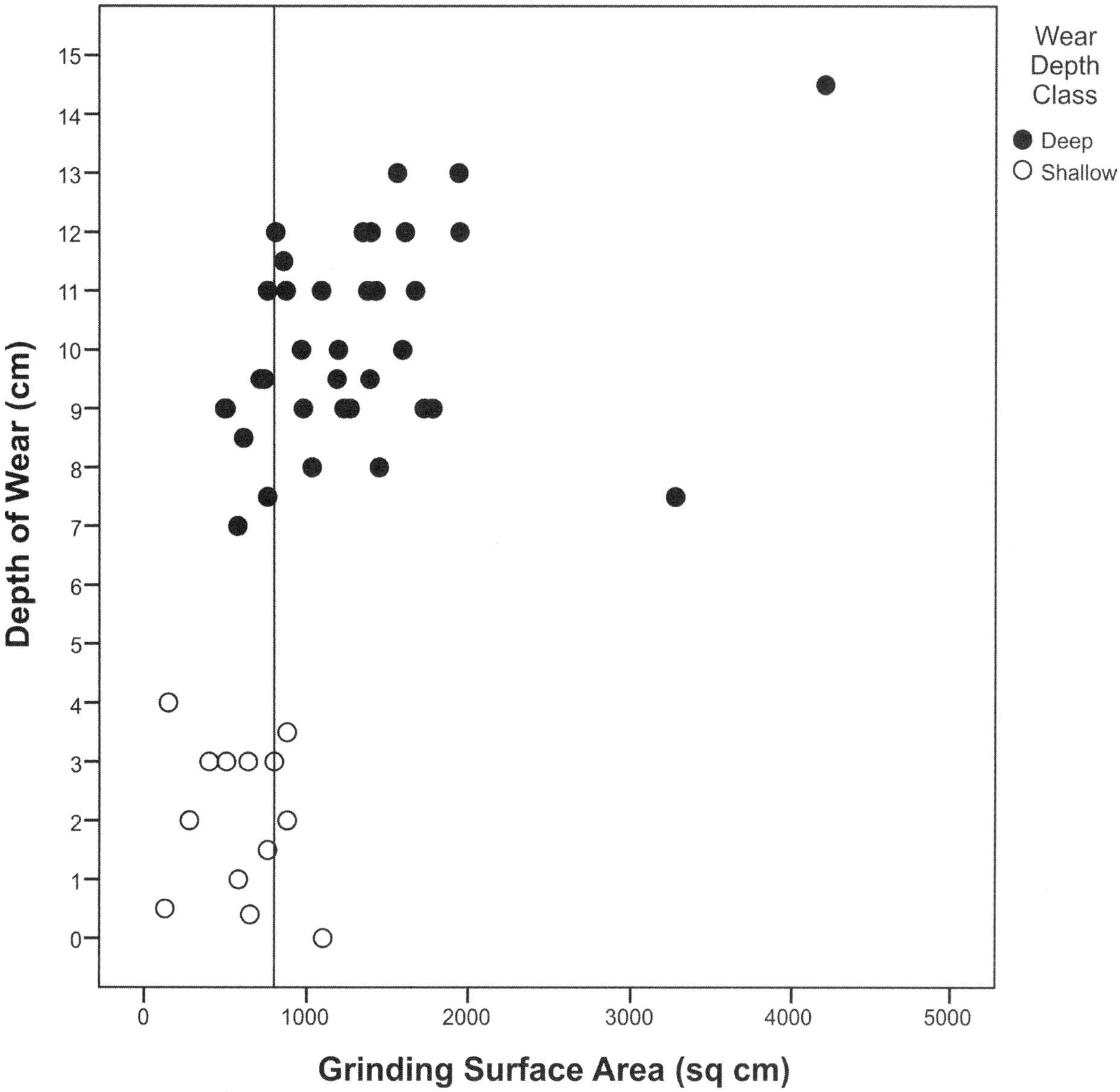

FIGURE 13.8. Scattergram of basin metate grinding surface area plotted against depth of wear for both the deep and shallow groups. Small grinding surface area strongly clusters with shallow wear. Vertical reference line marks 800 cm^2.

The correlation between grinding surface area and depth of wear exists not only between the shallow and deep groups, but also within the deep group. Considering only the 35 deeply worn basin metates, there is a weak but significant correlation between grinding surface area and grinding depth ($r^2 = .151$, $p = .021$).

There are a number of potential explanations for the depth of wear and grinding surface area patterns. The most straightforward is that grinding surface length and width tend to expand with grinding depth. As the use of the metate continues, more of the surface area is incorporated into the working surface. The 13 shallow basin metates were in early stages of their use lives when they were abandoned, while the deeply worn metates were advanced in their use lives. In some cases, the metates with shallow wear also had smaller grinding surface areas, perhaps resulting from an inefficient design. Users found them to be ineffective and stopped using them prematurely, or perhaps these metates served a different role. In contrast, the deep basin metates continued to be used since many become rotary querns as discussed below.

Basin Metate Rotary Querns

Holes are present at the bottoms of many of the complete basin metates and present an interesting question as to their manufacture and purpose. We consider four possibilities: 1) The metates were intentionally broken,

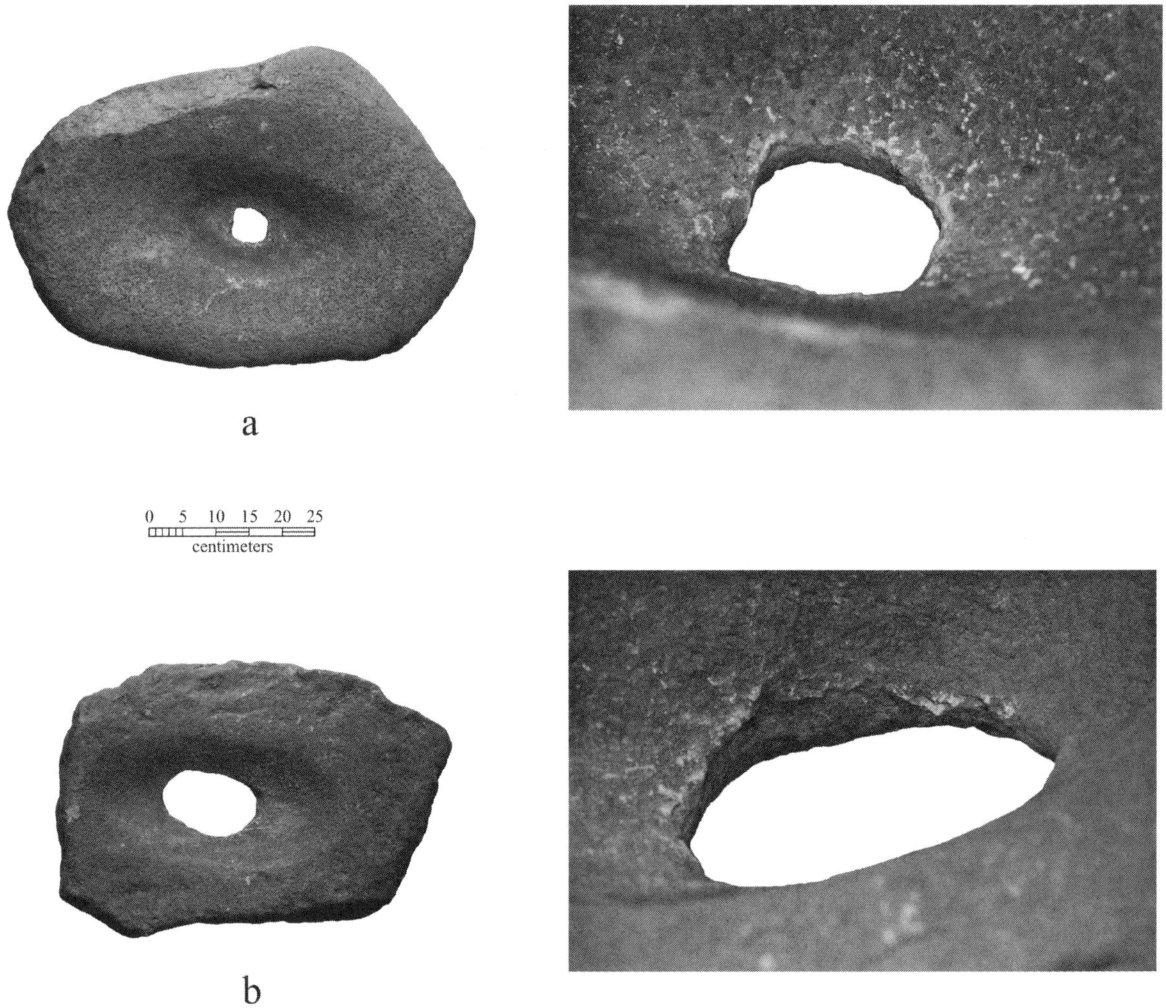

FIGURE 13.9. Manufactured openings in basin metates reused as rotary querns. Catalogue/bag numbers: (*a*) T175-31 (see Figure 13.6a). The opening is 6 cm by 5 cm, and the interior of the lip is rounded due to use wear. (*b*) sc735 (see Figure 13.6b). The shape of the hole is long and wide (16 cm by 9 cm) and is a rough geometric reflection of the plan of the grinding basin. The rounding of the interior lip is due to tool wear.

terminating their use lives or "killing" them (J. Adams 2002; Hayden 1969:154; Huckell and Huckell 1984:182). 2) The holes were an inadvertent byproduct of extensive use. Perhaps the grinding wore through the floor of the metate or the holes were accidently produced while pecking a thinned grinding surface. 3) The holes allowed the metate to become a hopper-mortar in a fashion similar to the bottomless baskets used with grinding slabs in California, as proposed by Sayles who found these tools to be a common element on Late Archaic period sites in southeastern Arizona (Sayles 1983:70; Sayles and Antevs 1941). 4) Finally, perhaps the holes were deliberately manufactured to form a tool analogous to the gyratory crushers found in the Sierra Pinacate of Sonora (Hayden 1969).

Holes are present in the bottom of 26.5% (13 out of 49 in detailed surface inventory) of the complete basin metates. The holes occur in the floors of the grinding basins and are usually oval, although a few circular ones do exist. Their mean size is 9.3 cm by 6.2 cm (Table 13.13). At the location of the holes, the floors of the metates are only 1 to 2 cm thick (Figure 13.9).

We quickly dismissed two of these ideas when we viewed the bottom of the metates and found multiple flake scars around the perimeters of the openings. Clearly, the users intentionally manufactured the holes with several blows to the already thinned grinding surfaces. In many cases, the interior rims of the holes are smoothly worn indicating that the basin metates continued to be utilized after the hole was made. Tools abraded and rounded the interior lips of the holes. This pattern eliminates the "killing" hypothesis. Manufactured holes with worn lips are inconsistent with intentional destruction of the metate or exhausted metate use life.

TABLE 13.13. Dimensions of Holes in the Floors of Basin Metates Used as Rotary Querns, $n = 13$.

Hole dimensions (cm)	Mean	Minimum	Maximum	Std Deviation	Valid *n*
Length	9.3	5.0	14.0	3.5	13
Width	6.2	4.0	12.0	2.2	13
Thickness	1.81	1.20	2.00	0.33	7

With regard to the third hypothesis, basin metates with holes are not analogous to California hopper mortars. The baskets attached to hopper-mortars have openings in their bottoms, but the hopper-mortars themselves do not. Among the Yokuts and Pomo for example, a conical or cylindrical basket with a hole in the bottom is placed on a slab nether stone in order to crush acorns with a pestle. The basket hopper prevents the product from spilling during crushing, and after repeated use the slab obtains a slight depression due to wear (Barrett 1952). If the basin metates with holes are analogous to baskets with holes, there would need to be yet another nether stone under the first which is difficult to envision. The oval shape of the holes and the wear around the rims are also inconsistent with this idea.

The gyratory crusher or rotary quern function is most consistent with the characteristics of the deep basin metates with holes. Hayden (1969) documented ancient gyratory crushers in the Sierra Pinacate of Sonora. Hayden suggested the gyratory crushers were used to process hard mesquite beans. The operators would place complete bean pods in the crusher and then rotate a pestle. The beans would be crushed into a fibrous mass and then drop through the gap that forms between the opening and pestle when the pestle is rotated. The purpose is to crush the pods and make it easier to pick out the hard seeds. The fibrous mass could then be ground in a metate. Hayden documented slab and deep forms of gyratory crushers. The deep form appears most similar to the Cerro Juanaqueña basin metate shape (Hayden 1969). The primary difference is that the Sierra Pinacate gyratory crushers tend to have a circular or conical mouth (frustum opening), more like a mortar, rather than the long oval opening in the Cerro Juanaqueña metates, so we use the term "rotary quern" rather than Hayden's more specific "gyratory crusher".

Several attributes of the Cerro Juanaqueña tools are consistent with the rotary quern interpretation. The oval holes are geometric parallels or reflections of the oval openings of the basin metates. We suspect that, like the Sierra Pinacate versions, a long wooden pestle was used. As the pestle was rotated around the walls of an oval metate, the hole served as a pivot point for the pestle. The pivoting end of the pestle caused the lower hole to wear in an oval form reflecting the oval shape of the metate's deep grinding basin (Figure 13.9, Items a and b).

In addition to holes, the Cerro Juanaqueña basin metates and Sierra Pinacate crushers share the following characteristics: use of basalt, worn rims of the pivot point holes, similar size of holes, and similar sized grinding surfaces. Also, in both the shapes of holes are geometric parallels of the upper use surfaces. Both are found in association with numerous other metates indicating they are an element in an extensive grinding tool assemblage.

Hayden (1969:159) observed about five similar basin metates with oval openings, both in the Sierra Pinacate and in collections attributed to the San Pedro phase. Similar to the Cerro Juanaqueña basin metate openings, his observations indicated that the basin metate gyratory crusher holes ranged from 10 by 13 cm to 13 by 18 cm. Unfortunately, Hayden provided no further elaboration on this metate form, which appears to be the same as the Cerro Juanaqueña form. Ironically, Hayden must not have felt they served a similar purpose as the gyratory crushers, since he indicated that he did not know their function. In contrast, we suspect the gyratory crushers and metates with holes served a similar role.

A gyratory crusher is designed as a tool with a singular purpose. In the Cerro Juanaqueña case, virtually exhausted basin metates were transformed into rotary querns. Instead of the circular motion of a gyratory crusher, the pestle was rotated in an oval fashion, so pressure would be placed against one of the sloping sides of the basin metate. The processed product would drop through the gaps created between the pestle and the hole when the pestle was rotated.

We think the pestle would have been carved from a hardwood. However, we identified a few crude stone pestles on the surface of Cerro Juanaqueña that have wear compatible with their use with gyratory crushers (see large pestles in Chapter 6). These could have been used occasionally, but we suspect that a wooden pestle would have been preferred.

Ground Stone Pollen Results

Cerro Juanaqueña pollen studies are fully reported in Chapter 10 and the results of the pollen washes from several ground stone tools are summarized here. The five samples include one metate, three manos, and one small mortar (stone bowl) examined by two different analysts. All were from excavated contexts and all were sealed in plastic bags following exposure. Suzanne Fish (Appendix 10.1) analyzed three pollen washes from Cerro Juanaqueña ground stone tools, and Linda Scott Cummings examined two manos (Appendix 10.3). We conducted some pollen washes in the field lab and sent the extract to Fish. Cummings processed the ground stone tools in her lab using procedures to extract and search for pollen, phytoliths, and starch. Cummings found pollen and phytoliths but not starch. She found that the total pollen concentration from the two mano washes varied from nearly 4,000 to more than 12,000 pollen per square cm of surface area.

Pollen was found on all five ground stone tools and included *Artemisia* (High spine), Compositae (Aster family), Low spine Compositae (Aster family), cheno-am, Poaceae (grass), Boerhaavia-type, *Sphaeralcea* (globemallow), *Kallstroemia* (poppy), Onagraceae (primrose), *Eriogonum* (buckwheat), *Larrea* (creosote bush), *Prosopis* (mesquite), *Ephedra* (e.g. Mormon tea), *Pinus* (pine), *Quercus* (oak), *Juniperus* (juniper), *Salix* (willow), *Alnus* (Alder), Umbelliferae, and Liguliflorae. This spectrum was similar to that recovered from modern floodplain agricultural fields, floodplain backhoe trenches, and Cerro Juanaqueña archaeological sediment samples, suggesting that the pollen on the ground stone tools derived from multiple sources, including the surrounding sediment. It was not simply a product of ground stone use. As discussed in Chapter 10, pine, alder, oak, and juniper are likely wind-transported as these species do not grow near the site but live at higher altitudes.

Cheno-am pollen was present in high levels on four of the five ground stone tools, and three of these included aggregates of cheno-am pollen. Typically, large clusters of pollen are not efficiently transported by wind and may indicate immediate introduction of the pollen from the plant, or possibly the grinding of cheno-ams that included immature floral parts. Cheno-am pollen aggregates were also found in the archaeological sediment samples, and both chenopod and amaranth flourished on the site at the time of our fieldwork. Chenopod and amaranth are well-known economic plants as described in Chapter 8. In Chapter 9, Gayle Fritz discusses the recovery of possibly domesticated amaranth seeds and non-domesticated chenopod seeds from the site.

Several taxa found at rare densities on these ground stone samples are not found in the floodplain samples or archaeological sediment samples, including *Euphorbia* type (spurge), *Simmondsia* (jojoba), *Cylindropuntia* (cholla), *Opuntia* (prickly pear), and *Cereus* type (hedgehog or related cacti). Several of these are succulent plants that typically grow on rocky hillslopes. Relatively rare cactus pollen is seldom recovered in samples from natural vegetation and these types may reflect resource residues. *Zea mays* pollen was recovered from one mano and is consistent with the evidence of the use of maize on the site and the arguments presented throughout this volume concerning the economic importance of maize.

The two manos that produced phytoliths yielded a high frequency of chloridoids, low frequency of festucoids, and panicoids on only one mano. Buliforms were present in low levels. None are suggestive of the presence of *Zea mays*. The chloridoids and panicoids typically represent C_4 grasses while the festucoids typically represent C_3 grasses. Further details are found in Chapter 10. The phytolith spectra is similar to that found in backhoe trench and archaeological sediment samples, and most likely the mano phytoliths derive from the surrounding sediments rather than from mano use.

In summary, the results of the pollen study are complicated by inputs of pollen and phytoliths from the surrounding sediment, but are consistent with ground stone use that included processing domestic and wild plants including maize, cheno-ams, and possibly succulents.

Maize Processing Technologies

Early Agricultural Period Maize

Endosperm type is a fundamental characteristic of maize that substantially affects the energy needed to grind maize, and three forms are relevant to this study—pop, flint, and flour. The earliest maize in the Southwest was a pop maize with an endosperm that is corneous, thick, hard, and translucent but surrounds a small proportion of flour endosperm (K. Adams 2015:Figure 1.2; K. Adams et al. 2006). Flint kernels also have a hard, translucent, corneous outer layer of endosperm, but the portion of flour endosperm which it envelopes is more substantial (K. Adams 2015:Figure 1.2; K. Adams et al. 2006). Kernels of flour maize contain a soft white endosperm with little of the corneous outer layer. The hard endosperm of dried pop and flint maize make these varieties difficult to grind, while flour maize is far easier to grind. The relatively flat grinding surfaces of Ceramic period mano-metate sets are well suited for reducing the soft flour maize. The timing of the emergence of flour maize is not well constrained but is thought to date to ca. AD 500 to 700, based on recovery of Maís de Ocho, an eight-rowed maize with flour kernels, from Tularosa Cave (K. Adams 1994:29, 2015:20; Cutler 1952:464–469; Martin et al. 1952:506). Indigenous farmers in historic and modern times typically plant several varieties of maize including those with different of endosperm types as each was valued for different reasons such as maturation and processing characteristics, as well as frost, drought, disease, fungus, or insect resistance (e.g. Castetter and Bell 1951:100–106; Pennington 1963; Rea 1997:347–352).

According to Karen Adams (Chapter 8), the charred maize kernels from Cerro Juanaqueña "…contained a mix of flint and flour endosperm…" and are "…generally consistent with a popcorn, having small, rounded kernels and flinty endosperm (with some lesser portion of the endosperm floury)." The Cerro Juanaqueña maize clearly does not possess a flour endosperm but is a pop or flint maize with a dominantly hard corneous endosperm and smaller proportions of floury endosperm, consistent with its Early Agricultural period age.

Maize Processing Alternatives

Transforming the hard, dried kernels of maize into something humans can consume is energetically costly (Barlow 2002; Diehl 1997; Kennett et al. 2006). In contrast, preparing maize in the green or milk stage involves low costs because the sugar in the kernels is in a chemical form that the human body can easily digest. Green corn requires little cooking and can be readily boiled or roasted. However, for storage, maize must be prepared, typically by a drying process. Doing so transforms the stored endosperm energy into polysaccharides that the human body cannot readily digest. This starch requires substantial preparation, involving some combination of cooking and grinding, in order to make it digestible (Braun 1983; Hard et al. 1996). Cooking chemically breaks the polysaccharide back to a more digestible state, such as a disaccharide or monosaccharide. Cooking by parching reduces the starch into an intermediate disaccharide while steeping in water for many hours gelatinizes the starch into a readily digestible monosaccharide. Soaking or cooking maize in an alkaline solution (typically achieved by the addition of wood ash) is called "nixtamalization." Nixtamalization not only gelatinizes the starch, but also helps enhance amino acids by making them more available to human digestion (Ellwood et al. 2013; Hard et al. 1996; Katz et al. 1974). With ceramics, nixtamalization was a relatively low cost method to cook dried maize for numerous societies in the New World (Katz et al. 1974). Using ceramic technology agriculturalists place a vessel containing maize and alkaline water on a fire to steep for many hours and the cooking pot requires little attention. In the absence of ceramics, placing heated stones in a water-tight container to cook dried maize may have been an alternative strategy (Ellwood et al. 2013), but it entails high labor costs (Thomas 2016). Other alternatives include popping corn or cooking hominy (J. Adams 2005; Diehl 1997). With regard to the latter, however, the labor demands of cooking dried maize in water in the absence of a functional ceramic technology may be underestimated, as stone boiling can be labor-intensive (Thomas 2016). Minimizing maize processing costs is critical for maize farming to be cost-effective (Barlow 2002; Diehl 1997).

Grinding dried maize into a flour or masa is fundamental to aiding digestibility; the particles have a higher ratio of surface area to volume so digestive acids and enzymes may attack them more effectively, but this effort is highly labor-intensive (Hard et al. 1996). Grinding maize to a fine flour frequently accompanies parching, while grinding the wet, cooked maize creates a masa. There are multiple combinations of cooking and grinding to convert dried maize into a digestible food, but all require the application of substantial energy. The elongated, deep basin metates at Cerro Juanaqueña led us to address the question of their relative efficiency for grinding maize.

Gonzalez's Maize Grinding Experiments

Gabriela Gonzalez (2014) conducted controlled experiments for her M.A. thesis to evaluate the relative grinding efficiency and mechanics of processing dried maize on a Cerro Juanaqueña basin metate. Gonzalez's (2014) controlled grinding experiments compared the relative efficiency of grinding pop, flint, and flour maize using a Cerro Juanaqueña basin metate and a manufactured trough metate. These experiments are described here in some detail as they are particularly relevant to understanding the Cerro Juanaqueña ground stone assemblage. The pop, flint, and flour maize were obtained from grow-out experiments conducted by Karen Adams and colleagues (K. Adams et al. 2006; Werth 2007). Gonzalez parched all maize kernels prior to grinding; parching makes grinding easier, reduces polysaccharides to disaccharides, and was a common recipe for Native American maize preparation when producing maize flour.

Gonzalez used two different mano and metate sets. One was a prehistoric elongated, deep basin metate from the surface of Cerro Juanaqueña (sc384, Figure 13.6, Item c) with a grinding surface area of 680 cm^2. The other was manufactured by a professional stone cutter in San Antonio, Texas, using modern tools to shape a trough metate with a rectangular flat grinding surface that was open only at one end. Both were made from the same local vesicular basalt from Cerro Juanaqueña. The stone cutter shaped the trough metate from a large, unmodified cobble we brought from the base of Cerro Juanaqueña. It was manufactured to have a grinding surface area similar to that of the basin metate (663 cm^2). The two one-hand manos used in the experiment were from Cerro Juanaqueña and were selected from the collections to have shapes compatible with

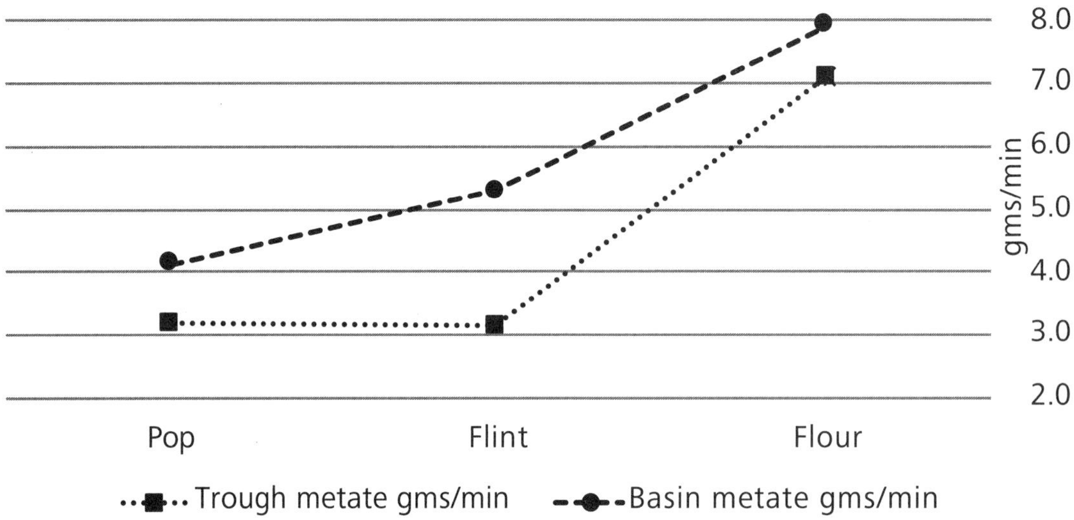

FIGURE 13.10. Maize and metate grinding experiment. Median grinding efficiency is expressed in g/min for six maize-metate pairs, *n* = 10 experiments for each pair, a total of 60 experiments. The basin metate is 68% more efficient than the trough metate when grinding flint variety of maize. The efficiency of grinding flour maize with either type of metate is 50% to 126% greater than that of grinding flint maize. When grinding flour maize there is little difference in efficiency between the basin and trough metates. This figure is modified from Gonzalez 2014:Figure 4.13.

TABLE 13.14. Median Times and Rates to Grind 150 g of Maize.[1]

	Trough Metate			Basin Metate			Percent Efficiency Increase Basin Compared to Trough
Maize	Minutes	IQR[2]	G/Min	Minutes	IQR	G/Min	
Pop	47.3	8.8	3.17	36.3	5.5	4.13	30.3%
Flint	47.7	12.4	3.15	28.3	4.3	5.30	68.3%
Flour	21.1	4.0	7.12	18.9	5.0	7.93	11.4%

[1] Each of six combinations of metate type and maize variety was repeated 10 times, for a total of 60 experiments (Gonzalez 2014:Table 4.1).
[2] Interquartile range.

that of their respective metate mate and to have similar grinding surface areas (mano used with basin metate = 120 cm^2, mano used with trough metate = 130 cm^2). The experiments controlled for material type and grinding surface area. They included two metate forms and three maize endosperm types, for a total of six different combinations. Each of the six possible metate forms and maize endosperm combinations was repeated 10 times for a total of 60 repetitions. Each repetition involved grinding 150 g of parched maize to a specified particle size. Gonzalez measured fineness by passing the ground product for each repetition through nested geological screens such that 80% of the ground maize particles were less than 1.0 mm but greater than 0.35 mm, approaching the fineness of modern Tarahumara pinole ground on a metate. Efficiency was measured as the time used to grind the maize as well as calories expended, excluding time for removing and measuring the flour.

Gonzales's results are given in Table 13.14 and are plotted on Figure 13.10 as median values reporting the time used to reduce kernels to flour for each of the six combinations along with the rate in grams/minute (g/min). Surprisingly, considering only the hard flint maize, the basin metate is 68% more efficient than the trough metate. Similarly, considering only hard pop maize, the basin metate is 30% more efficient than the trough metate. This demonstrates that the basin metate form is a more efficient design for grinding hard types of maize relative to the trough metate. Gonzalez (2014) showed that these efficiency increases were statistically significant using the independent samples median test.

Figure 13.10 shows that flour maize is ground with maximum efficiency, as expected. The most dramatic gains in efficiency are seen in the trough metate results. Flour maize was ground at 7.12 g/min reflecting a 126% increase over flint maize (3.15 g/min). In the basin metate, flour maize was ground at 7.93 g/min, reflecting a 50% increase in efficiency compared to that of flint maize (5.30 g/min). Surprisingly, the basin form was slightly more efficient grinding flour maize (7.93 g/min) than the trough metate (7.12 g/min), but this difference was not statistically significant (Gonzalez 2014). The 126% efficiency gain using the flat grinding surfaces of the experimental trough metate-mano suggests it is well designed to grind soft flour maize, but not effective in grinding hard maize. Larger grinding surface areas of trough metates would likely produce further efficiency gains while grinding soft maize.

The emergence of soft flour maize, presumably over a millennium after the occupation of Cerro Juanaqueña, would have revolutionized the productivity of maize farming due to its lowered cost of processing. These results are similar to results found in the food processing industry. There is a positive and direct correlation between endosperm hardness and energy needed to grind maize. Soft maize varieties require 50% to 67% less energy to grind than hard varieties (Li et al. 1996).

Finally, the basin metate and mano sets are an efficient grinding technology for processing hard, flinty maize because the concave-convex grinding surfaces facilitate substantial downward crushing force needed to reduce the corneous layer of the pop and flint maize, as well as the shearing force necessary for producing a fine particle size. Gonzalez reports that the initial minutes of grinding pop, flint, and flour maize on the trough metate progressed similarly. However, once the small proportion of floury endosperm in the pop and flint maize was ground, she observed that the remaining unground maize was made up of corneous particles, visible by their translucent appearance (Gonzalez 2014). Reducing those hard particles was presumably more time-consuming with the flat grinding surfaces of the trough metate and mano set. The flat grinding surface likely dilutes the potential downward crushing force across the surface area of the mano. There were no gains in efficiency between pop (3.17 g/min) and flint (3.15 g/min) maize on the trough metate, presumably because the high proportion of corneous endosperm limited efficiency gains. The corneous fraction became the limiting factor, despite the efficient reduction of the flour component of the endosperm. In contrast, the concave-convex design of the basin metate and mano sets can deliver greater crushing force to the large, hard particles of maize. This effort is facilitated by rocking and reciprocating the mano along the length of the metate to obtain both shearing force and an efficient design for processing quantities of maize. In contrast to the results on the trough metate, the basin metate demonstrated an increase in efficiency when comparing the grinding of pop maize (4.13 g/min) with the grinding of flint maize (5.30 g/min). Both the hard, corneous component and the soft flour component of the endosperm are readily ground. Apparently the efficiency gain was due to the higher proportion of soft endosperm in the flour maize. The elongated basin metates from Cerro Juanaqueña are an efficient tool for crushing and grinding hard pop or flint maize using a reciprocal grinding stroke.

Crushing and Grinding at Cerro Juanaqueña

The following model integrates our findings and addresses the Cerro Juanaqueña mano and metate assemblage as a technology designed to process substantial quantities of pop or flint maize without the use of ceramics. It links the demands of processing dried maize and small seeds with the attributes of the assemblage and grinding technology generally.

Flint or pop endosperm is harder in consistency and is therefore more difficult to grind. Flinty maize requires more crushing relative to grinding than floury maize. Parching not only reduces the starch chemically, but also enhances grindability, since parched kernels are more friable and crushing is more readily accomplished. Other experiments we conducted showed that parching popcorn reduced grinding time by 50% relative to unparched popcorn.

The elongated, deep basin metates and large one-hand manos are well adapted to the task of crushing and grinding substantial quantities of dried, flinty maize. As discussed above, the curvilinear shapes of mano and metate grinding surfaces are effective for the crushing and pounding action. In addition, the concave, basin metate and convex mano grinding surfaces provide an advantage over flat grinding surfaces as the basins help control the maize. The concave metate surface prevents maize kernels or seeds from scattering when pounded.

The small and large convex manos can deliver more downward force directly to a small kernel, while complementing the concave grinding basin. Many of the manos are convex both through their length and width. As the convex mano is reciprocated it is used with a rocking motion that increases pressure at the lowest portion of the basin. This rocking motion tends to maintain the mano's convexity in the lateral (width) dimension. Flat manos are likely used more frequently during the early stages of basin metate wear, while the grinding surface remains wide. The small convex manos may be used more towards the end of the metate's use life, as wear progresses and the grinding basin begins to develop a characteristic V shape. Flat manos can also be effective for grinding on the sides of the V-shaped grinding surface, but in this mode it is difficult to deliver as much downward force. The edges of flat manos are also useful for initial crushing of maize kernels. With grinding surface lengths much greater than widths, the elongated basin metates are designed primarily for reciprocal grinding. The relatively large grinding surfaces of the flat manos and large convex manos provide gains in efficiency for the grinding and regrinding that that is required in order to achieve a fine particle size. Metate depth of wear is a product of long use life in a setting where metates are reused for many years, such as a village.

One of the primary factors that condition the amount of energy needed to reduce material by grinding is the ratio between original size and final product size (Lowrison 1974). Since large maize kernels are being reduced to fine flour, more energy is required than grinding small seeds. Other experiments we conducted showed that an equivalent amount of parched amaranth seeds could be ground into a fine flour in a fraction of the time required to grind parched, flinty maize. The basin shape is particularly useful not only for controlling maize particles, but also for preventing the small round amaranth seeds from rolling off the metate during grinding.

The large, elongated basin metates with their compatible manos represent a transitional technology between Archaic style basin metates and typical trough metates. Elongated basin metates likely did not persist as the major tool form in the southwestern U.S. and northwestern Mexico once flour maize became available, although they are frequently found as a minor element in ground stone tool assemblages on Ceramic period sites. Increasing surface area likely offered dramatic increases in efficiency when processing maize (Diehl 1996; Hard et al. 1996; Lancaster 1984; Mauldin 1993b). However, the curvilinear grinding surfaces of basin metates and convex manos were limited in the degree to which grinding surface area could be increased while still maintaining the curvilinear grinding surface that was essential for crushing action. It is hard to imagine a two-hand mano that is convex in both length and width analogous in shape to a convex, one-hand mano. These engineering constraints limited the potential increases in grinding surface area of convex manos and concave metates. Increasing the grinding area of the planar surfaces of trough and slab metates and their manos was readily accomplished. Thus, with the introduction and widespread adoption of flour maize, manos with larger grinding surfaces provided increased efficiency, but with losses in their capacity for crushing.

Life Cycle of Metates and Manos at Cerro Juanaqueña

The occupants acquired the raw material, manufactured, used, and abandoned the basin metates at Cerro Juanaqueña. These metates had long use lives, certainly 20 years or more. Minimal investment appears to have been made in their manufacture, yet they were reused for over a generation. They were not intended for portability and were clearly made to become site furniture. Certainly, a basin metate with substantial wear was valued and remained in use in the household for decades. Once the working surface of the basin metate was formed, the grinding stone became an efficient tool for processing maize and seeds, and it was likely valued more than a new metate. When the floor of a basin metate wore thin, a hole was chipped in the bottom to convert the basin metate into a rotary quern.

Manos also showed little investment in manufacture and were largely made, used, and abandoned at the site. But unlike the metates, they had relatively short use lives. Most manos did not have heavy wear and were only used on one side. Little effort was invested in managing their wear through shaping or sharpening. Instead, most appeared to have been abandoned or lost before they were exhausted, and this may have been related to mobility. As families departed from the site they may have taken their manos with them, but not their metates. Different mano shapes had somewhat different performance characteristics. Different mano shapes were also needed to fit the evolving shape of the metate grinding surfaces.

Southwestern Contexts

The Cerro Juanaqueña mano and metate assemblage represents a specialized technology that is well suited for processing substantial amounts of dried, hard, flinty maize as well as wild seeds without the use of ceramics. As such, it represents a transitional technology between the classic Archaic basin metates used with small manos and the standard trough forms operated with large, flat manos used by later agriculturalists in the Southwest and northern Mexico.

If our suggestion is correct, elongated basin metates should be found in contexts where dried, flinty maize was an important element of the diet and where ceramic cooking technology was not established. Sayles (1945:50, 1983:69–70; Sayles and Antevs 1941:Plate XV) indicates that elongated, deep-basin metates are characteristic of the Late Archaic period but also occur in Early to Middle Archaic period sites, as well as in Early Ceramic period sites. He also notes the presence of this form with holes in the bottom, which he referred to as combination metates, mortars, and hopper-mortars. The Donaldson site in southeastern Arizona, dating to ca. 2500 BP (uncalibrated), produced deep, basin metates (Eddy 1958) including a single one in association with a burial (B. Huckell 1995:30, 61–64). A single, complete, large, elongated, deep basin metate with an oval hole was found at the Los Pozos site in the Tucson Basin in stratigraphic context indicating that it is Middle Archaic in age, perhaps dating between 2700 and 1900 cal BC (Gregory, 1999:28 and 47–48). J. Adams (1998:366) illustrates an elongated, deep basin metate from Cienega phase contexts, although she does not discuss these forms in particular. Two elongated, deep basin metates also occurred at the Early Ceramic period Houghton Road site in Tucson, and the illustrated shallow basin forms also have elongated grinding basins (B. Huckell 1998:120).

Elongated, deep basin metates are present in contexts consistent with expectations. That is, they are found in San Pedro and Cienega phase contexts in southern Arizona where maize was an important element in the diet and ceramic cooking technology was not yet established. However, they are apparently not common. Instead, shallow basin metates with elongated grinding surfaces are encountered more often. Were these elongated basin metates the tool of choice for grinding hard, flinty maize? The elongated basin provides a grinding surface suitable for reciprocal grinding, analogous to later trough metates. The basin form is certainly well suited for the crushing required to process both hard maize and wild seeds. Finally, the depth of the grinding surface is a product of years of accumulated use.

Substantial use of green corn which does not require extensive processing to be made edible would be consistent with the paucity of elongated, deep basin forms (J. Adams 2015b; J. Adams 2015c). Other alternatives include popping corn or cooking hominy (J. Adams 2005; Diehl 1997). With regard to the latter, however, the labor demands of cooking dried maize in water in the absence of a functional ceramic technology may be underestimated, as stone boiling can be highly labor-intensive (Thomas 2016). Future work on elongated, deep basin metates could include reporting the ratio of grinding surface length to width, further grinding experiments, and examination of metate fragments to better assess the frequency of these tools. Some examples of elongated, deep basin metates have been reported from Middle Archaic contexts and future research will have to evaluate their meaning. Elongated basin metates may be a tool increasingly selected for grinding and crushing quantities of dried, pop, or flint maize, as well as wild seeds. This design allows for an efficient reciprocal stroke, the curvilinear grinding surfaces are effective for crushing, and the basin constrains the product. Grinding surface depth increases with extended use.

Basin Metate Wear Rates and Population Estimates

The total grinding effort represented by all the worn basin metates at Cerro Juanaqueña is an indirect record of food production and the scale of the population that was processing food. In this section, we use the metate wear data to estimate the population that occupied the site. After estimating the total wear on all the basin metates on the surface of the site, we derive metate wear rates based on our experimental work. We finally estimate a series of parameters including hours of grinding per day, family size, length of occupation, and the amount of time per year people were grinding at the site to arrive at population estimates. This exercise is not meant to obtain a precise population value, but only to show that some population parameters are highly unlikely and others are more likely given the metate assemblage. This effort represents a revision of an earlier preliminary effort (Roney and Hard 2002).

The first step is to estimate the total wear represented by all the basin metates on the upper terraces Cerro Juanaqueña. We focus on the upper terraces as they represent the primary, San Pedro phase, occupation of the site. Two lower terraces had a Cienega phase occupation, which we want to exclude from the analysis. In fact, all 110 complete basin metates were from the upper terraces. There were 573 basin metate fragments on the upper terraces and only seven other fragments from the lower terraces.

We estimate that these 573 fragments are equivalent to a minimum of 92 complete metates based on volume. The median volume of the basin metate fragments is 7,714 cm^3 ($n = 103$, IQR [Interquartile Range] = 12,420 cm^3). The median volume of the complete metates is 47,940 cm^3 ($n = 46$, IQR = 41,322 cm^3). Therefore it takes on average the combined volume of 6.2 (47,940/7,714) fragments to equal one complete metate. The total metate fragments have an equivalent volume of a minimum of 92 complete basin metates (573/6.2). Summing 92 plus 110 complete basin metates produces a minimum estimate of 202 basin metates on the upper terrace. This is lower than our previous estimate (Roney and Hard 2002) as the median volume provides a better estimate than the mean because the fragment data are right-skewed.

We next need to estimate the total wear represented by these 202 metates. Recall the basin depth of wear was bimodally distributed and subdivided into deep and shallow groups. The deep metates make up 72.9% of the sample and average 101.6 mm deep. The shallow ones are 27.1% of the sample and average only 20.7 mm deep (Table 13.12). Extrapolating to the 202 metates, 147 (72.9%) would fall into the deep basin group and 55 into the shallow basin group.

Metates may have begun their use lives with a concavity. The concavity may have been manufactured or a large cobble with a natural concavity may have been selected. We suspect this initial condition because the presence of a concavity appears to be important in controlling the maize while it is being ground. Therefore, 10 mm will be deducted from each mean depth, so the deep group has 91.6 mm of wear and the shallow group only 10.7 mm.

In the earlier study we estimated that metates wore at an average rate of 5 mm/year (Roney and Hard 2002: 171–173) based on published ethnoarchaeological and experimental data that involved basalt metates used by maize agriculturalists in Mexico and Guatemala. Since then, we conducted a series of grinding experiments to better evaluate the rate at which Cerro Juanaqueña metates wear in order to enhance the accuracy of our population model, and these findings are described in Appendix 13.2.

The metate wear experiments resulted in a basin metate wear rate of .0034 mm/hr. If the level of maize use was high and metates were used for three hours per day every day of the year, then metate wear rates could have been as high as 3.7 mm/year (365 days/year × 3 hours/day × .0034 mm/hour). This is close to the original estimate of 5 mm per year or .0046 mm/hour.

The estimated rate of wear and the total assemblage of metates were combined to model the total person days of grinding required to produce the total wear. Once we estimate total person days we can use the known occupation span as determined by radiocarbon dates to estimate population size. We then consider a number of parameters that could affect the population size estimates with the following formula:

$P = T / (O \times S \times I)$ where

P = Average Annual Population during the Principal Span of Occupation

T = Total Number of Person Years of Occupation of Cerro Juanaqueña

O = Occupation Span

S = Seasonal Adjustment

I = Grinding Intensity

We estimate that on average a deep basin metate represents 24.7 years of use (91.57 mm/3.7 mm per year) for each of the estimated 147 metates in this group. Each metate in the shallow basin group represents only 2.9 years of use (10.69 mm/3.7 mm per year) for the estimated 55 metates in this group.

Combined, these metates represent about 3,790 years of use [(147 metates × 24.7 years per metate = 3,630.9 years) + (55 metates × 2.9 years of use = 159.5)]. If each metate was used by a family of five, using a common ethnographic average, we can convert years of wear to person years. The ground stone assemblage on the upper terraces at Cerro Juanaqueña reflects at least 18,952 person years of use. This number may sound high, but it is conservative. It only includes basin metates visible on the surface and does not include buried metates and metates removed from the site. Our approach for estimating the minimum number of metates is also conservative since it combines all basin metate fragments across the site. The assumption of 1 cm of manufacturing wear is also conservative as discussed previously. For these reasons, rounding down to an estimate of 18,500 person years is a moderate estimate of the intensity of occupation for the upper terraces at Cerro Juanaqueña.

The denominators on the right side of the equation represent modeled values of three parameters that directly impact population estimates: occupation span (O), seasonality (S), and intensity of grinding (I). Note that as any of these values decline the population estimate (P) increases proportionally. The variable O, occupation span, was derived from analysis of radiocarbon dates presented in Chapter 5. Using only the 13 dates from the upper terraces of Cerro Juanaqueña, the OxCal program estimates, at a 68% probability, that Cerro Juanaqueña's upper terrace occupation endured for as little as 30 years to no more than 230 years and 0 to 320 years at 95% probability (Fig 5.4, Table 5.2). Given many aspects of the site and the radiocarbon dating, the 68% probability range is the best estimate.

The variable S takes into account possible seasonal variation in occupation. We believe that Cerro Juanaqueña was occupied on a relatively sedentary basis as reviewed in the Ethnobotany and Conclusions Chapters (Chapters 8 and 15). However, quite a bit of room still remains for seasonal fluctuations in site population. We believe it is reasonable to model site occupation between six (S = .5) and 12 months each year (S = 1).

The variable I addresses the likelihood that the intensity of grinding activity may well have fluctuated on a seasonal basis and may have been less than the modeled three hours per day, 365 days of the year based on the grinding intensity of full-time agriculturalists (I = 1). The intensive grinding activity assumed by our discussion of metate wear may have only taken place during a few months each year or vary in other ways. During other seasons people may well have relied more on roots, greens, smaller and softer wild plant seeds, and meat. The characteristics of the ground stone assemblage, the proportion of maize recovered, and the overall scale of the occupation could not be supported without substantial maize production, suggesting that grinding intensity is unlikely to be much below 50%. The recovery of a wide range of wild plants also implies that grinding intensity was less than 100%. Therefore, I values likely range from .5 to .75. Note that lower S or I values result in greater population (P) estimates.

A group as small as 82 people could have generated

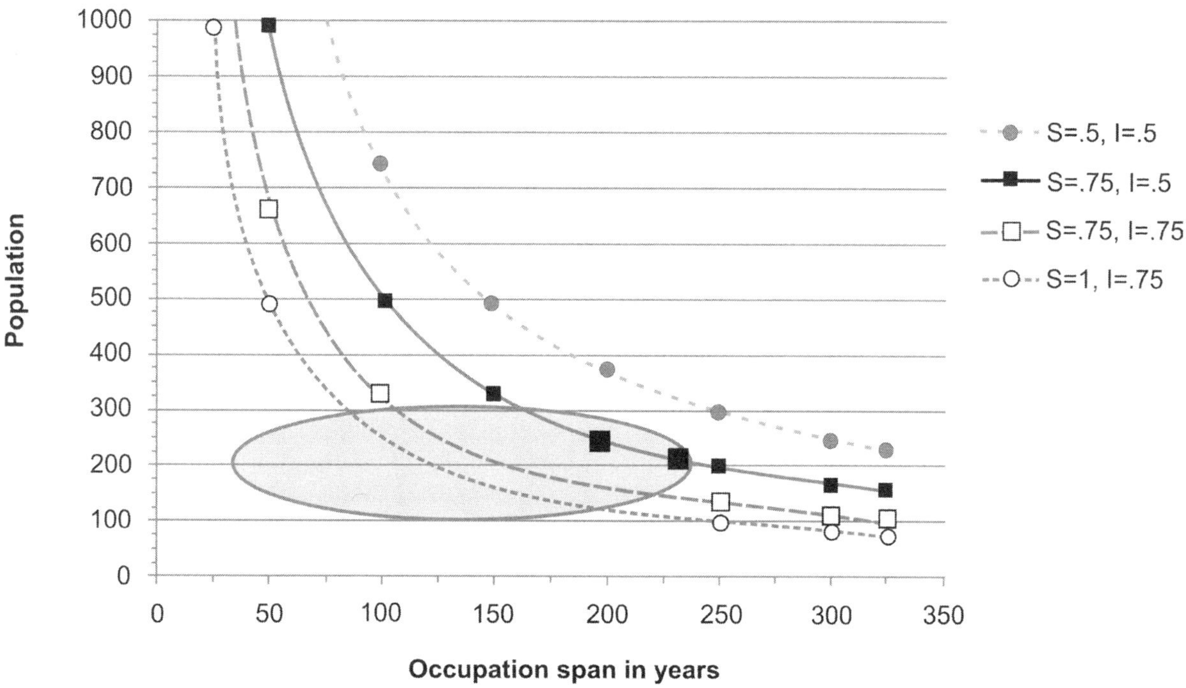

FIGURE 13.11. Model of population derived from total basin metate wear. The smoothed lines are derived from formula where S = Seasonal Adjustment and I = Grinding Intensity. The intersection of the lines and the ellipse defines likely scenarios. The two larger square markers on the S = .75, I = .5 line indicate our best estimate of population and length of occupation. The line where S = 1 models occupation of 12 months per year, S = .5 models occupation of six months per year. When I = 1, the model reflects grinding intensity equivalent to full-time agriculturalists.

the observed metate wear, but only if the site was occupied by a fully sedentary group for 300 years and only if their diet required grinding an intensity of .75 (S = 1, I = .75). This scenario begins to establish a lower limit for the size of the group that might have built and occupied Cerro Juanaqueña. At the other extreme, if the total metate wear could be accommodated by an occupation of only six months each year, for a span of occupation of 25 years, and a grinding intensity of .5, then the average population would need to have 2960 people! This unlikely possibility demonstrates that this approach systematically establishes upper and lower limits to the scale of the population.

Figure 13.11 illustrates what we consider to be four plausible solutions for this formula. One solution assumes that occupation was year round and that grinding intensity was 75% of that of assumed for full-time agriculturalists (S = 1.0, I = .75). Another possibility sets both seasonality and grinding intensity to .75 (S = .75, I = .75) and a third reduces grinding intensity to 50% of full-time (S = .75, I = .5). A fourth reduces both seasonality and grinding intensity to 50% (S = .5 and I = .5), but this scenario requires the highest population levels. Additional combinations with seasonality varying between .5 and 1.0 and grinding intensity varying between .5 and .75 are possible.

Figure 13.11 uses an ellipse to define the most likely scenarios given the sum of the archaeological data. The domain of the ellipse is set to an occupation span of 30 to 230 years based on the 68% probability span function. The range of the ellipse sets a minimum population of approximately 100 (107), which is the smallest population that the left-most curve models with an S = 1, I = .75, and a 230-year occupation span. We have set the maximum range of the ellipse at 300, simply because we think stable populations of greater than that level seem unlikely. The ellipse then encircles segments of three of the curves and eliminates the S = .5, I = .5 curve, as this scenario, given the occupation span, requires more than 300 people. Also note the absence of line segments in the left portion of the ellipse, indicating that occupation spans less than 100 years are unlikely given the other parameters. Assuming occupation spans of 100 to 230 years, relates to populations of 100 to 300 people living at the site.

Considering all the evidence, we favor the following estimate: the site was occupied continuously for about two centuries by a group of about 200 to 250 people who were involved in maize processing nine months of the year. They were involved in a level of hard seed and maize grinding that is about half the level of grinding utilized by full-time maize agriculturalists. This scenario is indicated by the two large square markers on the S = .75, I = .5 line.

Conclusions

The large assemblage of manos and metates found at Cerro Juanaqueña indicates that the site was the location of extensive domestic activity. We estimate that the metates reflect the accumulated activity of a couple of hundred people for about two centuries. The deep basin metates represent a technology that is ideal for crushing and grinding dried, hard, corneous maize into flour as well as wild seeds. The elongated shape of the grinding basin is well suited for reciprocal grinding. This design may represent a transitional morphology between typical Archaic basin metates and Ceramic period trough metates. Large, one-hand manos used with these metates also enhanced efficiency.

Comparisons across the region are needed to document the distribution of this technology in time and space and its association with early agricultural production. Depth of wear, elongated grinding surfaces, and large one-hand manos may be related to extensive flinty maize processing in Early Agricultural period contexts. Deeply worn metates are a product of long use life while elongated grinding surfaces are related to reciprocal grinding. The convex mano used against the concave surface of the metate facilitated the crushing and grinding needed to reduce the hard Early Agricultural period maize.

A few deep basin metates have been reported from Middle Archaic contexts in southern Arizona, but it is unknown if these are related to dried maize processing. Deep basin metates were not found at Ventana Cave or Picacho Reservoir, locations where Archaic period maize production is not expected. However deep, basin metates were present, but not common, on Tucson Basin Early Agricultural period sites although metate recovery was generally low at these sites. If elongated, deep basin metates are rare at intensively occupied Early Agricultural period locations, perhaps dried maize was not being extensively processed into flour and other maize processing strategies were being used.

In contrast, the Cerro Juanaqueña grinding slabs are expedient nether stones with little wear. Their small grinding surface would make them ineffective for repetitive grinding of volumes of food for routine consumption. They were likely used for grinding a wide range of products but involved tasks that did not involve substantial labor or volume.

The morphology of the manos is largely compatible with that of the basin metates. Three mano forms are present—large convex manos, small convex manos, and flat manos. A household may have used these forms at different times. The shape of the grinding basins of the metates was variable and changed as wear progressed, requiring manos of different shapes and sizes. Certainly the manos would have been incorporated into a wide range of grinding tasks, including use with the grinding slabs. Manos apparently had relatively short use lives as they tend not to be heavily worn or substantially modified in ways that would enhance their longevity. They were readily replaced given the inexhaustible supply of basalt cobbles found on the hill. Given their short use lives and the large number of basin metates, the frequency of manos at the site is lower than expected, leading to the suggestion that perhaps the residents were carrying them away from the site.

Future work on Early Agricultural period ground stone assemblages should emphasize systematic reporting of dimensional data and detailed morphological descriptions so comparisons across assemblages can be made. More controlled experimental work is needed to address the relative efficiency of varying mano and metate forms for processing maize and wild seeds. Alternative maize preparation recipes also need to be carefully considered. Dried maize requires some combination of cooking and grinding to make it suitable for consumption. These processes have finite physical constraints which provide a research avenue for reconstructing ancient economic practices are therefore worthy of further consideration.

Notes

1. The standard "one-hand" vs. "two-hand" mano classification is a misnomer since one-hand manos were frequently operated with both hands (Hard et al. 1996), but this traditional terminology is used here for convenience.
2. Hopper-mortars are deep basin metates that were reused as mortars and have a hole in the bottom.
3. Field records note a single large two-hand mano on a lower terrace found near Historic period sherds (Chapter 6).
4. Nether stones refer to any lower grinding stone and here refers to both the basin metates and grinding slabs.
5. Proximal refers to the edge closest to the grinder as the mano is positioned naturally on the metate.
6. Grinding surface area of all manos was calculated as length × width.
7. During the fieldwork we informally referred to the grinding slabs as slab metates but their size and wear patterns suggest otherwise.

CHAPTER 14

Chipped Stone Artifact Analysis

Bradley J. Vierra

Introduction

This chapter presents the results of the analysis of approximately 15,000 chipped stone artifacts recovered during the excavation of Cerro Juanaqueña. As previously described in this volume, the site is a unique place, a hilltop covered with over 550 terraces. It has been suggested that the labor investment in the construction of these features implies a degree of residential stability and labor organization beyond that of simple foragers. If so, how can studies of this chipped stone assemblage be used to evaluate whether or not Cerro Juanaqueña represents an early settled village? To answer this question we must understand several key aspects of technological organization involving the stone tool manufacturing process. These components include tool design, the selection of raw materials, stone tool production and maintenance, stone tool use, and the eventual discard of worn-out tools (Andrefsky 2009; Binford 1973, 1977, 1979; Hayden et al. 1996; Nelson 1991; Odell 2001; Torrence 2001; Torrence, ed. 1989). How might these various aspects of tool organization affect the structure of the archaeological record and provide insights into the past activities at Cerro Juanaqueña?

Recent research on Late Archaic stone tool technology has emphasized residential mobility as a possible explanation for variations in tool design. When I refer to *technology*, I am referring to all the tools and facilities needed to help a person survive. On the other hand, *mobility* is a positioning strategy that allows hunter-gatherers to distribute themselves across the landscape in order to procure critical resources. *Foraging strategies* refer to what foods people eat and *foraging tactics* to how people procure these foods. Foraging tactics have two organizational components—residential and logistical mobility. *Residential mobility* is when the whole group, including men, women, and children, move to a new residential location, whereas *logistical mobility* is when a specialized task group moves out from the residential base for the purpose of conducting a specific activity. The former tactic moves people to food and the latter tactic moves food to people. Any hunter-gatherer group uses a mixture of these two foraging tactics as a means of reducing the spatial and temporal incongruities in the distribution of resources. This mixture is conditioned by several factors including the structure of the environment, a dependence on storage, and regional demographic factors. Nonetheless, as hunter-gatherers shift from an economy based on foraging to agriculture, there is a decrease in the use of residential mobility and subsequent increase in the use of logistical mobility. The types of tools needed for the completion of foraging vs. task-group-oriented activities can be quite different (Binford 1980; Bleed 1986; Harwell and Kelly 1983; Keeley 1988; Kelly 1995; Kuhn 1989; Vierra 1995).

Stone tool design in North America is often characterized by a dichotomy of core reduction and bifacial tool production. That is, simple flake tools are generally associated with settled village communities, while an emphasis on the production of bifacial tools is more characteristic of mobile hunter-gatherers. In this case, *tools* are chipped stone artifacts that exhibit retouch and/or use-wear. Most rock types can be used for the production of simple flake tools since the sharp edge is used for a relatively short period of time and then discarded. However, higher quality materials easily worked by both percussion and pressure flaking techniques are required for the production of bifacial tools that are maintained over longer periods of time. Therefore, core reduction activities tend to be associated with the use of low quality materials like rhyolite, basalt, and quartzite that are available within the vicinity of the habitation site. In contrast, the production of bifacial tools is associated with the use of fine-grained materials like chert, chalcedony, and obsidian which often occur in restricted locations across the landscape and end up in a forager's campsite as a nonlocal rock type. Nonetheless, as was the case with residential vs. logistical mobility, North American stone tool technologies include a mix of both core reduction and bifacial tool production as a means of coping with the uncertainties of food procurement and processing (Andrefsky 1994; Bamforth 1986, 2002, 2003; Brady 2005; Buchanan 2002; Goodyear 1979; Johnson and Morrow 1987; Kelly and Todd 1988; Nelson 1991; Odell 1996a, 2001; Parry and Kelly 1987; Sullivan and Rosen 1985; Vierra 1990, 1993a).

The concept of residential mobility as a possible explanation for this technological variation often assumes that mobility limits the size and number of tools that a group can efficiently carry with them (Carr 1994; Ebert 1979; Kuhn 1994; Odell 1996b; Parry and Kelly 1987; Shott 1986a). For example, Parry and Kelly (1987) suggest that bifaces are portable tools that can also act as cores, something important for mobile groups with varying access to lithic materials; whereas, an expedient flake technology is sufficient for sedentary groups with access to locally available materials. This perspective has come under close scrutiny, especially in respect to the importance of bifacial cores for mobile groups. Counter arguments point out that flake production from platform cores can be as efficient as production from bifacial cores (Bamforth 2002, 2003; Eren et al. 2013; Jennings et al. 2010; Prasciunas 2007). Railey (2010) has also argued that the temporal shift from an emphasis on bifacial to core technology was a direct result of the replacement of dart with arrow technology and the use of a wide range of lithic materials suitable for the production of arrow points on smaller flake blanks. However, the use of flake blanks actually begins during the Late Archaic period prior to the introduction of the bow and arrow (Vierra 2013). Also, as Vierra and Foxx (2009) point out, Late Archaic residential sites contain a mix of core reduction and biface production activities, whereas, seasonal campsites emphasize biface production/maintenance.

By contrast, other studies of stone tool technology have stressed the importance of tool design, energetic efficiency, optimization, and risk reduction for explaining technological variation and long-term changes in technology (Beck 2008; Bleed 1986; Brantingham 2006; Jeske 1992; Nelson 1991; Odell 2001; Torrence 1983; Torrence, ed. 1989; Vierra 1995, 1997, 2013). As Schriever et al. (2011) point out, contrasting technologies used during the Pithouse period in the Mimbres area are presumably related to both subsistence and mobility. That is, agriculturally-dependent groups emphasized the use of informal tools, and groups more heavily engaged in foraging activities emphasized the use of formal tools. They suggest the former strategy reflects a maintainable and the latter a reliable technology due to continuous, vs. episodic use schedules, respectively. Gender has been explored only recently as a possible explanation for understanding stone tool technology (Arthur 2010; Vierra 2005).

Approximately 15,000 chipped stone artifacts were recovered during the excavation of Cerro Juanaqueña. The data derived from the analysis of these artifacts will be used to address three basic issues concerning the archaeological implications of residential site stability and labor organization. If Cerro Juanaqueña represents an early settled village site that is reliant on agriculture, then I would expect the chipped stone assemblage to: 1) be dominated by the reduction and use of locally available materials, 2) reflect the full range of core reduction and tool production/maintenance activities, and 3) be characterized by an increased emphasis on the production and use of simple flake tools. Lastly, also it could also be asked whether or not I can discern any archaeological evidence of changes in labor organization and the increasing role of logistical mobility.

The chapter is organized into five sections: Analytical Methods, Raw Material Selection, Lithic Reduction Strategies, Discussion (including Intrasite Comparisons, Intersite Comparisons with both other early assemblages and with assemblages from other time periods, and Technology and Labor Organization), and Summary and Conclusions which provide an overview of the analysis and concluding comments on the implications of the chipped stone assemblage for understanding the issues of residential stability and labor organization at Cerro Juanaqueña.

Analytical Methods

The by-products of stone tool manufacturing are some of the most ubiquitous remains present in the archaeological record. They represent a complicated process involving the acquisition of raw materials, tool production, tool use, and the subsequent discard of expended tools. The lithic analysis methods were designed to collect information necessary to address the research issues as presented in the previous section and in earlier chapters of this volume. This research therefore documents the various activities that are represented in the chipped stone assemblage at Cerro Juanaqueña. Through the recording of explicit artifact types, in conjunction with specific artifact attributes, the nature of the archaeological record is clearly defined and the raw data needed to reconstruct these past activities are generated (Andrefsky 1998, 2006; Odell 2004). Explicit definitions of artifact types and a discussion of the attributes recorded during this analysis are described in Appendix 14.1 (see electronic supplement). Analysis of the data presented in this chapter relies heavily on the chi-square statistical test, which is described in Appendix 14.2.

Of course, generalizations are made from a small excavated sample that is presumed to be representative of

TABLE 14.1. Chipped Stone Artifact Types Excavated from 26 Cerro Juanaqueña Features.[1]

		Artifact Type						
Feature	Location	Cores[2]	Macro Debitage	Micro Debitage	Retouched Tools	Other[3]	Thermal Items[4]	Total
R1A	Upper	3	228	101	3	0	0	335
T6	Upper	5	222	267	2	0	0	496
T10	Upper	1	72	62	2	0	0	137
R28	Upper	1	121	31	2	1	0	156
T97	Upper	0	189	78	3	0	0	270
T126	Upper	6	242	83	2	0	0	333
T163	Upper	5	1683	4303	30	0	0	6021
T167	Upper	13	568	199	15	1	2	798
T175	Upper	11	487	1133	11	2	1	1645
T222	Upper	2	77	189	1	0	0	269
R234	Upper	1	150	78	5	0	0	234
T239	Upper	0	67	92	1	0	0	160
R250	Upper	5	306	128	4	0	0	443
T273	Upper	2	348	361	8	1	3	723
R286	Upper	2	243	147	3	0	0	395
T287	Upper	3	819	617	23	0	2	1464
T290	Upper	1	85	42	1	0	0	129
T297	Upper	0	126	62	5	3	0	196
T537	Upper	4	143	32	1	1	0	181
T387	Lower	0	255	114	4	1	0	374
T413	Lower	0	24	16	0	0	0	40
T415	Lower	0	4	5	0	0	0	9
T463	Lower	0	6	6	0	0	0	12
T465	Lower	0	3	2	0	0	0	5
T487	Lower	6	276	157	4	1	1	445
T508	Lower	1	25	16	1	0	0	43
Total	*n*=26	72	6769	8321	131	11	9	15313

[1] Four of the 30 excavated features (see Chapter 3) yielded no chipped stone (R51, T195, R307, and BR1).
[2] Core types include split pebbles ($n = 2$), cobble unifaces ($n = 14$), and cores ($n = 56$).
[3] Other types include manuports ($n = 8$) and hammerstones ($n = 3$).
[4] Thermal items include fire-cracked rock (FCR) ($n = 3$) and pot lids ($n = 6$). Artifact totals (e.g., 15,304) referred to throughout text exclude these nine items.

the site as a whole. To this end, data derived from excavations, in-field analysis, and general surface artifact collections have been used to provide a wider picture of the activities at Cerro Juanaqueña. The excavations focused on terrace and rock ring features at the site. The site excavations were limited to certain sections of the features, but chipped stone items from all the levels of these sampled sections were analyzed. In addition, in-field analyses were conducted of cores and hammerstones found on the surface of the excavated features. These cores and hammerstones tended to be found along the edges of the terraces where they had been discarded. Because of this, these artifacts are underrepresented in the excavated sample. Lastly, systematic surface documentation was conducted of the entire site to collect, document, and map particular tool types, as described in Chapter 6. Artifacts that were collected during these activities are also considered in this chapter. Aspects of all three data sets—excavations, in-field analysis, and general surface collections—will be used to help understand the nature of the stone tool technology. In order to provide a more complete context for these observations, data was also collected on potential lithic raw material sources located within the Río Casas Grandes valley.

A total of 15,304 chipped stone artifacts were analyzed from 26 separate excavated features at Cerro Juanaqueña (Table 14.1).[1] Most of the excavated features are terraces ($n = 21$), with fewer rock rings ($n = 5$). The excavation collection is therefore biased towards artifacts recovered from the terraces ($n = 13{,}741$), vs. the rock

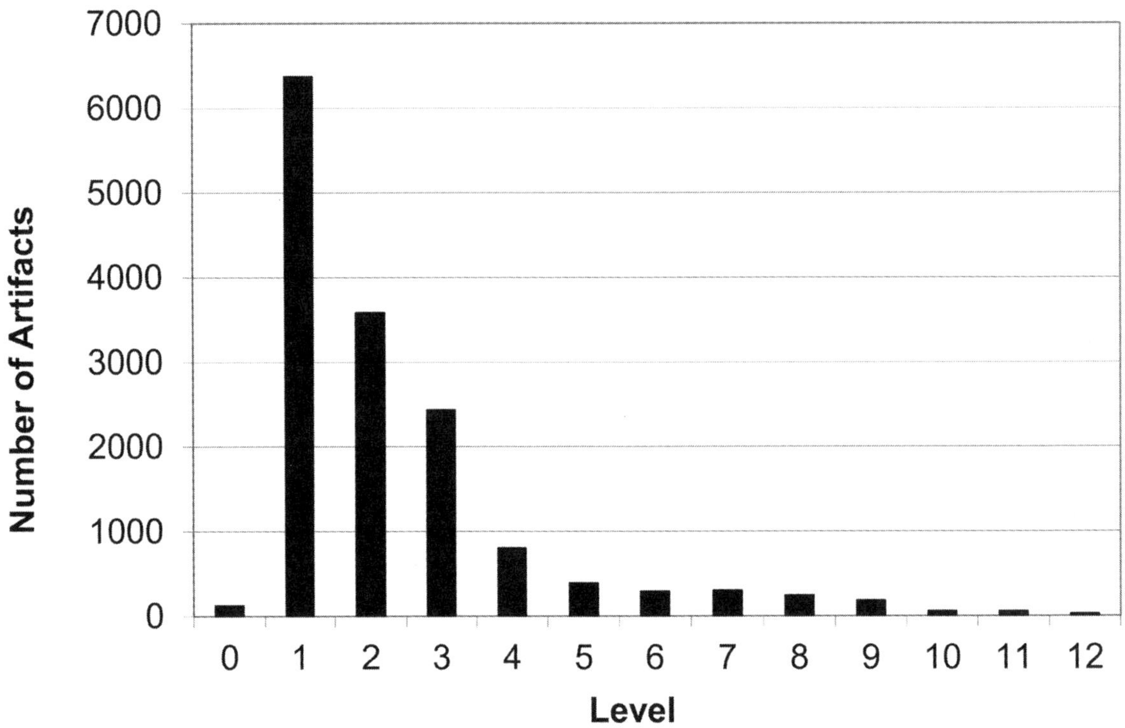

FIGURE 14.1. Number of excavated artifacts by level. Combined tabulation for all 26 excavated features at Cerro Juanaqueña that yielded chipped stone artifacts.

rings (n = 1,563). The collection is also biased towards features in the upper terrace zone (n = 19) that yielded a total of 14,377 artifacts, vs. features in the lower terrace zone (n = 7) with a sample of only 927 artifacts. This will affect intrasite comparisons between feature types and upper vs. lower terraces. In addition, samples from Cerro Juanaqueña will be used to compare with three other Late Archaic cerros de trincheras in the Río Casas valley (Cerro el Canelo, Cerro Torres, and Cerro Vidal), and two nearby Ceramic period sites. A sample of features was excavated at these other Late Archaic sites, and looters' backdirt from the two Ceramic period sites was screened in order to recover a sample of artifacts.

The site features range in size from about 25 to 300 m^2, with as few as five artifacts being recovered from one feature, and up to 6,021 from another. This latter feature is Terrace 163 which is located on the southeast edge of the upper level. The large number of artifacts from this feature includes 4,303 pieces of microdebitage. Although this terrace acted as a catch basin for sheet wash from this section of the site, less than 1% of this large collection of microdebitage was actually recovered from the surface of this feature. Instead most (73%) were recovered from levels 1 and 2 of Terrace 163. While the amount of microdebitage found in Terrace 163 was extraordinary, the vertical distribution of chipped stone was not. Figure 14.1 illustrates the distribution of artifacts by level for 26 features. As can be seen, the majority of the excavated site collection was recovered from levels 1–3, with level 1 containing the largest proportion.

Most of the data presented are derived from the analysis of the excavated samples. However, information is also presented on in-field analysis of cores and hammerstones and general site surface collections of tools. Some of these data are also discussed in Chapter 6.

Raw Material Selection

The first analysis of these data focuses on raw material selection. A variety of approaches can be used to obtain the lithic raw materials needed for the production of tools, involving both a procurement strategy and procurement tactic. *Procurement strategy* refers to the specific lithic material types selected for tool production. This information is readily available in the archaeological record as the varying proportions of worked lithic material types present. X-ray fluorescence (XRF) analysis of obsidian can also be used to identify specific sources. *Procurement tactic*, on the other hand, refers to the specific methods used to procure them.

Lithic raw materials can be obtained in three ways. An *embedded* tactic involves the collection of raw materials incidentally to subsistence-related movements (Binford 1977, 1979; Binford and Stone 1985). This practice typically involves the collection of materials for immediate replacement of exhausted tools. A *direct* tactic involves making a trip to the source location for the sole purpose of collecting raw materials (Binford 1979; Gould and Saggers 1981; Renfrew 1975). This practice typically involves the bulk acquisition of a raw material that is brought back to the residential site as cobbles, nodules, or prepared cores and cached for future use. An *indirect* tactic involves obtaining lithic materials from an intermediary. This practice usually involves some form of trade or exchange relationship. Trade items often consist of prepared cores, preforms, or specialized artifact forms (Earle and Ericson 1977; Renfrew 1975, 1977; Santley and Arnold 2005; Torrence 1986).

It is commonly assumed that Southwestern hunter-gatherers used an embedded procurement tactic and that the presence of nonlocal lithic materials is an indication of the area traversed by these groups (Shackley 1985, 1990:63; Vierra 1985, 1994a, 1994b; Vierra and Doleman 1994). A mixture of procurement tactics was probably used, but some may have been emphasized over the others (e.g., Brown 1990; Brown 1991; Vierra 1993a, 1993b, 1997; Young and Harry 1989). By contrast, agriculturalists in the Southwest could have obtained lithic raw materials using embedded, direct, or indirect procurement tactics (Arakawa et al. 2011; Brown 1990, 1991; Cameron 1984, 2001; Findlow and Bolognese 1980, 1982a, 1982b; Peterson et al. 1997; Shackley 2005b; Vierra 1993b, 1997; Walsh 1997, 1998; Young and Harry 1989).

Local Geological Sources

Local materials are defined as lithic materials that are obtainable within a 10-km (six-mile) catchment radius of the site. This is the typical foraging radius covered during daily activities around a hunter-gatherer residential site or the maximum distance traveled to tend agricultural fields (Binford 1983; Varien 1999). Nonlocal materials are those from outside this day-trip catchment.

Local bedrock geology in the vicinity of Cerro Juanaqueña has been mapped in detail by the Secretaría de Programación y Presupuesto (SPP) (1982, 1983a, 1983b, 1983c). An excellent description of the local bedrock geology is provided by Bautista and Goodell (1983), and Kettenbrink (1984) documents a stratigraphic section exposed near Janos, Chihuahua. The local geologic sequence reflects a series of volcanic eruptions, including multiple layers consisting of tuffaceous rocks, basalts, and rhyolite flow domes. Giant spherulites are present in one of the tuff units. Some of these spherulites are over a meter in diameter and filled with quartz crystals and chalcedony that sometimes form geodes. The site itself is situated on Quaternary basalts (Q[B]) including both aphanitic and vesicular varieties. At the base of the hill is the Río Casas Grandes valley. The alluvial sequence is described elsewhere (Chapter 12). The valley is bounded by both basalt and Tertiary rhyolite (T[R-Ta]) hills, including Cerro el Canelo. This formation contains reddish rhyolites that are both aphanitic and porphyritic with quartz phenocrysts. Located mainly along the southwest side of the valley is a Tertiary conglomerate (T[cg]) with a sand and gravel matrix mixed with volcanic rocks ranging from 1 to 50 cm in diameter. Lastly, a Paleozoic limestone (P[cz]) is exposed along the south side of the valley in the area of Sierra Escarate and Sierra del Chilicote. Gray chert nodules are present in this formation, and chert fragments can also be found in surface lag gravels sloping down towards the valley bottom.

As described by Nordt (Chapter 12), the Río Casas Grandes valley alluvial sequence contains multiple gravel lenses. Gravels are also present today in the stream beds of the Río Casas Grandes, Río San Pedro, and ephemeral streams draining the nearby hills. Another in-field analysis was conducted within each of these settings in order to identify potential raw material sources, raw material variation, and range in cobble size. Five 1-by-1-m surface units were analyzed in the Río Casas Grandes and two each in the Río San Pedro and a stream bed adjacent to the hill where Cerro Juanaqueña is located. The material type and maximum dimension were recorded for pebbles/cobbles larger than 5 cm in size. A total of 460 pebbles/cobbles were analyzed, mostly from the Río Casas Grandes. Eighty-two percent of these cobbles are rhyolite, with a little basalt (13%), chalcedony (3%), andesite (1%), conglomerate (0.1%), and vesicular basalt (0.9%). A chi-square analysis of a contingency table of sample location by rhyolite, basalt, and chalcedony materials only indicates significant difference in the distribution of these materials by location (chi-square = 85.7, $df = 4$, $p \leq 0.001$; $n = 452$, Appendix 14.3, electronic supplement). There appears to be relatively more basalt in the ephemeral stream channel vs. more rhyolite in both riverbeds. Indeed, the basalt boulders covering these hills could have also been used as a source of raw material. In contrast, the two rivers drain a much larger area that cuts through exposed rhyolite and basalt formations.

Information was also collected on nodule size by material type. Basalt cobbles are generally the largest (mean = 9.9 cm, $s = 3.4$), with rhyolite (mean = 8.6 cm, $s = 2.2$) and chalcedony (mean = 7.9 cm, $s = 2.4$) being somewhat smaller. The basalt and rhyolite cobbles are spheroidal in shape, whereas the chalcedony cobbles are more irregular or slightly tabular in shape with a bubbly surface appearance. The rhyolites exhibit a range in color variation that grades from gray to red to tan to yellow. Indeed, a single cobble can exhibit several of these colors. Additional information about color variation of lithic raw materials is provided in Appendix 14.4 (electronic supplement).

TABLE 14.2. Chipped Stone Artifact Type by Material Type From from Excavated Features.[1]

	Material Type						
Artifact type	Basalt	Rhyolite	Obsidian	Chalcedony	Chert	Other	Total
Cores							
Cores	2	30	1	17	6	0	56
Cobble uniface	0	13	0	0	1	0	14
Split pebble	0	0	2	0	0	0	2
Subtotal	**2 (3%)**	**43 (60%)**	**3 (4%)**	**17 (24%)**	**7 (9%)**	**0**	**72**
Debitage							
Debris	21	625	6	217	128	3	1000
Core flake	141	2894	47	942	645	6	4675
Biface flake	0	156	14	219	49	0	438
Und. Flake Frag.	3	257	17	236	68	0	581
Microdebitage	41	3269	170	4311	530	0	8321
Other deb.	2	31	4	25	11	2	75
Subtotal	**208 (1%)**	**7232 (48%)**	**258 (2%)**	**5950 (39%)**	**1431 (9%)**	**11(<1%)**	**15090**
Retouched Tools							
Retouched piece	0	12	0	7	4	0	23
Notch	0	8	1	4	1	0	14
Denticulate	0	3	0	0	0	0	3
Biface	0	19	2	21	11	0	53
Projectile point	0	14	1	4	7	0	26
Uniface	0	5	0	2	0	0	7
Drill	0	1	0	0	1	0	2
Perforator	0	2	0	0	0	0	2
Cruciform	0	0	0	1	0	0	1
Subtotal	**0**	**64 (49%)**	**4 (3%)**	**39 (30%)**	**24 (18%)**	**0**	**131**
Other Artifacts							
Hammerstone	2	1	0	0	0	0	3
Manuport	0	6	2	0	0	0	8
Subtotal	**2 (18%)**	**7 (64%)**	**2 (18%)**	**0**	**0**	**0**	**11**
Total	**212 (1%)**	**7346 (48%)**	**267 (2%)**	**6006 (39%)**	**1462 (10%)**	**11 (<1%)**	**15304**

[1] Row column subtotals with percentages are provided. Three FCR and six pot lids not included.

Raw Material Selection for Excavated Artifacts

The analysis of raw materials selected focuses first on the debitage and cores, then on the retouched tools, followed by other artifacts. The codes used for classifying material types are in Appendix 14.5. Next the x-ray fluorescence analysis of obsidian artifacts from both surface collections and excavations is presented. The general raw material analysis is presented by major rock type and not subtypes, since most of the basalt, rhyolite, chalcedony, and chert materials were similar to locally available materials. Table 14.2 presents the information on artifact type by material type for the total excavated chipped stone assemblage. A total of 15,090 pieces of debitage, 72 cores, 8 manuports, 3 hammerstones, and 131 retouched tools were analyzed from the excavated features at Cerro Juanaqueña, including those identified during in-field analysis.

Debitage and Cores

Analysis of the entire excavated debitage assemblage indicates that it is dominated by rhyolite (48%), with less chalcedony (39%), chert (10%) and other materials. The sample size is much smaller for the cores, but they represent a similar variation in material type as exhibited by the debitage.

While cores and debitage have similar proportions of material types, flake production is dominated by rhyolite and basalt while tool production relies more heavily on chalcedony and obsidian. The analysis of

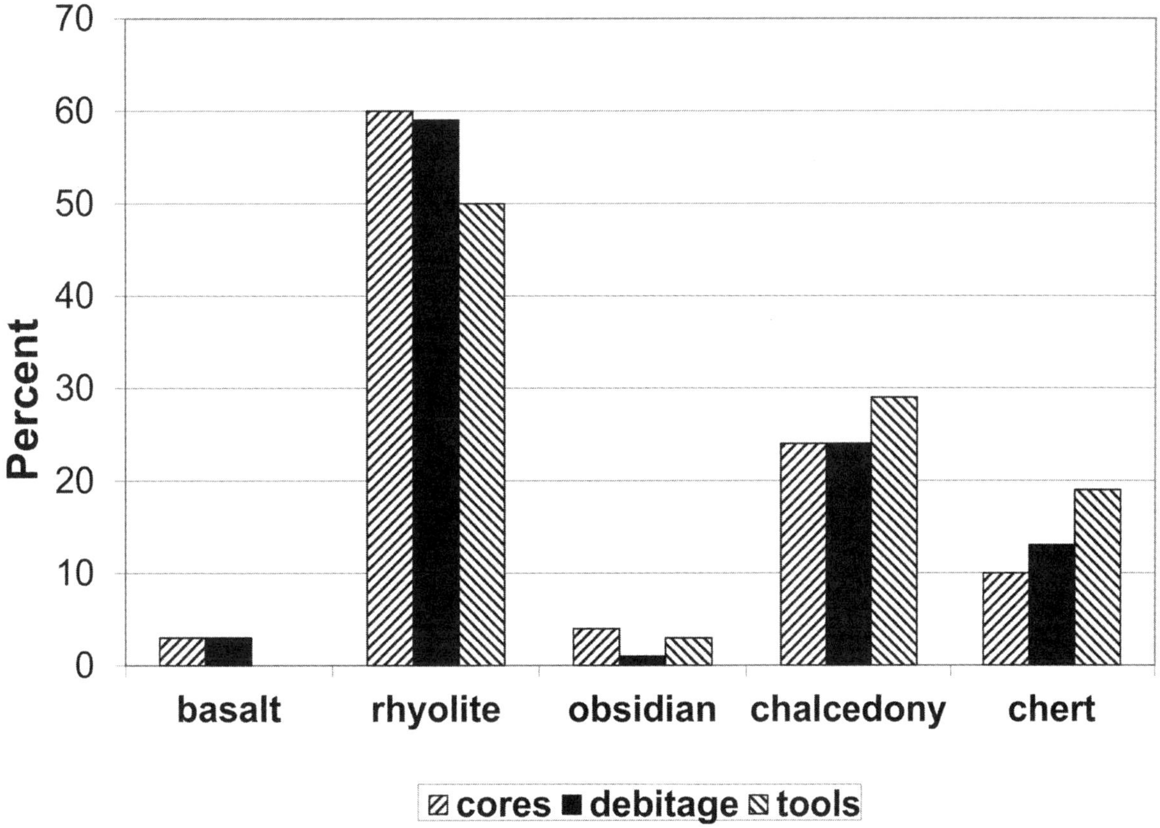

FIGURE 14.2. Bar chart showing lithic raw materials used for cores, debitage, and tools.

a contingency table of core flakes vs. biface flakes by material type shows this pattern (Appendix 14.6). An evaluation of the adjusted residuals indicates that there are relatively more rhyolite and basalt core flakes, vs. obsidian and chalcedony biface flakes (chi-square = 234.6, $df = 4$, $p \leq 0.001$, $n = 5{,}107$). A similar pattern is seen in Figure 14.2 considering the materials used for cores, debitage (excluding microdebitage), and retouched tools. As can be seen, retouched tool material is dominated by obsidian, chalcedony, and chert, while debitage material is dominated by rhyolite.

Assigning material types to material grain classifications shows that material quality is fundamental to its selection for different types of reduction. Obsidian and chalcedony are fine-grained materials, chert is both fine and medium-grained, rhyolite is both medium and coarse-grained, and basalt, along with the "other" materials, is mostly coarse-grained. Table 14.3 presents these data for debitage and shows how each material type is patterned with respect to its grain. Rhyolite debitage is dominated by medium and coarse grain varieties while chert debitage is dominated by medium grain material.

Material grain strongly impacts the selection of particular materials for the production of different artifacts. A chi-square analysis of core flakes and biface flakes by material grain indicates that significantly different materials were selected for the production of these artifacts (Appendix 14.7). Fine-grained materials were primarily selected for tool production vs. medium and coarse-grained materials for core reduction (chi-square = 272.5, $df = 2$, $p \leq 0.001$, $n = 5{,}113$). This supports the contention that higher quality materials are generally selected for tool production and lower quality materials for simple flake production.

The raw material analysis of local stream bed deposits indicates that many of the raw materials present at Cerro Juanaqueña could have been derived from this source. To evaluate this proposition, the nature of the exterior cortex was identified in the excavated debitage to determine possible sources (Appendix 14.8). The majority of the rhyolite (99%), chalcedony (97%), and chert (99%) debitage exhibit stream rolled (i.e., waterworn) cortex, indicating that they were obtained from secondary river gravels (e.g., the Río Casas Grandes). Basalt is roughly equally distributed between waterworn and a natural weathered or nodular cortex, reflecting acquisition from both river channels and nearby hills. The exact source is unclear for the chert cobbles exhibiting waterworn cortex; however, they could have been obtained elsewhere along the Río Casas Grandes,

TABLE 14.3. Excavated Debitage Material Type by Material Grain.[1]

Material Type	Fine	Medium	Coarse	Total
	Material Grain			
Basalt and Other[2]	0 **[−13.0]**	15 **[−11.5]**	206 **[36.8]**	221 (1.5%)
Rhyolite	165 **[−96.6]**	5434 **[71.6]**	1641 **[37.1]**	7240 (47.9%)
Obsidian	260 **[18.8]**	0 **[−14.7]**	0 **[−6.1]**	260 (1.7%)
Chalcedony	5951 **[114.6]**	0 **[−89.4]**	0 **[−37.2]**	5951 (39.4%)
Chert	89 **[−29.4]**	1330 **[38.3]**	16 **[−13.6]**	1435 (9.5%)
Total	6465 (42.8%)	6779 (44.9%)	1863 (12.3%)	15107

$\chi^2 = 15841.0$, $df = 8$, $p < 0.001$
[1] Top cell values are artifact counts. Bottom, bracketed values represent adjusted residuals. Significant ($p < 0.05$) values are shown in bold.
[2] "Other" Raw Material Type ($n = 11$) combined with Basalt ($n = 210$).

TABLE 14.4. Tool Group by Material Grain.[1]

Tool Group	Fine	Medium	Coarse	Total
	Material Grain			
Formal	35 [0.9]	46 [−1.1]	8 [0.4]	89 (67.9%)
Informal	13 [−0.9]	26 [1.1]	3 [−0.4]	42 (32.1%)
Total	48 (36.6%)	72 (55%)	11 (8.4%)	131

$\chi^2 = 1.2$, $df = 2$, $p = 0.55$.
No cell values are significant ($p < 0.05$).
[1] Top cell values are artifact counts. Bottom, bracketed values represent adjusted residuals.

where tributary drainages may have deposited chert derived from nearby limestone outcrops. For example, Rebnegger (2001:67) does note the presence of chert cobbles on the hills and in the gravels of the Río Casas Grandes near Colonia Juárez and Nuevo Casas Grandes. By contrast, the majority of the obsidian (93%) exhibit nodular cortex indicating that they were obtained from primary sources. However, most of the obsidian sources in northern and central Chihuahua are at least 10 million years old and have been subjected to long-term erosion and secondary deposition. So the exterior cortex may have hydrated, obscuring the original waterworn surface (M. Steven Shackley, personal communication 2004). Some very small nodules of obsidian were observed to the south near Cerro el Canelo, but these were too small for producing useable flakes or tools.

Retouched Tools

The excavated retouched tools were manufactured on rhyolite, chalcedony, and chert, with some obsidian. As previously noted, finer grained materials appear to have been selected for biface production vs. more coarse-grained materials for core reduction. A chi-square analysis was also conducted of a contingency table for tool group by material grain in order to determine if there were any selective preferences by tool group (Table 14.4). In this case, tool group was subdivided into informal and formal categories. Informal tools include retouched pieces, notches, denticulates, and perforators; whereas, formal tools include bifaces, projectile points, unifaces drills, and cruciforms. Various studies have shown that lower quality materials are generally selected for the production of informal tools, while higher quality ma-

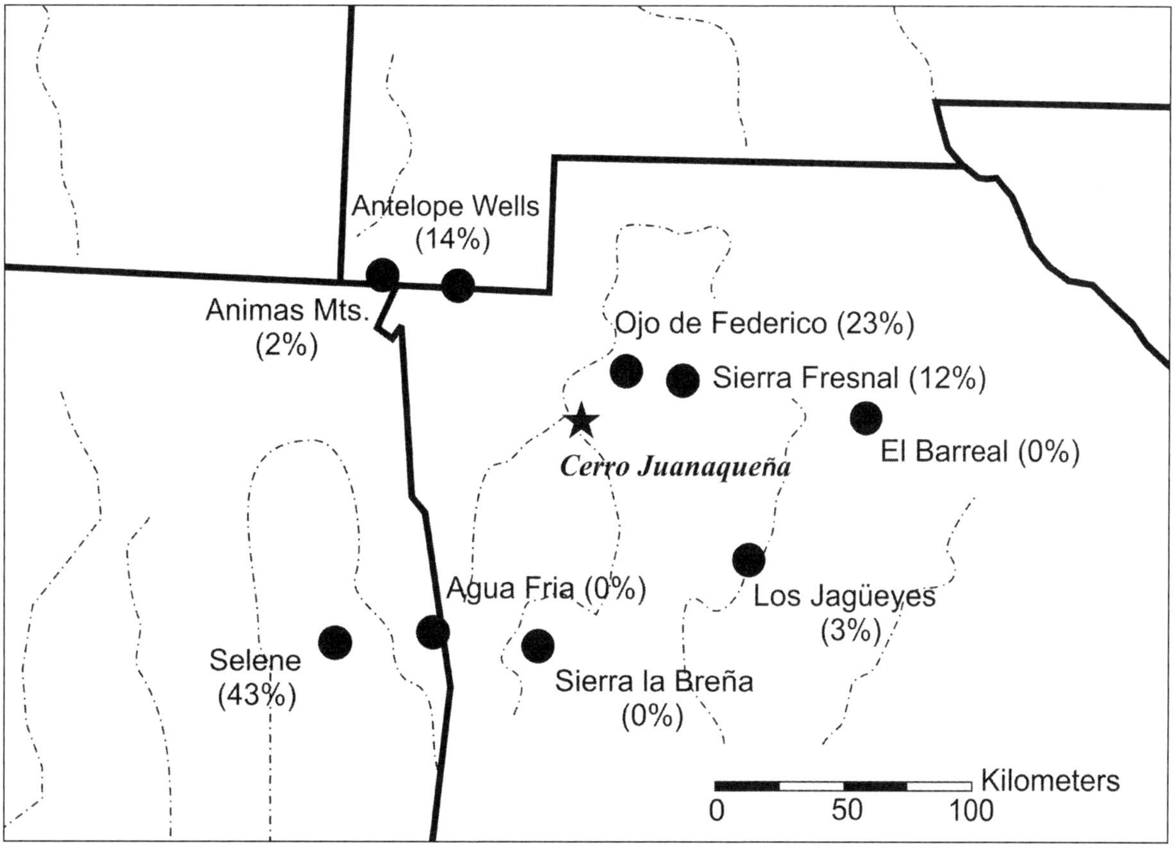

FIGURE 14.3. Sources of obsidian near Cerro Juanaqueña. Percentages reflect the proportion of analyzed obsidian from each source (*n*=89). Source could not be determined for 3% of the obsidian specimens.

terials are often used for the production of formal tools (Andrefsky 1994; Bamforth 1986; Vierra 1993a). The use of higher quality materials tends to increase tool use life through re-sharpening. Nonetheless, the results of this analysis indicate that there is no significant difference in the preference for material grain in the production of these tool groups. Perhaps the more coarse-grained materials used for core reduction were primarily blanks for flakes that were used without retouching and are therefore not included with the informal tools.

Other Artifacts

Only a few hammerstones and manuports were recovered during the excavations. The hammerstones are made of basalt and rhyolite, that is, materials that are more resistant to breakage. On the other hand, the manuports are unmodified cobbles of rhyolite and pebbles of obsidian that presumably represent materials to be used in core reduction, or else unused hammerstones.

X-ray Fluorescence Analysis of Obsidian Artifacts

As previously noted, with the exception of the obsidian, most of these materials are locally available in the Río Casas Grandes gravels. Indeed, practically all the cortex present on the debitage is waterworn cortex. Obsidian on the other hand, only comprises about 2% of the total Cerro Juanaqueña excavated assemblage, with 258 pieces of debitage, one core, two split pebbles, two unmodified pebbles, and four retouched tools being made of this material (Table 14.2). An x-ray fluorescence analysis was conducted by Steve Shackley of a sample of 99 obsidian artifacts including 89 from Cerro Juanaqueña, four from Cerro el Canelo, three from Cerro el Torres, and three from Cerro Vidal. The Cerro Juanaqueña artifacts are from both surface (55) and excavated contexts (34) and include 49 pieces of debitage, 10 bifaces, 17 projectile points, one core, and seven pebble manuports or nodules; tools were emphasized in the sample selection.

Six separate regional obsidian sources were identified (Figure 14.3). Selene, Sierra Fresnal, and Antelope Wells are associated with known primary sources while Ojo de Federico[2] and Los Jagüeyes and the provisional Animas Mountains are known only from secondary sources (Shackley 1998, 1999, 2005b) (Table 14.5). The Sierra Fresnal and Antelope Wells sources are documented to be widely dispersed in secondary settings over large areas. Geographically these six sources form three distinct geographic locations, one in northeast

TABLE 14.5. Cerro Juanaqueña XRF Analysis of Obsidian Sources.

Obsidian Source	Primary[1]/ Secondary	Subsurface	Surface	Total	Percent	Distance from Cerro Juanaqueña km	Direction from Cerro Juanaqueña	Geographic Cluster
Selene	Primary	22	16	38	42.7%	125	SW	northeast Sonora Sierra Madre
Ojo de Federico	Secondary	4	16	20	22.5%	30	NE	northwest Chihuahua Basin and Range
Sierra Fresnal	Primary	4	7	11	12.4%	52	NE	northwest Chihuahua Basin and Range
Los Jagüeyes	Secondary	1	2	3	3.4%	88	SE	northwest Chihuahua Basin and Range
Antelope Wells	Primary	0	12	12	13.5%	75	NW	New Mexico bootheel
Animas Mountains	Secondary	1	1	2	2.2%	100	NW	New Mexico bootheel
Indeterminate		2	1	3	3.4%			
Total				89	100.0%			

[1] Primary source has been identified.

Sonora in the Sierra Madre Occidental, a cluster of sources in the northwestern Chihuahua basin and range country, and a cluster of two sources in the New Mexico Bootheel.

All of these sources are Middle Tertiary (40–15 million years ago) in origin (Kibler et al. 2014; Shackley 2005b:16–20), but erosional processes over millions of years redistribute obsidian from its primary sources over unknown but finite areas. The Selene source is located in northeast Sonora in the Sierra Madre Occidental physiographic province. It is in the northward flowing upper Río Bavispe basin of northeastern Sonora, and its nodules may be distributed along its drainage although investigators could find little evidence of this (Kibler et al. 2014). The other sources are in the Basin and Range physiographic province of northern Chihuahua and southwestern New Mexico where ephemeral streams empty into interior basins. During the Pliocene and Pleistocene more of these basins were fluvially interconnected and backflushing may have been typical, in contrast to modern unidirectional flow (Shackley 1998, 1999). Such erosional processes redistributed obsidian substantial distances from primary sources making obsidian source provenance studies challenging since obsidian nodules may have been obtained from redeposited secondary sources as well as primary locations. The Sierra Fresnel, Ojo de Frederico, and Las Jagüeyes form a cluster of sources in northwest Chihuahua that includes portions of the Río Casas Grandes and Río Santa María watersheds. Antelope Wells and Animas Mountains sources form another geographic cluster in the New Mexico Bootheel. Future work will require sampling obsidian over large areas to better understand the secondary distribution of nodules and to continue to identify undiscovered sources (Dolan et al. 2017; Kibler et al. 2014; Shackley 2005b).

The Selene primary source derives from a Middle Tertiary rhyolite/perlite flow 125 km southwest of Cerro Juanaqueña in an oak savannah at 1400 to 1600 m elevation (Kibler et al. 2014). Originally, it was identified as Unknown A in our study (Shackley 1998, 1999). Kibler and his colleagues (2014) have identified its geographic location, and have named it the Selene source. Subsequently, Shackley (personal communication to Robert Hard, 2017) reanalyzed the XRF data and reported that all Cerro Juanaqueña Unknown A is Selene. Many Selene marekanites or nodules have excellent flaking properties and occur in densities up to 30 per m^2. The nodules trend 30 to 50 mm in diameter, and 50 to 80 mm nodules are present but uncommon. Therefore, Selene nodules are arguably larger in size than the Chihuahua Basin and Range sources (Kibler et al. 2014). These qualities likely account for the desirability of Selene obsidian at Cerro Juanaqueña as it is the most distant from the site yet the most commonly used.

The northwest Chihuahua Basin and Range cluster of sources include Sierra Fresnal, Ojo de Federico, and Los Jagüeyes. Sierra Fresnal is a primary source 50 km northeast of Cerro Juanaqueña where high densities of marekanites are associated with a series of heavily eroded and coalesced rhyolite domes (Shackley 1998). The erosional flows following the Middle Tertiary eruptive event are poorly known but likely include backflushes and other complex stream dynamics that spread Sierra Fresnal obsidian widely. It may be eroding on both the west and east sides of the mountain range which would carry it both into the Río Casas Grandes and ultimately into Laguna de Guzmán, as well as eastward into the neighboring Laguna el Fresnal basin. Sierra Fresnel nodules have been identified in Arroyo Seco channel gravels 100 km to the south of the Sierra Fresnel. This location is upstream from the primary source with respect to modern patterns of drainage and erosion. Sierra Fresnel nodules have also been found 70 km downstream of the Sierra Fresnal in the Río Casas Grandes alluvium near the New Mexico border. Thus Sierra Fresnal obsidian occurs both north and south of Cerro Juanaqueña over a distance of at least 130 km (Shackley 1998, 1999, 2005b). These data demonstrate that obsidian provenance research must consider both the primary sources as well as the finite distribution of secondary sources.

The Ojo de Federico obsidian is known only from a secondary source where nodules were found in the alluvium just west of Sierra Fresnal and only 30 km northeast of Cerro Juanaqueña. They may have originated from domes in the Sierra Fresnal or yet another source upstream in the Río Casas Grandes system. They have a different elemental signature from the Sierra Fresnal source and likely derived from a different magma source (Shackley 2005b:83).

The Los Jagüeyes secondary source was found along the Los Jagüeyes channel, a tributary of the Río Santa María 90 km southeast of Cerro Juanaqueña. This location is an archaeological site that not only contains Los Jagüeyes marekanites but also stone tools, including Early Agricultural period projectile points with metate fragments, but no ceramics (Shackley 2005b). Additional secondary sources are likely and the primary source is unknown. Sierra Fresnal nodules are also present at this location, and they may be either geological or cultural in origin (Shackley 2005b).

The Antelope Wells obsidian occurs in primary and secondary contexts over a large area in the New Mexico Bootheel in southcentral Hidalgo County, New Mexico, and also occurs within 15 to 20 km of adjacent Chihuahua and is about 75 km northwest of Cerro Juanaqueña (Shackley 2005b). The material is an excellent tool stone as it is free of phenocrysts. It was widely used in the Southwest and occurs as far west as Tucson (Findlow and Bolognese 1980; Shackley 2005b:57). Nodules range up to 100 mm in diameter but most are half that size.

Animas Mountains obsidian (known also as Sonora Unknown A) is a provisional source identified with a single geological nodule found in New Mexico between the Peloncillo and Animas Mountains near the international border in the New Mexico Bootheel (Shackley 2005b, 2017). Chemically distinct from Antelope Wells although from the same area, it was originally identified in the Cerro Juanaqueña study as Unknown B (Shackley 1999) but has now been confirmed as Animas Mountains obsidian (Shackley, personal communication to Robert Hard 2017). It has distinctive elemental signature with a relatively high concentration of Rb and very low Sr. It has also been found on archaeological sites in the Douglas, Arizona, area, and a single piece was discovered on the Casas Grandes outlier site 315 (Dolan et al. 2017; Shackley 2005b:85). Animas Mountains obsidian was the least used source at Cerro Juanaqueña.

The dominant source in the Cerro Juanaqueña XRF sample that includes both surface and excavated material (Table 14.5) is from the Selene source in northeastern Sonora ($n = 38$); the others, in declining order include: Ojo de Federico ($n = 20$), Antelope Wells ($n = 12$), Sierra Fresnal ($n = 11$), Los Jagüeyes ($n = 3$), and Animas Mountains ($n = 2$), plus indeterminates ($n = 3$).

All obsidian projectile points in the analyzed sample are from the surface. The nine dart points in the sample include Tularosa Corner-notched, San Pedro Corner-notched, and unidentifiable dart fragments (Chapter 6). Six dart points are made from Selene obsidian, two are from Ojo de Federico, and one is made from the Los Jagüeyes source (Table 14.6). Five small points which are assumed to be arrow points include two from the Ojo de Federico source, two from Antelope Wells, and one that yielded an indeterminate chemical signature. The three remaining point fragments could be either dart or arrow points and are represented by one from Antelope Wells and two from Selene. The Antelope Wells source is only represented by surface artifacts including the arrow points. Although a small sample, it may suggest later visitors to the site added that more distant source. Note that the discussion in Chapter 6 points out that bow and arrow technology may have been underway during the latter part of the Early Agricultural period (Sliva 1999a), an interval that includes the later occupation of Cerro Juanaqueña.

In summary, the three obsidian source geographic locations (northeast Sonora Sierra Madre Occidental, northwest Chihuahua Basin and Range, and New Mexico Bootheel) lie to southwest, eastward, and northwest from Cerro Juanaqueña, a catchment of roughly 30,000 km^2 (Table 14.5). It reflects use of the Basin and Range and the Sierra Madre provinces and presumably indicates the region that would have been exploited for other resources as well. Notably the Agua Fria and Sierra

TABLE 14.6. Obsidian Artifact Types and Sources.

	Biface	Core	Debitage	Dart Point	Small Point	Unknown Point Fragment	Nodule	Unknown	Total
Animas Mountains									
Count	0	0	2	0	0	0	0	0	2
% of Total	0.0%	0.0%	2.2%	0.0%	0.0%	0.0%	0.0%	0.0%	2.2%
Antelope Wells									
Count	0	0	6	0	2	1	3	0	12
% of Total	0.0%	0.0%	6.7%	0.0%	2.2%	1.1%	3.4%	0.0%	13.5%
Ojo de Federico									
Count	2	0	12	2	2	0	2	0	20
% of Total	2.2%	0.0%	13.5%	2.2%	2.2%	0.0%	2.2%	0.0%	22.5%
Los Jagüeyes									
Count	1	0	1	1	0	0	0	0	3
% of Total	1.1%	0.0%	1.1%	1.1%	0.0%	0.0%	0.0%	0.0%	3.4%
Sierra Fresnal									
Count	2	0	6	0	0	0	2	1	11
% of Total	2.2%	0.0%	6.7%	0.0%	0.0%	0.0%	2.2%	1.1%	12.4%
Selene									
Count	5	1	22	6	0	2	0	2	38
% of Total	5.6%	1.1%	24.7%	6.7%	0.0%	2.2%	0.0%	2.2%	42.7%
Indeterminate									
Count	0	0	0	0	1	0	0	2	3
% of Total	0.0%	0.0%	0.0%	0.0%	1.1%	0.0%	0.0%	2.2%	3.4%
Total									
Count	10	1	49	9	5	3	7	5	89
% of Total	11.2%	1.1%	55.1%	10.1%	5.6%	3.4%	7.9%	5.6%	100.0%

la Breña sources in the Sierra Madre Occidental, as well as the secondary El Barreal source[3] (Shackley 2005) in the northwest Chihuahua Basin and Range are within this catchment but do not appear to have been used.

The seven obsidian pieces from Cerro Torres, Cerro el Canelo, and Cerro Vidal that yielded valid XRF results indicate the same sources were being used at those sites as at Cerro Juanaqueña. Cerro Torres, which dates to the San Pedro phase, yielded two excavated obsidian artifacts that were submitted for XRF analysis. One is from the Ojo de Federico source and the other is from Antelope Wells, confirming that these sources were used during the period Cerro Juanaqueña was occupied. Two nodules found eroding out at the base of the San Pedro phase site of Cerro el Canelo are from Los Jagüeyes, as is a projectile point from the surface of the site, while another surface projectile point is from the Sierra Fresnal source. Cerro Vidal, which dates to the Cienega phase, produced one nodule from the Selene source.

We also attempted to identify correlations between chemical types and obsidian visual types. The visual types included translucent, banded semitranslucent, black opaque, semitranslucent with green edges, dark greenish gray, glossy gray, and dusky blue. Most of the obsidian was defined as the translucent type (54%), with the analysis revealing no clear correlation between visual type and obsidian source.

Vierra's (1994a, 1994b) studies of Archaic obsidian procurement ranges indicated that groups residing on the Colorado Plateau covered an area approximately 8,500 km^2 in size, in contrast to groups residing in the southern Basin and Range who ranged over an area covering about 38,000 km^2. This contrast presumably reflects the shorter distance between resource patches on the Colorado Plateau, vs. the homogenous structure of resources in the southern deserts (e.g., Wills 1988: 57). The procurement range represented at Cerro Juanaqueña appears to be similar to that exhibited by other Archaic groups residing in the southern Basin and

Range covering an area of approximately 30,000 km^2 (Vierra 2005). Possibly obsidian was collected during logistical foraging trips that involved hunting game at long distances from the residential site. This strategy would have increased the probability of encountering large game that might have otherwise been depleted in the area of the residence (Geib 2011:186; Grove 2010; Szuter and Bayham 1989).

Summary of Raw Material Analysis

The chipped stone assemblage from Cerro Juanaqueña is dominated by the use of locally available raw materials like rhyolite, chalcedony, and probably chert. Rhyolite cobbles were directly procured from nearby river gravels and cached at the site for core reduction. Obsidian, on the other hand, is the only nonlocal material present. It appears to have been brought to the site as small pebbles that could have been collected incidentally during subsistence-related activities. This tactic has been noted by Shackley (1990) and Vierra (1993a) for Southwestern Archaic groups (e.g., logistical hunting). Lastly, fine-grained materials were primarily selected for tool production, vs. medium and coarse-grained materials for core reduction.

Lithic Reduction Strategies

After the appropriate raw material has been selected, a variety of methods can be used to produce the desired finished product. The term *reduction trajectory* refers to a stage-like sequence of stone tool manufacture beginning with the preparation of a core and ending with the completion of a finished retouched tool (Bradley 1975; Chapman 1982; Collins 1975; Inizan et al. 1999; Van Peer 1992, 1993). For example, this process might consist of a cobble-flake or a prepared core-flake-bifacial tool trajectory. Again, a distinction should be made between reduction strategies and reduction tactics. *Reduction strategy* refers to the specific tool being produced, such as a biface or a simple flake, and *reduction tactic* refers to the manner in which the tool is produced, or the reduction trajectory. Although some raw material determinists propose that raw material availability conditions reduction strategy (Bamforth 1986), this is not the case. It is the foraging tactic—what you eat—and the procurement tactic—how you get it—that conditions the reduction strategy; whereas, it is the reduction tactic that is primarily affected by raw material availability. Nonetheless, a flint knapper produces a variety of by-products during the stone tool manufacturing process. These by-products provide the intricate details of the specific techniques used to produce stone tools (Andrefsky 2001, 2009; Shott 1994; Whittaker 1994).

The lithic reduction section of this chapter provides information about core reduction techniques, stages of reduction represented by the debitage, and evidences of tool production/maintenance. Descriptions of the variation in cores, hammerstones, debitage, and retouched tools are included.

In-field and Excavated Samples for Cores and Hammerstones

Only 56 prepared cores, 14 cobble unifaces, 2 split pebbles, and 3 hammerstones were recovered during the excavations (Table 14.2). Based on field surface observations, most of the cores and hammerstones are distributed in toss zones located along the edge of the terraces. Since excavation units were primarily located in the central area of these features, few of these items were recovered. Therefore, an in-field surface analysis was conducted of tools from these toss zones to increase the core and hammerstone sample sizes. A total of 45 prepared cores, 62 cobble unifaces, 20 hammerstones, two tested cobbles, two manuports, and an anvil were identified during the in-field surface analysis. These finds increased the prepared core sample size to 101, the number of cobble unifaces to 76, and hammerstones to 23. The sample size of all three artifact categories was dramatically grown.

The core reduction process primarily involved the use of a single-directional, single-face reduction technique on both cores and cobble unifaces (Table 14.7). In contrast, flakes were removed from single-faceted platforms on prepared cores, vs. simple cortical platforms on cobble unifaces. The presence of unmodified cobble manuports indicates that these raw materials were brought to the site for reduction. Once at the site, several reduction tactics were used to remove flakes.

Tactic #1 involves the removal of flakes from cortical platforms along a single-face of a cobble. Broad flakes with cortical platforms were removed from the center of the cobble and distinctive orange rind flakes from the lateral sides of the flaking surface. This technique was often used to reduce rhyolite cobble unifaces. *Tactic #2* involves the removal of flakes from platform cores that were produced by initially splitting the cobble roughly in half. The fresh break was used as a platform, or the cortical surface was removed adjacent to the fresh break, creating a single-faceted platform and single-face core. This tactic was mostly used to reduce both large rhyolite cobbles and small obsidian pebbles, producing flakes with faceted platforms. *Tactic #3* also involved the removal of flakes from platform cores, but the end of cobble was removed and used as the platform. Flakes with faceted platforms were removed down the long axis of the cobble. This tactic was useful for reducing smaller rhyolite, chert, and chalcedony cobbles. *Tactic #4* involves the use of large rhyolite flakes as cores. These flakes were sometimes the by-product of the other

TABLE 14.7. Core Type by Core Shape.[1]

Core Type	Core Shape									
	Single-face	Multi-face	Prismatic	Pyramidal	Discoidal	Bifacial	OSF, ODF[2]	Globular	Bipolar	Total
Single-directional	36	10	1	4	0	0	0	0	0	51 (61%)
Bidirectional	0	0	0	0	3	1	8	0	0	12 (14%)
Multidirectional	0	0	0	0	0	0	0	9	0	9 (11%)
Bipolar	0	0	0	0	0	0	0	0	1	1 (1%)
Flake core	7	4	0	0	0	0	0	0	0	11 (13%)
Total	43 (51%)	14 (17%)	1 (1%)	4 (5%)	3 (4%)	1 (1%)	8 (10%)	9 (11%)	1 (1%)	84

[1] Includes excavated ($n = 42$) and in-field surface ($n = 42$) samples.
[2] OSF is opposed same face and ODF is opposed different face.

TABLE 14.8. Core Type by Discard Condition.[1]

Core Type	Condition						
	Broken: Flaw	Broken: Cultural	Extensive Hinging	Extensive Battering	Exhausted	Still Useable	Total
Single-directional	0	1	8	1	17	24	51 (29%)
Bidirectional	0	0	4	0	4	4	12 (7%)
Multidirectional	0	0	1	0	4	4	9 (5%)
Bipolar	0	0	0	0	1	0	1 (1%)
Flake core	0	1	1	0	6	4	12 (7%)
Core fragment	3	13	0	0	0	0	16 (9%)
Cobble unifaces	1	5	10	0	14	46	76 (43%)
Total	4 (2%)	20 (11%)	24 (14%)	1 (1%)	46 (26%)	82 (46%)	177

[1] Excavated ($n = 70$) and In-field Surface Sample ($n = 107$).

reduction tactics, with the ventral surface primarily being used as the striking platform. At least one basalt flake core indicates that this technique was also used for boulder sources. Other reduction techniques like bifacial, discoidal, change-of-orientation, globular, and bipolar cores may represent attempts to increase core use life. On the other hand, pyramidal and prismatic cores appear to reflect attempts at producing more standardized elongated flakes or small blades.

Observations related to the reasons for core discard indicate that that single-directional cores and cobble unifaces were mostly still useable when discarded (Table 14.8). This finding contrasts with bidirectional and multidirectional cores that exhibit an equal mix of useable and exhausted condition when discarded. Although sample size is small, this pattern may indicate that varying reduction tactics were used to increase core use life prior to discard. The amount of cortex present decreases from single-directional to multidirectional types (Appendix 14.9). This information provides some preliminary support for changing core reduction tactics to increase core reduction use life.

Core weight was only monitored during lab analysis of the excavated items. Therefore, the sample size is small and variable, but information on core size also appears to support this proposition. That is, mean core weight decreases from single-directional to bidirectional to multidirectional cores (Table 14.9). On the other hand, mean flaking surface length is generally the same for all three of these core types, ranging from 28 to 30 mm, but is slightly longer for cobble unifaces. This pattern is indirectly monitored via mean core length in Table 14.9. It would appear that flake length was a limiting factor for core use life. Even flake cores exhibit a mean flaking surface length of about 26 mm, but are mostly exhausted like multidirectional cores. With the

TABLE 14.9. Excavated and In-field Surface Sample Core Type and Cobble Uniface Measurements.[1]

		Measurements			
Core Type		Minimum	Maximum	Mean	Std.
Single-directional (*n* = 47)	Length (mm)	11	65	34.8	11.8
	Width (mm)	19	69	45.8	12.5
	Thickness (mm)	15	65	39.0	10.2
	Weight (g)	4.7	238.7	85.5	55.5
Bidirectional (*n* = 11)	Length	20	48	36.0	11.1
	Width	15	62	41.0	15.9
	Thickness	17	55	36.4	12.2
	Weight	7.0	203.8	76.3	62.5
Multidirectional (*n* = 8)	Length	22	54	33.8	10.4
	Width	18	61	38.2	14.0
	Thickness	8	54	31.0	15.9
	Weight	2.5	153.9	67.4	58.3
Flake Core (*n* = 11)	Length	11	46	25.7	10.3
	Width	25	74	44.0	17.6
	Thickness	14	65	38.5	19.1
	Weight	4.4	211.3	55.7	56.5
Cobble Uniface (*n* = 69)	Length	12	155	42.0	19.2
	Width	29	95	63.4	12.5
	Thickness	29	93	54.6	14.1
	Weight	49.9	371.2	177.1	86.5

[1] Excludes four outliers from the single-directional category and a single outlier from the bidirectional, multidirectional, and cobble uniface categories.

exception of the flake cores, most other cores tend to look blocky, since they represent cobble fragments.

As previously noted, it appears that reduction tactics can vary by material type. That is, differing tactics are used to increase core use life for higher quality materials. For example, the proportion of cores made from higher quality materials (i.e., obsidian, chalcedony, and chert) increases from single-directional (22%, 11/51) to bidirectional (25%, 3/12) to multidirectional (44%, 4/9) (Appendix 14.10). This is another tentative pattern in need of further evaluation, since the sample size is small.

It is noteworthy that 30% (7/23) of the chalcedony cores are fragments, reflecting an emphasis on fully reducing this material type (Appendix 14.10). On the other hand, data on prepared platforms indicates a relatively similar investment in the reduction of the rhyolites (29%, 20/70) vs. the higher quality materials (21%, 6/29) (Appendix 14.11). This finding seems to indicate that irrespective of the material type, an attempt was made to prepare core platforms prior to flake removal.

In conclusion, the previous core analysis revealed a pattern of changing reduction tactics to increase core use life as reflected in the reason for core discard, cortex, and weight data. Most of the single-directional cores are still useable when discarded, whereas, the bidirectional and multidirectional cores are equally distributed between still useable and exhausted. The amount of cortex present decreases from single-directional to multidirectional types, and the mean core weight also decreases along this continuum. The proportion of cores made from higher quality materials (i.e., obsidian, chalcedony, and chert), increases from single-directional to bidirectional to multidirectional.

The hammerstones are primarily made of rhyolite, with less basalt, chalcedony, and andesite. The majority of these artifacts are battered on two or three loci, with a few battered only on one or four loci. The battering is mainly situated on a convex surface, with a few hammerstones exhibiting a battered ridge, flake scar surface and all over the artifact. Only two cores and five cobble unifaces exhibit damage that mostly represents reuse as chopping tools. On the other hand, no cobble bifaces were identified during the analysis that could reflect formal chopping tools.

Excavated Debitage

Debitage reflects the by-products of core reduction and tool production/maintenance activities, from the initial preparation of the core to the finishing of a retouched tool. As such it provides a much larger sample

TABLE 14.10. Excavated Whole Flake Reduction Stages.

Material	Primary	Secondary Cortical	Secondary Non-cortical	Bifacial	Total	Cortical: Non-cortical Ratio
Basalt	1	11	41	0	53	0.29
Rhyolite	57	582	370	49	1058	1.52
Obsidian	1	4	10	7	22	0.29
Chalcedony	9	99	236	70	414	0.35
Chert	9	80	124	20	233	0.61
Total	77 (4%)	776 (44%)	781 (44%)	146 (8%)	1780 (100%)	0.92

and more detailed picture of these processes at the site. The excavated debitage assemblage mostly consists of microdebitage (55.1%, 8321/15,090) and core flakes (31%, 4675/15,090), with some biface flakes (2.9%, 438/15,090), angular debris, undetermined flake fragments, core trimming flakes, and other debitage (Table 14.2). These artifacts indicate that a range of core reduction and tool production/maintenance activities occurred at the site. However, the details of this process require clarification.

We can begin by breaking up the reduction sequence into an arbitrary set of stages based on the presence/absence of cortex and debitage type. I define these stages somewhat differently that those used by some Southwestern researchers (e.g., Brown 1991; Parry 1987). Assuming that cobble raw materials are used, the outer cortex of the cobble is slowly removed during the reduction sequence. Primary reduction flakes refer to core flakes with 100% dorsal cortex. These flakes are removed early in the core reduction process. Secondary cortical reduction flakes consist of core flakes with a cortical platform and/or partial dorsal cortex, and secondary noncortical reduction flakes are core flakes with no cortex. Secondary reduction flakes are removed during the later stages of core reduction. Lastly, tertiary reduction solely consists of retouch or re-sharpening flakes that are removed during the tool manufacturing and maintenance process (e.g., biface flakes). The reduction of nodules at the site would therefore produce a relatively greater proportion of primary and secondary cortical flakes with cortical platforms; the reduction of prepared cores would produce relatively more secondary noncortical flakes and single-faceted platforms; and the production of retouched tools would result in more tertiary flakes and multifaceted platforms.

Table 14.10 summarizes the various stages of reduction represented by whole flakes. The debitage assemblage is composed of primary (4.3%, 77/1780), secondary cortical (43.5%), and secondary non-cortical (43.8%) core flakes, as well as biface flakes (8.2%), reflecting that all segments of the lithic reduction process are represented. The overall cortical: non-cortical ratio of 0.92 reflects a roughly equal balance between the earlier and later stages of core reduction with some tool production. However, there is some variation within specific material types. For example, the high cortical:non-cortical ratio for rhyolite reflects an emphasis on the earlier stages of core reduction with a ratio of 1.5. The ratios for basalt (0.3), obsidian (0.3), chalcedony (0.4), and chert (0.6) reflect an emphasis on the later stages of core reduction along with some tool production. However, since basalt flakes, with a ratio of 0.3, may have been removed from boulder and not cobble sources, the presence of cortex may not be an accurate measure of reduction.

Information on platform types also corresponds with the data on core reduction technique (Appendix 14.12). That is, most of the platforms on core flakes are single-faceted (54.3%, 1646/3031) or cortical (41.0%, 1242/3031). The large amount of cortex, in conjunction with the presence of cortical and single-faceted platforms, indicates that both cobble unifaces and prepared platform cores were reduced at the site. The rhyolite flakes show a similar mix of cortical (52.5%, 1004/1913) and single platforms (43.8%, 838/1913). In contrast, flakes from basalt, chalcedony, obsidian, and chert are each dominated (66 to 80%) by single platforms indicating relatively more reduction of these materials compared to rhyolite following trends seen in other analyses.

The information on cortex placement on core flakes reveals a similar pattern (Appendix 14.13). Chalcedony, obsidian, chert, and basalt flakes primarily lack cortex; whereas, rhyolite flakes exhibit a preponderance of cortical platforms. This finding again reflects an emphasis on the reduction of rhyolite cobble unifaces, vs. the use of chalcedony, obsidian, and chert in the later stages of core reduction and tool production.

Both the core and debitage data indicate that differing reduction trajectories are represented among the various material types. Rhyolite cobble unifaces may have primarily been used to produce simple flakes. Prepared chert cores were used for flakes and tool blanks (e.g., marginally retouched tools), and prepared chalcedony cores were used for tool blanks (e.g., bifacial tools). Indeed, there are significantly more chalcedony and obsidian flakes with prepared platforms, than rhyolite, chert, or basalt flakes (chi-square = 217.4, $df = 4$, $p \leq 0.001$, $n = 3424$; Appendix 14.14).

This indicates that a greater amount of platform preparation was being invested in the removal of flake blanks to be used in the production of retouched tools. The small obsidian sample seems to reflect reduction of prepared cores for flakes and tool blanks, but nodule size would presumably be a limiting factor. For example, only 23 (9%) of the 262 dart points analyzed from surface collections across the site are made of obsidian, vs. 6 of 13 (46%) of the arrow points. Carmichael (1986:99) presents a similar example.

A variety of factors can affect the breakage patterns of flakes, including material type, core reduction vs. tool production, tool use, and post-depositional processes (Mauldin and Amick 1989; McBrearty et al. 1998; Prentiss and Romansky 1989; Sullivan and Rosen 1985; Tomka 1989; Whittaker 1994). The majority of core flakes are broken (64.4%), with less whole flakes (Appendix 14.15). The broken flakes are roughly evenly divided between proximal ($n = 1{,}080$) and distal ($n = 1{,}222$) fragments, reflecting flakes that tended to break in half. Midsections ($n = 412$), lateral ($n = 292$), and undetermined ($n = 1$) fragments are underrepresented in the sample. A similar pattern was identified at the Santa Cruz Bend site, where most of the core flakes (63%) were also broken in half (R. Jane Sliva, personal communication 2013). Presumably the same range of processes was affecting both assemblages at these early agricultural village sites.

A chi-square analysis of a contingency table of material type by core flake condition found no significant differences in breakage patterns among the various material types (chi-square = 24.0, $df = 16$, $p = 0.89$, $n = 4668$; Appendix 14.15). The majority of the biface flakes are also broken (66.6%), with fewer whole flakes. However, unlike the core flakes, most of the broken biface flakes are proximal fragments ($n = 175$), with fewer midsection ($n = 36$), distal ($n = 78$), and lateral ($n = 3$) fragments. In contrast, the undetermined flake fragments are dominated by midsections ($n = 108$) and distal ($n = 306$) fragments, with fewer proximal ($n = 3$), lateral ($n = 1$), and undetermined ($n = 148$) sections. Undetermined flake fragments were often small pieces that were difficult to assign to a specific debitage category. The fact that most of the biface flakes are proximal sections and that the undetermined flake fragments are midsections and distal segments indicates that many of the latter fragments may represent portions of small biface retouch flakes. Overall, there appears to be no difference in breakage patterns by material type or core vs. biface flakes. So, this breakage pattern was not affected by the reduction of different material types.

The by-products of core reduction and tool production/maintenance can produce a variety of different sized flakes. Therefore, flake size can provide important information on reduction techniques or the potential flake blanks (Patterson 1982, 1990; Patterson and Sollberger 1978; Shott 1994; Sullivan and Rosen 1985) selected for the production of tools (Ahler 1989; Hayden 1979:144, 156; Stahle and Dunn 1982; Sullivan and Rosen 1985). How does the length of the flaking surface on cores compare with the size range of core flakes? That is, were specific sized flakes being used for the production of tools and are these items missing from the general debitage assemblage?

Mean whole core flake length (21.6 mm) is actually smaller than the mean maximum length for core flaking surfaces (28 to 30 mm). Length could only be determined on four blades, but they are consistently much longer than the core flakes with a mean of 31.5 mm (Appendix 14.16). A frequency graph of whole core flake length by material type is presented in Figure 14.4. Two distinct patterns are evident. Basalt and rhyolite primarily consist of larger flakes ranging from 11–30 mm in length, with some rhyolite flakes as large as 71 mm. In contrast, obsidian, chalcedony, and chert flakes are smaller, primarily ranging from 11–20 mm in length, with none larger than 48 mm. This finding corresponds with other core and debitage data that indicates that these higher quality materials are being more fully reduced than the basalts or rhyolites. On the other hand, basalt flakes exhibit a similar size distribution as that of rhyolite. This pattern differs from the cortex evidence. As previously noted, given that basalt is locally available in boulder outcrops and not as cobbles, cortex may not be an accurate indicator of reduction for this material.

The size distribution for whole core flakes ($n = 1655$) and bifaces ($n = 36$) is illustrated in Figure 14.5. The biface sample is derived from both excavated and surface-collected contexts. As can be seen, core flakes larger than 30 mm were being selected for biface production. The number of larger core flakes declined, because the larger flakes were being selected as blanks for manufacturing bifaces. These larger flakes were probably made of fine-grained materials, therefore biasing the core flake sample towards the smaller sizes. On the other hand, the presence of larger coarse-grained core flakes indicates that they were presumably being used as unretouched flake tools. Larger flakes are often selected for use since they are more easily held.

Burned artifacts can either represent items that were accidentally burned or intentionally heat-treated to increase the knappability of the raw material. As Hester and Collins (1974) have noted, heat-treatment was commonly used for the production of bifacial tools by residentially mobile groups in lithic resource poor areas of South Texas. Heat-treatment makes the stone less

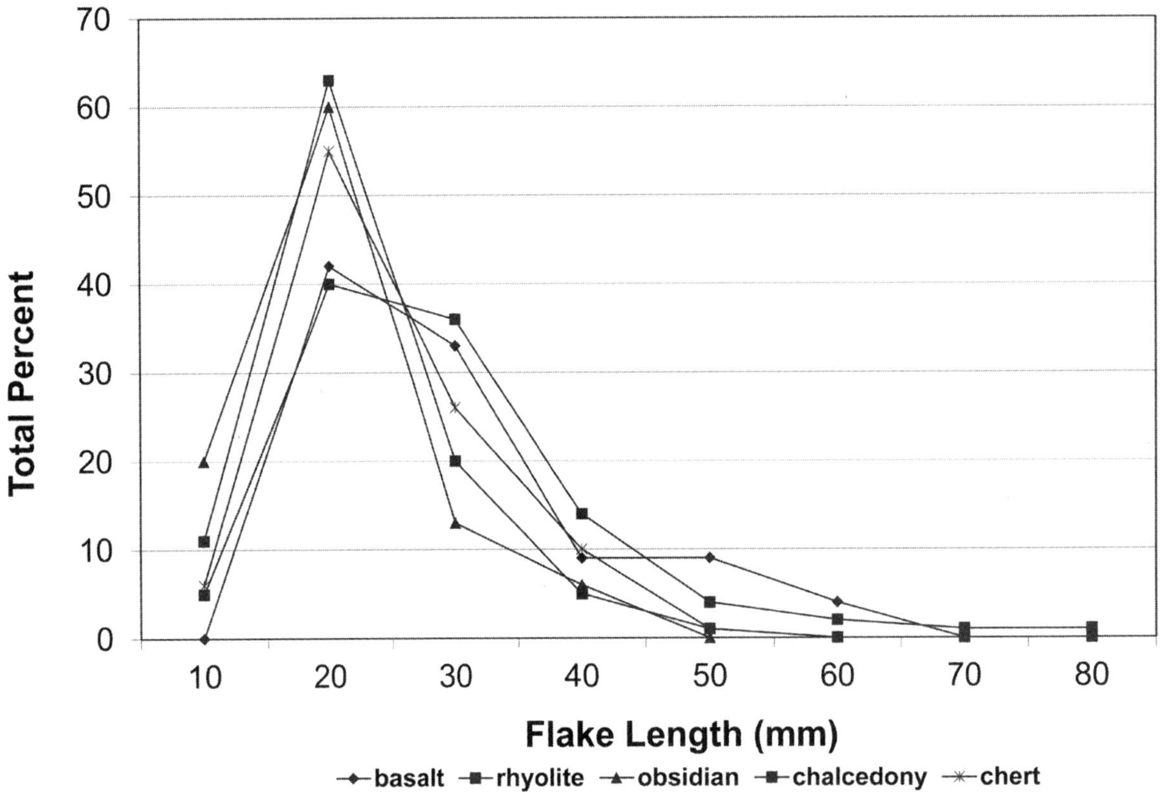

FIGURE 14.4. Frequency graph of whole core flake length by material type.

grainy, smoother in texture, more brittle, and therefore easier to flake (Whittaker 1994:72). Accidentally burned items are characterized by discoloration, crackling, and pot lids, whereas, intentionally heat-treated artifacts exhibit a glossy luster and greasy feel. My research found that medium-grained materials were being preferentially heat-treated for the production of bifacial tools at a Late Archaic campsite in South Texas (Vierra 1998). However, this does not appear to be the case at Cerro Juanaqueña where less than 1% of the debitage exhibits burning. Similar percentages are represented for both core and biface flakes. This finding contrasts with 24% of the debitage from the South Texas site. The few pieces of burned debitage at Cerro Juanaqueña appear to have been accidentally burned and not heat-treated, probably indicating that the site occupants were not as residentially mobile as some foraging groups in South Texas, and therefore, the supply of locally available raw materials was sufficient for both core reduction and tool production activities.

Several distinctive flake types also represent specific core reduction and tool production techniques. Besides biface flakes, the presence of an overstruck and a notching flake reflect biface and projectile point production. Three uniface flakes also represent the production of unifacial tools. Most core trimming flakes (n = 38) are from platform cores, but a few were removed to create platforms on cobble unifaces. These flakes, in addition to core tablets (n = 1) and opposing core flakes (n = 3), reflect three differing techniques that were used to remove step fractures adjacent to core platforms and/or to change the orientation of the core for flake removal. Another technique for removing steps from the core flaking surface involves *platform projection flakes*. These flakes are formed by retouching two adjacent notches along the platform edge, with the intermediate section forming a projection. This projection is then used as a platform that focuses the force of the blow towards the removal of isolated steps on the adjacent core flaking surface.

Other distinctive core reduction flakes include orange rind flakes with cortex around a portion of their lateral perimeter and circular or oval-shaped flakes with cortex around their total perimeter. The former are removed from cobble unifaces, and the latter represent the reduction of cigar-shaped cobbles. Orange rind flakes provide a natural cortex backing that makes the flake easy to hold while cutting or scraping. Bipolar flakes (n = 5) and possibly *pièces esquillèes* (n = 2) represent a bipolar core reduction technique (e.g. Flenniken 1981). Bipolar flakes are composed of rhyolite, obsidian, chalcedony, and chert, indicating that smaller pieces of these

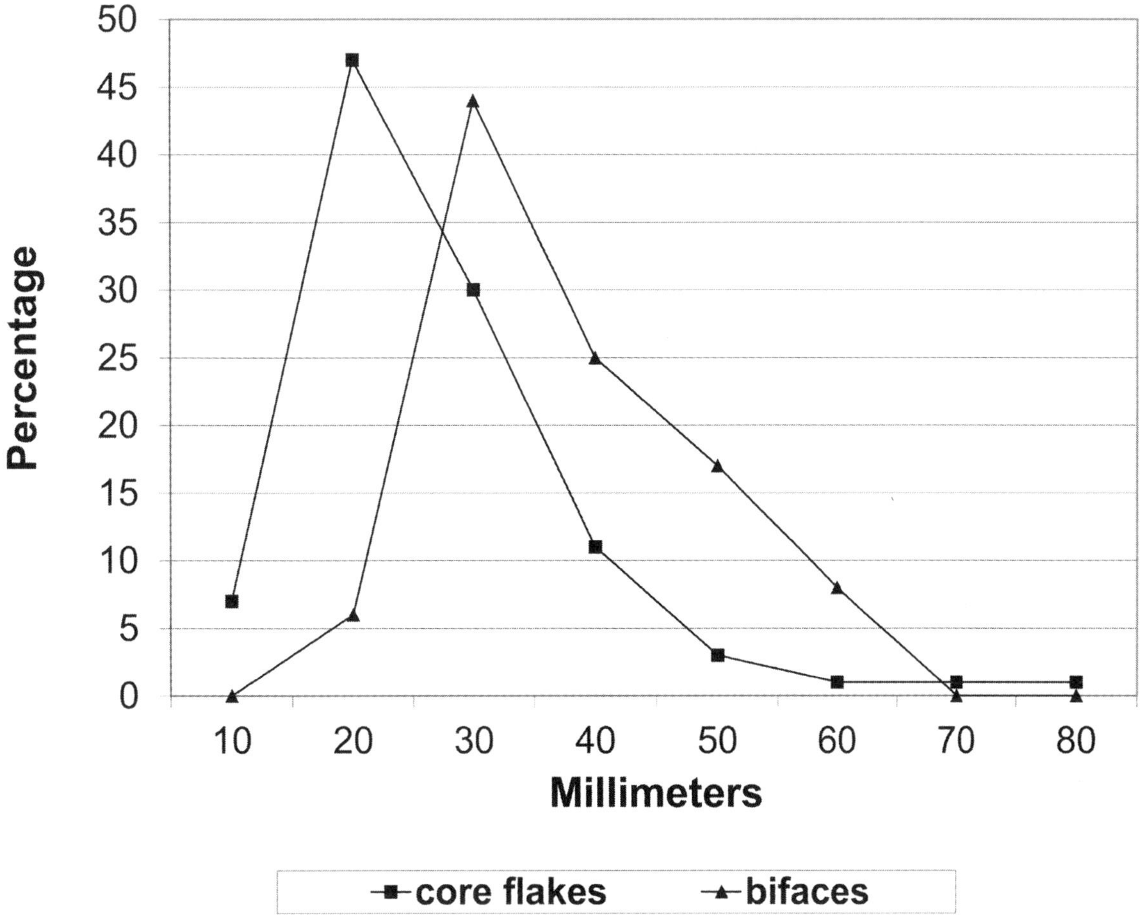

FIGURE 14.5. Size distribution for whole core flakes and bifaces.

materials were reduced using this technique. The *pièces esquillèes* are also made of rhyolite and obsidian.

Lastly, two flake types do not directly reflect core reduction or tool production/maintenance activities. Hammerstone flakes (*n* = 11) are incidentally detached from hammerstones during percussion. They correspond with the presence of several hammerstones in the excavated assemblage. On the other hand, ground stone flakes (*n* = 4) with flat ground facets on their dorsal surfaces are rather unique. These are mostly made from basalt and could have been removed from metates, possibly for breaking holes through the bottom of the grinding trough, or from manos that were recycled into small stone mortars or bowls.

Excavated Retouched Tools From Excavations

Retouch is both a technique used to shape a tool and to extend its use life (Inizan et al. 1999; Kelly 1988; Shott 1986b). At Cerro Juanaqueña the excavated retouched tools (*n* = 131) are dominated by bifaces, with fewer projectile points and retouched pieces (Table 14.11).

Informal tools like retouched pieces, notches, and denticulates make up about a third (40/131, 31%) of the excavated retouched tool assemblage (Table 14.2). Informal tools are tools that are only marginally retouched and reflect short use lives. This contrasts with formal tools like bifaces and projectile points, which compose most of the assemblage (60%). These tools are facially retouched and reflect longer use lives. In addition, seven unifaces, two drills, two perforators, and a cruciform are also represented. Given the few retouched tools collected from the excavations, the following discussion will primarily consist of simple artifact descriptions and possible tool production processes. Information on additional retouched tools collected during the general site reconnaissance is included in a later section of this chapter. Chapter 6 also presents further analyses of the tools.

Informal Tools

The retouched pieces, notches, and denticulates were mostly made on large thick flakes. They exhibit steep, unidirectionally retouched edges with a mean of about 63 degrees. These edges were produced using a hard

TABLE 14.11. Condition of Retouched Tools from Excavation.

	Condition						
Tool Type	Whole	Proximal	Midsection	Distal	Lateral	Und.	Total
Retouched Piece	10	3	4	3	3	0	23 (18%)
Notch	9	0	1	4	0	0	14 (11%)
Denticulate	2	0	1	0	0	0	3 (2%)
Perforator	2	0	0	0	0	0	2 (2%)
Biface	7	14	12	17	1	2	53 (40%)
Projectile Point	6	9	7	4	0	0	26 (20%)
Uniface	2	1	0	4	0	0	7 (5%)
Drill	1	0	0	1	0	0	2 (2%)
Cruciform	1	0	0	0	0	0	1 (1%)
Total	40 (31%)	27 (21%)	25 (19%)	33 (25%)	4 (3%)	2 (2%)	131

TABLE 14.12. Measurements of Informal Tools from Excavations.

		Measurements				
Tool Type		*n*	Minimum	Maximum	Mean	Standard Deviation
Retouched Pieces	Length (mm)	11	20	48	33.3	9.4
	Thickness (mm)	13	2	20	9.6	5.7
	Edge angle	23	40	80	63.4	10.7
Notches	Length	9	13	42	28.6	10.4
	Thickness	10	4	18	9.7	4.7
	Edge angle	13	50	80	67.3	9.7
Denticulates	Length	2	27	43	35.0	11.3
	Thickness	2	5	14	9.5	6.3
	Edge angle	3	55	70	63.3	7.6
Perforator	Length	2	20	28	24.0	5.6
	Thickness	2	5	9	7.0	2.8
	Edge angle	2	70	70	70.0	0.0

hammer percussion technique. Indeed, most of the notches were created with a single blow. Information on informal tool measurements are provided in Table 14.12. The two perforators are shorter and thinner, but have a steeper edge angle. The points on these tools were produced by unidirectionally retouching the distal of the flake. The perforators presumably functioned as expedient drills. The informal tools are about evenly divided between whole and fragments and all probably had short use lives, with some having been broken during manufacturing (Table 14.11).

Bifaces

There are 53 bifaces in the excavated collection and some can be separated into an early, middle, and late-stage reduction sequence based on the width:thickness ratio and edge angle (Callahan 1979:10, 18). Width:thickness ratios can be calculated for 15 bifaces including seven whole artifacts and eight fragments. Early stage bifaces ($n = 5$) exhibit width:thickness ratios of 2.1 to 2.4, with edge angles ranging from 65 to 75 degrees. Whereas, middle-stage bifaces ($n = 6$) have ratios of 3.1 to 4.3 and edge angles of 45 to 65 degrees, and late-stage bifaces ($n = 4$) have ratios from 5.1 to 6.0 and edge angles of 35 to 50 degrees. The roughly even division among stages of reduction indicate that the full range of biface production activities was occurring at the site.

More detail can be extracted from the seven whole bifaces. One is an ovate-shaped tool that was produced on a large core flake. The blank is oriented so that a single-faceted platform is situated at the tip of the biface. A width-thickness ratio of 2.1 is consistent with an early-

TABLE 14.13. Projectile Point Measurements and Descriptive Data for 17 Excavated Specimens.[1]

Feature	Bag	Material	Cond.	Overall Length	Blade Length	Neck Width	Stem Length	Stem Width	Thickness	Wt. (g)	Haft type	Blade shape	Base shape
10	25	rhyolite	whole	42.9	29.1	13.1	13.8	17.7	5.3	4.0	sn	st	conv
97	3	rhyolite	whole	51.6	40.3	14.3	11.3	19.4	6.2	7.3	sn	st	conv
163	69	rhyolite	proximal	—	—	—	—	20.4	5.0	2.3	tri	st	st
163	86	chert	proximal	—	—	—	—	13.9	6.9	4.5	stem	st	st
163	106	rhyolite	proximal	—	—	11.9	8.2	13.2	4.9	2.6	stem	—	st
167	12	chert	whole	63.3	—	—	—	36.0	7.4	16.2	tri	st	st
167	54	rhyolite	proximal	—	—	12.9	8.8	17.8	5.5	3.0	sn	st	st
175	77	chert	whole	46.8	—	—	—	34.5	6.2	9.5	tri	st	st
273	9	rhyolite	proximal	—	—	13.9	9.7	13.7	6.9	4.5	stem	st	st
273	13	rhyolite	whole	32.8	23.1	11.3	9.7	14.8	4.4	2.7	sn	conv	st
287	2	chalcedony	whole	35.6	28.2	11.6	7.4	—	5.0	4.0	stem	st	st
287	8	rhyolite	midsection	—	—	—	—	—	5.3	4.5	—	st	st
287	40	rhyolite	whole	38.0	28.5	13.5	9.5	15.2	6.7	4.2	stem	st	conv
287	44	rhyolite	midsection	—	—	13.2	—	—	9.5	7.2	stem	st	—
297	35	chert	midsection	—	—	—	—	—	4.3	3.9	bn	conv	—
387	26	rhyolite	midsection	—	—	11.0	—	—	5.1	3.9	stem	st	—
487	114	rhyolite	whole	51.8	37.7	16.6	14.1	24.2	7.3	7.7	sn	st	st

[1] Dimensions are in mm. Side-notched (sn), triangular (tri), stemmed (stem), basal-notch (bn), convex (conv), and straight (st). Nine other excavated projectile points were too fragmentary to list.

stage biface, as is an edge angle of 75 degrees. Two others are also early-stage bifaces. One is ovoid-shaped and the other is lanceolate-shaped. These are smaller than the first biface, being about 30 mm vs. 64 mm long. They have a width-thickness ratio of 2.1 and 2.2, and edge angles of 65 and 70 degrees. The latter biface exhibits steep alternate beveled retouch, rather than bidirectional retouch. A single middle-stage lanceolate-shaped biface is represented, with a width-thickness ratio of 3.4 and edge angle of 45 degrees. The remaining three complete bifaces are all triangular-shaped late-stage bifaces with width-thickness ratios of about 5.0 and edge angles of 45 to 50 degrees.

The majority (46/53) of the bifaces from excavations are fragments and most appear to have been broken during manufacturing. Only two could be classified as lanceolate-shaped and one as ovoid-shaped. The fragments are relatively evenly distributed by portion, including proximal ($n = 14$), midsection ($n = 12$), and distal ($n = 17$) fragments (Table 14.11). There is only a single lateral and two undetermined fragments. It is common for the proximal and distal fragments to exhibit transverse snap breaks, vs. the midsections which often represent wedge-shaped sections with twisting (perverse) oblique breaks. The former breakage pattern is most likely the result of manufacturing activities, but could also be due to use or post-depositional breakage; whereas, the latter breakage pattern is the result of manufacturing error (e.g., see Callahan 1979; Crabtree 1972; Johnson 1979).

Projectile Points

The 26 excavated projectile points exhibit a variety of shapes, mostly consisting of stemmed ($n = 7$), and side-notched ($n = 5$) forms, with a few triangular-shaped points ($n = 3$) and a basal-notched point ($n = 1$). Point shape could not be determined for 10 fragments. However, most of the projectile points were broken during manufacturing and use activities (Table 14.13). Some of the proximal fragments represent points with broken tips that were discarded after use. Most of these are burinated, and one was re-sharpened before being discarded. Other proximal fragments consist mainly of the base and appear to have been rejected after a snap break that occurred during manufacturing. At least one triangular-shaped point base exhibits both hafting polish and basal grinding indicating that it was discarded after use. The midsections represent both manufacture and use-breakage, whereas the tips were probably broken during manufacturing. Although most of the points were retouched using a simple invasive bidirectional technique, three points do exhibit steep beveled retouched edges. The latter consists of two stemmed and a side-notched point with edge angles of 70 degrees.

Other Retouched Tools

The seven excavated unifaces are similar to the retouched pieces, except that they exhibit invasive and not marginal retouch. That is, they are roughly shaped, made on thick flakes, but with steeper retouched edge

TABLE 14.14. General Surface Collection Artifact Type by Material Type.

	Material Type					
Artifact Type	Basalt	Rhyolite	Obsidian	Chalcedony	Chert	Total
Cores						
Core	1	8	20	6	3	38
Tested material	0	0	2	1	0	3
Cobble uniface	0	1	0	0	0	1
Split pebble	0	0	4	0	0	4
Retouched Tools						
Notch	0	2	0	0	0	2
Biface	0	61	7	47	80	195
Projectile point	4	72	23	63	100	262
Uniface	0	6	0	0	5	11
Drill	0	3	1	5	6	15
Cruciform	8	2	0	1	3	14
Total	13	155	57	123	197	545

angles. The mean thickness for these tools is 10.4 mm ($s = 2.8$) with a mean edge angle of 75.8 degrees ($s = 8.0$). They are most often retouched along the distal portion of the flake. All but two of these artifacts are broken. One fragment exhibits a perverse break, indicating that it was broken during manufacturing. One of the whole artifacts is a formal endscraper that is 55 mm long, 15 mm thick, and has an edge angle of 80 degrees.

The two drills are quite similar. Both exhibit relatively blunt tips that are heavily worn, including rounding and polish. One of the drill tips has been re-sharpened. The blades are 11–13 mm wide, 5–7 mm thick, and roughly diamond-shaped in cross section. The single whole drill is 39 mm long.

A single possible cruciform was identified. The item consists of a bidirectionally retouched flake with four lobes. One of the lobes is broken, possibly during manufacturing. It is 26 mm long, 23 mm wide, and 4 mm thick, with an edge angle of 65 degrees. The item is presumably unfinished, so it is unclear as to whether it actually represents a cruciform or some other retouched artifact such as a drill.

Summary of Reduction Strategies

The chipped stone assemblage from Cerro Juanaqueña appears to be dominated by the use of locally available materials that were mostly in the form of river cobbles brought to the site from nearby stream deposits for reduction. Nonlocal obsidian artifacts represent a small proportion of the assemblage, with small obsidian pebbles possibly being procured during subsistence-related activities. Fine-grained materials were preferentially selected for the production of retouched tools, vs. medium and coarse materials for core reduction. Several different reduction tactics were used to produce flakes from cores, with higher quality materials being more fully reduced. The full range of core reduction and tool production/maintenance activities is represented at the site, including primary, secondary, and tertiary stages of debitage reduction, as well as early, middle, and late-stage biface production. Larger core flakes were being selected as blanks for the production of bifaces. The projectile points also reflect manufacturing and use-related breakage patterns. However, the assemblage is not characterized by an increased emphasis on the production and use of simple flake tools, but rather a balance between core reduction and bifacial tool production/maintenance activities. This pattern is represented by both the debitage and the dominance of formal retouched tools in the assemblage.

Discussion

General Surface Artifact Collection

A total of 545 chipped stone artifacts were collected during a general surface reconnaissance across Cerro Juanaqueña as described in Chapter 6. These include 38 cores, 1 cobble uniface, 195 bifaces, 262 projectile points, and 42 other retouched items (Table 14.14).[4] This collection provides a larger picture within which to view the smaller sample obtained for many of these artifacts from the excavations. Note, however, that the general surface collection strategy was not the same for all types of artifacts. For projectile points, bifaces, drills, and cruciforms, an attempt was made to find and collect all specimens exposed on the surface. Chert unifaces resembling endscrapers were also collected whenever they were recognized. Other types of chipped stone artifacts, including cores, hammerstones, notches, and

TABLE 14.15. General Surface Collection Retouched Tool Type by Condition.[1]

	Condition					
Tool Type	Whole	Proximal	Midsection	Distal	Lateral	Total
Notch	2	0	0	0	0	2 (<1%)
Biface	23	53	41	63	15	195 (40%)
Projectile Point	82	117	27	18	18	262 (54%)
Uniface	9	2	0	0	0	11 (2%)
Drill	2	4	7	2	0	15 (3%)
Total	118 (24%)	176 (36%)	75 (15%)	83 (17%)	33 (7%)	485

[1] Fourteen cruciforms are not included.

unshaped unifaces made on thick rhyolite flakes were collected opportunistically. Relative frequencies of the collected examples of these latter artifact types are not meaningful in terms of prehistoric behavior.

The large number of projectile points present on the surface of Cerro Juanaqueña would seem to reinforce the notion of competition and warfare; however, it could alternatively simply reflect the importance of hunting. Are most of the discarded items the by-products of manufacturing dart points, or is the production of bifacial knives also an important activity at the site? One-hundred and ninety-five bifaces were collected from the surface of the site. As was the case with the previously discussed excavated sample, many of the surface bifaces are also broken and have the same distribution of fragment types (Table 14.15). Most of the breaks are lateral (transverse) snap breaks, with a few oblique snap and oblique perverse breaks. Use-related breaks are primarily present on some of the tip and midsection fragments, including obvious polish, impact scars, and burination.

The bifaces can be subdivided into early, middle, and late reduction stages as previously defined. Data on these stages are provided in Table 14.16. Again, all the reduction stages are represented in the biface collection, increasing in number from early to late stages. Tabulation of biface shape by reduction stage shows some significant differences (Table 14.16). That is, there are significantly more ovate early stage bifaces and more triangular late stage bifaces than expected by chance alone. In contrast, there are relatively equal numbers of early to late stage lanceolate-shaped bifaces, and these proportions are similar to what chance alone would predict. Most of the undetermined fragments were broken during the latter stages of reduction. It appears that the reduction trajectory for bifaces on the site involved the initial production of ovate-shaped bifaces that were later worked into both triangular and lanceolate shapes. At least some of the lanceolate-shaped bifaces were probably also worked into triangular shapes. For example, if the base of a lanceolate-shaped biface was broken during manufacturing process, it could easily be reworked into a triangular form. Nonetheless, flake blanks

TABLE 14.16. General Surface Collection Biface Shape by Reduction Stage.[1]

Biface Shape	Stage 1	Stage 2	Stage 3	Total
Ovate	13	4	1	18
	[3.8]	[–0.9]	**[–2.9]**	(23.4%)
Lanceolate	13	14	10	37
	[0.0]	[1.2]	[–1.2]	(48.1%)
Triangular	1	6	15	22
	[–3.5]	[–0.5]	**[4.0]**	(28.6%)
Total	27	24	26	77
	(35.1%)	(31.2%)	(33.8%)	

$\chi^2 = 26.4$, $df = 4$, $p < 0.001$.
[1] Top cell values are artifact counts. Bottom, bracketed values represent adjusted residuals. Significant ($p < 0.05$) values are shown in bold.

were primarily used for the production of all these bifaces. Measurements of the bifaces by reduction stage are found in Appendix 14.17.

Projectile points ($n = 262$) comprise the largest data set from the surface collections (Chapter 6). Most of the points can be classified as stemmed ($n = 99$; 37.8%), with fewer corner-notched ($n = 62$), side-notched (43), basal-notched ($n = 7$), contracting ($n = 6$), triangular ($n = 3$), and indeterminate fragments ($n = 42$). Table 14.17 presents the information on hafting type by base shape. It appears that stemmed, corner-notched, side-notched, and triangular points primarily exhibit straight bases with fewer convex-shaped.

Metric data for the projectile points are provided in Table 14.18. Separating the metric data by hafting type, some minor differences are apparent (Table 14.19). That is, overall length and thickness is similar for stemmed, corner-notched, and side-notched points. However, side-notched points tend to have slightly shorter blades with longer and wider stems. Corner-notched points have narrower stems and contracting-stemmed points have narrower neck widths.

A histogram of overall point length does discriminate three size classes: small (18–29 mm), medium (30–35 mm), and large (36–51 mm) (Figure 14.6). Most of the

TABLE 14.17. General Surface Collection Projectile Point Hafting Type by Base Shape.

	Base Shape					
Hafting Type	Indeterminate	Concave	Contracting	Convex	Straight	Total
Indeterminate	35	0	0	1	6	42
Basal-notched	0	1	0	2	4	7
Corner-notched	4	4	0	18	36	62
Contracting	0	0	0	6	0	6
Side-notched	2	0	0	10	31	43
Stemmed	5	5	1	26	62	99
Triangular	0	0	0	1	2	3
Total	46	10	1	64	141	262

TABLE 14.18. General Surface Collection Projectile Point Measurements.

Measurement	*n*	Minimum	Maximum	Mean	Standard Deviation
Overall length	84	18.3	53.7	36.1	7.6
Blade length	96	9.0	53.7	27.5	7.9
Neck width	198	5.3	19.1	12.5	2.4
Stem length	193	3.8	15.6	9.5	1.8
Stem width	189	6.4	35.3	14.8	3.6
Thickness	247	2.3	8.7	5.7	1.0
Edge angle	199	20	70	53.4	7.3

TABLE 14.19. Surface Collection Projectile Point Metric Data for Hafting Types.

	Overall Length		Blade Length		Neck Width		Stem Length		Stem Width		Thickness	
Haft Type	*n*	Mean (std)	*n*	Mean (std)	*n*	Mean (std)	*n*	Mean (std)	*n*	Mean (std)	*n*	Mean (std)
Stemmed	36	36.4 (8.5)	38	28.1 (9.1)	87	12.6 (2.8)	82	9.3 (1.8)	85	13.9 (3.0)	97	5.9 (1.0)
Corner-notched	24	36.4 (6.8)	26	27.5 (6.5)	52	12.3 (1.6)	55	9.2 (1.4)	55	9.2 (1.4)	60	5.5 (1.0)
Side-notched	17	35.2 (8.2)	17	24.9 (7.4)	39	13.3 (2.3)	39	10.4 (1.7)	36	17.1 (3.0)	42	5.8 (0.9)
Basal-notched	2	31.8 (2.2)	2	23.7 (2.2)	6	11.1 (0.5)	6	7.9 (1.2)	6	11.3 (0.6)	7	4.8 (0.6)
Contracting	4	36.5 (4.2)	4	30.8 (6.9)	6	10.1 (2.2)	6	8.5 (1.7)	6	9.1 (1.9)	6	6.2 (1.1)
Triangular	1	43.9	1	28.4	0		0		3	25.6 (9.7)	3	6.8 (1.6)

stemmed points are large, while corner-notched points are medium and side-notched actually vary from large to small in size (Table 14.20). A histogram of neck width aids in discriminating possible arrow from dart points (Figure 14.7). The overall graph illustrates a bell-shaped curve with a modal distribution at 12–13 mm. However, the left tail of the distribution appears to reflect small arrow points with neck widths from 5.3–8.7 mm. These points exhibit an overall length of 24.9–32.7 mm, and consist of 10 stemmed, one side-notched, one corner-notched, and one contracting stemmed type ($n = 13$). The smallest dart point appears to be a contracting stemmed form with a neck width of 8.8 mm. The presence of these arrow points also indicates that the site was reused by later populations. As discussed in Chap-

TABLE 14.20. General Surface Collection Projectile Point Hafting Type by Size.

	Size		
Hafting type	Small	Medium	Large
Stemmed	8	9	19
Corner-notched	1	15	8
Side-notched	4	7	6
Basal-notched	0	2	0
Contracting	0	2	2
Triangular	0	0	1

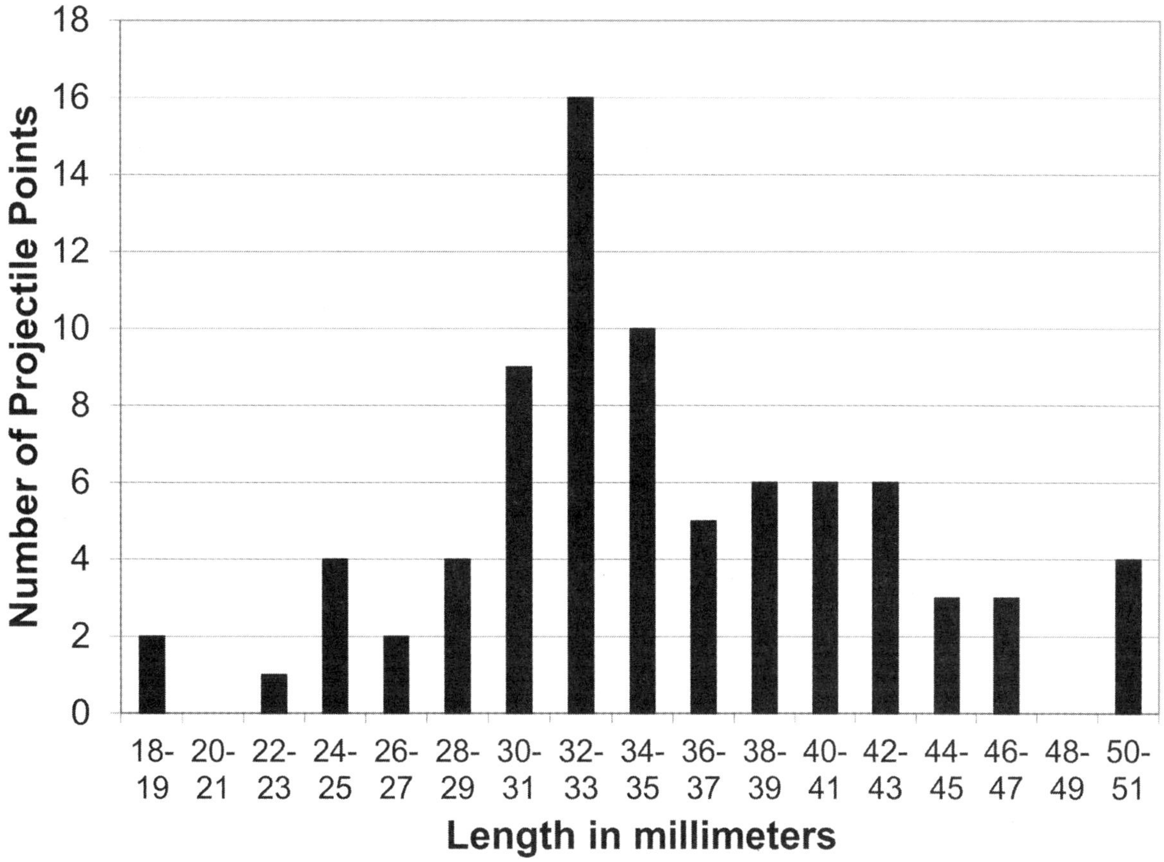

FIGURE 14.6. Histogram showing lengths of projectile points from general surface collection.

ter 6, these may belong to the Ceramic period or the second occupation of Cerro Juanaqueña, as Sliva (1999a) suggested bow and arrow technology was introduced during the later part of the Early Agricultural period.

So how do the projectile points relate to the biface preforms? The late stage bifaces are similar to the projectile points, having mean thicknesses of 5.3 vs. 5.7 mm and edge angles of 50.2 vs. 53.4 degrees (Appendix 14.17 and Table 14.18). It seems likely that many of the lanceolate and triangular-shaped bifaces were acting as dart preforms. The lanceolate-shape could have been used to produce stemmed and side-notched point types and the triangular-shape for corner-notched, basal-notched, and triangular point types. That is, the lanceolate-shape was more conducive to manufacturing larger points and the triangular-shape for medium and smaller-sized points. Nonetheless, at least some of the lanceolate-shaped bifaces were also being used as knives due to the presence of broken tips and midsections with obvious use-wear.

It therefore appears that biface preforms were being worked into both projectile points and generalized bifaces. Generalized bifaces compose 39% (51/131) and projectile points 21% (28/131) of the excavated retouched tool assemblage (Table 14.2). This compares with Sliva's (2001) comparative study of Late Archaic village sites in the Tucson Basin where projectile points comprise 23%, but generalized bifaces only consist of 25% of the retouched tool assemblages. So, there actually appears to be a relatively greater number of generalized bifaces present in the excavated sample at Cerro Juanaqueña than the average village site in the Tucson Basin.

An important aspect of bifacial technologies is the ability to increase tool use life through re-sharpening (Kelly 1988; Shott 1986b). This practicality is especially important for residentially mobile groups traveling through lithic-resource poor areas where re-sharpening and beveling is often used to conserve lithic materials (Sollberger 1971; Wiant and Hasesn 1985). As previously noted, heat-treatment was another tactic for extending tool-use but was not one used by the occupants at Cerro Juanaqueña. Only one dart point and three bifaces were identified with obvious heat-treatment. Beveling was only identified on seven stemmed projectile points in the surface collection, with edge angles of about 65–70 degrees. However, it is noteworthy that beveling does seem to be primarily restricted to a specific point form. On the other hand, re-sharpening was used on a few points with impact fractures to increase the tool use life. Overall, these observations show that little was done at

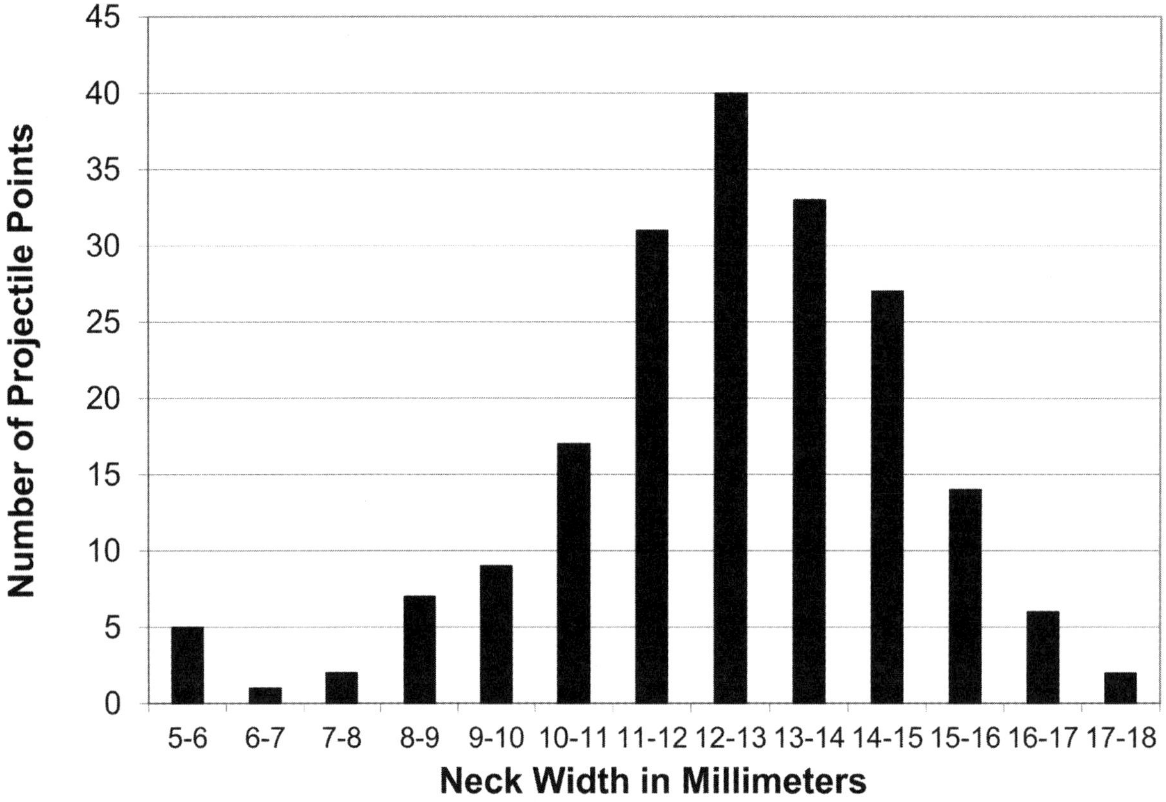

FIGURE 14.7. Histogram showing neck width of projectile points from general surface collection.

the site to increase tool use life by re-sharpening, beveling, or heat-treatment.

Only a few other retouched tool types were collected from the surface of the site. These consist of two notches, 11 unifaces, and 15 drills. The notches represent both single blow and retouched types with edge angles of 80 degrees. The unifaces can be separated into two groups. The first are primarily made on large, thick (13–21 mm) rhyolite flakes that are unifacially retouched along a lateral edge with a natural cortical backing. The second are smaller, thinner (8–12 mm) chert flakes that resemble endscrapers. The natural cortical backing on the larger flakes would make them easier to hold while in use, while some of the endscrapers exhibit hafting polish along the high points of the flake scar arrises. Edge angles range from 60 to 75 degrees, with some rounding and polish being observed along the edge of both uniface types. Although unifaces are an important retouched tool in many Late Archaic assemblages along the Borderlands, they are not at Cerro Juanaqueña (e.g., Hester 1995; Sliva 1998, 2001; Vierra 1999a). Indeed, unifaces only account for 2% (11/499) of the retouched tools collected on the surface of the site, and 5% (7/131) in the excavation assemblage, respectively.

Data from a Late Archaic village site in the Tucson Basin (Santa Cruz Bend site) is illustrative. Sliva (2001) identified mostly bifaces (46%), unifaces (33%), and unidirectionally retouched flakes (20%) for these sites. It is unclear what these unifaces were used for, but some preliminary studies indicate that woodworking may be one possible tool function for the larger flake unifaces (see Hayden [1979], Hester et al. [1973], and Gould et al. [1971] for ethnoarchaeological examples). However, the smaller unifaces may have also been used for other activities (e.g., hide processing).

The 15 drills from surface contexts (Table 14.14) are generally similar to the two recovered from excavation contexts. However, the surface drills can be separated into two groups. The first group is long and narrow with a diamond-shaped cross section (n = 12). They have a mean width of 12.7 mm (n = 7, s = 3.2) and thickness of 5.2 mm (n = 12, s = 0.9). These drills also exhibit heavily worn and rounded tips and edges. In contrast, the second group is relatively triangular-shaped with a wide base and does not exhibit obvious use-wear (n = 3). They exhibit a mean width of 18.5 mm (n = 2, s = 12.0) and thickness of 6 mm (n = 3, s = 2.6). It could be that the first group was used to drill the hole through stone pipes and the second group for various domestic activities (see Chapter 6).

Most of the 14 cruciforms from surface contexts appear to be rejects from the initial stage of production.

TABLE 14.21. Excavated Tool Type by Rock Feature.[1]

Tool Type	Rock Ring	Terrace	Total
Informal	11 **[2.8]**	31 **[−2.8]**	42 (32.1%)
Formal	7 **[−2.8]**	82 **[2.8]**	89 (67.9%)
Total	18 (13.7%)	113 (86.3%)	131

$\chi^2 = 8.1$, $df = 1$, $p = 0.004$.

[1] Top cell values are artifact counts. Bottom, bracketed values represent adjusted residuals. All values are significant ($p < 0.05$) and are shown in bold.

These primarily consist of thick basalt flakes that have been worked into the distinctive "X" shape with hard hammer percussion. However, the four artifacts made of chert and chalcedony have been retouched. Two of the 8 basalt cruciforms are highly polished and reflect a finished item. Further discussion of the cruciforms and other artifacts from the surface collections are found in Chapter 6.

Intrasite Comparisons of Excavated Features

Two distinctive forms of rock features have been identified at Cerro Juanaqueña—rock rings and terraces (see Chapter 3). Yet it is unclear whether these distinctively-shaped features had similar or different functions. To help answer this question, a comparative analysis can be conducted on the debitage assemblages excavated from these features to determine how similar these assemblages are. A comparison of material type by feature type does indicate that significantly more rhyolite and chert is represented in the rock rings, vs. more chalcedony than expected in the terraces (chi-square = 85.12, $df = 4$, $p \leq 0.001$, $n = 15{,}095$; Appendix 14.18). These differences are also represented in a comparison of debitage types, with relatively more core flakes being present within the rock rings, vs. more debris in the terraces (chi-square = 29.9, $df = 2$, $p \leq 0.001$, $n = 6{,}113$; Appendix 14.19). Although the cells for the biface flakes do not have a significant value, relatively more of them are present on the terraces. This finding is important because a comparison of the presence of retouched tool types by feature reveals that relatively more informal tools are found within the rock rings, vs. formal tools on the terraces (Table 14.21). These patterns in assemblage variability seem to indicate that the production and use of informal tools was a more important activity within the rock rings. In contrast, the production and use of formal tools was presumably a more important activity on the terraces.

Of course, rock features are not limited to the upper area of the site. There are also terraces located on the lower slopes of Cerro Juanaqueña. Could the lower terraces have been used differently than the upper terraces? A comparison of debitage assemblages again reveals some significant differences between these two sections of the site. It appears that there is relatively more basalt and rhyolite present in the lower terraces vs. chalcedony and chert in the upper terraces (Appendix 14.20). There is also relatively more debris on the upper terraces and more biface flakes on the lower terraces (Appendix 14.21). These results appear contradictory. That is, the debitage indicates that core reduction is a more prevalent activity on the upper terraces, vs. the production/maintenance of bifaces on the lower terraces. Higher quality materials that are generally associated with the production of bifaces are more prevalent on the upper terraces, and lower quality materials that are often used for core reduction are more common on the lower terraces. However, one possible solution to this conundrum is that the lower terraces have greater access to rhyolite from stream deposits and fine to medium grain rhyolite is being preferentially used on the lower terraces for biface production. Note that over one-third of biface flakes are indeed made from rhyolite at Cerro Juanaqueña (Table 14.2).

Since some of the lower terraces are associated with a later occupation at the site, could these be contributing to the differences between the upper and lower terraces? It seems unlikely, but this issue cannot be addressed with the samples at hand. Only two lower terraces (T415 and T463) date to the late occupation, and these excavations yielded only a small sample of chipped stone artifacts.

Intersite Comparisons with Other Late Archaic Cerro de Trincheras Site Assemblages

Test excavations were conducted at three other Late Archaic cerros de trincheras sites along the Río Casas Grandes valley: Cerro el Canelo, Cerro los Torres, and Cerro Vidal (Chapter 4). How do the excavated debitage assemblages from these sites compare to Cerro Juanaqueña? Tabulations in Tables 14.22, 14.23, and 14.24 provide excavated data on debitage type by material type for these three sites. The sample size is obviously much larger for Cerro el Canelo ($n = 3{,}327$) when compared to the other two sites (Cerro los Torres, $n = 304$; Cerro Vidal, $n = 263$), but all three are much smaller than the Cerro Juanaqueña sample.

Given the disparity in debitage sample sizes between Cerro Juanaqueña and these three sites, I decided to select an assemblage from a single terrace feature at Cerro Juanaqueña that exhibited a similar range in debitage types as the overall site, but that had a sample size of about 300 artifacts—that is, a similar sample size as the two smaller assemblages. Cerro Juanaqueña Terrace 97

TABLE 14.22. Cerro el Canelo Excavated Debitage Type by Material Type.

	Material Type						
Debitage Type	Basalt	Rhyolite	Obsidian	Sandstone	Chalcedony	Chert	Total
Debris	25	37	0	1	56	7	126
Core flake	33	302	3	0	300	60	698
Biface flake	0	56	0	0	169	16	241
Micro-debitage	10	548	7	0	1470	98	2133
Und. Flake fragment	0	40	0	0	76	12	128
Other	0	1	0	0	0	0	1
Total	68	984	10	1	2071	193	3327

TABLE 14.23. Cerro los Torres Excavated Debitage Type by Material Type.

	Material Type						
Debitage Type	Basalt	Rhyolite	Andesite	Obsidian	Chalcedony	Chert	Total
Debris	0	9	0	0	2	1	12
Core flake	2	163	1	1	28	27	222
Biface flake	0	1	0	0	3	1	5
Micro-debitage	0	31	0	0	10	3	44
Und. Flake fragment	1	7	0	1	4	2	15
Other	0	3	0	0	3	0	6
Total	3	214	1	2	50	34	304

TABLE 14.24. Cerro Vidal Excavated Debitage Type by Material Type.

	Material Type						
Debitage Type	Basalt	Rhyolite	Andesite	Obsidian	Chalcedony	Chert	Total
Debris	6	4	0	0	0	3	13
Core flake	7	69	4	0	14	26	120
Biface flake	0	0	0	0	2	1	3
Micro-debitage	4	34	1	7	41	31	118
Und. Flake fragment	0	3	0	2	2	1	8
Other	0	1	0	0	0	0	1
Total	17	111	5	9	59	62	263

met these requirements with a sample size of 267 pieces of debitage.

A chi-square analysis of a contingency table of debitage material type for the sites reveals some significant differences (Appendix 14.22). There is significantly more rhyolite and chert in the Terrace 97, Cerro los Torres and Cerro Vidal assemblages relative to the Cerro el Canelo assemblage (chi-square = 483.9, $df = 9$, $p =< 0.001$, $n =$ 4,132). This pattern is driven by Cerro el Canelo having significantly more chalcedony than the other sites. As previously noted in the material selection section, both a white chalcedony and red jasper were observed in a rhyolite formation located near Cerro el Canelo. Therefore, the prevalence of use of this material at the site is probably due to its local availability.

These site assemblage differences are also represented in a comparison of debitage types between the sites. A chi-square analysis shown in Appendix 14.23 indicates that significantly more core flakes are represented in the Cerro Juanaqueña Terrace 97, Cerro los Torres, and Cerro Vidal assemblages, vs. relatively more pieces of angular debris and biface flakes in the Cerro el Canelo assemblage (chi-square = 126.2, $df = 6$, $p \leq 0.001$, $n = 1{,}601$). However, chalcedony biface flakes ($n = 169$)

TABLE 14.25. Viejo Period Site, Debitage Type by Material Type.

	Material Type					
Debitage Type	Basalt	Rhyolite	Andesite	Chalcedony	Chert	Total
Debris	2	26	0	4	1	33
Core flake	3	59	1	4	3	70
Micro-debitage	0	50	0	12	0	62
Und. Flake frag	0	1	0	0	0	1
Other	0	1	0	0	0	1
Total	5	137	1	20	4	167

TABLE 14.26. Medio Period Site, Debitage Type by Material Type.

	Material Type						
Debitage Type	Basalt	Rhyolite	Andesite	Obsidian	Chalcedony	Chert	Total
Debris	1	32	0	0	0	2	35
Core flake	2	101	6	1	16	8	134
Micro-debitage	7	78	2	1	31	0	119
Und. Flake frag	0	0	0	0	2	0	2
Other	0	1	0	0	0		1
Total	10	212	8	2	49	10	291

and chalcedony microdebitage ($n = 1{,}470$) comprise a large proportion of the Cerro el Canelo debitage assemblage ($n = 2{,}133$). Therefore, smaller items like angular debris and biface flakes may also be overrepresented, possibly due to access to chalcedony as well as specific site formation processes that are not represented in the other excavated samples, or to differences in site function.

Intersite Comparisons with Medio and Viejo Phase Site Assemblages

How does the chipped stone assemblage at Cerro Juanaqueña compare to later Ceramic period sites? If the Cerro Juanaqueña chipped stone assemblage reflects a settled village community, then it should resemble these later assemblages. To evaluate this question, samples were collected from Viejo (Sitio Lucero) and Medio (Sitio Antonio) phase sites located on the Río Casas Grandes floodplain less than two km from Cerro Juanaqueña (see Chapter 4). Because of their proximity, access to raw materials is the same for all three sites. Both Ceramic period sites had been looted, and debitage samples were recovered by screening looters' backdirt. A total of 167 and 291 pieces of debitage were recovered from the Viejo and Medio phase sites, respectively (Tables 14.25 and 14.26). Given the differences in debitage samples sizes between Cerro Juanaqueña and the two Ceramic period sites, I decided to use Cerro Juanaqueña Terrace 97 for comparisons again.

TABLE 14.27. Site Debitage Assemblage by Material Type.[1]

	Material Type			
Assemblage	Basalt	Rhyolite	Chalcedony	Chert
T97[2]	8	118	99	41
	−0.3	−9.2	**6.6**	**5.9**
Viejo	5	137	20	4
	−0.2	**5.3**	−4.0	−2.9
Medio	10	212	49	10
	0.4	**4.5**	−3.1	−3.4

$\chi^2 = 97.5$, $df = 6$, $p < 0.001$.

[1] Top value in cell represents artifact count and bottom value represents adjusted residuals. Significant ($p < 0.05$) positive values are shown in bold.

[2] Cerro Juanaqueña.

A review of lithic material types indicates that there are some significant differences between the assemblages (Table 14.27). Terrace 97 contains significantly more chalcedony and chert while both Viejo and Medio Ceramic period assemblages contain significantly more rhyolite despite virtually identical access to raw materials. As previously discussed, these differences in lithic raw material selection generally represent biases in tool production vs. core reduction activities. Indeed, a comparison of debitage types by assemblage indicates that there are significantly more core and biface flakes in the Terrace 97 assemblage, vs. more angular debris in the Viejo period assemblage (Table 14.28). Although the

Table 14.28. Site Assemblage by Debitage Type.[1]

Assemblage	Debris	Core flake	Biface flake	Total
	Debitage Type			
T97[2]	12	141	8	161
	−4.5	**3.1**	**3.7**	
Viejo	33	**70**	0	103
	4.1	**−3.4**	−1.6	
Medio	35	134	0	169
	1.0	−0.2	−2.3	
Total	80	345	8	433

$\chi^2 = 37.8$, $df = 4$, $p < 0.001$.
[1] Top value in cell represents artifact count, and bottom value represents adjusted residuals. Significant ($p < 0.05$) positive values are shown in bold.
[2] Cerro Juanaqueña.

Table 14.29. Site Assemblage by Platform Type.[1]

Assemblage	Cortical	Single-faceted	Multi-faceted
	Platform Type		
T97[2]	30	51	9
	−4.6	**3.2**	**3.4**
Viejo	31	21	0
	1.2	−0.5	−1.7
Medio	61	30	1
	3.5	−2.7	−1.9

$\chi^2 = 27.7$, $df = 4$, $p < 0.001$.
[1] Top value in cell represents artifact count and bottom value represents adjusted residuals. Significant ($p < 0.05$) positive values are shown in bold.
[2] Cerro Juanaqueña.

table illustrates that the Medio period assemblage also contains relatively more angular debris than the terrace assemblage, the adjusted residual for this cell is not significant at the 0.05 level.

Lastly, a comparison of platform types by assemblage also indicates some significant differences. That is, the Cerro Juanaqueña Terrace 97 assemblage contains significantly more single-faceted and multifaceted platforms while there are significantly more cortical platforms in the Medio site assemblage (Table 14.29). Although the table illustrates that the Viejo assemblage also contains relatively more cortical platforms than the terrace assemblage, the adjusted residual for this cell is also not significant at the 0.05 level.

Overall, the Terrace 97 assemblage reflects a greater emphasis on the production of bifacial tools made from chalcedony and chert platform cores. This contrasts with the Ceramic period assemblages that emphasize the production of flakes, which were detached from unprepared cortical platforms on rhyolite cobble unifaces. This pattern is similar to the findings of Miller (1995) and Rebnegger (2001) studies of Viejo and Medio period site assemblages in the area around Paquime. These site assemblages also emphasize the use of locally available materials for core reduction.

Technology and Labor Organization at Cerro Juanaqueña

The results presented in this chapter reinforce my previous research showing that the chipped stone assemblage from Cerro Juanaqueña was similar to other Late Archaic habitation sites, but contrasted with an emphasis on core reduction at later Pueblo structural sites (Vierra 1999b, 2005). That is, this regional comparison revealed three general chipped stone assemblage groups: 1) the quarry and the Pueblo structural site show an emphasis on core reduction activities; 2) the pithouse and Late Archaic habitation sites reflect a mix of core reduction and biface production/maintenance; and 3) the Late Archaic summer campsites have a greater emphasis on biface production/maintenance (also see Vierra 1990; Vierra and Foxx 2009). Solely based on the chipped stone assemblage, the artifacts from Cerro Juanaqueña seem to represent a technology oriented towards a mixed foraging and agricultural economy.

The Cerro Juanaqueña assemblage reflects a mix of the full range of core reduction, tool production, and tool maintenance activities. A combination of informal and formal tools, mostly consisting of bifaces and projectile points with fewer retouched pieces and notches, is present. The lithic assemblage is dominated by the use of rhyolite and chalcedony which are locally available from river gravels; whereas, obsidian appears to be the only nonlocal material reduced at the site. The formal tools are primarily associated with the terraces when contrasted with the rock ring features which contain relatively more informal tools.

As Binford (1980:13) pointed out, technology does not solely consist of tools but also includes labor. Indeed, technological organization is a direct reflection of corporate labor group structure and economic organization. The chipped stone component of a site cannot be studied in isolation from other technological components, if a full and accurate picture of the past is to be gained. These other components include heavy-duty tools, milling equipment, ceramics, storage facilities, architecture, and agricultural features.

Marsha Ogilvie's (2000, 2005) work is important for understanding this point. Her biological research identified differences in femur bone structure for Southwestern foragers vs. agriculturalists. However, her study of the Tucson Basin Late Archaic populations revealed an important pattern. That is, the males resembled foragers while the females resembled agriculturalists.

Watson and Stoll (2013) examined femoral morphology among Early Agricultural period populations from the Tucson Basin and northern Sonora and they found a similar pattern—males were engaged in more mobility than females, including travel over greater distances and rougher terrain. This finding implies that important changes in labor organization were occurring during the Late Archaic/Early Agricultural period. Would we not also expect these changes to be reflected in the technology being employed by men vs. women (e.g., see Crown 2000; Roth 2006)? If so, then my study of the chipped stone technology at Cerro Juanaqueña may be providing a biased view of the site.

Anyone walking over the Cerro Juanaqueña is immediately impressed by two things: the large number of terraces and the presence of numerous basin-shaped metates that have extremely deep grinding basins. Both of these aspects of technology would seem to suggest a high degree of residential stability. Chapter 13 provides a detailed discussion of grinding technology and emphasizes that the heavy wear on the basin metates is related to the extent of food-processing of maize and other plants conducted at Cerro Juanaqueña. The wear patterns are a result of crushing and reciprocal grinding that is particularly effective at processing substantial quantities of flinty maize. The Cerro Juanaqueña basin metates are argued to be a transitional grinding technology between that of Archaic foragers and full-time agriculturalists. This transitional status is also reflected in the grinding surface area of the manos which is intermediate in size between that of foragers and full-time farmers.

A similar pattern is found when I compare the size of Cerro Juanaqueña manos with those of Basketmaker and Pueblo manos from the northwest New Mexico. Considering mano grinding surface length, the Cerro Juanaqueña manos are situated in between those of Basketmaker III and Pueblo I (Vierra 2005:Figure 7.10). I have suggested elsewhere that the Pueblo I period is characterized by a heavy dependence on maize agriculture (Vierra 1994c). Clearly Cerro Juanaqueña manos also reflect a diet becoming more reliant on domesticated plants (e.g., maize and perhaps amaranth).

I suggest that the differences between chipped and ground stone assemblages seem to support the contention that the sexual division of labor possibly represented at the Tucson Basin sites may also be represented at Cerro Juanaqueña. The changing roles played by men and women in these early villages may not only be affecting the chipped vs. ground stone assemblages, but also the relationship between tool production vs. core reduction activities. Spatial analyses conducted of the site indicate that there are more projectile points present in non-terrace contexts (e.g., cleared areas), cores on terraces, and metate fragments on rubble berms where they had been discarded near terraces (Appendix 6.2). Cores may have also been discarded in toss zones located along the edges of the terraces. As previously noted, relatively more projectile points and bifaces are represented among the surface collections, vs. the excavated sample of retouched tools. Nonetheless, whole metates are often located on or near terraces.

The association of grinding and core reduction activities on residential terraces may reflect the increased role of women in simple flake production and tool use. If so, then the increasing presence of core reduction activities is not a simple effect of residential stability, but rather changes in labor organization that are associated with increases in occupation duration and a reliance on agriculture. That is, women were possibly spending more time at the villages and less time foraging. Nonetheless, the importance of flake blanks for the production of projectile points also appears to be an important factor affecting Late Archaic core technology. As recently suggested by Vierra (2013), the growing variability in Late Archaic projectile point design may be related to increasing diet breadth and an attempt to raise hunting success rates by focusing on specific target species. Therefore, the production of flakes becomes important in respect to both female and male activities at these early villages.

Summary and Conclusions

A basic question was asked at the beginning of this chapter: how can studies of the chipped stone assemblage be used to evaluate whether or not Cerro Juanaqueña represents an early settled village? To answer this question, I needed information on several key aspects of technological organization involving the stone tool manufacturing process. These components included tool design, the selection of raw materials, stone tool production and maintenance, stone tool use, and the eventual discard of worn-out tools. The chapter was then organized around this perspective by collecting information on core vs. biface technologies, as well as their relationship to labor organization and residential site stability.

Several research issues were also identified at the beginning of this chapter. These concerns were based on the assumption that if Cerro Juanaqueña represented an early village reliant on agriculture, then I would expect the chipped stone assemblage to: 1) be dominated by the reduction and use of locally available materials, 2) reflect the full range of core reduction and tool production/maintenance activities, and 3) be characterized by an increased emphasis on the production and use of simple flake tools. Data on the analysis of approximately 15,000 chipped stone artifacts was used to address these issues.

This analysis indicates that the chipped stone assemblage from Cerro Juanaqueña does emphasize the reduction of locally available materials like rhyolite, chalcedony, chert, and basalt. These raw materials were directly procured from local river gravels or outcrops and brought to the site for reduction. On the other hand, obsidian is the only material that does not appear to be locally available and could have been obtained incidentally to subsistence-related activities. X-ray fluorescence analysis indicated that obsidian samples were from three distinct geographic clusters: northeastern Sonora Sierra Madre (Selene source), northwest Chihuahua Basin and Range (Sierra Fresnal, Ojo de Federico, Los Jagüeyes), and the New Mexico Bootheel cluster (Antelope Wells and Animas Mountains). These sources are 30 to 125 km distant from Cerro Juanaqueña in virtually all directions, reflecting a geographically wide subsistence region.

The analysis also determined that the full range in core reduction and tool production activities are represented at the site. A variety of core reduction tactics were used, depending on the material type. Cores made on higher quality material were more fully reduced than those made of lower quality materials. Indeed, the evidence shows that different reduction tactics were used for chalcedony, chert, and obsidian, vs. rhyolite and basalt. A greater emphasis was placed on the fine-grained materials for tool production and medium-coarse grained materials for core reduction. Formal tools like bifaces and projectile points dominate the retouched tool assemblage, with fewer informal tools consisting of retouched pieces, notches, and denticulates. The presence of early, middle, and late stage bifaces also attests to the fact that tool production was an important activity at the site. Many of these bifaces probably represent preforms for the production of dart points.

Lastly, the suggestion that the Cerro Juanaqueña chipped stone assemblage should emphasize the production and use of simple flake tools is certainly not the case. A comparison of this assemblage with other cerros de trincheras sites in the area revealed that they all reflect a mix of core reduction and tool production/maintenance activities. In contrast, comparisons between Cerro Juanaqueña and local Medio and Viejo phase village sites indicate that the latter assemblages are dominated by core reduction activities.

The analytical results presented in this chapter reinforce my previous research showing that the chipped stone assemblage from Cerro Juanaqueña was similar to other Late Archaic habitation sites, but that it contrasted with later Ceramic period sites that emphasize core reduction and also with Late Archaic campsites that emphasize biface production (Vierra 1999b, 2005). On the other hand, a suggestion was made that the differences represented between chipped and ground stone assemblages at Cerro Juanaqueña seem to indicate that important changes in sexual division of labor were occurring at the site. This transition would include a greater distinction between the roles played by men vs. women in respect to an increasing dependence on agriculture and subsequent changes in early village organization (Ogilvie 2005; Vierra 2005; Watson and Stoll 2013). And a greater distinction between the roles played by men vs. women may be observed in respect to increasing diet breadth, foraging, agricultural activities, and subsequent changes in early village social organization. That is, women were dividing their time between foraging and agricultural activities; whereas, foraging continued to be an important activity for men, including traveling greater distances to hunt game. The result was the increasing importance of a flake-based technology for manufacturing simple flake tools and projectile points by women and men.

Notes

1. All artifact totals exclude three fire-cracked rocks (FCR) and six pot lids as indicated in Table 14.1.
2. Also known as the Lago (Ojo) Frederico obsidian source.
3. Also known as the Lago (El) Barreal obsidian source.
4. Differences in counts between Tables 6.1 and 14.14 are largely because there were tools from the early phases of the project for which we only had documentation and not the collections. Table 6.1 includes these, while Table 14.4 does not.

CHAPTER 15

Conclusions

Introduction

In this book, we have emphasized direct archaeological observations and have made relatively limited interpretations from these data. For the most part, our interpretations have been concerned with the specific human behaviors that are reflected in the archaeological record of northwestern Chihuahua between about 1300 cal BC and cal AD 1. In this final chapter, we summarize these findings briefly and then consider some of their wider implications.

Dating

When we began our fieldwork, the age of the cerros de trincheras was a major issue. Many similar sites had been described in the basin and range country of northwestern Mexico and the southwestern United States, but all were thought to date to late prehistoric times, after 1100 BC. No one suspected that massive hilltop constructions could date to a time before the advent of pottery. Chapter 5 describes the results of 23 radiocarbon dates from Cerro Juanaqueña. Almost all of these are AMS dates from annual or short-lived species. The samples came from a wide range of materials, including maize, wild plants, and even small fish remains. Despite this diversity, they tell a remarkably consistent story: the most intensive occupation of Cerro Juanaqueña was underway by 1300 cal BC and ended by 1100 cal BC, with a second, much less intensive occupation between 400 cal BC and cal AD 1. These dates are a conservative rounding of the Oxcal boundary function described in Table 5.2. We consider the temporal boundaries in Table 5.2 to be the most precise estimates of the occupation dates. Both of these occupations were relatively brief. The early, most intensive occupation likely lasted no more than 220 years and may have been as brief as 30 years, based on the range of radiocarbon dates.

We had not expected such a restricted range of radiocarbon dates at Cerro Juanaqueña, but we were even more surprised as radiocarbon dating results began to come from the three other cerros de trincheras that we tested. Dates from Cerro el Canelo and Cerro los Torres fell precisely within the range of the early, intensive occupation at Cerro Juanaqueña, while the single date from Cerro Vidal corresponds to the later, less intensive occupation at Cerro Juanaqueña.

These results have several important implications. First, the unprecedented labor investments represented at these sites were not made over long spans of time. Instead, the constructions came about during brief, intensive occupations. Second, the artifact assemblages and other archaeological materials documented here are not palimpsests, the result of multiple unrelated occupations distributed over a long period of time. Instead most were deposited during two discrete periods of occupation. This result is surprising, but it is being borne out by evidence of brief, discrete occupations at other Early Agricultural period cerros de trincheras, such as Tumamoc Hill in Tucson and Round Mountain along the Gila River in southeastern Arizona (Cajigas et al. 2016; Roney et al. 2016). Finally, residential occupation of the cerros de trincheras cannot be modeled in normative terms as a behavior characteristic of particular groups throughout the Early Agricultural period. Instead, it is a residential strategy that was adopted during specific, brief intervals of time and then discontinued. This cycle of occupation and abandonment was repeated at various times during the Early Agricultural period.

Residential Occupation

As we began our investigations in northwestern Chihuahua, other archaeologists working in southern Arizona were beginning to report residential structures on cerros de trincheras in much later time periods and to stress residential occupation as a consistent theme (Downum 1986, 1993; Downum et al. 1994). Our findings extend this observation and expand upon it.

Multiple lines of evidence lead to this conclusion. In Chapters 3 and 4, we describe numerous rock rings found on the terraces, which we interpret as domiciles and storage structures. Chapters 6 and 14 describe a wide range of artifacts, tools that reflect all of the activities which might be expected archaeologically on residential sites. Chapter 13, in particular, describes the phenomenal assemblage of ground stone from Cerro Juanaqueña. This assemblage reflects intensive food processing and is even more impressive considering the brevity of its use. In Chapters 8, 9, and 11, Karen R. Adams, J. Kevin Hanselka, Gayle J. Fritz, and Kari M. Schmidt describe the abundant and varied record of plant and animal remains used by the inhabitants of Cerro Juanaqueña. These materials were recovered from rich midden deposits, which also reflect intensive residential occupation. Similar evidence was found at the other cerros de trincheras which we probed, including Cerro el Canelo, Cerro los Torres, and Cerro Vidal. These observations leave little doubt that many of the terraces at these four sites were used as living surfaces and house platforms.

Subsistence

Even before we began excavations at Cerro Juanaqueña, we suspected that agriculture played an important role in the economy of the people who lived there. The archaeobotanical analyses completed by Adams and Hanselka (Chapter 8) certainly support the inference that maize played an important role in the diet. They show that maize was by far the most common recovered taxon. It was present in 62% of the 122 non-sterile flotation samples and was found on 15 of the 17 terraces which were tested.

Chenopod/amaranth (cheno-ams) were found in 32% of non-sterile flotation samples. Regarding amaranth, Fritz, in Chapter 9, was able to show that "the relatively high ubiquity of amaranth seeds with seed coat thicknesses well within the cultigen range indicates that grain amaranth may have been part of the early agricultural system in the Río Casas Grandes valley." Cultigen amaranth could have made a difference by adding diversity and, given its drought-resistant properties, reducing risk. Although wild *Cucurbita* was identified, domesticated varieties were not. The absence of squash is puzzling and may imply diffusion processes independent of those of amaranth and maize.

The inhabitants of Cerro Juanaqueña and other cerros de trincheras in northwestern Chihuahua were clearly successful farmers, but they also continued to make extensive use of a broad repertoire of wild plants as well, including 19 different taxa. However, aside from cheno-ams, the overall ubiquity of wild plants was low (1% to 5% of non-sterile samples). These wild plants were primarily forbs (chenopod, milkvetch, a non-domesticated cucurbit, spurge, sunflower, groundcherry, portulaca, chia, globemallow, horse carpetweed, purslane), with one or two grasses (lovegrass and an unknown grass), a few shrubs (lemonade berry, mesquite, and an unknown legume), one succulent (barrel cactus), a tree (walnut), and a riparian plant (bulrush).

Cerro Juanaqueña also yielded a rich faunal assemblage. Kari M. Schmidt in Chapter 11 analyzed a total of 33,173 bones from Cerro Juanaqueña, including 17,799 from excavation and 15,374 from flotation samples. Sixty-five percent of the 2,558 identified faunal remains from excavations at Cerro Juanaqueña consist of jackrabbit. Another 11% consist of cottontail rabbit and indeterminate rabbits. Only 5% of the identified sample consists of artiodactyl remains, including bighorn sheep, pronghorn, mule deer, and a single possible bison. The remaining 19% of the identified specimens in the assemblage include small fish, venomous and nonvenomous snakes, lizards, terrestrial turtles, indeterminate turtles, spadefoot toads, quail, duck, perching birds, other birds, coyote, badger, squirrels, and other rodents.

The faunal assemblage indicates heavy use of jackrabbits relative to cottontails. In addition, the use of all leporids is high relative to all artiodactyls. Both patterns are somewhat typical of low elevation, southern Southwestern Archaic sites, and therefore do not necessarily indicate overhunting of large species. Other aspects of the faunal assemblage were surprising. The recovery of 88 small fish bones from the flotation samples and another six from excavation samples offered additional insight on the subsistence focus. These included both cyprinids (chubs, redheads, and dace) and catostomids (suckers), species that would have been smaller than 100 mm in length. An extrapolation of these flotation data to unfloated deposits suggests that if the entire excavation sample had been subjected to flotation, fish would have been exceeded only by leporids and rodents in number of identified samples. Clearly fish were an important element in the diet. Very low frequencies of freshwater mollusk fragments were found, suggesting they were occasionally used as well.

Excavations at other cerros de trincheras in northwestern Chihuahua, Cerro el Canelo, Cerro los Torres, and Cerro Vidal were much more limited in extent and yielded less evidence related to subsistence. Nevertheless, early maize was found at all three of these sites, and other faunal and botanical remains are consistent with those described here from Cerro Juanaqueña.

Farming Strategy

Subsistence remains described above show that hunting and gathering wild plants continued to be important, but that corn (*Zea mays*) was the most ubiquitous dietary item. Clearly maize farming was a critical economic activity, yet the early cerros de trincheras of northwestern Chihuahua are located in Semidesert Grassland, where rainfall is hardly sufficient for dryland farming (Chapter 2). How can this be? We believe that it is no coincidence that all 14 cerros de trincheras that we documented are within two kilometers of a major floodplain, either along the Río Casas Grandes or the Río Santa Maria. Indeed, most of the sites are directly adjacent to the floodplain and close to river reach boundaries where subsurface water is forced to the surface.

Studies completed by Lee C. Nordt (Chapter 12) show that when Cerro Juanaqueña was occupied the Río Casas Grandes was characterized by a wide, stable floodplain, aggrading deposits, low velocity overbank flooding with rare high-magnitude destructive flooding, and deep alluvial sediments with good moisture retention properties. In addition, paleoenvironmental studies summarized in Chapter 2 suggest that climatic conditions may have been slightly more mesic than today. These conditions were excellent for agriculture and provided a niche where maize farming likely required lower investment, produced higher yields, and proffered lower risks relative to many other environments in the southwestern United States and northwestern Mexico.

Today all farming in the vicinity of the documented cerros de trincheras relies on irrigation, drawing seasonal water from the principal watercourses or else water which is available year-round from wells. In the recent past, farmers immediately below Cerro Juanaqueña irrigated a large area of the floodplain using year-round water from Ojo San Pedro, a major spring which used to exist about 1.5 km upstream from the site. Recent investigations in Tucson, Arizona, show that irrigation technology focused on floodplains was established by 1500 cal BC (Nials 2015), raising the possibility that similar technologies were used in northwestern Chihuahua. Our own investigations on the floodplain of the Río Casas Grandes did not discover any direct evidence of irrigation, leaving open the possibility that early agriculture relied on dry farming on the floodplain, natural overbank flooding, or less sophisticated irrigation technology (Nabhan 1979a). However, the floodplain surface at 1300 cal BC is deeply buried today, and it is still possible that irrigation systems similar to those found near Tucson remain undiscovered in this setting.

Seasonality

One of the most important aspects of a transition from foraging to farming economies is residential mobility. In arid environments such as northwestern Chihuahua, residential mobility is one of the most fundamental strategies used by foragers to cope with changing suites of subsistence resources. Although the inhabitants of Cerro Juanaqueña made appreciable use of wild game and non-domesticated plants, by the time the site was occupied we believe that agriculture was thoroughly entrenched in their economic systems. By this we mean that farming activities intersected with most aspects of life and labor, including residential mobility (MacWilliams et al. 2008).

Farming tethered the occupants of Cerro Juanaqueña to the Río Casas Grandes valley, because they had to remain nearby in order to cultivate, protect, harvest, dry, store, and process maize. It is well established that agriculture does not mandate year-round sedentism. Nonetheless it does anchor populations to fields and winter stores (e.g. Hard and Merrill 1992). Tasks related to field preparation, planting, tending, and harvesting span many months of the growing season from early spring through late fall. The labor demands of the agricultural cycle are frequently managed by people residing in the vicinity of fields and their stored harvests for a major portion, if not all, of the year to minimize travel time to fields, and this activity, of course, facilitates guarding fields.

Indicators for season of use in the archaeological record are rare and often ambiguous. Nevertheless, limited evidence from Cerro Juanaqueña does provide support for year-round occupation (contra Roth 2016a:126). One of the non-domesticated economic plants, chia, is a cool season, springtime resource. As described above, the emphasis on cultigens, including maize and perhaps amaranth, indicates a substantial presence in late spring, summer, and early fall. Another economic plant, lovegrass, is a grain that ripens during late summer and fall. However, because most economically important plants are dormant during late fall, winter, and early spring, there are no good botanical indicators for these seasons.

Additional information about seasonality and duration of occupation can be inferred from Schmidt's leporid data (Chapter 11). Young rabbits are abundant in the summer-to-fall season. Therefore, if the site was only occupied during the growing season we would expect a high frequency of immature leporids. Instead, immature leporids make up a very small percentage of the identified remains. These data allow us to say with some confidence that the occupation included a large proportion of the nongrowing season. When combined with the ethnobotanical data, evidence for occupation through most of the year emerges, with plant reproductive parts from a range of taxa representing the spring, summer, and fall seasons complementing the leporid age distribution and indicating that the site was occupied for much of the year, perhaps on a sedentary or semi-sedentary basis. By semi-sedentary, we mean that some or all of the population could have departed the site for a few weeks or months of the year, returning to the site to remain for most of the year.

Other traditional, if less precise archaeological indicators of sedentary behavior also occur at Cerro Juanaqueña and other cerros de trincheras. These findings include abundant midden deposits, tools discarded in all stages of manufacture and use, and human bone, which was present in limited quantities (Chapter 11). The domiciles and likely storage structures found on virtually all of the cerros de trincheras include superstructures made of perishable materials, but also permanent stone foundations, sometimes several courses in height. These features did not result incidentally from clearing cobbles off of a natural surface. Instead, most were deliberately constructed on surfaces which are frequently themselves artificially constructed. These observations imply more than brief, seasonal occupation.

Scale of Construction and Overall Organization

Cerro Juanaqueña is among the most ambitious early building projects in North America. As described in Chapter 3, it includes eight kilometers of constructed terrace walls and approximately 100 rock rings. Its construction involved moving about 53,000 metric tons of material, and we estimate that its construction required over 30 person years of labor (Plate 3). Residential constructions of this scale were not known to be present in the Southwest for another 2,000 years.

We believe that the upper terrace complex at Cerro Juanaqueña was built rapidly by a large labor force, and that its construction implies a degree of planning and organized leadership not usually associated with the Early Agricultural period. The alternate hypothesis is that the terrace complex grew gradually, with individual terraces being added a few at a time as more terrace space was needed. If the latter was the case, we would expect the overall organization of the complex to be haphazard, resembling the middle terraces described in Chapter 3 and illustrated in Figure 3.2. Instead the individual terraces are often joined to one another to form long macrofeatures which follow contours around the hillslope. There is little or no unused or unmodified space within the complex.

Perimeter berms are one of the most distinctive features of the Early Agricultural period cerros de trincheras in northwestern Chihuahua. These features define the exterior boundaries of the upper terrace complex at Cerro Juanaqueña (Plate 2), as well as the exterior boundaries of Cerro el Canelo, Cerro los Torres, and Cerro Vidal (Plates 7, 10 and 13). At Cerro Juanaqueña the perimeter berm is a macrofeature extending around the northern, eastern, and southern edges of the upper terrace complex. It consists of 22 adjoined terraces forming a 400 m long alignment. Where this macrofeature runs perpendicular to the hill slope it resembles other macrofeatures composed of linked individual terraces, but on the northern and southern sides of the site, it runs parallel to the slope and has the form of a large cobble berm, rather than a terrace.

The other three cerros de trincheras, which have been chronometrically dated to the Early Agricultural period, are smaller than Cerro Juanaqueña, but they are very impressive constructions, nevertheless. Cerro el Canelo includes 3,900 meters of terrace wall, and likely required 14 person years to construct, while Cerro los Torres and Cerro Vidal each include between 2,300 and 2,200 meters of terraces and would have each required about eight person years to build.

Aggregation

Equifinality is a problem that sometimes bedevils archaeological interpretation. In the case of Cerro Juanaqueña, we believe that its impressive archaeological signature resulted from intensive occupation by a large number of people over a relatively short period of time. However, a similar outcome could result from other processes. For example, camp groups composed of several extended families might have reoccupied this location repeatedly, perhaps on an annual basis, over a long period of time. This issue is a point of contention in southern Arizona, where large concentrations of Early Agricultural period pithouses are interpreted by some as aggregated villages, and by others as palimpsests of smaller-scale occupations (e.g. Gregory and Diehl 2002; Roth 2016b; Schurr and Gregory 2002; Whittlesey 2016). The palimpsest model is plausible in southern Arizona because use lives of pithouses are limited. As some pithouses are abandoned, other new ones are built, and the result over time can be an impressive number of individual structures. However, only a fraction may have ever been in use at the same time.

For Cerro Juanaqueña, and probably for other Early Agricultural period cerros de trincheras, several lines of evidence converge to suggest strongly that these sites represent sizeable population aggregations, rather than palimpsests of small-scale occupations. The perimeter berm at Cerro Juanaqueña is one argument against growth through accretion. This distinctive feature is coherent around much of the upper terrace complex, and within this perimeter virtually all space is taken up by terrace walls and surfaces. Its coherence implies that it was planned and built all at one time. The area that it encloses and the amount of labor involved in its construction imply a relatively large number of people. Similar features are present at the other Early Agricultural period cerros de trincheras in northwestern Chihuahua.

The brief spans of occupation demonstrated at Cerro Juanaqueña and inferred at Cerro el Canelo, Cerro los Torres, and Cerro Vidal further reinforce the notion that these sites did not grow through gradual accretion. The constructions on these sites represent large labor investments made during a tightly bracketed time frame. Cerro Juanaqueña, in particular, also exhibits a large number of features and a high density of cultural material built and deposited in a short period of time. These observations are also not consistent with accretionary small-scale occupations. They most likely resulted from brief but intensive occupation by an aggregated population.

Basin Metates and Population

One of the many intriguing aspects of Cerro Juanaqueña is the ground stone assemblage, particularly the numerous, heavily worn basin metates. There are 110 complete basin metates and 573 fragments on the surface of the upper terraces. In Chapter 13, we developed a model which estimates population of the site based on the amount of measured wear on these metates. The total wear represented by these metates offers evidence of the intensity of the ancient occupation that is the combined effect of the total number of people and length of occupation. We also modeled different combinations of population, grinding effort, and years of occupation that might account for the total wear represented on these metates, measured by the depth of the grinding surface. What does this ground stone assemblage have to say about the nature of the occupation of Cerro Juanaqueña? By integrating various classes of data, we sought to estimate the general magnitude of the population of Cerro Juanaqueña through examination of metate wear (Chapter 13).

We made inferences regarding population by documenting the wear on all Cerro Juanaqueña basin metates. Using experimentally derived metate wear rates, we estimated that 18,500 person-years of use accounted for the total wear on the basin metates.

Returning to the results of the radiocarbon analysis, the estimate of the duration of occupation was between 30 and 230 years at 68% probability. The model considers the proportion of year that the site may have been occupied as well as the grinding intensity.

There are a number of ways to frame these results, including demonstrating that some upper and lower limits on the nature of the occupation are unlikely. For example, occupations of less than 100 years are probably untenable as the average population required to account for the metate wear would be more than 300 people. Populations of less than 100 people would require year-round occupations that lasted more than 250 years, a combination that also seems unlikely. The model that appears most consistent with all other evidence suggests that 200 to 250 people, occupied the site for about 200 years, assuming the site was occupied for nine months of the year. Grinding intensity of maize and other seeds was perhaps 50% of that of a full-time agriculturalist population. Although this population estimate seems high, Jonathan Mabry et al (2008b) arrived at comparable population levels for the Early Agricultural period settlement of Las Capas in the Tucson Basin.

Why Do People Choose to Live on Hilltops?

In the sections above, we have summarized a series of archaeological observations and have developed a number of inferences from them. We have shown that Cerro Juanaqueña and at least three other cerros de trincheras in northwestern Chihuahua were built and occupied during relatively brief intervals within the Early Agricultural period. We have argued that they hosted intensive residential occupations and that they represent substantial population aggregations. The subsistence economy of their inhabitants was based on maize agriculture practiced on riverine floodplains, along with wild game and locally available non-domesticated plant foods. We have also argued that these populations were significantly less mobile than their predecessors may have been. In all of this, we have avoided the question of why people live on hilltops in the first place.

Hilltop modifications made prior to occupation entailed unprecedented labor investments, and actual occupation of these settings significantly increased the energetic costs of many daily activities. Moreover, in comparison to traditional village locations on terraces or floodplains, the hilltop settings are more exposed to natural hazards such as wind, rain, lightning, and harsh summer temperatures.

Our approach to this problem began with a cross-cultural study that encompassed 186 cultures from around the world (Chapter 7; Hard and Roney 2007). From this sample, we identified 40 reasonably well-documented cases in which people customarily built residential settlements on elevated locations. The central question we sought to answer was "Why do people choose to live on elevated landforms?" As we researched each ethnographic case we classified all of the explanations provided in the ethnographies as to why hilltops were chosen as a settlement location and found that investigators provided one to five explanations for each case.

The most common explanation given for occupation of elevated landforms was defense, which was a factor in 65% of the cases (n=26), and the second most common circumstance was occupation in a rugged landscape where hilltop residential occupation was the only viable option (35%, n=14). Together these two factors, defense and limited options, accounted for 85% of the cases.[1]

The next most common reason given for living on elevated landforms was environmental context, but this reason was the sole one given in only three cases. Two of these involved avoidance of wet soils or drainage, and the third was a case where people established temporary residential sites near honey trees which were located on hills. Proximity to agricultural land was the sole reason given for living on elevated land forms in only two cases. These observations suggest that explanations grounded in environmental context, as we have used the term, or in proximity to agricultural lands are unlikely to be especially useful in explaining residential occupation of isolated hilltops in northwestern Chihuahua during the Early Agricultural period.

Monumentality and symbolic reasons were given only in combination with other explanations; there were no examples where either of these was cited as the primary reason that people lived on hilltops. This analysis showed that multiple factors motivate people to live on hilltops, but it also showed that some explanations are much more likely than others.

Based on this sample, we propose the following: hilltop residential living is costly as it involves increased transport costs and can be understood in terms of a relatively small number of explanations. Most people of the world choose to live on elevated landforms only when there is an absence of lowland settlement space or for the defensive qualities of hilltops. Occasionally special circumstances do occur and people elect hill sites for other reasons which may include access to agricultural land or to avoid low-lying wet conditions. Notably no examples were found of people occupying hills solely for incidental reasons such as improved health, for the view, for microclimatic conditions, or as seasonal aggregation locations. Also, there are no cases of people either initially occupying hills or continuing to live on them solely for symbolic reasons. In our sample, symbolic justification always co-occurs with other conditions.

In most cases where defense is a factor, groups abandon their hills when hostilities cease, although symbolic significance, along with other reasons, may sometimes motivate people to continue living on an elevated landform after the defensive need has passed. We could not find any examples of people occupying hills solely for their monumental characteristics or structures. It is not unusual for monumental construction to occur as element of hilltop locations, but these structures are always a product of complex political organizations and usually the settlements, and sometimes the monuments themselves, have defensive qualities. We do not suggest that this ethnographic sample is a complete inventory of all hilltop residential use either in the past or ethnographically. But we do suggest that all alternative explanations of hilltop residential use do not have equal probability of being correct. With these thoughts in mind, we now turn to archaeological explanations.

The Role of Cerros de Trincheras

Over the years a number of different interpretations of cerros de trincheras have been offered. In this section we will review some of these with specific reference to the four sites we radiocarbon dated. All four of these hilltop sites provided evidence of extensive residential occupation, yet the hilltops are not obvious choices for settlement locations. To us, there are two central questions: "Why did people choose to live in these settings?" and "Why did they invest so much labor in their construction?"

Agricultural Hypothesis

Some of the earliest discussions of cerros de trincheras in Sonora and southern Arizona proposed that at least some were built for agricultural purposes (Huntington 1914:69). More recently Downum and others (1994) proposed that some terraces on late prehistoric cerros de trincheras in southern Arizona may have been designed to take advantage of microenvironmental characteristics of the hillsides for agriculture. Is it possible that hilltop villages were established so that people could live and work adjacent to critical terraced agricultural plots in order to better care for and protect their plantings?

When we began our investigations the agricultural hypothesis seemed quite viable as a general explanation for cerros de trincheras, but today few subscribe to this position. There are a number of reasons to reject interpretation of cerros de trincheras as complexes of agricultural terraces (Chapter 7). For one thing, they do not resemble later agricultural terraces in either setting or design. They were built in areas that are marginal at best for dry land farming, and they do not exhibit design characteristics that would concentrate the little moisture that is available in this setting. More importantly, they represent a huge labor investment that would have yielded a very slender return, even under ideal conditions. We cannot exclude the possibility that domesticated or even non-domesticated plants may have been grown on some terraces, but we do not believe that agriculture was ever more than an incidental use of the terraces built on the early cerros de trincheras in northwestern Chihuahua.

Seasonal Aggregation

Some have proposed that Early Agricultural period cerros de trincheras may have been built as places of seasonal aggregation by peoples whose subsistence was based on hunting and gathering supplemented by limited agriculture (Roth 2016a:126). Seasonal aggregation is a residence pattern observed ethnographically among mobile hunting and gathering peoples around the world, especially in marginal environments such as the southwestern United States and northwestern Mexico. Limited and widely dispersed subsistence resources require that the population spend much of the year in small, widely scattered, and often highly mobile settlements, consisting of a single family or a small group of families. However, these small groups may aggregate seasonally for a few weeks or even for longer periods of time, based upon a seasonally abundant or communally harvested resource, such as fish, antelope, or rabbits. It was once thought that the initial role of maize in the Southwest might have been to underwrite such gatherings. Seasonal aggregations were important times characterized by ritual and intensified social interaction. Barbara Roth (2016a:126) points to specialized artifacts such as pipes and small mortars or trays as possible indications of such activities. To this, we might add the large rock ring at Cerro el Canelo (Chapter 4; Plate 8), which may have provided a setting for communal ritual or social activities, and the roasting features near the summit of Cerro Juanaqueña, which could have been used for communal cooking. Although inconvenient, under this scenario the hilltop settings would certainly have provided highly visible settings for such gatherings.

However, for Cerro Juanaqueña and the other three cerros de trincheras which we tested, several lines of evidence contradict an interpretation of these sites as locations of seasonal aggregation. First, multiple lines of evidence indicate intensive use of the hilltops. These include deeply worn metates, dense midden deposits, abundant plant and faunal remains, and a wide range of artifacts and features. Under the seasonal aggregation model, such remains might accumulate over a very long span of use. However, radiocarbon evidence shows that the span of use at Cerro Juanaqueña was limited to a few centuries, and perhaps even less. This finding is not consistent with a model of seasonal aggregation. Second, the labor investments and the nature of the constructions are not consistent with seasonal aggregation. Their layout suggests planning and rapid construction, rather than haphazard growth through accretion, and their brief span of use does not accommodate accretionary growth. Third, biological evidence does not support a model proposing that occupation was limited to a particular season. Instead, there are indications of use in all four seasons.

Environmental Context

Another explanation which we have termed "environmental context" is actually a group of loosely defined possibilities. The best summary of this position is the statement, "People live on hills for all sorts of reasons." Although not explicitly formulated in the literature, these explanations are informally articulated in conversations with friends and colleagues. These uses include idiosyncratic possibilities such as the hill was occupied for: the view, health benefits, favorable microclimate, insect avoidance, game monitoring, flood avoidance, and so forth. However, the cross-cultural study described above implies that "environmental context" is unlikely to be the primary reason that people chose a 140-meter high hill as a place of permanent residence. In the ethnographic sample reasons like these are sometimes cited, but almost always in combination with other explanations which are much more common and arguably more powerful.

Over the years, no specific models of this nature have been advanced. In our opinion, it will be difficult to formulate a model based on environmental context which accounts for widespread residential use of hilltops over very brief periods of time.

Symbolic Context

We also considered separately the possibility that hilltop locations were chosen for residential occupation solely for reasons of symbolic context. Under this hypothesis, an elevated location may hold symbolic meaning only for its occupants (as opposed to signaling power, dominance, or some other concept communicated to a wider audience). Many peoples in the southwestern United States and northwestern Mexico revere mountain peaks and other elevated places. Shrines are often found on isolated, elevated landforms, and sometimes religious buildings or community centers occupy these settings as well (Fish and Fish 2007). In our global ethnographic sample hilltops are often used for specialized purposes, many of them with symbolic meaning (Chapter 7). Some have suggested that symbolic context alone might have provided sufficient motivation for people to live in settings like Cerro Juanaqueña, Cerro el Canelo, Cerro los Torres, and Cerro Vidal.

To us, the strongest objection to this hypothesis is the principal of least cost. It is one thing to climb a mountain in order to worship in a sacred place. However, in order to live in such a place, inhabitants would have to incur substantial costs on a daily basis, impinging on virtually all aspects of daily life. Absent other powerful motivations, there are few precedents for human behavior of this sort. In the sample of ethnographic cases which we considered, we found only three cases where symbolic context was identified as a factor in settlement location, and it was never the only reason given for occupying isolated, elevated landforms.

Monumentality

Here we use the term "monumentality" to describe constructions which far exceed the functional requirements of their intended purpose (Trigger 1990). Symbolic context and monumentality are often intertwined.

However, the distinction drawn here is that monumentality is intended to impress others (either outsiders or subgroups within a society) by signaling power, dominance, or some other concept. Emil Haury may have been the first to raise this point, suggesting that late prehistoric cerros de trincheras in Sonora and southern Arizona could represent a northern modification of the terraced pyramid so widely known in Mesoamerica (Haury 1976:348). This interpretation is perhaps best developed in recent literature by Christian Downum (2007). Writing about some of the larger, Hohokam-era cerros de trincheras in southern Arizona, Downum suggests that they may have been "rather aggressive political and symbolic statements, designed to send a message to other groups regarding territorial control and political power.... they were instead built as highly visible monuments commemorating the power (and implied threat of resistance or retaliation) represented by a cooperating village unit or other population aggregate" (Downum 2007:134).

There is no denying that many of the cerros de trincheras in northwestern Chihuahua have a monumental character, and the concept of monumentality provides a useful framework for interpreting important aspects of their construction. However, it does not explain residential occupation and reoccupation of these places. Our cross-cultural ethnographic sample included 11 examples of monumental constructions coinciding with hilltop residential settlements. However, in all of these examples, other factors were important determinants of the settlement location. Defense was cited in nine of these 11 examples.

Defense

The very earliest account of cerros de trincheras in Sonora dates to the mid-eighteenth century and includes a description of their use as fortifications (Pfefferkorn 1949). Since that time, many investigators have inferred that these sites were defensive refuges or fortified villages (Lumholtz 1912; McGee 1895, 1898; Sauer and Brand 1931). The approach we used to evaluate the hypothesis that the Río Casas Grandes cerros de trincheras were built for defense relied on LeBlanc's (1999) characteristics of settlements involved in warfare in the Southwest, coupled with several lines of evidence derived from our cross-cultural study. These are not post hoc arguments developed to explain selected aspects of the cerros de trincheras. Instead they are characteristics that we can stipulate in advance, and if they are present they suggest warfare or the threat of warfare existed.

Sites involved in warfare tend to located on defensible landforms. Defensible characteristics include high elevated locations, defined through the ethnographic study as landforms rising 60 to 540 meters (200 to 2000 feet) above surrounding landforms. Other characteristics of defensive settings detailed in Chapter 7 include isolated landforms and locations with steep approaches. All of the cerros de trincheras are either isolated or are elements of small isolated ranges, and all have steep sides. The lowest site is Cerro los Torres, an 80-meter high hill, and the other three are situated on hills that rise between 120 and 160 meters above the surrounding terrain. The hills offer views in 360 degrees, and feature and artifact distributions from Cerro Juanaqueña indicate that residential occupation occurred on all sides of the hill such that any potential attacker could be readily detected.

The sites show strong evidence of planning and a defensive layout, another of LeBlanc's criteria. The four sites reviewed here contain perimeter berms and other linear groups of features. The berms would have offered protection to defenders standing behind them, and because of their height and the insecure footing that they offer, they would have impeded attackers. Within the sites a maze of successive macrofeatures composed of adjacent terraces would have provided alternative lines of defense. LeBlanc argues that larger communities have a significant military advantage over smaller neighbors. Cerro Juanaqueña may have had on the order of 200 occupants, certainly a large population. Extrapolating, the other three cerros de trincheras may have included as few as 50 or as many as 150 inhabitants.

When small, dispersed communities are abandoned, large defensive sites are often built immediately thereafter using techniques particularly suited to rapid construction, another characteristic of defensive sites (LeBlanc 1999). The tight series of 23 radiocarbon dates from Cerro Juanaqueña suggest that its principal period of occupation lasted only about 200 years or less. The perimeter berms at the four sites imply that the sites were planned to accommodate definite spatial needs and were constructed in coordinated episodes. Large portions of these sites did not grow through accretionary, uncoordinated construction episodes.

We do know that line-of-site communication between many of the cerros de trincheras in northwestern Chihuahua is possible, consistent with an expectation of site intervisibility for sites involved in conflict. Figure 7.1 shows that the visual linkages among the fourteen cerros de trincheras sites form three, or perhaps four separate intervisible groups.

The radiocarbon dates from four cerros de trincheras suggest that that the use of these sites was brief and synchronous throughout the region. During the principal period of use in the San Pedro phase, between 1300 and 1100 cal BC, Cerro Juanaqueña, Cerro Torres, and Cerro de Canelo were occupied. A second, more limited occupation, that is not as precisely dated, occurred during the Cienega phase, between 400 cal BC and cal AD 1, of at least two lower terraces at Cerro Juanaqueña and Cerro Vidal. These data suggest that within northwestern Chihuahua the residential occupation of cerros de trincheras was limited to brief episodes which were synchronous throughout the region. Residential reoccupation did not reoccur on these hilltops during the ensuing two millennia, despite the sometimes dense occupations of the Río Casas Grandes valley bottomlands.

LeBlanc and other authors often consider poorly treated human remains as evidence of warfare. At Cerro Juanaqueña we excavated only 1/10th of 1% of the cultural deposits (37.8 cu meters excavated; 31,000 cu m total), and we did not encounter any formal burials. However, three different proveniences on the site did yield 13 small fragments of human bone, and two of the human bone fragments from two different proveniences were burned (see Schmidt's Chapter 11). These bones may reflect poorly treated human remains, although other explanations for them are equally as likely.

Several of LeBlanc's expectations are either not met or cannot be addressed using data from northwestern Chihuahua. None of the sites have protected domestic water supplies, a characteristic which is largely precluded by their hilltop location. However, all are adjacent to the floodplain of the Río Casas Grandes, with relatively easy access to water. LeBlanc also suggests that smaller sites needed more defensive capabilities than larger sites, hypothesizing that smaller sites are more likely to be located in highly defensible settings. We have found no evidence of this. Indeed, Cerro Juanaqueña is almost certainly the largest Late Archaic site in the region and it is among the most defensible. LeBlanc does state that during intervals of extreme warfare small sites may cease to exist. Instead, everyone resides in some form of larger, fortified community. Given the brief span of occupation at Cerro Juanaqueña, and the apparently synchronous use of cerros de trincheras in this region, this interpretation may well be the case during two distinct episodes in the Late Archaic period. Other criteria that cannot be satisfactorily addressed on the basis of current information pertain to site distributions. The cerros de trincheras in northwestern Chihuahua do form a distinct cluster as LeBlanc predicts, but settlement pattern data from the Late Archaic period are incomplete.

The large number of dart points from Cerro Juanaqueña contrasts dramatically with artifact assemblages from many other Late Archaic sites in the Southwest. They accompany a faunal assemblage which is dominated by rabbits, a prey species most often taken by clubs, throwing sticks, and nets rather than by projectiles. These considerations raise the possibility that the artifact assemblage at Cerro Juanaqueña, and perhaps some of the other cerros de trincheras as well, reflects armament for raiding, warfare, and defense. The spatial distribution of projectile points is also different from all other classes of tools, being more evenly distributed across all sectors of the site (Appendix 6.2).

The sum total of observations is difficult to reconcile with the alternative explanations of cerros de trincheras, that is, the agriculture and symbolic models or other explanations of cerros de trincheras. People likely were willing to incur the daily inconvenience and costs of hilltop residence only when compelled to do so during periods of particularly intense raiding and warfare. Nor do we see evidence that agricultural use was combined with a defensive role. Rather we argue Cerro Juanaqueña was a large residential, defensive settlement and that once the need for a defensive posture passed the site was abandoned, only to be briefly reused in a similar fashion about a millennium later.

We did not encounter human burials in northwestern Chihuahua. However, Watson and Phelps (2016) have recently summarized interments of 172 individuals from the Early Agricultural period in the Tucson Basin and La Playa in northern Sonora. They found that 16.3% of these showed evidence of violent trauma, defined as injury from projectile points and blunt force. These observations imply that violent conflict was prevalent during the Early Agricultural period. We believe that the settlement patterns exemplified by Cerro Juanaqueña, Cerro el Canelo, Cerro los Torres, Cerro Vidal, and Tumamoc Hill in Tucson imply that this occupation was not simply a result of interpersonal or interfamily conflict. Instead it was almost certainly organized at the level of entire communities.

Floodplain Dynamics and Site Abandonments

The paleoclimatic literature summarized in Chapter 2 and geomorphologic studies presented in Chapter 12 suggest that people involved in agriculture settled at this location at virtually an ideal point in time for farming in the Río Casas Grandes valley, a situation paralleling conditions in southern Arizona where changes in stream behavior created aggrading floodplains suitable for farming (B. Huckell 1996b). During the two millennia or so that preceded the occupation of Cerro Juanaqueña, high magnitude floods dominated evolution of the floodplain, erosional events had been lowering much of the floodplain, and river channels were deeply entrenched relative to later times. The Río Casas Grandes flowed with mixed-load meander belts and deposited Nordt's Unit D between 3420 and 1620 cal BC under relatively mesic conditions. A key change in stream behavior occurred around 1300 cal BC, concurrent with a reduction of large magnitude flooding throughout the North American Southwest and in northwest Chihuahua. Along the Río Casas Grandes a brief period of landscape stability allowed for the development of the Trincheras paleosol between 1620 and 1130 cal BC. At the same time the occupation of Cerro Juanaqueña was underway 1300 to 1100 cal BC, and the occupants likely farmed the Trincheras soil (Figure 12.5). Samples from this paleosol yielded stable isotope, pollen, and phytolith data (Chapter 10). This setting supported a mixed desert grassland with disturbance plants, habitats with damp soils, and locations with continual moisture, further evidence that agricultural fields located on the Trincheras soil would have been productive.

At this time a moist rainfall regime continued, but high magnitude floods were significantly less common. The river carried gravels, silts, and clays in a meandering stream. Monsoonal overbank floods spread slow-moving water over much of the floodplain, delivering water to fields but not destroying plants. Unit D stratigraphy showed that sediments grade up into fine-grained overbank deposits capped by the weakly developed Trincheras soil (Chapter 12). Silts and clays filled old meander scars yielding deep, arable fields that probably required little preparation prior to planting.

An erosional episode that Nordt refers to as E4 followed the formation of the Trincheras paleosol but preceded the next depositional episode. The E4 erosional event was bracketed at 1130 to 790 cal BC and correlated with the abandonment of Cerro Juanaqueña at 1100 cal BC. This erosional event removed a large portion of the Unit D floodplain along with much of the Trincheras paleosol. Examination of Figure 12.4B shows that channel downcutting associated with E4 removed most of Unit D. Similar downcutting occurred at the same time in the Río San Pedro valley (Figures 12.3B and 12.3C). The E4 erosional episode correlates with a spike in Southwestern-wide flood alluvial frequencies that occurred at 1200 to 1000 cal BC (Harden et al. 2010). At about the same time, a notable increase in monsoonal circulation occurred 1100 to 900 cal BC (Poore et al. 2003).

Assuming that the inhabitants of Cerro Juanaqueña farmed on the Unit D floodplain and that our backhoe trenches are representative of the area, this erosional event would have substantially removed a large segment of the productive farmland that was critical to the Cerro Juanaqueña economy. With only fragments of Unit D remaining, only the Oaxaca Terrace (Unit A, Figure 12.10) and the Petra paleosol (Unit C) offered other options for alluvial terrace farming. Both were higher in elevation, far smaller in area, and had lower agricultural potential (Chapter 12). We suggest that erosional event E4 degraded the agricultural potential and altered the cultural and ecological dynamics such that maintaining a defensive aggregated posture on the top of Cerro Juanaqueña was no longer a viable strategy. The sequence of formation and loss of productive agricultural land may offer insight into the suite of conditions that led to the settlement's success and its abandonment.

Subsequently, this pattern of floodplain deposition, occupation, and erosion repeated itself. A long depositional regime following 790 cal BC created Unit E. Unit E, in turn, was capped by the Janos paleosol that formed between about 390 cal BC and cal AD 230 (Chapter 12). The second, minor occupation of Cerro Juanaqueña and the occupation of Cerro Vidal occurred sometime between 400 cal BC and cal AD 1 correlating with the formation of the Janos paleosol. Another erosional event, E5, truncated the Janos paleosol sometime between cal AD 20 and cal AD 520. Southwestern flood frequencies sharply increased after 50 cal BC (Harden et al. 2010). The apparent correlations between the formation of the Trincheras and Janos paleosols with both occupations of the cerros de trincheras indicate that conditions of landscape stability created opportunities for farming on productive floodplain terraces. The correlations between the two erosional events (E4 and E5) and both abandonments suggest that the erosional processes degraded the agricultural potential of the floodplain. The correlations between paleosol formations and site occupations, and between erosional events and site abandonments, provide an ecological context for understanding the cerros de trincheras. We suspect that the geomorphological and paleoclimatic processes coalesced to form particularly productive agricultural niches along the river reach boundaries. Once those ecological niches were degraded, continued farming of the floodplain and occupation of the hilltops in a defensive posture was no longer a viable strategy.

Why Adopt Farming?

These processes beg the larger question of why foragers become farmers, a topic of ample consideration (e.g. Barker 2009; Bellwood 2005; Binford 2001; Kennett and Winterhalder 2006; Mabry 2005a; Matson 1991; Smith 2015; Wills 1988). One approach to this complex issue is identifying the tradeoffs that push or pull foragers to adopt farming. Push factors include population increases that reduce the relative availability of resources or limit mobility, or climatic deterioration which would force hunters and gatherers toward the adoption of farming. Such processes have been described as models of necessity (Minnis 1992). Pull variables would include highly productive ecological settings which would attract hunters and gatherers to reduce residential mobility and reliance on hunting and gathering in favor of increased investment in agriculture, and these processes are described as models of opportunity (Minnis 1992). Human behavioral ecology considers these tradeoffs simultaneously in the form of net return rates. In one study we used this approach to model foraging and farming the Río Casas Grandes valley versus that of the Jornada region and argued that maize under productive conditions would offer higher return rates than foraging alone after consideration of the differences in the wild plant resources of each region (Hard and Roney 2005). Explanations of the adoption and spread of agriculture must consider both the opportunities and costs involved in reducing foraging and increasing farming.

In this volume we emphasize the fertility of the ancient Río Casas Grandes as a pull factor. Parallel conditions were present at multiple locations in the International Four Corners region, including the Tucson Basin, La Playa, and the Cienega Valley (B. Huckell 1995, 1996b). The evidence is consistent that floodplains provide opportunities that attract regional foragers to adopt farming in these productive niches. In our studies we also did not find evidence of environmental deterioration that would have pushed foragers to farm in the Río Casas Grandes valley. At face value, the data at hand support a model of opportunity; that is, foragers are pulled toward productive farming locations. This pattern is also consistent with recent niche construction approaches in which productive locations are occupied, modified, and defended (Boivin et al. 2016; Smith 2012, 2015; Zeder 2012, 2015). Under this view there is no necessity to adopt farming. Hunters and gatherers are attracted to productive locations to reduce the costs of mobility, to reduce uncertainty, and to benefit from the increased productivity of farming.

However, the project did not attempt to collect evidence of changing Archaic period demography across the landscape to evaluate evidence for population packing or population resource imbalances. We know little about the preceding Middle Archaic period populations in the immediate region. We do know that archaeological survey in the Río Casas Grandes valley and Río San Pedro valley found few Archaic period sites (Whalen and Minnis 2001). On the other hand, large numbers of Archaic projectile points, including those from the Middle Archaic period, as well as burned rock middens and ground stone artifacts, have been found in lowland desert, playa, and dune settings up to 150 km to the east, north, and south of the Río Casas Grandes (Phelps 1998). The New Mexico Bootheel region and adjacent Arizona also yielded many Middle Archaic period projectile points found in lowland, non-riverine settings (Formby 1986; Knight 2003). In the adjacent Jornada Mogollon region of southern New Mexico, west Texas and northern Chihuahua, 200 km to the northeast (Figure 2.3, Location 9), decades of cultural resource management research on military installations have yielded far better data (Miller 2018). There the hunter-gatherer archaeological record of the early Middle Archaic period (Tularosa phase, 4050 to 2550 cal BC) is sparse suggesting populations were low. However, in the following late Middle Archaic period (Keystone phase, 2550 to 1450 cal BC) there is substantial evidence of population growth. The number of sites with contracting stemmed points rise, and there is dramatic increase in the frequency of radiocarbon dates relative to the previous period (Miller 2018). The frequencies of late Middle Archaic earth ovens, residential base camps, shallow pithouses, hearths, and pit features all sharply increase relative to the previous period. Late Middle Archaic sites are found in multiple settings including basins, playa margins, alluvial fans, and foothills (Miller 2018). If the Jornada Mogollon late Middle Archaic record parallels that of adjacent northwest Chihuahua, we cannot eliminate the possibility that hunter-gatherer populations were also increasing in the lowland regions east, north, and south of Cerro Juanaqueña prior to its occupation.

A preliminary model for further evaluation suggests we cannot assume that the attractiveness of the floodplain was the sufficient condition that accounts for the emergence of Cerro Juanaqueña. While floodplain conditions were becoming more productive for farming that may not have been necessarily true of the productivity and predictability of wild resources in surrounding lowlands. Monsoons were at an intermediate level 3250 to 1550 cal BC and sharply declined from 1550 to 1350 cal BC (Barron et al. 2012; Poore et al. 2003; Poore et al. 2005). Could the decrease in monsoons have reduced the productivity and predictability of lowland, nonriverine resources? At the same time, enhanced winter moisture from ENSO activity and reduced flooding boosted potential agricultural productivity of floodplains. If elevated hunter gatherer populations during the late Middle Archaic period (2550 to 1450 cal BC) were exploiting lowland playa and basin resources as the Jornada region data suggest and reductions in summer monsoons following 1550 cal BC impacted lowland productivity, farming in the productive river valleys may have been an attractive option. Thus diverse hunter and gatherer groups from the surrounding regions were pulled and pushed into this newly opened agricultural niche by 1300 cal BC, creating a setting for the formation of new social units and beginning some of the earliest agricultural societies in northwest Chihuahua.

Regional Perspective

When thinking of long-term and fundamental cultural change, our training and background lead us to think first of environmental factors. In Chapter 2, we reviewed a large body of paleoenvironmental literature pertinent to the basin and range country of the southwestern United States and northwestern Mexico. This review suggested that the cool, wet climate of the late Pleistocene and early Holocene gave way to drying conditions in the early part of the middle Holocene, between 7500 and 5500 cal BC. This interval was harsh in the region, and human use of the basin and range country of the borderland region may have been quite restricted during the early middle Holocene.

However, during the late part of the middle Holocene, around 5000 cal BC, the North American Monsoon (NAM) narrowed to its modern distribution, ending the dry conditions of the early middle Holocene and bringing growing season moisture from both the Gulf of Mexico and the Gulf of California. This climactic change linked southwestern Mexico, northwestern Mexico, and the southwestern United States in a similar rainfall regime. The cold, dry winters of the late middle Holocene, coupled with wet summers, allowed C_4 grasslands to flourish in the Southwest, but limited the spread of C_3 desert shrubs into the Chihuahuan Desert (Holmgren et al. 2007). By 3700 to 2500 cal BC, milder winter temperatures and a decrease in winter freezes, coupled with ongoing monsoonal rains, allowed characteristic Chihuahuan Desert C_3 shrubs such as *Acacia greggi*, *Fouquieria splendens*, and *Larrea tridentata*, as well as other plants derived from tropical species, to expand northward.

Zea mays was among the tropical plants which spread into the southwestern United States and northwestern Mexico at this time. Because maize requires human intervention to propagate, we can infer that it was introduced by prehistoric peoples. It may have been present in the Tucson Basin as early as 3700 cal BC (Vint 2015a:448); certainly by 2200 cal BC it was present in multiple settings in the monsoonal regions of the Southwest (Merrill et al. 2009). This may have been a time of incipient cultivation, when maize was used locally in different ways by people whose lifeways were still fundamentally those of hunter-gatherers. Maize was useful, but not yet critical for subsistence. In the basin and range country of the southern Southwest and northwestern Mexico evidence of human populations becomes increasingly common at this time, as discussed in the preceding section.

In addition to facilitating changes in the distribution of plant species, changes in climate were also altering the landscape itself. During the late middle Holocene there may have been extended periods when monsoons were 50 to 100% stronger than today, and riverine settings in the Southwest experienced much more frequent high-magnitude flooding events. A pronounced peak in bedrock flood events throughout the Southwest occurred between about 5000 cal BC and 3000 cal BC (Harden et al. 2010). In the Tucson Basin, most of the alluvium was stripped from valley bottoms during this interval, and the Santa Cruz River was a braided stream flowing over coarse cobbles (Gregory and Nials 2005:43). Similar conditions may have prevailed at about this time in other riverine systems, including the Río Casas Grandes (Chapter 12). Between 4000 and 3000 cal BC, high-magnitude flooding became notably less common. Since that time, alluvial cycles of erosion and deposition continue to occur, but for several millennia now, deposition has prevailed over erosion. Over time, this process has created the lush and diverse, oasis-like niches which we now associate with these riverine settings, as well as soils and topographic conditions which facilitated early agriculture (Huckell 1996b). We do not argue that the natural evolution of this productive niche resulted in intensive early agricultural adaptations in the basin and range country of the southern Southwest, but it was almost certainly a necessary condition. It is important to remember that these settings were once more restricted in extent, and that they have developed gradually over a long period of time. At about 2200 cal BC, the El Nino-Southern Oscillation (ENSO) began to bring wet winters to the Southwest, with a notable increase at 1200 cal BC (Onken et al. 2017). This change, too, may have facilitated the development of early agriculture in the southern Southwest by enhancing moisture availability and recharging ground water.

Beginning about 1500 cal BC, subsistence practices in a large portion of the southern Southwest were focused on floodplains in riverine settings, and people were beginning to make substantial investments in agriculture. We see this first in construction of canals along the Santa Cruz River in the Tucson Basin, associated with radiocarbon dates near 1500 cal BC (Thiel and Mabry, 2006). Not too long afterwards, around 1300 cal BC, people built impressive hilltop residential complexes adjacent to the Río Casas Grandes floodplain at Cerro Juanaqueña, Cerro el Canelo, Cerro los Torres, and perhaps other locations. Additional canal systems were built in the Tucson Basin beginning around 1220 cal BC (Vint 2015b). The inhabitants of La Playa, Sonora, also farmed a reach boundary along the floodplain of the Río Boquillas during the San Pedro phase, and by the Cienega phase, if not earlier, they had constructed substantial irrigation systems. These projects required planning and multifamily coordination, implying a degree of leadership and social organization not documented in earlier times. They demonstrate both a motivation and capacity for collective action.

In tandem with the appearance of these large-scale construction projects, we find a suite of new artifact classes, several of which have no direct or obvious connection to subsistence or manufacturing activities. For example, although use of pigments is documented in the Tucson Basin at about 2100 cal BC, beginning around 1500 cal BC, its use becomes more widespread and more intense. Evidence of red pigments have been found in mortuary contexts, and at least one specially shaped lapstone used to process pigment has been found at this date in the Tucson Basin (J. Adams 2006:9.6). In northwestern Chihuahua, small, shallow, carefully shaped mortars and small compatible pestles may have been used for this purpose, beginning around 1300 cal BC (Chapter 6). Subsequently, equivalent although stylistically distinct artifacts become an integral element of San Pedro phase assemblages from the Tucson Basin, dated 1200 cal BC to 800 cal BC.

Stone pipes are another category of artifacts that are well represented in both northwestern Chihuahua and the Tucson Basin, along with heavily worn chipped stone drills used to manufacture their bores. According to Alan Ferg (1998), the earliest such artifacts reported from the Southwest are from the Upper Cave at Ventana Cave, dated by artifact associations to between 3000 BC and 1200 BC. In southern Arizona, northern Sonora, and northwestern Chihuahua, stone pipes are relatively common throughout the Early Agricultural period. Other unusual artifacts that are found consistently in Early Agricultural period assemblages of this region include chipped stone cruciforms and, more rarely, small spindle whorls made of ground stone.

We interpret at least some of these as social artifacts. That is, they were items which played important roles in social, symbolic, and ideological realms rather than having been employed directly in subsistence or manufacturing activities. They were used, at least in part, to facilitate ritual and other activities related to group identity, social integration, and development of group solidarity. Based on current evidence, they appear rather suddenly in the context of population aggregation, relatively permanent settlements, and large-scale construction projects. We argue here that danger of armed conflict was another factor driving rapid social change and shaping material culture at this time.

The spatial and temporal patterns that we have outlined here are part of a regional phenomenon that Jonathan Mabry (2008b) describes as the "San Pedro complex." We understand this as an adaptation that developed rapidly beginning around 1500 cal BC, emphasizing floodplain agriculture, especially near reach boundaries with persistent moisture such as cienegas. This adaptation appears almost synchronously at various places in the International Four Corners region (Figure 2.2). At present it is best known from the Tucson Basin in southern Arizona, from the site of La Playa in northern Sonora, and from the cerros de trincheras in northwestern Chihuahua described in this volume. The Cienega Creek sites in southern Arizona represent similar adaptations (B. Huckell 1995) but may represent smaller populations. We expect that additional Early Agricultural period pithouse settlements, similar to those in the Tucson Basin, will eventually be recognized along riverine floodplains in northwestern Chihuahua, as well as elsewhere in northern Sonora and southern Arizona. Occasional construction and residential occupation of cerros de trincheras are an integral part of this complex, and we predict that Early Agricultural period occupations will eventually be found on other cerros de trincheras in Sonora and Arizona.

The origins of the San Pedro complex are still hotly debated. Some argue for gradual, incremental change among peoples who continued to rely primarily on nondomesticated plants and animals for their subsistence needs. Others maintain that the transition from incipient agriculture to agricultural dependence occurred rapidly, and some argue that it occurred through population replacement. Under this latter model, fully committed agriculturalists migrated northwards into the southern Southwest and either displaced or absorbed earlier inhabitants who were less reliant on agriculture.

The model of gradual, incremental change fits what we know of the period of incipient agriculture, from 2100 cal BC or earlier until about 1500 cal BC. However, as we read the evidence, fundamental changes between 1500 cal BC and 1200 cal BC occurred with startling rapidity. They are evident in the archaeological record as a series of interrelated shifts in subsistence technology (irrigation), settlement patterns (focus on floodplains, residential occupation of cerros de trincheras, reduced residential mobility), architecture (integrative structures), social behavior (use of pigments and unusual artifacts), scale of organization (large-scale labor projects), and warfare (osteological evidence, residential use of cerros de trincheras). Based on current evidence, these elements are largely absent prior to 1500 cal BC, and yet the pattern appears to be fully developed by the time Cerro Juanaqueña and other sites in northwestern Chihuahua were occupied, 1300 BC to 1100 BC.

Sudden appearance of entire suites of new traits in the archaeological record, such as the San Pedro complex, can indicate migration, and we did take this possibility seriously. We spent five field seasons searching for antecedents along the eastern slopes of the Sierra Madre of central and southern Chihuahua (MacWilliams et al. 2008), and others have surveyed the similar settings on the western slopes. The amount of archaeological work which has been done in this part of Mexico is miniscule in comparison to that done in the southwestern United States. However, results to date have failed to discover major Early Agricultural period developments south of the borderlands. In 2009 we marshalled linguistic, genetic, and archaeological evidence to argue against the hypothesis that large-scale migrations of Proto-Uto-Aztecans brought agriculture to northwestern Mexico and the southwestern United States (Merrill et al. 2009).

Subsequently Merrill (2012) reconstructed the Proto-Southern Uto-Aztecan speech community of foragers to have been living in northern Sonora on the west side of the Sierra Madre, between 30 and 32 degrees latitude before 4000 years ago when they acquired maize. This location places them in the vicinity of the Tucson Basin earliest maize dates and the Early Agricultural period site of La Playa, Sonora. This research implies that the Proto-Southern Uto-Aztecan speech community could have been an important contributor to emergence of agricultural communities in the International Four Corners region.

Short distance migrations and ethnogenesis may have played important roles in the emergence of these agricultural communities. The introduction of agriculture opens new ecological niches that multiple societies may fill as farming became the technological and economic basis of change (Binford 2001). As groups occupied this new agricultural niche, it provided a setting for the formation of new ethnic identities. Ethnogenesis represents an alternative view that has received less attention in discussions of the adoption and spread of agriculture. Rather than emphasizing stability and lineal descent, this perspective stresses blending of cultural elements with disparate origins, resulting in formation of cultural systems which are fundamentally different from their various antecedents (Moore 1994). Blanton (2015) refers to this process as "ethnic construction" and views it as a means of overcoming the problem of collective action by establishing socially purposed cooperative groups. To us, this description seems apt for processes that were underway around 1500 cal BC in the southern Southwest. People were organizing in new ways in order to provide for common defense; to undertake large-scale, community level construction projects; and to provide for cooperative management of agricultural lands as they occupied new agricultural floodplain niches.

The San Pedro complex is a coherent archaeological pattern found over a large area of the southern Southwest. Consistencies in settlement patterns and material culture, as well as common evolutionary trajectories, imply a high degree of interaction throughout the region. However, differences among subareas within the larger San Pedro complex area imply local autonomy and may also indicate that the San Pedro complex was a multiethnic entity. For example, the incipient pottery that includes figurines, fired-clay bowls, and ceramic pipes seem to be limited to the Tucson Basin, and there are stylistic differences in small stone mortars or trays. Cruciforms are found throughout the region but are far more common in Sonora and Chihuahua, and certain projectile point types such as Cortaro and Empire points have more restricted distributions within the region.

Floodplain agriculture is at the heart of the San Pedro complex. Although people continued to exploit wild plants and animals, their hunting and gathering activities were substantially tethered to agricultural lands along the river systems. We believe that competition between these regionally-based groups, as well as rapidly changing attitudes toward land tenure and territoriality, led to at least sporadic warfare throughout the Early Agricultural period in the southern Southwest. Additional cultural instabilities likely arose from the natural cycles of erosion and alluviation, which alternately facilitated and impeded agricultural production. Despite these sources of instability, at a regional level this cultural system managed to perpetuate itself in a recognizable form, with strong continuity in both settlement patterns and material culture through the San Pedro and Cienega phases, lasting until cal AD 50 or later.

Notes

1. Percentages do not sum to 100% as many cases had more than one explanation.

References

Abrams, Elliot M.

1989 Architecture and Energy: An Evolutionary Perspective. In *Archaeological Method and Theory, Vol. 1*, edited by Michael B. Schiffer, pp. 47–87. University of Arizona Press, Tucson.

1994 *How the Maya Built Their World: Energetics and Ancient Architecture.* University of Texas Press, Austin.

Adams, Jenny L.

1998 Ground Stone Artifacts. In *Archaeological Investigations of Early Village Sites in the Middle Santa Cruz Valley: Analysis and Synthesis, Part 1*, edited by Jonathan B. Mabry, pp. 357–422 Anthropological Papers No. 19. Center for Desert Archaeology, Tucson.

1999 Refocusing the Role of Food-Grinding Tools as Correlates for Subsistence Strategies in the U.S. Southwest. *American Antiquity* 64:475–498.

2001 The Ground Stone Assemblage. In *Excavations in the Santa Cruz River Floodplain: The Early Agricultural Period Component at Los Pozos*, edited by David A. Gregory, pp. 107–134 Anthropological Papers No. 21. Center for Desert Archaeology, Tucson.

2002 *Ground Stone Analysis: A Technological Approach.* University of Utah Press, Salt Lake City.

2005 Early Agricultural Period Grinding Technology. In *Material Cultures and Lifeways of Early Agricultural Communities in Southern Arizona*, edited by R. Jane Sliva, pp. 99–120. Anthropological Papers 35. Center for Desert Archaeology, Tucson.

2006 Ground Stone Artifacts. In *Rio Nuevo Archaeology Program, 2000–2003: Investigations at the San Agustin Mission and Mission Gardens, Tucson Presidio, Tucson Pressed Brick Company, and Clearwater Site*, edited by J. Homer Thiel and Jonathan B. Mabry, pp. 9.1–9.32. Technical Report 2004-11. Center for Desert Archaeology, Tucson.

2015a Building Models for Understanding How Maize Was Incorporated into Diet of Early Agriculturalists. In *Las Capas Archaeological Project: Ground Stone and Maize Processing Experiments*, edited by Jenny L. Adams, J. Skeldon Rychener, and A. J. Denoyer, pp. 19–42. Technical Report 2014-02. Desert Archaeology, Inc., Tucson.

2015b Exploring Early Agricultural Technological Traditions at Las Capas with Experiments. Paper presented at the 80th Annual Meeting of the Society of American Archaeology, San Francisco.

2015c San Pedro Phase Grinding Technology at Las Capas, AZ AA:12:111 (ASM). In *Implements of Change: Tools, Subsistence, and the Built Environment of Las Capas, an Early Agricultural Irrigation Community in Southern Arizona*, edited by James M. Vint, pp. 95–159. Anthropological Papers 51. Desert Archaeology, Inc., Tucson.

Adams, Karen R.

1980 *Pollen, Parched Seeds and Prehistory: A Pilot Investigation of Prehistoric Plant Remains from Salmon Ruin, a Chacoan Pueblo in Northwestern New Mexico.* Contributions in Anthropology 9. Eastern New Mexico University, Portales, New Mexico.

1988 *The Ethnobotany and Phenology of Plants in and Adjacent to Two Riparian Habitats in Southeastern Arizona.* PhD Dissertation, Department of Ecology and Evolutionary Biology, University of Arizona, Tucson. University Microfilms, Ann Arbor.

1994 A Regional Synthesis of *Zea mays* in the Prehistoric American Southwest. In *Corn and Culture in the Prehistoric New World*, edited by Sissel Johannessen and Christine A. Hastorf, pp. 273–302. Westview Press, Boulder, Colorado.

1998 Plant Remains Recovered from Juanaqueña, a Late Archaic (3000 BP) Trincheras Site in Northern Chihuahua. Manuscript on file, Department of Anthropology, University of Texas at San Antonio.

1999 Plant Remains Recovered in 1998 from Cerro Juanaqueña, Cerro Vidal, and Cerro Los Torres, Trincheras Sites in Northern Chihuahua. Manuscript on file, Department of Anthropology, University of Texas at San Antonio.

2002 Subsistence and Wood Use in Northwestern Mexico: Plant Remains from Late Archaic Cerros de Trincheras Sites in Chihuahua. Manuscript on file, Department of Anthropology, University of Texas at San Antonio.

2004 Anthropogenic Ecology of the North American Southwest. In *People and Plants in Ancient Western North America*, edited by Paul E. Minnis, pp. 167–204. Smithsonian Books, Washington, DC.

2015 The Archaeology and Agronomy of Ancient Maize (*Zea mays* L.). In *Traditional Arid Lands Agriculture: Understanding the Past for the Future*, edited by Scott Ingram and Robert C. Hunt, pp. 15–53. University of Arizona Press, Tucson.

Adams, Karen R. and Vorsila L. Bohrer

1998 Archaeobotanical Indicators of Seasonality: Examples from Arid Southwestern United States. In *Seasonality and Sedentism: Archaeological Perspectives from Old and New World Sites*, edited by Thomas R. Rocek and Bar-Ofer Yosef, pp. 129–138.

Peabody Museum Bulletin 6. Peabody Museum of Archaeology and Ethnology, Harvard University, Cambridge, Massachussetts.

Adams, Karen R. and Vandy E. Bowyer

2002 Sustainable Landscape: Thirteenth Century Food and Fuel Use in the Sand Canyon Locality. In *Seeking the Center Place: Archaeology and Ancient Communities in the Mesa Verde Region*, edited by Mark D. Varien and Richard H. Wilshusen, pp. 123–142. University of Utah Press, Salt Lake City.

Adams, Karen R. and Suzanne K. Fish

2006 Southwest Plants. In *Environment, Origins, and Population*, edited by Douglas H. Ubelaker, pp. 292–312. Handbook of North American Indians Vol. 3, William C. Sturtevant, general editor. Smithsonian Institution, Washington, DC.

Adams, Karen R., Cathryn M. Meegan, Scott G. Ortman, R. Emerson Howell, Lindsay C. Werth, Deborah A. Muenchrath, Michael K. O'Neill, and Candice A. C. Gardner

2006 MAÍS (Maize of American Indigenous Societies) Southwest: Ear Descriptions and Traits That Distinguish 27 Morphologically Distinct Groups of 123 Historic USDA Maize (*Zea mays* L. spp. *mays*) Accessions and Data Relevant to Archaeological Subsistence Models. Arizona State University, Tempe.

Adams, Karen R., Deborah A. Muenchrath, and Dylan M. Schwindt

1999 Moisture Effects on the Morphology of Ears, Cobs and Kernels of a South-Western U.S. Maize (*Zea Mays L.*) Cultivar, and Implications for the Interpretation of Archaeological Maize. *Journal of Archaeological Science* 26:483–496.

Ahler, Stanley A.

1989 Mass Analysis of Flaking Debris: Studying the Forest Rather Than the Trees. In *Alternative Approaches to Lithic Analysis*, edited by Donald O. Henry and George Odell, pp. 85–118. Archaeological Papers No. 1. American Anthropological Association.

Alipayou, Daniel, Jerry L. Holechek, Raul Valdez, Ackim Tembo, Lewis Saiwana, Michael Rusco, and Manual Cardenas

1993 Range Condition Influence on Chihuahuan Desert Cattle and Jackrabbit Diets. *Journal of Range Management* 46:296-301.

Alley, R. B., P. A. Mayewski, T. Sowers, M. Stuiver, K. C. Taylor, and P. U. Clark

1997 Holocene Climatic Instability: A Prominent, Widespread Event 8200 Yr Ago. *Geology* 25:483–486.

Alvarez de Williams, Anita

1983 Cocopa. In *Southwest*, edited by Alfonso Ortiz, pp. 99–112. Handbook of North American Indians Vol. 10, William C. Sturtevant, general editor. Smithsonian Institution, Washington, DC.

Anderson, Edgar

1952 *Plants, Man, and Life*. University of California Press, Berkeley.

Anderson, Sydney

1972 *Mammals of Chihuahua: Taxonomy and Distribution*. Bulletin of the American Museum of Natural History, Vol. 148, article 2. American Museum of Natural History, New York.

Andrefsky, William, Jr.

1994 Raw-Material Availability and the Organization of Technology. *American Antiquity* 59:21–34.

1998 *Lithics: Macroscopic Approaches to Analysis*. Cambridge University Press, Cambridge.

2001 *Lithic Debitage: Context, Form, Meaning*. University of Utah Press, Salt Lake City.

2006 Experimental and Archaeological Verification of an Index of Retouch for Hafted Bifaces. *American Antiquity* 71:743–757.

2009 The Analysis of Stone Tool Procurement, Production, and Maintenance. *Journal of Archaeological Research* 17:65–103.

Antevs, Ernst

1948 Climatic Changes and Pre-White Man. In *The Great Basin: With Emphasis on Glacial and Postglacial Times*, pp. 168–191. Bulletin of the University of Utah Vol. 38(20), Biological Series, Vol. X. The University of Utah, Salt Lake City.

Anyon, Roger and Steven A. LeBlanc

1984 *The Galaz Ruin*. University of New Mexico Press, Albuquerque.

Arakawa, Fumiyasu, Scott G. Ortman, M. Steven Shackley, and Andrew I. Duff

2011 Obsidian Evidence of Interaction and Migration from the Mesa Verde Region, Southwest Colorado. *American Antiquity* 76:774–796.

Arizona Game and Fish Department

2012 *Anodonta Californiensis*. Unpublished Abstract Compiled and edited by the Heritage Data Management System, Arizona Game and Fish Department, Phoenix.

Arnon, Nancy S. and W. W. Hill

1979 Santa Clara Pueblo. In *Southwest*, edited by Alfonso Ortiz, pp. 296–307. Handbook of North American Indians, Vol. 9, William C. Sturtevant, general editor. Smithsonian Institution, Washington, DC.

Arthur, Kathryn Weedman

2010 Feminine Knowledge and Skill Reconsidered: Women and Flaked Stone Tools. *American Anthropologist* 112:228–243.

Asmerom, Yemane, Victor Polyak, Stephen Burns, and Jessica Rassmussen

2007 Solar Forcing of Holocene Climate: New Insights from a Speleothem Record, Southwestern United States. *Geology* 35:1–4.

Assad Hunter, Cristi

2001 Vertebrate Faunal Remains. In *The Grewe Archaeological Research Project, Volume 2, Material Culture Part II: Stone, Shell, and Bone Artifacts and Biological Remains*, edited by Douglas B. Craig, pp. 647–679. Anthropological Papers No. 99-1. Northland Research, Inc., Phoenix.

Bamforth, Douglas B.

1986 Technological Efficiency and Tool Curation. *American Antiquity* 51:38–50.

2002 High-Tech Foragers? Folsom and Later Paleoindian

Technology on the Great Plains. *Journal of World Prehistory* 16:55–98.
2003 Rethinking the Role of Bifacial Technology in Paleoindian Adaptations on the Great Plains. In *Multiple Approaches to the Study of Bifacial Technologies*, edited by M. Soressi and H.L. Dibble, pp. 209–228. University Museum Monograph 115. Museum of Archaeology and Anthropology, University of Pennsylvania, Philadelphia.

Barker, Graeme
2009 *The Agricultural Revolution in Prehistory: Why Did Foragers Become Farmers?* Oxford University Press, Oxford.

Barlett, Peggy F.
1980 *Agricultural Decision Making: Anthropological Contributions to Rural Development.* Academic Press, New York.

Barlow, K. Renee
2002 Predicting Maize Agriculture among the Fremont: An Economic Comparison of Farming and Foraging in the American Southwest. *American Antiquity* 67:65–88.

Barneby, Rupert C.
1964 *Atlas of North American Astragalus.* Memoirs of the New York Botanical Garden 13. 2 vols. New York Botanical Garden, New York.

Barrett, Samuel A.
1952 *Material Aspects of Pomo Culture*, Part One. Bulletin of the Public Museum of the City of Milwaukee, Vol. 20. Board of Trustees, Public Museum of the City of Milwaukee, Milwaukee, Wisconsin. Electronic Document, http://ehrafworldcultures.yale.edu/document?id=ns18-023, accessed November 26, 2017.

Barron, John A. and Lesleigh Anderson
2011 Enhanced Late Holocene ENSO/PDO Expression Along the Margins of the Eastern North Pacific. *Quaternary International* 235:3–12.

Barron, John A., S. E. Metcalfe, and J. A. Addison
2012 Response of the North American Monsoon to Regional Changes in Ocean Surface Temperature. *Paleoceanography* 27:1–17.

Bartlett, John Russell
1854 [1965] *Personal Narrative of Explorations and Incidents in Texas, New Mexico, California, Sonora, and Chihuahua.* D. Appleton and Company, New York. The Rio Grande Press, Inc., Chicago.

Bautista, Jesus and Philip Goodell
1983 Bimodal Volcanism in the Colonia Mexico-Highway 10 Area, Chihuahua, Mexico. In *Geology and Mineral Resources of North-Central Chihuahua. Guidebook for the 1983 Field Conference*, pp. 225–231. El Paso Geological Society, El Paso.

Bayham, Frank E.
1982 *A Diachronic Analysis of Prehistoric Animal Exploitation at Ventana Cave.* PhD dissertation, Department of Anthropology, Arizona State University, Tempe. University Microfilms, Ann Arbor.

Bayham, Frank E. and Pamela Hatch
1985 Archaeofaunal Remains from the New River Area. In *Hohokam Settlement and Economic System in the Central New River Drainage, Arizona*, edited by David E. Doyel and Mark D. Elson, pp. 405–433. Soil Systems Publications in Anthropology No. 4, Phoenix.

Beck, R. Kelly
2008 Transport Distance and Debitage Assemblage Diversity: An Application of the Field Processing Model to Southern Utah Toolstone Procurement Sites. *American Antiquity* 73:759–780.

Bellwood, Peter
1997 Prehistoric Cultural Explanations for Widespread Language Families. In *Archaeology and Linguistics: Aboriginal Australia in Global Perspective*, edited by Patrick McConvell and Nicholas Evans, pp. 123–134. Oxford University Press, Melbourne.
2005 *First Farmers: The Origins of Agricultural Societies.* Blackwell Publishing, Malden, Massachusetts.

Benz, Bruce F. and Carlos Alvarez del Castillo
n.d. Racial Identification of Maize from the Late Prehistoric Site of Cuarenta Casas, Chihuahua. Manuscript on file, Department of Botany, University of Wisconsin, Madison.

Berry, Claudia F. and Michael S. Berry
1986 Chronological and Conceptual Models of the Southwestern Archaic. In *Anthropology of the Desert West: Essays in Honor of Jesse D. Jennings*, edited by Carol J. Condie and Donald D. Fowler, pp. 253–327. Anthropological Papers 110. University of Utah, Salt Lake City.

Berry, Michael S.
1982 *Time, Space and Transition in Anasazi Prehistory.* University of Utah Press, Salt Lake City.

Binford, Lewis R.
1962 Archaeology as Anthropology. *American Antiquity* 28:217–225.
1973 Interassemblage Variability: The Mousterian and the 'Functional Argument.' In *The Explanation of Cultural Change: Models in Prehistory*, edited by C. Renfrew, pp. 227–254. Duckworth, London.
1977 Forty-Seven Trips. In *Stone Tools as Cultural Markers*, edited by R. Wright, pp. 24–36. Australian Institute of Aboriginal Studies, Canberra, Australia.
1979 Organization and Formation Processes: Looking at Curated Technologies. *Journal of Anthropological Research* 35:255–273.
1980 Willow Smoke and Dogs' Tails: Hunter-Gatherer Settlement Systems and Archaeological Site Formation. *American Antiquity* 45:4–20.
1983 *In Pursuit of the Past.* Thames and Hudson, London.
2001 *Constructing Frames of Reference: An Analytical Method for Archaeological Theory Building Using Hunter-Gatherer and Environmental Data Sets.* University of California Press, Berkeley.

Binford, Lewis R. and Nancy M. Stone
1985 "Righteous Rocks" and Richard Gould: Some Observations on Misguided "Debate." *American Antiquity* 50:151–153.

Blackmer, A.
2000 Bioavailability of Major Essential Elements. In *Handbook of Soil Science*, edited by M. Sumner, pp. D3–D18. CRC Press, Boca Raton, Florida.

Blanton, Richard E.
2015 Theories of Ethnicity and the Dynamics of Ethnic Change in Multiethnic Societies. *Proceedings of the National Academy of Sciences of the United States of America* 112:9176–9181.

Bleed, Peter
1986 The Optimal Design of Hunting Weapons: Maintainability or Reliability. *American Antiquity* 51:737–747.

Bohrer, Vorsila L.
1962 Nature and Interpretation of Ethnobotanical Materials from Tonto National Monument. In *Archaeological Studies of Tonto National Monument, Arizona*, edited by Charles R. Steen, Loyd M. Pierson, Vorsila L. Bohrer, and Kate Peck Kent, pp. 75–114. Technical Series 2. Southwestern Monuments Association, Globe, Arizona.
1973 Tentative List of Utilized Plant Remains from Fresnal Shelter, January 1972. In *Technical Manual: 1973 Survey of the Tularosa Basin*, pp. 211–218. Human Systems Research, Tularosa, New Mexico.
1987 The Plant Remains from La Ciudad, a Hohokam Site in Phoenix. In *La Ciudad, Specialized Studies in the Economy, Environment, and Culture of La Ciudad*, edited by JoAnn E. Kisselburg, Glen E. Rice, and Brenda L. Shears, pp. 67–202. Anthropological Field Studies 20. Arizona State University, Tempe, Arizona.
2007 *Preceramic Subsistence in Two Rock Shelters in Fresnal Canyon, South Central New Mexico*. Archaeological Series 199. Arizona State Museum, The University of Arizona, Tucson.

Bohrer, Vorsila L. and Karen R. Adams
1977 *Ethnobotanical Techniques and Approaches at Salmon Ruin, New Mexico*. Contributions in Anthropology 8, No. 1. Eastern New Mexico University, Portales.

Boivin, N. L., M. A. Zeder, D. Q. Fuller, A. Crowther, G. Larson, J. M. Erlandson, T. Denham, and M. D. Petraglia
2016 Ecological Consequences of Human Niche Construction: Examining Long-Term Anthropogenic Shaping of Global Species Distributions. *Proceedings of the National Academy of Sciences of the United States of America* 113:6388–6396.

Bowers, Janice E. and Steven P. McLaughlin
1982 Plant Species Diversity in Arizona. *Madroño* 29: 227–233.
1987 Flora and Vegetation of the Rincon Mountains, Pima County, Arizona. *Desert Plants* 8:51–94.
1996 Flora of the Huachuca Mountains, a Botanically Rich and Historically Significant Sky Island in Cochise County, Arizona. *Journal of the Arizona-Nevada Academy of Science* 29:66–107.

Bradley, Bruce
1975 Lithic Reduction Sequences. In *Lithic Technology: Making and Using Stone Tools*, edited by E. Swanson, pp. 5–13. Mouton Publishers, The Hague.

Brady, Ryan T.
2005 Assessing the Role of Bifaces in Hunter-Gatherer Toolkits of Eastern California. *Proceedings of the Society for California Archaeology* 18:238–246.

Brand, Donald D.
1937 *The Natural Landscape of Northwestern Chihuahua*. University of New Mexico Press, Albuquerque.

Brantingham, P. Jeffrey
2006 Measuring Forager Mobility. *Current Anthropology* 47:435–459.

Braun, David P.
1983 Pots as Tools. In *Archaeological Hammers and Theories*, edited by James A. Moore and Arthur S. Keene, pp. 107–134. Academic Press, New York.

Bronk Ramsey, Christopher
2001 Development of the Radiocarbon Calibration Program Oxcal. *Radiocarbon* 43:355–363.
2005 Oxcal 3.10. Electronic Document, https://c14.arch.ox.ac.uk/oxcal/OxCal.html, accessed June 8, 2010.
2017 Oxcal 4.3. Electronic Document, https://c14.arch.ox.ac.uk/oxcal/OxCal.html, accessed April 15, 2017. University of Oxford.

Bronson, F. H. and O. W. Tiemeier
1958 Reproduction and Age Distribution of Black-Tailed Jackrabbits in Kansas. *Journal of Wildlife Management* 22:409–414.

Brown, Cecil H.
2008 A Lack of Linguistic Evidence for Domesticated Sunflower in Pre-Columbian Mesoamerica. *Proceedings of the National Academy of Sciences of the United States of America* 105:E47.

Brown, Coleen F. and Paul R. Krausman
2003 Habitat Characteristics of 3 Leporid Species in Southeastern Arizona. *The Journal of Wildlife Management*:83–89.

Brown, David E.
1982a Chihuahuan Desertscrub. In *Biotic Communities of the American Southwest-United States and Mexico*, edited by David E. Brown, pp. 169–179. Special Issue of *Desert Plants*, Vol 4. The University of Arizona for the Boyce Thompson Southwestern Arboretum, Superior, Arizona.
1982b Madrean Evergreen Woodland. In *Biotic Communities of the American Southwest-United States and Mexico*, edited by David E. Brown, pp. 59–65. Special Issue of *Desert Plants*, Vol 4. The University of Arizona for the Boyce Thompson Southwestern Arboretum, Superior, Arizona.
1982c Plains and Great Basin Grassland. In *Biotic Communities of the American Southwest-United States and Mexico*, edited by David E. Brown, pp. 115–121. Special Issue of *Desert Plants*, Vol 4. The University of Arizona for the Boyce Thompson Southwestern Arboretum, Superior, Arizona.
1982d Semidesert Grassland. In *Biotic Communities of the American Southwest-United States and Mexico*, edited by David E. Brown, pp. 123–131. Special Issue of *Desert Plants*, Vol 4. The University of Arizona

for the Boyce Thompson Southwestern Arboretum, Superior, Arizona.
Brown, David E. (editor)
1982 *Biotic Communities of the American Southwest-United States and Mexico*. Special Issue of *Desert Plants*, Vol 4. The University of Arizona for the Boyce Thompson Southwestern Arboretum, Superior, Arizona.
1994 *Biotic Communities: Southwestern United States and Northwestern Mexico*. University of Utah Press, Salt Lake City.
Brown, David E. and Charles H. Lowe
1980 *Biotic Communities of the Southwest*. Thematic Map, 1:1,000,000. General Technical Report RM-78. Rocky Mountain Forest and Range Experiment Station, Forest Service, U.S. Department of Agriculture.
Brown, David E. and Charles P. Pase
1982 Rocky Mountain (Petran) and Madrean Montane Conifer Forests. In *Biotic Communities of the American Southwest-United States and Mexico*, edited by David E. Brown, pp. 43–48. Special Issue of *Desert Plants*, Vol 4. The University of Arizona for the Boyce Thompson Southwestern Arboretum, Superior, Arizona.
Brown, Gary M.
1990 Specialized Lithic Exchange and Production on Anderson Mesa. In *Technological Change in the Chavez Pass Region, North-Central Arizona*, edited by G. Brown, pp. 173–206. Anthropological Research Papers No. 41. Arizona State University, Tempe.
1991 Embedded and Direct Lithic Procurement Strategies on Anderson Mesa. *Kiva* 56:359–384.
Bruno, Maria C.
2006 A Morphological Approach to Documenting *Chenopodium* Domesticates in the Andes. In *Documenting Domestication: New Genetic and Archaeological Paradigms*, edited by Melinda A. Zeder, Daniel G. Bradley, Eve Emshwiller, and Bruce D. Smith, pp. 32–45. University of California Press, Berkeley.
Buchanan, Briggs
2002 Folsom Lithic Procurement, Tool Use, and Replacement at the Lake Theo Site, Texas. *The Plains Anthropologist* 47(181):121–146.
Buck, Brenda J. and H. Curtis Monger
1999 Stable Isotopes and Soil-Geomorphology as Indicators of Holocene Climate Change, Northern Chihuahuan Desert. *Journal of Arid Environments* 43:357–373.
Bull, W. B.
1997 Discontinuous Ephemeral Streams. *Geomorphology* 19:227–276.
Bunn, Henry T.
1993 Bone Assemblages at Base Camps: A Further Consideration of Carcass Transport and Bone Destruction by the Hadza. In *From Bones to Behavior: Ethnoarchaeological and Experimental Contributions to the Interpretation of Faunal Remains*, edited by Jean Hudson, pp. 156–168. Occasional Paper No. 21. Center for Archaeological Investigations, Southern Illinois University, Carbondale.
Burt, William H. and Richard P. Grossenheider
1976 *A Field Guide to the Mammals*. 3rd ed. Peterson Field Guides. Houghton-Mifflin Co., Boston.
Butzer, K. W., J. T. Abbott, C. D. Frederick, P. H. Lehman, C. E. Cordova, and J. F. Oswald
2008 Soil-Geomorphology and "Wet" Cycles in the Holocene Record of North-Central Mexico. *Geomorphology* 101:237–277.
Cajigas, Rachel
2016 Geoarchaeological Investigation of the Early Agricultural Period Irrigation Canals at the La Playa Site in Sonora, Mexico. Paper presented at the Geological Society of America Annual Meeting, Denver.
2018 The Early Agricultural Period at La Playa, Mexico, a Geoarchaeological Investigation. Paper presented at the 83rd Annual Meeting of the Society for American Archaeology, Washington, DC.
Cajigas, Rachel, James T. Watson, and Todd Pitezel
2016 Anthropogenic Influences on Terrace Soil Development at Tumamoc Hill. Poster presented at the 15th Biennial Southwest Symposium, University of Arizona, January 14–16. Tucson, Arizona.
Callahan, Eric
1979 The Basics of Biface Knapping in the Eastern Fluted Point Tradition: A Manual for Flintknapping and Lithic Analysis. *Archaeology of Eastern North America* 7:1–180.
Callahan, Errett
1981 *Pamunkey Housebuilding: An Experimental Study of Late Woodland Construction Technology in the Powhatan Confederacy*, PhD dissertation, Department of Anthropology, Catholic University of America. Washington, DC. ProQuest Dissertations & Theses Global.
Cameron, Catherine M.
1984 A Regional View of Chipped Stone Raw Material Use in Chaco Canyon, New Mexico. In *Recent Research on Chaco Prehistory*, edited by W. James Judge and John D. Schelberg, pp. 137–152. Reports of the Chaco Center No. 8. Division of Cultural Research, National Park Service, Albuquerque.
2001 Pink Chert, Projectile Points, and the Chacoan Regional System. *American Antiquity* 66:79–101.
Campbell, Thomas N.
1983 Coahuiltecans and Their Neighbors. In *Southwest*, edited by Alfonso Ortiz, pp. 343–358. Handbook of North American Indians, Vol. 10, William C. Sturtevant, general editor. Smithsonian Institution, Washington, DC.
Cannon, Michael D.
1999 A Mathematical Model of the Effects of Screen Size on Zooarchaeological Relative Abundance Measures. *Journal of Archaeological Science* 26:205–214.
Cannon, Michael D. and Michael A. Etnier
1997 Report on the Analysis of Faunal Materials from the Reconocimiento Regional Paquimé, 1996. Report

on file, Department of Anthropology, University of Tulsa, Oklahoma.

Carmichael, David L.
1986 *Archeological Survey in the Southern Tularosa Basin, New Mexico.* Publications in Anthropology No. 10. Centennial Museum, University of Texas at El Paso, El Paso.

Carpenter, John, Guadalupe Sanchez, James T. Watson, and Elisa Villalpando
2015 The La Playa Archaeological Project: Binational Interdisciplinary Research on Long-Term Human Adaptation in the Sonoran Desert. *Journal of the Southwest* 57:213–264.

Carr, Philip J.
1994 *The Organization of North American Prehistoric Chipped Stone Tool Technologies.* Archaeological Series 7. International Monographs in Prehistory, Ann Arbor.

Castetter, Edward F. and Willis H. Bell
1951 *Yuman Indian Agriculture.* University of New Mexico Press, Albuquerque.

Castiglia, Peter J.
2002 *Late Quaternary Climate History of the Pluvial Lake Palomas System, Northern Chihuahua, Mexico.* Master's thesis, Department of Earth and Planetary Sciences, University of New Mexico. Albuquerque.

Castiglia, Peter J. and Peter J. Fawcett
2006 Large Holocene Lakes and Climate Change in the Chihuahuan Desert. *Geology* 34:113–116.

Cavazos, T., A. C. Comrie, and D. M. Liverman
2002 Intraseasonal Variability Associated with Wet Monsoons in Southeast Arizona. *Journal of Climate* 15:2477–2490.

Ceballos González, Gerardo J., Ana Davidson, Rurik List, Jesús Pacheco, Patricia Manzano-Fischer, Georgina Santos-Barrera, and Juan Cruzado
2010 Rapid Decline of a Grassland System and Its Ecological and Conservation Implications. *PLoS One* 5:e8562.

Ceballos González, Gerardo J. and Rurik List
2011 *Inventario de Especies de Vertebrados Para Apoyar la Creación del Ordenamiento Ecológico y la Reserva de la Biósfera Janos, Chihuahua.* Informe Final SNIB-CONABIO Proyecto No. FQ016. Instituto de Ecología, Universidad Nacional Autónoma de México, México, D.F.

Chapin, Nicholas M.
n.d. Archaic Period Projectile Point Types in Northwestern New Mexico, Unpublished poster on file at Bureau of Land Management, Rio Puerco Field Office, Albuquerque.
2017 *Oshara Revisited: The Archaic Period in Northern New Mexico.* Anthropological Papers No. 10. University of New Mexico, Maxwell Museum of Anthropology, Albuquerque.

Chapman, Richard C.
1982 Dating Archaeological Sites with Lithic Manufacturing Debris. In *Inventory Survey of the Lower Hidden Mountain Floodpool, Lower Rio Puerco Drainage, Central New Mexico*, edited by P. Eidenbach, pp. 235–286. Human Systems Research, Las Cruces.

Cockrum, E. Lendell
1960 *The Recent Mammals of Arizona: Their Taxonomy and Distribution.* University of Arizona Press, Tucson.

Cohen, Alan and Dale Serjeantson
1996 *A Manual for the Identification of Bird Bones from Archaeological Sites, Revised Edition.* Archetype Publications, London.

COHMAP Members
1988 Climatic Changes of the Last 18,000 Years: Observations and Model Simulations. *Science* 241:1043–1052.

Cole, Sarah M.
2012 Population Dynamics and Warfare in the Central Mesa Verde Region. In *Emergence and Collapse of Early Villages: Models of Central Mesa Verde Archaeology*, edited by Timothy A. Kohler and Mark D. Varien, pp. 230–252. University of California Press, Berkeley.

Collins, Michael B.
1975 Lithic Technology: Making and Using Stone Tools: Lithic Technology as a Means of Processual Inference. In *Lithic Technology: Making and Using Stone Tools*, edited by E. Swanson, pp. 15–34. Mouton Publishers, The Hague.

Coltrain, Joan B., Joel C. Janetski, and Shawn W. Carlyle
2007 The Stable- and Radio-Isotope Chemistry of Western Basketmaker Burials: Implications for Early Pueblo Diets and Origins. *American Antiquity* 72:301–321.

Comisión Nacional de Áreas Naturales Protegidas
2006 Flora Registrada en El Área de la Reserva Propuesta, Anexo 1. Estudio Previo Justificativo Para El Establecimiento del Área Natural Protegida. Reserva de la Biosfera Janos, Chihuahua, México.

Conroy, Jessica L., Jonathan T. Overpeck, Julia E. Cole, Timothy M. Shanahan, and Miriam Steinitz-Kannan
2008 Holocene Changes in Eastern Tropical Pacific Climate Inferred from a Galápagos Lake Sediment Record. *Quaternary Science Reviews* 27:1166–1180.

Copeland, Audrey, Jay Quad, James T. Watson, Brett T. McLaurin, and Elisa Villalpando
2012 Stratigraphy and Geochronology of La Playa Archaeological Site, Sonora, Mexico. *Journal of Archaeological Science* 39:2934–2944.

Cornejo, Dennis O., Linda S. Leigh, Richard S. Felger, and Charles F. Hutchinson
1982 Utilization of Mesquite in the Sonoran Desert: Past and Future. Paper presented at the Symposium on Mesquite Utilization, Texas Tech University, Lubbock, Texas.

Correll, Donovan Stewart and Marshall Conring Johnston
1970 *Manual of the Vascular Plants of Texas.* Texas Research Foundation, Renner, Texas.

Crabtree, Don E.
1972 *An Introduction to Flintworking.* Occasional Papers of the Idaho State Museum, No. 28, Pocatello.

Craig, Douglas B., James P. Holmlund, and Jeffrey J. Clark
1998 Labor Investment and Organization in Platform

Mound Construction: A Case Study from the Tonto Basin of Central Arizona. *Journal of Field Archaeology* 25:245–259.

Crown, Patricia L.
2000 Women's Role in Changing Cuisine. In *Women and Men in the Prehispanic Southwest: Labor, Power and Prestige*, edited by Patricia L. Crown, pp. 221–266. School of American Research Press, Santa Fe.

Crumrine, N. Ross and M. Louise Crumrine
1967 Ancient and Modern Mayo Fishing Practices. *Kiva* 33:25–33.

Cruz Antillón, Rafael, Robert D. Leonard, Timothy D. Maxwell, Todd L. VanPool, Marcel J. Harmon, Christine S. VanPool, David A. Hyndman, and Sidney S. Brandwein
2004 Galeana, Villa Ahumada, and Casa Chica: Diverse Sites in the Casas Grandes Region. In *Surveying the Archaeology of Northwest Mexico*, edited by Gillian E. Newell and Emiliano Gallaga, pp. 149–175. University of Utah Press, Salt Lake City.

Currie, Pat O. and D. L. Goodwin
1966 Consumption of Forage by Black-Tailed Jackrabbits on Salt-Desert Ranges of Utah. *The Journal of Wildlife Management* 30:304–311.

Curtin, Leonard S. M.
1984 *By the Prophet of the Earth: Ethnobotany of the Pima.* University of Arizona Press, Tucson.

Cutler, Hugh C.
1952 A Survey of the Plant Remains of Tularosa Cave. In *Mogollon Cultural Continuity and Change: The Stratigraphic Analysis of Tularosa and Cordova Caves*, edited by Paul S. Martin, John B. Rinaldo, Elaine Bluhm, Hugh C. Cutler, and Roger Grange Jr., pp. 446–480. Fieldiana: Anthropology 40. Field Museum of Natural History, Chicago.
1960 Cultivated Plant Remains from Waterfall Cave, Chihuahua. *American Antiquity* 26:277–279.

Cutler, Hugh C. and Leonard W. Blake
1974 Corn from Casas Grandes. In *Casas Grandes, a Fallen Trading Center of the Gran Chichimeca*, Vol. 8, edited by Charles C. Di Peso, John B. Rinaldo, and Gloria J. Fenner, pp. 308–314. Amerind Foundation, Dragoon, Arizona.

Dabo, Sira-Mady
1980 *Botanical Composition of Black-Tailed Jackrabbit Diets on Semidesert Rangeland.* Master's thesis, Department of Animal and Range Sciences, New Mexico State University. Las Cruces, New Mexico.

Damp, Jonathan E., Stephen A. Hall, and Susan J. Smith
2002 Early Irrigation on the Colorado Plateau near Zuni Pueblo, New Mexico. *American Antiquity* 67: 665–676.

Davenport, Harbert
1924 The Expedition of Pánfilo de Narváez by Gonzalo Fernández Oviedo y Valdez. *The Southwestern Historical Quarterly* 27:276–304.

Davis, Charles A., Joel A. Medlin, and James P. Griffing
1975 *Abundance of Black-Tailed Jackrabbits, Desert Cottontail Rabbits, and Coyotes in Southeastern New Mexico.* Research Report 293. Agricultural Experiment Station, New Mexico State University, Las Cruces.

Davis, Owen K.
1994 Pollen Analysis of Borderland Cienegas. Report Submitted to Richard P. Young, The Nature Conservency, Arizona Field Office. Contract Number HQ/AZ-920815-1. Manuscript on file, Tucson, Arizona.

Dean, Jeffrey S. and John C. Ravesloot
1993 The Chronology of Cultural Interaction in the Gran Chichimeca. In *Culture and Contact: Charles C. Di Peso's Gran Chichimeca*, edited by Anne I. Woosley and John C. Ravesloot, pp. 83–103. Amerind Foundation and University of New Mexico Press, Albuquerque.

Dean, Rebecca M.
2003 *People, Pests, and Prey: The Emergence of Agricultural Economies in the Desert Southwest.* PhD dissertation, Department of Anthropology, University of Arizona. Tucson.
2007 The Lagomorph Index: Rethinking Rabbit Ratios in Hohokam Sites. *Kiva* 73:7–30.

Dean, Rebecca M. and Joseph E. Beaver
2017 Jackrabbit Age Profiles as an Indicator of Hunting Behavior: An Example from Southern Arizona. *Kiva* 83:162–182.

DeNiro, Michael J.
1985 Postmortem Preservation and Alteration of *in vivo* Bone Collagen Isotope Ratios in Relation to Paleodietary Reconstruction. *Nature* 317:806–809.

Desmond, Martha J.
2004 Habitat Associations and Co-Occurances of Chihuahuan Desert Hares (*Lepus californicus and L. callotis*). *The American Midland Naturalist* 151:414–419.

DeVeaux, Jennie S. and Eugene B. Schultz, Jr.
1985 Development of Buffalo Gourd (*Cucurbita foetidissima*) as a Semiaridland Starch and Oil Crop. *Economic Botany* 39:454–472.

Díaz, Sara C., Matthew D. Therrell, David W. Stahle, and Malcolm K. Cleaveland
2002 Chihuahua (Mexico) Winter-Spring Precipitation Reconstructed from Tree-Rings, 1647–1992. *Climate Research* 22:237–244.

Dick, Herbert W.
1965 *Bat Cave.* Monograph 27. School of American Research, Santa Fe.

Dick-Peddie, William A.
1993 *New Mexico Vegetation Past Present and Future.* University of New Mexico Press, Albuquerque.

Diehl, Michael W.
1996 The Intensity of Maize Processing and Production in Upland Mogollon Pithouse Villages AD 200–1000. *American Antiquity* 61:102–115.
1997 Rational Behavior, the Adoption of Agriculture, and the Organization of Subsistence During the Late Archaic Period in the Greater Tucson Basin. In *Rediscovering Darwin: Evolutionary Theory and Archaeological Explanation*, edited by C. M. Barton and G.A. Clark, pp. 251–266. Archaeological Papers 7. American Anthropological Association.

1999a Macrobotanical Remains. In *Prehistoric Uses of a Developing Floodplain: Archaeological Investigations on the East Bank of the Santa Cruz River at A-Mountain*, edited by Jonathan B. Mabry, Michael W. Lindeman, and Helga Wöcherl, pp. 81–87. Center for Desert Archaeology, Tucson.

1999b Paleobotanical Remains. In *Excavations in the Santa Cruz River Floodplain: The Middle Archaic Component at Los Pozos*, edited by David A. Gregory, pp. 49–59. Anthropological Papers 20. Center for Desert Archaeology, Tucson.

2005a Epilogue: "Farmaging" During the Early Agricultural Period. In *Subsistence and Resource Use Strategies of Early Agricultural Communities in Southern Arizona*, edited by Michael W. Diehl, pp. 181–184. Anthropological Papers No. 34. Center for Desert Archaeology, Tucson.

2005b Morphological Observations on Recently Recovered Early Agricultural Period Maize Cob Fragments from Southern Arizona. *American Antiquity* 70: 361–375.

2005c When Corn Was Not Yet King. In *Subsistence and Resource Use Strategies of Early Agricultural Communities in Southern Arizona*, edited by Michael W. Diehl, pp. 1–18. Anthropological Papers 34. Center for Desert Archaeology, Tucson.

Diehl, Michael W. and Owen K. Davis

2016 The Short, Unhappy Use Lives of Early Agricultural Period "Food Storage Pits" at the Las Capas Site, Southern Arizona. *American Antiquity* 81:333–344.

Dinerstein, Eric, David Olson, Jennifer Atchley, Colby Loucks, Salvador Contreras-Balderas, Robin Abell, Eduardo Inigo, Ernesto Enkerlin, Christopher Williams, and Guillermo Castilleja (editors)

2000 Ecoregion-Based Conservation in the Chihuahuan Desert: A Biological Assessment. World Wildlife Fund. Electronic document, https://www.researchgate.net/publication/264419963, accessed March 28, 2019.

DiPeso, Charles C.

1974 *Casas Grandes, a Fallen Trading Center of the Gran Chichimeca, Vol. 2.* Amerind Foundation Publications No. 9. Northland Press, Flagstaff, Arizona.

DiPeso, Charles C., J. B. Rinaldo, and Gloria J. Fenner

1974 *Casas Grandes: A Fallen Trading Center of the Gran Chichimeca.* 8 vols. Amerind Foundation Publications No. 9. Northland Press, Flagstaff, Arizona.

Doebley, John F.

1984 "Seeds" of Wild Grasses: A Major Food of Southwestern Indians. *Economic Botany* 38:52–64.

1990 Molecular Evidence and the Evolution of Maize. In *New Perspectives on the Origin and Evolution of New World Domesticated Plants*, edited by Peter K. Bretting, pp. 6–28. Supplement to Economic Botany 44. Allen Press, Lawrence, Kansas.

Dolan, Sean G., Michael E. Whalen, Paul E. Minnis, and M. Steven Shackley

2017 Obsidian in the Casas Grandes World: Procurement, Exchange, and Interaction in Chihuahua, Mexico, CE 1200–1450. *Journal of Archaeological Science: Reports* 11:555–567.

Donkin, R. A.

1979 *Agricultural Terracing in the Aboriginal New World.* Viking Fund Publications in Anthropology No. 56. University of Arizona Press, Tucson.

Doolittle, William Emery

2000 *Cultivated Landscapes of Native North America.* Oxford Geographical and Environmental Studies. Oxford University Press, New York.

Douglas, M. W., R. A. Maddox, K. Howard, and S. Reyes

1993 The Mexican Monsoon. *Journal of Climate* 6: 1665–1677.

Downum, Christian E.

1986 The Occupational Use of Hill Space in the Tucson Basin: Evidence from Linda Vista Hill. *Kiva* 41: 219–232.

1993 *Between Desert and River: Hohokam Settlement and Land Use in the Los Robles Community.* Anthropological Papers No. 57. University of Arizona, Tucson.

2007 Cerros de Trincheras in Southern Arizona, Review and Current Status of the Debate. In *Trincheras Sites in Time, Space, and Society*, edited by Suzanne K. Fish, Paul R. Fish, and María Elisa Villalpando C., pp. 101–136. Amerind Studies in Archaeology 1, John Ware, general editor. University of Arizona Press, Tucson.

Downum, Christian E., Paul R. Fish, and Suzanne K. Fish

1994 Refining the Role of Cerros de Trincheras in Southern Arizona Settlement. *Kiva* 59:271–296.

Driver, Jonathan C. and David Maxwell

2013 Bison Death Assemblages and the Interpretation of Human Hunting Behaviour. *Quaternary International* 297:100–109.

Dunne, T. and L. Leopold

1972 *Water in Environmental Planning.* W. H. Freeman and Company, New York.

Earle, Timothy E. and John E. Ericson

1977 *Exchange Systems in Prehistory.* Academic Press, New York.

Ebert, James I.

1979 An Ethnoarchaeological Approach to Reassessing the Meaning of Variability in Stone Tool Assemblages. In *Ethnoarchaeology: Implications of Ethnography for Archaeology*, edited by Carol Kramer, pp. 59–74. Columbia University Press, New York.

Eddy, Frank W.

1958 *A Sequence of Cultural and Alluvial Deposits in the Cienega Creek Basin, Southeastern Arizona.* Master's thesis, Department of Anthropology, University of Arizona, Tucson.

Ellwood, Emily C., M. Paul Scott, William D. Lipe, R. G. Matson, and John G. Jones

2013 Stone-Boiling Maize with Limestone: Experimental Results and Implications for Nutrition among SE Utah Preceramic Groups. *Journal of Archaeological Science* 40:35–44.

Ely, L. L.

1997 Response of Extreme Floods in the Southwestern

United States to Climatic Variations in the Late Holocene. *Geomorphology* 19:175–201.

Ely, L. L., Y. Enzel, V. R. Baker, and D. R. Cayan
1993 A 5000-Year Record of Extreme Floods and Climate Change in the Southwestern United States. *Science* 262:410–412.

Elyea, Janette
1995 *Macbeth Data Recovery Program Final Report.* Office of Contract Archeology, University of New Mexico, Albuquerque.

Ember, Carol R. and Melvin Ember
2014 Violence in the Ethnographic Record: Results of Cross-Cultural Research on War and Aggression. In *Troubled Times: Violence and Warfare in the Past,* edited by Debra L. Martin and David W. Frayer, pp. 1–20. Routledge, New York.

Erasmus, Charles J.
1977 Monument Building: Some Field Experiments. In *Experimental Archaeology,* edited by Daniel Ingersoll, John E. Yellen, and William McDonald, pp. 52–78. Columbia University Press, New York.

Eren, Metin I., Fernando Diez-Martin, and Manuel Dominguez-Rodrigo
2013 An Empirical Test of the Relative Frequency of Bipolar Reduction in Beds VI, V, and III at Mumba Rockshelter, Tanzania: Implications for the East African Middle to Late Stone Age Transition. *Journal of Archaeological Science* 40:248–256.

Ezzo, Joseph A. and Deaver L. William
1998 *Watering the Desert: Late Archaic Farming at the Costello-King Site.* Technical Series 68. Statistical Research, Inc., Tucson.

Fatehi, Mohammad
1986 *Comparative Seasonal Food Habits of Black-Tailed Jackrabbits and Cattle on Semidesert Rangeland.* PhD Dissertation, Department of Animal and Range Sciences, New Mexico State University. Las Cruces, New Mexico.

Ferg, Alan
1998 Rare Stone, Fired Clay, Bone, and Shell Artifacts. In *Archaeological Investigations of Early Village Sites in the Middle Santa Cruz Valley,* edited by Jonathan B. Mabry, pp. 545–654. Anthropological Papers 19, pt. II. Center for Desert Archaeology, Tucson.

Findley, James S., Arthur H. Harris, Don E. Wilson, and Clyde Jones
1975 *Mammals of New Mexico.* University of New Mexico Press, Albuquerque.

Findlow, Frank J. and Marisa Bolognese
1980 An Initial Examination of Prehistoric Obsidian Exchange in Hidalgo County, New Mexico. *Kiva* 45:227–251.
1982a A Preliminary Analysis of Prehistoric Obsidian Use within the Mogollon Area. In *Mogollon Archaeology: Proceedings of the 1980 Mogollon Conference,* edited by Patrick H. Beckett, pp. 297–316. Acoma Press, Ramona, California.
1982b Regional Modeling of Obsidian Procurement in the American Southwest. In *Contexts for Prehistoric Exchange,* edited by J. E. Ericson and T. K. Earle, pp. 53–81. Academic Press, New York.

Fish, Paul R., Suzanne K. Fish, Austin Long, and Charles Miksicek
1986 Early Corn Remains from Tumamoc Hill, Southern Arizona. *American Antiquity* 51:563–572.

Fish, Paul R., Suzanne K. Fish, and M. Elisa Villalpando
2007a Introduction. In *Trincheras Sites in Time, Space, and Society,* edited by Suzanne K. Fish, Paul R. Fish, and M. Elisa Villalpando, pp. 1–9. Amerind Studies in Archaeology 1, John Ware, general editor. University of Arizona Press, Tucson.
2007b *Trincheras Sites in Time, Space, and Society.* Amerind Studies in Archaeology 1. John Ware, general editor. University of Arizona Press, Tucson.

Fish, Suzanne K. and Paul R. Fish
2007 Regional Heartlands and Transregional Trends. In *Trincheras Sites in Time, Space, and Society,* edited by Suzanne K. Fish, Paul R. Fish, and M. Elisa Villalpando, pp. 165–194. Amerind Studies in Archaeology 1, John Ware, general editor. University of Arizona Press, Tucson.

Fish, Suzanne K., Paul R. Fish, Gary Christopherson, and Todd A. Pitezel
2011 Two Villages on Tumamoc Hill. *Journal of Arizona Archaeology* 1:185–196.

Fish, Suzanne K., Paul R. Fish, and Christian Downum
1984 Hohokam Terraces and Agricultural Production in the Tucson Basin. In *Prehstoric Agricultural Strategies in the Southwest,* edited by Suzanne K. Fish and Paul R. Fish, pp. 55–71. Anthropological Research Papers No. 33. Arizona State University, Tempe.

Fish, Suzanne K., Paul R. Fish, Todd Pitezel, Gary Christopherson, James T. Watson, and Phillip O. Leckman
2013 Emerging Settlement Differentiation in Preceramic and Early Hohokam Villages on Tumamoc Hill. In *New Perspectives on the Rock Art and Prehistoric Settlement Organization of Tumamoc Hill, Tucson, Arizona,* edited by Gayle Harrison Hartmann and Peter C. Boyle, pp. 1–22. Archaeological Series 208. Arizona State Museum, University of Arizona, Tucson.

Flenniken, J. Jeffrey
1981 *Replicative Systems Analysis: A Model Applied to the Vein Quartz Artifacts from the Hoko River Site.* Laboratory of Anthropology Reports of Investigation 59. Washington State University, Pullman.

Flinders, Jerran T. and John A. Crawford
1977 Composistion and Degradation of Jackrabbit and Cottontail Fecal Pellets, Texas High Plains. *Journal of Range Management* 30:217–230.

Flinders, Jerran T. and Richard M. Hansen
1972 *Diets and Habitats of Jackrabbits in Northeastern Colorado.* Range Science Department Science Series 12. Colorado State University, Fort Collins.

Folk, R. L.
1980 *Petrology of Sedimentary Rocks.* Hemphill Publishing Company, Austin, Texas.

Fontana, Bernard L.
1983 Pima and Papago: Introduction. In *Southwest*, edited by Alfonso Ortiz, pp. 125–136. Handbook of North American Indians Vol. 10, William C. Sturtevant, general editor. Smithsonian Institution, Washington DC.

Fontana, Bernard L., J. Cameron Greenleaf, and Donnelly D. Cassidy
1959 A Fortified Arizona Mountain. *Kiva* 25:41–52.

Formby, Donald E.
1986 Pinto-Gypsum Complex Projectile Points from Arizona and New Mexico. *Kiva* 51:99–127.

Foster, Michael S. and Shirley Gorenstein (editors)
2000 *Greater Mesoamerica: The Archaeology of West and Northwest Mexico*. University of Utah Press, Salt Lake City.

Fournier, Patricia
1997 Mexican Ceramic Analysis. In *A Presidio Community on the Rio Grande: Phase III Testing and Historical Research at San Elizario, Texas*, 2 vols, edited by Bradley J. Vierra, June-el Piper, and Richard C. Chapman, pp. 199–255. University of New Mexico, Office of Contract Archeology, Albuquerque.

Freeman, Andrea Kelly Lee
1997 *Middle to Late Holocene Stream Dynamics of the Santa Cruz River, Tucson, Arizona: Implications for Human Settlement, the Transition to Agriculture and Archaeological Site Preservation*. PhD dissertation, Department of Anthropology, The University of Arizona. ProQuest Dissertations Publishing, Ann Arbor.

French, Norman R., Ray McBride, and Jack Detmer
1965 Fertility and Population Density of the Black-Tailed Jackrabbit. *The Journal of Wildlife Management* 29:14–26.

Fritz, Gayle J.
1984 Identification of Cultigen Amaranth and Chenopod from Rockshelter Sites in Northwest Arkansas. *American Antiquity* 49:558–572.
2007 Pigweeds for the Ancestors: Cultural Identities and Archaeobotanical Identification Methods. In *The Archaeology of Food and Identity*, edited by Katheryn C. Twiss, pp. 288–307. Occasional Paper 34. Center for Archaeological Investigations, Southern Illinois University, Carbondale.

Fritz, Gayle J., Karen R. Adams, Glen E. Rice, and John L. Czarzasty
2009 Evidence for Domesticated Amaranth from a Sedentary Period Hohokam House Floor at Las Canopas. *Kiva* 74:393–419.

Fritz, Gayle J. and Bruce D. Smith
1988 Old Collections and New Technology: Documenting the Domestication of Chenopodium in Eastern North America. *Midcontinental Journal of Archaeology* 13:3–27.

Gallaga Murrieta, Emiliano, Jesús Emmanuel Moreno Alvarado, and Marisela Guzmán Aguirre
2016 *Informe Final Del Salvamento Del Rancho Santa Maria, Municipio de Galeana, Chihuahua*. Instituto Nacional de Antropología e História, Centro INAH Chihuahua, Ciudad Chihuahua, Chihuahua.

García, Enriqueta, Rosalía Vidal, Luz Ma. Tamayo, Teresa Reyna, Rubén Sánchez, Margarita Soto, and Enrique Soto
1973 *Precipitación y Probabilidad de Lluvia en la República Mexicana y Su Evaluación: Chihuahua*. Climas 17. Instituto de Geografía, U.N.A.M. y Comisión de Estudios del Territoria Nacional, México, D.F.

Geib, Phil R.
2011 *Foragers and Farmers of the Northern Kayenta Region: Excavations Along the Navajo Mountain Road*. University of Utah Press, Salt Lake City.

Gerald, Rex E.
1957 *A Historic House Excavation near Janos, Northwest Chihuahua, Mexico*. Master's thesis, Department of Anthropology, University of Pennsylvania. Philadelphia.
1990 Report on a U.T. El Paso Mini-Grant to Investigate Prehistoric Fortifications in a Primitive State in the Casas Grandes Area of Chihuahua. *The Artifact* 28:59–64.

Gerhard, Peter
1982 *The North Frontier of New Spain*. Princeton University Press, Princeton.

Gilbert, B. Miles
1993 *Mammalian Osteology*. Missouri Archaeological Society, Columbia.

Gillespie, William B.
1989 Vertebrate Faunal Remains. In *Hohokam Archaeology Along Phase B of the Tucson Aqueduct, Central Arizona Project: Vol. 4, Small Sites and Specialized Reports*, edited by Jon S. Czaplicki and John C. Ravesloot, pp. 85–98. Archaeological Series No. 178. Arizona State Museum, University of Arizona, Tucson.
1994 Vertebrate Faunal Remains from Five Schuk Toak Sites. In *Archaeological Studies of the Avra Valley, Arizona: Excavations in the Schuk Toak District: Vol. 2, Scientific Studies and Interpretations*, edited by Allen Dart, pp. 215–242. Anthropological Papers 16. Center for Desert Archaeology, Tucson.

Glass, Margaret
1984 Faunal Remains from Hohokam Sites in the Rosemont Area, Northern Santa Rita Mountains. In *Hohokam Habitation Sites in the Northern Santa Rita Mountains*, edited by Allen Ferg, Kenneth C. Rozen, William L. Deaver, Martyn D. Tagg, David A. Phillips, Jr., and David A. Gregory, pp. 823–916. Archaeological Series 147. Arizona State Museum, University of Arizona, Tucson.

Gobalet, Kenneth W.
2002 *Fish Remains from Cerro Juanaqueña, Chihuahua, Mexico*, Manuscript on file, Center for Archaeological Research, University of Texas at San Antonio.

Gonzalez, Gabriela P.
2014 *Understanding the Maize Grinding Technology at*

Cerro Juanaqueña, Chihuahua, Mexico. Master's thesis, Department of Anthropology, University of Texas at San Antonio. Proquest Dissertations Publishing, Ann Arbor.

Goodyear, Albert C.

1979 *A Hypothesis for the Use of Crypocrystaline Raw Materials among Paleoindian Groups of North America.* Research Manuscript Series, No. 156. Institute of Archaeology and Anthropology, University of South Carolina, Columbia.

Gould, Richard A., Dorothy A. Koster, and Ann H. L. Sontz

1971 The Lithic Assemblage of the Western Desert Aborigines of Australia. *American Antiquity* 36:149–169.

Gould, Richard A. and Sherry Saggers

1981 Lithic Procurement in Central Australia: A Closer Look at Binford's Idea of Embeddedness in Archaeology. *American Antiquity* 50:117–136.

Grayson, Donald K.

1984 *Quantitative Zooarchaeology: Topics in the Analysis of Archaeological Faunas.* Academic Press, New York.

Greer, John W.

1969 Micro-Eccentrics from the Four Corners Region of the Southwest. *Kiva* 34:251–126.

Gregory, David A.

1999a Data Integration and Synthesis. In *Excavations in the Santa Cruz River Floodplain: The Middle Archaic Component at Los Pozos*, edited by David A. Gregory, pp. 85–123. Anthropological Papers 20. Center for Desert Archaeology, Tucson.

2001 An Evaluation of Early Agricultural Period Chronology in the Tucson Basin. In *Excavations in the Santa Cruz River Floodplain: The Early Agricultural Period Component at Los Pozos*, edited by David A. Gregory, pp. 237–254. Anthropological Papers No. 21. Center for Desert Archaeology, Tucson.

Gregory, David A. (editor)

1999b *Excavations in the Santa Cruz River Floodplain: The Middle Archaic Component at Los Pozos.* Anthropological Papers 20. Center for Desert Archaeology, Tucson.

Gregory, David A. and Sam W. Baar IV

1999 Stratigraphy, Chronology, and Characteristics of the Natural and Cultural Deposits. In *Excavations in the Santa Cruz River Floodplain: The Middle Archaic Component at Los Pozos*, edited by David A. Gregory, pp. 13–31. Anthropologhical Papers 20. Center for Desert Archaeology, Tucson.

Gregory, David A. and Michael W. Diehl

2002 Duration, Continuity, and Intensity of Occupation at a Late Cienega Phase Settlement in the Santa Cruz River Floodplain. In *Traditions, Transitions, and Technologies, Themes in Southwestern Archaeology*, edited by Sarah H. Schlanger, pp. 200–223. University Press of Colorado, Boulder.

Gregory, David A. and Fred L. Nials

2005 The Environmental Context of Early Agricultural Period Occupations in the Tucson Basin. In *Subsistence and Resource Use Strategies of Early Agricultural Communities in Southern Arizona*, edited by Michael W. Diehl, pp. 19–71. Anthropological Papers 34. Center for Desert Archaeology, Tucson, Arizona.

Gremillion, Kristen J., Loukas Barton, and Dolores R. Piperno

2014 Particularism and the Retreat from Theory in the Archaeology of Agricultural Origins. *Proceedings of the National Academy of Sciences of the United States of America* 111:6171–6177.

Griffen, William B.

1983 Southern Periphery: East. In *Southwest*, edited by Alfonso Ortiz, pp. 329–342. Handbook of North American Indians, Vol. 10, William C. Sturtevant, general editor. Smithsonian Institution, Washington DC.

Grissino-Mayer, Henri D.

1996 A 2129-Year Reconstruction of Precipitation for Northwestern New Mexico, USA. In *Tree Rings, Environment, and Humanity: Proceedings of the International Conference*, edited by Jeffrey S. Dean, David M. Meko, and Thomas W. Swetnam, pp. 191–204. Radiocarbon, Department of Geosciences, University of Arizona, Tucson.

Grove, Matt

2010 Logistical Mobility Reduces Subsistence Risk in Hunting Economies. *Journal of Archaeological Science* 37:1913–1921.

Haas, Jonathan

1990 Warfare and the Evolution of Tribal Polities in the Prehistoric Southwest. In *The Anthropology of War*, edited by Jonathan Haas, pp. 171–189. Cambridge University Press, New York.

Haas, Jonathan and Winifred Creamer

1996 The Role of Warfare in the Pueblo III Period. In *The Prehistoric Pueblo World, A.D. 1150–1350*, edited by Michael A. Adler, pp. 205–213. University of Arizona Press, Tucson.

Hack, John T.

1942 *The Changing Physical Environment of the Hopi Indians of Arizona.* Papers of the Peabody Museum of American Archaeology and Ethnography Vol. 35, No. 1. Harvard University, Cambridge.

Hall, Stephen A.

2018 Paleoenvironments of the American Southwest. In *The Archaic Southwest: Foragers in an Arid Land*, edited by Bradley J. Vierra, pp. 16–18. University of Utah Press, Salt Lake City.

Hanselka, J. Kevin

2000 *Late Archaic Plant Use and Early Agriculture in Northwestern Chihuahua, Mexico: Insights from Cerros de Trincheras Sites.* Master's thesis, Division of Behavioral and Cultural Sciences, University of Texas at San Antonio.

2018 A Pan-Regional Overview of Archaic Agriculture in the Southwest. In *The Archaic Southwest: Foragers in an Arid Land*, edited by Bradley J. Vierra, pp. 269–295. University of Utah Press, Salt Lake City.

Hard, Robert J.
1983 A Model for Prehistoric Land Use, Ft. Bliss, Texas. *American Society for Conservation Archaeology: Proceedings*:41–51.
1990 Agricultural Dependence in the Mountain Mogollon. In *Perspectives on Southwestern Prehistory*, edited by Paul E. Minnis and Charles L. Redman, pp 135–149. Westview Press, Boulder.

Hard, Robert J. and M. Anne Katzenberg
2011 Stable Isotope Study of Hunter-Gatherer-Fisher Diet, Mobility, and Intensification on the the Texas Gulf Coastal Plain. *American Antiquity* 76:709–751.

Hard, Robert J., Raymond P. Mauldin, and Gerry R. Raymond
1996 Mano Size, Stable Carbon Isotope Ratios, and Macrobotanical Remains as Multiple Lines of Evidence of Maize Dependence in the American Southwest. *Journal of Archaeological Method and Theory* 3:253–318.

Hard, Robert J. and William L. Merrill
1992 Mobile Agriculturalists and the Emergence of Sedentism: Perspectives from Northern Mexico. *American Anthropologist* 94:601–620.

Hard, Robert J., William L. Merrill, A. C. MacWilliams, John R. Roney, Jacob C. Freeman, and Karen R. Adams
2015 Rainfed Farming and Settlement Aggregation: Reflections from Chihuahua, Mexico. In *Traditional Arid Lands Agriculture: Understanding the Past for the Future*, edited by Scott Ingram and Robert C. Hunt, pp. 237–272. University of Arizona Press, Tucson.

Hard, Robert J. and John R. Roney
1998a A Massive Terraced Village Complex in Chihuahua, Mexico, 3000 Years before Present. *Science* 279: 1661–1664.
1998b *Una Investigación Arqueológica de Los Sitios Cerros con Trincheras del Arcáico Tardío en Chihuahua, México: Las Investigaciones de Campo de 1997*. Informe al Consejo de Arqueología, Instituto Nacional de Antropología E Historia. Center for Archaeological Research, University of Texas at San Antonio and included as an electronic appendix to Chapter 3.
1999a *An Archaeological Investigation of Late Archaic Cerros de Trincheras Sites in Chihuahua, Mexico: Results of the 1998 Investigations*. Special Report 25. Center for Archaeological Research, University of Texas at San Antonio and included as an electronic appendix to Chapter 3.
1999b *Una Investigación Arqueológica de Los Sitios Cerros con Trincheras del Arcáico Tardio en Chihuahua, México: Las Investigaciones de Campo de 1998*. Special Report 25. Center for Archaeological Research, The University of Texas at San Antonio.
2004 Late Archaic Period Hilltop Settlements in Northwestern Chihuahua, Mexico. In *Identity, Feasting, and the Archaeology of the Greater Southwest: Proceedings of the 2002 Southwest Symposium*, edited by Barbara J. Mills, pp. 276–294. University Press of Colorado, Boulder.
2005 The Transition to Farming on the Rio Casas Grandes and in the Southern Jornada Mogollon Region. In *The Late Archaic across the Borderlands: From Foraging to Farming*, edited by Bradley J. Vierra, pp. 141–186. University of Texas Press, Austin.
2007 Cerros de Trincheras in Northwestern Chihuahua: The Arguments for Defense. In *Trincheras Sites in Time, Space, and Society*, edited by Suzanne K. Fish, Paul R. Fish, and Elisa M. Villalpando, pp. 11–52. University of Arizona Press, Tucson.

Hard, Robert J., John R. Roney, A. C. MacWilliams, Karen Adams, and Mary Whisenhunt
2016a Round Mountain and Other Cerros de Trincheras of the Upper Gila River, Arizona. Poster presented at the 15th Biennial Southwest Symposium, Tucson, Arizona.

Hard, Robert J., John R. Roney, A. C. MacWilliams, Mary Whisenhunt, and Mark Willis
2016b Early Agricultural Period Cerros de Trincheras on the Upper Gila River, Arizona. Paper presented at the 81st Annual Meeting of the Society for American Archaeology, Orlando, Florida.

Hard, Robert J., John R. Roney, A.C. MacWilliams, Mary Whisenhunt, and Karen R. Adams
2018 The Sanchez Site: An Early Pithouse and Early Agricultural Period Cerro de Trincheras on the Upper Gila River, Arizona. Paper presented at the 20th Biennial Mogollon Archaeology Conference, Las Cruces, New Mexico.

Hard, Robert J., José E. Zapata, Bruce K. Moses, and John R. Roney
1999 Terrace Construction in Northern Chihuahua, Mexico: 1150 B.C. and Modern Experiments. *Journal of Field Archaeology* 26:129–146.

Hard, Robert J., José E. Zapata, and John R. Roney
2001 *Una Investigación Arqueológica de Los Sitios Cerros con Trincheras del Arcaico Tardío en Chihuahua, México*. Special Report 27S. Center for Archaeological Research, University Texas at San Antonio and included as an electronic appendix to Chapter 3.

Harden, T., M. G. Macklin, and V. R. Baker
2010 Holocene Flood Histories in South-Western USA. *Earth Surface Processes and Landforms* 35: 707–716.

Harter, Abigail V., Keith A. Gardner, Daniel Falush, David L. Lentz, Robert A. Bye, and Loren H. Riesberg
2004 Origin of Extant Domesticated Sunflowers in Eastern North America. *Nature* 430:201–205.

Harwell, H. and M. Kelly
1983 Maricopa. In *Southwest*, edited by A. Ortiz, pp. 71–85. Handbook of North American Indians, Vol. 10, William C. Sturtevant, general editor. Smithsonian Institution, Washington, DC.

Haury, Emil W.
1950 *The Stratigraphy and Archaeology of Ventana Cave*. University of New Mexico Press and University of Arizona Press, Albuquerque and Tucson.
1962 The Greater American Southwest. In *Courses Toward Urban Life*, edited by R. Braidwood and G. Willey,

pp 106–131. Viking Fund Publications in Anthropology 32. New York.
1976 *The Hohokam, Desert Farmers and Craftsmen, Excavations at Snaketown, 1964–1965*. University of Arizona Press, Tucson.
1983 Concluding Remarks. In *The Cochise Cultural Sequence in Southeastern Arizona*, by E. B. Sayles, pp. 158–166. Anthropological Papers of the University of Arizona 42. University of Arizona Press, Tucson.

Haury, Emil W. and Lisa W. Huckell
1993 A Prehistoric Cotton Cache from the Pinaleño Mountains, Arizona. *Kiva* 59:95–145.

Hawley, John W.
1969 Notes on the Geomorphology and Late Cenozoic Geology of Northwestern Chihuahua. In *Guidebook of the Border Region: Twentieth Field Conference, October 23, 24, and 25, 1969*, edited by Diego A. Córdoba, Sherman A. Wengerd, and John Shomaker, pp. 131–142. Guidebook 20. New Mexico Geological Society, Socorro, New Mexico.

Hayden, Brian
1979 *Paleolithic Reflections: Lithic Technology and Ethnographic Excavations among Australian Aborigines*. Australian Institute of Aboriginal Studies, Canberra.

Hayden, Brian, Nora Franco, and Jim Spafford
1996 Evaluating Lithic Strategies and Design Criteria. In *Stone Tools: Theoretical Insights into Human Prehistory*, edited by George Odell, pp. 9–50. Plenum Press, New York.

Hayden, Julian D.
1969 Gyratory Crushers of the Sierra Pinacate, Sonora. *American Antiquity* 34:154–161.

Hayden, Page
1966 Food Habits of Black-Tailed Jack Rabbits in Southern Nevada. *Journal of Mammalogy* 47:42–46.

Heerwagen, Arnold
1956 Mixed Prairie in New Mexico. In *Grasslands of the Great Plains: Their Nature and Use*, edited by John E. Weaver and F. W. Albertson, pp. 284–300. Johnson Publishing, Lincoln, Nebraska.

Heidke, James M.
2005 Early Agricultural Period Pottery from Las Capas and Los Pozos. In *Material Cultures and Lifeways of Early Agricultural Communities in Southern Arizona*, edited by R. Jane Sliva, pp. 171–205. Anthropological Papers No. 35. Center for Desert Archaeology, Tucson.
2006 Native American Pottery. In *Rio Nuevo Archaeology, 2000–2003: Investigations at the San Agustin Mission and Mission Gardens, Tucson Presidio, Tucson Pressed Brick Company, and Clearwater Site*, edited by J. Homer Thiel and Jonathan B. Mabry, pp. 7.1–7.86. Technical Report No. 2004-11. Desert Archaeology, Inc., Tucson.
2015 San Pedro and Early Cienega Phase Clay Objects and Hohokam Pottery from Las Capas, AZ AA:12:111 (ASM): Dating, Technology, Use, and Discard. In *Implements of Change: Tools, Subsistence, and the Built Environment of Las Capas, an Early Agricultural Irrigation Community in Southern Arizona*, edited by James M. Vint, pp. 161–226. Anthropological Papers No. 51. Desert Archaeology, Inc., Tucson.

Heiser, Charles B.
2001 About Sunflowers. *Economic Botany* 55:470–473.
2008a How Old Is the Sunflower in Mexico? *Proceedings of the National Academy of Sciences of the United States of America* 105:E48.
2008b The Sunflower (*Helianthus annuus*) in Mexico: Further Evidence for a North American Domestication. *Genetic Resources and Crop Evolution* 55:9–13.

Heizer, Robert F. and Martin A. Baumhoff
1962 *Prehistoric Rock Art of Nevada and Eastern California*. University of California Press, Berkeley.

Helmke, P.
2000 The Chemical Composition of Soils. In *Handbook of Soil Science*, edited by M. Sumner, pp. B3–B24. CRC Press, Boca Raton, Florida.

Herold, Laurance C.
1965 *Trincheras and Physical Environment Along the Río Gavilan, Chihuahua, Mexico*. Publications in Geography 651. University of Denver, Colorado.

Hester, Thomas R.
1995 The Prehistory of South Texas. *Bulletin of the Texas Archeological Society* 66:427–459.

Hester, Thomas R. and Michael B. Collins
1974 Evidence of Heat Treating of Southern Texas Projectile Points. *Bulletin of the Texas Archeological Society* 45:119–224.

Hester, Thomas R., Delbert Gilbow, and Alan D. Albee
1973 Functional Analysis of "Clear Fork" Artifacts from the Rio Grande Plain of Texas. *American Antiquity* 38:90–96.

Hill, Jane H.
2001 Proto-Uto-Aztecan: A Community of Cultivators in Central Mexico? *American Anthropologist* 103: 913–934.

Hillson, Simon
2005 *Teeth*. Cambridge University Press.

Hoagland, Donald B.
1992 Feeding Ecology of an Insular Population of the Black-Tailed Jackrabbit (*Lepus californicus*) in the Gulf of California. *The Southwestern Naturalist* 37:280–286.

Hodgetts, Lisa M.
1996 Faunal Evidence from El Zurdo. *Kiva* 62:149–170.

Hoffman, M. T., C. D. James, G. I. H. Kerley, and W. G. Whitford
1993 Rabbit Herbivory and Its Effect on Cladode, Flower and Fruit Production of *Opuntia violacea* Var. *macrocentra* (Cactaceae) in the Northern Chihuahuan Desert, New Mexico. *The Southwestern Naturalist* 38:309–315.

Holmgren, Camille A., Jodi Norris, and Julio L. Betancourt
2007 Inferences About Winter Temperatures and Summer Rains from the Late Quaternary Record of C4 Perennial Grasses and C3 Desert Shrubs in the Northern Chihuahuan Desert. *Journal of Quaternary Science* 22:141–161.

Holmgren, Camille A., M. Cristina Peñalba, Kate Aasen Rylander, and Julio L. Betancourt
2003 A 16,000 ^{14}C Yr B.P. Packrat Midden Series from the USA-Mexico Borderlands. *Quaternary Research* 60:319–329.
Hoover, J.W.
1941 Cerros de Trincheras of the Arizona Papagueria. *The Geographical Review* 31:228–239.
Howard, William and Thomas M. Griffiths
1966 *Trinchera Distribution in the Sierra Madre Occidental, Mexico*. Publications in Geography 66-1. University of Denver, Colorado.
Hruska, T., D. Toledo, R. Sierra-Corona, and V. Solis-Gracia
2017 Social-Ecological Dynamics of Change and Restoration Attempts in the Chihuahuan Desert Grasslands of Janos Biosphere Reserve, Mexico. *Plant Ecology* 218:67–80.
Huang, Y., F.A. Street-Perrott, S.E. Metcalfe, M. Brenner, M. Moreland, and K.H. Freeman
2001 Climate Change as the Dominant Control on Glacial-Interglacial Variations in C-3 and C-4 Plant Abundance. *Science* 293:1647–1651.
Huber, Edgar K.
2005 Early Maize at the Old Corn Site (LA 137258). In *Fence Lake Project: Archaeological Data Recovery in the New Mexico Transportation Corridor and First Five Year Permit Area, Fence Lake Coal Mine Project, Catron County, New Mexico. Volume 4: Synthetic Studies and Summary*, edited by Edgar K. Huber and Carla R. Van West, pp. 36.31–36.33. Technical Series 84. Statistical Research, Inc., Tucson.
Huckell, Bruce B.
1988 Late Archaic Archaeology of the Tucson Basin: A Status Report. In *Recent Research on Tucson Basin Prehistory: Proceedings of the Second Tucson Basin Conference*, edited by William H. Doelle and Paul R. Fish, pp. 57–80. Anthropological Papers No. 10. Institute for American Research, Tucson.
1990 *Late Preceramic Farmer-Foragers in Southeastern Arizona: A Cultural and Ecological Consideration of the Spread of Agriculture into the Arid Southwestern United States*. PhD dissertation, Department of Arid Lands Resource Sciences, University of Arizona, Tucson.
1995 *Of Marshes and Maize: Preceramic Agricultural Settlements in the Cienega Valley, Southeastern Arizona*. Anthropological Papers of the University of Arizona No. 59. University of Arizona Press, Tucson.
1996a The Archaic Prehistory of the North American Southwest. *Journal of World Prehistory* 10:305–373.
1996b Middle to Late Holocene Stream Behavior and the Transition to Agriculture in Southeastern Arizona. In *Early Formative Adaptations in the Southern Southwest*, edited by Barbara J. Roth, pp. 27–36. Prehistory Press, Madison.
1998 The Ground Stone Collection. In *Early Farmers of the Sonoran Desert: Arhaeological Investigations at the Houghton Road Site, Tucson, Arizona*, edited by Richard Ciolek-Torello, pp. 119–126. Technical Series 72. Statistical Research, Inc., Tucson.
Huckell, Bruce B. and Lisa W. Huckell
1984 Excavations at Milagro, a Late Archaic Site in the Eastern Tucson Basin. Manuscript on file, Cultural Resource Management Division, Arizona State Museum, University of Arizona, Tucson.
Huckell, Lisa W.
1995 Farming and Foraging in the Cienega Valley: Early Agricultural Period Paleoethnobotany. In *Of Marshes and Maize: Preceramic Agricultural Settlements in the Cienega Valley, Southeastern Arizona*, edited by Bruce B. Huckell, pp. 74–97. University of Arizona Press, Tucson.
1998 Appendix B, Paleoethnobotany. In *Early Farmers of the Sonoran Desert: Archaeological Investigations at the Houghton Road Site, Tucson Arizona*, edited by Richard Ciolek-Torello, pp. 327–344. Technical Series 72. Statistical Research, Inc., Tucson.
2002 Paleoethnobotany. In *Tonto Creek Archaeological Project: Artifact and Environmental Analyses: Vol. 2: Stone Tool and Subsistence Studies*, edited by Jeffery J. Clark, pp. 643–710. Anthropological Papers 23. Center for Desert Archaeology, Tucson.
2005 Early Maize from LA 137258, Catron County, New Mexico. In *Fence Lake Project: Archaeological Data Recovery in the New Mexico Transportation Corridor and First Five-Year Permit Area, Fence Lake Coal Mine Project, Catron County, New Mexico. Volume 3: Environmental Studies*, edited by Edgar K. Huber and Carla R. Van West, pp. 30.31–30.17. Technical Series 84. Statistical Research, Inc., Tucson.
2006 Ancient Maize in the American Southwest: What Does It Look Like and What Can It Tell Us? In *Histories of Maize: Multidisciplinary Approaches to the Prehistory, Linguistics, Biogeography, Domestication, and Evolution of Maize*, edited by John E. Staller, Robert H. Tykot, and Bruce F. Benz, pp. 97–107. Elsevier, Amsterdam.
Huckleberry, Gary A.
1995 Archaeological Implications of Late Holocene Channel Changes on the Middle Gila River, Arizona. *Geoarchaeology* 10:159–182.
1999 Prehistoric Flooding and Its Effect on Indigenous Agriculture in the Northern Sonoran Desert, U.S.A. *Journal of Arid Lands Studies* 9:277–284.
2015 Defining the Envrionmental Context of Indigenous Agriculture in the Southwest. In *Traditional Arid Lands Agriculture: Understanding the Past for the Future*, edited by Scott E. Ingram and Robert C. Hunt, pp. 89–130. University of Arizona Press, Tucson.
Huckleberry, Gary A., Charlotte Beck, George T. Jones, Amy Holmes, Michael Cannon, Stephanie Livingston, and Jack M. Broughton
2001 Terminal Pleistocene/Early Holocene Environmental Change at the Sunshine Locality, North-Central Nevada, U.S.A. *Quaternary Research* 55:303–312.

Huntington, Ellsworth
1914 *The Climatic Factor as Illustrated in Arid America.* Carnegie Institution of Washington, Washington, DC.

Inizan, M. L., M. Reduron-Ballinger, H. Roche, and J. Tixier
1999 Technology and Terminology of Knapped Stone. In *Préhistorique de la Pierre Taillée* 5. Cercle de Recherches et d'Etudes Préhistoriques, Nanterre, France.

James, Steven R.
1989 Archaeofaunal Remains from the Tempe Outer Loop Corridor. In *Prehistoric Agricultural Activities on the Lehi-Mesa Terrace: Excavations at La Cuenca del Sedimento*, edited by T. Kathleen Henderson, pp. 303–333. Northland Research, Inc., Flagstaff, Arizona.

Jennings, Thomas A., Charlotte D. Pevny, and William A. Dickens
2010 A Biface and Blade Core Efficiency Experiment: Implications for Early Paleoindian Technological Organization. *Journal of Archaeological Science* 37:2155–2164.

Jeske, Robert J.
1992 Energetic Efficiency and Lithic Technology: An Upper Mississippian Example. *American Antiquity* 57:467–481.

Jewell, Peter A. (editor)
1963 *The Experimental Earthwork on Overton Down Wiltshire 1960: An Account of the Construction of an Earthwork to Investigate by Experiment the Way in Which Archaeological Structures Are Denuded and Buried.* British Association for the Advancement of Science, London.

Johannessen, Sissel and Christine A. Hastorf
1994 *Corn and Culture in the Prehistoric New World.* Westview Press, Boulder.

Johnson, Alfred E.
1963 The Trincheras Culture of Northern Sonora. *American Antiquity* 29:174–186.

Johnson, Jay K.
1979 Archaic Biface Manufacture: Production Failures, a Chronicle of the Misbegotten. *Lithic Technology* 8:25–35.

Johnson, Jay K. and Carol A. Morrow
1987 *The Organization of Core Technology.* Westview Press, Boulder.

Johnson, Randal D. and Jay E. Anderson
1984 Diets of Black-Tailed Jack Rabbits in Relation to Population Density and Vegetation. *Journal of Range Management* 37:79–83.

Justice, Noel D.
2002 *Stone Age Spear and Arrow Points of the Southwestern United States.* Indiana University Press, Bloomington.

Kaplan, Lawrence
1963 Archaeoethnobotany of Cordova Cave, New Mexico. *Economic Botany* 17:350–359.

Katz, S. H., M. L. Hediger, and L. A. Valleroy
1974 Traditional Maize Processing Techniques in the New World. *Science* 184:765–773.

Katzenberg, M. Anne
2008 Stable Isotope Analysis: A Tool for Studying Past Diet, Demography, and Life History. In *Biological Anthropology of the Human Skeleton*, edited by M. Anne Katzenburg and Shelley R. Saunders, pp. 413–441. 2nd ed. Wiley-Liss, New York.

Katzer, Keith
1993 A Geomorphic Evaluation of the Agricultural Potential of Cerros de Trincheras. In *Between Desert and River: Hohokam Settlement and Land Use in the Los Robles Community*, edited by Christian Downum. Anthropological Papers of the University of Arizona 57. University of Arizona Press, Tucson.

Kearney, Thomas H. and Robert H. Peebles
1960 *Arizona Flora.* University of California Press, Berkeley.

Keeley, Lawrence H.
1988 Hunter-Gatherer Complexity and "Population Pressure": A Cross-Cultural Analysis. *Journal of Anthropological Archaeology* 7:373–411.
1995 Protoagricultural Practices among Hunter-Gatherers. In *Last Hunters First Farmers: New Perspectives on the Prehistoric Transition to Agriculture*, edited by Douglas T. Price and Anne Birgitte Gebauer, pp. 243–272. School of American Research, Santa Fe, New Mexico.
1996 *War Before Civilization.* Oxford University Press, New York.

Keeley, Lawrence H., Marisa Fontana, and Russell Quick
2007 Baffles and Bastions: The Universal Features of Fortifications. *Journal of Archaeological Research* 15:55–95.

Kelley, J. Charles
1952 The Historic Indian Pueblos of La Junta de Los Rios. *New Mexico Historical Review* 27:257–295.
1953 The Historic Indian Pueblos of La Junta de Los Rios. *New Mexico Historical Review* 28:21–51.

Kelly, Robert L.
1988 Three Sides of a Biface. *American Antiquity* 53: 717–734.
1992 Mobility/Sedentism: Concepts, Archaeological Measures, and Effects. *Annual Review of Anthropology* 21:43–66.
1995 *The Foraging Spectrum: Diversity in Hunter-Gatherer Lifeways.* Smithsonian Institution Press, Washington, DC.

Kelly, Robert L. and Lawrence C. Todd
1988 Coming into the Country: Early Paleoindian Hunting and Mobility. *American Antiquity* 53:231–244.

Kemp, Leonard, Cynthia Munoz, Raymond Mauldin, and Robert Hard
2013 Preliminary Results of Stable Carbon Isotopes of Modern and Prehistoric Leporid Bone Collagen from the Jornada Mogollon Region. In *Advances in Jornada Mogollon Archaeology: Proceedings from the 17th Jornada Mogollon Conference*, edited by Todd L. VanPool, Elizabeth M. McCarthy, and Christine S. VanPool, pp. 197–210. El Paso Museum of Archaeology, El Paso, Texas.

Kennett, Douglas J. and Bruce Winterhalder (editors)
2006 *Behavioral Ecology and the Transition to Agriculture.* University of California Press, Berkeley.

Kennett, Douglas J., Barbara Voorhies, and Dean Martorana
2006 An Ecological Model for the Origins of Maize-Based Food Production on the Pacific Coast of Southern Mexico. In *Behavioral Ecology and the Transition to Agriculture*, edited by Douglas J. Kennett and Bruce Winterhalder, pp. 103–136. University of California Press, Berkeley.

Kent, Susan
1993 Variability in Faunal Assemblages: The Influence of Hunting Skill, Sharing, Dogs and Mode of Cooking on Faunal Remains at a Sedentary Kalahari Community. *Journal of Anthropological Archaeology* 12:323–385.

Kettenbrink, Edwin C., Jr. (editor)
1984 *Geology and Petroleum Potential of Chihuahua, Mexico: West Texas Geological Society Field Trip Guidebook, 1984.* Publication 84–80. West Texas Geological Society, Midland.

Kibler, Karl W., Héctor R. Hinojosa-Prieto, M. Steven Shackley, and Héctor J. Hinojosa-García
2014 The Selene Obsidian Source (Formerly Sonora Unknown B) of the Upper Río Bavispe Basin, Sonora, Mexico. *Kiva* 80:168–192.

Klein, Richard G. and Kathryn Cruz-Uribe
1984 *The Analysis of Animal Bones from Archaeological Sites.* University of Chicago Press, Chicago.

Knight, Brian D.
2003 *Early Hunter-Gatherer Occupations in the Southern Playas and Animas Valleys, Hidlago County, New Mexico.* Master's thesis, Department of Anthropology, New Mexico State University. Las Cruces.

Kowalewski, Stephen A.
2007 Concluding Observations: Perspectives from the Hill Towns of Oaxaca. In *Trincheras Sites in Time, Space, and Society*, edited by Suzanne K. Fish, Paul R. Fish, and M. Elisa Villalpando, pp. 247–267. Amerind Studies in Archaeology 1, John Ware, general editor. University of Arizona Press, Tucson.

Krider, P. Reed
1998 Paleoclimatic Significance of Late Quaternary Lacustrine and Alluvial Stratigraphy, Animas Valley, New Mexico. *Quaternary Research* 50:283–289.

Kuhn, Steven L.
1989 Hunter-Gatherer Foraging Organization and Strategies of Artifact Replacement and Discard. In *Experiments in Lithic Technology*, edited by D. Amick and R. Mauldin, pp. 33–47. BAR International Series 528. British Archaeological Reports, Oxford, England.
1994 A Formal Approach to the Design and Assembly of Mobile Toolkits. *American Antiquity* 59:426–442.

Kwiatkowski, Scott M.
1989 The Paleoethnobotany of the Grand Canal Ruins: Results from Flotation, Macrobotanical and Wood Charcoal Analysis. In *Archaeological Investigations at the Grand Canal Ruins: A Classic Period Site in Phoenix, Arizona*, edited by Douglas R. Mitchell, pp. 497–558. Publications in Archaeology 12. Soil Systems, Phoenix.

Lambert, Patricia M.
2002 The Archaeology of War: A North American Perspective. *Journal of Archaeological Research* 10:207–241.

Lancaster, James
1983 *An Analysis of Manos and Metates from the Mimbres Valley Valley, New Mexico.* Master's thesis, Department of Anthropology, University of New Mexico, Albuquerque.
1984 Groundstone Artifacts. In *The Galaz Ruin: A Prehistoric Mimbres Village in Southwestern New Mexico*, edited by Roger Anyon and Stephen A. LeBlanc, pp. 247–262. University of New Mexico Press, Albuquerque.

Langlie, BrieAnna S., Natalie G. Mueller, Robert N. Spengler, and Gayle J. Fritz
2014 Agricultural Origins from the Ground Up: Archaeological Approaches to Plant Domestication. *American Journal of Botany* 101:1601–1617.

Laumbach, Karl W.(editor)
2013 *Cañada Alamosa Project: A Preliminary Report on the 2011 Research Season.* Human Systems Research Report No. 2011–20. Human Systems Research, Inc., Las Cruces, New Mexico.

Lawrence, Barbara
1951 *Post-Cranial Skeletal Characters of Deer, Pronghorn, and Sheep-Goat with Notes on Bos and Bison.* Papers of the Peabody Museum of Archaeology and Ethnology Vol. 35, No. 3. Harvard University, Cambridge, Massachusetts.

LeBlanc, Steven A.
1976 Archeological Recording Systems. *Journal of Field Archaeology* 3:159–168.
1999 *Prehistoric Warfare in the American Southwest.* University of Utah Press, Salt Lake City.
2002 Conflict and Language Dispersal: Issues and a New World Example. In *Examining the Farming/Language Dispersal Hypothesis*, edited by Peter Bellwood and Colin Renfrew, pp. 357–365. MacDonald Institute for Archaeological Research, University of Cambridge, Cambridge, England.
2008 The Case for an Early Farmer Migration into the Greater American Southwest. In *Archaeology without Borders: Contact, Commerce, and Change in the U.S. Southwest and Northwestern Mexico*, edited by Laurie D. Webster, Maxine E. McBrinn, and Eduardo Gamboa C., pp. 107–142. University Press of Colorado, Boulder.

Lee-Thorp, J.A.
2008 On Isotopes and Old Bones. *Archaeometry* 50: 925–950.

Lee, Y.K. and John D. Speth
2004 Rabbit Hunting by Farmers at the Henderson Site, New Mexico. In *Life on the Periphery: Economic Change in Late Prehistoric Southeastern New Mexico*, edited by John D. Speth, pp. 225–277. Memoir 37. Museum of Anthropology, University of Michigan, Ann Arbor.

Legler, Robert P.
1970 *Habitat Preference of the Desert Cottontail with Additional Notes on the Blacktailed Jackrabbit.* Master's thesis, Department of Biology, New Mexico State University, Las Cruces.

Lekson, Stephen H.
2002 War in the Southwest, War in the World. *American Antiquity* 67:607–624.

Lentz, David L., Mary E. D. Pohl, José Luis Alvarado, Somayeh Tarighat, and Robert Bye
2008 Sunflower (*Helianthus annuus* L.) as a Pre-Columbian Domesticate in Mexico. *Proceedings of the National Academy of Sciences of the United States of America* 105:6232–6237.

Lentz, David L., Mary E. D. Pohl, Kevin O. Pope, and Andrew R. Wyatt
2001 Prehistoric Sunflower (*Helianthus annuus* L.) Domestication in Mexico. *Economic Botany* 55: 370–376.

Lesueur, Hardeman David
1945 *The Ecology of the Vegetation of Chihuahua, Mexico, North of Parallel Twenty-Eight.* University of Texas Publication No. 4521. University of Texas, Austin.

Li, P. X. P., A. K. Hardacre, O. H. Campanella, and K. J. Kirkpatrick
1996 Determination of Endosperm Characteristics of 38 Corn Hybrids Using the Stenvert Hardiness Test. *Cereal Chemistry* 73:466–471.

Lightfoot, David C., Ana D. Davidson, Christopher M. McGlone, and Dana G. Parker
2010 Rabbit Abundance Relative to Rainfall and Plan Production in Northern Chihuahuan Desert Grassland and Shrubland Habitats. *Western North American Naturalist* 70:490–499.

List, Rurik, Jesús Pacheco, Eduardo Ponce, Rodrigo Sierra-Corona, and Gerardo Ceballos
2010 The Janos Biosphere Reserve, Northern Mexico. *International Journal of Wilderness* 15:35–41.

Loendorf, Chris and Glen E. Rice
2004 *Projectile Point Typology, Gila River Indian Community, Arizona.* Anthropological Research Papers 2. Gila River Indian Community, Sacaton, Arizona.

Long, Austin, Bruce F. Benz, Douglas J. Donahue, A. J. T. Jull, and L. J. Toolin
1989 First Direct AMS Dates on Early Maize from Tehuacan, Mexico. *Radiocarbon* 31:1035–1040.

Lowrison, G. C.
1974 *Crushing and Grinding: The Size Reduction of Solid Materials.* Butterworths, London.

Lumholtz, Carl
1912 *New Trails in Mexico.* Charles Scribner's Sons, New York.

Lyman, R. Lee
2008 *Quantitative Paleozoology.* Cambridge University Press, New York.

Mabry, Jonathan B.
1998 Conclusion. In *Archaeological Investigations of Early Village Sites in the Middle Santa Cruz Valley: Analysis and Synthesis, Part II*, edited by Jonathan B. Mabry, pp. 757–791. Anthropological Papers No. 19. Center for Desert Archaeology, Tucson.
2002 The Role of Irrigation in the Transition to Agriculture and Sedentism in the Southwest: A Risk Management Model. In *Traditions, Transitions, and Technologies: Themes in Southwestern Archaeology*, edited by Sarah H. Schlanger, pp. 178–199. University Press of Colorado, Boulder.
2005a Diversity in Early Southwestern Farming and Optimization Models of Transitions to Agriculture. In *Subsistence and Resource Use Strategies of Early Agricultural Communities in Southern Arizona*, edited by Michael W. Diehl, pp. 113–152. Anthropological Papers No. 34. Center for Desert Archaeology, Tucson.
2005b Reading the Traces of Early Farming Villages. In *Material Cultures and Lifeways of Early Agricultural Communities in Southern Arizona*, edited by R. Jane Sliva, pp. 1–17. Anthropological Papers No. 35. Center for Desert Archaeology, Tucson.
2006 Chronology. In *Rio Nuevo Archaeology Program, 2000–2003: Investigations at the San Agustin Mission and Mission Gardens, Tucson Presidio, Tucson Pressed Brick Company, and Clearwater Site*, edited by J. Homer Thiel and Jonathan B. Mabry, pp. 67–93. Technical Report 2004-11. Center for Desert Archaeology, Tucson.
2008a Chronology. In *Las Capas: Early Irrigation and Sedentism in a Southwestern Floodplain*, edited by Jonathan B. Mabry, pp. 55–76. Anthropological Papers 28. Center for Desert Archaeology, Tucson.
2008b Introduction. In *Las Capas: Early Irrigation and Sedentism in a Southwestern Floodplain*, edited by Jonathan B. Mabry, pp. 1–33. Anthropological Papers 28. Center for Desert Archaeology, Tucson.
2008c Irrigation, Short-Term Sedentism, and Corporate Organization During the San Pedro Phase. In *Las Capas: Early Irrigation and Sedentism in a Southwestern Floodplain*, edited by Jonathan B. Mabry, pp. 257–277. Anthropological Papers 28. Center for Desert Archaeology, Tucson.

Mabry, Jonathan B. (editor)
1998 *Archaeological Investigations of Early Village Sites in the Middle Santa Cruz Valley: Analysis and Synthesis, Parts I and II.* Anthropological Papers No. 19. Center for Desert Archaeology, Tucson.
2008 *Las Capas: Early Irrigation and Sedentism in a Southwestern Floodplain.* Anthropological Papers 28. Center for Desert Archaeology, Tucson.

Mabry, Jonathan B., John P. Carpenter, and Guadalupe Sanchez
2008a Archaeological Models of Early Uto-Aztecan Prehistory in the Arizona-Sonora Borderlands. In *Archaeology without Borders: Contact, Commerce, and Change in the U.S. Southwest and Northwestern Mexico*, edited by Laurie D. Webster and Maxine E. McBrinn, pp. 155–183. University Press of Colorado, Boulder.

Mabry, Jonathan B. and William E. Doolittle
2008 Modeling the Early Agricultural Frontier in the Desert Borderlands. In *Archaeology without Borders: Contact, Commerce, and Change in the U.S. Southwest and Northwestern Mexico*, edited by Laurie D. Webster and Maxine E. McBrinn, pp. 55–70. University Press of Colorado, Boulder.

Mabry, Jonathan B., James P. Holmlund, Fred L. Nials, and Manuel R. Palacios-Fest
2008b Modeling Canal Characteristics and Trends. In *Las Capas: Early Irrigation and Sedentism in a Southwestern Floodplain*, edited by Jonathan B. Mabry, pp. 235–247. Anthropological Papers 28. Center for Desert Archaeology, Tucson.

Mabry, Jonathan B. and Michelle N. Stevens
2014 Beyond the Cochise Culture: New Views of Archaic Foragers and Early Farmers in Southeastern Arizona. In *Between Mimbres and Hohokam: Exploring the Archaeology and History of Southeastern Arizona and Southwestern New Mexico*, edited by Henry D. Wallace, pp. 115–164. Archaeology Southwest, Tucson.

Mabry, Jonathan B., Deborah L. Swartz, Helga Wöcherl, Jeffery J. Clark, Gavin H. Archer, and Michael W. Lindeman
1997 *Archaeological Investigations of Early Village Sites in the Middle Santa Cruz Valley: Description of the Santa Cruz Bend, Square Hearth, Stone Pipe, and Canal Sites*. Anthropological Papers 18. Center for Desert Archaeology, Tucson.

MacNeish, Richard S.
1958 *Preliminary Archaeological Investigations in the Sierra de Tamaulipas*. Transactions of the American Philosophical Society 48(6). American Philosophical Society, Philadelphia.
1993 *Preliminary Investigations of the Archaic in the Region of Las Cruces, New Mexico*. Historic and Natural Resources Report 9. Cultural Resources Management Branch, Directorate of Environment, United States Army Air Defense Artillery Center, Fort Bliss.

MacWilliams, A. C., Robert J. Hard, John R. Roney, Karen R. Adams, and William L. Merrill
2008 The Setting of Early Agriculture in Southern Chihuahua. In *Archaeology without Borders: Contact, Commerce, and Change in the U.S. Southwest and Northwestern Mexico*, edited by Laurie D. Webster and Maxine E. McBrinn, pp. 35–54. University Press of Colorado, Boulder.

Mallouf, Robert J.
1985 *A Synthesis of Eastern Trans-Pecos Prehistory*. Master's thesis, Department of Anthropology, University of Texas at Austin.

Mangelsdorf, Paul C. and Robert H. Lister
1956 Archaeological Evidence on the Evolution of Maize in Northwestern Mexico. *Harvard University, Botanical Museum Leaflets* 17:151–177.

Mangelsdorf, Paul C., Richard S. MacNeish, and W. Galinat
1956 Archaeological Evidence on the Diffusion and Evolution of Maize in Northeastern Mexico. *Harvard University, Botanical Museum Leaflet* 17:125–150.

Marin, Ariadna I., Lucina Hernandez, and John W. Laundre
2003 Predation Risk and Food Quantity in the Selection of Habitat by Black-Tailed Jackrabbit (*Lepus Californicus*): An Optimal Foraging Approach. *Journal of Arid Environments* 55:101–110.

Martin, Alexander C. and William D. Barkley
1961 *Seed Identification Manual*. University of California Press, Berkeley.

Martin, Paul Sydney and John B. Rinaldo
1947 *The SU Site: Excavations at a Mogollon Village Western New Mexico, Third Season 1946*. Anthropological Series 32, No. 3. Field Museum of Natural History, Chicago.

Martin, Paul Sydney, John B. Rinaldo, Elaine Bluhm, Hugh C. Cutler, and Roger Grange, Jr.
1952 *Mogollon Cultural Continuity and Change: The Stratigraphic Analysis of Tularosa and Cordova Caves*. Fieldiana: Anthropology 40. Chicago Natural History Museum, Chicago.

Matson, R. G.
1991 *The Origins of Southwestern Agriculture*. University of Arizona Press, Tucson.
2002 The Spread of Maize Agriculture into the U.S. Southwest. In *Examining the Farming/Language Dispersal Hypothesis*, edited by Peter Bellwood and Colin Renfrew, pp. 341–356. MacDonald Institute for Archaeological Research, University of Cambridge, Cambridge.

Matson, R. G. and Brian Chisholm
1991 Basketmaker II Subsistence: Carbon Isotopes and Other Dietary Indicators from Cedar Mesa, Utah. *American Antiquity* 56:444–459.

Matthews, J. A.
1985 Radiocarbon Dating of Surface and Buried Soils: Principles, Problems and Prospects. In *Geomorphology and Soils*, edited by K. S. Richards, R. R. Arnett, and S. Ellis, pp. 269–288. George Allen and Unwin, London.

Mauldin, Raymond P.
1993a Exploring the Potential Variation in Jackrabbit Carbon Isotopic Signatures as an Indicator of Seasonality on Archaeological Sites in the Chihuahuan Desert. Manuscript on file, Archaeological Studies Program, Environmental Management Office, Fort Bliss, Texas.
1993b The Relationship between Ground Stone and Agricultural Intensification in Western New Mexico. *Kiva* 58:317–330.
1995 *Groping for the Past: Investigating Archaeological Patterns across Space and Time in the Southern Southwestern United States*. PhD dissertation, Department of Anthropology, The University of New Mexico. Albuquerque.

Mauldin, Raymond P. and Daniel S. Amick
1989 Investigating Patterning from Experimental Bifacial Core Reduction. In *Experiments in Lithic Technology*, edited by Daniel Amick and Raymond P.

Mauldin, pp. 67–99. BAR International Series 528. British Archaeological Reports, Oxford.

McAuliffe, Joseph R. and Thomas R. Van Devender

1998 A 22,000-Year Record of Vegetation Change in the North-Central Sonoran Desert. *Palaeogeography, Palaeoclimatology, Palaeoecology* 141:253–275.

McBrearty, Sally, Laura Bishop, Thomas Plummer, Robert Dewar, and Nicholas Conard

1998 Tools Underfoot: Human Trampling as an Agent of Lithic Artifact Edge Modification. *American Antiquity* 63:108–130.

McBrinn, Maxine

2005 *Social Identities among Archaic Mobile Hunters and Gatherers in the American Southwest.* Arizona State Museum Archaeological Series 197. University of Arizona, Tucson.

McCarthy, Elizabeth M., Gordon F. M. Rakita, Christine S. VanPool, and Todd L. VanPool

2013 Faunal Analysis and Environmental Reconstruction of the 76 Draw Site, New Mexico. In *Advances in Jornada Mogollon Archaeology: Proceedings from the 17th Jornada Mogollon Conference*, edited by Todd L. VanPool, Elizabeth M. McCarthy, and Christine S. VanPool, pp. 185–196. El Paso Museum of Archaeology, El Paso, Texas.

McClung de Tapia, Emily, Javier González Vázquez, Judith Zurita Noguera, and Emilio Ibarra Morales

1996 *La Domesticación Prehispánica de Amaranthus spp.* Cuadernos de Trabajo IIa. Universidad Nacional Autónoma de México, Instituto de Investigaciones Antropológicas, México, D.F.

McGee, William John

1895 The Beginning of Agriculture. *The American Anthropologist* 8:350–375.

1898 Papgueria. *National Geographic Magazine* 9:345–371.

McGuire, Randall H.

1998 Cerro de Trincheras: A Prehispanic Terraced Town in Sonora, Mexico. *Archaeology in Tucson* 12:1–5.

McGuire, Randall H., Maria Elisa Villalpando C., James P. Holmlund, and Maria O'Donovan

1993 Cerro de Trincheras Mapping Project: Final Technical Report to the National Geographic Society for Grant #4454-91. Manuscript on file.

McGuire, Randall H. and M. Elisa Villalpando

2007 Excavations at Cerro de Trincheras. In *Trincheras Sites in Time, Space, and Society*, edited by Suzanne K. Fish, Paul R. Fish, and M. Elisa Villalpando, pp. 137–164. Amerind Studies in Archaeology 1, John Ware, general editor. University of Arizona Press, Tucson.

2015 War and Defense on Cerros de Trincheras in Sonora, México. *American Antiquity* 80:429–450.

McLaughlin, Steven P.

1986 Floristic Analysis of the Southwestern United States. *Great Basin Naturalist* 46:46–65.

1989 Natural Floristic Areas of the Western United States. *Journal of Biogeography* 16:239–248.

1992 Are Floristic Areas Hierarchically Arranged? *Journal of Biogeography* 19:21–32.

1994 Floristic Plant Geography—the Classification of Floristic Areas and Floristic Elements. *Progress in Physical Geography* 18:185–208.

2007 *Tundra to Tropics: The Floristic Plant Geography of North America.* Botanical Research Institute of Texas, Fort Worth, Texas.

McLaughlin, Steven P. and Janice E. Bowers

1982 Effects of Wildfire on a Sonoran Desert Plant Community. *Ecology* 63:246–248.

McLaurin, Brett T., Aileen C. Elliott, James T. Watson, and Maria Elisa Villalpando Canchola

2012 Quaternary Stratigraphy of the La Playa Archaeological Site (SON F:10:3), Northern Sonora, Mexico. In *Reconstructing Human-Landscape Interactions, Vol. 1 Interpreting Desert and Fluvial Environments*, edited by Brett T. McLaurin, Aileen C. Elliott, and N. Torres, pp. 3–20. Springer Briefs in Earth System Science. Springer, Berlin.

McNamee, Calla

2012 *Soil Phytolith Assemblages of the American Southwest: The Use of Historical Ecology in Taphonomic Studies.* PhD dissertation, Department of Archaeology, University of Calgary. Calgary, Alberta.

Medlin, J. A.

1974 *Abundance of Black-Tailed Jackrabbits, Desert Cottontail Rabbits, and Coyotes in Southeastern New Mexico.* Master's thesis, New Mexico State University. Las Cruces.

Mehringer, Peter J.

1967 Pollen Analysis and the Alluvial Chronology. *Kiva* 32:96–101.

Mehringer, Peter J., Paul S. Martin, and C. Vance Haynes Jr.

1967 Murray Springs, a Mid-Postglacial Pollen Record from Southern Arizona. *American Journal of Science* 265:786–797.

Melgoza, Alicia

2005 Base de Datos de Las Plantas del Municipio de Janos, Chihuahua. Instituto Nacional de Investigaciones Forestales, Agricolas, Pecuarias. Manuscript on file, Department of Anthropology, University of Texas at San Antonio, San Antonio, Texas.

Meltzer, David J., Robert D. Leonard, and Susan K. Stratton

1992 The Relationship between Sample Size and Diversity in Archaeological Assemblages. *Journal of Archaeological Science* 19:375–387.

Mendiola Galván, Francisco

2002 *El Arte Rupestre en Chihuahua, Expresión Cultural de Nómadas y Sedentarios en El Norte de México.* Instituto Chihuahuense de la Cultura and Instituto Nacional de Antropología e Historia, Chihuahua and México, D.F.

Menking, Kirsten M. and Roger Y. Anderson

2003 Contributions of La Niña and El Niño to Middle Holocene Drought and Late Holocene Moisture in the American Southwest. *Geology* 31:937–940.

Merrill, William L.

2012 The Historical Linguistics of Uto-Aztecan Agriculture. *Anthropological Linguistics* 54:203–260.

Merrill, William L., Robert J. Hard, Jonathan B. Mabry, Gayle J. Fritz, Karen R. Adams, John R. Roney, and A. C. MacWilliams
2009 The Diffusion of Maize to the Southwestern United States and Its Impact. *Proceedings of the National Academy of Sciences of the United States of America* 106:21019–21026.
Merrill, William L. and Celia López González
2007 Humans and Other Mammals in Prehispanic Chihuahua. In *Human and Faunal Relationships Reviewed: An Archaeozoological Approach*, edited by Eduardo Corona-M. and Joaquín Arroyo-Cabrales, pp. 53–62. BAR International Series 1627. Archaeopress, Oxford.
Metcalfe, S. E.
1997 Palaeolimnological Records of Climate Change in Mexico—Frustrating Past, Promising Future? *Quaternary International* 43–4:111–116.
Metcalfe, S. E., A. Bimpson, A. J. Courtice, S. L. Ohara, and D. M. Taylor
1997 Climate Change at the Monsoon/Westerly Boundary in Northern Mexico. *Journal of Paleolimnology* 17:155–171.
Miksicek, Charles H.
1987 Late Sedentary–Early Classic Period Hohokam Agriculture: Plant Remains from the Marana Community Complex. In *Studies in the Hohokam Community of Marana*, edited by Glenn E. Rice, pp. 197–216. Office of Cultural Resource Management, Department of Anthropology, Arizona State University, Tempe.
Miller, Christopher L.
1995 *Chipped Stone Analysis from Northwestern Chihuahua, Mexico*. Master's thesis, Department of Anthropology, University of Tulsa, Oklahoma.
Miller, Myles R.
2018 Archaic Transitions and Transformations in the Jornada Mogollon Region of Southern New Mexico and Western Texas. In *The Archaic Southwest: Foragers in an Arid Land*, edited by Bradley J. Vierra, pp. 119–144. University of Utah Press, Salt Lake City.
Minnis, Paul E.
1981 Seeds in Archaeological Sites: Sources and Some Interpretive Problems. *American Antiquity* 46:143–152.
1985 Domesticating Plants and People in the Greater Southwest. In *Prehistoric Food Production in North America*, edited by Richard I. Ford, pp. 309–339. Anthropological Papers 75. Museum of Anthropology, University of Michigan.
1992 Earliest Plant Cultivation in the Desert Borderlands of North America. In *The Origins of Agriculture: An International Perspective*, edited by C. Wesley Cowan and Patty Jo Watson, pp. 121–141. Smithsonian Institution Press, Washington, DC.
Minor, Rick
1975 *The Pit-and-Groove Petroglyph Style in Southern California*. Ethnic Technology Notes 15. San Diego Museum of Man, San Diego, California.
Monks, Gregory G.
1981 Seasonality Studies. In *Advances in Archaeological Method and Theory, Vol. 4*, edited by Michael B. Schiffer, pp. 177–240. Academic Press, New York.
Moore, John H.
1994 Putting Anthropology Back Together Again: The Ethnogenetic Critique of Cladistic Theory. *American Anthropologist*, n.s. 96:925–948.
Moran-Zenteno, D.
1994 *Geology of the Mexican Republic*. Studies in Geology 39. American Association of Petroleum Geologists.
Morris, Donald H.
1990 Changes in Groundstone Following the Introduction of Maize into the American Southwest. *Journal Of Anthropological Research* 46:177–194.
Muenchrath, Deborah A.
1995 *Productivity, Morphology, Phenology, and Physiology of a Desert-Adapted Native American Maize (Zea mays L.) Cultivar*. PhD dissertation, Department of Agronomy, Iowa State University, Ames.
Muenchrath, Deborah A., M. Kuratomi, Jonathan A. Sandor, and Jeffrey A. Homberg
2002 Observational Study of Maize Production Systems of Zuni Farmers in Semiarid New Mexico. *Journal of Ethnobiology* 22:1–33.
Muenchrath, Deborah A. and Ricardo J. Salvador
1995 Maize Productivity and Agroecology: Effects of Environment and Agricultural Practices on the Biology of Maize. In *Soil, Water, Biology, and Belief in Prehistoric and Traditional Southwestern Agriculture*, edited by H. Wolcott Toll, pp. 303–334. New Mexico Archaeological Council, Albuquerque, New Mexico.
Muenscher, Walter Conrad
1939 *Poisonous Plants of the United States*. The MacMillan Company, New York.
Muñiz, Arturo, M.
1998 The Mobile Faunas: Reliable Seasonal Indicators for Archaeozoologists? In *Seasonality and Sedentism: Archaeological Perspectives from Old and New World Sites*, edited by Thomas R. Rocek and Ofer Bar Yosef, pp. 25–40. Peabody Museum of Archaeology and Ethnology, Harvard University, Boston.
Munoz, Cynthia M., Raymond Mauldin, Debajyoti Paul, and Leonard Kemp
2011 Monitoring Paleovegetation Shifts through Stable Carbon Isotope Variability in Archaeologically Recovered Leporids. *Texas Journal of Science* 62:113–140.
Nabhan, Gary P.
1979a The Ecology of Floodwater Farming in Arid Southwestern North America. *Agro-Ecosystems* 5:245–255.
1979b Grain Amaranth and Other Rare Crops of Northern Mexico. *Dry Country News* 3:20–22.
National Research Council
1984 *Amaranth: Modern Prospects for an Ancient Crop*. National Academy Press, Washington, DC.
Nelson, Ben A.
2007 Crafting of Places: Mesoamerican Monumentality in Cerros de Trincheras and Other Hilltop Sites. In

Trincheras Sites in Time, Space, and Society, edited by Suzanne K. Fish, Paul R. Fish, and M. Elisa Villalpando, pp. 230–246. Amerind Studies in Archaeology 1, John Ware, general editor. University of Arizona Press, Tucson.

Nelson, Margaret C.

1991 The Study of Technological Organization. In *Archaeological Method and Theory, Vol. 3*, edited by Michael B. Schiffer, pp. 57–100. University of Arizona Press, Tucson.

Nelson, T., J. L. Holechek, R. Valdez, and M. Cardenas

1997 Wildlife Numbers on Late and Mid Seral Chihuahuan Desert Rangelands. *Journal of Range Management* 50:593–599.

Nials, Fred L.

2015 Life on the Floodplain: The Promise and Perils of Prehistoric Irrigation Agriculture at Las Capas, AZ:12:111 (ASM). In *Implements of Change: Tools, Subsistence, and the Built Environment of Las Capas, an Early Agricultural Irrigation Community in Southern Arizona*, edited by James M. Vint, pp. 419–468. Anthropological Papers 51. Desert Archaeology, Inc., Tucson.

Nials, Fred L., David A. Gregory, and J. Brett Hill

2011 The Stream Reach Concept and the Macro-Scale Study of Riverine Agriculture in Arid and Semiarid Environments. *Geoarchaeology* 26:724–761.

Nissen, Karen M. and Eric W. Ritter

1986 Cupped Rock Art in North-Central California: Hypothesis Regarding Age and Social/Ecological Context. In *American Indian Rock Art, Volume 11*, edited by William D. Hyder, Helen Crotty, Kay Sanger, and Frank G. Bock, pp. 59–75. American Rock Art Research Association, El Toro, California.

Nordt, Lee C.

2001 Stable Carbon and Oxygen Isotopes in Soils: Applications for Archaeological Research. In *Earth Sciences and Archaeology*, edited by Paul Goldberg, Vance T. Holliday, and C. Reid Ferring, pp. 419–445. Kluwer/Plenum Publishers, New York.

2003 Late Quaternary Fluvial Landscape Evolution in Desert Grasslands of Northern Chihuahua, Mexico. *Geological Society of American Bulletin* 115:596–606.

Nordt, Lee C., Thomas W. Boutton, Charles T. Hallmark, and Michael R. Waters

1994 Late Quaternary Vegetation and Climate Changes in Central Texas Based on the Isotopic Composition of Organic Carbon. *Quaternary Research* 41:109–120.

Nordt, Lee C., Thomas W. Boutton, Rolfe D. Mandel, and John S. Jacob

2002 C_4 Plant Productivity and Climate-CO_2 Variations in South-Central Texas during the Late Quaternary. *Quaternary Research* 58:182–188.

Nordt, Lee C., J. Von Fisher, and L. L. Tieszen

2008 Coherent Changes in Relative C4 Plant Productivity and Climate During the Late Quaternary in the North American Great Plains. *Quaternary Science Reviews* 27:1600–1611.

O'Donovan, Maria

1997 *Confronting Archaeological Enigmas: Cerro de Trincheras, Cerros de Trincheras and Monumentality.* PhD dissertation, State University of New York at Binghamton.

2002 *New Perspectives on Site Function and Scale of Cerro de Trincheras, Sonora, Mexico: The 1991 Surface Survey.* Arizona State Museum Archaeological Series 195. University of Arizona, Tucson.

2004 The Role of Cerro de Trincheras Sites in the Northwest Mexican and the American Southwest Landscape. In *Surveying the Archaeology of Northwest Mexico*, edited by Gillian E. Newell and Emiliano Gallaga, pp. 27–45. University of Utah Press, Salt Lake City.

O'Rourke, Charles Edward (editor)

1940 *General Engineering Handbook.* McGraw-Hill Book Company, Inc., New York and London.

Odell, George H.

1996a Economizing Behavior and the Concept of "Curation". In *Stone Tools: Theoretical Insights into Human Prehistory*, edited by George Odell, pp. 51–80. Plenum Press, New York.

1996b *Stone Tools and Mobility in the Illinois Valley.* Archaeological Series No. 10. International Monographs in Prehistory, Ann Arbor.

2001 Stone Tool Research at the End of the Millennium: Classification, Function, and Behavior. *Journal of Archaeological Research* 9:45–100.

2004 *Lithic Analysis. Manuals in Archaeological Method, Theory and Technique.* Kluwer/Academic Press, New York.

Ogilvie, Marsha D.

2000 *A Biological Reconstruction of Mobility Patterns at the Foraging to Farming Transition in the American Southwest.* PhD dissertation, Department of Anthropology, University of New Mexico. University Microfilms, Ann Arbor.

2005 A Biological Reconstruction of Mobility in Late Archaic Populations. In *The Late Archaic across the Borderlands*, edited by Bradley J. Vierra, pp. 84–112. University of Texas Press, Austin.

Olsen, Stanley J.

1964 *Mammal Remains from Archaeological Sites, Part I, Southeastern and Southwestern United States.* Papers of the Peabody Museum of Archaeology and Ethnology Vol. 56, No. 1. Harvard University, Cambridge.

1968 *Fish, Amphibian and Reptile Remains from Archaeological Site, Part I, Southeastern and Southwestern United States.* Papers of the Peabody Museum of Archaeology and Ethnology Vol. 56, No. 2. Harvard University, Cambridge.

Olson, R. and D. Sander

1988 Corn Production. In *Corn and Corn Improvement*, edited by G. Sprague and J. Dudley, pp. 638–686. Agronomy Monograph 18. American Society of Agronomy, Madison, Wisconsin.

Onken, Jill, Susan J. Smith, Manuel R. Palacios-Fest, and Karen R. Adams
2017 Late Holocene Hydroclimatic Change at Cienega Amarilla, West-Central New Mexico, USA. *Quaternary Research* 87:227–245.
Opler, Morris E.
1935 A Note on the Cultural Affiliations of Northern Mexican Nomads. *American Anthropologist* 37:702–706.
Ortega-Ramírez, José R.
1995 Los Paleoambientes Holocénicos de la Laguna de Babícora, Chihuahua, México. *Geofisica Internacional* 34:107–116.
Ortega-Ramírez, José R., Alfonso Valiente-Banuet, Jaime Urrutia-Fucugauchi, Carlos A. Mortera-Gutiérrez, and Guillermo Alvarado-Valdez
1998 Paleoclimatic Changes During the Late Pleistocene-Holocene in Laguna Babícora, near the Chihuahuan Desert, México. *Canadian Journal of Earth Sciences/Revue Canadienne des Sciences de la Terre* 35:1168–1179.
Otterbein, Keith F.
1999 A History of Research on Warfare in Anthropology. *American Anthropologist* 101:794–805.
Palacios-Fest, Manuel R., Ana Luisa Carreño, José R. Ortega-Ramírez, and Guillermo Alvarado-Valdéz
2002 A Paleoenvironmental Reconstruction of Laguna Babícora, Chihuahua, Mexico Based on Ostracode Paleoecology and Trace Element Shell Chemistry. *Journal of Paleolimnology* 27:185–206.
Parmenter, R. R. and Thomas R. Van Devender
1995 Diversity, Spatial Variability, and Functional Roles of Vertebrates in the Desert Grassland. In *The Desert Grassland*, edited by Mitchel P. McClaran and Thomas R. Van Devender, pp. 196–229. University of Arizona Press, Tucson.
Parry, William J.
1987 Sources of Chipped Stone Materials. In *Prehistoric Stone Technology on Northern Black Mesa, Arizona*, edited by William J. Parry and Andrew L. Christenson. Occasional Paper No.12. Center for Archaeological Investigations, Southern Illinois University, Carbondale.
Parry, William J. and Robert L. Kelly
1987 Expedient Core Technology and Sedentism. In *The Organization of Core Technology*, edited by Jay K. Johnson and Carol A. Morrow, pp. 285–304. Westview Press, Boulder.
Patterson, Leland W.
1982 Replication and Classification of Large Size Lithic Debitage. *Lithic Technology* 11:50–58.
1990 Characteristics of Bifacial-Reduction Flake-Size Distribution. *American Antiquity* 55:550–558.
Patterson, Leland W. and J. B. Sollberger
1978 Replication and Classification of Small Size Lithic Debitage. *Plains Anthropologist* 23:103–112.
Peebles, Christopher S. and Susan M. Kus
1977 Some Archaeological Correlates of Ranked Societies. *American Antiquity* 42:421–448.
Pennington, Campbell W.
1963 *The Tarahumar of Mexico: Their Environment and Material Culture*. University of Utah Press, Salt Lake City.
1983 Tarahumara. In *Southwest*, edited by Alfonso Ortiz, pp. 276–289. Handbook of North American Indians, Vol. 10, William C. Sturtevant, general editor. Smithsonian Institution, Washington, DC.
Peregrine, Peter N.
2004 Cross-Cultural Approaches in Archaeology: Comparative Ethnology, Comparative Archaeology, and Archaeoethnology. *Journal of Archaeological Research* 12:281–309.
Peterson, Jane, Douglas R. Mitchell, and Steven Shackley
1997 The Social and Economic Contexts of Lithic Procurement: Obsidian from Classic-Period Hohokam Sites. *American Antiquity* 62:231–259.
Pfefferkorn, Ignaz
1949 *Sonora, a Description of the Province*. Translated by Theodore E. Treutlein. University of New Mexico Press, Albuquerque.
Phelps, Alan L.
1966 *Cruciform: An Unusual Artifact of the El Paso Southwest*. Special Report No. 5. El Paso Archaeological Society, El Paso.
1998 An Inventory of Prehistoric Native American Sites in Northwestern Chihuahua. *The Artifact* 36:1–176.
Piperno, Dolores R. and Kent V. Flannery
2001 The Earliest Archaeological Maize (*Zea mays* L.) from Highland Mexico: New Accelerator Mass Spectometry Dates and the Implications. *Proceedings of the National Academy of Sciences of the United States of America* 98:2101–2103.
Piperno, Dolores R., Anthony J. Ranere, Irene Holst, Jose Iriarte, and Ruth Dickau
2009 Starch Grain and Phytolith Evidence for Early Ninth Millennium B.P. Maize from the Central Balsas River Valley, Mexico. *Proceedings of the National Academy of Sciences of the United States of America* 106:5019–5024.
Pohl, Mary D., Kevin O. Pope, John G. Jones, John S. Jacob, Dolores R. Piperno, Susan D. deFrance, David L. Lentz, John A. Gifford, Marie E. Danforth, and J. Kathryn Josserand
1996 Early Agriculture in the Maya Lowlands. *Latin American Antiquity* 7:355–372.
Polyak, Victor J. and Yemane Asmerom
2001 Late Holocene Climate and Cultural Changes in the Southwestern United States. *Science* 294:148–151.
Poore, R. Z., H. J. Dowsett, S. Verardo, and T. M. Quinn
2003 Millennial- to Century-Scale Variability in Gulf of Mexico Holocene Climate Records. *Paleoceanography* 18:1048.
Poore, R. Z., M. J. Pavich, and H. D. Grissino-Mayer
2005 Record of the North American Southwest Monsoon from Gulf of Mexico Sediment Cores. *Geology* 33:209–212.

Pope, Kevin O., Mary D. Pohl, John G. Jones, David L. Lentz, Christopher von Nagy, Francisco J. Vega, and Irvy R. Quitmyer
2001 Origin and Environmental Setting of Early Agriculture in the Lowlands of Mesoamerica. *Science* 292:1370–1373.
Powell, A. Michael
1994 *Grasses of the Trans-Pecos and Adjacent Areas.* University of Texas Press, Austin.
1998 *Trees and Shrubs of the Trans-Pecos and Adjacent Areas.* University of Texas Press, Austin.
Prasciunas, Mary M.
2007 Bifacial Cores and Flake Production Efficiency: An Experimental Test of Technological Assumptions. *American Antiquity* 72:334–348.
Prentiss, W. C. and E. J. Romansky
1989 Experimental Evaluation of Sullivan and Rosen's Debitage Typology. In *Experiments in Lithic Technology*, edited by Daniel S. Amick and Raymond P. Mauldin, pp. 89–100. BAR International Series 528. British Archaeological Reports, Oxford.
Quirt-Booth, Tina
1996 *Examination of Sinagua Faunal Exploitation in Prehistoric Northern Arizona.* Master's thesis, Department of Anthropology, Northern Arizona University, Flagstaff.
Rafferty, Janet E.
1985 The Archaeological Record on Sedentariness: Recognition, Development, and Implications. In *Advances in Archaeological Method and Theory, Vol. 8*, edited by Michael B. Schiffer, pp. 113–156. Academic Press, New York.
Railey, Jim A.
2010 Reduced Mobility or the Bow and Arrow? Another Look at "Expedient" Technologies and Sedentism. *American Antiquity* 75:259–286.
Rawls, W.
1983 Estimating Soil Bulk Density from Particle Size Analysis and Organic Matter Content. *Soil Science* 135:123–125.
Rea, Amadeo M.
1997 *At the Desert's Green Edge: An Ethnobotany of the Gila River Pima.* University of Arizona Press, Tucson.
Rebnegger, Karin J.
2001 *Lithic Technology, Craft Production and Site Variation in the Casas Grandes Region.* Master's thesis, Department of Anthropology, University of Oklahoma, Norman.
Reimer, Paula J, Edouard Bard, Alex Bayliss, J. Warren Beck, Paul G. Blackwell, Christopher Bronk Ramsey, Caitlin E. Buck, Hai Cheng, R. Lawrence Edwards, and Michael Friedrich
2013 IntCal13 and Marine13 Radiocarbon Age Calibration Curves 0–50,000 Years Cal BP. *Radiocarbon* 55: 1869–1887.
Reimer, Paula J., Mike G. L. Baillie, Edouard Bard, Alex Bayliss, J. Warren Beck, Chanda J. H. Bertrand, Paul G. Blackwell, Caitlin E. Buck, George S. Burr, Kirsten B. Cutler, Paul E. Damon, R. Lawrence Edwards, Richard G. Fairbanks, Michael Friedrich, Thomas P. Guilderson, Konrad A. Hughen, Bernd Kromer, Gerry McCormac, Sturt Manning, Christopher Bronk Ramsey, Ron W. Reimer, Sabine Remmele, John R. Southon, Minze Stuiver, Sahra Talamo, F. W. Taylor, Johannes van der Plicht, and Constanze E. Weyhenmeyer
2004 Intcal04 Terrestrial Radiocarbon Age Calibration, 0–26 Cal Kyr BP. *Radiocarbon* 46:1029–1058.
Reiseberg, Loren and John M. Burke
2008 Molecular Evidence and the Origin of the Domesticated Sunflower. *Proceedings of the National Academy of Sciences of the United States of America* 105:E46.
Renfrew, Colin
1975 Trade as Action at a Distance: Questions of Integration and Communication. In *Ancient Civilization and Trade*, edited by Jeremy A. Sabloff and C. C. Lamberg-Karlovsky, pp. 3–59. University of New Mexico Press, Albuquerque.
1977 Alternative Models for Exchange and Spatial Distribution. In *Exchange Systems in Prehistory*, edited by Timothy K. Earle and Jonathon E. Ericson, pp. 71–90. Academic Press, New York.
Retallack, Gregory J.
2001 *Soils of the Past.* Blackwell Science, Oxford.
Rhode, David
1988 Measurement of Archaeological Diversity and the Sample-Size Effect. *American Antiquity* 53: 708–716.
Rice, Glen E. and Steven A. LeBlanc
2001 *Deadly Landscapes: Case Studies in Prehistoric Southwestern Warfare.* University of Utah Press, Salt Lake City.
Robertson, Kenneth L.
1981 The Genera of Amaranthaceae in the Southeastern United States. *Journal of the Arnold Arboretum* 62: 267–314.
Robinson, Richard W. and D. S. Decker-Walters
1997 *Cucurbits.* CAB International, New York.
Roney, John R.
1996a Cerro Juanaqueña: A Late Archaic Cerro de Trincheras in Northwestern Chihuahua. Paper presented at the Conference on the Archaic Prehistory of the North American Southwest, Albuquerque, New Mexico.
1996b Late Archaic Cerros de Trincheras in Northwestern Chihuahua. Paper presented at the 61st Annual Meeting of the Society for American Archeology, New Orleans.
1999 Canador Peak, an Early Pithouse Period Cerro de Trincheras in Southwestern New Mexico. In *La Frontera: Papers in Honor of Patrick H. Beckett*, edited by Meliha S. Duran and David T. Kirkpatrick, pp. 173–184. Papers of the Archaeological Society of New Mexico 25. The Archaeological Society of New Mexico, Albuquerque.

Roney, John R. and Robert J. Hard
1998 A Note on the Temporal and Geographic Distribution of Cerros de Trincheras. Paper presented at the 63rd Annual Meeting of the Society for American Archaeology, Seattle.
2000 *Una Investigación Arqueológica de Los Sitios Cerros con Trincheras del Arcaico Tardío en Chihuahua, México: Las Investigaciones de Campo de 1999*. Special Report 26-S. Center for Archaeological Research, University of Texas at San Antonio, and included as an electronic appendix to Chapter 3.
2002 Early Agriculture in Northwestern Chihuahua. In *Traditions, Transitions, and Technologies: Themes in Southwestern Archaeology in the Year 2000*, edited by Sarah Schlanger, pp. 163–180. University Press of Colorado, Boulder.

Roney, John R., Robert J. Hard, A.C. MacWilliams, and Mary Whisenhunt
2014 Initial Investigations at the Round Mountain Cerro de Trincheras Site in Southeastern Arizona. Paper presented at the 18th Biennial Mogollon Archaeology Conference, Las Cruces, New Mexico.
2015a Preliminary Field Report for 2014 Archeological Investigations at Round Mountain (AZ CC:4:61 (ASM)) near Duncan, Arizona. Unpublished report on file Bureau of Land Management, Safford Field Office. Safford, Arizona.
2015b Recent Test Excavations at an Early Agricultural Period Cerro de Trincheras Site on the Upper Gila River, Arizona. Paper presented at the 80th Annual Meeting of the Society for American Archaeology, April 15–19, San Francisco, California.

Roney, John R., Robert J. Hard, Mary Wisenhunt, and Ashley Jones
2016 Preliminary Results of the 2015 Archaeological Survey and Test Excavations of the Duncan and York Valleys. Report on file, Bureau of Land Management, Safford Field Office, Arizona.

Roscoe, Paul
2008 Settlement Fortification in Village and 'Tribal' Society: Evidence from Contact-Era New Guinea. *Journal of Anthropological Archaeology* 27:507–519.

Rosenswig, Robert M.
2015 A Mosaic of Adaptation: The Archaeological Record for Mesoamerica's Archaic Period. *Journal of Archaeological Research* 23:115–162.

Rosenswig, Robert M. and Richard L. Burger
2012 Considering Early New World Monumentality. In *Early New World Monumentality*, edited by Richard L. Burger and Robert M. Rosenswig, pp. 3–22. University Press of Florida, Gainesville.

Roth, Barbara J.
1989 *Late Archaic Settlement and Subsistence in the Tucson Basin*. PhD dissertation, Department of Anthropology, University of Arizona, Tucson.
2006 The Role of Gender in the Adoption of Agriculture in the Southern Southwest. *Journal of Anthropological Research* 62:513–538.
2016a *Agricultural Beginnings in the American Southwest*. Rowman & Littlefield, Lanham, Maryland.
2016b Were They Sedentary and Does It Matter? Early Farmers in the Tucson Basin. In *Late Holocene Research on Foragers and Farmers in the Desert West*, edited by Barbara J. Roth and Maxine E. McBrinn, pp. 108–135. University of Utah Press, Salt Lake City.

Roth, Barbara J. and Bruce B. Huckell
1992 Cortaro Points and the Archaic of Southern Arizona. *Kiva* 57:353–370.

Russo, Michael
1998 Measuring Sedentism with Fauna: Archaic Cultures Along the Southwest Florida Coast. In *Seasonality and Sedentism: Archaeological Perspectives from Old and New World Sites*, edited by Thomas R. Rocek and Yosef-Ofer Bar, pp. 25–40. Peabody Museum of Archaeology and Ethnology, Harvard University.

Rust, William F., III and Barbara Leyden
1994 Evidence of Maize Use at Early and Middle Preclassic La Venta Olmec Sites. In *Corn and Culture in the Prehistoric New World*, edited by Sissel Johannessen and Christine A. Hastorf, pp. 181–201. Westview Press, Boulder, Colorado.

Rzedowski, Jerzy and M. Laura Huerta
1981 *Vegetación de México*. Editorial Limusa, México.

Santley, Robert S. and Phillip J. Arnold, III
2005 The Obsidian Trade to the Tuxtlas Region and Its Implications for the Prehistory of Southern Veracruz, Mexico. *Ancient Mesoamerica* 16:179–194.

Sassaman, Kenneth E. and Asa R. Randall
2012 Shell Mounds of the Middle St. Johns Basin, Northeast Florida. In *Early New World Monumentality*, edited by Richard L. Burger and Robert M. Rosenswig, pp. 53–77. University Press of Florida, Gainesville.

Sauer, Carl O. and Donald Brand
1931 *Prehistoric Settlements of Sonora with Special Reference to Cerros de Trincheras*. University of California Publications in Geography 5. University of California Press, Berkeley.

Sauer, Johathan D.
1950 The Grain Amaranths: A Survey of Their History and Classification. *Annals of the Missouri Botanical Garden* 37:561–632.
1967 The Grain Amaranths and Their Relatives: A Revised Taxonomic and Geographic Survey. *Annals of the Missouri Botanical Garden* 54:103–137.
1969 Identity of Archaeological Grain Amaranths from the Valley of Tehuacan, Puebla, Mexico. *American Antiquity* 34:80–81.
1993 *Historical Geography of Crop Plants: A Select Roster*. CRC Press, Boca Raton, Florida.

Saunders, Joe
2012 Early Mounds in the Lower Mississippi Valley. In *Early New World Monumentality*, edited by Richard L. Burger and Robert M. Rosenswig, pp. 25–52. University Press of Florida, Gainesville.

Sayles, Edwin B.
1945 *The San Simon Branch: Excavations at Cave Creek and in the San Simon Valley, I: Material Culture*. Medallion Papers 34. Gila Pueblo, Globe, Arizona.
1983 Archaeological Analysis. In *The Cochise Cultural Se-*

quence in Southeastern Arizona, edited by Edwin B. Sayles. Anthropological Papers of the University of Arizona, Vol. 42, pp. 58–81. University of Arizona Press, Tucson.

Sayles, Edwin B. and Ernst V. Antevs
1941 *The Cochise Culture.* Medallion Papers 29. Gila Pueblo, Globe, Arizona.

Schaafsma, Polly
1997 *Rock Art Sites in Chihuahua, Mexico.* Archaeology Notes 171. Museum of New Mexico, Office of Archaeological Studies, Santa Fe, New Mexico.
2000 *Warrior, Shield and Star.* Western Edge Press, Santa Fe.

Schiffer, Michael B.
1986 Radiocarbon Dating and the "Old Wood" Problem: The Case of the Hohokam Chronology. *Journal Of Archaeological Science* 13:13–30.

Schmidly, David J.
1994 *The Mammals of Texas.* Texas Parks and Wildlife and University of Texas Press, Austin.

Schmidt, Kari M.
1998 *Logistical Mobility and Faunal Remains: A Diachronic Study from the San Simon Valley of Southeastern Arizona.* Master's thesis, Department of Anthropology, University of Oklahoma, Norman.
1999 The Five Feature Site (AZ CC:7:55[ASM]): Evidence for a Prehistoric Rabbit Drive in Southeastern Arizona. *Kiva* 65:103–124.
2000 1999 Excavations: A Report on the Faunal Remains from Cerro Juanaqueña, Cerro El Canelo, and the Floodplain Site, Chihuahua, Mexico. Manuscript on file, Department of Anthropology, University of Texas at San Antonio.
2001 Faunal Remains from the 1997 Test Escavations at McEuen Cave (AZ W:13:6 ASM), Fishook Wilderness, Gila Mountains, South-Central Arizona. Manuscript on file.
2005 Faunal Remains for Sites 204 and 242, Chihuahua, Mexico. Manuscript on file, Department of Anthropology, University of Tulsa.
2008 *An Assessment of Settlement and Subsistence in Emergent Agricultural Economies in the Tucson Basin, United States, and Chihuahua, Mexico.* PhD dissertation, Department of Anthropology, University of New Mexico. Albuquerque.

Schmidt, Kari M. and Jennifer E. Nisengard
2001 Los Restos de Fauna. In *Una Investigación Arqueológica de Los Sitios Cerros con Trincheras del Arcaico Tardío en Chihuahua, México*, edited by Robert J. Hard, José E. Zapata, and John R. Roney, pp. 31–36. Special Report 27-S. Center for Archaeological Research, University of Texas at San Antonio.

Schmidt, Robert H., Jr.
1973 *A Geographical Survey of Chihuahua.* Southwestern Studies Monograph 37. University of Texas at El Paso, Texas Western Press, El Paso.

Schmidt, Robert H., Jr. and Rex E. Gerald
1988 The Distribution of Conservation-Type Water-Control Systems in the Northern Sierra Madre Occidental. *Kiva* 53:165–179.

Schott, Amy M.
2017 Site Formation Processes and Depositional Environment of a Fine-Grained Alluvial Floodplain at La Playa Archaeology Site, Sonora, Mexico. *Geoarchaeology* 32:283–301.

Schriever, Bernard A., Matthew Taliaferro, and Barbara J. Roth
2011 A Tale of Two Sites: Functional Site Differentiation and Lithic Technology During the Late Pithouse Period in the Mimbres Area of Southwestern New Mexico. *Journal of Anthropological Archaeology* 30:102–115.

Schumm, S. A. and G. R. Brackenridge
1987 River Responses. In *North American and Adjacent Oceans During the Last Deglaciation*, edited by W. F. Ruddiman and H. E. Wright Jr., pp. 221–240. The Geology of North America, Vol. K-3. Geological Society of America, Boulder.

Schurr, Mark R. and David A. Gregory
2002 Fluoride Dating of Faunal Materials by Ion-Selective Electrode: High Resolution Relative Dating at an Early Agricultural Period Site in the Tucson Basin. *American Antiquity* 67:281–299.

Schwatka, Frederick
1895 *In the Land of Cave and Cliff Dwellers.* (New Edition). Educational Publishing Company, Boston.

Sealy, Judith C.
1986 *Stable Carbon Isotopes and Prehistoric Diets in the Southwest Cape Province, South Africa.* BAR International Series 293. British Archaeological Reports, Oxford.

Servicio Meteorológico y Geográfico
1997 Altura de la Lluvia, Termperatura Media, Estado de Chihuahua, Estación Janos, 1961–1993. Manuscript on file.

Sezate, Adam
2017 *Pipe Construction Techniques in the Pre-Columbian Southwest.* Poster presented at the 82nd Annual Meeting of the Society for American Archaeology, Vancouver, British Columbia.

Shackley, M. Steven
1985 Lithic Raw Material Procurement and Hunter-Gatherer Mobility Strategies in the Southwest Archaic. Paper presented at the 50th Annual Meeting of the Society for American Archaeology, Denver, Colorado.
1990 *Early Hunter-Gatherer Procurement Ranges in the Southwest: Evidence from Obsidian Geochemistry and Lithic Technology.* PhD dissertation, Department of Anthropology, Arizona State University. University Microfilms, Ann Arbor.
1998 *Source Provenance of Archaeological Obsidian from Cerro Juanaqueña, Cerro Vidal and Cerro Los Torres, Northwestern Chihuahua.* Archaeological XRF Laboratory, University of California, Berkeley. Report on file, Department of Anthropology, University of Texas at San Antonio.
1999 *Source Provenance of Archaeological Obsidian from the 1999 Season at Cerro Juanaqueña and Cerro El Canelo, Northwestern Chihuahua.* Archaeological

XRF Laboratory, University of California, Berkeley. Report on file, Department of Anthropology, University of Texas at San Antonio.

2005a Chronometry and Geochemistry at McEuen Cave: The Radiocarbon and Obsidian Geochemical Data. Paper presented at the 70th Annual Meeting of the Society for American Archaeology, Salt Lake City.

2005b *Obsidian: Geology and Archaeology in the North American Southwest.* University of Arizona Press, Tucson.

2017 Sources of Archaeological Obsidian in the Greater American Southwest. Electronic Document, http://swxrflab.net/swobsrcs.htm, accessed May 18, 2017.

Shaffer, Brian S.

1992 Interpretation of Gopher Remains from Southwestern Archaeological Assemblages. *American Antiquity* 57:683–691.

Shaffer, Brian S. and Julia Sanchez

1994 Comparison of ⅛" and ¼" Mesh Recovery of Controlled Samples of Small-to-Medium Animals. *American Antiquity* 59:525–530.

Shaw, Robert H.

1988 Climate Requirement. In *Corn and Corn Improvement*, edited by G. F. Sprague and J. W. Dudley, pp. 609–638. Agronomy Monograph 18. 3rd ed. American Society of Agronomy, Madison, Wisconsin.

Shott, Michael J.

1986a Technological Organization and Settlement Mobility. *Journal of Anthropological Research* 42:15–52.

1986b On Tool-Class Use Lives and the Formation of Archaeological Assemblages. *American Antiquity* 54:9–30.

1994 Size and Form in the Analysis of Flake Debris: Review and Recent Approaches. *Journal of Archaeological Method and Theory* 1:69–110.

Simmons, Alan H.

1986 New Evidence for the Early Use of Cultigens in the American Southwest. *American Antiquity* 51:73–89.

Sliva, R. Jane

1997 *An Introduction to the Study and Analysis of Flaked Stone Artifacts and Lithic Technology.* Center for Desert Archaeology, Tucson.

1998 Flaked Stone Artifacts. In *Archaeological Investigations of Early Village Sites in the Middle Santa Cruz Valley (Part I)*, edited by Jonathan B. Mabry, pp. 299–356. Anthropological Papers No. 19. Center for Desert Archaeology, Tucson.

1999a Cienega Points and Late Archaic Period Chronology in the Southern Southwest. *Kiva* 64:339–367.

1999b Flaked Stone Artifacts. In *Excavations in the Santa Cruz River Floodplain: The Middle Archaic Component at Los Pozos*, edited by David A. Gregory, pp. 33–45. Anthropological Papers No. 20. Center for Desert Archaeology, Tucson.

2001 Flaked Stone Artifacts. In *Excavations in the Santa Cruz Floodplain: The Early Agricultural Period Component at Los Pozos*, edited by David A Gregory, pp. 91–106. Anthropological Papers No. 21. Center for Desert Archaeology, Tucson.

2005 Developments in Flaked Stone Technology During the Transition to Agriculture. In *Material Cultures and Lifeways of Early Agricultural Communities in Southern Arizona*, edited by R. Jane Sliva, pp. 47–98 Anthropological Papers No. 35. Center for Desert Archaeology, Tucson.

2006 Flaked Stone. In *Rio Nuevo Archaeology Program, 2000–2003: Investigations at the San Agustin Mission and Mission Gardens, Tucson Presidio, Tucson Pressed Brick Company, and Clearwater Site*, edited by J. Homer Thiel and Jonathan B. Mabry, pp. 10.11–10.17. Technical Report No. 2004-11. Center for Desert Archaeology, Tucson.

2009 Common Middle Archaic and Early Agricultural Period Points in Southern Arizona. *Archaeology Southwest* 23:supplemental information.

2015 *Projectile Points of the Early Agricultural Southwest: Typology, Migration, and Social Dynamics from the Sonoran Desert to the Colorado Plateau.* Archaeology Southwest, Tucson.

Smiley, Francis E.

1994 The Agricultural Transition in the Northern Southwest: Patterns in the Current Chronometric Data. *Kiva* 60:165–189.

Smith, Bruce D.

1984 Chenopodium as a Prehistoric Domesticate in Eastern North America: Evidence from Russell Cave, Alabama. *Science* 226:165–167.

1985 *Chenopodium berlandieri* ssp. *jonesianum*: Evidence for a Hopewellian Domesticate from Ash Cave, Ohio. *Southeastern Archaeology* 4:107–133.

1997 The Initial Domestication of *Cucurbita pepo* in the Americas 10,000 Years Ago. *Science* 276:932–934.

1998 *The Emergence of Agriculture.* Scientific American Library, New York.

2001 Documenting Plant Domestication: The Consilience of Biological and Archaeological Approaches. *The Proceedings of the National Academy of Sciences of the United States of America* 98:1324–1326.

2005 Reassessing Coxcatlan Cave and the Early History of Domesticated Plants in Mesoamerica. *The Proceedings of the National Academy of Sciences of the United States of America* 102:9438–9445.

2006 Eastern North America as an Independent Center of Plant Domestication. *Proceedings of the National Academy of Sciences of the United States of America* 103:12223–12228.

2008 Winnowing the Archaeological Evidence for Domesticated Sunflower in Pre-Columbian Mesoamerica. *Proceedings of the National Academy of Sciences of the United States of America* 105:E45.

2012 A Cultural Niche Construction Theory of Initial Domestication. *Biological Theory* 6:260–271.

2015 A Comparison of Niche Construction Theory and Diet Breadth Models as Explanatory Frameworks for the Initial Domestication of Plants and Animals. *Journal of Archaeological Research* 23: 215–262.

Smith, Grant D. and Michael McFaul

1997 Paleoenvironmental and Geoarchaeologic Implications of Late Quaternary Sediments and Paleosols: North-central to Southwestern San Juan Basin, New Mexico. *Geomorphology* 21:107–138.

Smith, Graham W.
1990 Home Range and Activity Patterns of Black-tailed Jackrabbits. *Great Basin Naturalist* 50:249–256.

Smith, M. L. and R. R. Miller
1986 The Evolution of Rio Grande Basin as Inferred from Its Fish Fauna. In *The Zoogeography of North American Freshwater Fishes*, edited by C. H. Hocutt and O. E. Wiley, pp. 457–485. John Wiley and Sons, New York.

Smith, Steve, Raymond Mauldin, Cynthia M. Munoz, Robert J. Hard, Debajyoti Paul, Grzegorz Skrzypek, Patricio Villanueva, and Leonard Kemp
2014 Exploring the Use of Stable Carbon Isotope Ratios in Short-Lived Leporids for Local Paleoecological Reconstruction. *Open Journal of Archaeometry* 2:5306:81–87.

Soil Survey Division Staff
1993 *Soil Survey Manual.* U.S. Department of Agriculture Handbook 18. U.S. Government Printing Office, Washington, DC.

Sollberger, J. B.
1971 A Technological Study of Beveled Knives. *Plains Anthropologist* 16:209–218.

Somerville, Andrew D., Andrew W. Froehle, and Margaret J. Schoeninger
2018 Environmental Influences on Rabbit and Hare Bone Isotope Abundances: Implications for Paleoenvironmental Research. *Palaeogeography, Palaeoclimatology, Palaeoecology* 497:91–104.

Somerville, Andrew D., Nawa Sugiyama, Linda R. Manzanilla, and Margaret J. Schoeninger
2016 Animal Management at the Ancient Metropolis of Teotihuacan, Mexico: Stable Isotope Analysis of Leporid (Cottontail and Jackrabbit) Bone Mineral. *PLoS One* 11.
2017 Leporid Management and Specialized Food Production at Teotihuacan: Stable Isotope Data from Cottontail and Jackrabbit Bone Collagen. *Archaeological and Anthropological Sciences* 9:83–97.

Spencer, Joseph E. and Gary A. Hale
1961 The Origin, Nature, and Distribution of Agricultural Terracing. *Pacific Viewpoint* 2:1–40.

Spier, Leslie
1970 *Yuman Tribes of the Gila River.* Cooper Square Publishers, Inc., New York.

SPP, Secretaría de Programación y Presupuesto
1980 *Ciudad Juárez H13-1, Carta Uso del Suelo y Vegetación.* Thematic Map, 1:250,000. SPP, México, D.F.
1981a *Agua Prieta H12-3, Carta Uso del Suelo y Vegetación.* Thematic Map, 1:250,000. SPP, México, D.F.
1981b *Nuevo Casas Grandes, H13-4, Carta Uso del Suelo y Vegetación.* Thematic Map, 1:250,000. SPP, México, D.F.
1982 *Agua Prieta, H12-3, Carta Geológica.* Thematic Map, 1:250,000. SPP, México, D.F.
1983a *Ciudad Juárez H13-1, Carta Geológica.* Thematic Map, 1:250,000. SPP, México, D.F.
1983b *Nacozari H12-6, Carta Geológica.* Thematic Map, 1:250,000. SPP, México, D.F.
1983c *Nuevo Casas Grandes H13-4, Carta Geológica.* Thematic Map, 1:250,000. SPP, México, D.F.
1985 *Nacozari H12-6, Carta Uso del Suelo y Vegetación.* Thematic Map, 1:250,000. SPP, México, D.F.

Stacy, Valeria Kay Pheriba
1974 *Cerros de Trincheras in the Arizona Papagueria.* PhD dissertation, Department of Anthropology, University of Arizona. Tucson.

Stafford, Thomas W., P. E. Hare, Lloyd Currie, A. J. T. Jull, and Douglas J. Donahue
1991 Accelerator Radiocarbon Dating at the Molecular Level. *Journal of Archaeological Science* 18:35–72.

Stahle, David W. and J. E. Dunn
1982 *An Analysis and Application of the Size Distribution of Waste Flakes from Biface Reduction.* Technical Paper No. 2. Arkansas Archaeological Survey, Fayetteville.

Stark, Barbara L.
1981 The Rise of Sedentary Life. In *Handbook of Middle American Indians, Supplement 1: Archaeology*, edited by V.R. Bricker and J.A. Sabloff, pp. 345–372. University of Texas Press, Austin.

Stewart, Kenneth M.
1957 Mohave Fishing. *Masterkey* 31:198–203.
1983 Mohave. In *Southwest*, edited by Alfonso Ortiz, pp. 55–71. Handbook of North American Indians, Vol. 10, William C. Sturtevant, general editor. Smithsonian Institution, Washington, DC.

Stoney, Stephen A.
1993 Cupule Art in Southern Nevada: An Analysis. In *American Indian Rock Art, Volume 19*, edited by Frank G. Bock, pp. 49–56. American Rock Art Research Association, San Miguel, California.

Struever, Stuart
1968 Flotation Techniques for the Recovery of Small-Scale Archaeological Remains. *American Antiquity* 33:353–362.

Stuiver, M. and P. J. Reimer
1993 Extended C-14 Data Base and Revised Calib 3.0 C-14 Age Calibration Program. *Radiocarbon* 35:215–230.

Sullivan, Alan and Steven Rosen
1985 Debitage Analysis and Archaeological Interpretation. *American Antiquity* 50:755–799.

Swanson, Eric R., Kirt A. Kempter, Fred W. McDowell, and Wililiam C. McIntosh
2006 Major Ignimbrites and Volcanic Centers of the Copper Canyon Area: A View into the Core of Mexico's Sierra Madre Occidental. *Geosphere* 2:125–141.

Swanson, Eric R. and Fred W. McDowell
1984 Calderas of the Sierra Madre Occidental Volcanic Field Western Mexico. *Journal of Geophysical Research* 89:8787–8799.

Szuter, Christine R.
1991 *Hunting by Prehistoric Horticulturalists in the American Southwest.* Garland Publishing, New York.

Szuter, Christine R. and Frank E. Bayham
1989 Sedentism and Animal Procurement among Desert Horticulturalists of the North American Southwest.

In *Farmers as Hunters: The Implications of Sedentism*, edited by Susan Kent, pp. 80–95. Cambridge University Press, Cambridge.

1996 Faunal Exploitation During the Late Archaic and Early Ceramic/Pioneer Periods in South-Central Arizona. In *Early Formative Adaptations in the Southern Southwest*, edited by Barbara J. Roth, pp. 65–72. Prehistory Press, Madison.

Tagg, Martyn D.

1996 Early Cultigens from Fresnal Shelter, Southeastern New Mexico. *American Antiquity* 61:311–324.

Taylor, R.E.

1997 Radiocarbon Dating. In *Cronometric Dating in Archaeology*, edited by R.E. Taylor and Martin J. Aitken, pp. 65–96. Advances in Archaeological and Museum Science 2. Plenum, New York.

Teague, Lynn S.

1998 *Textiles in Southwestern Prehistory*. University of New Mexico Press, Albuquerque.

Thiel, J. Homer

1998 Faunal Remains. In *Archaeological Investigations of Early Village Sites in the Middle Santa Cruz Valley: Analysis and Synthesis, Part I*, edited by Jonathan B. Mabry, pp. 165–208. Anthropological Papers No. 19. Center for Desert Archaeology, Tucson, Arizona.

Thiel, J. Homer and Jonathan B. Mabry (editors)

2006 *Rio Nuevo Archaeology, 2000–2003: Investigations at the San Agustin Mission and Mission Gardens, Tucson Presidio, Tucson Pressed Brick Company, and Clearwater Site*. Technical Report No. 2004–11. Center for Desert Archaeology, Tucson.

Thomas, Andrea M.

2016 *The Evaluation of the Labor Costs of Stone Boiling Dried Maize During the Early Agricultural Period in the Southwest*. Master's thesis, Department of Anthropology, University of Texas at San Antonio. ProQuest Dissertations and Theses, Ann Arbor.

Tomka, Steven A.

1989 Differentiating Lithic Reduction Techniques. In *Experiments in Lithic Technology*, edited by Daniel S. Amick and Raymond P. Mauldin, pp. 137–161. BAR International Series 528. British Archaeological Reports, Oxford.

Torrence, Robin

1983 Time Budgeting and Hunter-Gatherer Technology. In *Hunter-Gatherer Economy in Prehistory: A European Perspective*, edited by G.N. Bailey, pp. 11–22. Cambridge University Press, Cambridge.

1986 *Production and Exchange of Stone Tools: Prehistoric Obsidian in the Aegean*. Cambridge University Press, Cambridge.

2001 Hunter-Gatherer Technology: Macro-and Microscale Approaches. In *Hunter-Gatherers: An Interdisciplinary Perspective*, edited by, Robert H. Layton, Catherine Panter-Brick and Peter Rowley-Conwy, pp. 73–98. Cambridge University Press, Cambridge.

Torrence, Robin (editor)

1989 *Time, Energy and Stone Tools*. Cambridge University Press, Cambridge.

Trigger, Bruce G.

1990 Monumental Architecture: A Thermodynamic Explanation of Symbolic Behavior. *World Archaeology* 22:119–132.

Tucker, Jonathan B.

1986 Amaranth: The Once and Future Crop. *Bioscience* 36:9–13.

Turnbow, Christopher A.

1997 Projectile Points as Chronological Indicators. In *OLE*, Vol. 2, edited by John C. Acklen, , pp. 161–230. Public Service Company of New Mexico, Albuquerque.

2000 *A Highway through Time: Archaeological Investigations Along NM 90, in Grant and Hidalgo Counties, New Mexico* Technical Report 20003. 2 vols. New Mexico State Highway and Transportation Department, Albuquerque.

Turner, B.L., II

1983a Constructional Inputs for Major Agrosystems of the Ancient Maya. In *Drained Field Agriculture in Central and South America, Proceedings from the International Congress of Americanists, Manchester 1982*, edited by J.P. Darch, pp. 11–25. BAR International Series 189. British Archaeological Reports, Oxford.

1983b *Once Beneath the Forest: Prehistoric Terracing in the Rio Bec Region of the Maya Lowlands*. Dellplain Latin American Studies No. 13. Westview Press, Boulder.

Turner, Ellen Sue and Thomas R. Hester

1995 *A Field Guide to Stone Artifacts of Texas Indians*. 2nd ed. Gulf Publishing Co., Houston.

Ugan, Andrew and Joan Coltrain

2011 Variation in Collagen Stable Nitrogen Values in Black-Tailed Jackrabbits (*Lepus californicus*) in Relation to Small-Scale Differences in Climate, Soil, and Topography. *Journal of Archaeological Science* 38:1417–1429.

United Nations Economic Commission for Asia and the Far East

1961 *Earthmoving by Manual Labour and Machines: Report and Discussion Papers of the Working Party on Earthmoving Operations*. United Nations, Bangkok.

Van Auken, Oscar W.

2000 Shrub Invasions of North American Semiarid Grasslands. *Annual Review of Ecologicial Systems* 31:197–215.

Van Devender, Thomas R.

1990 Late Quaternary Vegetation and Climate of the Chihuahuan Desert, United States and Mexico. In *Packrat Middens: The Last 40,000 Years of Biotic Change*, edited by Julio L. Betancourt, Thomas R. Van Devender, and Paul S. Martin, pp. 104–133. University of Arizona Press, Tucson.

1995 Desert Grassland History: Changing Climates, Evolution, Biogeography, and Community Dynamics. In *The Desert Grassland*, edited by M.P. McClaran and Thomas R. Van Devender, pp. 68–99. University of Arizona Press, Tucson.

Van Devender, Thomas R. and Richard D. Worthington
1977 The Herpetofauna of Howell's Ridge Cave and the Paleoecology of the Northwestern Chihuahuan Desert. In *Transactions of the Symposium on the Biological Resources of the Chihuahuan Desert Region: United States and Mexico*, edited by Roland H. Wauer and David H. Riskind, pp. 85–106. Transactions and Proceedings Series No. 3. National Park Service, U.S. Department of the Interior, Washington, DC.

Van Peer, Philip
1992 *The Levallois Reduction Strategy*. Monographs in World Archaeology No. 13. Prehistory Press, Madison.
1993 Technological Variation and Subsistence Strategies: Explaining Changes in Stone Tool Technology. Paper presented at the The Fifth Occasional Anasazi Symposium, Farmington, New Mexico.

VanPool, Christine S.
2003 The Shaman-Priests of the Casas Grandes Region, Chihuahua, Mexico. *American Antiquity* 68:696–717.

Varien, Mark
1999 *Sedentism and Mobility in a Social Landscape: Mesa Verde & Beyond*. University of Arizona Press, Tucson.

Vierra, Bradley J.
1985 Archaic Hunter-Gatherer Mobility Strategies in Northwestern New Mexico. Paper presented at the 50th Annual Meeting of the Society for American Archaeology, Denver, Colorado.
1990 Archaic Hunter-Gatherer Archaeology in Northwestern New Mexico. In *Perspectives on Southwestern Prehistory*, edited by Paul E. Minnis and Charles L. Redman, pp. 57–67. Westview Press, Boulder.
1993a Explaining Long-Term Changes in Lithic Procurement and Reduction Strategies. In *Architectural Studies, Lithic Analyses and Ancillary Studies*, edited by Bradley J. Vierra, Tim W. Burchett, Kenneth L. Brown, Marie E. Brown, Paul T. Kay, and Carl J. Phagan, pp. 139–381. Across the Colorado Plateau: Anthropological Studies for the Transwestern Pipeline Expansion Project, Vol. 17, Parts 1 and 2. Office of Contract Archaeology and Maxwell Museum of Anthropology, Albuquerque.
1993b Technological Variation and Subsistence Strategies: Explaining Changes in Stone Tool Technology. Paper presented at the Fifth Occasional Anasazi Symposium, Farmington, New Mexico.
1994a Archaic Hunter-Gatherer Mobility Patterns in the American Southwest. In *Excavations and Interpretations of Aceramic and Archaic Sites*, edited by Tim W. Burchett, Bradley J. Vierra, and Kenneth L. Brown, pp. 385–398. Across the Colorado Plateau: Anthropological Studies for the Transwestern Pipeline Expansion Project, Vol. 14. Office of Contract Archeology, University of New Mexico, Albuquerque.
1994b Archaic Hunter-Gatherer Mobility Strategies in the Northwestern New Mexico. In *Archaic Hunter-Gatherer Archaeology in the American Southwest*, edited by B. Vierra, pp. 121–154. Contributions in Anthropology 13(1), Phillip H. Shelley, general editor. Eastern New Mexico University, Portales.
1994c A Study of Basketmaker II Lithic Technology. In *Excavations and Interpretations of Aceramic and Archaic Sites*, edited by Tim W. Burchett, Bradley J. Vierra, and Kenneth L. Brown, pp. 413–441. Across the Colorado Plateau: Anthropological Studies for the Transwestern Pipeline Expansion Project Vol. 14. Office of Contract Archeology, University of New Mexico, Albuquerque.
1995 *Subsistence and Stone Tool Technology: An Old World Perspective*. Anthropological Research Papers No. 47. Department of Anthropology, Arizona State University, Tempe.
1997 A Study of Stone Tool Technology in the Fort Wingate Area. In *Cycles of Closure: A Cultural Resource Inventory of Fort Wingate Depot Activity, New Mexico*, edited by Jeanne A. Schutt and Richard C. Chapman, pp. 105–148. Office of Contract Archaeology, University of New Mexico, Albuquerque.
1998 *41MV120: A Stratified Late Archaic Site in Maverick County, Texas*. Archaeological Survey Report No. 251. Center for Archaeological Research, University of Texas at San Antonio.
1999a Analysis and Synthesis. In *Assessment of 431 Archeological Sites in Doña Ana Range Firing Groups C, D, G, H, I, and J, Fort Bliss, Texas and New Mexico*, edited by Bradley J. Vierra, Paul Lukowski, Grant D. Smith, Sue M. Sitton, Bruce Boeke, Valerie Provencio, Mark T. Sechrist, Amy Sullivan, Jeffery Johnson, Scott Wilcox, and Martha Yduarte. Historic and National Resources Report No. 97-11. 2 vols. TRC Mariah Associates, El Paso.
1999b Late Archaic Stone Tool Technology across the Borderlands. Paper presented at the 64th Annual Meeting of the Society for American Archaeology, Chicago.
2005 Late Archaic Stone Tool Technology across the Borderlands. In *The Late Archaic Across the Borderlands*, edited by Bradley J. Vierra, pp. 187–218. University of Texas Press, Austin.
2013 Archaic Foraging Technology and Land Use in the Northern Rio Grande, New Mexico. In *From Mountaintop to Valley Bottom: Understanding Past Land Use in the Northern Rio Grande Valley, New Mexico*, edited by Bradley J. Vierra, pp. 145–160. University of Utah Press, Salt Lake City.

Vierra, Bradley J. and William H. Doleman
1994 The Organization of Archaic Settlement-Subsistence Systems in the Northern Southwest. In *Archaic Hunter-Gatherer Archaeology in the American Southwest*, edited by Bradley J. Vierra, pp. 76–102. Contributions in Anthropology 13(1), Phillip H. Shelley, general editor. Eastern New Mexico University, Portales.

Vierra, Bradley J. and Teralene Foxx
2009 Archaic Upland Resource Use: The View from the Pajarito Plateau. In *Between the Mountains, Beyond*

the Mountains: Papers in Honor of Paul R. Williams, edited by Emily Brown, Karen Armstrong, and David A. Brugge, pp. 153–166. Archaeological Society of New Mexico, Albuquerque.

Vint, James M.

2015a Time and Place in the Early Agricultural Period Santa Cruz Valley. In *Implements of Change: Tools, Subsistence, and the Built Environment of Las Capas, an Early Agricultural Irrigation Community in Southern Arizona*, edited by James M. Vint, pp. 469–499. Anthropological Papers 51. Archaeology Southwest, Tucson.

2015b Las Capas, AZ AA:12:111 (ASM), Introduced: Background, Chronology, and Research Orientation. In *Implements of Change: Tools, Subsistence, and the Built Environment of Las Capas, an Early Agricultural Irrigation Community in Southern Arizona*, edited by James M. Vint, pp. 1–31. Anthropological Papers 51. Archaeology Southwest, Tucson.

Vint, James M. (editor)

2015 *Implements of Change: Tools, Subsistence, and the Built Environment of Las Capas, an Early Agricultural Irrigation Community in Southern Arizona.* Anthropological Papers 51. Archaeology Southwest, Tucson.

Vint, James M. and Barbara J. Mills

2016 Niches, Networks and the Pathways to the Forager-to-Farmer Transition in the US Southwest/NorthWest Mexico. In *The Origins of Food Production/Los Orígenes de la producción de alimentos* pp. 264–281. United Nations Educational, Scientific and Cultural Organization, Mexico Office, Mexico City.

Virden-Lange, Christine H.

2015 Shell Material from Las Capas, AZ AA:12:111 (ASM). In *Implements of Change: Tools, Subsistence, and the Built Environment of Las Capas, an Early Agricultural Irrigation Community in Southern Arizona*, edited by James M. Vint, pp. 227–260. Anthropological Papers No. 51. Archaeology Southwest, Tucson.

Von Fisher, Joseph C., Larry L. Tieszen, and David S. Schimel

2008 Climate Controls on C_3 Vs. C_4 Productivity in North American Grasslands from Carbon Isotope Composition of Soil Organic Matter. *Global Change Biology* 14:1–15.

Vorhies, Charles T. and Walter P. Taylor

1933 *The Life Histories and Ecology of Jack Rabbits, Lepus alleni and Lepus californicus ssp., in Relation to Grazing in Arizona.* Technical Bulletin 49. Agricultural Experiment Station, College of Agriculture, University of Arizona, Tucson.

Wallace, Henry D.

1983 The Mortars, Petroglyphs, and Trincheras on Rillito Peak. *Kiva* 48:137–246.

Wallace, Henry D. and William H. Doelle

2001 Classic Period Warfare in Southern Arizona. In *Deadly Landscapes: Case Studies in Prehistoric Southwestern Warfare*, edited by Glen E. Rice and Steven A. LeBlanc, pp. 239–287. University of Utah Press, Salt Lake City.

Wallace, Henry D., Paul R. Fish, and Suzanne K. Fish

2007 Tumamoc Hill and the Early Pioneer Period Occupation of the Tucson Basin. In *Trincheras Sites in Time, Space, and Society*, edited by Suzanne K. Fish, Paul R. Fish, and M. Elisa Villalpando, pp. 53–99. University of Arizona Press, Tucson.

Wallace, Laurel T. (editor)

1998 *Ormand Village: Final Report on the 1965–1966 Excavation.* Archaeology Notes No. 229. Office of Archaeological Studies, Museum of New Mexico, Santa Fe.

Wallace, William J.

1955 Mohave Fishing Equipment and Methods. *Anthropological Quarterly* 28:87–94.

Walsh, Michael R.

1997 *Lines in the Sand: Competition and Territoriality in the Northern Rio Grande AD 1150–1325.* PhD dissertation, Department of Anthropology, University of California, Los Angeles. University Microfilms, Ann Arbor.

1998 Lines in the Sand: Competition and Stone Selection on the Pajarito Plateau, New Mexico. *American Antiquity* 63:573–593.

Wansi, Tchouassi, Rex D. Pieper, Reldon F. Beck, and Leigh W. Murray

1992 Botanical Content of Black-Tailed Jackrabbit Diets on Semidesert Rangeland. *Great Basin Naturalist* 52:300–308.

Waters, Jennifer A.

1998 Vertebrate Faunal Remains from the Wetlands Site. In *Archaeological Investigations at the Wetlands Site, AZ AA:12:90 (ASM)*, edited by Andrea K. L. Freeman, pp. 205–228. Technical Report No. 97-5. Center for Desert Archaeology, Tucson.

2000 Vertebrate Fauna from Sunset Mesa. In *Excavations at Sunset Mesa Ruin*, edited by Michael W. Lindeman, pp. 177–194. Technical Report 2000-02. Desert Archaeology, Inc, Tucson.

Waters, Jennifer A., Stephanie M. Reyes, and Sarah E. Wolff

2015 Vertebrate Faunal Remains Recovered from Las Capas, AZ AA:12:111 (ASM). In *Implements of Change: Tools, Subsistence, and the Built Environment of Las Capas, an Early Agricultural Irrigation Community in Southern Arizona*, edited by James M. Vint, pp. 261–304. Anthropological Papers No. 51. Archaeology Southwest, Tucson.

Waters, Michael R.

1989 Late Quaternary Lacustrine History and Paleoclimatic Significance of Pluvial Lake Cochise, Southeastern Arizona. *Quaternary Research* 32:1–11.

1992 *Principles of Geoarchaeology: A North American Perspective.* University of Arizona Press, Tucson.

Waters, Michael R. and C. Vance Haynes

2001 Late Quaternary Arroyo Formation and Climate Change in the American Southwest. *Geology* 29: 399–402.

Waters, Michael R. and John C. Ravesloot

2000 Late Quaternary Geology of the Middle Gila River, Gila River Indian Reservation, Arizona. *Quaternary Research* 54:49–57.

Watson, James T. and Danielle O. Phelps
2016 Violence and Perimortem Signaling among Early Irrigation Communities in the Sonoran Desert. *Current Anthropology* 57:586–609.
Watson, James T. and Marijke Stoll
2013 Gendered Logistic Mobility among the Earliest Farmers in the Sonoran Desert. *Latin American Antiquity* 24:433–450.
Webster, Monica
2001 *Prehistoric Diet and Human Adaptation in West Central Chihuahua, Mexico.* Master's thesis, Department of Archaeology, University of Calgary, Canada.
Werth, Lindsay C.
2007 *Characterization and Classification of Native American Maize Landraces from the Southwestern United States.* Master's thesis, Department of Agronomy, Iowa State University. Ames.
Whalen, Michael E.
1994a Moving out of the Archaic on the Edge of the Southwest. *American Antiquity* 59:606–621.
1994b *Turquoise Ridge and Late Prehistoric Residential Mobility in the Desert Mogollon Region.* University of Utah Press, Salt Lake City.
Whalen, Michael E. and Paul E. Minnis
2001 *Casas Grandes and Its Hinterland.* University of Arizona Press, Tucson.
2009 *The Neighbors of Casas Grandes: Excavating Medio Period Communities of Northwest Chihuahua, Mexico.* University of Arizona Press, Tucson.
Whisenhunt, Mary E., Robert J. Hard, and John R. Roney
2016 Cerros de Trincheras on the Upper Gila River, Arizona: Implications for Defense. Paper presented at the South Texas Archaeological Association Quarterly Meeting, San Antonio.
White, Theodore E.
1953 A Method of Calculating the Dietary Percentages of Various Food Animals Utilized by Aboriginal Peoples. *American Antiquity* 18:396–398.
Whittaker, John. C.
1994 *Flinknapping: Making and Understanding Stone Tools.* University of Texas Press, Austin, Texas.
Whittlesey, Stephanie M.
2016 Deconstructing the Early Agricultural Period in Southern Arizona. In *Late Holocene Research on Foragers and Farmers in the Desert West*, edited by Barbara J. Roth and Maxine E. McBrinn, pp. 78–107. University of Utah Press, Salt Lake City.
Whittlesey, Stephanie M. and Richard Ciolek-Torrello
1996 The Archaic-Formative Transition in the Tucson Basin. In *Early Formative Adaptations in the Southern Southwest*, edited by Barbara J. Roth, pp. 49–64. Prehistory Press, Madison.
Whittlesey, Stephanie M., S. Jerome Hesse, and Michael S. Foster (editors)
2010 *Recurrent Sedentism and the Making of Place: Archaeological Investigations at Las Capas, a Preceramic Period Farming Community in the Tucson Basin, Southern Arizona.* SWCA Cultural Resources Report 07556. SWCA Environmental Consultants, Tucson, Arizona.
Wiant, Michael D. and Howard Hasesn
1985 The Role of Lithic Resource Availability and Accessibility in the Organization of Technology. In *Lithic Resource Procurement: Proceedings from the Second Conference on Prehistoric Chert Exploitation*, edited by Susan Vehic, pp. 101–104. Center for Archaeological Investigations Occasional Paper 4. Southern Illinois University, Carbondale.
Wilcox, David R.
1979 Warfare Implications of Dry-Laid Masonry Walls on Tumamoc Hill. *Kiva* 45:15–38.
Wilcox, David R. and Jonathan Haas
1994 The Scream of the Butterfly: Competition and Conflict in the Prehistoric Southwest. In *Themes in Southwest Prehistory*, edited by George J. Gumerman, pp. 211–238. School of American Research, Santa Fe.
Wilcox, David R., Stephen M. Larson, Bruce W. Masse, Gayle H. Hartmann, and Alan Ferg
1979 A Summary of Conclusions and Recommendations of the Tumamoc Hill Survey. *Kiva* 45:187–195.
Wilken, Gene C.
1976 Traditional Slope Management: An Analytical Approach. In *Hill Lands: Proceedings of an International Symposium*, edited by J. Luchok, J. D. Cawthorn, and M. J. Breslin, pp. 416–421. West Virginia University/West Virginia Forage and Grassland Council, Morgantown, West Virginia.
1987 *Good Farmers: Traditional Agricultural Resource Management in Mexico and Central America.* University of California Press, Berkeley.
Wilkens, David E. and Donald R. Currey
1999 Radiocarbon Chronology and $\delta^{13}C$ Analysis of Mid-to Late-Holocene Aeolian Environments, Guadalupe Mountains National Park, Texas, USA. *The Holocene* 9:363–371.
Wills, Wirt H.
1988 *Early Prehistoric Agriculture in the American Southwest.* School of American Research Press, Santa Fe.
Wills, Wirt H. and Bruce B. Huckell
1994 Economic Implications of Changing Land-Use Patterns in the Late Archaic. In *Themes in Southwest Prehistory*, edited by George J. Gumerman, pp. 33–52. School of American Research, Santa Fe.
Wöcherl, Helga
1997 *The Role of Faunal Resources in Subsistence Practices During the Transition to Sedentism and Agriculture in Southeastern Arizona.* PhD dissertation, Department of Anthropology, University of Arizona, Tucson. Microfilms International, Ann Arbor.
2001 Faunal Remains. In *Excavations in the Santa Cruz River Floodplain: The Early Agricultural Period Component at Los Pozos*, edited by David A. Gregory, pp. 209–236. Anthropological Papers No. 21. Center for Desert Archaeology, Tucson.
Wooton, E. O. and Paul C. Standley
1915 *Flora of New Mexico.* Contributions from the United States National Herbarium 19. Washington Government Printing Office, Washington, DC.
Yale University Library: Social Science Research Services
2004 Economic Growth Center Digital Library: Mexico

Collection. Electronic Document, http://ssrs.yale .edu/egcdl/mxdl/index.jsp, accessed June 15, 2012.

York, J. C. and William A. Dick-Peddie

1969 Vegetation Changes in Southern New Mexico During the Past Hundred Years. In *Arid Lands in Perspective*, edited by William G. McGinnies and Bram J. Goldman, pp. 157–199. University of Arizona Press, Tucson.

Young, Lisa C.

1996 *Mobility and Farmers: The Pithouse-to-Pueblo Transition in Northern Arizona*. PhD dissertation, Department of Anthropology, University of Arizona, Tucson.

Young, Lisa C. and Karen G. Harry

1989 A Preliminary Analysis of Temporal Changes in the Homol'ovi IV Assemblage. *Kiva* 54:273–284.

Zeder, Melinda A.

2001 A Metrical Analysis of a Collection of Modern Goats (*Capra hircus aegargus* and *C. h. hircus*) from Iran and Iraq: Implications for the Study of Caprine Domestication. *Journal of Archaeological Science* 28:61–79.

2012 The Broad Spectrum Revolution at 40: Resource Diversity, Intensification, and an Alternative to Optimal Foraging Explanations. *Journal of Anthropological Archaeology* 31:241–264.

2015 Core Questions in Domestication Research. *Proceedings of the National Academy of Sciences of the United States of America* 112:3191–3198.

Contributors

Karen R. Adams
Archaeobotanical Consultant
2387 E. Beverly Dr.
Tucson, AZ 85716-5716
agave@dakotacom.net

Gayle J. Fritz
Professor Emerita
Washington University in St. Louis
602 Ostle Drive
Collinsville IL 62234
gjfritz@wustl.edu

J. Kevin Hanselka
Archaeologist, Archaeobotanist
Texas Department of Transportation
1227 Covina Ct.
Allen, Texas 75013
khanselk@yahoo.com

Robert J. Hard
Professor of Anthropology
University of Texas at San Antonio
One UTSA Circle
San Antonio, TX 78249
robert.hard@utsa.edu

Lee C. Nordt
Professor, Dean
Department of Geosciences
College of Arts and Sciences
Baylor University
Waco, TX 76798-7344
Lee_Nordt@baylor.edu

John R. Roney
Archaeologist
Colinas Cultural Resource Consulting
Bureau of Land Management (retired)
316 Osuna Rd NE, Suite 102
Albuquerque, NM 87107
jroney.ccrc@yahoo.com

Kari M. Schmidt
Faunal Analyst
4037 Windward Dr.
Tega Cay, SC 29708
schmidt.karim@gmail.com

Bradley J. Vierra
Adjunct Professor
New Mexico State University
8913 Cherry Hills Road, NE
Albuquerque, NM 87111
bvierra53@gmail.com

Index

Page numbers printed in *italic* type refer to figures, illustrations, maps, or tables.